Geography: Realms, Regions, and Concepts

Seventeenth Edition

About the Cover

On September 18, 2015, this large group of mostly Syrian refugees reached far western Turkey, following a long and arduous journey from their war-torn homeland. Walking along an expressway near the town of Edirne, they approached the border with Greece. If they could get across they would be inside the European Union (EU)—but by this time, their chances were slim. During 2015 alone, there were an estimated two million clandestine border crossings and national authorities, including the Greek government, were increasingly clamping down on illegal entries and hardening their borders. A few months later in the spring of 2016, the EU and Turkey would reach an agreement whereby the latter would no longer allow refugees to cross into neighboring Greece and Bulgaria. It is impossible to know if these particular refugees made it to their preferred destination, remained in Turkey, or were deported to their home countries.

JAN NIJMAN

Distinguished University Professor, Geosciences,
Georgia State University
Professor of Geography, University of Amsterdam

PETER O. MULLER

Senior Professor of Geography and
Regional Studies, University of Miami

H.J. DE BLIJ

The late John A. Hannah Professor of Geography,
Michigan State University

Geography: Realms, Regions, and Concepts

Seventeenth Edition

WITH

ANN-MARGARET ESNARD

Professor of Regional and Urban Planning, Andrew Young School of Policy Studies,
Georgia State University

RICHARD GRANT

Professor of Geography and Regional Studies, University of Miami

MICHAEL SHIN

Associate Professor of Geography, University of California, Los Angeles

WILEY

VICE PRESIDENT AND DIRECTOR	Petra Recter
EXECUTIVE EDITOR	Jessica Fiorillo
SPONSORING EDITOR	Jennifer Yee
ASSOCIATE DEVELOPMENT EDITOR	Mallory Fryc
MARKET SOLUTIONS ASSISTANTS	Lindsey Myers, Esther Kamar
SENIOR MARKETING MANAGER	Alan Halfen
PRODUCTION EDITOR	John Du Val
SENIOR PHOTO EDITOR	Billy Ray
SENIOR DESIGNER	Wendy Lai

Front cover image: Osman Orsal/Reuters/Zuma Press
Back cover image: Sam Edwards/Getty Images

This book was set in 9.5 Source Sans Pro by Cenveo® Publisher Services and printed and bound by LSC Comm-Kendallville. This book is printed on acid-free paper.

Founded in 1807, John Wiley & Sons, Inc. has been a valued source of knowledge and understanding for more than 200 years, helping people around the world meet their needs and fulfill their aspirations. Our company is built on a foundation of principles that include responsibility to the communities we serve and where we live and work. In 2008, we launched a Corporate Citizenship Initiative, a global effort to address the environmental, social, economic, and ethical challenges we face in our business. Among the issues we are addressing are carbon impact, paper specifications and procurement, ethical conduct within our business and among our vendors, and community and charitable support. For more information, please visit our Web site: www.wiley.com/go/citizenship.

ePUB ISBN: 978-1-119-30189-9

Printed in the United States of America

The inside back cover will contain printing identification and country of origin if omitted from this page. In addition, if the ISBN on the back cover differs from the ISBN on this page, the one on the back cover is correct.

10 9 8 7 6 5 4 3 2 1

This Seventeenth Edition of Regions *is dedicated to the memory of Harm de Blij, our long-time co-author, mentor, and friend. Harm revolutionized Geography textbook writing in the United States, and he was without equal in communicating geographic knowledge and awareness to wider audiences. His legacy will live on in our books.*

This is the Seventeenth Edition of a book that, since its first appearance in 1971, has consistently broken new ground in the interpretation and teaching of world regional geography. For more than four decades, *Geography: Realms, Regions, and Concepts* has reported (and sometimes anticipated) trends in the discipline of Geography and developments in the world at large. *REGIONS*, as the book has come to be called, has explained the contemporary world's geographic realms (the largest regional entities on the face of the Earth) and their natural environments and human dimensions. In the process, this book has become an introduction to Geography itself, the discipline that links the study of human societies and natural environments. We look at the ways people have organized their living space, adapted to changing social as well as environmental circumstances, and continue to confront forces largely beyond their control ranging from globalization to climate change.

This book was the first to introduce an approach to Geography that meshes theoretical concepts with regional realities. This approach to the integrated treatment of regions and concepts is a constant throughout all editions, but over the years we have adapted to the changing world around us. The evolving regional content of the chapters, and their organization, reflect the dynamic nature of the world's geography; the changing (and growing number of) concepts mirror the progress of the discipline; and the ongoing introduction of new (digital) features reflects the instructional possibilities of new technologies.

As always, we put a premium on top-quality, original maps to accompany our regional analyses, and we rely, in large part, on our own field observations. Much of our time and research for this book involves field visits and investigations in many of the world's realms. This is expressed most concretely in the numerous field notes and photographs supplied by the authors throughout the book. For the geographer, there is no substitute for *being there*.

Geographic Knowledge, Global Awareness

The book before you is an information highway to geographic knowledge and global awareness. Universities and colleges prepare their students to become critical thinkers, creative problem solvers, and, increasingly, global citizens. If we are now living in a "global village," then students must get to know the world as they know their own local areas. But while the influence and interests of the United States reach across the globe, it is no secret that geographic illiteracy at home remains widespread. Much of what students encounter in this book is of immediate, practical value to them—as citizens, as consumers, as travelers, as voters, as job-seekers. North America is a geographic realm marked by ever-intensifying global connections and relationships. Those involvements require countless, often instantaneous decisions, whether they relate to international business, media, affairs of state, disaster relief, and many other spheres. Such decisions must be based on the best possible knowledge of the world beyond our borders. That understanding can be gained by studying the layout of our world, its environments, societies, traditions, resources, policies, development strategies, and other properties—in short, its **regional geography**.

The New Author Team

REGIONS carries the lifelong imprint of Harm J. de Blij, our founding author, mentor, faculty colleague, and cherished friend. Harm's passing in 2014 brought some necessary changes, but this book will always radiate his passionate belief we all share: that geography is fundamental to our understanding of the world and to our efforts to shape and sustain it. Until a few weeks before his death, Harm was heavily involved in discussions about the newly formed author team and about the ongoing digital transformation of this book. And he was highly supportive of the planned changes for this latest edition.

Jan Nijman and Peter Muller, Harm's long-time coauthors, are the lead authors of this revision. They are joined by contributing authors Ann-Margaret Esnard, Richard Grant, and Michael Shin. Ann-Margaret is a Professor at the Andrew Young School of Policy Studies at Georgia State University. With her international expertise in regional planning (especially international disaster planning and management), she is ideally positioned to contribute a new and important feature entitled *Regional Planning Cases*. Next, Richard Grant is responsible for the revision of the chapter on Subsaharan Africa. A Professor of Geography and Regional Studies at the University of Miami, he is one of the discipline's foremost authorities on the geography of contemporary Subsaharan Africa. Richard has more than two decades of field research experience in Southern and West Africa and is affiliated with universities in Ghana and South Africa. Michael Shin contributes two of the new text features: *Technology & Geography* and *Map Analysis*. He also provides general expertise and support regarding the use of digital geographic data and cutting-edge methods of interactive learning. Michael is a highly valued teacher of GIS and possesses the rare ability to combine high-level technical skills with in-depth thematic knowledge (in regional political geography). We are delighted to have Ann-Margaret, Richard, and Michael on our author team to enrich the book with their collective expertise and insights.

REGIONS: Print or Digital?

The print version of this book is also available in digital form and we—both authors and publisher—have invested substantially in both formats. We are now well underway in preparing for a future in which users will increasingly opt for the digital version because it facilitates greater interactive learning and incorporates an ever-expanding array of online connections to monitor real-time changes in the world around us. At this time, we want to be sure that you have a choice and that both products, print and digital, are at the leading edge of world regional geography. This revision of the print book has, in fact, benefited notably from our work on the digital version; for instance, the new *Map Analysis* feature in print (elaborated below) is a direct outcome of these efforts. We are convinced that you will be pleased with this new edition, and we also encourage you to explore the digital version to see which option best suits your needs and the needs of your students.

New Features:
Map Analysis

Our book is well known for the quality and currency of its maps, indispensable tools for understanding and resolving geographic questions. We have now gone a step further and added a new feature in each chapter that actively draws students "inside" the map, challenging them to utilize and develop their analytical skills in map interpretation. These new maps typically present the spatial distribution of one or more variables tied to concepts highlighted in that chapter of the text; the accompanying map caption provides pertinent information and asks one or two open-ended analytical questions, usually related to the mapped spatial patterns and their possible explanations. Since the questions are open-ended and we do not supply any answers, they are well suited for classroom discussion or short writing assignments.

For example, the South Asia chapter discusses the concept of megacities, with a *Map Analysis* focused on comparing the size of the urbanized areas centered by Dhaka, Bangladesh and Los Angeles, California (see Box 8-5). The questions posed to student readers concern their thoughts as to why the areal sizes are so different (L.A. is about 17 times larger than Dhaka), even though their population sizes are roughly equal (ca. 15 million). The answers involve notions of density that are, in turn, related to factors such as mobility, modes of transportation, housing conditions, and income levels.

We know how challenging it can be to engage students with maps in a meaningful way in an introductory course, especially in getting them to *read* maps. We believe the new *Map Analysis* feature offers a helpful mechanism for achieving one of the central learning objectives in any regional geography course.

If you are interested in exploring the digital version of this book, be sure to examine the new *Geographic Information Analysis* modules—the online counterparts to the *Map Analysis* feature in this print book. They are specifically designed to advance geographical knowledge, to increase spatial awareness, and to engage students with interactive, GIS-based spatial analysis. A set of questions—both multiple choice and discussion-oriented—accompanies each of the modules to facilitate student learning, progress, and assessment.

Technology & Geography

Technological advances have always influenced geography, especially when they involve improvements in transportation and communication—from railroads to the telegraph, from air travel to the Internet. This new feature highlights technologies that either have a direct impact on the regions under discussion, or technologies that add significantly to the toolkit of geographers as they investigate these regions. Examples of the first category of technologies include nuclear icebreakers in the Arctic (Chapter 5), driverless cars in North America (Chapter 1), and desalination methods to produce drinking water in the Arabian Peninsula (Chapter 6). Examples of technologies that allow geographers to conduct new kinds of regional analyses include satellite-based remote sensing in the measurement of deforestation in South America (Chapter 3), and *LiDAR* in the mapping of archeological sites in Cambodia (Chapter 10). This is an important new feature because it underscores the dynamic nature of geography in response to technological change. Make no mistake: geography is anything but static. Regions change, and so do the techniques used by geographers at work–especially during today's transformative digital age.

Regional Planning Case

The third new feature in this Seventeenth Edition is the *Regional Planning Case*. Geography is often defined in terms of the interaction between humans and their environment—where the environment influences social behavior *and* where human activities can reshape that environment. Regional planning, as practiced by sovereign states, subnational governments, and other agencies around the world, represents a fundamental effort to shape or reshape the environment. For that reason, regional planning is an intrinsic part of geography and commonly undertaken by professionals who have had geographic training. With this new feature we pay special attention to regional planning projects in each realm that have major geographic ramifications. A planning case can cover an international region (e.g., the China-Pakistan Economic Corridor), a national region (e.g., Saudi Arabia's Economic Cities), or a single urban area (e.g., earthquake recovery in Christchurch, New Zealand). In each instance, we explain the planning goals, the scope and scale of the plan, the challenges and consequences, and likely outcomes. The importance of this new feature lies in the recognition of the practitioner's perspective: as geographers we not only seek to better understand the world around us—we are also engaged in applying our knowledge to actively change it.

A New Organizational Structure

Organizationally, this Seventeenth Edition differs from its predecessor in four ways. First, it is both shorter and more concise. We are keenly aware of the persistent challenge to instructors and students of World Regional Geography trying to cover all of the world's realms in a single course. We have judiciously condensed the text by about 10 percent without compromising the explanation of the most essential regional topics and concepts. Indeed, notwithstanding this reduction in overall length, we have expanded the number of concepts while adding the new features described above.

The second difference involves the reordering of the chapters that follow the Introduction. Previous editions began with Europe, followed by Russia, then turned to the Americas, and so on. Of course, there is no self-evident sequence for the presentation of the world's realms. Users should always feel free to choose their own sequence, and we have maintained the flexibility of the book to fit every possible ordering scheme. Nonetheless, we decided to make a major change and begin the regional chapters in this edition with North America—because that is where most of us are based; because, overall, it presents the most familiar terrain; and because it is from here that we view and develop perspectives on the rest of the world. Starting with North America, then, is not to privilege that realm but rather to acknowledge our own vantage point for this survey of world regions. Following North America (Chapter 1), the succession of chapters is largely based on proximity: the second regional chapter covers Middle America; the third focuses on South America. From there, we move to Europe (Chapter 4); next to Russia/Central Asia (5); and then on to North Africa/Southwest Asia (6), which is followed by Subsaharan Africa (7). The final five chapters—South Asia (8), East Asia (9), Southeast Asia (10), the Austral Realm (11), and the Pacific Realm/Polar Futures (12)—follow the same order used in the Sixteenth Edition. It might therefore be useful to closely examine this new organization for the purposes of course planning.

Third, the world's ongoing geographic transformations have led us to reconsider some critical boundaries that separate certain realms. In this edition, we pay more attention to *transition zones* between realms in general—and we also saw the need to redraw a few boundaries. The chapter on the Russian realm now includes the additional region of Central Asia, which as

Turkestan was previously allocated to the North Africa/Southwest Asia realm. At the same time, we have shifted Afghanistan to the South Asian realm from its previous positioning within the North Africa/Southwest Asia realm. The reasons for these changes are elaborated in the appropriate chapters, but they reflect the changing regional dynamics that operate in various parts of the world. Whereas the world's geography is still anchored to a relatively stable set of realms, at a finer scale—often along the edges of realms—pieces of the regional puzzle are shifting.

Finally, the Seventeenth Edition restores the use of single chapters for each world geographic realm. Thus we have moved away from subdividing realms into A and B chapters, a result of our streamlining the text. Internally, however, each regional chapter preserves our practice of focusing first on large-scale, realmwide geographies, and then turning to the more detailed coverage of each realm's constituent regions, often breaking them down further as needed into subregions and/or individual countries.

Highlights of Newly Added Content

Since the Sixteenth Edition of this book appeared in 2014, the world has undergone some momentous changes, all of which have important geographic dimensions. Some of the most profound changes of the mid-2010s involve the European, Russian, East Asian, and North African/Southwest Asian realms. Elsewhere, too, geographies are being reshaped by geopolitical, environmental, economic, and/or social forces—from the Arctic Basin to Myanmar and from the Panama Canal to the South China Sea.

Occasionally, changes are so rapid and transformative that they force us to rethink the fundamentals of particular realms, regions, and countries—or even of the global system that binds them together. In the mid-2010s, cogent examples include the massive refugee drama that engulfed Europe (see cover photo and caption); Russia's revanchist policies in its Near Abroad (especially in Ukraine); China's relentlessly expanding global presence; and, of course, the deepening conflicts in North Africa/Southwest Asia (e.g., Syria). This Seventeenth Edition not only contains myriad updates but also, where necessary, provides reappraisals of the fundamental nature of the world's ever-changing geography.

No summary can adequately encapsulate everything this new edition contains, but here are chapter-by-chapter highlights of the new content:

The **Introduction**, as usual, provides the foundation and sets the stage for the regional chapters, discussing a broad range of basic geographic concepts along with a set of cornerstone world maps. It includes a new section on the concepts of sustainable development and inclusive development, and an expanded section on global refugee and migration flows. The newly featured *Map Analyses* focus on the critical relationship between economic growth and sustainable development, and on the global geography of capital punishment. The new *Technology & Geography* feature explains the mapping of carbon dioxide in the Earth's atmosphere.

Chapter 1, North America, has a new section and map on the U.S.-Mexican border that simultaneously functions as a sharp boundary and a transition zone. Also look for new text on high-technology clusters and a new map on U.S. agriculture. The *Regional Planning Case* deals with the upgrading of the Port of Savannah in anticipation of larger ("neo-Panamax") ships coming through the enlarged Panama Canal, and the *Technology & Geography* feature discusses the future of driverless cars. Look for *Map Analyses* on urban inequality and presidential elections.

Chapter 2, Middle America, notes the beginning of a new era in United States-Cuba relations, the implications of the enlarged Panama Canal that opened in 2016, and the apparent diversification—from exclusively Mexico to incorporating the Caribbean—of U.S.-bound routes for smuggled cocaine and other illicit drugs from South America. The *Map Analyses* concentrate on the geography of drug-related homicides in Mexico and on the role of connectivity in the economic development of the Central America. The planning of a new interocean canal across Nicaragua that began in 2016, underwritten by China, represents one of the biggest and most expensive planning projects the world has ever seen—if indeed it can be built.

Chapter 3, South America, contains a new section on the waning of the commodity boom that affects so much of this realm, and a section on Colombia's recovery from years of civil strife and economic near-paralysis (including the remarkable revival of the city of Medellín). The *Map Analyses* in this chapter treat the geographies of inequality and the geography of wine. Look for a new map of the intensifying economic development of the Amazon Basin.

Chapter 4, Europe, discusses the human drama of the refugee crisis and the enormous challenges it poses for Europe, the EU, and individual European countries. Look for several new maps on refugee flows and asylum-seekers across the realm, and for a new section on Brexit, the rather unexpected outcome of a 2016 referendum in the United Kingdom to leave the EU. The Ukrainian crisis, too, receives considerable attention and is the focus of one of the *Map Analyses* in this chapter. Also look for new material on Islamic jihadist terrorism in Belgium and France. The *Technology & Geography* feature explains the basics of GIS, and the *Regional Planning Case* explores Amsterdam's new subway, in a city that lies below sea level.

Chapter 5, Russia/Central Asia, emphasizes the increasingly revanchist policies of the Putin regime toward the "Near Abroad" (extending from Ukraine to the Transcaucasian Transition Zone to Central Asia). There is a new section and map on Russia's power as a leading oil and gas supplier of Europe, and a major new section has been added on Central Asia—where Russia and China compete for influence, and the Chinese are busily planning their "New Silk Road" connection to Europe

through Russia's "backyard." Kazakhstan's new (and affluent) capital of Astana is showcased in the *Regional Planning Case*, and the *Map Analyses* highlight Russia's population decline as well as the intricate political geographies of Transcaucasia.

Chapter 6, North Africa/Southwest Asia (NASWA), takes a close look at the anticlimactic aftermath of the 2011 "Arab Spring" movement and the dislocation and turmoil it triggered in various regions and countries from Syria to Somalia, from Libya to Yemen. There are several new maps: one details the surging trans-Mediterranean flow of undocumented immigrants from NASWA to Europe and the refugee crisis it spawned in Southern Europe; another shows the territorial conflicts involving ISIS (Islamic State) across Syria and Iraq through mid-2016. Note, too, that this chapter now ends with the regional discussion of the crucial African Transition Zone (previously part of the chapter on Subsaharan Africa). One of the *Map Analyses* focuses on the relationship between economic prosperity and democracy.

Chapter 7, Subsaharan Africa, bears the imprint of our new contributing author, Africa specialist Richard Grant. It contains several new field notes, and documents the Ebola crisis of 2014-2015. The *Map Analysis* feature in this chapter addresses the challenges of communication and infrastructure in this enormous postcolonial realm. There are new maps on patterns of economic growth and the spatial distribution of Chinese investments across the realm, both linked to an analysis of the heavy reliance on raw materials. Also look for new sections on "mobile money" in Subsaharan Africa and plans to create an ultramodern urban hub (Eko-Atlantic City) amidst the chaotic sprawl on the outskirts of Lagos, Nigeria.

Chapter 8, South Asia, now incorporates Afghanistan as a transition zone on the realm's western flank. That new section includes a map of the reconfigured West region of South Asia, highlighting the critical connections between Afghanistan and Pakistan—that, in turn, are crucial to relations between Pakistan and India. The *Technology & Geography* feature explores the use of U.S. military drones in far-off conflict zones. Another new section discusses extreme weather events in India and Bangladesh in the context of ongoing global climate change. In the *Regional Planning Case*, see the new map showing the China-Pakistan Economic Corridor. One of the *Map Analyses* explains changing patterns in the gender imbalance across India. Also look for new material on Indian Ocean geopolitics, population dynamics, and megacities.

Chapter 9, East Asia, introduces many new field notes and maps to reflect China's westward march of both economic development and urbanization (also the focus of one of the *Map Analyses* in this chapter). Look for a new section on China's massive water diversion project to help quench the thirst of the urban north, and for an updated discussion of geopolitical controversies in the East China Sea. See the new maps that compare China's rapid expansion of its high-speed rail (HSR) system with the almost nonexistent development of HSR in the United States. The second *Map Analysis* in this chapter spotlights the lack of economic development in North Korea.

Chapter 10, Southeast Asia, contains new sections and maps on the Mekong River Basin and the astonishing proliferation of dams along the main stream and its tributaries; on the regional impacts of Myanmar's liberalization (vis-à-vis India, Thailand, and China); on the geopolitical confrontation over Chinese maritime territorial claims in the South China Sea; on China's singular dominance in this realm's international trade; and on the intensifying persecution of Myanmar's Muslim minority—the Rohingyas—thousands of whom have fled the country in a desperate search for safer havens in Muslim Malaysia and Indonesia. The *Regional Planning Case* focuses on the development of the new port city of Dawei on the Indian Ocean that could have a major impact on Myanmar's position in Southeast Asia.

Chapter 11, Austral Realm, includes an expanded section on Australia's environmental challenges, especially with regard to climate change and growing water shortages. Look for a new map that shows changes in precipitation and the distribution of desalination plants along segments of the country's coastline. The *Map Analysis* in this chapter addresses the geography of Australia's Aboriginal population.

Chapter 12, The Pacific Realm and Polar Futures, in addition to new field notes and photos, contains an updated section on the accelerated melting of the Arctic ice cap and its consequences for competing territorial claims in the Arctic Basin. The *Map Analysis* centers on rival claims by Russia and Denmark, and on the implications of the shrinking ice cap for the navigational potential of both the Northeast and Northwest Passages. Also see new boxes on the Tsunami Warning System of the Pacific Ocean (with accompanying map) and adaptation to rising sea level in low-island Kiribati.

Ancillaries

A broad spectrum of print and electronic ancillaries are available to support instructors:

Test Bank—Prepared by Travis Bradshaw of Liberty University. Includes over 1200 multiple-choice, true/false, and fill-in-the-blank short-answer questions.

PowerPoint Lecture Slides—Prepared by Michele Barnaby of Pittsburg State University. These slides highlight key chapter topics to help reinforce students' grasp of essential concepts.

PowerPoint Slides with Text Images—Images, maps, and figures from the text are available in PowerPoint format. Instructors may use these images to customize their presentations and to provide additional visual support for quizzes and exams.

Clicker Questions—Prepared by Tama Nunnelley of the University of North Alabama. A bank of questions is available for instructors who utilize personal-response-system technology in their courses.

Instructor's Media Guide—Includes information about the various media resources (videos, animations, interactive maps) available for use with this text and offers tips and suggestions on how to use these resources in your course, in conjunction with WileyPLUS Learning Space and Orion.

Acknowledgments

Over the 46 years since the publication of the First Edition of *Geography: Realms, Regions, and Concepts*, we have been fortunate to receive advice and assistance from literally thousands of people. One of the rewards associated with the publication of a book of this kind is the steady stream of correspondence and other feedback it generates. Geographers, economists, political scientists, education specialists, and others have written us, often with fascinating enclosures. We still make it a point to respond personally to every such email, and our editors have communicated with many of our correspondents as well. Moreover, we have considered every suggestion made, and many who wrote or transmitted their reactions through other channels will see their recommendations in print in this edition.

Student Response

A major part of the correspondence we receive comes from student readers. We would like to take this opportunity to extend our deep appreciation to the several million students around the world who have studied from our books. In particular, we thank the students from more than 150 different colleges across the United States who took the time to send us their opinions. Students told us they found the maps and graphics attractive and functional. We have not only enhanced the map program with exhaustive updating but have added a number of new maps to this Seventeenth Edition as well as making significant changes in many others. Generally, students have told us that they found the pedagogical devices quite useful. We have kept the study aids the students cited as effective: a boxed list of each chapter's key concepts, ideas, and terms (numbered for quick reference in the text itself); a box that summarizes each realm's major geographic qualities; a chapter-end box of thought questions entitled Points to Ponder: and an extensive and still-expanding Glossary.

Faculty Feedback

In developing the Seventeenth Edition, we are indebted to the following people, including several faculty colleagues, for advising us on numerous matters:

Thomas L. Bell, *University of Tennessee, Knoxville*
Jason B. Greenberg, *Sullivan University (Kentucky)*
Margaret Gripshover, *Western Kentucky University*
Chris Hanson, *University of Miami*
J. Miguel Kanai, *University of Sheffield (UK)*
Max Lu, *Kansas State University*

Ian Maclachlan, *University of Lethbridge (Alberta)*
Imelda Moïse, *University of Miami*
Veronica Mormino, *William Rainey Harper College (Illinois)*
Clifton W. Pannell, *University of Georgia*
Bimal K. Paul, *Kansas State University*
Elvira Maria Restrepo, *University of Miami*
Shouraseni Sen Roy, *University of Miami*
Ira Sheskin, *University of Miami*
Justin Stoler, *University of Miami*
Diana Ter-Ghazaryan, *University of Miami*

In addition, many colleagues and other scholars from around the world assisted us with earlier editions, and their contributions continue to grace the pages of this book. Among them are:

James P. Allen, *California State University, Northridge*
Stephen S. Birdsall, *University of North Carolina*
Bilal Butt, *University of Wisconsin-Madison*
Leo Dillon, *U.S. Department of State*
J. Douglas Eyre, *University of North Carolina*
Charles Fahrer, *Georgia College and State University*
Fang Yong-Ming, *Shanghai, China*
Edward J. Fernald, *Florida State University*
Ray Henkel, *Arizona State University*
Richard C. Jones, *University of Texas at San Antonio*
Gil Latz, *Indiana University-Purdue University, Indianapolis*
Ian Maclachlan, *University of Lethbridge (Alberta)*
Melinda S. Meade, *University of North Carolina*
Henry N. Michael, *Temple University (Pennsylvania)*
David B. Miller, *National Geographic Society*
Eugene J. Palka, *U.S. Military Academy (New York)*
Charles Pirtle, *Georgetown University (D.C.)*
J. R. Victor Prescott, *University of Melbourne (Australia)*
Mika Roinila, *SUNY College at New Paltz*
Rinku Roy Chowdhury, *Clark University (Massachusetts)*
John D. Stephens, *University of Washington*
George E. Stuart, *North Carolina*
Canute Vander Meer, *University of Vermont*
Jack Weatherford, *Macalester College*
Henry T. Wright, *University of Michigan*

Faculty members from a large number of North American colleges and universities continue to supply us with vital feedback and much-appreciated advice. Our publishers arranged several feedback sessions, and we are most grateful to the following professors for showing us where the text could be strengthened and made more precise:

Martin Arford, *Saginaw Valley State University (Michigan)*
Donna Arkowski, *Pikes Peak Community College (Colorado)*
Greg Atkinson, *Tarleton State University (Texas)*
Christopher Badurek, *Appalachian State University (North Carolina)*
Denis Bekaert, *Middle Tennessee State University*
Donald J. Berg, *South Dakota State University*
Jill (Alice) Black, *Missouri State University*
Kathleen Braden, *Seattle Pacific University*
David Cochran, *University of Southern Mississippi*
Joseph Cook, *Wayne County Community College (Michigan)*
Deborah Corcoran, *Southwest Missouri State University*
Marcelo Cruz, *University of Wisconsin at Green Bay*
William V. Davidson, *Louisiana State University*

Larry Scott Deaner, *Kansas State University*
Jeff de Grave, *University of Wisconsin, Eau Claire*
Jason Dittmer, *Georgia Southern University*
James Doerner, *University of Northern Colorado*
Steven Driever, *University of Missouri-Kansas City*
Elizabeth Dudley-Murphy, *University of Utah*
Dennis Ehrhardt, *University of Louisiana-Lafayette*
Bryant Evans, *Houston Community College*
William Flynn, *Oklahoma State University*
William Forbes, *Stephen F. Austin State University (Texas)*
Bill Foreman, *Oklahoma City Community College*
Eric Fournier, *Samford University (Alabama)*
Gary A. Fuller, *University of Hawai'i*
Randy Gabrys Alexson, *University of Wisconsin-Superior*
William Garbarino, *Community College of Allegheny County (Pennsylvania)*
Hari Garbharran, *Middle Tennessee State University*
Chad Garick, *Jones County Junior College (Mississippi)*
Jon Goss, *University of Hawai'i*
Debra Graham, *Messiah College (Pennsyluania)*
Sara Harris, *Neosho County Community College (Kansas)*
John Havir, *Ashland Community & Technical College (Kentucky)*
John Hickey, *Inver Hills Community College (Minnesota)*
Shirlena Huang, *National University of Singapore*
Ingrid Johnson, *Towson State University (Maryland)*
Kris Jones, *Saddleback College (California)*
Uwe Kackstaetter, *Front Range Community College, Westminster (Colorado)*
Cub Kahn, *Oregon State University*
Thomas Karwoski, *Anne Arundel Community College (Maryland)*
Robert Kerr, *University of Central Oklahoma*
Eric Keys, *University of Florida*
Jack Kinworthy, *Concordia University-Nebraska*
Marti Klein, *Saddleback College (California)*
Christopher Laingen, *Kansas State University*
Heidi Lannon, *Santa Fe College (Florida)*
Unna Lassiter, *Stephen F. Austin State University (Texas)*
Richard Lisichenko, *Fort Hays State University (Kansas)*
Catherine Lockwood, *Chadron State College (Nebraska)*
George Lonberger, *Georgia Perimeter College*
Claude Major, *Stratford Career Institute (Quebec)*
Steve Matchak, *Salem State College (Massachusetts)*
Chris Mayda, *Eastern Michigan University*
Trina Medley, *Oklahoma City Community College*
Dalton W. Miller, Jr., *Mississippi State University*
Ernando F. Minghine, *Wayne State University (Michigan)*
Tom Mueller, *California University (Pennsylvania)*
Irene Naesse, *Orange Coast College (California)*
Valiant C. Norman, *Lexington Community College (Kentucky)*
Richard Olmo, *University of Guam*
J. L. Pasztor, *Delta College (Michigan)*
Iwona Petruczynik, *Mercyhurst College (Pennsylvania)*
Paul E. Phillips, *Fort Hays State University (Kansas)*
Rosann Poltrone, *Arapahoe Community College (Colorado)*
Jeff Popke, *East Carolina University*
David Privette, *Central Piedmont Community College (North Carolina)*
Joel Quam, *College of DuPage (Illinois)*
Rhonda Reagan, *Blinn College (Texas)*
Paul Rollinson, *Missouri State University*
A. L. Rydant, *Keene State College (New Hampshire)*

Justin Scheidt, *Ferris State University (Michigan)*
Kathleen Schroeder, *Appalachian State University (North Carolina)*
Nancy Shirley, *Southern Connecticut State University*
Dmitrii Sidorov, *California State University, Long Beach*
Dean Sinclair, *Northwestern State University (Louisiana)*
Richard Slease, *Oakland, North Carolina*
Susan Slowey, *Blinn College (Texas)*
Dean B. Stone, *Scott Community College (Iowa)*
Jamie Strickland, *University of North Carolina at Charlotte*
Ruthine Tidwell, *Florida Community College at Jacksonville*
Irina Vakulenko, *Collin County Community College (Texas)*
Cathy Weidman, *Austin, Texas*
Kirk White, *York College of Pennsylvania*
Thomas Whitmore, *University of North Carolina*
Keith Yearman, *College of DuPage (Illinois)*
Laura Zeeman, *Red Rocks Community College (Colorado)*
Yu Zho, *Vassar College (New York)*

Personal Appreciation

We are privileged to work with a team of professionals at John Wiley & Sons that is unsurpassed in the college textbook publishing industry. As authors we are aware of these talents on a daily basis during the crucial production stage, masterminded by the skillful coordination and leadership of our Sponsoring Editor, Jennifer Yee. Others who played a leading role in this process were Senior Photo Editor Billy Ray, Senior Designer Wendy Lai, and cartographers Don Larson and Terry Bush of Mapping Specialists, Ltd. in Fitchburg, Wisconsin. We are also most fortunate to be guided in our work by Executive Editor Jessica Fiorillo, whose many talents have positively impacted every aspect of this durable publication enterprise; Jess is not only a delight to work with, but has proven to be a wise and bold leader as *Regions* continues to evolve to meet the needs of higher education's new digital era. We also salute the efforts of Marketing Manager Christine Kushner (and welcome her successor, Alan Halfen), Production Editor John Du Val, Associate Development Editor Mallory Fryc, and Market Solutions Assistants Lindsey Myers and Esther Kamar. Beyond this immediate circle, we acknowledge the support and encouragement we receive from many others at Wiley, most notably Vice President and Director, Petra Recter.

We also want to express our continuing appreciation and admiration for the outstanding efforts of Jeanine Furino, one of the most gifted production editors we have ever worked with. Her organizational skills, extraordinary knowledge, technical prowess, unerring professional instincts, and marvelous ability to work smoothly with every collaborator, truly make her an indispensable team leader in the demanding process of fashioning this attractive volume out of a mountain of text and graphic files, email attachments, design layouts, and seemingly endless rounds of preliminary pages. We also are much in Jeanine's debt for continuing to usher us into the fast-changing digital age of book publishing.

Finally, above all else, we express our gratitude to our families for yet again seeing us through the constantly challenging schedule of creating this latest edition of *Regions*.

JAN NIJMAN
Atlanta. Georgia

PETER O. MULLER
Coral Gables, Florida

September 6, 2016

Brief Contents

Contents

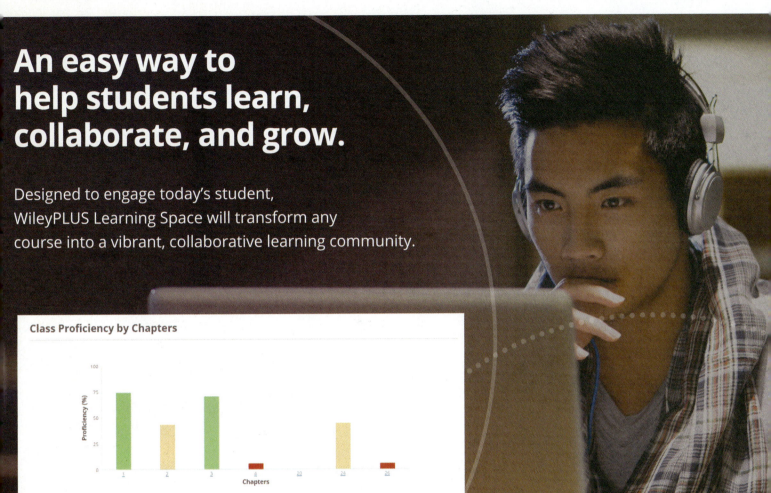

Geography: Realms, Regions, and Concepts

Seventeenth Edition

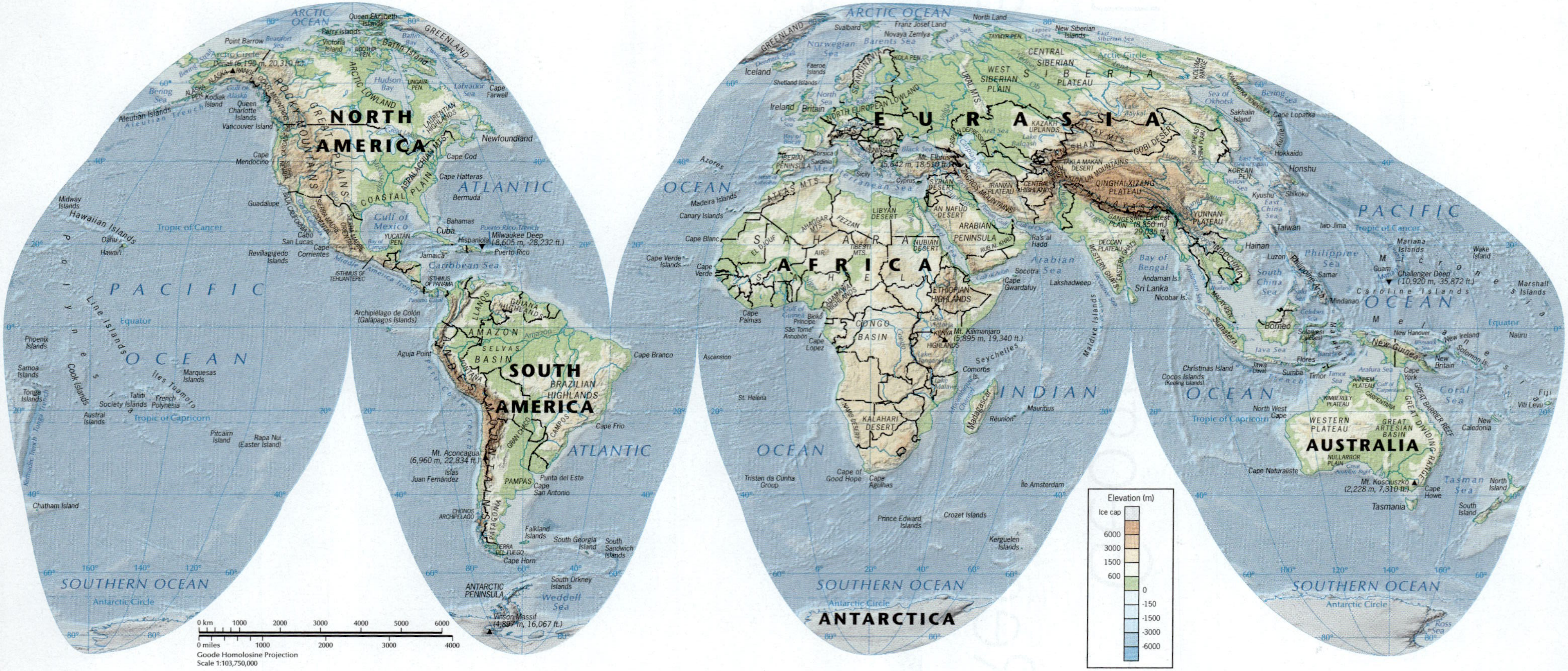

FIGURE G-1

Nature and culture, entwined in geography. The Tuscany region in the northwestern corner of peninsular Italy contains some of the most elegant cultural landscapes in the world with its rolling hills, contoured grain fields, olive groves, and vineyards.

© Jan Nijman

World Regional Geography: A Global Perspective

IN THIS CHAPTER

- The power of maps
- The spatial order of the world
- Persistent cultural diversity
- Geopolitics and the specter of terrorism
- Technology and geography
- The challenge of sustainable development
- Widening global inequalities

CONCEPTS, IDEAS, AND TERMS

Mental maps [1]	Continental drift [14]	European state model [27]
Spatial perspective [2]	Tectonic plate [15]	Geopolitics [28]
Scale [3]	Pacific Ring of Fire [16]	Development [29]
Geographic realm [4]	Climate [17]	Sustainable development [30]
Transition zone [5]	Holocene [18]	Inclusive development [31]
Region [6]	Global climate change [19]	Global core [32]
Absolute location [7]	Geospatial data [20]	Global periphery [33]
Relative location [8]	Population distribution [21]	Core-periphery relationship [34]
Formal region [9]	Urbanization [22]	Core area [35]
Spatial system [10]	Megacity [23]	Globalization [36]
Hinterland [11]	Cultural landscape [24]	Migration [37]
Functional region [12]	State [25]	
Natural landscape [13]	Sovereignty [26]	

What are your expectations as you open this book? You have signed up for a course that will take you around the world to try to understand how it functions today. You will discover how interesting and challenging the discipline of geography is. We hope that this course, and this book, will open new vistas, bring new perspectives, raise your awareness of place, and help you navigate our increasingly complex and often daunting world.

You could not have chosen a better time to study geography. The world is changing on many fronts and is doing so ever faster in response to the rapid advancement of communication technologies. The world truly is "shrinking," and the likelihood has never been greater that your professional career will be taking you to places far from home. As globalization steadily advances, geography becomes ever more important. Moreover, the United States has the world's most globalized economy, with political influence to match. Intentionally or not, the United States is affecting nations and peoples, lives and livelihoods, from pole to pole. That power confers on Americans a responsibility to learn as much as they can about those places, nations, and livelihoods, so that the decisions of their government representatives are well informed. But in this respect, the United States is no superpower. Geographic literacy is a measure of international comprehension and awareness, and Americans' geographic literacy ranks low. That is a liability, for both the United States and the rest of the world, because such geographic fogginess afflicts not only voters but also the representatives they elect, from local school boards to the U.S. Congress.

A World on Maps

Just a casual glance at the pages that follow reveals a difference between this and other textbooks: there are almost as many maps as there are pages. Geography is more closely identified with maps than any other discipline, and we urge you to give as much (or more!) attention to the maps in this book as you do to the text. It is often said that a picture is worth a thousand words, and the same applies to maps. When we write "see Figure XX," we really mean it … and we hope that you will get into the habit. We humans are territorial creatures, and the boundaries that fence off our 200 or so countries reflect our divisive ways. Other, less visible borders—between religions, languages, wealth, and poverty—partition our planet as well. When political and cultural boundaries are at odds, there is nothing like a map to summarize the circumstances. Just look, for example, at the map of the African Transition Zone in Chapter 6: this corridor's turbulence and challenges are steeped in geography.

Maps in Our Minds

All of us carry in our minds maps of what psychologists call our *activity space*: the apartment building or house we live in, the streets nearby, the way to school or workplace, the general layout of our hometown or city. You will know what lane to use when you turn into a shopping mall, or where to park at the movie theater. You can probably draw from memory a pretty good map of your hometown.

These **mental maps** [1] allow you to navigate your activity space with efficiency, predictability, and safety. When you arrived as a first-year student on a college or university campus, a new mental map would have started forming. At first you needed a GPS, online, or hard-copy map to find your way around, but soon you dispensed with that because your mental map became sufficient. And it will continue to improve as your activity space expands.

If a well-formed mental map is useful for decisions in daily life, then an adequate mental map is surely indispensable when it comes to decision making in the wider world. You can give yourself an interesting test. Choose some part of the world beyond North America in which you have an interest or about which you have a strong opinion—for example, Israel, Iran, Russia, North Korea, or China. On a blank piece of paper, draw a map that reflects your impression of the regional layout there: the country, its neighbors, its internal divisions, major cities, water bodies, and so forth. That is your mental map of the place. Put it away for future reference, and try it again at the end of this course. You will have proof of your improved mental-map inventory.

Mapping Revolutions

The maps in this book show larger and smaller parts of the world in various contexts. Some depict political configurations; others display ethnic, cultural, economic, or environmental features. **Cartography** (the making of maps) has undergone a dramatic and continuing technological revolution. Earth-orbiting satellites equipped with remote-sensing technology (special on-board sensors and imaging instruments) transmit remotely sensed information to computers on the surface, recording the expansion of deserts, the shrinking of glaciers, the depletion of forests, the growth of cities, and myriad other geographic phenomena. Earthbound computers possess ever-expanding capabilities not only to organize this information but also to display it graphically. This allows geographers to develop a **geographic information system (GIS)**, bringing geospatial data to a monitor's screen that would have taken months to assemble just three decades ago.

A parallel map revolution is embodied in the astounding proliferation of navigation systems in cars and on mobile phones. Smartphones enable the use of maps on the go, and many of us, in the developed world at least, have already become dependent on them to traverse cities, find a store or restaurant, even to move around shopping malls. Whereas the personalized maps on our smartphones allow us to navigate more efficiently, the maps in this book are aimed at better understanding the world and its constituent parts.

Satellites—even spy satellites—cannot record everything that occurs on the Earth's surface. Sometimes the borders between ethnic groups or cultural sectors can be discerned by satellites—for example, in changing house types or religious shrines—but this kind of information tends to require on-the-ground verification through field research and reporting. No satellite view of Iraq could show you the distribution of Sunni and Shi'ite Muslim adherents. Many of the boundaries you see on the maps in this book cannot be observed from space because long stretches are not even marked on the ground. So the maps you will be "reading" here have their continued uses: they summarize complex situations and allow us to begin forming durable mental maps of the areas they represent.

There is one other point we should make that is especially important when it comes to world maps: never forget that the world is a sphere, and to project it onto a two-dimensional flat surface must necessarily entail some very significant distortions. Try peeling an orange and flattening the entire peel on a surface—you will have to tear it up and try to stretch it in places to get the job done. Take a look at **Figure G-1** and note how the Atlantic Ocean and other segments of the planetary surface are interrupted. You can produce a map like this in many different ways, but you will always end up distorting things. When studying world maps, there is nothing like having a globe at hand to remind you of our three-dimensional reality.

Geography's Perspective

Geography is sometimes described as the most interdisciplinary of disciplines. That is a testimonial to geography's historic linkages to many other fields, ranging from geology to economics and from sociology to environmental science. And, as has been the case so often in the past, geography is in the lead because interdisciplinary studies and research are now more prevalent than ever.

A Spatial Viewpoint

Most disciplines focus on one key theme: economics is about money; political science is about power; psychology is about the mind; biology is about life forms. Geography is about the explanation of space on the Earth's surface. More specifically, geographers are concerned with the organization of ***terrestrial space***. Social space (cities, buildings, political boundaries, etc.) as well as natural space (climates, terrain, vegetation, etc.) are not randomly configured. Instead, there generally prevails a particular order, regularity, even predictability about the ways in which space is organized. Sometimes it is the deliberate work of human beings, and sometimes it is the work of nature, but both produce specific patterns. Geographers consider these spatial patterns and processes as not only interesting but also crucial to how we live and how we organize our societies.

The **spatial perspective [2]** has defined geography from its beginning.

Environment and Society

There is another connection that binds geography and has done so for centuries: an interest in the relationships between human societies and the natural environment. Geography lies at the intersection of the social and natural sciences and integrates perspectives from both, being the only discipline to do so explicitly. This perspective comes into play frequently: environmental modification is in the news on a daily basis in the form of worldwide climate change, but this current surge of global warming is only the latest phase of endless atmospheric and ecological fluctuation. Geographers are involved in understanding current environmental issues not only by considering climate change in the context of the past, but also by looking carefully at the implications of global warming for human societies. Geographers are acutely aware that human beings will always be part of nature, no matter how far technology advances.

More generally, think of this relationship between humans and their environment as a two-way street. On one hand, human beings have always had a transformative effect on their natural surroundings, from the burning of forests to the creation of settlements. On the other hand, humans have always been heavily dependent on the natural environment, their individual and collective behaviors very much a product of it. There are so many examples that it is hard to know where to begin or when to end: we eat what nature provides, and traditional diets vary regionally; rivers allow us to navigate and connect with other peoples—or they serve as natural boundaries like the Rio Grande; wars are fought over access to water or seaports; landlocked countries seem to have different cultures from those of islands; and so on.

Spatial Patterns

Geographers, therefore, need to be conversant with the location and distribution of salient features on the Earth's surface. This includes the natural (physical) world, simplified in Figure G-1, as well as the human world; our inquiry will view these in temporal (historical) as well as spatial perspective. The spatial structure of cities, the layout of farms and fields, the networks of transportation, the configurations of rivers, the patterns of climate—all these form part of our investigation. As you will find, geography employs a comprehensive spatial vocabulary with meaningful terms such as area, distance, direction, clustering, proximity, accessibility, and many others that we will encounter in the pages ahead. For geographers, some of these terms have more specific definitions than is generally assumed. There is a difference, for example, between *area* (surface) and *region*, between *boundary* and *frontier*, and between *place* and *location*. Sometimes, what at first may seem to be simple ideas turn out to be rather complex concepts.

NASA Earth Observatory/NOAANGDC

This map is an assemblage of nighttime satellite images that show the dominance of electric lighting in certain parts of the world and darkness in others. What does this spatial distribution suggest about the geography of development?

Scale and Scope

One very prominent term in the geographic vocabulary is **scale** [3]. Whenever a map is created, it represents all or part of the Earth's surface at a certain level of detail. Obviously, Figure G-1 displays a very low level of detail; it is little more than a general impression of the distribution of land and water as well as lower and higher elevations on our planet's surface. A limited number of prominent features such as the Himalayas and the Sahara are named, but not the Pyrenees Mountains or the Nile Delta. At the bottom of the map you can see that one inch at this scale must represent about 1650 miles of the real world, leaving the cartographer little scope to insert information.

A map such as Figure G-1 is called a *small-scale* map because the ratio between map distance and real-world distance, expressed as a fraction, is very small at 1:103,750,000. Increase that fraction (i.e., zoom in), and you can represent less territory—but also enhance the amount of detail the map can exhibit. In Figure G-2, note how the fraction increases from the smallest (1:103,000,000) to the largest (1:1,000,000). Montreal, Canada, is just a dot on Map A but an urban area on Map D. Does this mean that world maps like Figure G-1 are less useful than larger-scale maps? It all depends on the purpose of the map. In this chapter, we often use world maps to show global distributions as we set the stage for the more detailed discussions to follow. In later chapters, the scale tends to become larger as we focus on smaller areas, even on individual countries and cities. But whenever you read a map, be aware of the scale because it is a guide to its utility.

The importance of the scale concept is not confined to maps. Scale plays a fundamental role in geographic research and in the ways we think about geographic problems—scale in terms of *level of analysis*. This is sometimes referred to as *operational scale*, the scale at which social or natural processes operate or play out. For instance, if you want to investigate the geographic concentration of wealth in the United States, you can do so at a range of scales: within a neighborhood, a city, a county, a State,* or at the national level. You choose the scale that is the most appropriate for your purpose, but it is not always that straightforward. Suppose you had to study patterns of ethnic segregation: what do you think would be the most relevant scale(s)?

In this book, our main purpose is to understand the geography of the world at large and how it works, and so, inevitably, we must deal with broad spatial entities. Our focus is on the world's realms and on the main regions within those realms, and in most cases we will have to forego analyses at a finer scale. For our purposes, it is the big picture that matters most.

World Geographic Realms

Ours is a globalized, interconnected world, a world of international trade and travel, migration and movement, tourism and television, financial flows and Internet traffic. It is a world that, in some contexts, has taken on the properties of a "global village"—but that village still has its neighborhoods. Their names are Europe, South America, Southeast Asia, and others familiar to us all.

---

* Throughout this book we will capitalize State when this term refers to an administrative division of a country: for example, the U.S. State of Ohio or the Australian State of New South Wales. Since this term is also synonymous with country (e.g., the state of Brazil), we use the lower case when referring to such a national state.

EFFECT OF SCALE

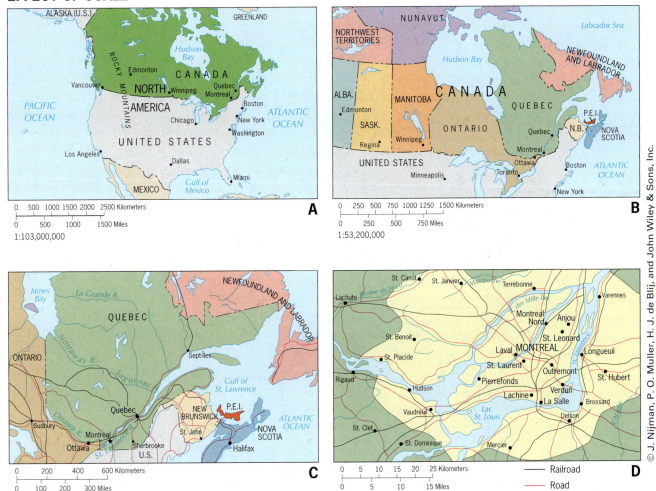

© J. Nijman, P. O. Muller, H. J. de Blij, and John Wiley & Sons, Inc.

Like the neighborhoods of a city or town, these global neighborhoods may not have sharply defined borders, but their persistence, after tens of thousands of years of human dispersal, is beyond doubt. We call such global neighborhoods **geographic realms [4]**. Each of the realms possesses a particular combination of environmental, cultural, and organizational properties.

These blended, characteristic qualities are imprinted on the landscape, giving each realm its own traditional attributes and social settings. As we come to understand the human and environmental makeup of these geographic realms, we learn not only where they are located but also why they are located where they are (a central question in geography), how they are constituted, and what their future is likely to be in our fast-changing world. **Figure G-3**, therefore, forms the overall framework for our investigation in this book.

Criteria for Geographic Realms

The existence and identification of world geographic realms depend on a combination of factors. Our world consists of a highly complex and variable environment of large and small continents, enormous oceans and countless waterways, innumerable islands, diverse habitats and cultures, and intricate political geographies. What constitutes a realm depends on the circumstances, but we can still identify three main sets of criteria that apply to all realms:

- **Physical and Human** Geographic realms are based on sets of spatial criteria. They are the largest units into which the inhabited world can be divided. The criteria on which such a broad regionalization is based include both physical (that is, natural) and human (or social) yardsticks. For instance, South America is a geographic realm because physically it is a continent and culturally it comprises comparable societies. The realm called South Asia, on the other hand, lies on a Eurasian landmass shared by several other geographic realms; high mountains, wide deserts, and dense forests combine with a distinctive social fabric to create this well-defined realm centered on India.

- **Functional** Geographic realms are the result of the interaction of human societies and natural environments, a *functional* interaction revealed by farms, mines, fishing ports, transport routes, dams, bridges, villages, and countless other

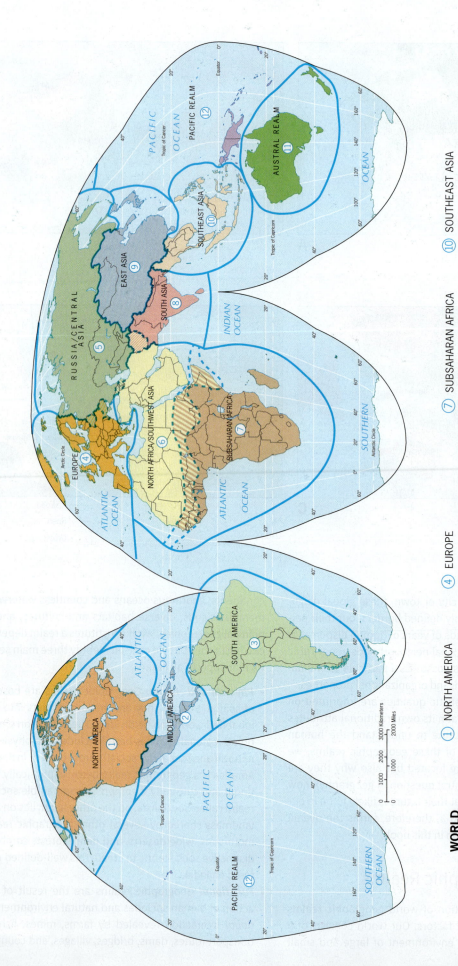

WORLD GEOGRAPHIC REALMS

① NORTH AMERICA
② MIDDLE AMERICA
③ SOUTH AMERICA

④ EUROPE
⑤ RUSSIA/CENTRAL ASIA
⑥ NORTH AFRICA/SOUTHWEST ASIA

⑦ SUBSAHARAN AFRICA
⑧ SOUTH ASIA
⑨ EAST ASIA

⑩ SOUTHEAST ASIA
⑪ AUSTRAL REALM
⑫ PACIFIC REALM

© J. Nijman, P. O. Muller, and John Wiley & Sons, Inc.

FIGURE G-3

features that mark the landscape. According to this criterion, Antarctica is a continent but not a geographic realm.

- ***Historical*** Geographic realms represent the most comprehensive and encompassing definition of the great clusters of humankind in the world today. China lies at the heart of such a cluster, as does India. Most of Africa constitutes a geographic realm from the southern margin of the Sahara (an Arabic word for desert) to the Cape of Good Hope and from its Atlantic to its Indian Ocean shores. Geographic realms are generally defined on the basis of physical features, but they are also the product of historical evolution.

Figure G-3 displays the 12 world geographic realms based on these criteria. As we will show in greater detail later, waters, deserts, and mountains as well as cultural and political shifts mark the borders of these realms. We shall discuss the positioning of these boundaries as we examine each realm.

Delineating Realms: Boundaries and Transition Zones

Oceans and seas are the most common natural boundaries of the world's realms, such as the South Atlantic to Subsaharan Africa's west or the North Atlantic to North America's east. But where two geographic realms meet, **transition zones [5]**, not sharp boundaries, often mark their contacts.

We need only remind ourselves of the border zone between the geographic realm in which most of us live, North America, and the adjacent realm of Middle America. The line in Figure G-3 coincides with the boundary between Mexico and the United States, crosses the Gulf of Mexico, and then separates Florida from Cuba and the Bahamas. But Hispanic influences are strong in North America north of this boundary, and the U.S. economic influence is strong south of it. The line, therefore, represents an ever-changing zone of regional interaction. Again, there are many ties between South Florida and the Bahamas, but the Bahamas resemble a Caribbean more than a North American society. On the other hand, metropolitan Miami has so many ethnic Cuban residents that it is sometimes referred to as the second-largest Cuban city after Havana.

In Africa, the transition zone from Subsaharan to North Africa is so wide and well defined that we have put it on the world map; elsewhere, transition zones tend to be narrower and less easily represented. In the second decade of this century, such countries as Belarus (between Europe and Russia/Central Asia) and Afghanistan (between Southwest Asia and South Asia) lie in inter-realm transition zones. Remember, over much (though not all) of their length, borders between realms are zones of regional change.

Transition zones are fascinating spaces: it is almost as if they rebel against a clear ordering of the world's geography. They remind us that the world is a restless and contested place with shifting boundaries and changing geographic fortunes. They challenge the geographer and, above all, they underscore just how complex the study of geography is.

Geographic Realms: Dynamic Entities

Had we drawn Figure G-3 before Columbus made his voyages beginning in 1492 (and assuming we had the relevant geographical knowledge), the map would have looked different: indigenous states and peoples would have determined the boundaries in the Americas; Australia and New Guinea would have constituted a single realm, and New Zealand would have been part of the Pacific Realm. The colonization, Europeanization, and Westernization of the world changed that map dramatically. Since World War II, the world map has been redrawn as a result of decolonization and the rise and then demise of the Cold War. That Cold War division between western and eastern Europe has now given way to far-reaching European integration across that geographic realm. Realms and regions are dynamic entities, and their geographies are always subject to change.

Two Types of Realms

The world's geographic realms can be divided into two categories. The first are *monocentric* realms that are dominated by a single major political entity, in terms of territory and/or population. North America (United States), Middle America (Mexico), East Asia (China), South Asia (India), Russia/Central Asia, and the Austral Realm (Australia) are all monocentric realms. They are, in their entirety, heavily influenced by the presence of that single country. It is as if the realm is organized around them.

The second type of realm is *polycentric* in nature. In these, the appearance, functioning, and organization of the realm are dispersed among a number of more or less equally influential regions or countries. Europe, North Africa/Southwest Asia, Subsaharan Africa, and the Pacific Realm all fall into this category. Polycentric realms can be more volatile in some ways, their development determined by the sum of many different parts.

Two of the world's realms are a bit more difficult to categorize. Southeast Asia is a dynamic realm that contains almost a dozen countries, some of them regarded as emerging economies. Arguably, Indonesia is becoming the most influential power, but it would be premature to label this a monocentric realm. The other realm that seems to fall in between is South America. Here it is Brazil that has the biggest population and increasingly the largest and most influential economy. South America, more emphatically than Southeast Asia, may be moving toward a monocentric spatial organization.

Of course, some of the dominant powers in the monocentric realms influence events beyond their realm and demonstrate a truly global reach. The United States has dominated world events (though not unchallenged) since World War II, but in recent decades it has had to make way for newly emergent powers such as Japan and Germany. Nowadays, China exercises a major influence not just across East and Southeast Asia but around the globe. Our discussion of the various realms will give due consideration to the influence of global trends and outside powers.

Regions Within Realms

The compartmentalization of the world into geographic realms establishes a broad global framework, but for our purposes a more refined level of spatial classification is needed. This brings us to an important organizing concept in geography: the **region** [6]. To establish regions within geographic realms, we need more specific criteria.

Let us use the North American Realm to demonstrate the regional idea. When we refer to a part of the United States or Canada (e.g., the South, the Midwest, or the Prairie Provinces), we employ a regional concept—not scientifically but as part of everyday communication. We reveal our perception of local or distant space as well as our mental image of the region we are describing.

But what exactly is, for example, the Midwest? How would you draw this region on the North American map? Regions seem easy to imagine and describe, but they can be difficult to outline with precision on a map. You might suggest that, broadly, the Midwest lies between the Ohio River and the Rocky Mountains. Or you might use the borders of States: certain States are part of this region, others are not. Alternatively, you could use agriculture as the principal criterion: the Midwest is where corn and/or soybeans occupy a certain percentage of the farmland. Each method results in a different delimitation; a Midwest based on physical boundaries is different from a Midwest based on State borders or farm production. Therein lies an important principle: regions are devices that allow us to make spatial generalizations, and they are based on criteria to help us construct them. If you were studying the geography of presidential-election politics, then a Midwest region defined by State boundaries would make sense. If you were studying agricultural distributions, you would need a different definition.

Criteria for Regions

Given these different dimensions of the same region, we can identify properties that all regions have in common:

- **Area** To begin with, all regions have *area*. This observation would seem obvious, but there is more to this idea than meets the eye. Regions may be intellectual constructs, but they are not abstractions: they exist in the real world, and they occupy space on the Earth's surface.
- **Boundaries** It follows that regions have *boundaries*. Occasionally, nature itself draws sharp dividing lines, for instance, along the crest of a mountain range or along a coast. More often, regional boundaries are not self-evident, and we must determine them using criteria that we establish for that purpose. For example, to define a cultural region, we may decide that only areas where more than 50 percent of the population belong to a particular religion or speak a certain language qualify to be part of that region.
- **Location** All regions also possess *location*. Often the name of a region contains a locational clue, as in Amazon Basin or

Indochina (a region of Southeast Asia lying between India and China). Geographers refer to the **absolute location** [7] of a place or region by providing the latitudinal and longitudinal extent of the region with respect to the Earth's grid coordinates. A more useful measure is a region's **relative location** [8], that is, its location with reference to other regions. Again, the names of certain regions reveal aspects of their relative locations, as in *Mainland* Southeast Asia and *Equatorial* Africa.

- **Homogeneity** Many regions are marked by a certain *homogeneity* or sameness. Homogeneity may lie in a region's cultural properties, its physical characteristics, or both. Siberia, a vast region of northeastern Russia, is marked by a sparse human population that resides in widely scattered small settlements of similar form, frigid climates, extensive areas of permafrost (permanently frozen subsoil), and cold-adapted vegetation. This dominant uniformity makes it one of Russia's natural and cultural regions, extending from the Ural Mountains in the west to the Pacific Ocean in the east. When regions display a measurable and often visible internal homogeneity, they are called **formal regions [9]**. But not all formal regions are visibly uniform. For instance, a region may be delimited by the area in which, say, 90 percent or more of the people speak a particular language. This cannot be seen in the landscape, but the region is a reality, and we can use that criterion to draw its boundaries accurately. It, too, is a formal region.
- **Regions as Systems** Other regions are marked not by their internal sameness but by their functional integration—that is, by the way they work. These regions are defined as **spatial systems [10]** and are formed by the areal extent of the activities that define them. Take the case of a large city with its surrounding zone of suburbs, urban-fringe countryside, satellite towns, and farms. The city supplies goods and services to this encircling zone, and it buys farm products and other commodities from it. The city is the heart, the **core** of this region, and we call the surrounding zone of interaction the city's **hinterland [11]**. But the city's influence wanes along the outer **periphery** of that hinterland, and there lies the boundary of the functional region of which the city is the focus. A **functional region [12]**, therefore, is usually forged by a structured, urban-centered system of interaction. It has a core and a periphery.

Interconnections

Even if we can easily demarcate a particular region and even if its boundaries are sharp, that does not mean it is isolated from other parts of the realm or even the world. All human-geographic regions are more or less interconnected, being linked to other regions. As we shall see, globalization is forging ongoing integration and connections among regions around the world. Trade, migration, education, television, computer linkages, and other interactions sometimes blur regional identities. Interestingly, globalization tends to have a seemingly paradoxical effect: in some ways, regions and places become more alike, more homogeneous (think of certain consumption patterns), but in other respects the contrasts can become stronger (for example, a reassertion of ethnic or religious identities).

The Physical Setting

Natural Landscapes

The landmasses of Planet Earth present a jumble of **natural landscapes [13]** ranging from rugged mountain chains to smooth coastal plains (Fig. G-1). Certain continents are readily linked with a dominant physical feature—for instance, North America and its Rocky Mountains, South America with its Andes and Amazon Basin, Europe with its Alps, Asia with its Himalaya Mountains and massive river basins, and Africa with its Sahara and Rift Valley complex. Physical features have long influenced human activity and movement. Mountain ranges form barriers to movement but have also channeled the spread of agricultural and technological innovations around them. Large deserts similarly form barriers as do rivers, although rivers also facilitate accessibility and connectivity between people. River basins in Asia still contain several of the planet's largest population concentrations: the advantages of fertile soils and ample water supplies that first enabled clustered human settlement now sustain hundreds of millions in crowded South and East Asia.

As we study each of the world's geographic realms, we will find that physical landscapes continue to play significant roles in this modern world. That is one reason why the study of world regional geography is so important: it puts the human map in environmental as well as regional perspective (see, for instance, **Box G-1**).

Geology and Natural Hazards

Our planet may be 4.5 billion years old, but it is far from placid. As you read this chapter, Earth tremors are shaking the still-thin crust on which we live, volcanoes are erupting, storms are raging. Even the continents themselves are moving measurably, pulling apart in some areas, colliding in others. Hundreds of thousands of human lives are lost to natural calamities of this sort in almost every decade.

More than a century ago, the German geographer Alfred Wegener used spatial analysis to explain something that is obvious even from a small-scale map like Figure G-1: the apparent jigsaw-like fit of the landmasses, especially across the South Atlantic Ocean. His theory of **continental drift [14]** held that the landmasses on the map were actually pieces of a supercontinent, *Pangaea*, that existed hundreds of millions of years ago. Today we know that the continents are "rafts" of relatively light rock that rest on slabs of heavier rock known as **tectonic plates [15]** whose movement is propelled by giant circulation cells in the red-hot magma below (when this

---

Box G-1 From the Field Notes …

"On the descent into Tibet's Lhasa Gongga Airport, I have a great view of the Yarlung Zangbo Valley, its braided stream channels gently flowing toward the distant east. The Yarlung Zangbo is the highest major river on Earth, running from the Tibetan Plateau into northeastern India where it joins the mighty Brahmaputra River that continues on to Bangladesh where it empties into the Indian Ocean. It was mid-October and the water levels were low. The landing strip of the airport can be seen in the center-right of the photo, on the south bank. The airport is quite far from Lhasa, the Tibetan capital, located about 62 kilometers (40 mi) to its southwest. Despite major road and tunnel construction, it is still more than an hour's drive. The airport had to be built away from the city and in this widest part of the valley because it allows the easiest landings and takeoffs in this especially rugged terrain. It lies at 3700 meters above sea level (12,100 ft), one of the highest airports in the world."

©Jan Nijman

molten magma reaches the surface through volcanic vents, it is called lava).

When moving tectonic plates collide, earthquakes and volcanic eruptions result, and the physical landscape is thrown into spectacular relief. **Figure G-4** shows the outlines of the tectonic plates and the coincidence of earthquakes and volcanic eruptions. The Earth's largest ocean is almost completely encircled by active volcanoes and earthquake epicenters. Appropriately, this geologic frame is known as the **Pacific Ring of Fire [16]**. In comparison, Russia, Europe, Africa, and Australia are relatively safe; in other realms, the risks are far greater in one sector than in others (western as opposed to eastern North and South America, for instance). As we shall discover, for certain parts of the world the activity mapped in Figure G-4 presents a clear and present danger. Some of the world's largest cities, such as Tokyo and Mexico City, lie in zones that are highly vulnerable to sudden disaster—as indeed occurred with Japan's huge 2011 earthquake and tsunamis not very far north of Tokyo.

Climate

The prevailing **climate [17]** constitutes a key factor in the geography of realms and regions (in fact, some regions are essentially defined by climate). But climates change: those dominating in certain regions today may not have prevailed there several thousand years ago. Thus any map of climate, including the maps in this chapter, is but a still-picture of our constantly changing world.

Climatic conditions have swung back and forth for as long as the Earth has had an atmosphere. Imagine this: just 18,000 years ago, great ice sheets had spread all the way south to the Ohio River Valley, covering most of the Midwest; this was the peak of a glaciation that had lasted about 100,000 years (**Fig. G-5**). The Antarctic Ice Sheet was bigger than ever, and even in the tropics great mountain glaciers pushed down valleys and onto plateaus. But then the **Holocene [18]** warming began about 10,000 years ago, the continental and mountain glaciers receded, and ecological zones that had been squeezed between the advancing ice sheets spread north and south. In Europe, particularly, living space expanded and human numbers grew.

Global Climate Change   Today we are living in an era of **global climate change [19]**, particularly natural global warming that has been accelerated by anthropogenic (human-source) causes. Since the Industrial Revolution, we have been emitting gases that have enhanced nature's **greenhouse effect** whereby the sun's radiation becomes trapped in the Earth's atmosphere. This is leading to a series of atmospheric changes, especially the overall warming of the planet (see **Box G-2**). One important international organization of experts, the Intergovernmental Panel on Climate Change (IPCC), predicts an increase of about 2°C (3.6°F) during the course of the twenty-first century for the world at large, but with significant regional variability (e.g., more at higher latitudes, less at lower latitudes). Precipitation patterns are predicted to become

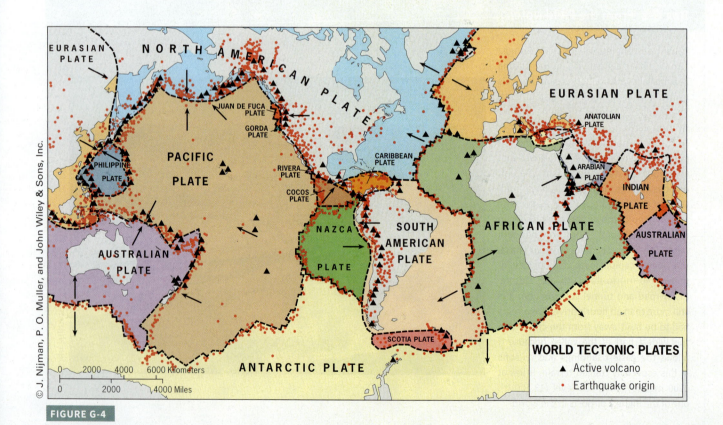

© J. Nijman, P. O. Muller, and John Wiley & Sons, Inc.

FIGURE G-4

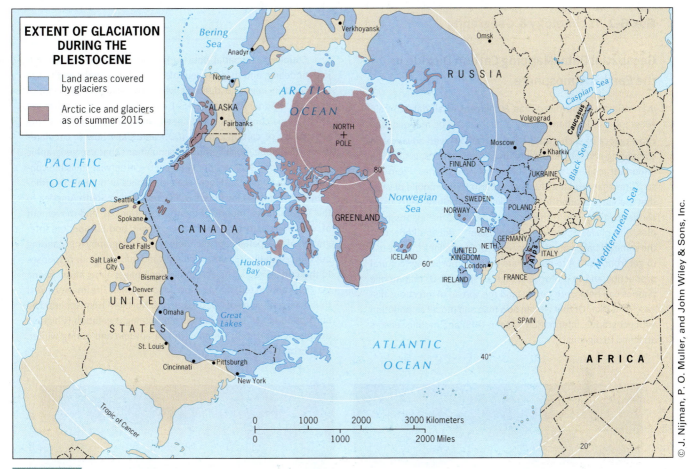

EXTENT OF GLACIATION DURING THE PLEISTOCENE

- Land areas covered by glaciers
- Arctic ice and glaciers as of summer 2015

FIGURE G-5

more variable, particularly in regions where they are already seasonal.

This change in temperature may seem small but is expected to have significant impacts on global climate patterns, agricultural zones, and the quality of human lives. The full ramifications are not known, but scenarios are being modeled so that societies can confront the changes that are coming. Leaders of some countries are more skeptical than others, and some have already made greater adjustments than others. One of the most significant consequences of global climate change is that the icecap atop the Arctic Ocean is melting faster than even recent models predicted, with environmental and geopolitical implications. We pick up this issue in Chapters 5 and 12. Elsewhere, the effects of global warming are evident as well, even in equatorial zones (see **Box G-3**).

Climate Regions

We have just learned how variable climate can be, but in a human lifetime most of us see relatively little evidence of this variability—although this may change as a result of global warming. We talk about the weather (the immediate state of the atmosphere) in a certain place at a given time, but as a technical term *climate* defines the aggregate, total record of weather conditions at a place, or in a region, over the entire period during which records have been kept. Figure G-6 may appear rather complicated, but even a glance at this map

shows its utility. Devised long ago by Wladimir Köppen and later modified by Rudolf Geiger, it represents climatic regions through a combination of colors and letter symbols. In the legend, note that the **A** climates (rose, gold, and peach) are equatorial and tropical; the **B** climates (tan, yellow) are dry; the **C** climates (shades of green) are temperate, that is, moderate and neither hot nor cold; the **D** climates (shades of purple) are cold; the **E** climates (blue) are frigid; and the **H** climates (gray) prevail in highlands like the Himalayas/Tibetan Plateau and the Andes.

A good way to get a sense of how this map works is to start with a climate with which you are familiar. If you live in the southeastern United States (the large green **C** area extending from near the Great Lakes to Florida), you would feel at home—climatically—in the green area in southeastern South America, in southeastern Australia, and in southeastern China. If you live in the **B** zones of the U.S. Southwest, either in the tan (Texas) or the yellow (Arizona) area, conditions in much of Australia, southwestern Africa, or the Middle East would be familiar to you. This can even be taken to the city scale. If you enjoy the climate of San Francisco, you will rediscover it in Santiago (Chile), Cape Town (South Africa), Athens (Greece), and Perth (Australia). Details will differ, of course—higher wind incidence or maybe lower overall humidity—but, in general, the similarities will be greater than such differences. And if you live in Canada, you would be prepared for much of Russia and northeastern China.

Box G-2 Technology & Geography

Geospatial Data: Mapping Carbon Dioxide in the Earth's Atmosphere

It is estimated that more than 80 percent of all data contain geographic or locational information that can be mapped. The convergence of such **geospatial data [20]** with advances in computing and technology are driving innovations, generating important insights, and facilitating new discoveries. From the satellites that orbit and survey the Earth to the mobile devices that we carry around in our pockets, technology helps us to connect with, define, reshape, and understand the regions, peoples, and places of the world. For instance, did you know that the same global positioning systems (GPS) satellites that we use to get from point A to point B are used to improve agricultural production and to protect endangered species? Or that mobile phones can help us to understand the spread of infectious diseases like Zika? Or that we can now map and model air pollution with geographic information systems (GIS) and mobile sensors attached to pigeons?

Coupling geospatial data with recent advances in computing, modeling, and mapping enables us to learn about and to see the world in new and different ways. The National Aeronautics and Space Administration (NASA) compiles an array of data on atmospheric conditions, wind speed, and pollution to model the seasonality of carbon dioxide (CO_2) and show how it circulates across the globe. Such computer simulations that use geographic data and information as inputs reveal the otherwise invisible continental-, regional-, and local-scale dynamics of the Earth's atmosphere. From the daily pulses of CO_2 released in South America to the seasonal (i.e., month to month) variations in, and movement of, emissions across North America, China, and Europe, the integration of geospatial data, technology, and geography is expanding our understanding of climate change and informing our responses to this critical issue. The accompanying images show atmospheric CO_2 (indicated in red) during winter and summer. The images reveal that most emissions of CO_2 occur in the Northern Hemisphere. The absorption of CO_2 by the Earth's blossoming vegetation is highest during the (Northern) summer; in late autumn and during winter, CO_2 absorption is lowest.

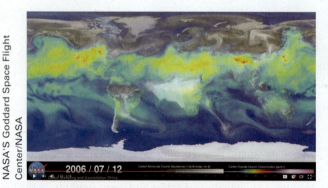
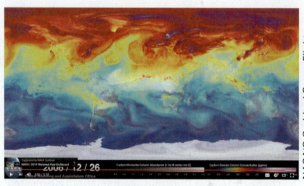

NASA'S Goddard Space Flight Center/NASA

NASA'S Goddard Space Flight Center/NASA

Carbon dioxide in the Earth's atmosphere during the Northern-Hemisphere summer (left) and winter (right). CO_2 is shown in red.
Source: NASA: http://svs.gsfc.nasa.gov/goto?11719

Even at this relatively small geographic scale, Figure G-6 can tell us a great deal about entire climate regions. For instance, among the equatorial/tropical (**Af**) climates, the areas with the darkest (rose) shade get the most rain and it falls year-round, so that this is where tropical rainforests still stand. But in the peach-colored (**Aw**) areas, the rainfall regime is subject to dry seasons, and the vegetation reflects this by its very name: savanna. Much more open space and more widely scattered stands of trees prevail in these tropical grasslands. And a third kind of regime is signified by the **m** after the **A**: for *monsoon*, the annual copious rainy season on which the lives of hundreds of millions of people still depend.

Let us look at the world's climatic regions in greater detail.

Humid Equatorial (A) Climates

The humid equatorial, or tropical, climates are characterized by high temperatures

all year and by heavy precipitation. In the **Af** subtype, the rainfall arrives in substantial amounts every month; but in the **Am** areas, the arrival of the annual wet **monsoon** (the Arabic word for "season" [see Chapter 8]) marks a sudden enormous increase in precipitation. The **Af** subtype is named after the vegetation that develops there—the tropical rainforest. The **Am** subtype, prevailing in part of peninsular India, in a coastal area of West Africa, and in sections of Southeast Asia, is appropriately referred to as the monsoon climate. A third tropical climate, the **savanna (Aw)**, has a wider daily and annual temperature range and a more sharply seasonal distribution of rainfall.

Savanna rainfall totals tend to be lower than those in the rainforest zone, and savanna seasonality is often expressed in a "double maximum." Each year produces two periods of increased rainfall separated by pronounced dry spells. In many savanna zones, inhabitants refer to the "long rains" and the "short rains"

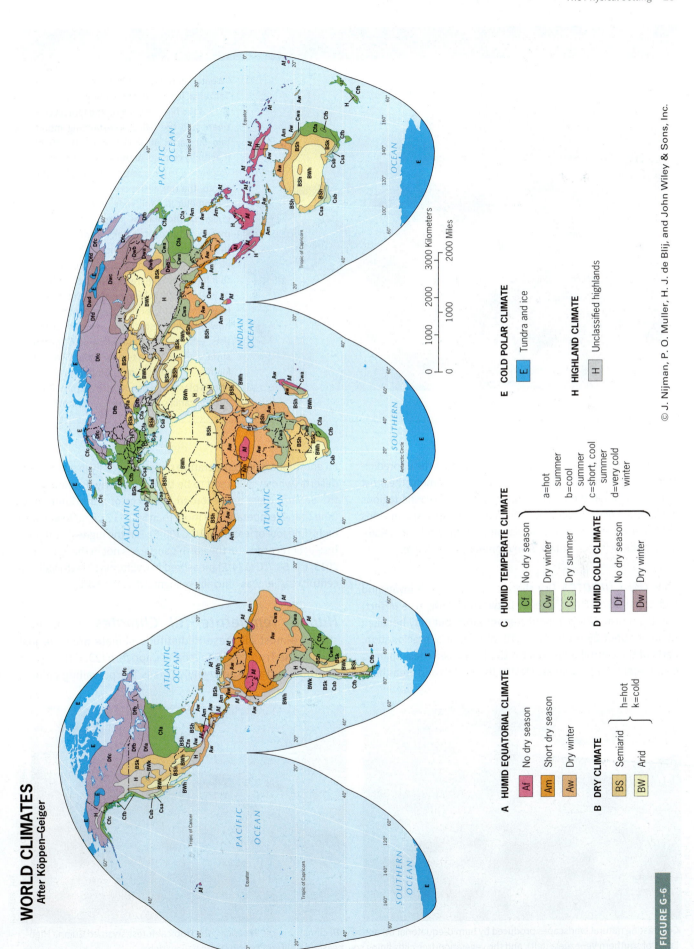

WORLD CLIMATES
After Köppen–Geiger

A HUMID EQUATORIAL CLIMATE

Af	No dry season
Am	Short dry season
Aw	Dry winter

B DRY CLIMATE

BS	Semiarid
BW	Arid

h=hot
k=cold

C HUMID TEMPERATE CLIMATE

Cf	No dry season
Cw	Dry winter
Cs	Dry summer

a=hot summer
b=cool summer
c=short, cool summer
d=very cold winter

D HUMID COLD CLIMATE

Df	No dry season
Dw	Dry winter

E COLD POLAR CLIMATE

E	Tundra and ice

H HIGHLAND CLIMATE

H	Unclassified highlands

0   1000   2000   3000 Kilometers
0   1000   2000 Miles

FIGURE G-6

Box G-3 From the Field Notes ...

© Jan Nijman

"What a sight! Mount Kilimanjaro in northern Tanzania, just south of the equator (see Fig. G-1), is the highest peak on the African continent and the tallest freestanding mountain on Earth; at 5895 meters (19,340 ft), it actually consists of three closely grouped volcanoes. To get a sense of the scale of this photo, the rim of the outer crater is about 1 kilometer (0.6 mi) wide. In the local language of the Maasai people, Kilimanjaro means "source of water," attributable to the streams fed by rain and snow at higher elevations—and, importantly, the glaciers that have existed here for tens of thousands of years. Nonetheless, on this 23rd day of November 2014, it is frightening to observe how little of these glaciers remain. Scientists estimate that this shrunken glacial surface is now down to only about 2 square kilometers (0.8 sq mi). Global warming affects places and people all over the world, and Tanzania's Maasai are not exempt."

to identify those seasons; a persistent problem is the unpredictability of the rain's arrival. Savanna soils are not among the most fertile, and when the rains fail hunger looms. Savanna regions are far more densely peopled than rainforest areas, and millions of residents of the savanna subsist on what they cultivate. Rainfall variability is their principal environmental challenge.

Dry (B) Climates

Dry climates occur in both lower and higher latitudes. The difference between the **BW** (true **desert**) and the moister **BS** (semiarid **steppe**) varies but may be taken to lie at about 25 centimeters (10 in) of annual precipitation. Parts of the central Sahara in North Africa receive less than 10 centimeters (4 in) of rainfall. Most of the world's arid areas have

an enormous daily temperature range, especially in subtropical deserts (whose soils tend to be thin and poorly developed). In the Sahara, there are recorded instances of a maximum daytime shade temperature of more than 50°C (122°F) followed by a nighttime low of less than 10°C (50°F). But the highest temperature ever recorded on the Earth's surface is not in the Sahara: in 2013, the 56.7°C (134°F) measured in California's Death Valley a century earlier was officially recognized as the hottest.

Humid Temperate (C) Climates

As the map shows, almost all these midlatitude climate areas lie just beyond the Tropics of Cancer and Capricorn (23.5° North and South latitude, respectively). This is the prevailing climate

Jacques Jangoux / Getty Images, Inc.

Francesco Riccardo Iacomino / Getty Images, Inc.

Contrasting natural landscapes produced by humid-equatorial (**A**) and dry (**B**) climates in the Americas: the tropical-forest-swathed Guiana Highlands of southern Venezuela (left), and the vegetation-free cliffs lining southeastern California's arid Death Valley (right).

in the southeastern United States from Kentucky to central Florida, on North America's west coast, in most of Europe and the Mediterranean, in southern Brazil and northern Argentina, in coastal South Africa, in eastern Australia, and in eastern China and southern Japan. None of these areas suffers climatic extremes or severity, but the winters can be cold, especially away from water bodies that moderate temperatures. These areas lie midway between the winterless equatorial climates and the summerless polar zones. Fertile and productive soils have developed under this regime, as we will note in our discussion of the North American and European realms.

The humid temperate climates range from moist, as along the densely forested coasts of Oregon, Washington, and British Columbia, to relatively dry, as in the so-called Mediterranean (dry-summer) areas that include not only coastal southern Europe and northwestern Africa but also the southwestern tips of Australia and Africa, central Chile, and Southern California. In these Mediterranean environments, the scrubby, moisture-preserving vegetation creates a natural landscape different from that of richly green western Europe.

Humid Cold (D) Climates

The humid cold (or "snow") climates may be called the continental climates, for they seem to develop in the interior of large landmasses, as in the heart of Eurasia or North America. No equivalent land areas at similar latitudes exist in the Southern Hemisphere; consequently, no **D** climates occur there. Great annual temperature ranges mark these humid continental climates, and cold winters and relatively cool summers are the rule. In a **Dfa** climate, for instance, the warmest summer month (July) may average as high as 21°C (70°F), but the coldest month (January) might average only 11°C (12°F). Total precipitation, much of it snow, is not high, ranging from about 75 centimeters (30 in) to a steppe-like 25 centimeters (10 in). Compensating for this paucity of precipitation are cool temperatures that inhibit the loss of moisture from evaporation and evapotranspiration (moisture loss to the atmosphere from soils and plants).

Some of the world's most productive soils lie in areas under humid cold climates, including the U.S. Midwest, parts of southwestern Russia and Ukraine, and northeastern China. The winter dormancy (when all water is frozen) and the accumulation of plant debris during the fall balance the soil-forming and enriching processes. The soil differentiates into well-defined, nutrient-rich layers, and substantial organic humus accumulates. Even where the annual precipitation is light, this environment sustains extensive coniferous forests.

Cold Polar (E) and Highland (H) Climates

Cold polar (**E**) climates are differentiated into true icecap conditions, where permanent ice and snow keep vegetation from gaining a foothold, and the tundra, which may have average temperatures above freezing up to four months of the year. Like rainforest, savanna, and steppe, the term **tundra** is vegetative as well as climatic, and the boundary between the **D** and **E** climates in

Figure G-6 corresponds closely to that between the northern coniferous forests and the tundra.

Finally, the **H** climates—the unclassified highlands mapped in gray (Fig. G-6)—resemble the **E** climates. High elevations and the complex topography of major mountain systems often produce near-Arctic climates above the tree line, even in the lowest latitudes such as the equatorial section of the high Andes of South America.

Here is an important qualification to keep in mind when studying Figure G-6: this map is also a still-picture of a changing scene, a single frame from an unspooling film. Global climate never stops changing, and less than a century from now those changes may compel revisions to this map. If current predictions about rising sea levels turn out to be accurate, we may even have to start redrawing some familiar coastlines.

You will find larger-scale maps of climate in several of the regional chapters that follow, but it is always useful to refer back to this Köppen-Geiger map whenever the historical or economic geography of a region or country is under discussion. This world climatic map reflects agricultural opportunities and limitations as well as climatic regimes, and as such helps explain some enduring patterns of human distribution on our planet. We turn next to this crucial topic.

Realms of Population

Earlier we noted that population numbers by themselves do not define geographic realms or regions. Population distributions, and the functioning society that gives them common ground, are more significant criteria. That is why we can identify one geographic realm (the Austral) with less than 30 million people and another (South Asia) with more than 1.7 billion inhabitants. Neither population numbers nor territorial size alone can delimit a geographic realm. Nevertheless, the map of global population distribution shows some major clusters that are part of certain realms (Fig. G-7).

Before we examine these clusters in some detail, remember that the world's human population now rounds off at 7.4 billion (see Appendix B, available online, for a detailed breakdown)—confined to the landmasses that constitute less than 30 percent of our planet's surface, much of which is arid desert, inhospitable mountain terrain, or frigid tundra. (And keep in mind that Fig. G-7 is yet another still-picture of a continually changing scene as the rapid growth of humankind continues.) After thousands of years of slow growth, world population during the nineteenth and twentieth centuries grew at an increasing rate. That rate has recently been slowing down, even imploding in some parts of the world. But consider this: it took about 17 centuries following the birth of Christ for the world to add 250 million people to its numbers; now we are adding 250 million about every three years. While the *rate* of population growth has declined in some parts of the world, in absolute terms the global population continues to grow apace and is expected to reach just under 10 billion by 2050.

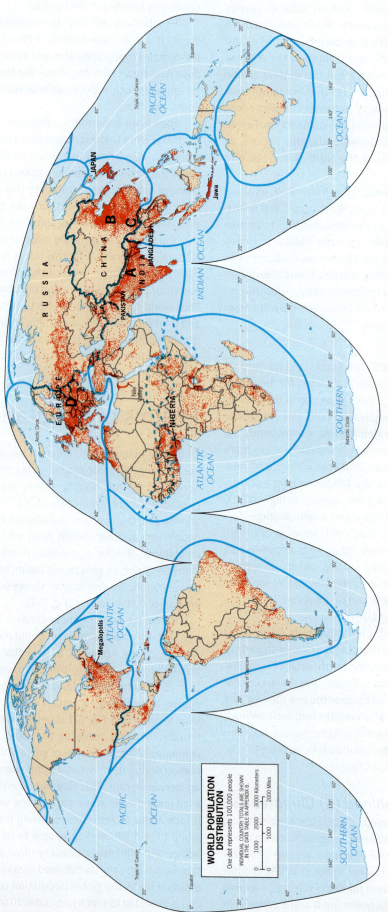

© J. Nijman, P. O. Muller, H. J. de Blij, and John Wiley & Sons, Inc.

WORLD POPULATION DISTRIBUTION

One dot represents 100,000 people

INDIVIDUAL COUNTRY TOTALS ARE SHOWN IN THE DATA TABLE IN APPENDIX B.

FIGURE G-7

This raises the important question as to whether there are limits to the Earth's carrying capacity—will there be enough food to go around? That question has become more and more pressing over the past decade due to rapidly rising food prices resulting from increased demand in China and India, a general dietary shift from grains to meat and vegetables, and the use of agricultural resources for the production of biofuels. The actual increase of population is only part of the problem; our growing appetite for certain products is another. And we are not just talking about food—think drinking water, fossil fuels, and minerals as well. Therefore, it seems inconceivable that almost 10 billion people by mid-century could be consuming at the rate we do today in the developed world.

Major Population Clusters

One way to present an overview of the location of people on the planet is to create a map of **population distribution [21]** (Fig. G-7). As you can see in the map's legend, each dot represents 100,000 people, and the clustering of large numbers of them in certain areas, as well as the near-emptiness of others, is readily evident. By the way, there is a technical difference between population distribution and *population density*, which is another way of showing where people live. Density maps reveal the number of persons per unit area.

- **South Asia** The *South Asia* population cluster lies centered on India and includes its populous neighbors, Pakistan and Bangladesh. This huge agglomeration of humanity focuses on the wide plain of the Ganges River (**A** in Fig. G-7). South Asia recently became the world's largest population cluster, overtaking East Asia earlier in this decade.
- **East Asia** The *East Asia* population cluster is centered on eastern China and includes the Pacific-facing Asian coastal zone from the Korean Peninsula to Vietnam. Not long ago, we would have reported this as a dominantly rural, farming population, but rapid economic growth and burgeoning urbanization have altered the picture. In China's interior river basins of the Huang (Yellow) and Chang/Yangzi (**B** and **C** on the map), and in the Sichuan Basin between these two letters, most of the people remain farmers. But the booming cities of coastal, and increasingly interior, China are attracting millions of new inhabitants, and in 2011 the Chinese urban population surpassed the 50 percent milestone.
- **Europe** The third-ranking population cluster, *Europe*, also lies on the Eurasian landmass but at the opposite end from China. The European cluster, including western Russia, counts more than 700 million inhabitants, which puts it in a class with the two larger Eurasian concentrations—but there the similarity ends. In Europe, the key to the linear, east-west orientation of the axis of population (**D** in Fig. G-7) is not a fertile river basin but a zone of raw materials for industry. Europe is among the world's most highly urbanized and industrialized realms, its human agglomeration sustained by factories and offices rather than paddies and pastures.

The three world population concentrations just discussed (South Asia, East Asia, and Europe) account for about 4 billion of the world's 7.4 billion people. No other cluster comes close to these numbers. The next-ranking cluster, eastern North America, is only about one-quarter the size of the smallest Eurasian concentrations. As in Europe, the population in this zone is concentrated in major metropolitan complexes; the rural areas are now relatively sparsely settled.

Toward an Urban World

Geographic realms and regions display varying levels of **urbanization [22]**, the proportion of the total population residing in cities and towns (which globally stands at 55 percent today). Some regions are urbanizing far more rapidly than others, a phenomenon we will investigate as we examine each realm. For now, it is important to be aware of the distinction between urbanization level and urban growth rates. The urbanization level refers to the proportion of a country's or region's total population living in cities (however defined by national governments). The urban growth rate, on the other hand, concerns the growth of the urban population over time, usually on an annual basis. So, for example, the U.S. urbanization level is 83 percent, whereas its yearly urban growth rate is around 1 percent; for Tanzania, the urbanization level is far less, just over 30 percent, but the urban growth rate is a much higher 5.5 percent. It makes sense that urbanization levels and growth rates are often inversely related: once the percentage of the population already living in cities reaches a certain level, urban growth rates will subside, as has been the case for the United States. Alternatively, when urbanization levels are low, there are ample opportunities for rapid urban growth.

Thus the fastest urban growth rates tend to occur in realms and regions in which a large proportion of the population resides in rural areas, as in Subsaharan Africa as well as South and East Asia. It is also worth noting—in these realms where urbanization levels remain low and urban growth rates are high—that a large number of **megacities [23]** have emerged over the past three decades (see **Box G-4**). These huge urban agglomerations are defined as containing populations of greater than 10 million, and in 2016, 36 of them could be found on the world map. Among the latter, only two are in North America (New York and Los Angeles), whereas 7 are located in South Asia and no less than 6 in China alone.

Realms of Culture

Imagine yourself aboard a riverboat on Africa's White Nile, heading upstream (south) from Khartoum, Sudan. The desert sky is blue, the heat is searing. You pass by villages that look much the same: low, square, or rectangular dwellings, some recently whitewashed, others gray, with flat roofs, wooden doors, and small windows. The minaret of a modest mosque may rise above the houses, and there may be a small central

Box G-4 From the Field Notes …

"After a chat with the manager of a high-rise hotel in Mumbai (formerly Bombay), Zach Woodward and I are allowed access to the roof in early evening to take shots of the city's mushrooming skyline. Zach is an undergraduate photography major who accompanied me on this field research trip, and he has a sharp eye for both depth and scenic drama. In the foreground, waterfront-lining Marine Drive, also known as 'The Queen's Necklace,' faces the Arabian Sea as the setting sun shines faintly on some of the low-rise Art Deco apartment blocks that date back to the 1930s. We are at the southern tip of the Mumbai Peninsula, from where the British ruled their South Asian empire during colonial times. Today, the megacity has expanded far beyond this slender peninsula onto the surrounding mainland, and the metropolitan-area population exceeds 23 million. Looking northward, Mumbai's unrelenting densification presents itself as the proverbial concrete jungle."

© Jan. Nijman

square. There is very little vegetation; here and there a hardy palm tree stands in a courtyard. People on the paths wear long white or colored robes and headgear. A few goats lie in the shade, and children are playing on the dusty ground.

All of this is part of Sudan's rural **cultural landscape [24]**, the distinctive attributes of a society imprinted on its portion of the world's physical stage. The cultural landscape concept was initially articulated in the 1920s by a University of California geographer named Carl Sauer, who stated that "a cultural landscape is fashioned from a natural landscape by a culture group"

and that "culture is the agent; the natural environment the medium." What this means is that people, starting with their physical environment and using their culture as their agency, construct a landscape layered with forms such as buildings, gardens, and roads, and is also distinguished by modes of dress, aromas of food, and sounds of music.

Continue your journey southward on the Nile, and you will soon witness a remarkable transition. Quite suddenly, the square, solid-walled, flat-roofed houses of Sudan give way to the round, wattle-and-thatch, conical-roofed dwellings of

PATRICK BAZ / Getty Images, Inc.

Cultura Travel/Philip Lee Harvey /Getty Images, Inc.

Contrasting settlement landscapes in Sudan. To the left, the town of Umm Dawban just north of the capital, Khartoum, which stands in sharp contrast with those in South Sudan (right, the village of Nyaro just south of the Sudan-South Sudan border).

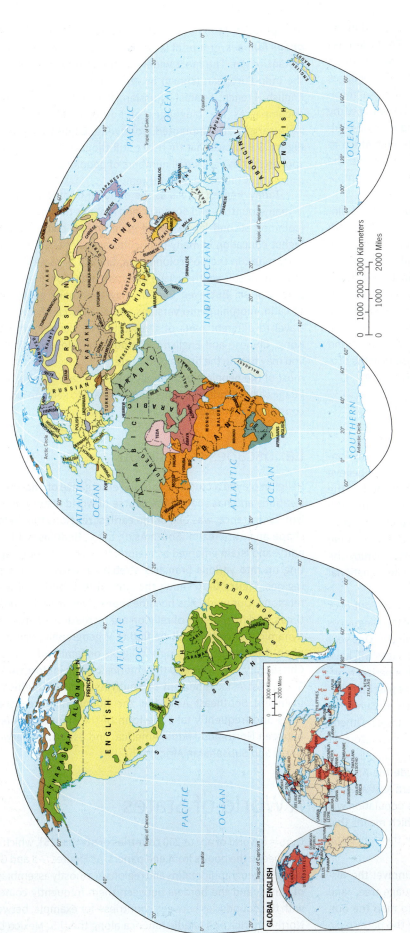

LANGUAGE FAMILIES OF THE WORLD
Majority Speakers

INDO-EUROPEAN

AFRO-ASIATIC

NIGER-CONGO

SAHARAN

SUDANIC

KHOISAN

URALIC

ALTAIC

SINO-TIBETAN

JAPANESE AND KOREAN

DRAVIDIAN

AUSTRO-ASIATIC

AUSTRONESIAN

TRANS-NEW GUINEA AND AUSTRALIAN

AMERINDIAN

OTHERS

UNPOPULATED AREAS

Modified from Hammond World Atlas, 1977.

© J. Nijman, P. O. Muller, H. J. de Blij, and John Wiley & Sons, Inc.

FIGURE G-8

South Sudan. You may note that clouds have appeared in the sky: it rains more here, and flat roofs will not do. The desert has yielded to green. Vegetation, natural as well as planted, grows between houses, flanking even the narrow paths. The villages seem less orderly, more varied. You'll see women, often in colorful dresses, carrying water in jars on their heads from a nearby well or the river to their huts. You have traveled from one cultural landscape into another, from Arabized, Islamic Africa to animist/Christian Africa. More broadly, you have crossed the boundary between two geographic realms (see the accompanying pair of photos).

Although no geographic realm, is marked by only a single cultural landscape, these landscapes can help define both realms and regions. The cultural landscape of the high-rise North American city with its far-reaching suburbs differs from that of Europe; the meticulously organized, terraced paddies of Southeast Asia are unlike anything to be found in the rural cultural landscape of neighboring Australia. Important variations of cultural landscapes also occur within geographic realms, such as between highly urbanized and dominantly rural (and more traditional) areas, and they can be quite useful in the process of delineating several of the world's regions as well.

The Geography of Language

Language is the essence of culture. People tend to feel passionately about their mother tongue, especially when they believe it is threatened in some way. In the United States today, the English Only movement reflects many people's fears that the primacy of English as the national language is under threat as a result of immigration. As we will see in later chapters, some governments try to suppress the languages (and thereby the cultures) of minorities in mistaken attempts to enforce national unity, provoking violent reactions.

In fact, many languages emerge, thrive, and die out over time, and linguists estimate that the number of lost languages is in the tens of thousands—a process that continues. One year from the day you read this, about 25 more languages will have become extinct, leaving no trace. Just in North America, more than 100 native languages have been lost over the past half-century. Some major languages of the past, such as Sumerian and Etruscan, have left fragments in later languages. Others, like Sanskrit and Latin, live on in their modern successors. At present, about 6900 languages remain, half of them classified by linguists as endangered ("hot spots" include the Amazon Basin, the Andes to its west, Siberia, and northern Australia). By the end of this century, the bulk of the world's population will be speaking just a few hundred languages, which means that many millions will no longer be able to speak their ancestral mother tongues.

Scholars have tried for many years to unravel the historic roots and branches of the global "language tree," and their debates continue. Geographers trying to map the outcome of this research keep having to modify the pattern, so you should take **Figure G-8** as a work in progress, not the

final product. At minimum, there are some 15 so-called **language families**, groups of languages with a shared but usually distant origin. The most widely distributed language family, the Indo-European (shown in yellow on the map), includes English, French, Spanish, Russian, Persian, and Hindi. This encompasses the languages of European colonizers that were carried and implanted worldwide, English most of all. **Figure G-9** shows that there are still thousands of languages spoken, but by very small numbers of people; their continued existence is therefore under threat.

Today, English serves as the national or official language of many countries and outposts, and remains the **lingua franca** (common second language) of government, commerce, and higher education in many multicultural societies (see Fig. G-8 inset map). In the postcolonial era, English became the chief medium of still another wave of ascendancy now in progress: globalization. But even English may eventually go the way of Latin, morphing into versions you already can hear as you travel to different parts of the world. For example, in Hong Kong, Chinese and English are producing a local "Chinglish" you may hear in the first taxi you enter. In Lagos, Nigeria, where most of the people are culturally and ethnically Yoruba, a language called "Yorlish" is emerging. It is hard to keep up with the constant evolution of language.

Landscapes of Religion

Religion played a crucial role in the emergence of ancient civilizations and has shaped the course of world history. Hinduism, for instance, was one of the earliest religions that helped shape an entire realm (South Asia). Later, Buddhism, Christianity, and Islam emerged as major belief systems, often splitting up into various branches stretching across realms and regions. **Figure G-10** shows the current distribution of world religions. Our world has become more complicated in recent times, and its patterns of religion are increasingly diffuse and dynamic. But today, still, we find that geographic realms are often dominated by a single religion or family of religions: Christianity in Europe and the Americas; Islam in North Africa/Southwest Asia; Hinduism in South Asia; and Buddhism in mainland Southeast Asia. But the boundaries are usually fuzzy and frequently take the form of transition zones (e.g., between North and Middle America, and especially between North and Subsaharan Africa).

A World of States

Ours is a world of about 200 countries or **states [25]**, which are mapped in the book's front end papers. As Figures G-3 and G-11 suggest, geographic realms and regions are mostly assemblages of states, and the borders between them frequently coincide with the boundaries between countries—for example, between North America and Middle America along the U.S.-Mexico border. It is also possible for a realm boundary to cut across a state

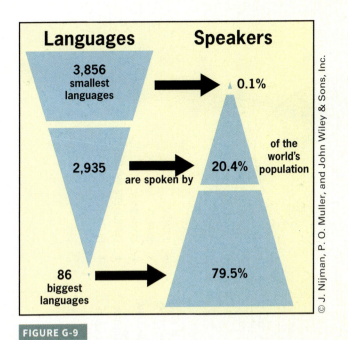

Languages

3,856 smallest languages

2,935

are spoken by

86 biggest languages

Speakers

0.1%

20.4%

of the world's population

79.5%

© J. Nijman, P. O. Muller, and John Wiley & Sons, Inc.

FIGURE G-9

conterminous, so that a ***nation-state*** would enclose an ethnically and culturally homogeneous people within a national boundary. That was never truly the case (even France, the "model of models," had its minorities), and today the ideal state is defined as a clearly and legally defined territory inhabited by a citizenry governed from a capital city by a representative government. As we shall discover in Chapter 4, not even in Europe itself are all governments truly representative. Nonetheless, the European state model has, for better or worse, been adopted throughout the world.

Although the modern state is a longstanding phenomenon, many states experience external and internal pressures and challenges. For instance, European states are challenged "from above" by the European Union as member-states transfer some of their power to "Brussels" (EU's headquarters city), mainly because they believe it will be to their economic advantage. States can also be challenged "from below" by ethnic minorities, such as the Kurds in southeastern Turkey or by regional secessionist movements such as the Catalonians in northeastern Spain.

(Indonesia is one), so that state and realm boundaries do not always concur (see the African Transition Zone in Fig. G-3).

The political territorial organization of the world within a system of states is based on the notion of **sovereignty [26]**. It is a concept from international law, which means that the government of a state reigns supreme within its borders. Normally, states recognize each other's sovereignty, but this becomes a matter of contention in times of conflict and war. Think, for example, of the ongoing conflict in which Russia is challenging Ukraine's sovereignty in that country's eastern provinces next to the Russian border. Or think of China's refusal to even consider the possible sovereignty of Taiwan.

The European Origins of the Modern State

In the long course of human history, the modern state is a relatively recent invention, and so is the international system of which it forms the cornerstone. The modern state emerged from other varieties of politico-territorial organization that date back to the earliest complex civilizations. In their studies of ancient history, scholars sometimes use the term *polity* or *proto-state* to indicate the difference. Ever since agricultural surpluses enabled the growth of large and prosperous towns, this was accompanied by the more sophisticated and centralized exercise of power and political organization. From these origins, the earliest states took shape.

It was not until the seventeenth century that European rulers and governments began to negotiate treaties that defined the state in international law. That is why the modern state is quite often described as based on the **European state model [27]**, with definitions of nationality and sovereignty. Often, the model assumed that state and nation were ideally

Power and Geopolitics

As the term **geopolitics [28]** implies, geography plays a major role in the politics of the state, in the international relations among states within realms and regions, and even in warfare and terrorism. Geography does not, of course, explain everything, but it is certainly an essential dimension in the understanding of world politics. A state's power and influence are related to its size, access to natural resources, location vis-à-vis other states, trading routes, access to waterways, and the like. More than ever today, news reports about conflicts in any part of the world—whether on television, the website and hard-copy pages of leading newspapers, or in newsmagazines—are accompanied by maps crucial to the explanation of these clashes.

Even though the role of geography in political affairs can be regarded as fundamental, it certainly does not remain static. Relative geopolitical location itself is subject to change, especially in response to new technologies. For example, states and regions once deemed impregnable because they could not easily be reached by enemies (such as the United States prior to the age of aviation), can swiftly see their invulnerability disappear. It is not hard to imagine how the arrival of long-range missiles, state-of-the-art spy-satellite technology, advances in geospatial and mapping systems, and other high-tech military innovations such as drones can hugely impact a country's strategic planning and positioning.

The territorial functioning and integrity of many states in this new century have also been challenged by the emergence, proliferation, and growing sophistication of terrorist groups. Today's most heavily contested zones of conflict are found in the North Africa/Southwest Asian Realm (e.g., Syria, Iraq, Yemen) where longstanding sectarian disputes have been complicated by the rise of violent, radical organizations such

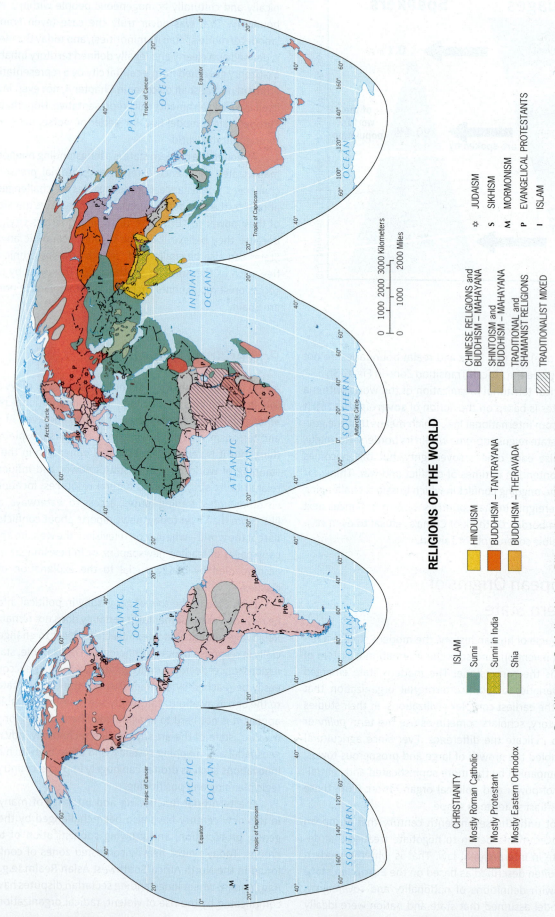

RELIGIONS OF THE WORLD

HINDUISM

BUDDHISM – TANTRAYANA

BUDDHISM – THERAVADA

CHINESE RELIGIONS and BUDDHISM – MAHAYANA

SHINTOISM and BUDDHISM – MAHAYANA

TRADITIONAL and SHAMANIST RELIGIONS

TRADITIONALIST MIXED

ISLAM
- Sunni
- Sunni in India
- Shia

CHRISTIANITY
- Mostly Roman Catholic
- Mostly Protestant
- Mostly Eastern Orthodox

✡ JUDAISM

S SIKHISM

M MORMONISM

P EVANGELICAL PROTESTANTS

I ISLAM

© J. Nijman, P. O. Muller, H. J. de Blij, and John Wiley & Sons, Inc.
Adapted from E. H. Foubeg et al., *Human Geography*, 9e, based on several data sources.

FIGURE G-10

The *hajj* is the yearly pilgrimage of Muslims to the holy city of Mecca in Saudi Arabia. The pilgrimage is referred to as the fifth "pillar" of Islam, the obligation of every able-bodied Muslim to worship Allah in this holiest of sites at least once in their lifetime. This is the Grand Mosque of Mecca on November 17, 2010, as more than two million Muslim pilgrims launched into the final rituals of this largest religious pilgrimage in the world.

as Al Qaeda and ISIS. Increasingly, these threats are also being felt far beyond that realm, in places such as western Europe, Russia, and West Africa. You will learn about the geographic dimensions of these and other key conflicts in the chapters ahead. For now, it is important to realize that the world map of states is far from stable in certain realms and regions, and that states are not the only major geopolitical actors on the global scene.

Geographies of Development

Finally, as we prepare for our study of world regional geography, it is all too clear that realms, regions, and states do not enjoy the same level of prosperity. The field of **economic geography** focuses on spatial aspects of the ways people make their living, and deals with patterns of production, distribution, and consumption of goods and services (see, for instance, **Box G-5**). As with all else in this world, these patterns reveal significant variation. Individual states report the nature and value of their imports and exports, farm and factory output, and many other economic data to the United Nations and other international agencies. From such information, economic geographers can measure the comparative well-being of the world's countries. The concept of **development [29]** is used to gauge a population's economic, social, and institutional growth and overall well-being.

As this definition indicates, the meaning of development is not confined to economics but can be expressed, for example, in access to schools or medical care, democratic institutions, environmental stewardship, social inclusiveness, political freedom, and the quality of governance. Development, therefore, is a broad theme that attracts attention not only from economic geographers but also from geographers who take a more comprehensive approach. In recent years, the dual notions of **sustainable development [30]** and **inclusive development [31]**

Box G-5 From the Field Notes ...

"As I sit enjoying breathtaking views of the spectacular Andean highlands of southern Peru, a farmer slowly comes up the side of the hill with her donkeys. She tells me that her name is Carmen and that the donkeys are carrying hay grass to market in the nearby town of Chinchero. The hay will feed guinea pigs that many rural and small-town Peruvian families keep in their home ... to eat. Guinea pigs are said to be quite tasty, and they are much cheaper than beef as well as a good source of protein. Carmen's family owns a small plot of land on one of these hillsides, and the donkeys are her main means of transportation. Although incomes for small farmers here are low and life is simple, she exudes an air of confidence and pride; she is at home here, and I am on her turf. I'm now thinking about trying guinea pig before I depart...."

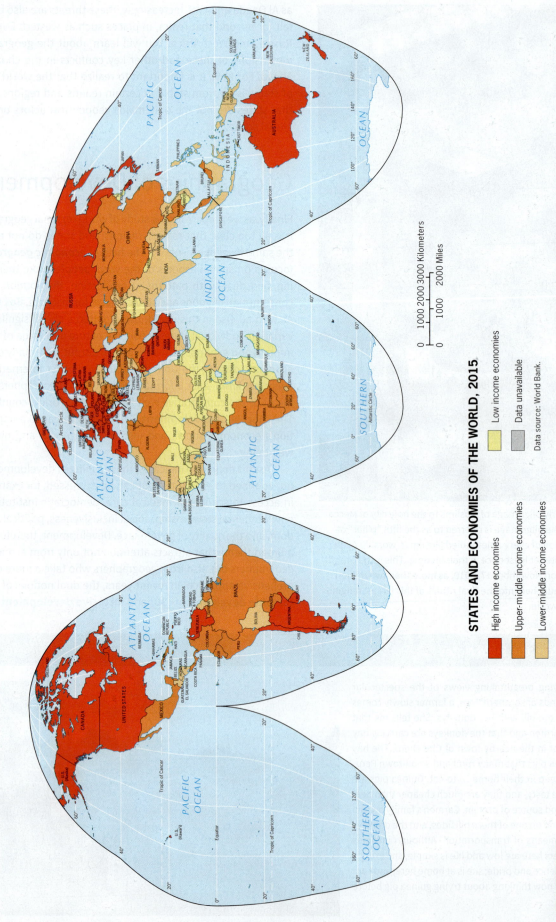

STATES AND ECONOMIES OF THE WORLD, 2015

- High income economies
- Upper-middle income economies
- Lower-middle income economies
- Low income economies
- Data unavailable

Data source: World Bank.

© J. Nijman, P. O. Muller, H. J. de Blij, and John Wiley & Sons, Inc.

FIGURE G-11

have emerged as particularly important themes. The first focuses on the long-term viability of economic patterns in relation to the physical environment and natural resources—think of pollution, or finite resources like fossil fuels. The second emphasizes the extent of equal economic (and social) opportunities for different population groups, especially minorities and the poor.

Development in Spatial Perspective

Various schemes to group the world's states into economic-geographic categories have come and gone, and others will probably arise in the future. For our purposes, the classification scheme used by the World Bank (one of the agencies that monitor economic conditions across the globe) is the most effective. It sorts countries into four categories based on the success of their economies: (1) high-income, (2) upper-middle income, (3) lower-middle income, and (4) low-income. These categories, when mapped, display important regional patterns (see **Fig. G-11**). Compare this map to our global framework (Fig. G-3), and you can see the role of economic geography in the layout of the world's geographic realms. Also evident are regional contrasts within realms—for instance, between Brazil and its western neighbors, between South Africa and most of the rest of Subsaharan Africa, and between west and east in Europe. Thus the geographic scale at which we consider development is of critical importance.

A Complex, Core-Periphery World

The economic success of human societies on the Earth's surface tends to be concentrated in certain areas while bypassing others. This was true in ancient and more recent historical times, and it is true today. The contemporary world economy—an integrated, international spatial system—is structured within the framework of a dominant **global core [32]** and subordinate **global periphery [33]**. During the nineteenth century, the core more or less coincided with western Europe, controlling as it did vast areas of the world through its empires. In the twentieth century, the core expanded first to North America and then grew to include Japan, Australia, and New Zealand. By the 1980s, Hong Kong, Singapore, Taiwan, and South Korea had become part of the core as well. And as the twenty-first century opened, the newest entrant stepped forward: Pacific-fronting China. The regions constituting this still-evolving global core are mapped in tan in **Figure G-12**.

Note that the global core is not a contiguous area but rather consists of four dominant regions: North America, Europe, eastern East Asia, and Australia/New Zealand. Moreover, Singapore (not shown in Fig. G-12) is often considered to be part of the Asian core region as well. Sometimes, the world outside the core is divided into a *periphery* and *semi-periphery*, where the latter comprises countries that occupy somewhat

of a middle position between affluence and poverty, between dominance and dependency. Examples of these semi-periphery countries today would be Brazil, South Africa, and Saudi Arabia.

It is also important to realize that **core-periphery relationships [34]** are not limited to the international scale. Countries as well can and do exhibit such patterns at the national scale. China is a particularly good example: its coastal provinces form the core, while the interior and westernmost reaches of the country comprise China's periphery. Except for a few special cases, all countries contain **core areas [35]**. These national cores are most often anchored by the country's capital and/or largest city: Paris (France), Tokyo (Japan), Buenos Aires (Argentina), and Bangkok (Thailand) are only a few notable examples. Larger countries may have more than one core area, such as Australia with its eastern and western coastal cores and intervening periphery. And uneven development also exists at the local scale, especially marking the complex socioeconomic landscapes of urban areas.

The world continues to exhibit major differences in productivity and well-being. One of the most intriguing economic-spatial outcomes of globalization is that a growing number of countries have accelerated their development—but this growth is often confined to specific city-regions, whereas the rest of the country remains quite poor. Figure G-12, therefore, is provisional: it conveys a broad impression of the global core, but we realize that it is not really a continuous geographic space (and neither is the global periphery). The core-periphery concept is most useful in thinking about the organization of space as well as the geographies of power and dependence. But it is difficult to map precisely, not least because it operates simultaneously at multiple scales. For instance, the United States constitutes part of the global core; its northeastern seaboard anchors the national core area; and New York's Midtown and Lower Manhattan form the core of its largest city.

We also need to add a cautionary note about spatially aggregated statistics. The concept of development, as measured by data that reflect totals and averages for entire national populations, entails certain pitfalls that we should be aware of. When a state's economy is growing as a whole, and even when it is "booming" relative to other states, this does not automatically mean that every citizen is better off and the income of every worker is rising. Averages have a way of concealing regional variability and local stagnation. In very large states such as India and China, it is useful to assess regional, provincial, and even local economic data to discover to what extent the whole country is sharing in "development." In the case of India, we should know that the State of Maharashtra (containing the megacity of Mumbai) is far ahead of most other States when it comes to sharing in the national economy. In China, the coastal provinces of the Pacific Rim still overshadow those of the west. Thus national-level statistics can conceal as much as they reveal, and this is why *inclusive development* is becoming a leading policy goal around the world.

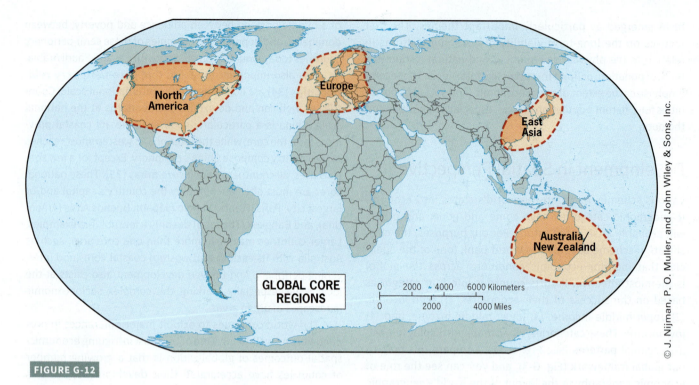

North America

Europe

East Asia

Australia/ New Zealand

GLOBAL CORE REGIONS

0	2000	4000	6000 Kilometers
0	2000		4000 Miles

© J. Nijman, P. O. Muller, and John Wiley & Sons, Inc.

FIGURE G-12

Globalization

Globalization [36] is fundamentally a geographical process in which spatial relations—economic, cultural, political—shift to the macro-scale. What this means is that what happens in one place has repercussions in places ever more distant—thereby integrating the entire world into a "global village," linked ever more tightly by recent advances in communications and transport technologies. Globalization now comes into our homes via television, computers, and smartphones; news today has never traveled faster, and sometimes even government leaders turn to the Internet on their personal electronic devices to get the

---

Box G-6 From the Field Notes …

"Would you like to float in an 'infinity pool' atop a 57-story skyscraper that provides amazing views? You can in Singapore. During a 2012 visit, I had spotted this futuristic-looking tower across the water from downtown and thought it would be worth a visit. It turned out to be the massive Marina Bay Sands Hotel (developed by the Sands Corporation of Las Vegas) that opened for business in 2010. Singapore is one of the best examples of a city propelled into prominence by the forces of globalization. Located strategically on one of the world's busiest shipping lanes (the Strait of Malacca) and possessing a fine harbor, this minuscule city-state was bound to benefit from expanding world trade. It now boasts the busiest transshipment port on Earth, and the city has also become a shopping magnet for elites all across the Southeast Asian realm. Like most 'world-cities,' Singapore is a hub of both production and consumption."

© Jan Nijman

latest reports on international events. Note that some places are (much) more globalized and connected than others. Global cities, such as New York, Paris, or Singapore, are major nodes in worldwide networks of production, consumption, trade, and travel (see **Box G-6**).

Globalization, however, is not all that new. The second half of the nineteenth century, for instance, also witnessed major advances in the intensification of global interdependence. It was particularly affected by such new technologies as the steamship, the railway, and the telegraph, which subsequently were followed by first-generation motor vehicles and aircraft. But with today's newest technologies, the global village is achieving an unprecedented level of long-distance interconnectivity. Accordingly, geography and our knowledge of the world's realms and regions have become more important than ever for understanding and navigating this transforming macro-spatial reality.

Globalization and Environmental Challenges

Globalization plays out in various spheres, from the environmental to the cultural to the economic. One of today's most pressing environmental issues is global warming, which threatens every corner of our planet. It is clear that the problem must be addressed internationally, yet it is far from easy to agree on mitigation strategies. Some countries are bigger polluters than others, some have more resources than others, and some are more developed than others. How to divide the burdens?

Countries from around the world have been meeting for many years at United Nations Climate Change Conferences, the most recent one held in Paris in late 2015. The main purpose is to reach agreement on collective reductions in the emission of greenhouse gases so as to limit global warming—but progress has been slow. **Figure G-13** provides some insight into the relationship between economic growth and environmental impact. Often, national economic growth comes at a price—but in the end all of us pay (see **Box G-7**).

Globalization and Population Movements

Culturally, too, the world is coming closer together. This is apparent in the global spreading of new fashions, music, foods, and other innovations. In a more problematic way, it is also expressed in global migration flows. In the past, **migration [37]** was relatively uncommon because most people were rooted in their home environment, where they lived out their entire lives. When residential relocation did occur, it used to be one-way, with people migrating from one place to another and then staying put. But over the past quarter-century, migration flows have intensified, in part because people increasingly possess better knowledge about opportunities elsewhere. Moreover, it is now much easier to travel back and forth, which allows migrants to maintain close ties with their countries of origin. Not surprisingly, as the number of highly mobile ***transnational migrants*** has risen, these transients have become major players in the spreading of cultures around the world. Examples include Algerians in Paris,

Box G-7 MAP ANALYSIS

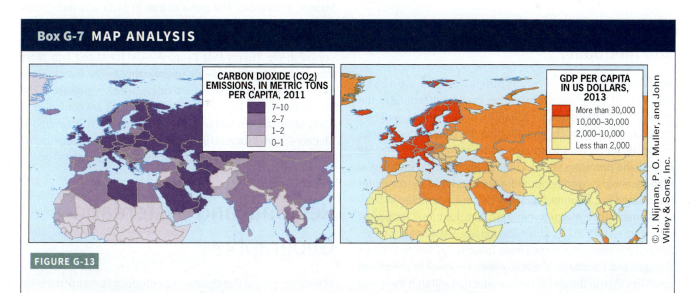

FIGURE G-13

Maps can reveal patterns that are difficult to discern using only statistics, tables, or graphs. Furthermore, making comparisons between maps also enables us to identify and understand the relationships and interconnections of our world. The maps above allow us to compare global patterns of economic development, measured as per capita GDP (right), with patterns of per-capita carbon dioxide (CO_2) emissions (left). CO_2 emissions are linked to

the greenhouse effect and global climate change. Reading these maps, how would you describe the general relationship between GDP and CO_2 emissions per capita? What is the relevance to notions of sustainable development?

Access your WileyPLUS Learning Space course to interact with a dynamic version of this map and to engage with online map exercises and questions.

Haitians in Montreal, Cubans in Miami, Mexicans in Los Angeles, Indians in Singapore, and Indonesians in Sydney.

Yet it is also important to keep in mind that people's mobility is often constrained because certain parts of this highly uneven world are so much better off than others. High-income countries exert the strongest pull on migrants—but all too often they cannot gain access. Millions of workers aspire to leave the poverty of the global periphery and seek a better life somewhere in the core. Trying to get there, every year many of them perish in the waters of the Mediterranean, the Caribbean, and the Atlantic. Others risk their lives at the barriers that encircle the global core as if it were a gated community—from the "security fence" between Mexico and the United States to the walls that guard Israel's safety to the razor wire that encircles Spain's tiny outposts on North Africa's shore.

Globalization refers to economics and politics as much as it does to culture. Think of clothing fashions, food, or popular music. It is sometimes argued that the spread of ideas is fundamental to all forms of globalization because it underscores how people think and what they value most. This, in turn, influences what they eat, how they vote, and their aspirations in life. Clearly, such aspirations play a part in the accelerating flows of migrants and refugees mentioned above. But globalization does not mean homogenization and the world remains highly diverse, especially in its cultural and political dimensions. One fascinating illustration of this diversity is the acceptance or rejection of capital punishment (see **Box G-8**). In the United States, to be sure, the death penalty continues to be a controversial and divisive issue, and in its adherence to capital punishment the country stands apart from its neighbors, Canada and Mexico.

Economic Globalization, Growth, and Inequality

Most economists, politicians, and businesspeople have a favorable view of economic globalization because it facilitates the expansion of international capitalism, open markets, and free trade. In theory, such globalization stimulates commerce, breaks down barriers to foreign trade, brings jobs to remote places, and promotes social, cultural, and political exchanges. High-technology workers in India are employed by computer firms based in California. Japanese cars are assembled in Thailand. American footwear is made in China. Fast-food restaurant chains spread standards of service and hygiene as well as familiar (and standardized) menus from Tokyo to Tel Aviv to Tijuana. If wages and standards of employment are lower in peripheral countries than in the global core, production will shift there and the development gap will shrink. Everybody wins. In general, the **gross national income (GNI)**[†] of countries that increase their foreign trade (and thereby become more "globalized")

[†] Gross national income (GNI) is the total income earned from all goods and services produced by the citizens of a country, within or outside of its borders, during a calendar year. *Per capita GNI* is a widely used indicator of the variation of spendable income around the globe.

An informal refugee camp near the Greek town of Idomeni in March 2016 at the border with Macedonia. These refugees are part of a massive stream of migrants fleeing civil wars in Syria and Iraq. Their hope of reaching the heart of Europe from Greece via the Balkan Peninsula was thwarted when the Macedonians abruptly blocked this route by hardening their border with a hastily built fence.

rises, whereas the GNI of those with diminished trading activity actually declines.

But there is another, more complicated issue: the liberalization and opening up of national economies to global markets is often accompanied by widening inequalities. In other words, uneven development within countries becomes more pronounced. As noted earlier, this imbalance is obvious in a country like China, whose economy has grown quite rapidly over the past two decades; much of this growth occurred in its Pacific coastal zone, not in the interior of the country, and income differentials became ever wider. The same is true in India and most other **emerging markets**. Moreover, this phenomenon is not confined to developing economies. In his widely read 2013 book entitled *Capital in the Twenty-First Century*—now translated into more than 30 languages worldwide (globalization!)—the French economist Thomas Piketty showed that inequality is growing in the United States and Europe as well. His findings are confirmed in many other studies. Inequality in the United States is covered in Chapter 1; that discussion underscores why a regional approach is vital to interpreting this process at both national and global scales.

Regional and Systematic Geographies

At the beginning of this chapter, we introduced a map of the great geographic realms of the world (Fig. G-3). We then addressed the task of dividing these realms into regions, and we used criteria ranging from physical geography to economic geography. The result is **Figure G-15**. On this map, note that we display not only the world geographic realms but also the regions into which they subdivide. The numbers in the legend reveal the order in which the realms and regions are discussed, starting with North America (1) and ending with the Pacific Realm (12).

Box G-8 MAP ANALYSIS

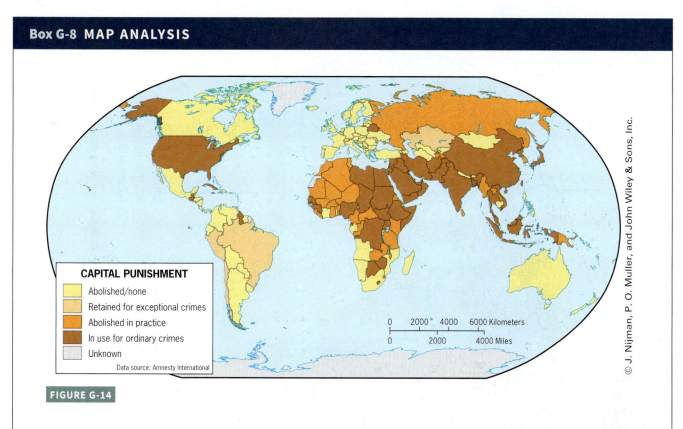

CAPITAL PUNISHMENT

- Abolished/none
- Retained for exceptional crimes
- Abolished in practice
- In use for ordinary crimes
- Unknown

Data source: Amnesty International

© J. Nijman, P. O. Muller, and John Wiley & Sons, Inc.

FIGURE G-14

In many parts of the world, but not all, the death penalty is a contentious matter. Whereas the number of executions, globally, has generally been declining in recent years, wide disparities remain in terms of the legality of capital punishment, the circumstances in which it is deemed to be warranted, and number of executions. The grisly nature of executions varies as well, from hanging to shooting, from beheading to lethal injection. Figure G-14 shows the disparities across the world and from one realm to another (though, in several cases, there is also variation within realms). Observe the patterns and consider what might explain the variation. Compare this map to Figures G-10 and G-11. Which of them shows a stronger relationship to capital punishment—economics or culture (i.e., income or religion)?

Access your WileyPLUS Learning Space course to interact with a dynamic version of this map and to engage with online map exercises and questions.

As this introductory chapter demonstrates, this book does not merely offer a series of descriptions of places and areas. We combine the study of realms and regions with a look at geography's ideas and concepts—the notions, generalizations, and basic theories that make the discipline what it is. We continue this approach in the chapters ahead so that we will become better acquainted with the world and with geography as a science. By now you are aware that geography is a wide-ranging, multifaceted discipline. It is often described as a social science, but that is only half the story: geography, in fact, straddles the divide between the social and the physical sciences. Many of the ideas and concepts you will encounter involve the multiple interactions between human societies and natural environments.

Regional geography allows us to view the world in an all-encompassing way. As we have seen, regional geography integrates information from many sources to create an overall image of our divided world. Those sources are not random: they represent topical or *systematic geography*. Research in the systematic fields of geography makes our world-scale generalizations possible. As **Figure G-16** shows, these systematic fields connect closely to those of other disciplines. Cultural geography, for example, is allied with anthropology; it is the spatial perspective that distinguishes cultural geography. Economic geography focuses on the spatial dimensions of economic activity; political geography concentrates on the spatial imprints of political behavior. Other systematic fields include historical, medical, behavioral, environmental, and urban geography. We will also draw on information from biogeography, marine geography, population geography, geomorphology, and climatology (as we did earlier in this chapter).

These systematic fields of geography are so named because their approach is thematic, not regional. Take the geographic study of cities, urban geography. Urbanization is a worldwide process, and urban geographers can identify certain human activities that all cities in the world exhibit in one form or another. But cities also display regional properties. The typical American city is quite distinct from, say, the European or the Chinese city. Regional geography, therefore, borrows from the systematic field of urban geography, but it injects this regional perspective. In the following chapters, we call upon these systematic fields

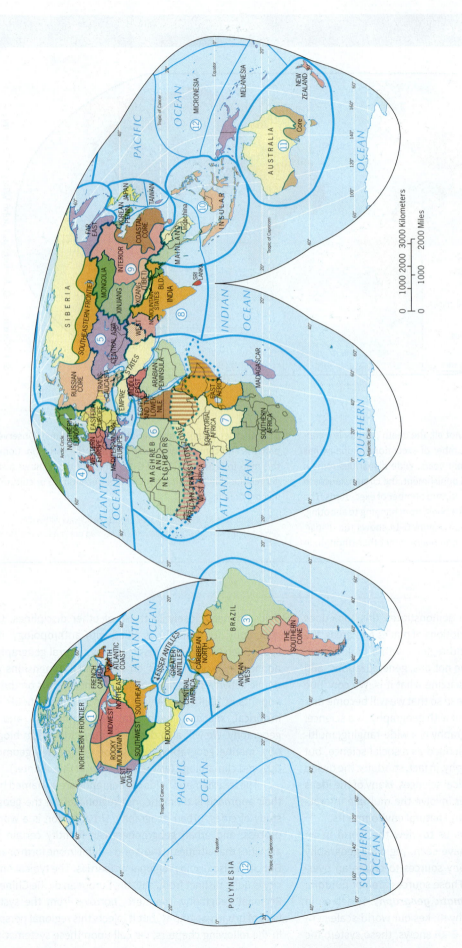

WORLD GEOGRAPHIC REALMS AND THEIR CONSTITUENT REGIONS

① NORTH AMERICA
② MIDDLE AMERICA
③ SOUTH AMERICA

④ EUROPE
⑤ RUSSIA/CENTRAL ASIA
⑥ NORTH AFRICA/SOUTHWEST ASIA

⑦ SUBSAHARAN AFRICA
⑧ SOUTH ASIA
⑨ EAST ASIA

⑩ SOUTHEAST ASIA
⑪ AUSTRAL REALM
⑫ PACIFIC REALM

© J. Nijman, P. O. Muller, and John Wiley & Sons, Inc.

FIGURE G-15

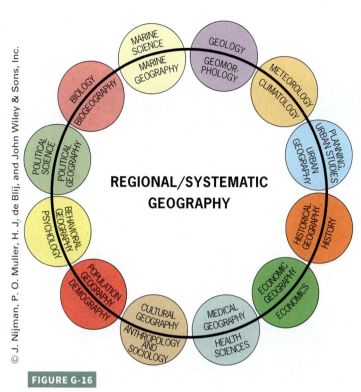

REGIONAL/SYSTEMATIC GEOGRAPHY

FIGURE G-16

to provide you with a better understanding of the world's realms and regions. As a result, you will gain insights into the discipline of geography as well as the regions we investigate.

We will begin our investigation of the world's realms and regions in North America. This is, after all, where most of you were born and raised—and it is from this geographic realm that your expanding worldview has evolved.

Points to Ponder

- According to some measures, China in 2015 surpassed the United States and became the world's biggest national economy.
- The global human population in 2017 is 7.4 billion and is predicted to reach just under billion by 2050. Will there be enough food and water to sustain everyone?
- Global warming is expected to cause a global rise in sea level of at least 10 centimeters (4 in) by the end of this century. Which cities are most vulnerable?
- Fast-forwarding globalization: the number of smartphones in the world surpassed the 2 billion mark in 2016; most of these devices have map navigation capabilities.

FIGURE 1-1

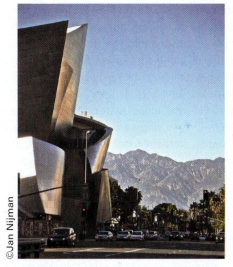

CHAPTER 1

©Jan Nijman

©Jan Nijman

Q Where were these pictures taken?

The North American Realm

IN THIS CHAPTER

- The U.S.-Mexico borderland
- North America's high-tech regions
- Changing geographies of inequality
- East Coast ports prepare for Panamax ships
- The immigration debate
- The geography of U.S. presidential elections
- China's impact on the realm's Pacific coast

CONCEPTS, IDEAS, AND TERMS

Borderland	**[1]**	Information economy	**[16]**
Transition zone	**[2]**	GPS (Global Positioning System)	**[17]**
Physiographic region	**[3]**	Gentrification	**[18]**
Continentality	**[4]**	Neighborhood effect	**[19]**
Rain shadow effect	**[5]**	Residential geography	**[20]**
Federation	**[6]**	Sunbelt	**[21]**
Aquifer	**[7]**	Migration	**[22]**
Fossil fuel	**[8]**	Electoral geography	**[23]**
Urban system	**[9]**	Melting pot	**[24]**
American Manufacturing Belt	**[10]**	First Nations	**[25]**
Distribution center	**[11]**	World-City	**[26]**
Intermodal connections	**[12]**	Technopole	**[27]**
Outer city	**[13]**	Pacific Rim	**[28]**
Deindustrialization	**[14]**	Tar sands	**[29]**
Central business district (CBD)	**[15]**	Boreal forest	**[30]**

In our survey of the world's geography, we first turn to the Western Hemisphere. It contains the two great interconnected continents that separate the Atlantic and Pacific oceans and extend, very nearly, from pole to pole, flanked by numerous islands large and small, indented by gulfs and bays of historic and economic import, and endowed with an enormous range of natural resources. Two continents—North and South—form the Americas, but three geographic realms blanket them (North, Middle, and South). In the context of physical (natural) geography, North America—from Canada's Ellesmere Island in the far north to the Isthmus of Panama in the south—is a continent. In terms of modern human geography, the northern continent is subdivided into the North American and Middle American realms along a transition zone marked by a political as well as a physical boundary between the United States and Mexico (Fig. 1-1; Box 1-2). In Texas, from the Gulf of Mexico upstream to El Paso/Ciudad Juárez, the Rio Grande forms this border. From El Paso westward to the Pacific Ocean, straight-line boundaries, hardened by fences and walls, separate the North from the Middle. We begin in North America.

Defining the Realm

North America is constituted by two of the world's most highly advanced countries in virtually every measure of social and economic development. Blessed by an almost unlimited range of natural resources and bonded by trade as well as culture, Canada and the United States are locked in a mutually productive embrace that is reflected by economic statistics: in an average recent year, about two-thirds of Canadian exports go to the United States, and just over half of Canada's imports come from its southern neighbor. For the United States, Canada is not quite as dominant, but it is still its leading export market and the second-biggest origin of imports (after China).

This realm is also defined on the basis of broad cultural traits, ranging from urban landscapes and a penchant for mobility to religious beliefs, language, and political persuasions (also see Box 1-1). Both countries rank among the world's most highly urbanized: nothing symbolizes the North American city as strongly as the skyscrapered panoramas of New York, Chicago, and Toronto—or the vast, beltway-connected suburban expanses of Los Angeles, Washington-Baltimore, and Houston. North Americans are also the most mobile people on Earth. Each year about one out of every nine individuals in the United States changes residence, a proportion that has declined lately but still leads the world.

Despite noteworthy multilingualism, in most areas English is the *lingua franca* of both countries. And notwithstanding all the religions followed by minorities, the great majority of both Canadians and Americans are Christians (whose U.S. total has dropped sharply since 2007). Moreover, both states are stable democracies, and both have federal systems of government.

Driving from Michigan across southern Ontario to upstate New York does not present a sharp contrast in cultural landscapes. Americans visiting Canadian cities (and vice versa) find themselves in mostly familiar settings. Canadian cities have fewer impoverished neighborhoods, no ethnic ghettos (although low-income ethnic districts do exist), lower crime rates, and, in general, better public transportation; but rush

---

Box 1-1 Major Geographic Features of North America

1. North America encompasses two of the world's biggest states territorially (Canada is the second-largest in size; the United States is third).

2. Both Canada and the United States are federal states. Canada's is divided into ten provinces and three territories. The United States consists of 50 States, the Commonwealth of Puerto Rico, and a number of island territories under U.S. jurisdiction in the Caribbean Sea and the Pacific Ocean.

3. Both Canada and the United States are plural societies. Canada's pluralism is most strongly expressed in regional bilingualism. In the United States, divisions occur largely along ethnic, racial, and income lines.

4. North America's population, not large by international standards, is one of the world's most highly urbanized and mobile, its enduring growth largely propelled by continuing waves of immigration.

5. By world standards, this is a rich realm where high incomes and high rates of consumption prevail. North America possesses a highly diversified resource base, but nonrenewable fuel and mineral deposits are consumed prodigiously.

6. North America has long been home to one of the world's great manufacturing complexes that generated an intricately connected, mature urban system. Over the past few decades, North America has come to rely increasingly on a highly advanced information economy and high-technology industries.

7. The two countries heavily depend on each other for supplies of critical raw materials (e.g., Canada is a leading source of U.S. energy imports) and have long been each other's primary trading partners.

8. The U.S.-Mexico border separates the North American realm from Middle America; it is a sharp boundary in some respects but in other ways a broadening transition zone.

9. Continued immigration and high transnational mobility make for an exceptionally diverse multicultural realm—but increasingly, immigration is politically contested.

Box 1-2 The U.S-Mexico Borderland: Boundary or Transition Zone?

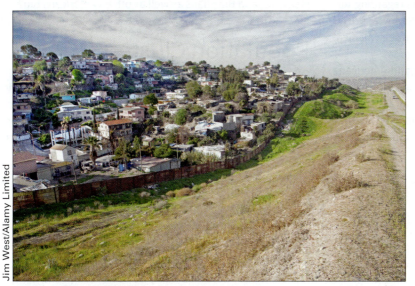

A stretch of the fortified fence along the Mexican border, suggesting a sharp boundary between the United States and Mexico. This is a ground-level view of the improvised wall that slashes westward, between Tijuana, Mexico, and the southernmost suburbs of San Diego, California, as it makes its way toward the Pacific shore on the horizon. The poverty-stricken cityscape of this eastern part of Tijuana extends almost to the rusty barrier itself, while development on the U.S. side has kept its distance from this well-patrolled segment of the border.

Most of what is now the southwestern United States used to belong to Mexico. In the so-called Treaty of Guadalupe Hidalgo (1848), Mexico ceded the territory to the United States. Today, that history is still reflected here in the presence of substantial numbers of Mexicans, Mexican Americans, and Americans of Mexican descent. Therefore, in terms of cultural geography, the U.S.-Mexico borderland [1] should be viewed as a broad transition zone [2] marked by a decidedly Mexican imprint on the landscape on both sides of the actual boundary. This duality extends into other spheres as well. The infusion of Mexican ethnicity profoundly shapes the demography of this quadrant of the country (see Fig. 1-2). And in the twenty-first century, the economies of the United States and Mexico are becoming ever more closely linked as their longstanding trade partnership intensifies—fully two-thirds of Mexican exports are now being shipped to its northern neighbor.

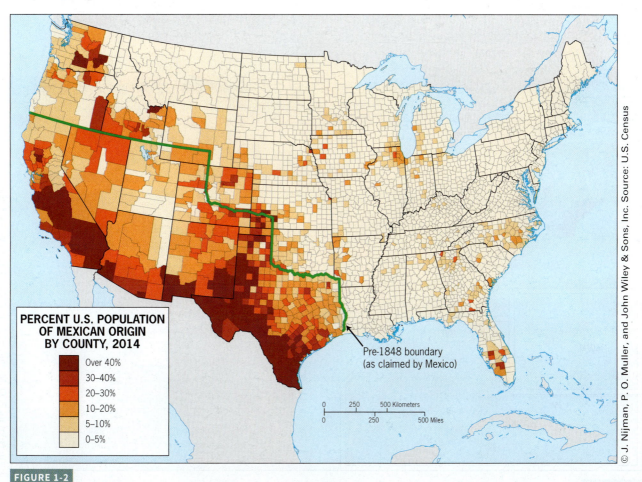

PERCENT U.S. POPULATION OF MEXICAN ORIGIN BY COUNTY, 2014

Over 40%
30–40%
20–30%
10–20%
5–10%
0–5%

Pre-1848 boundary (as claimed by Mexico)

© J. Nijman, P. O. Muller, and John Wiley & Sons, Inc. Source: U.S. Census

FIGURE 1-2

hour in Toronto very much resembles rush hour in Chicago. The boundary between the two states remains porous, even after the events of 9/11 and the recurring alerts regarding terrorism. Americans living near the border shop for lower-priced medicines in Canada; Canadians who can afford it seek medical treatment in the United States.

None of this is to suggest that Canadians and Americans don't have their differences—they do. Most of these differences may seem subtle to people from outside the realm, but they are important nonetheless, particularly for Canadians who at times feel dominated by Americans whose political or cultural values they do not always share. You can even argue over names: if this entire hemisphere is called the Americas, aren't Canadians, Mexicans, and Brazilians as well as others Americans too?

Population Clusters

Although Canada and the United States share many historical, cultural, and economic qualities, they also differ in significant ways and thereby diversify this realm. The United States, somewhat smaller territorially than Canada, occupies the heart of

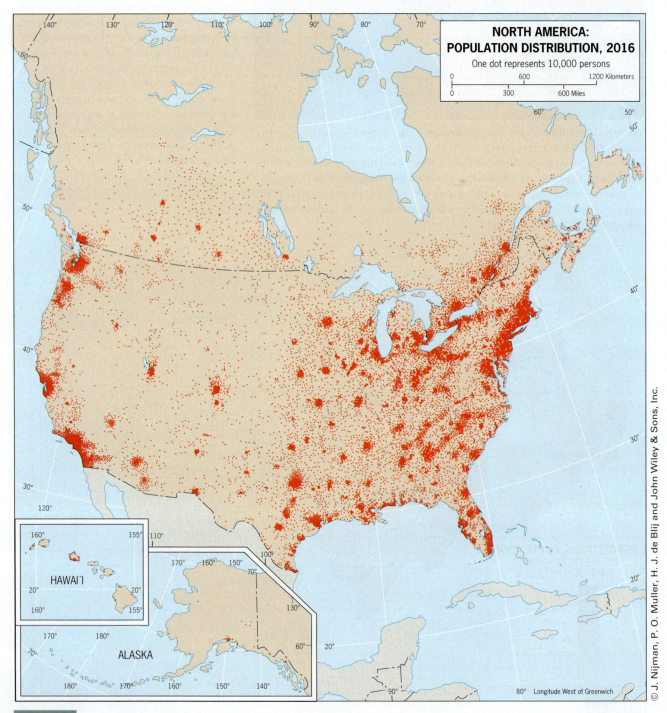

NORTH AMERICA: POPULATION DISTRIBUTION, 2016

One dot represents 10,000 persons

HAWAI'I

ALASKA

Longitude West of Greenwich

© J. Nijman, P. O. Muller, H. J. de Blij and John Wiley & Sons, Inc.

FIGURE 1-3

Box 1-3 Major Cities of the Realm 2016

	Population metro area, 2016 (in millions)
New York	20.7
Los Angeles	15.1
Chicago	9.2
Toronto, Canada	6.6
Dallas-Ft. Worth	6.3
Houston	6.0
San Francisco	6.0
Miami-Ft. Lauderdale-West Palm Beach	5.8
Philadelphia	5.6
Atlanta	5.1
Washington, D.C.	5.0
Boston	4.5
Phoenix	4.3
Detroit	3.7
Montreal, Canada	3.6
Seattle	3.5
San Diego	3.1
Minneapolis-St. Paul	2.8
Vancouver, Canada	2.3
Ottawa, Canada	1.0

The population of the United States, which reached 325 million in 2017, is growing 25 percent faster than that of Canada, and the 400 million mark may be surpassed as soon as 2050. This is an unusually high rate of growth for a high-income economy, resulting from a combination of natural increase (which leads today) and substantial immigration (on a pace to lead after 2030).

Although Canada's overall growth rate is significantly lower than that of the United States, immigration contributes proportionally even more to this increase than in the United States. With 36.3 million residents, Canada, like the United States, has been relatively open to legal immigration, and as a result both societies exhibit a high degree of cultural diversity in ancestral and traditional backgrounds. Indeed, Canada recognizes two official languages, English and French (the United States does not even designate English as such); and by virtue of its membership in the British Commonwealth, East and South Asians form a larger sector of Canada's population than Asians and Pacific Islanders do in the United States. In contrast, cultural diversity in the United States reflects large Hispanic (17.4 percent), African American (13.2 percent), and Asian (6.3 percent) minorities as well as a comprehensive range of other ethnic backgrounds.

Robust urbanization, substantial immigration, and cultural diversity are defining properties of the human geography of the North American realm. But before going further, we must also become familiar with the physical stage on which the human drama is unfolding.

North America's Physical Geography

Physiographic Regions

One of the distinguishing properties of the North American landmass, all of which lies on the North American tectonic plate, is its remarkable variegation into regional physical landscapes. So well defined are many of these landscape regions that we use their names in everyday parlance—for example, when we say that we flew over the Rocky Mountains or drove across the Great Plains or hiked in the Appalachians. These landscapes are called **physiographic regions [3]**, and nowhere else in the hemisphere is their diversity greater than to the north of the Rio Grande Valley (**Fig. 1-4**).

In the Far West, the Pacific Mountains extend all the way from Southern California through coastal Canada to Alaska. In the western interior, the Rocky Mountains form a continental backbone from central Alaska to New Mexico. Around the Great Lakes, the low-relief landscapes of the Interior Lowlands and the Great Plains to the west are shared by Canada and the United States, and the international boundary even divides the Great Lakes. In the east, the Appalachian Mountains (which, as the cross-sectional inset shows, is no

the North American continent and, as a result, encompasses a greater environmental range. The U.S. population is dispersed across most of the country, forming major concentrations along both the (north-south trending) Atlantic and Pacific coasts (**Fig. 1-3**). The overwhelming majority of Canadians, on the other hand, live in an interrupted east-west corridor that extends across southern Canada, mostly within 300 kilometers (200 mi) of the U.S. border. The United States, again unlike Canada, is a fragmented country in that the broad peninsula of Alaska is part of it (offshore Hawai'i, however, belongs in the Pacific Realm).

Figure 1-3 reveals that the great majority of both the U.S. and Canadian population reside to the east of a north-south line drawn down the center of the realm, reflecting historic core-area development in the east as well as the later and still-continuing shift to the west, and, in the United States, also to the south. Certainly this map shows the urban agglomerations that dominate the distribution of North America's population: you can easily identify cities such as Toronto, Chicago, Denver, Dallas-Fort Worth, San Francisco, and Vancouver (see data displayed in **Box 1-3**). Overall, nearly 85 percent of the realm's population is concentrated in cities and towns, a higher proportion even than Europe's.

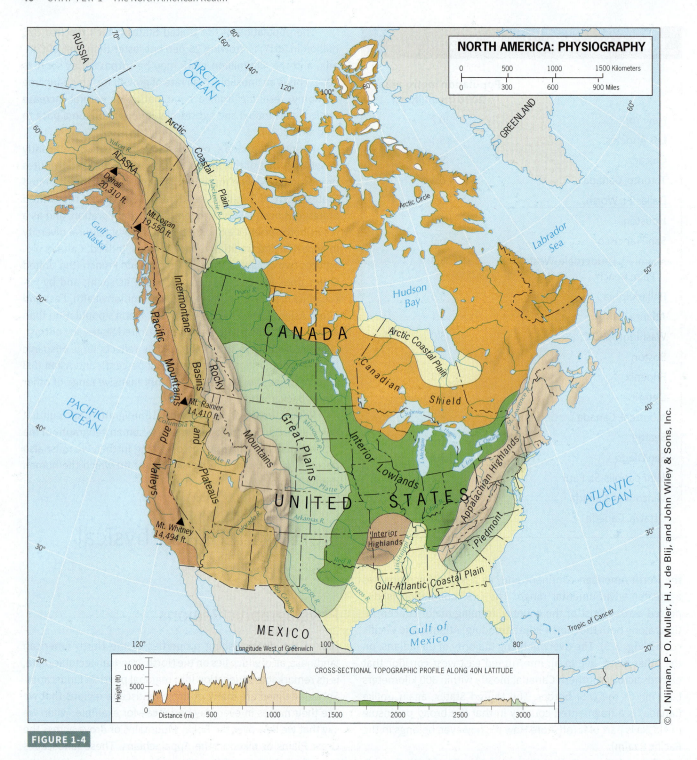

NORTH AMERICA: PHYSIOGRAPHY

CROSS-SECTIONAL TOPOGRAPHIC PROFILE ALONG 40N LATITUDE

FIGURE 1-4

match for the Rockies) form a corridor of ridges, valleys, and plateaus that represent a familiar topography from Alabama and Georgia to Nova Scotia and Newfoundland. If there is a major physiographic province that belongs to only one of the realm's two countries, it is the Canadian Shield, scoured bare by the Pleistocene glaciers that deposited their pulverized rocks as fertile soil in the U.S. Midwest, that is, in the Interior Lowlands.

Climate

This diversity of landscapes is matched by a variety of climates. Take another look at Figure G-6, and you can see the lineaments of landscape mirrored in the contrasts of climate. North America may not have it all—there are no areas of true tropical environment in the North American realm to speak of except at the southern tip of Florida—but it does exhibit a great deal

© B.A.E. Inc. / Alamy Inc

Superstorm Sandy, the powerful remnant of a Caribbean hurricane of the same name, will forever be remembered along the Mid-Atlantic seaboard of the northeastern United States (and far beyond) as the second-costliest natural disaster in American history, through mid-2016. It struck with a vengeance at high tide on the night of October 28–29, 2012, unleashing sustained winds of up to 130 kilometers per hour (80 mph) and a 2.6-meter (9-ft) storm surge of seawater. The brunt of Sandy's landfall was absorbed by the low-lying northern New Jersey shore. The heaviest wind and flooding damage occurred on the narrow offshore barrier islands that "guard" and parallel the mainland. This aerial view encompasses the oceanside resort town of Mantoloking on the well-known barrier island that terminates a few miles south at the Jersey Shore's popular Island Beach State Park. The inset photo was taken a few months prior to Sandy, and the most dramatic landscape change is the new inlet at the end of the bridge (center of main photo) that breached the island between the Atlantic Ocean (right) and Barnegat Bay (left), the lagoon that separates Mantoloking from Brick Township on the mainland.

of variation. North America contains moist coastal zones and arid interiors, well-watered plains, and even bone-dry deserts. On the world map, *Cf* and *Df* climates are especially good for commercial farming; note how large North America's share of these environments is.

Temperature-wise, this map leaves no doubt: the farther north you go, the colder it gets. And even though coastal areas derive moderation from warmer offshore waters, the

rigors of **continentality [4]** (inland climate environment remote from ameliorating maritime influences) set in not far from the coast. Hot summers, frigid winters, and limited precipitation make these higher-latitude continental interiors difficult places to make it on the land. Figure G-6 has much to do with the southerly concentration of the Canadian people as well as the lower population densities that mark the realm's interior.

In the west, especially in the United States, you can see the effects of the Pacific Mountains on inland areas. Moisture-laden air arrives from the ocean, the mountain wall forces the air upward, cools it, condenses the moisture in it, and produces rain—the rain for which Seattle and Portland as well as other cities of the Pacific Northwest are (in)famous. By the time the air crosses the mountains and descends on the landward side, most of the moisture has been extracted from it, and the forests of the ocean-facing side quickly yield to scrub and brush. This **rain shadow effect [5]** extends all the way across the Great Plains; North America does not turn moist again until the Gulf of Mexico sends humid tropical air northward into the eastern interior via the Mississippi Basin.

In a very general way, therefore, and not including the coastal strips along the Pacific, nature divides North America into an arid west and a humid east. Again the population map reveals more than a hint of this: draw a line approximately from Lake Winnipeg to the mouth of the Rio Grande, and look at the contrast between the (comparatively) humid east and drier west when it comes to population density. Water is a critical part of this story.

Between the Rocky Mountains and the Appalachians, North America lies open to air masses from the frigid north and tropical south. In winter, southward-plunging polar fronts send frosty, bone-dry air masses deep into the heart of the realm, turning even places like Memphis and Atlanta into iceboxes; in summer, hot and humid tropical air surges northward from the Gulf of Mexico, giving Chicago and Toronto a taste of the tropics. Such air masses clash in low-pressure systems along weather fronts loaded with lightning, thunder, and, frequently, dangerously destructive tornadoes. And the summer heat brings an additional threat to the Gulf-Atlantic Coastal Plain (Fig. 1-4): hurricanes capable of inflicting catastrophic devastation on low-lying areas (see photo). These tropical cyclones also prune natural vegetation, replenish groundwater reservoirs, fill natural lakes, and flush coastal channels.

Great Lakes and Great Rivers

Two great drainage systems lie between the Rockies and the Appalachians: (1) the five Great Lakes that drain into the St. Lawrence River and the Atlantic Ocean, and (2) the mighty Mississippi-Missouri river system that carries water from a vast interior watershed to the Gulf of Mexico, where the Mississippi forms one of the world's major **deltas**. Both natural systems have been modified by human engineering. In the case of the St. Lawrence Seaway, a series of locks and canals has created a direct shipping route, via the Great Lakes and their outlet, from the Midwest to the Atlantic. The Mississippi has been fortified by artificial levees that, while failing to contain the worst of flooding, have enabled farmers to cultivate the most fertile of American soils.

Native Americans and European Settlement

When the first Europeans set foot on North American soil, the continent was inhabited by millions of people whose ancestors had reached the Americas from Asia via Alaska, and possibly also across the Pacific, more than 14,000 years earlier (and perhaps as long as 30,000 years ago). In search of Asia, the Europeans misnamed them Indians. These Native Americans or **First Nations**—as they are now called, respectively, in the United States and Canada—were organized into hundreds of nations with a rich mosaic of languages and a great diversity of cultures.

The eastern nations were the first to bear the brunt of the European invasion. By the end of the eighteenth century, ruthless and land-hungry settlers had driven most of the native peoples living along the Atlantic and Gulf coasts from their homes and lands, initiating a westward push that was to devastate indigenous society. Today, less than 3 million Native Americans remain in the United States, and about one-third that number in Canada. In the United States they are left with only about 4 percent of the national territory in the form of mostly impoverished reservations. Aboriginal peoples today are better off territorially in Canada where they hold titles to large tracts of land, especially in the northern sectors of British Columbia and Quebec (but keep in mind that northern Canada is nearly empty to begin with and that environmental conditions there can be quite harsh).

The realm's current population geography is rooted in the colonial era of the seventeenth and eighteenth centuries, which was dominated by Britain and France (**Fig. 1-5**). The French sought mainly to organize a lucrative fur-trading network, while the British established settlements along the coast of what is now the eastern seaboard of the United States (the oldest, Jamestown, was founded in tidewater Virginia more than four centuries ago).

Cultural Foundations

The modern American creed, if one can be identified, is sometimes idealistically characterized as exhibiting an adventurous drive, a liking for things new, a willingness to move, a sense of individualism, an aggressive pursuit of goals and ambition, a need for societal acceptance, and a firm sense of destiny. These qualities are not unique to modern U.S. culture, of course, but in combination they seem to have created a particular and

pervasive mindset that is reflected in many ways in this geographic realm. Generally speaking, these tendencies tend to translate into an intense pursuit of educational and other goals to fulfill high expectations concerning upward socioeconomic mobility (at least among the aspiring classes). Thus the *American Dream*, whether a genuine prospect or an ideological fabrication, is very much an expression of this set of cultural values, and it is closely entwined with the character of North America as a land of immigrants.

Facilitating these aspirations is **language**. None of these goals would be within reach for so many were it not for the use of English throughout most of the realm. In Europe, by contrast, language inhibits the mobility so routinely practiced by Americans: workers moving from Poland to Ireland, for example, found themselves at a disadvantage in competition with immigrants from Ghana or Sri Lanka. In North America, a worker from Arkansas would not even consider language to be an issue when applying for a job in Toronto.

Also reflecting the cultural values cited above is the role of **religion**, which sets American (more so than Canadian) culture apart from much of the rest of the Western world. The overwhelming majority of North Americans express a belief in God, and a (declining) majority regularly attend church, in contrast to much of Europe. Two out of three people in the United States say they adhere to a religion, and three out of ten say they attend a religious service once a week (more than in Iran!). Religious observance is a virtual litmus test for political leaders; no other developed country prints "In God We Trust" on its currency.

Figure 1-6 shows the mosaic of Christian faiths that blankets the realm, but a map at this scale can only suggest the broadest outlines of what is a much more intricate pattern. Protestant denominations are estimated to number in the tens of thousands in North America, and no map can capture them all. Southern (and other) Baptists form the majority across the U.S. Southeast from Texas to Virginia, Lutherans in the Upper Midwest and northern Great Plains, Methodists in a belt across the Lower Midwest, and Mormons in the interior West centered on Utah. Roman Catholicism prevails in most of Canada as well as the U.S. Northeast and Southwest, where ethnic Irish and Italian adherents form majorities in the Northeast and Hispanics in the Southwest. Behind these patterns lie histories of proselytism, migration, conflict, and competition, but tolerance of diverse religious (even nonreligious) views and practices is a hallmark of this realm.

The Federal Map of North America

The two states that constitute North America may have arrived at their administrative frameworks with different objectives and at different rates of speed, but the result is unmistakably similar: their internal political geographies are dominated

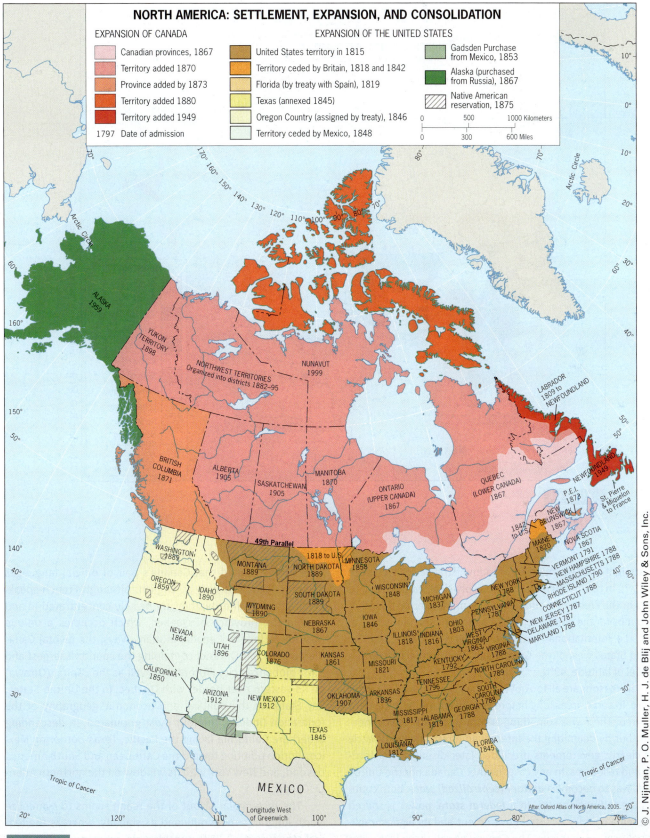

NORTH AMERICA: SETTLEMENT, EXPANSION, AND CONSOLIDATION

EXPANSION OF CANADA

- Canadian provinces, 1867
- Territory added 1870
- Province added by 1873
- Territory added 1880
- Territory added 1949
- 1797 Date of admission

EXPANSION OF THE UNITED STATES

- United States territory in 1815
- Territory ceded by Britain, 1818 and 1842
- Florida (by treaty with Spain), 1819
- Texas (annexed 1845)
- Oregon Country (assigned by treaty), 1846
- Territory ceded by Mexico, 1848
- Gadsden Purchase from Mexico, 1853
- Alaska (purchased from Russia), 1867
- Native American reservation, 1875

0   500   1000 Kilometers
0   300   600 Miles

FIGURE 1-5

© J. Nijman, P. O. Muller, H. J. de Blij and John Wiley & Sons, Inc.

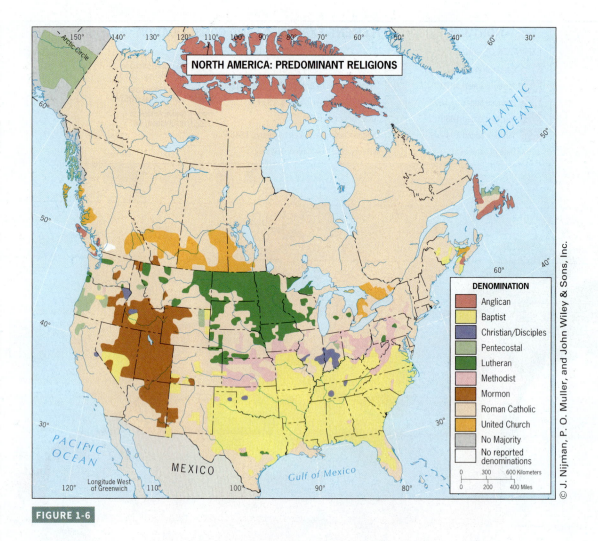

NORTH AMERICA: PREDOMINANT RELIGIONS

DENOMINATION

- Anglican
- Baptist
- Christian/Disciples
- Pentecostal
- Lutheran
- Methodist
- Mormon
- Roman Catholic
- United Church
- No Majority
- No reported denominations

© J. Nijman, P. O. Muller, and John Wiley & Sons, Inc.

FIGURE 1-6

by straight-line boundaries of administrative convenience (Fig. 1-7). In comparatively few places, physical features such as rivers or the crests of mountain ranges mark internal boundaries, but by far the greater length of boundaries is ruler-straight. Even the international boundary between Canada and the United States west of the Great Lakes mostly coincides with a parallel of latitude, 49°N.

The reasons are all too obvious: the framework was laid out before, in some areas long before, significant white settlement occurred. There was something arbitrary about these delineations, but by delimiting the internal administrative units early and clearly, governments precluded later disputes over territory and resources. In any case, neither Canada nor the United States was to become a **unitary (centralized) state**: both countries are **federations [6]**. In a **federal state**, power is shared between a country's central government and its highest-ranking political subdivisions. The Canadians call these first-order subdivisions provinces, whereas in the United States they are, as the country's name implies, States.

It is no trivial matter to take a careful look at some of the provinces and States that compartmentalize this realm because some are of greater consequence than others. We have already noted that Canada's pair of official languages includes French as well as English. Although French-speakers are in the minority, they are heavily concentrated in and around the province of Quebec, which occupies a unique position in federal Canada. In many ways, the pivotal Canadian province is Ontario, containing the country's largest and most globalized city (Toronto) as well as the capital (Ottawa) on its eastern river border with Quebec. In the United States, key States in the political and economic geography of the federation (indeed, the four most populated, in descending order) are California facing the Pacific, Texas bordering Mexico, Florida pointing toward the Caribbean and South America beyond, and New York with its window on the northern Atlantic Ocean.

The rectangular layout of the realm seems to symbolize North America's modernity and its stability, whereas the federal structure itself (with considerable autonomy at the State or provincial level) reflects the culture of freedom and independence. If federalism works within these highly advanced economies, why shouldn't the rest of the world emulate it? Federalism does indeed have its assets, but it also carries some significant liabilities.

FIGURE 1-7

The Distribution of Natural Resources

One of these liabilities has to do with the variable allotment of natural resources among the States and provinces. In Canada, Alberta is favored with massive energy reserves (and even more in the offing), but its provincial government is not always eager to see the federal administration take that wealth to assist less affluent provinces. In the United States, when oil prices are high, Texas benefits while California's budget suffers. So it is important to compare the map of natural resources (**Fig. 1-8**) with the map of States and provinces. Keep in mind, though, that the world market price of raw materials, including oil, is notoriously volatile: in mid-2016, oil prices were less than one-third of what they were only two years earlier, with a massive impact on producing and consuming regions.

Water certainly is a natural resource, and North America as a realm is comparatively well supplied with it despite concerns over long-term prospects in the American Southwest and the Far West, where some States depend on sources in other States. Another concern focuses on lowering the water tables in some of North America's most crucial **aquifers [7]** (underground water reserves) in which a combination of overuse and decreasing replenishment foretells the rise of supply problems.

North America is endowed with abundant reserves of **minerals** that are mainly found in three zones: the Canadian Shield north of the Great Lakes, the Appalachian Mountains in the East, and the mountain ranges of the West. The Canadian Shield contains substantial iron ore, nickel, copper, gold, uranium, and diamonds. The Appalachians yield lead, zinc, and iron ore. And the western mountain zone has significant deposits of copper, lead, zinc, molybdenum, uranium, silver, and gold.

In terms of **fossil fuel [8]** energy resources (oil, natural gas, and coal), North America is also quite well endowed. Nonetheless, the voracious demand of the United States, the highest in the world, cannot be met by domestic supplies alone and generates significant imports. Figure 1-8 displays the distribution of oil, natural gas, and coal deposits.

The leading **oil**-production areas lie (1) along and offshore from the Gulf Coast, where the floor of the Gulf of Mexico is yielding a growing share of the output; (2) in the Midcontinent District, from western Texas to eastern Kansas; and (3) along Alaska's North Slope facing the Arctic Ocean. An important development is taking place in Canada's northeastern Alberta, where oil is being drawn from deposits of *tar sands* in the vicinity of Fort McMurray. The process is expensive and can reward investors only when the price of oil is comparatively high (in 2016, tar-sand oil was selling at a loss), but the reserves of these oilfields may exceed those of Saudi Arabia.

The distribution of **natural gas** reserves resembles that of oilfields because petroleum and natural gas tend to be found in similar geological formations (i.e., the floors of ancient shallow seas). What the map cannot reveal is the volume of production, in which this realm leads the world (Russia and Iran lead in proven reserves). That output has risen significantly in recent decades, as natural gas has become the fuel of choice for electricity generation in North America. Driving this expansion has been the widespread application of hydraulic fracturing (**fracking**) technology. By injecting pressurized fluids into deeply buried shale rocks to create fractures, vast quantities of trapped gas can be extracted, resulting in surging supplies and dropping prices. (Fracking is also widely used to extract oil from shale formations.)

The **coal** reserves of North America, perhaps the largest on the planet, are found in Appalachia, beneath the Great Plains of the United States as well as Canada, and in the southern Midwest, among other places. These reserves guarantee an adequate supply for centuries to come, although coal has become a less desirable fuel owing to its polluting emissions (especially carbon dioxide) that contribute to global warming.

Given the relative scarcity of oil and the geopolitical disadvantages of having to import almost one-quarter of what is needed in the United States (24 percent in 2015, the lowest percentage since 1970), you might expect the use of nuclear energy to be more widespread, especially because it is relatively cheap to produce. The United States has just over 60 nuclear power plants that provide roughly one-fifth of all electrical energy (Canada has five plants that produce about 15 percent of the country's electricity). But ever since the notorious radiation leak and near-calamity at Pennsylvania's Three Mile Island in 1979, the fear of nuclear accidents has been pervasive. In fact, the federal government did not approve new nuclear plants again until 2012, and the first such facility opened in Tennessee in mid-2016.

Urbanization and the Spatial Economy

Industrial Cities

When the Industrial Revolution (which originated in England around 1800) crossed the Atlantic to America in the 1870s, it took hold so successfully and advanced so robustly that only 50 years later North America was surpassing Europe as the world's mightiest industrial stronghold. Industrialization progressed in tandem with urbanization as manufacturing plants in need of large labor supplies were built in and near cities. This, in turn, propelled rural-urban migration (including the migration of African Americans from the South to northern cities during the first half of the twentieth century) and the growth of individual cities. At a broader scale, a *system* of new cities emerged that specialized in the collection and processing of raw materials and the distribution of manufactured products. This **urban system [9]** was interconnected through a steadily expanding network of railroads, themselves a product of the Industrial Revolution.

As new technologies and innovations emerged, and as specializations such as Detroit's automobile industry strengthened,

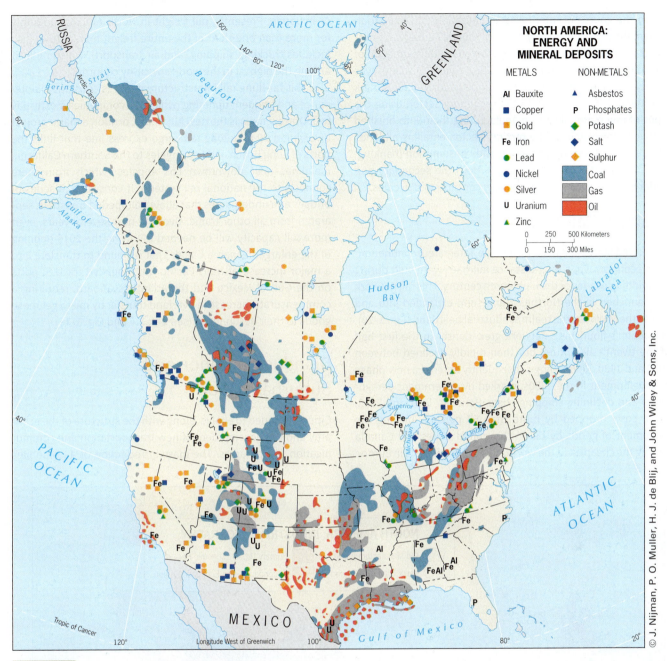

FIGURE 1-8

American Manufacturing Belt [10] (which included southern Ontario) swiftly evolved into the cornerstone of the North American Core (**Fig. 1-9**). This core area, which soon became the world's most productive and important, contained the majority of the realm's industrial activity and leading cities, including New York, Chicago, Toronto, and Pittsburgh (the "steel city").

Wind farms have been spreading to many locales in North America as the clean-energy campaign to generate electricity from sources other than fossil fuels accelerates. This wind farm is located in the San Gorgonio Pass (near Palm Springs), one of the windiest places in Southern California. The tallest of the 2,700 turbines here reach 53 meters (160 ft). This facility generates enough electricity to power about 300,000 homes.

In the course of interlinked urbanization and industrialization, the spatial economy of North America underwent profound changes. The ***primary*** sector, involving the extraction of raw materials from nature (agriculture, mining, fishing), was rapidly mechanized, and the workforce in this sector shrank considerably. Employment in the ***secondary*** sector, using the input of raw materials and manufacturing them into finished products, grew rapidly. The ***tertiary*** sector, entailing all kinds of services to support production and consumption (banking, retail, transport), expanded as well. Both the secondary and tertiary sectors were overwhelmingly concentrated in cities.

Realm of Railroads

The United States first became a single, integrated, continental-scale economy when long-distance railroads were built during the middle decades of the nineteenth century. This rail network was launched in the Northeast and soon expanded into and across the Midwest, the realm's industrial heartland (**Fig. 1-10**). The realm's flourishing network grew denser in the first half of the twentieth century, but then rapidly declined between 1960 and 1980. Railroading subsequently entered a new phase of development in the 1980s, marked by government deregulation and declining shipping costs. Between 1981 and 2012, those costs dropped by more than 50 percent, making the movement of goods by rail in the United States and Canada among the cheapest in the world, half of what it costs in Europe

or in Japan. Presently, rail freight in North America accounts for more than one-third of the entire freight market (that also includes trucking, shipping, and air transport), which ranks it among the highest in the world's wealthy countries.

This far-flung rail infrastructure requires constant maintenance and frequent updating (which rail companies mostly pay for). One example is the new Alameda Corridor, a major project involving a 20-mile-long rail cargo express line that links the ports of Long Beach and Los Angeles to the Southern California mainlines near downtown Los Angeles (which from there connects to the national rail system). It consists of a series of bridges and underpasses that allow unimpeded transport, separated from all other traffic. Over the next several years, even more rail capacity will be needed following the 2016 opening of the enlarged Panama Canal. That is bound to translate into a major increase of trans-Pacific cargo headed for U.S. ports on the Gulf of Mexico and the Atlantic seaboard, requiring a commensurate increase in overland transfer by rail to get these goods to their destinations (see **Box 1.4** and **Fig. 1-11**).

Deindustrialization and Suburbanization

Cities continued to evolve, along with the structure of the economy and the development of new transportation and communication technologies. The mass introduction of automobiles

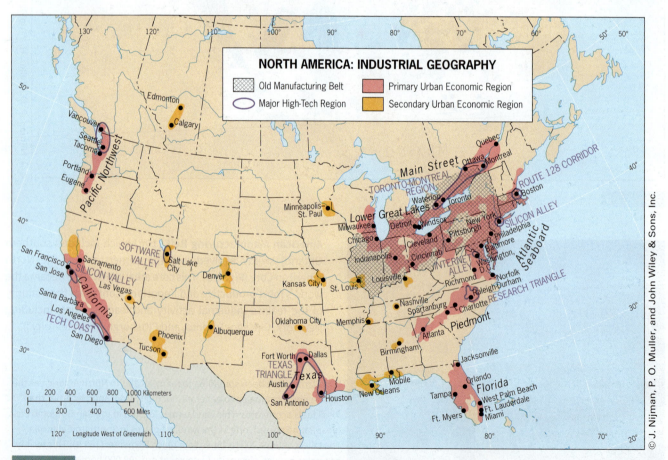

FIGURE 1-9

© J. Nijman, P. O. Muller, and John Wiley & Sons, Inc.

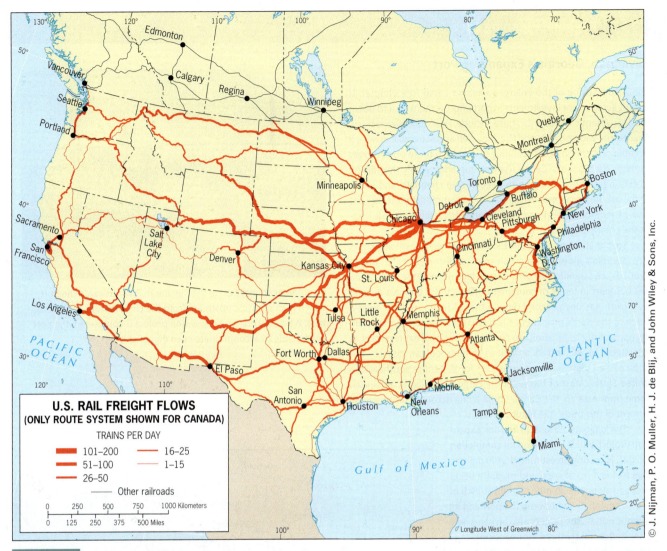

FIGURE 1-10

and the accompanying construction of a large-scale highway system, especially after World War II, had two major effects on cities. First, it resulted in much higher interconnectedness and mobility among cities. Second, it propelled a process of *sub-urbanization*—the transformation from compact city to widely dispersed metropolis through the evolution of residential suburbia into a complete **outer city [13]** with its own clusters of office businesses and industries, sports and entertainment complexes, and myriad other amenities. As the newly urbanized suburbs increasingly captured major economic activities, many central cities saw their comparative status diminish.

Importantly, this transformation coincided with the **dein-dustrialization [14]** that devastated American cities in the 1970s, manifested in the loss of manufacturing jobs as a result of automation and the relocation of production to countries with lower wages. The suburban outer cities now became the primary destination of intrametropolitan migrants and of economic activity and employment in the service sector. Meanwhile, the **central business district (CBD) [15]** of the central city lost its role as the dominant metropolitan core,

with many of its functions now reduced to serving the less affluent populations of the central city's close-in neighborhoods. The fate of Detroit (see photo) is often cited as exemplifying this sequence of events, but to a certain degree all major U.S. cities suffered.

The Information Economy and City Regions

Since the 1980s, a number of North American cities have rebounded from the destructive effects of deindustrialization and unemployment in conjunction with the arrival of the **infor-mation economy [16]**. Many northern cities had lost population and experienced "hollowing out" for nearly two decades, but for a few lucky ones the tide had turned by the end of the twentieth century. Employment in the tertiary sector started to grow rapidly, especially in high technology, producer services (such as consulting, advertising, and accounting), finance, research and development, and the like. The geography of the

Box 1-4   Regional Planning Case

Savannah, Georgia's Expanding Port

The 2016 opening of the enlarged Panama Canal (see Chapter 2) has triggered major investments in port improvement up and down the U.S. Atlantic seaboard. These include New York–New Jersey, Charleston, Jacksonville, and Miami as well as Savannah, because a sizeable proportion of the shipping to and from these ports moves through the Panama Canal. Typical upgrades entail the raising of existing bridges, dredging to deepen ship channels, and state-of-the-art terminal modernization to accommodate the bigger, new-generation ("Neo-Panamax") container vessels that now transit the enlarged Canal.

Savannah, home to about 135,000, is mostly known as a small city (Georgia's oldest) with a rich colonial history, but that image overlooks the significance of its port. In fact, it is now the fourth busiest U.S. port in terms of container traffic, and for the past decade its growth has been second only to that of New York–New Jersey. Much of the port's prominence is based on Savannah's accessibility, particularly its proximity to such major regional expressways as Interstates 16, 20, and 95 as well as to metropolitan Atlanta—the unparalleled **distribution center [11]** for the entire southeastern United States. Most of the seaborne cargoes arriving at Savannah come from Asia via the Canal. China alone accounts for nearly 30 percent of this trade (see **Figure 1-11**).

The ongoing Savannah Harbor Expansion Project, which is targeted for completion in 2021, is costing an estimated $700 million. Five feet (1.5 m) of sand and mud are being dredged along the 30-mile (50-km) stretch of the Savannah River between the port and its Atlantic outlet. The project also involves new and better **intermodal connections [12]** (mainly, ship to truck or train), port terminal upgrades, and new regional warehouse distribution facilities.

The Expansion Project's overall goals are to accelerate the local growth of container traffic and to solidify Savannah's position as one of the realm's leading ports. But first, two challenges must be confronted. One is that the country's Pacific-coast ports can already accommodate larger container ships, and may therefore be able to absorb even more of the U.S. oceanic trade with Asia. The other is that transshipment activities (transfer of goods to smaller ships that service ports with shallower harbors) are expected to grow at favored locations in the Caribbean Basin, which could divert cargo that would otherwise flow through America's east-coast ports. Time will tell if these port upgrading and infrastructural investments prove to be successful. Today, however, it is clear that Savannah's immediate economic future is heavily staked on its becoming a winner in the intense competition to capture more of the Asian container trade in the dawning Neo-Panamax era.

A container ship glides up the Savannah River towards the port, which is located just upstream from the city's famous historic district. When the ongoing expansion project is completed, the port will be able to accommodate even bigger ships.

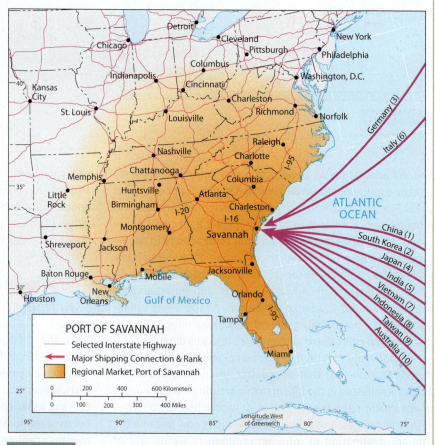

FIGURE 1-11

An abandoned house in Detroit's Brush Park neighborhood, barely 2 miles (3 km) north of downtown, sits by itself during the summer of 2014. The poster child of a city ravaged by deindustrialization, the Motor City has followed a downward spiral in population, workforce, and tax base since the 1960s. In 2013, Detroit was forced to file for bankruptcy, by far the biggest American city ever to do so. Some say that Brush Park now possesses "regrowth potential" thanks to its low real estate values, but to date few developers have taken notice.

information economy is quite different from that of the old manufacturing industries and is still not fully understood. Some of it concentrates in established CBDs such as Midtown Manhattan; other economic activities center around important hubs of the Internet infrastructure to benefit from maximum bandwidth (e.g., North Dallas); and still other activities cluster in suburban areas containing large pools of highly skilled workers, notably Silicon Valley, metropolitan Seattle, and North Carolina's Research Triangle. The largest and best known high-technology clusters in North America are mapped in Figure 1-9.

Northern California's **Silicon Valley**—the world's leading hub of computer research and development as well as the headquarters of the U.S. microprocessor industry—exquisitely illustrates the locational dynamics of this sector of the spatial economy. Proximity to Stanford, a world-class research university; adjacent to cosmopolitan San Francisco; the major local concentration of highly educated and skilled workers; a strong, thriving business culture; ample local investment capital; high-quality housing; and a scenic environment with good weather—all combine to make Silicon Valley the prototype for similar developments now springing up around the globe. Under different names (e.g., *technopolis* is used in Brazil, France, and Japan; *science park* in China, Taiwan, and South Korea), these ultramodern, campus-like complexes symbolize the digital economy just as the smoke-belching factory embodied the industrial age of the past (see **Box 1-5**).

The Polycentric Metropolis

Fly into any large American metropolitan region, and you will see the high-rises of **suburban downtowns** encircling the old central-city CBD, some of them boasting their own impressive skylines. Thus the overall structure of the modern North American metropolis has become polycentric, resembling a pepperoni pizza in its general form (**Fig. 1-12**). The traditional CBD still tends to be situated at the center, much of its former cross-traffic diverted by beltways; but the outer city's CBD-scale nodes are both ultramodern and thriving.

Efforts to attract businesses and higher-income residents back to the old CBD often entail the construction of multiple-use high-rises that displace low-income residents, producing conflicts and lawsuits. The revitalization and upgrading—or **gentrification [18]**—of crumbling downtown-area neighborhoods increases real estate values as well as taxes, which tends to drive lower-income, longtime residents from their homes. The rejuvenation of old CBDs and their immediate surroundings is occurring in numerous urban areas, from Manhattan's Lower West Side to central Seattle; but the growth and development of large interconnected metropolitan regions, anchored by multiple urban centers, is certain to continue.

Urban Geographies of Inequality

Inequality has always been a feature of the United States, more so than of Canada. One could argue that differences in income and well-being are part and parcel of all free (market) societies, where rewards are based on initiative, risk, talent, and hard work, all of which tend to vary among the population. But the particular organization of the economy and of the political system play a very important part—and this explains, for example, why Canada is generally less unequal. Think, for instance, of the role of progressive taxes, college tuition fees, minimum wages, rent subsidies, or healthcare—all matters of government intervention even in countries that generally embrace the free market. Of course, there are plenty of other countries that exhibit significant inequalities (e.g., South Africa, Brazil, Russia, and others), but among "Western" countries, the United States is generally considered highly unequal.

Inequality is often not only a matter of economics or class. It is also influenced by gender, ethnicity, race, age, marital status, and immigration category. For example, an elderly, single, immigrant woman is much more likely to be poor than a married, middle-aged, white, U.S.-born man. In addition, and this is where geography comes in, a person's fortunes are strongly influenced by the neighborhood in which she or he is born and raised or is trying to earn a living. And inequality does not only refer to income: it is also reflected in health conditions, access to transportation, educational achievement, and even longevity. One's neighborhood tends to have an impact on one's social network, contacts, social identity, frame of mind, aspirations, and life chances. Overall, this is known as the so-called **neighborhood effect [19]**.

U.S. cities, in particular, have always been characterized by neighborhoods that are distinct in terms of social composition, income level, and race/ethnicity. The economic well-being of a residential neighborhood can be quickly ascertained by examining housing prices: the more expensive an area the greater the earnings of its households. Yet a close look at high-end or lower-end residential areas, especially in U.S. cities, often also

Box 1-5 Technology & Geography

GPS, Sensors, and Self-Driving Cars

No single geospatial technology is more ubiquitous than global positioning systems or GPS. **GPS [17]** refers to the global network of satellites, sensors, and devices that are used to define absolute location, to determine elevation, and to calculate transportation routes. GPS satellites orbiting the Earth continuously transmit signals that are received by devices like our mobile phones to calculate precise locations. The signals are often enhanced by ground-based instruments such as cell-phone towers with well-known locations. Coupled with algorithms, digital maps, and other sensors, GPS is revolutionizing commerce, industry, and transportation.

Originally developed by the U.S. Department of Defense for military purposes, GPS is now one of the most widely used and applied geospatial technologies. In addition to providing directions on how to get from point A to point B, GPS technology serves as the foundation for autonomous or *self-driving vehicles*. Google's well-known self-driving car project is built on digital maps created with GPS, which are then enhanced by real-time data collected from sensors mounted on the cars themselves. According to Google, self-driving cars have already logged 1.5 million miles (2.4 million km) on the roads of California, Texas, and Washington State. Trials are expanding to other States in order to test performance under different conditions.

Once considered the stuff of science fiction, it is now envisioned that self-driving vehicles will become common within the next two decades. The ramifications for urban planning, social behavior, transportation systems, and regional geography could be enormous.

NOAH BERGER/Getty Images, Inc.

ELIJAH NOUVELAGE/Newscom

Left, a self-driving car traverses a parking lot at Google's headquarters in Mountain View, California, in January 2016. On the right, video captured by a Google self-driving car in late 2015: the vehicle's real-time sensing of its surroundings is visualized and coupled with the same street scene on film (lower left corner).

reveals correlations with race and ethnicity (see **Box 1-6** and **Fig. 1-13**).

Neighborhoods are subject to change, and occasionally they can move "up" as in the case of a number of central cities discussed above. Often, older residents are displaced, reinvestment and redevelopment take off, and gentrification gathers momentum. For the displaced residents, such improvement is obviously unwelcome, but their concerns mostly carry little weight. In other cases, growing unemployment and declining income can result in a neighborhood losing its vitality. The **Great Recession** of 2008–2011, often referred to as the housing mortgage crisis, negatively impacted hundreds of middle- and lower-income neighborhoods—deepening inequality across urban America's mosaic of residential landscapes.

Changing Geographies of Inequality and Economic Opportunity

Two major legacies of the Great Recession are the exacerbation of growing economic inequality and the reshaping of the map of economic opportunity. In the late 2010s, the longstanding gap in personal income continues to widen as the distance between affluence and poverty grows while the ranks of the intervening middle class steadily shrink. In the United States today, the number of families living in middle-income neighborhoods is more than one-third less than it was in 1970; during that same half-century, residential areas housing the wealthy *and* the poor have doubled in size.

Over the past two decades across North America, the most lucrative employment opportunities have shifted from manufacturing, construction, and other eroding middle-class job sectors into the burgeoning healthcare, technology, informatics, and energy sectors that require advanced levels of education and training. As a result, the pay-scale spectrum is beginning to resemble the bimodal pattern of the income-based residential fabric—expanding lower and upper extremes at the expense of a contracting middle.

Not surprisingly, the States with greatest inequality are those with the highest levels of high-tech and information-based employment such as California, Texas, New York, Connecticut, and Massachusetts. And those inequalities are the

THE MULTINODAL AMERICAN METROPOLIS

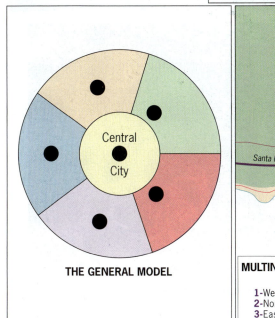

THE GENERAL MODEL

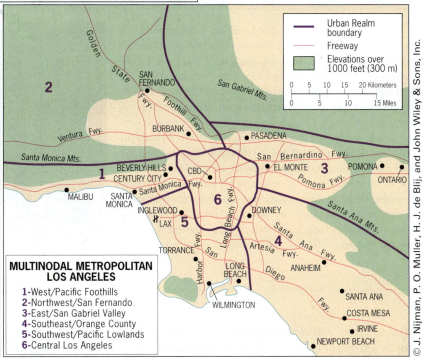

MULTINODAL METROPOLITAN LOS ANGELES

1-West/Pacific Foothills
2-Northwest/San Fernando
3-East/San Gabriel Valley
4-Southeast/Orange County
5-Southwest/Pacific Lowlands
6-Central Los Angeles

Legend:
- Urban Realm boundary
- Freeway
- Elevations over 1000 feet (300 m)

© J. Nijman, P. O. Muller, H. J. de Blij, and John Wiley & Sons, Inc.

A   **B**

FIGURE 1-12

Box 1-6 MAP ANALYSIS

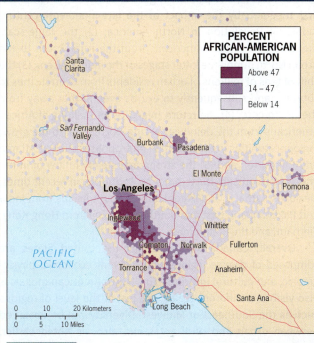

PERCENT AFRICAN-AMERICAN POPULATION
- Above 47
- 14 – 47
- Below 14

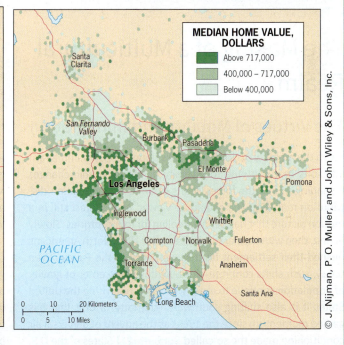

MEDIAN HOME VALUE, DOLLARS
- Above 717,000
- 400,000 – 717,000
- Below 400,000

© J. Nijman, P. O. Muller, and John Wiley & Sons, Inc.

FIGURE 1-13

North America is sometimes described as a melting pot, but an analysis of the internal spatial distribution of metropolitan populations in terms of income, race, or ethnicity suggests otherwise. The maps in Figure 1-13 allow you to compare the **residential geography [20]** of African Americans (left) with the distribution of home values (right) across Greater Los Angeles. How would you describe the distribution of African Americans, and how does it relate to patterns of home values? What does it reveal about the nature of inequality in U.S. urban regions?

Access your WileyPLUS Learning Space course to interact with a dynamic version of this map and to engage with online map exercises and questions.

Box 1-7 From the Field Notes …

Driving along San Francisco's 5th Street on a late afternoon in March 2016, I pass by northern California's largest homeless shelter, the St. Vincent de Paul Center. It is nearly 4pm, and there is a line of about 60 or 70 people waiting for the doors to open (see photo). Right now, San Francisco is the most expensive U. S. city to live in, with 2016 home prices averaging $1.1 million. This year, the city has also witnessed a record 2120 home evictions, mostly because people living in much more modest housing cannot afford to pay the higher rent or because landlords sell the entire rental building to a developer who turns it into private owner-occupied housing units. San Francisco now has some 6700 homeless people, and this shelter serves 400 to 500 persons daily, with separate accommodations for men and women. They can have a bed for the night, a meal, shower, counseling, and basic medical help. The shelter closes from 9am to 4pm.

© Jan Nijmans

most conspicuous within metropolitan areas. The *average* price of an apartment in Midtown or Lower Manhattan in 2016 was $1.9 million, while 21 percent of New York City's population was living below the poverty line (less than $25,000 per year for a family of four). Such extreme contrasts are found in many other cities, with San Francisco topping the list (see **Box 1-7**).

The Making of a Multicultural Realm

The Virtues of Mobility and Immigration

Given the mobility of North America's population, the immigrant infusions that continue to diversify it, the economic changes reshaping it, and the forces acting upon it, the population distribution map (Fig. 1-3) should be viewed as the latest still in a motion picture that has been unreeling for four centuries. Slowly at first, then with accelerating speed after 1800, North Americans pushed their settlement frontier westward to the Pacific. Even today, such shifts continue. Not only does the center of gravity of population continue to move westward, but within the United States it is also shifting southward—the latter a drift that has gained momentum since the 1960s as the advent of universal air conditioning made the so-called **Sunbelt [21]** States of the U.S. southern tier ever more attractive to internal migrants.

The current population map is the still-changing product of numerous forces. For centuries, North America has attracted a pulsating influx of immigrants who, in the faster-growing United States, were rapidly assimilated into the societal mainstream. Throughout the realm, people have arranged themselves geographically to maximize their proximity to evolving economic opportunities, and they have shown little resistance to relocating as these opportunities successively favored new locales.

During the past century, such transforming forces have generated a number of major **migrations [22]**, of which the still-continuing shift to the west and south is only the latest. Five others were: (1) the persistent growth of metropolitan areas, first triggered by the impact of the late-nineteenth-century Industrial Revolution in North America; (2) the large-scale movement of African Americans from the rural South to the urban North during the latter stages of the industrial era; (3) the shift of tens of millions of urban residents from central cities to suburbs, and subsequently to exurbs even farther away from the urban core; (4) the return migration of millions of African Americans from the deindustrializing North back to the growing opportunities in the South (in metropolises such as Atlanta and Charlotte); and (5) the strong and steady inflow of immigrants from outside North America, including, in recent times, Mexicans, Cubans, and other Latinos; South Asians from India and Pakistan; and East and Southeast Asians from Hong Kong, Vietnam, and the Philippines.

Overall, Canada and especially the United States are composed of diverse regions with different economic, physiographic, and cultural imprints. This North American mosaic is also visible at different scales, not just at the level of regions such as the Southeast or Midwest. Diversity can be observed between urban and rural areas and also, as noted earlier, between neighborhoods in the same city. Interestingly, these geographies are critical, and always have been, to political preferences from the local to the federal level. Political candidates, particularly in presidential elections, can sometimes be seen to shift their rhetoric depending on the region or State where they are campaigning, and support for one candidate or the other can vary significantly from place to place (see **Box 1-8** and **Fig. 1-14**).

Box 1-8 MAP ANALYSIS

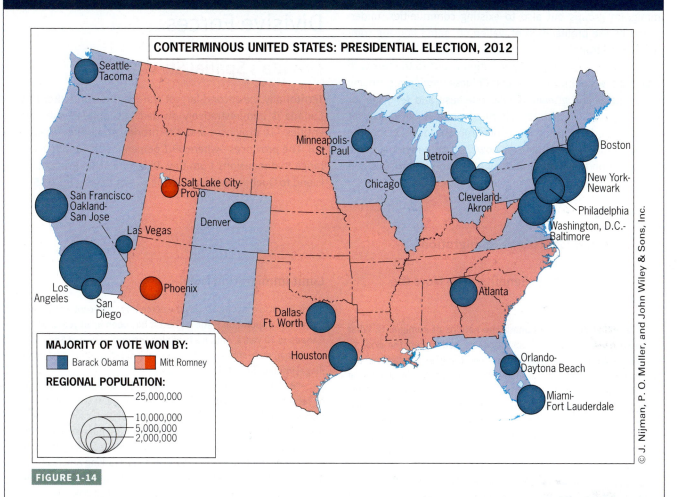

CONTERMINOUS UNITED STATES: PRESIDENTIAL ELECTION, 2012

MAJORITY OF VOTE WON BY:
Barack Obama   Mitt Romney

REGIONAL POPULATION:
25,000,000
10,000,000
5,000,000
2,000,000

© J. Nijman, P. O. Muller, and John Wiley & Sons, Inc.

FIGURE 1-14

Geography plays a most significant role in politics, elections, and democracy. Important insights into politics and elections can be gained by mapping election results. Figure 1-13 shows the presidential election results of 2012, between Barack Obama and Mitt Romney. In the United States, presidential elections are mostly framed as a competition for votes between Democratic or "blue" states versus Republican or "red" states. The importance of **electoral geography [23]** is evident in broad regional patterns (e.g., West Coast versus the South) and also at the finer scale of urban-rural contrasts. How would you describe these main regional and urban-rural patterns? Inside which two States is the urban-rural divide clearest? How do these 2012 patterns compare to the result of the 2016 presidential election?

Access your WileyPLUS Learning Space course to interact with a dynamic version of this map and to engage with online map exercises and questions.

The Challenge of Multiculturalism

Growing diversity, especially at a time of globalization and **transnationalism**, comes with challenges. The sheer dimensions of immigration into the United States, from virtually all corners of the world, creates an increasingly complex ethnic and cultural mosaic that tests **melting pot [24]** assertions at every geographic scale. Today there are more people of African descent living in the United States than in Kenya. The number of Hispanic residents in America is nearing half the population of Mexico. Miami is the second-biggest Cuban city after Havana, and Montreal is the largest French-speaking city in the world after Paris. In short, there are sufficient immigrant numbers for them to create durable societies within the overall national fabric. The challenge is to ensure that the great majority of the immigrants become full participants in that larger society.

Such are the numbers, and so diverse are the cultural traditions, that the melting pot that America once was is today morphing into something else. Spatially, the United States is now completing its transformation into a **mosaic culture**, an increasingly heterogeneous complex of separate, more or less uniform "tiles" whose residents spend less time than ever interacting and "melting."' This new pattern applies not only to new immigrant groups but also to existing communities, underscored by the proliferation of walled-off, gated housing complexes in metropolitan areas across the realm.

In Canada, the melting-pot notion is put to the test with large numbers of (mostly affluent) Chinese immigrants arriving in Vancouver and elsewhere who purchase and renovate (or frequently replace) traditional homes in long-stable Vancouver neighborhoods, arousing the ire of numerous locals. Across the realm, multicultural tendencies now vary on a regional basis, depending on the subtle interplay of economics, existing cultural patterns, immigration, and local politics (see **Box 1-9**).

Regionalism in Canada: Divisive Forces

Canada's Spatial Structure

Territorially the second-largest state in the world after Russia, Canada is administratively divided into only 13 subnational entities—10 provinces and 3 territories (the United States has 50). The provinces—where 99.7 percent of all Canadians live—range in dimensions from tiny, Delaware-sized Prince Edward Island to Quebec, nearly twice the size of Texas (**Fig. 1-15**).

Box 1-9 Regional Issue: Immigration

Immigration Brings Benefits—The More the Merrier!

"The United States and Canada are nations of immigrants. What would have happened if our forebears had closed the door to America after they arrived and stopped the Irish, the Italians, the eastern Europeans, and so many other nationalities from entering this country? Now, we're arguing over Latinos, Asians, Russians, Muslims, you name it. Fact is, newcomers have always been viewed negatively by most of those who came before them. When Irish Catholics began arriving in the 1830s, the Protestants already here accused them of assigning their loyalty to some Italian pope rather than to their new country, but Irish Catholics soon proved to be pretty good Americans. Sound familiar? Muslims can be very good Americans too. Why do you think they want to be here? And America's immigrants have always been a primary engine of growth because they are a self-selected group of people with initiative and work ethic. They become part of the world's most dynamic economy and make it more dynamic still.

"My ancestors came from Sweden in the 1800s, and the head of the family was an architect from Stockholm. I work here in western Michigan as an urban planner. People who want to limit immigration seem to think that only the least educated workers flood into the United States and Canada, depriving the less-skilled among us of jobs and causing hardship for citizens. But in fact America attracts skilled and highly educated as well as unskilled immigrants, and they all make contributions. The highly educated foreigners, including doctors and technologically skilled workers, are quickly absorbed into the workforce; you're very likely to have been treated by a physician from India or a dentist from South Africa. The unskilled workers take jobs we're not willing to perform at the wages offered.

"And our own population is aging, which is why we need the infusion of younger people that immigration brings with it. We don't want to become like Japan or some European countries, where they won't have the younger working people to pay the taxes needed to support the social security system. And few people realize that today, in 2016, the percentage of immigrants in the United States is actually less than it was in the late nineteenth century, on the eve of this country achieving greatness. We need more, not fewer, immigrants."

Limit Immigration Now!

"Well, it is also a fact that today the percentage of recent immigrants in the U.S. population is the highest it has been in 80 years, and in Canada in 70 years. America is adding the population of San Diego every year, over and above the natural increase, and not counting illegal immigration. This can't go on.

"By 2020, almost one-sixth of the U.S. population will consist of recent immigrants. At least one-third of them will not have a high school diploma. They will need housing, education, medical treatment, and other social services that put a huge strain on the budgets of the States they enter. The jobs they're looking for often aren't there, and then they start displacing working Americans by accepting lower wages. It's easy for the elite to pontificate about how great immigration is for the American melting pot, but they're not the ones affected on a daily basis. Immigration is a problem for the working people. We see company jobs disappearing across the border to Mexico, and at the same time we have Mexicans arriving here by the hundreds of thousands, legally and illegally, and more jobs are taken away.

"Don't talk to me about how immigration now will pay social security bills later. I know a thing or two about this because I'm an accountant here in Los Angeles, and I can calculate as well as the next guy in Washington. Those fiscal planners seem to forget that immigrants grow older and will need social security too.

"Comparisons with immigration to this country in historical times don't hold up because the world has changed—if you hadn't noticed. Millions of poor migrants from war-torn countries are trying to get into Europe, and the same goes for the United States. Migrants or refugees—it is hard to tell the difference anymore—from Syria, Morocco, Libya, Iraq, Sudan, and a host of impoverished and destabilized African countries want to make their way into our country as well. You do realize that we are targeted by terrorist organizations such as ISIS and that they are very likely to be among those so-called refugees?

"We live in a harsh, globalized world in which security is precarious and we have to adjust. Tighten immigration rules, I say; set up a sophisticated system for background checks and selection, and send a message to the world that this country is hard to get into. For our prosperity and for our security."

As in the United States, the smallest provinces lie in the northeast. Canadians call these four the Atlantic Provinces (or simply Atlantic Canada): Prince Edward Island, Nova Scotia, New Brunswick, and the mainland-island province named Newfoundland and Labrador. Gigantic, mainly *Francophone* (French-speaking) Quebec and populous, heavily urbanized Ontario, both flanking Hudson Bay, form the heart of the country. Most of western Canada (which is what Canadians call everything west of Ontario) is organized into the three Prairie Provinces: Manitoba, Saskatchewan, and Alberta. In the Far West, beyond the Canadian Rockies and extending to and along the Pacific Ocean, lies the tenth province, British Columbia.

Of the three territories in Canada's Arctic North, the Yukon is the smallest—but nothing is small in Arctic Canada. With the Northwest Territories and recently created Nunavut, the three territories cover almost 40 percent of Canada's total area.

The population, however, is extremely small: only about 115,000 people inhabit this vast, frigid frontier zone.

Canada's population is clustered in a discontinuous ribbon, roughly 300 kilometers (200 mi) wide, nestled against the U.S. border and along ocean shores, whereas most of the country's energy reserves lie far to the north of this corridor. As such, the maps of Canada's human settlement (Figs. 1-3 and 1-15) bear out this generalization and emphasize the significance of environment. People choose to live in the more agreeable areas to the south, but territorial control of abundant northern resources is highly important. This population pattern creates cross-border affinities with major American cities in several places: Toronto-Buffalo (see **Box 1-10**), Windsor-Detroit, and Vancouver-Seattle. Very little of this occurs in the three Prairie Provinces of the Canadian West, which for the most part adjoin sparsely settled North Dakota and Montana.

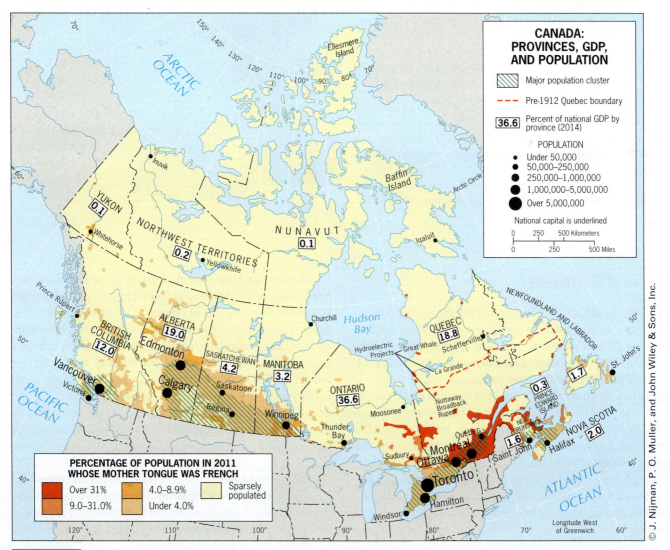

© J. Nijman, P. O. Muller, and John Wiley & Sons, Inc.

FIGURE 1-15

Box 1-10 Among the Realm's Great Cities: Toronto

Toronto, capital of Ontario and Canada's largest metropolis (more than six million), is the historic heart of English-speaking Canada. The landscape of much of its center is dominated by exquisite Victorian-era architecture and surrounds a healthy downtown that is one of Canada's leading economic centers. Landmarks abound in this CBD, including the Sky Dome stadium (now called Rogers Centre), the famous City Hall with its facing pair of curved high-rises, and mast-like CN Tower, which at 553 meters (1815 ft) is the world's sixth-tallest freestanding structure. Toronto also is a major port and industrial complex, whose facilities line the shore of Lake Ontario.

Livability is one of the first labels Torontonians apply to their city, which has retained more of its middle class than central cities of its size in the United States. *Diversity* is another leading characteristic because this is North America's richest urban ethnic mosaic. Among the largest of more than a dozen thriving ethnic communities are those dominated by Italians, Portuguese, Chinese, Greeks, and Ukrainians; overall, Toronto now includes residents from about 170 countries who speak more than 100 languages; the immigrant inflow continues steadily, with those born outside Canada constituting fully half of the city's population (in all the world only two other million-plus cities exhibit a higher percentage).

Originally, the city's advantageous location on the northwestern shore of Lake Ontario was a function of its centrality in what has always been Canada's most populated region, its access to regional waterways, and its relative proximity to the U.S. core area. Subsequently, it was the advent of the railroad during the late-nineteenth-century industrialization era that solidified Toronto's position as Canada's leading city. Today, the Greater Toronto area produces about one-fifth of Canada's entire economic output and is home to nearly half of the country's corporate headquarters.

Toronto has functioned successfully in recent decades, thanks to a metropolitan government structure that fostered central city–suburban cooperation. But that relationship is increasingly stressed as the outer city gains a critical mass of population, economic activity, and political clout. Externally, the city has become the centerpiece of the Main Street conurbation (see Fig. 1-9), which extends for 1000 kilometers (650 mi) from Windsor-Detroit in the southwest through Toronto and Montreal to Quebec City in the northeast.

© J. Nijman, P. O. Muller, H. de Blij, and John Wiley & Sons, Inc.

Cultural Contrasts

Canada's capital, Ottawa, is symbolically located astride the Ottawa River, which marks the boundary between English-speaking Ontario and French-speaking Quebec. Though comparatively small, Canada's population is markedly divided by culture and tradition, and this division has a pronounced regional expression. Linguistically, 75 percent of Canada's citizens speak English as their mother tongue, and 23 percent speak French (18 percent of the population claim to be bilingual); the remaining 2 percent speak neither English nor French. The spatial concentration of more than 90 percent of the country's Francophones in Quebec accentuates this division (**Fig. 1-15**), and the province overwhelmingly continues to function as the historic, traditional, and emotional focus of French culture in Canada.

Over the past half-century, a strong nationalist movement has emerged in Quebec, and at times it has demanded outright separation from the rest of Canada. This ethnolinguistic division continues to be the litmus test for Canada's federal system, and the issue continues to pose a latent threat to the country's future national unity. The historical roots of the matter are related to French-British competition since the seventeenth century in which Britain gained the upper hand. To this day, the British monarch remains the official head of state (as Canada remains part of the British Commonwealth), represented by the governor general. It is one of the issues that riles Quebec nationalists.

There does seem to be an overall decline in support for Quebec's independence, however. This is partly due to a number of laws that have been enacted to reassure the primacy of Québécois culture in the province (a 2006 federal parliamentary resolution effectively recognizes the Québécois as a *distinct nation*). In addition, there appears to be a growing realization among the French-speaking people that continued integration within Canada combined with a high degree of autonomy is the best possible option. But the struggle for a maximum degree of autonomy continues, along with accompanying political and ethnic tensions.

The Ascendancy of Indigenous Peoples

On the wider Canadian scene, the culture-based Quebec struggle has stirred ethnic consciousness among the country's 1.4 million native peoples—dominated by First Nations [25], but also including Inuit and Métis. They, too, have received a sympathetic hearing in Ottawa, and in recent years breakthroughs have been achieved with the creation of Nunavut as well as local treaties for limited self-government in northern British Columbia.

A foremost concern among these peoples—who constitute 3.9 percent of Canada's population—is that their aboriginal rights be protected by the federal government against the provinces of which they are a part. This is especially true for the largest First Nation of Quebec, the Cree, whose historic domain covers the northern half of the province. As Figure 1-15 shows, the territory of the Cree (north of the red dashed line) is no unproductive wilderness: it contains the James Bay Hydroelectric Project, a massive scheme of dikes and dams that has transformed much of northwestern Quebec and generates hydroelectric power for a huge market within and outside the province.

Regionalism and Ethnicity in the United States

In Canada, indigenous peoples were able to hold on to sparsely populated areas in the north. By contrast, in the United States the relentless push westward by European settlers and the U.S. government in effect dispossessed—and decimated—the original inhabitants. In the end, they were confined to some 300 reservations located mainly in the western half of the country (Fig. 1-16, lower-right map). As noted earlier, there are fewer than 3 million Native Americans left today, with the Navajo and Cherokee being the largest surviving nations. Whereas in Canada First Nations hold significant territorial power, in the United States Native Americans are in a far weaker position to challenge the federal government.

Within U.S. society, more so than in Canada, the social fabric and ethnic relations are determined by past and present waves of immigration. In certain cases, immigration has helped forge a particular regional geography. This was especially true in the case of the hundreds of thousands of Africans who were forcibly brought to the United States in bondage from the seventeenth to the mid-nineteenth centuries and put to work as slaves on plantations in the American South.

Slavery was abolished in 1865, but racial segregation persisted in a variety of legal forms until the civil rights movement crested in the 1960s. In violation of the Constitution, a wide range of State and local Jim Crow laws were enacted to codify the repression of African Americans. This legacy is still visible in the present-day geography of race in the United States, with blacks regionally concentrated in the Southeast, from eastern Texas to Maryland (Fig. 1-16, upper-right map). We noted earlier that substantial numbers of blacks migrated north to work in the industrial centers of the American Manufacturing Belt during the second quarter of the twentieth century; major African American population clusters persist in those northern cities, but they are not clearly apparent at the scale of this national map.

Race continues to be a sensitive issue in the United States, more so than in Canada, because of this terrible history, enduring rifts between blacks and whites, continuing deprivation among African Americans, and the recurrent racial injustices impeding the nation's evolution into a truly postracial society. Emblematic of this challenged transition were the violent 2014–2015 confrontations that erupted first in the St. Louis suburb of Ferguson after the fatal shooting of an unarmed black teenager by a white police officer (who was later exonerated), and then inner-city Baltimore where another unarmed black died violently while in police custody. Frustration about the systematic maltreatment of African Americans by police gave way to the "Black Lives Matter" movement, a campaign that spread across the nation during the mid-2010s.

Another highly visible regional expression of ethnic migration is the large concentration of Hispanics in the southwestern United States (Fig. 1-16, upper-left map; see also Fig. 1-2). Since 1980, the number of Hispanics in the United States has more than tripled to surpass 55 million, forming the nation's biggest ethnic minority—and more than one-sixth of the country's total population. For some Anglophone Americans who fear the erosion of the status of English in their communities, language remains an emotional issue. Indeed, Hispanics are now dispersing and increasing their presence in many more parts of the United States, ranging from the rural areas of the Southeast and Midwest to thousands of urban communities from Massachusetts to Washington State to the small towns that still remain on the Great Plains.

Finally, the Asian immigrant population, whose numbers have risen over the past few decades, is geographically the most agglomerated of the leading ethnic minorities (Fig. 1-16, lower-left map). Ethnic Chinese remain the most numerous among the nearly 20 million Americans who identify themselves as Asian, but this diverse minority also includes Japanese, Filipinos, Koreans, and Vietnamese as well as a sizeable cohort of South Asian Indians. Whatever their origins, the great majority have remained in California, where many have achieved economic success and upward mobility.

DISTRIBUTION OF MAJOR ETHNIC GROUPS, CONTERMINOUS UNITED STATES, 2014

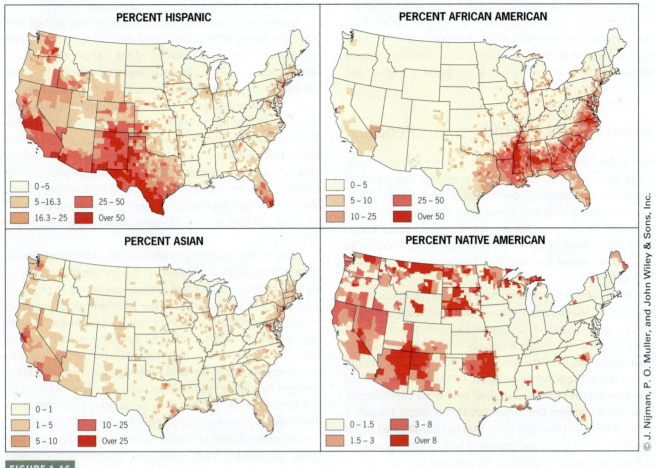

PERCENT HISPANIC

0 – 5
5 – 16.3
16.3 – 25
25 – 50
Over 50

PERCENT AFRICAN AMERICAN

0 – 5
5 – 10
10 – 25
25 – 50
Over 50

PERCENT ASIAN

0 – 1
1 – 5
5 – 10
10 – 25
Over 25

PERCENT NATIVE AMERICAN

0 – 1.5
1.5 – 3
3 – 8
Over 8

FIGURE 1-16

© J. Nijman, P. O. Muller, and John Wiley & Sons, Inc.

Regions of North America

Mountain ranges, coastal zones, deserts, low-lying plains, and a range of climatic environments all demand different kinds of human adaptation. Successive waves of immigrants brought with them a variety of cultures; sometimes the differences were subtle, but sometimes they were quite stark. As **Figure 1-17** shows, a majority of this realm's regions include both countries regardless of the international boundary. We can regionalize North America's geography by combining functional and formal principles. The resulting map of nine regions is by no means the only solution, but for our general purposes these boundaries are the most pertinent, and the regional distinctions they facilitate will be helpful in understanding the overall fabric of the realm.

Region The Northeast (1)

The historic core region of the North American realm integrates what have long been the most prominent parts of the United States and Canada (Fig. 1-17). The region contains the largest cities and the federal capitals of both countries, the leading financial markets, the largest number of corporate headquarters, dominant media centers, prestigious universities, cutting-edge research complexes, and the busiest airports and intercity expressways. Moreover, both the U.S. and Canadian portions of the North American core still contain roughly one-third of their respective national populations. Political and business decisions, investments, and many other commitments made here affect not just North America but the world at large.

Nonetheless, the dominance of the Northeast has been declining. As was true of the American Manufacturing Belt—whose eastern anchor it forms—it once commanded unmatched supremacy. But deindustrialization and the subsequent emergence of the information economy have eroded that dominance as competitors to the south and west continue to siphon away some of its key functions. An example is eastern Michigan, or parts of Ohio, which have experienced long-term economic decline as jobs in the automobile industry have steadily disappeared over the past four decades.

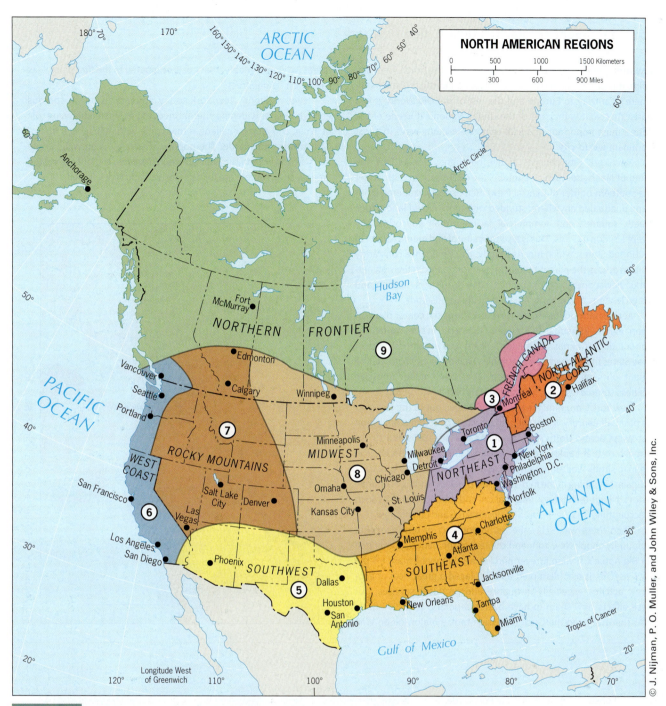

NORTH AMERICAN REGIONS

ARCTIC OCEAN

PACIFIC OCEAN

ATLANTIC OCEAN

Hudson Bay

Gulf of Mexico

NORTHERN FRONTIER

WEST COAST

ROCKY MOUNTAINS

MIDWEST

NORTHEAST

FRENCH CANADA

NORTH ATLANTIC COAST

SOUTHEAST

SOUTHWEST

Anchorage · Fort McMurray · Edmonton · Vancouver · Seattle · Portland · Calgary · Winnipeg · San Francisco · Salt Lake City · Las Vegas · Denver · Los Angeles · San Diego · Phoenix · Dallas · Houston · San Antonio · New Orleans · Minneapolis · Milwaukee · Detroit · Chicago · Omaha · Kansas City · St. Louis · Memphis · Atlanta · Charlotte · Jacksonville · Tampa · Miami · Toronto · Montréal · Halifax · Boston · New York · Philadelphia · Washington, D.C. · Norfolk

Arctic Circle · Tropic of Cancer

Longitude West of Greenwich

© J. Nijman, P. O. Muller, and John Wiley & Sons, Inc.

FIGURE 1-17

Region North Atlantic Coast (2)

When you travel north from Boston into New Hampshire, Vermont, and Maine, you move across one of North America's historic culture hearths whose identity has remained strong for four centuries. Even though Massachusetts, Connecticut, and Rhode Island—traditional New England States—share these qualities, they have become part of the Northeast and became oriented to the New York urban region, so that on a functional as well as formal basis, the North Atlantic Coast extends from the northern border of Massachusetts to Newfoundland, thereby incorporating all four Canadian Atlantic Provinces (Fig. 1-17).

Difficult environments, a maritime orientation, limited resources, and a persistent rural character have combined to slow economic development here. Primary industries—fishing, logging, and some farming—are mainstays, although recreation and tourism have boosted the regional economy in recent decades. Upper New England experiences the spillover effect of metropolitan Boston's prosperity to some degree, but not enough to ensure steady and sustained economic growth. Atlantic Canada

Box 1-11 Among the Realm's Great Cities: New York

New York is much more than the largest city of the North American realm. It is one of the most famous places on Earth; it is the hemisphere's gateway to Europe and the rest of the Old World; it is a tourist mecca; it is the seat of the United Nations; it is one of the globe's most important financial centers—in many ways its most prominent **world-city**.

New York City consists of five boroughs, centered by the island of Manhattan, which contains the CBD south of 59th Street (the southern border of Central Park). Here is a skyscrapered landscape unequaled anywhere, studded with more fabled landmarks, streets, squares, and commercial facilities than any other city except London or Paris. This is also the cultural and media capital of the United States, which means that the city's influence constantly radiates across the planet thanks to New York-based television networks, newspapers and magazines, book publishers, fashion and design leaders, and artistic and new-media trendsetters.

At the metropolitan scale, New York forms the center of a vast urban region—250 square kilometers (150 sq mi) in size, containing a population of almost 21 million—which sprawls across parts of three States in the heart of Megalopolis. That outer city has become a giant in its own right with its population of 13 million, massive business complexes, and flourishing suburban downtowns.

New York's prominence has certainly been put to the test over the past half-century. It was hit hard by waves of deindustrialization that swept across North America during the late 1960s and 1970s, so much so that bankruptcy was declared in 1975. The population began to shrink while inner-city decay advanced. But New York took off again, beginning in the mid-1980s, led by its expanding economic sectors in high-technology, information, and producer services such as accounting, finance, insurance, advertising, and consulting.

Then there was 9/11—the horrendous act of terrorism that profoundly shook New York and the rest of the nation. It changed the Big Apple's skyline forever and left a permanent mark on the city's psyche. It also disrupted and displaced a great deal of economic activity, especially in finance and producer services, but these effects now seem to have been temporary. The 2007 fiscal crisis that triggered the Great Recession took a major toll on Wall Street (where, arguably, that crisis originated in the first place)—as well as on "Main Street."

Ever resilient, New York today appears to be prosperous enough, despite widening internal inequalities, its steadily rising cost of living, and all the challenges that come with managing a metropolis of this magnitude. The population overall is growing, crime rates are lower than they have been in decades, and once decaying areas such as Harlem are being gentrified and redeveloped.

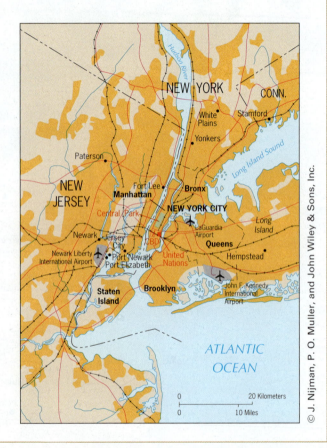

© J. Nijman, P. O. Muller, and John Wiley & Sons, Inc.

Lobster boats and wharf at Port Clyde, Maine. The cold, clean waters off Maine's rocky coast are ideal for lobster harvesting, and the industry is interwoven with the culture of the area. The annual catch exceeds 100 million pounds, and thanks to successful regulation it has been remarkably sustainable. Catching lobster eggs or egg-bearing females is prohibited; the only legal way to fish is by using traps (no nets); there is a limit of 800 traps per fisher, and they must be designed to allow smaller lobsters to escape. The combined lobster industry of the North Atlantic Region amounts to about $4 billion annually.

© Daniel Dempster Photography/Alamy Stock Photo

has endured economic hard times as well in the recent past. For example, overfishing has seriously depleted fish stocks. Alternate opportunities focus on this subregion's spectacular scenery and the tourism it attracts. Economic prospects have also been boosted in Canada's poorest province, Newfoundland and Labrador, by the discovery of significant offshore oil reserves. Production is now well underway, and in the 2010s additional energy deposits have been discovered both on and around Newfoundland Island.

Region French Canada (3)

Francophone Canada constitutes the inhabited southern portion of Quebec, focused on the St. Lawrence Valley from near Montreal northeastward to the river's mouth (Fig. 1-17). It also includes the French-speaking Acadians who reside in neighboring New Brunswick (Fig. 1-15). The French cultural imprint on the region's cities and towns is matched in farming zones by the narrow, rectangular *long lots* perpendicular to the river, also of French origin.

The economy of French Canada was historically less industrial and more rural than the adjacent Northeast region (and Toronto far outranks Montreal here). Nevertheless, Montreal's economic role has advanced since 2000, with information technology, telecommunications, and biopharmaceuticals all growing in importance. In some ways, French Canada is a richly endowed part of North America. Northern Quebec produces so much hydroelectricity that it exports the surplus to Ontario and to the northeastern United States. Most of this power is generated via the James Bay Project, a series of dams and power stations on the eastern side of Hudson Bay along the La Grande River (Fig. 1-15). Quebec is faring so well in the energy-production sphere that the recent discovery of natural gas in the St. Lawrence Basin was met with tempered enthusiasm. Many Québécois feel they have no need for gas: unlike hydropower, it is a nonrenewable resource that would boost carbon emissions and the chances of widespread water pollution.

© Chris Cheadle/Alamy Stock Photo

Montreal's, Rue Bonsecours. The old center of the city is marked by an unmistakably French cultural landscape. The restaurant and hotel (the brick building to the right) have been here since the early eighteenth century, predating the British rule of Canada.

Region The Southeast (4)

For more than a century following the U.S. Civil War (1861–1865), the Southeast (Fig. 1-17) remained in economic stagnation and cultural isolation from the rest of the country. During the 1970s, however, things changed so fast that a New South was born almost overnight. The Sunbelt migration stream drove people and enterprises into the long-dormant cities. Northeast-based companies looking for headquarters or branch offices found Atlanta, Charlotte, Tampa, Miami-Fort Lauderdale, and other urban areas both economical and attractive, swiftly turning them into boomtowns. Racial segregation had been dismantled in the wake of the civil rights movement. A new social order was matched by new facilities ranging from airports (Atlanta's quickly became one of the world's busiest) to theme parks open to all. Soon the nation was watching Atlanta-based CNN on television, vacationing at Orlando's Walt Disney World, and monitoring space flights originating at Cape Canaveral.

The geography of development in the South, however, is rather uneven. Although many cities and certain agricultural areas have benefited, others have not; the South contains some of the realm's poorest rural areas, and the widening gap between rich and poor here is greater than in any other region in the realm. The "Silicon Alley" corridor of Northern Virginia's Washington suburbs, central North Carolina's Research Triangle, and metropolitan Atlanta's corporate campuses are locales that typify the New South; but Appalachia and rural Mississippi still represent the Old South, where depressed farming areas and stagnant small industries inhibit change.

Another, relatively small part of this region warrants special attention. South Florida used to be a particularly remote and nearly unpopulated corner of the United States. Before the twentieth century, it was too far from the northeastern core of the country (and even from closer urban centers such as Atlanta) to warrant development; the heat, humidity, and mosquitoes posed a formidable challenge to prospective settlers; and there were no natural resources to exploit. It was not until the 1960s that this subregion of the Southeast became fully integrated into the national economy and urban system—thanks to the widespread introduction of air conditioning and affordable air travel, not to mention the mass immigration of Cubans fleeing Fidel Castro's communist regime (see **Box 1-12**). Today, Miami has evolved into a **world-city [26]**, a globally linked, southward-oriented conduit between North America and the rest of the hemisphere (**Fig. 1-18**).

Region The Southwest (5)

As Figure 1-17 shows, by our definition the Southwest begins in East Texas and extends all the way to the eastern edge of Southern California. New Mexico, in the middle of the Southwest region, is the least developed economically and ranks low on many U.S. social indices, but its environmental and cultural attractions benefit Albuquerque and Santa Fe, with Los Alamos a famous name

Box 1-12   From the Field Notes …

© Jan Nijman

"Domino Park lies at the center of Little Havana, perhaps the most iconic neighborhood in Greater Miami. After Fidel Castro came to power in 1959, thousands of Cubans fled the nearby island and resettled here. Today, over one million Cubans reside in the metropolitan area, the biggest concentration of any single ethnic group in the United States. Miami would easily qualify as the second-largest city in Cuba. Domino Park is situated on Southwest 8th Street, known locally as 'Calle Ocho.' It is only a short walk from the Bay of Pigs Monument that commemorates the fallen soldiers of the failed invasion of Cuba by exiles in 1961. Taking a stroll along Calle Ocho, you can stop at a cafe for a *cortadito*, watch cigars being made by hand, or get fitted for a guayabera shirt. Domino Park recently opened to women (!), but it is mostly older men who come here to play games, talk politics, and meet with friends—just as they would in Cuba."

in research-and-development circles. To the west, Arizona's technologically transformed desert now harbors two large, coalescing metropolises—Phoenix and Tucson. Growth continues, but rising concerns about climate change add to the uncertainties of maintaining water supplies already stretched to the limit.

Texas leads this region in most respects. Its economy, once all but fully dependent on oil and natural gas, has been restructured and diversified so that today the Dallas-Fort Worth–Houston–Austin–San Antonio triangle has become one of the world's most productive **technopoles [27]**—state-of-the-art, high-technology industrial complexes—especially Austin at the heart of it. This subregion is also a hub of international trade and the northern anchor of a North American Free Trade Agreement (NAFTA)-generated transnational growth corridor that extends into Mexico as far south as Monterrey.

The huge State of Texas, larger than France, is surpassed in areal size only by Alaska and is also the second most populous U.S. State after California (28 and 39 million, respectively). Parts of the State are quite affluent, with much of its wealth still emanating from the energy sector: Texas claims about a quarter of all U.S. oil reserves and nearly a third of its natural gas deposits (though world prices have tumbled since 2014 and this had a dampening effect on the economy). The Lone Star State has also become a leader, along with California, in the generation of wind energy, with wind turbines increasingly dotting the sparsely settled landscape of immense West Texas.

Region   The West Coast (6)

The Southwest meets the Far West near metropolitan San Diego on California's border with Mexico, and from there the region we call

© J. Nijman, P. O. Muller, and John Wiley & Sons, Inc.

WORLD–CITY CONNECTIONS: THE CASE OF MIAMI

● Cities with largest number of headquarters of corporations conducting business in Miami

● Cities with largest number of branches of Miami-based corporations

FIGURE 1-18

© Suzanne Cordeiro / Corbis Images

Austin, Texas—highly regarded for its burgeoning technology-led economy—is one of the fastest growing American cities in the 2010s. As this photo of the 2014 Austin City Limits Music Festival shows, the Lone Star State's capital also occupies a prominent position on the U.S. popular culture scene. Austin is a product of the postindustrial Sunbelt boom of the past half-century, an anchor of the new high-tech industrial corridor that stretches from the Dallas-Fort Worth Metroplex in the north to Greater San Antonio in South Texas (see Fig. 1-9). In addition to its function as a far-reaching employment magnet for young people, Austin is considered to be one of the most livable cities in the nation.

the West Coast extends all the way north into Canada where Vancouver forms the northern anchor in southwestern British Columbia (Fig. 1-17). We include almost all of California in this region but (for environmental as well as economic reasons) only the western portions of Oregon and Washington State. This is a particularly important part of North America and increasingly serves as a counterweight to the historic core area in the U.S. Northeast.

The West Coast is a major economic region not just in this realm but in the global economy as well. It includes major cities such as San Francisco, San Diego, Portland, Seattle, and Los Angeles (see Boxes 1-13, 1-14, and 1-15). And it contains America's most populous State, California, whose economy ranks among the world's ten largest *by country*. This region also encompasses one of the realm's most productive agricultural areas in California's Central Valley, magnificent scenery, and a culturally diverse population drawn by its long-term economic growth and pleasant living conditions.

Our regional definition is heavily based on economic considerations and intensifying trade connections across the Pacific Ocean. This part of North America is deeply involved in the development of countries in eastern Asia. In the postwar era, Japan's success had a salutary impact here, but what has since taken place in China, South Korea, Taiwan, Singapore, and other Asia-Pacific economies has created unprecedented opportunities. The term **Pacific Rim [28]** has come into use to describe the discontinuous regions surrounding the great Pacific Ocean that have experienced spectacular economic growth and progress over the past four decades: not only coastal China and various parts of East and Southeast Asia, but also Australia, South America's Chile, and the western shores of Canada and the conterminous United States.

The Pacific Rim is therefore a classic example of a **functional region**, with economic activity in the form of capital flows, raw-material movements, and trading linkages that generate urbanization, industrialization, and labor migration. As part of this process, human landscapes from Sydney to Santiago are being transformed within a 32,000-kilometer (20,000-mi) corridor that rings the globe's largest body of water.

Region The Rocky Mountains Region (7)

Where the forests of the West Coast yield to the scrub of the rain shadow on the inland slopes of the mountain wall that parallels the West Coast, and the terrain turns into intermontane ("between the mountains") basins and plateaus lies a region stretching eastward from the Sierra Nevada and Cascades to the Rockies, encompassing segments of southern Alberta and British Columbia, eastern Washington State and Oregon, all of Nevada, Utah, and Idaho plus most of Montana, Wyoming, and Colorado. As Figure 1-17 shows, Edmonton, Calgary, and Denver are situated along the eastern edge of the Rocky Mountains Region, and Salt Lake City, anchoring the Software Valley technopole and symbolizing the new high-tech era, lies at the heart of it.

Remoteness, dryness, and sparse population typified this region for a very long time. It became the redoubt of the Mormon faith and a place known for boom-and-bust cycles of mining, logging, and, where possible, livestock raising. However, in recent decades advances in communications and transportation technologies, sunny climates, wide-open spaces, lower costs of living, and growing job opportunities have combined to form effective pull factors for myriad outsiders. Every time an earthquake struck in California, eastward migration got a boost. And during the most recent ten-year period, millions of Californians emigrated from the overcrowded and overpriced Pacific coast to the inland high desert.

Development in the Rocky Mountains Region centers on its widely dispersed urban areas, which are making this one of the realm's fastest-growing regions (albeit from a low base). Here thousands of new high-technology manufacturing and specialized service jobs have propelled a two-decade-long influx, slowed but not reversed by the Great Recession. Metropolitan Las Vegas typifies this accelerated growth trajectory. Far more than a gambling and entertainment magnet that draws some

Box 1-13   Among the Realm's Great Cities: Los Angeles

Los Angeles, the City of Angels, became part of the United States in 1848 at the end of the Mexican-American War. The Mexican influence has always remained, and today Hispanics outnumber Whites in the metropolitan population of just over 15 million. But Los Angeles also became a quintessential *American* city: far away from the Northeast with its colonial European influences; cradle of the iconic American film industry (Hollywood); and symbolic for the American automobile city, the freeways, and urban sprawl.

The plane-window view during the descent into Los Angeles International Airport, almost always across the heart of the metropolis, gives a good feel for the immensity of this urban landscape. It not only fills the huge natural amphitheater known as the Los Angeles Basin, but it also oozes into adjoining coastal strips, mountain-fringed valleys and foothills, and even the margins of the Mojave Desert more than 90 kilometers (60 mi) inland. This quintessential spread city, of course, could only have materialized in the automobile age. Most of it has been built rapidly since 1920, propelled by the swift postwar expansion of a high-speed freeway network unsurpassed anywhere in metropolitan America. In the process, Greater Los Angeles became so widely dispersed that today it has reorganized within a sprawling, multinodal geographic framework (see Fig. 1-12).

The metropolis as a whole constitutes North America's second-largest urban agglomeration and forms the southern anchor of the huge California megalopolis that parallels the Pacific coast from San Francisco southward to the Mexican border. It also is the West Coast's leading trade, manufacturing, and financial center. In the global arena, Los Angeles is the eastern Pacific Rim's dominant city and the origin of the greatest number of transoceanic flights and sailings to Asia.

In quite a few ways, due to globalization, the Asian influence on Los Angeles has surpassed that of Mexico. More than a quarter of Angelinos were born in Asia, and the bulk of port trade is trans-Pacific—hence with Asia. At its largest university, UCLA, about one-third of all undergraduate students are now Asian.

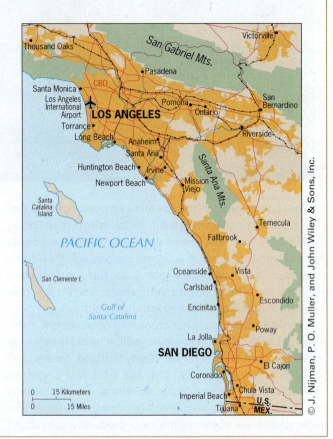

© J. Nijman, P. O. Muller, and John Wiley & Sons, Inc.

Box 1-14   From the Field Notes …

"On this street corner in central Los Angeles, four different cultures converge. Los Angeles is one of the most ethnically diverse metropolitan areas in the United States and in the world. More than a third of its population was born abroad, and more than half speak a language other than (or in addition to) English. The largest non-native ethnic group are Hispanics (mainly Mexicans and Central Americans), but there also are many Asians, especially Chinese and Koreans. Cesar E. Chavez, after whom this avenue is named, was a charismatic Mexican American farm worker turned union organizer and became nationally known as a leading civil rights activist. His birthday, March 31, is a State holiday in both California and Texas—and an optional holiday in Arizona (!). Chavez Avenue is a major east-west thoroughfare just north of downtown Los Angeles, and even a short stroll along it offers a sampling of global cultures."

© Jan Nijman

San Francisco is one of North America's most iconic cities, known for its liberal political climate, its unmatched Bay and natural harbor, the Golden Gate Bridge, Mediterranean climate, prestigious universities, infamous Alcatraz Island, and a location at the northern tip of the peninsula that makes for a high population density.

The city was named after Saint Francis in 1776 by Spanish colonialists, and it became part of Mexico in 1821. Like Los Angeles, it was ceded to the United States after the Mexican-American War in 1848. The city grew quickly as a result of the California Gold Rush (1848–1855), a wild time even for the Wild West, and more growth ensued with the arrival of the transcontinental railroad in 1869, connecting San Francisco to the rest of the U.S. The city became an important military (naval) base in the late nineteenth century and through World War II, but a number of facilities have now been closed or relocated to San Diego or Seattle. Key remaining facilities, to the far northeast of the city (not shown on the map) are Concord Naval Base and Travis Air Force Base.

Today, San Francisco is very much a part of an extended urban region of 6 million that runs south across the peninsula and the Valley, and east and northeast across the Bay. It is a huge region that includes the main urban centers of San Jose (now with a bigger population than San Francisco itself) and Oakland but also suburban communities such as Sausalito or Menlo Park and smaller cities like Berkeley, Richmond, or Redwood City. The entire urban complex is now referred to as the *Bay Area*, and another classic example of the twenty-first century polycentric urban region.

The Bay Area is also very much a **functional region**, with major highways and railroads running along the bay on both sides, connecting the port and airports (located, respectively, west and east of the bay), and facilitating daily commuters. *Silicon Valley* is essential to the region's economy, and as noted earlier is the prototype of high-tech clusters that contain major universities, concentrations of corporate upstarts as well as big business, eager venture capital, and attractive living environments for high-salaried workers. The concentration of high-tech work in the Valley, along with gentrification in the City of San Francisco, means that today the commuter flow out of San Francisco to the south is greater than the inflow. And, at a larger scale, the region is tightly linked to the Pacific Rim, as underscored by the presence of a large Asian population.

The Bay Area is a highly varied region with extremely expensive real estate and pockets of wealthy residents, but also contains deprived areas, sometimes in close proximity to the affluent enclaves (e.g., East Palo Alto). The City of San Francisco has become a contested arena between high-income gentrifiers and lower-income renters (see Box 1-7). This conflict attracted national attention when protesters in the City blocked the corporate buses of Google and other high-tech companies that transport these new city residents to and from work in the Valley.

40 million visitors annually, Las Vegas attracts in-migrants because of its history of job creation, relatively cheap land and low-cost housing, moderate taxes, and sunny if seasonally hot weather. Although the pace of development sharply leveled off during the late 2000s, by 2015 things were back on track (see photo).

Region The Midwest (8)

This vast North American heartland extends from interior Canada to the borders of the Southeast and Southwest (Fig. 1-17). The region contains many noteworthy cities, including Chicago, Kansas City, Omaha, Minneapolis, and Winnipeg; nonetheless, agriculture is the dominant story here. As the map of U.S. farmlands (**Fig. 1-19**) shows, this is the nation's breadbasket. This is the land of the beef- and pork-producing Corn Belt (*Meat Belt* would be a more accurate label); of the mighty soybean, America's and the world's most rapidly expanding crop of the past half-century; and of spring wheat in the Dakotas and Canada's Prairie Provinces as well as winter wheat in Kansas. Indeed, the cities and towns of this region share histories of food processing, packing, and marketing.

This is also the scene of the struggle for survival of the family farm as the unrelenting incursion of large-scale corporate farming threatens a longstanding way of life. Even a cursory look at demographic statistics makes it clear why many States in the Midwest are losing population or gaining far fewer than the national average. The Great Plains, the western component

© trekkerimages /Alamy Stock Photo

The sprawling Las Vegas suburb of Summerlin. In recent decades, Las Vegas has been one of the fastest growing metro areas in the United States, and it also became the epicenter of the mortgage meltdown that accompanied the Great Recession, which lasted from 2008 until about 2011. By 2012, nearly 70 percent of all mortgaged homes were "underwater," meaning the market value of the house was less than the mortgage balance (that number had declined to about 25 percent by 2016). The housing crisis was concentrated, here and elsewhere, in relatively new suburbs, where homes had come onto the market just before the bubble burst in 2008. For many, that crisis left a painful legacy that is felt to this day.

of this region, is especially hard hit: younger people and those better off are leaving, whereas older and less affluent residents stay behind. Villages and small towns continue to die, and the notion of abandoning certain areas to return them to their natural state is being seriously discussed.

But things are not the same everywhere. The Midwest may not have as prominent a presence on the national scene as New York or California, and the region is not as wealthy, but with abundant coal and oil deposits (in Wyoming and North Dakota, respectively) as well as South Dakota's booming financial sector, unemployment rates here rank among the lowest in the nation. Buoyed by an economy largely driven by agriculture and mining, and with fossil-fuel production a leading growth industry, the effects of the Great Recession were minimal and short-lived in the Midwest.

Region The Northern Frontier (9)

Figure 1-17 leaves no doubt as to the dimensions of this final North American region: it is by far the largest of the realm, covering just about 90 percent of Canada and all of the biggest U.S. State, Alaska. Not only does this region include the northern parts of seven of Canada's provinces, but it also comprises the Yukon and Northwest Territories as well as recently established Nunavut. The very sparsely populated Northern Frontier remains a land of isolated settlement based on the exploitation of newly discovered resources. The Canadian Shield, underlying most of the eastern half of the region, is a rich storehouse of mineral resources, including metallic ores such as nickel, uranium, copper, gold, silver, lead, and zinc. The Yukon and Northwest Territories have proven especially bountiful, with the mining of gold and especially diamonds (Canada currently ranks third in the world); at the opposite, eastern edge of the Shield, near Voisey's Bay on the central coast of Labrador, the largest body of high-grade nickel ore ever discovered is now open for extraction.

The Province of Alberta experienced one of the most spectacular economic booms in the realm's history. Recent estimates indicate that Alberta's oil reserves, dominated by the Athabasca **tar sands [29]** (see **Fig. 1-20**), rank third in the world behind only those of Venezuela and Saudi Arabia. In the early twenty-first century, Alberta became Canada's fastest-growing province, and Calgary its fastest-expanding major city. The boom makes Calgary a magnet for business and highly skilled workers; the city's skyline has been spectacularly transformed, comparable to what has recently occurred on the Arabian Peninsula. But much depends on the price of oil, which has dropped significantly in the past few years through mid-2016.

Productive locations in the Northern Frontier constitute a far-flung network of mines, oil and gas installations, pulp mills, and hydropower stations that have spawned hundreds of small settlements as well as thousands of kilometers of interconnecting

© J. Nijman, P. O. Muller, and John Wiley & Sons, Inc.

CANADA

MEXICO

CONTERMINOUS UNITED STATES: PERCENT LAND IN FARMS BY COUNTY, 2012

- 90% or more
- 70–89%
- 50–69%
- 30–49%
- 10–29%
- Less than 10%

0   250   500 Kilometers
0   250   500 Miles

FIGURE 1-19   United States: Percent Land in Farms by County

Alexandra Kobalenko/Corbis Images

The Inuit hamlet of Pangnirtung (pop. 1425) on Baffin Island, located just south of the Arctic Circle (see Fig. 1-20). Baffin Island is larger in area than California, and has rugged, mountainous terrain that is snow-covered most of the year. These fjords were shaped by inland glaciers reaching the coast, just as in Norway.

transportation and communications lines. Inevitably, these activities have infringed on the lands of indigenous peoples without the preparation of treaties or agreements, which has led to recent negotiations between the government and leaders of First Nations over resource development in the Northern Frontier. But essentially this gigantic region remains a frontier in the truest sense of that term.

Alaska's regional geography differs from the rest of the Northern Frontier in that the State contains several urban settlements and an incipient core area in the coastal zone around Anchorage. The population of Alaska (ca. 750,000) accounts for more than

one-third of the region's total. Here also lies the only metropolis of any size, Anchorage (population: 300,000), a key node on the aviation network that links the realm to Pacific Asia. The State's internal communications (air and surface) are also better developed.

Finally, Nunavut merits special mention. Created in 1999, this newest Territory of Canada is the outcome of a major aboriginal land claim agreement between the *Inuit* people (formerly called Eskimos) and the federal government, and encompasses all of Canada's eastern Arctic as far north as Ellesmere Island (Figs. 1-15, 1-20), an area far larger than any other province or territory. With around one-fifth of Canada's total area, Nunavut—which means "our land"—contains only about 37,000 residents, of whom 80 percent are Inuit.

Climate change is likely to affect the Northern Frontier in as-yet-unforeseeable ways. As the map shows, this region extends all the way north to the Arctic Ocean; it therefore includes waters that, with continued global warming and associated contraction of the Arctic icecap, will open (possibly year-round) nearshore waterways that would dramatically alter global shipping routes and intercontinental distances—a topic discussed in some detail in Chapter 12. The question of ownership of these waters—and the submerged seafloor beneath them, coming within reach of exploitation—will be a matter of significant international contention in the decades ahead.

In 2010, Canada solidified its reputation as an environmentally conscious nation through the signing of an unprecedented agreement. The Canadian Boreal Forest Agreement involves 21 logging firms and 9 environmental organizations as well as the Canadian government. As shown in Figure 1-20, it imposes a moratorium on logging and any other development within a huge zone of 290,000 square kilometers (112,000 sq mi). Greenpeace, which originated in Canada, has called it the biggest agreement to date that preserves nature. The word "boreal" is derived from the Latin word for north; **boreal forests [30]**—identical to the Russian *taiga* (snowforest) discussed in Chapter 5—are dominated by dense stands of coniferous (cone-bearing) needleleaf trees such as spruce, fir, and pine. The boreal forests of the Northern Frontier also constitute the habitat of such endangered wildlife species as the caribou, lynx, American black bears, and wolverines.

Points to Ponder

- Whereas more than 70 percent of Canada's population resides within 200 kilometers (125 mi) of the northern U.S. border, only about 12 percent of the Mexican population lives that close to the southern U.S. border.

- The United States does not have an official language. Should it?

- Thanks to the Pacific Rim trade, California's ports are among the busiest in the United States. How will they be affected by the newly enlarged Panama Canal that opened in 2016?

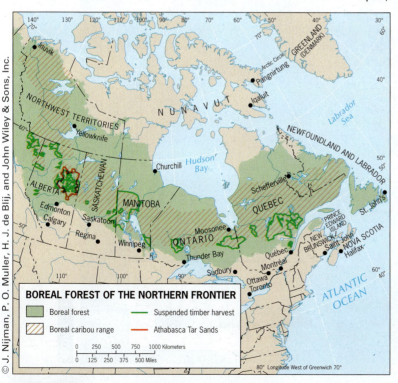

© J. Nijman, P. O. Muller, H. J. de Blij, and John Wiley & Sons, Inc.

BOREAL FOREST OF THE NORTHERN FRONTIER

- Boreal forest
- Boreal caribou range
- Suspended timber harvest
- Athabasca Tar Sands

0 250 500 750 1000 Kilometers
0 125 250 375 500 Miles

FIGURE 1-20

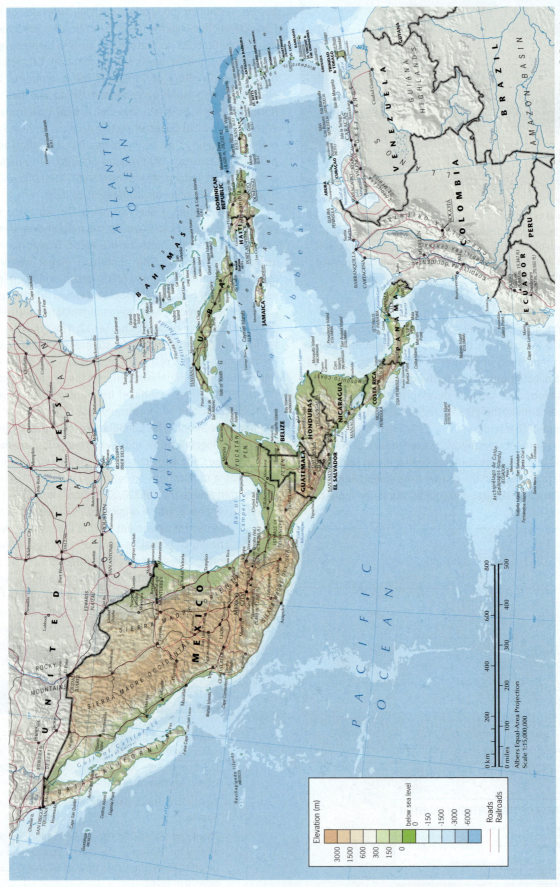

FIGURE 2-1

CHAPTER 2

© Jan Nijman

© Jan Nijman

Q Where were these pictures taken?

The Middle American Realm

IN THIS CHAPTER

- The enlarged Panama Canal opens for business
- Chinese influence behind Nicaragua Canal plans
- Drug trafficking: Shifting from Mexico to the Caribbean?
- Organized crime and violence in Central America
- Is small beautiful? The predicament of tiny island-nations

CONCEPTS, IDEAS, AND TERMS

Look at a world or hemispheric map, and it is obvious that the Americas comprise two landmasses: North America extending from Alaska to Panama and South America from Colombia to Argentina. But here we are reminded that continents and geographic realms do not necessarily coincide.

Between North America and South America lies the small but important geographic realm known as Middle America. Consisting of an impediment-filled mainland corridor and myriad Caribbean islands, Middle America is a highly fragmented realm.

Defining the Realm

From **Figure 2-1** it is clear that Middle America is much wider than it is long. The distance from Baja California to Barbados is about 6000 kilometers (3800 mi), but from the latitude of Tijuana to Panama City it is only half that distance. In terms of total area, Middle America is the second smallest of the world's geographic realms. As the map shows, the dominant state of this realm is Mexico, larger than all its other countries and territories combined (see also **Box 2-1**).

A Realm of High Densities and Cultural Diversity

Middle America may be a rather small geographic realm by global standards, but comparatively it is densely peopled. Its population passed the 200-million milestone in 2011, more than half of it residing in Mexico alone, and the rate of natural increase (at 1.3 percent) remains above the world average. **Figure 2-2** shows Mexico's populous interior core area quite clearly, but note that the population in Guatemala and Nicaragua tends to cluster toward the Pacific rather than the Caribbean coast. Among the islands, this map reveals how crowded Hispaniola (containing Haiti and the Dominican Republic) is, but at this scale we cannot clearly discern the pattern on the smaller islands of the eastern Caribbean and the Bahamas, some of which are also densely populated. In many Middle American countries, both on the mainland and the islands, large **primate cities [1]** dominate the landscape, from Mexico City (20 million) to Managua (1 million), which both contain about one-sixth of their national populations (see **Box 2-2**).

What Middle America lacks in size it makes up in physiographic and cultural diversity. This is a realm of soaring volcanoes and spectacular shorelines, of lush tropical forests and barren deserts, of windswept plateaus and scenic islands. It holds the architectural and technological legacies of ancient **indigenous [2]** civilizations. Today it is a mosaic of immigrant cultures from Africa, Europe, and elsewhere, richly reflected in music and the visual arts. Material poverty, however, is endemic: Haiti is often cited as the poorest country in the Americas; Nicaragua, on the mainland, is almost as badly off. As we will discover, a combination of factors has produced a distinctive but challenged realm between North and South America.

The "Latin America" Misnomer

Sometimes you will see Middle and South America referred to in combination as "Latin" America, alluding to their prevailing Spanish-Portuguese heritage. At the very least, this is an imprecise regional designation, much as Anglo America, a term once commonly used for North America, was also inappropriate. Such culturally based terminologies reflect historic power and dominance, and they tend to make outsiders of those people they do not represent. Thus you will not find the cultural landscape particularly 'Latin' in the Bahamas, Jamaica, Trinidad, Belize, or lengthy stretches of Guatemala and Mexico.

Box 2-1 Major Geographic Features of Middle America

1. Middle America is a relatively small realm consisting of eight mainland countries from Mexico to Panama and all the islands of the Caribbean Basin to the east.

2. Middle America's mainland constitutes a crucial barrier between Atlantic and Pacific waters. In physiographic terms, this is a land bridge connecting the continental landmasses of North and South America.

3. Middle America's cultural geography is complex. Various African and European influences dominate the Caribbean, whereas Spanish and indigenous traditions survive on the mainland.

4. Middle America is a realm of intense cultural and political fragmentation. The presence of many small, insular, and remote countries poses major challenges to economic development.

5. The realm's northern political boundary is in certain respects the axis of a dynamic borderland and transition zone.

6. In terms of area, population, and economic potential, Mexico leads the realm.

7. Many of the countries in this realm find themselves relatively isolated from the rest of the hemisphere.

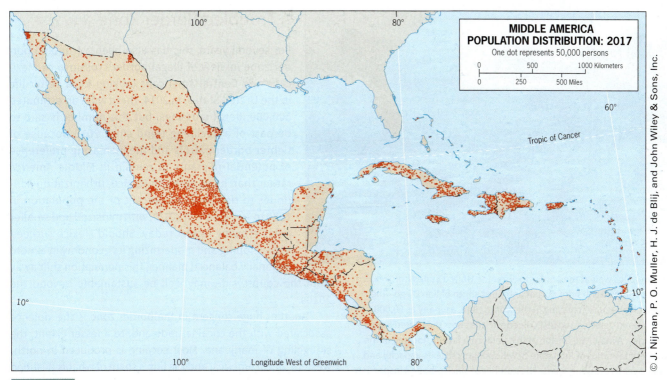

FIGURE 2-2

But is Middle America sufficiently different from either North or South America to merit distinction as a separate realm? Consider this: North America encompasses just a pair of states, and the entire continent of South America has only 12 states

Box 2-2 MAJOR CITIES OF THE REALM, 2016

	Population Metro Area (millions)
Mexico City, Mexico	20.2
Guadalajara, Mexico	4.7
Monterrey, Mexico	4.2
Santo Domingo, Dominican Rep.	3.6
Port-au-Prince, Haiti	2.7
Guatemala City, Guatemala	2.4
San Juan, Puerto Rico	2.1
Puebla, Mexico	2.1
Havana, Cuba	2.0
San José, Costa Rica	1.7
Panama City, Panama	1.5
San Salvador, El Salvador	1.5
Tijuana, Mexico	1.4
Ciudad Juárez, Mexico	1.4
Tegucigalpa, Honduras	1.2
Managua, Nicaragua	1.0

(plus France's dependency on the northeast coast). But far smaller Middle America, as we define it, incorporates more than three dozen political entities, including several dependencies (or quasi-dependencies) of the United Kingdom, the Netherlands, and France as well as a few constituent territories of the United States. Therefore, unlike South America, Middle America is a multilingual patchwork of independent states, territories in political transition, and residual colonial dependencies, with strong continuing ties to the United States and non-Iberian Europe. Middle America is defined in large measure by its vivid multicultural geography.

The Realm's Northern Boundary

The 3169-kilometer (1969-mi) land border between North America and Middle America is the longest in the world separating an affluent realm from a relatively poor one. The U.S.-Mexican boundary crosses half the continent from the Pacific to the Gulf of Mexico, yet Mexican cultural influences penetrate deeply into the southwestern States and American impacts reach far inside Mexico. To some Mexicans, the border is a constant reminder of territory lost to the United States in historic conflicts; to many Americans, it is a symbol of economic contrasts and illegal immigration. Along the Mexican side, the effects of **NAFTA [3]**—the North American Free Trade Agreement between Canada, the United States, and Mexico that went into effect in 1994—have transformed Mexico's economic geography.

David R. Frazier Photolibrary, Inc./Alamy

Automobiles wait at the border to enter the United States from Mexico at the Tijuana–San Ysidro port of entry just south of San Diego—the busiest land border crossing in the world. Here, as many as 50,000 northbound vehicles and 25,000 pedestrians enter the United States every day. A major expansion of this California facility, located at the southern terminus of Interstate-5, will soon increase its capacity and reduce vehicle waiting times. The number of Mexico-bound inspection lanes recently jumped from 11 to 27 in 2015; the U.S.-bound lanes will expand even more dramatically from 34 in 2012 to 72 by 2019.

A Troubled Border Zone

For several years, this has also been a troubled border zone in view of illegal border crossings by Mexicans and Central Americans looking for a better life in the United States. In 2014, there were an estimated 500,000 illegal crossings, but this number is said to be part of a declining trend since 2010 because of tighter border patrols and the increasing preference of potential migrants to remain in Middle America rather than head north. In Mexico, urbanization continues to rise (from 66 percent of the population in 1980 to 79 percent in 2015), with midsized and smaller cities growing fastest today. Should Mexico succeed in developing and modernizing its economy in a more regionally balanced manner, this population shift away from the countryside may well be sustainable beyond the 2010s.

Another huge concern in the border zone is the violence associated with the cocaine trade and, to a lesser extent, the smuggling of marijuana. Most cocaine is produced in northwestern South America, and the lion's share of what is destined

Although the two realms are separated by a sharp political (and economic) boundary, this is a good example of a dynamic **borderland [4]** marked by considerable interaction, wherein the boundary can be more broadly viewed as the axis of a transition zone between North and Middle America. The implementation of NAFTA sparked an economic boom as Mexico became part of a free-trade zone and market now encompassing some 490 million people. A major beneficiary was the strip of Mexican land that ran along the boundary with the United States. The resulting boom changed urban landscapes all along the emerging border zone—but it could not, of course, close the massive economic gap between the two sides (see photo).

Under NAFTA, factories based in Mexico can assemble imported, duty-free raw materials and components into finished products, which are then exported back into the U.S. market. Logically, most of these factories, called *maquiladoras*[5], are located as close to the U.S. border as possible. Thereby, manufacturing employment in the cities and towns along that border, from Tijuana on the Pacific to Matamoros at the mouth of the Rio Grande, expanded rapidly. Mexico now provides roughly 15 percent of all U.S. imports annually, the third-largest source of U.S. imports after Canada and China. China in particular has been increasing its share more rapidly, even though it faces tariffs that NAFTA-member Mexico (and Canada) does not. This puts pressure on Mexico to improve its competitiveness in order to build an even closer economic relationship with the United States and Canada.

Courtesy of John Wark/WarkPhotography

The boundary between Mexico and the United States displays some stunning cross-border contrasts. In this scene, the crowded Mexican town of San Luis Rio Colorado ends at a tall metal fence, beyond which the Arizona desert takes over. This photo, therefore, looks eastward; south is to the right. Like other Mexican towns pressed against the border, San Luis (the shorthand name used by locals) grew rapidly and chaotically in response to the opening of several *maquiladoras*, assembly plants fabricating products for export to the United States.

for the U.S. market passes through Mexico (see Fig. 2-12). Much of the marijuana consumed within the United States is grown in western Mexico and similarly finds its way across the border. Contraband narcotics are mainly carried in trucks and cars, often imaginatively disguised as regular cargo (e.g., stuffed in teddy bears, inside hollowed-out furniture, or simply in boxes labeled as anything from candy to engine parts). It is the sheer volume of this illicit cross-border flow that makes it virtually impossible for authorities to conduct sufficient inspections. Unquestionably, the infrastructure of cities and industrial areas on the Mexican side of the border further facilitates the operation of the drug trade.

The city of Ciudad Juárez seems ideally situated to handle this clandestine activity. It is located directly on the international boundary, and together with El Paso, Texas, on the U.S. side, forms a single metropolis — even though the two cities are separated by a continuous artificial barrier running across this entire urban area (**Fig. 2-3**). This map also shows the four ports of entry, which account for about 25 million border crossings every year. Juárez has grown quite rapidly and exhibits numerous manufacturing districts as well as a sprawling mosaic of neighborhoods, ranging from clusters of upper-income *colonias* to the slums that house the poorest of the new migrants from the south. With a plentiful supply of poor people willing to take the risk of driving or carrying drugs across the border, Juárez also has dozens of *maquiladoras* that can be bribed to store, conceal, and transport the contraband north.

In 2006, the U.S. government, responding both to homeland security concerns (potential terrorists may be entering the country via Mexico) and the tidal wave of cocaine smuggling, initiated the construction of a fortified fence along the entire length of the Mexican border. The very idea seems at odds with the ideals of closer economic and political cooperation, and many security experts questioned its efficacy. By 2011, however, with about one-third of the border hardened by fences and walls, the project to build a "virtual fence" to fill the remaining gaps had become such a boondoggle that it was terminated. Nevertheless, imposing physical barriers are now concentrated around all the main crossing points, and it remains unclear how the longer-term economic health of the border zone they divide will be affected.

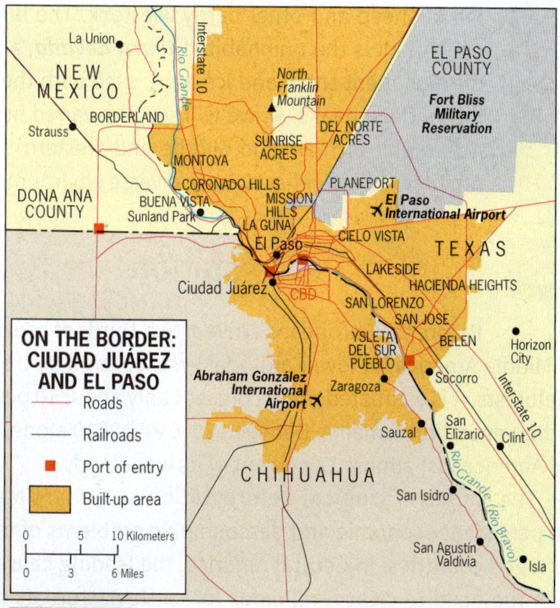

Physical Geography of Middle America

A Land Bridge

The funnel-shaped mainland, a 4800-kilometer (3000-mi) connection between North and South America, is wide enough in the north to contain two major mountain chains and a vast interior plateau, but it narrows to a slim 65-kilometer (40-mi) ribbon of land in Panama. Here this strip of land — or *isthmus*— bends eastward so that Panama's orientation is east-west. Accordingly, mainland Middle America exemplifies what physical geographers call a **land bridge [6]**, an isthmian link between continents.

If you examine a globe, you can see some other present and former land bridges: Egypt's Sinai Peninsula between Asia and Africa, the (now-broken) Bering land bridge between northeasternmost Asia and Alaska, and the shallow waters between New Guinea and Australia. Such land bridges, though temporary features in geologic time, have played crucial roles in the dispersal of animals and humans across the planet. But even though mainland Middle America forms a land bridge, its internal fragmentation has always inhibited movement. Mountain ranges, swampy coastlands, and dense rainforests make contact and interaction difficult.

Island Chains

As shown in Figure 2-1, the approximately 7000 islands of the Caribbean Sea stretch in a lengthy arc from Cuba and the Bahamas eastward and then southward to Trinidad, with numerous outliers outside (such as Barbados) and inside (e.g., the Cayman Islands) the main chain. The four large islands in the Caribbean's east-west segment—Cuba, Hispaniola (containing Haiti and the Dominican Republic), Puerto Rico, and Jamaica—are called the ***Greater Antilles***, and all the remaining smaller islands constitute the ***Lesser Antilles***. The entire Antillean **archipelago [7]** (island chain) consists of the crests and peaks of mountain chains that rise from the floor of the Caribbean, the result of collisions between the Caribbean Plate and its neighbors (Fig. G-4). Some of these crests are relatively stable, but elsewhere they contain active volcanoes, and almost everywhere in this realm earthquakes are an ever-present danger—in the islands as well as on the mainland. Add to this constant hazard the realm's seasonal exposure to Atlantic/Caribbean hurricanes, and it amounts to some of the highest-risk real estate on Earth.

ON THE BORDER: CIUDAD JUÁREZ AND EL PASO

— Roads
— Railroads
■ Port of entry
▨ Built-up area

0   5   10 Kilometers
0   3   6 Miles

© J. Nijman, P. O. Muller, H. J. de Blij, and John Wiley & Sons, Inc.

FIGURE 2-3

Dangerous Terrain and Skies

The danger from below is dramatically illustrated in the Haitian, western half of Hispaniola, which is laced with geologic fault lines associated with the nearby boundary that separates the North American and Caribbean plates (see Fig. 2-16, inset map). On January 12, 2010, Haiti was devastated by a massive earthquake; the epicenter of that 7.0 temblor was located just outside Port-au-Prince and virtually destroyed this teeming, impoverished capital city of nearly 3 million. At least 300,000 people died, and within a week more than a million had fled to the countryside. It was the worst natural disaster in Haitian history but certainly not the first (see photo).

The environmental hazard from above comes in the form of hurricanes, powerful tropical cyclones that annually threaten the Caribbean Basin and its surrounding coastlines. The eastern half of Middle America is one of the most hurricane-prone areas in the world. One of the key conditions for the formation of hurricanes is very warm ocean water, which further heats the hot moist air rising above it to "fuel" the evolving storm. The prolonged Atlantic/Caribbean hurricane season extends from June 1 to December 1, with the greatest number of tropical cyclones occurring in August and September when seawater reaches its highest temperatures. Most storms travel in a westerly direction from their low-latitude spawning ground off the coast of West Africa, steered across the Atlantic by the trade winds to reach the Caribbean Sea. Once there, many of these cyclones follow similar routes within **Hurricane Alley [8]**, whose wide axis lies along all of the Greater Antilles and then broadens to include southern Florida, Mexico's Yucatán Peninsula, and all of the Gulf of Mexico. On average, every season sees the development of four to eight major hurricanes, and rarely does a year go by without a destructive landfall on at least one of these densely populated areas.

Altitudinal Zonation of Environments

Continental Middle America and the western margin of South America are areas of high relief and strong environmental contrasts. Even though settlers have always favored temperate intermontane basins and valleys, people also cluster in hot tropical lowlands as well as high plateaus just below the snow line in South America's Andes Mountains. In each of these zones, distinct local climates, soils, vegetation, crops, domestic animals, and modes of life prevail. Such **altitudinal zones [9]** (diagrammed in **Fig. 2-4**) are known by specific names as if they were regions with distinguishing properties—as in reality they are.

The lowest of these vertical zones, from sea level to 750 meters (2500 ft), is known as the ***tierra caliente***, the "hot land" of the coastal plains and low-lying interior basins where tropical agriculture predominates. Above this zone lie the tropical highlands containing Middle and South America's largest population clusters, the ***tierra templada*** of temperate land reaching up to about 1800 meters (6000 ft). Temperatures here are cooler; prominent among the commercial crops is coffee, while corn (maize) and wheat are the staple grains. Still higher, from approximately 1800 to 3600 meters (6000 to nearly 12,000 ft), is the ***tierra fría***, the cold country of the higher Andes where hardy crops such as potatoes and barley are mainstays. Above the tree line, which marks the upper limit of the *tierra fría*, lies the ***tierra helada***; this fourth altitudinal zone, extending from about 3600 to 4500 meters (12,000 to 15,000 ft), is so cold and barren that it can support only the grazing of sheep and other hardy livestock. The highest zone of all is the uninhabited ***tierra nevada***, a zone of permanent snow and ice associated with the loftiest Andean peaks. As we will see, the varied human geography of mainland Middle and western South America closely reflects these diverse environments.

Demonstrators in Port-au-Prince, Haiti's capital, march during a recent protest to demand improvement of their living conditions. Recovery since the catastrophic earthquake of 2010 has been painfully slow. On the sixth anniversary of the disaster in 2016, Haiti continued to be plagued by recurrent outbreaks of cholera and other serious diseases, hundreds of thousands of inadequately sheltered homeless citizens, and the relentlessly crushing effects of a stagnant, still-battered economy. With the rebuilding effort all but stalled, this overwhelmingly vulnerable country was facing its seventh hurricane season since the quake: how much longer could its fortunate string of uneventful storm seasons hold out?

Tropical Deforestation

Before the Europeans arrived, two-thirds of continental Middle America (at lower altitudes) was covered by tropical rainforests. It is estimated that at present only about 10 percent of this vegetation remains. Between 2000 and 2010 alone, Central America lost almost 12 percent of its woodlands.

The causes of **tropical deforestation [10]** are related to the persistent economic and demographic problems of disadvantaged countries. In Central America, the leading cause has been the need to clear rural lands for cattle pasture as many countries, especially Costa Rica, became meat producers and

Meters | Feet

TIERRA NEVADA		
TIERRA HELADA	livestock grazing	
TIERRA FRÍA	potatoes barley wheat dairying	
TIERRA TEMPLADA	coffee maize (corn) wheat vegetables	
TIERRA CALIENTE	bananas sugarcane rice other tropical crops	

4,500 — 15,000
3,600 — 12,000
1,800 — 6,000
750 — 2,500
Sea Level — Sea Level

Snow Line

Tree Line

ALTITUDINAL ZONATION

FIGURE 2-4

exporters. Because tropical soils are so nutrient-poor, newly deforested areas are able to function as pastures for only a few years at most. These fields are then abandoned for other freshly cut lands and quickly become a ravaged landscape (see photo). Without the protection of binding tree roots, local soil erosion and flooding immediately become problems, affecting still-productive areas nearby. A second cause of deforestation is the rapid logging of tropical woodlands as the timber industry increasingly turns from the exhausted forests of the midlatitudes to harvest the rich tree resources of the equatorial zones, in response to accelerating global demands for housing, paper, and furniture. A third major contributing factor is related to the region's population explosion: as more and more peasants are required to extract a subsistence from inferior lands, they have no choice but to cut down the remaining forest for both firewood and additional crop-raising space, and their intrusion prevents the trees from regenerating.

Cultural Geography

Mesoamerican Legacy

Mainland Middle America was the scene of the emergence of major ancient civilizations. Here lay one of the world's true **culture hearths [11]**, a source area from which new ideas radiated outward and whose population could expand and make significant material as well as intellectual progress. Agricultural specialization, urbanization, trade, and transportation networks developed, and writing, science, art, and other spheres of achievement saw major advances. Anthropologists refer to the Middle American culture hearth as **Mesoamerica**, which extended southeast from the vicinity of

what is now Mexico City to central Nicaragua. Its development is particularly remarkable because it occurred in highly different geographic environments, each presenting obstacles that had to be overcome in order to unify and integrate large territories. First, in the low-lying tropical plains of what is now northern Guatemala, Belize, and Mexico's Yucatán Peninsula, and perhaps simultaneously in Guatemala's highlands to the south, the Maya civilization arose more than 3000 years ago. Later, in the fourteenth century AD, far to the northwest on the high plateau in central Mexico, the Aztecs founded a major civilization

Despite scattered attempts to reverse this landscape scourge, deforestation continues to afflict Central America. Its worst effects plague the steeper slopes of interior highlands, as here in western Panama. In the wake of recent deforestation, the land near the top of this hill has already begun to erode because in the absence of binding tree roots the copious tropical rains are making short work of the unprotected topsoil.

centered on Tenochtitlán (present-day Mexico City) in the Valley of Mexico. Some of the greatest contributions of Mesoamerica's indigenous peoples came from the agricultural sphere and included the domestication of corn (maize), the sweet potato, cacao beans (the raw material of chocolate), and tobacco.

Spanish Conquest

Spain's defeat of the Aztecs in the early sixteenth century opened the door to Spanish penetration and supremacy. The Spaniards were ruthless colonizers but not more so than other European powers that subjugated other cultures. The Spaniards enslaved the native population and were determined to destroy the strength of their society. Biology accomplished what ruthlessness could not have achieved in so short a time: diseases introduced by the Spaniards and the slaves they imported from Africa killed millions of indigenous people.

Middle America's cultural landscape was drastically modified. Unlike the indigenous peoples, who had utilized stone as their main building material, the Spaniards employed great quantities of wood and used charcoal for heating, cooking, and smelting metal. The onslaught on the forests was immediate, and rings of deforestation swiftly expanded around the colonizers' towns. The Spaniards introduced large numbers of cattle and sheep, and brought over their own crops (notably wheat) and farming equipment. Soon large wheatfields began to encroach upon the small plots of corn that the native people cultivated.

The Spaniards' most far-reaching cultural changes derived from their traditions as town dwellers. The indigenous people were moved off their land into nucleated villages and towns that the Spaniards established and laid out. In these settlements, the Spaniards could exercise the kind of rule and administration to which they were accustomed. The internal focus of each Spanish town was the central **plaza** or market square, around which both the local church and government buildings were located. The surrounding street pattern was deliberately laid out in gridiron form. Each town was located near what was thought to be good agricultural land (which often was not so good), so that the indigenous people could venture out each day and work in the fields. Packed tightly into these towns and villages, they came face to face with Spanish culture. Here they (forcibly) learned the Europeans' Roman Catholic religion and Spanish language, and they paid their taxes and tribute to a new master. Many of Middle America's leading cities still bear this Spanish imprint.

Collision of Cultures

But Middle America is not Spain, and its fragmented cultural mosaic reflects the collision of indigenous, Spanish, and other European influences. Indeed, in more remote areas in southeastern Mexico and interior Guatemala, indigenous societies survived, and to this day native languages prevail over Spanish (see Fig. 2-10).

In Middle America outside Mexico, only Panama, with its twin attractions of interoceanic transit and gold deposits, became an early focus of Spanish activity. From there, following the Pacific-fronting side of the isthmus, Spanish influence radiated northwestward through Central America and into Mexico. The major arena of international competition in Middle America, however, lay not on the Pacific side but on the islands and coasts of the Caribbean Sea. Here the British gained a foothold on the mainland, controlling a narrow coastal strip that extended southeast from Yucatán to what is now Costa Rica. As the colonial-era map (**Fig. 2-5**) shows, in the Caribbean the Spaniards faced not only the British but also the French and the Dutch, all interested in the lucrative sugar trade, all searching for instant wealth, and all seeking to expand their empires.

Much later, after centuries of European colonial rivalry in the Caribbean Basin, the United States entered the picture and made its influence felt in the coastal areas of the mainland, not through colonial conquest but through the introduction of widespread, large-scale, banana-plantation agriculture. The effects of these plantations were as far-reaching as the impact of colonialism on the Caribbean islands. Because the diseases the Europeans had introduced were most rampant in these hot humid lowlands (as well as the Caribbean islands to the east), the indigenous population that survived was too small to provide a sufficient workforce. This labor shortage was quickly remedied through the trans-Atlantic slave trade from Africa that transformed the population composition of the Caribbean Basin.

The cultural variegation of the Caribbean Basin is especially striking, and it is hardly an arena of exclusive Hispanic cultural heritage. For example, Cuba's southern neighbor, Jamaica (population 2.8 million, mostly of African ancestry), has a legacy of British involvement, while to the east in Haiti (10.9 million, overwhelmingly of African ancestry) the strongest imprints have been African and French. The Lesser Antilles also exhibit great cultural diversity. There are the (once Danish) U.S. Virgin Islands; French Guadeloupe and Martinique; a group of British-influenced islands, including Barbados, St. Lucia, and Trinidad and Tobago; and the Dutch St. Maarten (shared with the French) as well as the now-autonomous A-B-C islands of the former Netherlands Antilles—Aruba, Bonaire, and Curaçao—off the northwestern Venezuelan coast.

Political and Economic Fragmentation

Independence

Independence movements stirred Middle America at an early stage. On the mainland, insurrections against Spanish authority (beginning in 1810) achieved independence for Mexico by 1821 and for the Central American republics by the end of the 1820s, resulting in the creation of eight different countries. The United States, concerned over European designs in the realm,

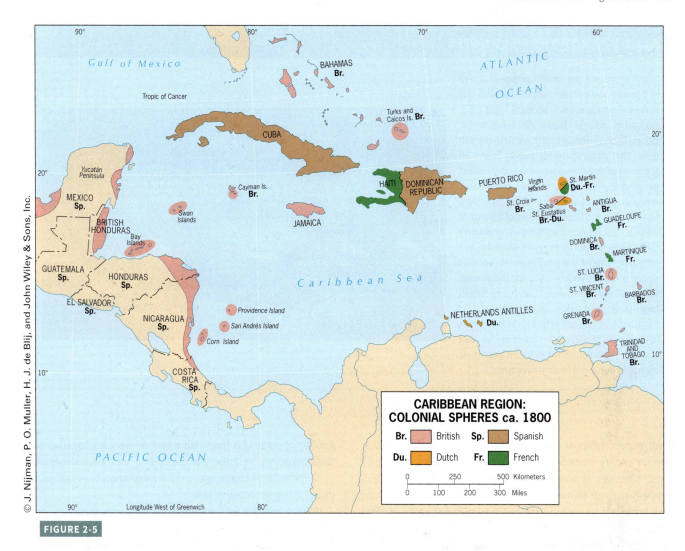

**CARIBBEAN REGION:
COLONIAL SPHERES ca. 1800**

Br.	British	**Sp.**	Spanish
Du.	Dutch	**Fr.**	French

FIGURE 2-5

proclaimed the Monroe Doctrine in 1823 to deter any European power from reasserting its authority in the newly independent republics or from further expanding its existing domains.

By the end of the nineteenth century, the United States itself had become a major force in Middle America. The Spanish-American War of 1898 made Cuba independent and placed Puerto Rico under the U.S. flag; soon thereafter, the Americans were in Panama constructing the Panama Canal across their newly acquired Canal Zone. Meanwhile, with U.S. corporations propelling a boom based on massive banana plantations, the Central American republics had become colonies of the United States in all but name.

Independence came to the Caribbean Basin in fits and starts. African-Caribbean Jamaica as well as Trinidad and Tobago, where the British had brought in a large South Asian population, attained full sovereignty from the United Kingdom in 1962; other British islands (among them Barbados, St. Vincent, and Dominica) became independent later on. France, however, retains Martinique and Guadeloupe as *Overseas Départements* of the French Republic, and the Dutch islands (see **Box 2-3**) are at various stages of autonomy. No less than 33 states are found on the political map of the Caribbean Basin today.

Regional Contrasts

There are some important contrasts, socially and economically, between the Middle American mainland on the one hand and the Caribbean coasts and islands on the other. The mainland is dominated by Spanish as well as indigenous influences and also includes **mestizo [12]** sectors where the two ancestries mixed. Agriculture on the mainland is often organized around the **hacienda [13]**, a Spanish institution through which landowners possessed a domain whose productivity they might never push to its limits; the very possession of such a vast estate brought with it social prestige and a comfortable lifestyle. Native workers lived on the land—which may once have been their land—and had plots where they could grow their own subsistence crops. All this is written as though it is mostly in the past, but the legacy of the hacienda system, with its inefficient use of land and labor, still exists throughout mainland Middle America.

The Caribbean coasts (from Belize southward) and the Caribbean islands possess a very different cultural heritage, based on a fusion of European and African influences. Here, **plantations [14]** long dominated the economies. The plantation, in contrast to the hacienda, is all about efficiency and profit.

Box 2-3   FROM THE FIELD NOTES ...

"Driving around the small Caribbean island of Bonaire and coming face to face with this surreal landscape, I first thought I had experienced a mirage. These glistening white, perfectly cone-shaped hills are actually salt piles. This small (formerly Dutch) island off the coast of Venezuela possesses the perfect geographic conditions for salt production: it has a series of salt-water inlets, it is very hot and dry, and lies in the zone of persistent trade winds. Remember that salt, nowadays taken for granted in every household, has long been one of the world's most precious spices and was widely used for the preservation of meat and fish. The Dutch began large-scale production on salt 'plantations' here in the 1620s; today this local industry is in the hands of the Antilles International Salt Company and continues to be an important source of Bonaire's foreign revenues."

© Jan Nijman

Foreign ownership and investment is the norm, as is production for export. Most plantations grow only a single crop, be it sugar, bananas, or coffee. Much of the labor is seasonal, being needed in large numbers mainly during the harvest period; such labor has been imported because of the scarcity of indigenous workers. With its "factory-in-the-field" operations, the plantation enables far more efficient use of land and labor than the hacienda. Profit and wealth, rather than social prestige, are the dominant motives for the plantation's establishment, operation, and perpetuation.

Connections Matter

The unusual layout of the Middle American realm offers an opportunity to consider the spatial dimensions of development. Geographical fragmentation and territorial size play their parts but so does *accessibility* in the form of **connectivity [15]** in both the Caribbean Basin and on the mainland. **Figure 2-6** in **Box 2-4** provides a closer look at the relationship between development and connectivity.

Is Small Beautiful?

As noted, the Middle American realm is exceptional in the modest size of its territorial extent and population. But the number of countries is considerable, and they tend to be quite small. Caribbean islands often invoke images of paradise: beautiful scenery, tropical drinks, and shiny blue waters. But the economic realities are almost invariably harsh. The limited land area and small population of most islands (those of the Lesser Antilles average less than half a million people), together with their insularity, relative inaccessibility, and low connectivity, pose formidable challenges that are common to **small-island developing economies [16]**.

What are the consequences of these geographical properties for economic development? First, natural resources are frequently limited, which requires a heavy reliance on imports made more expensive by added transport costs. Second, the cost of government is relatively high per capita: even the smallest population will require services such as schools, hospitals, and waste disposal. Third, these specialized services must often be brought in from elsewhere. And fourth, local production cannot really benefit from **economies of scale [17]** (unit-cost savings resulting from large-scale production). Consequently, local producers can be put out of business by cheaper imports, thereby driving up unemployment.

Given the Caribbean Basin's limited economic options, does the tourist industry provide better opportunities? The resort areas, scenic treasures, and historic locales of Caribbean America attract more than 25 million visitors annually, with about half of them traveling on Florida-based cruise ships. But Caribbean tourism also has serious drawbacks (**Box 2-5**). The invasion of poor communities by affluent tourists contributes to rising local resentment, which is fueled by the glaring contrasts of shiny new hotels towering over substandard housing, or luxury liners gliding past poverty-stricken villages. At the same time, tourism can debase local culture, which often is adapted to suit the visitors' tastes at hotel-staged "culture" shows. In addition, the cruise industry tends to monopolize revenues (accommodations, meals, and entertainment), with relatively few dollars flowing into

Box 2-4 MAP ANALYSIS

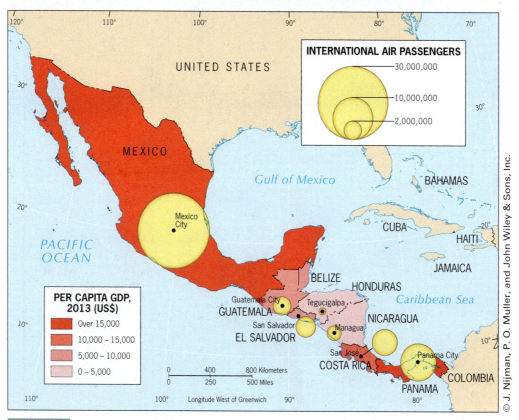

INTERNATIONAL AIR PASSENGERS

30,000,000
10,000,000
2,000,000

PER CAPITA GDP, 2013 (US$)

Over 15,000
10,000 – 15,000
5,000 – 10,000
0 – 5,000

© J. Nijman, P. O. Muller, and John Wiley & Sons, Inc.

FIGURE 2-6 Location, Connectivity, and Economic Development in Mainland Middle America, 2013

This map shows annual GDP per capita as well as the number of international air passengers in 2013. Mapping these variables allows us to analyze them in spatial context and compare their locational patterns. What seems to be the relationship between development and connectivity? What are the key geographic factors that shape connectivity within Panama and Mexico?

Access your WileyPLUS Learning Space course to interact with a dynamic version of this map and to engage with online map exercises and questions.

the local economy. Finally, even though tourism does generate considerable income in the Caribbean, the intervention of island governments and multinational corporations can often remove opportunities from local entrepreneurs in favor of large operators and major resorts.

The Push for Regional Integration

Another challenge for the countries of the Middle American realm is to foster greater **economic integration [18]** (the benefits of forging new international partnerships). Many of the countries on the mainland as well as in the Caribbean are poorly connected within the realm and are heavily dependent on major outside countries, particularly the United States. For the large majority of these countries, the United States is their primary trading partner. Consider this: of all the trade that engages Middle American countries, less than 10 percent occurs within the realm. And less than 1 percent takes place

between the Caribbean Basin and the Middle American mainland.

Over the years, efforts have been made to advance economic integration and convert this realm into more of a *functional region*. There are presently two trade organizations in Middle America (in addition to Mexico's membership in NAFTA). **CARICOM** (Caribbean Community), established in 1989, consists of 15 full members, including nearby Guyana and Suriname in South America. To a certain degree, CARICOM emulates the European Union (EU), and in 2009 it even introduced a common passport; but unlike the EU, this organization has had only a superficial economic impact. The more influential **CAFTA-DR** (Central American Free Trade Agreement) includes the United States, five Central American countries, and the Caribbean's Dominican Republic (hence the "DR" suffix). But this organization seems to be a mixed blessing: it may increase access to U.S. markets and lead to cheaper imports, yet it also reinforces the United States' dominant position at the expense of greater intra-realm integration.

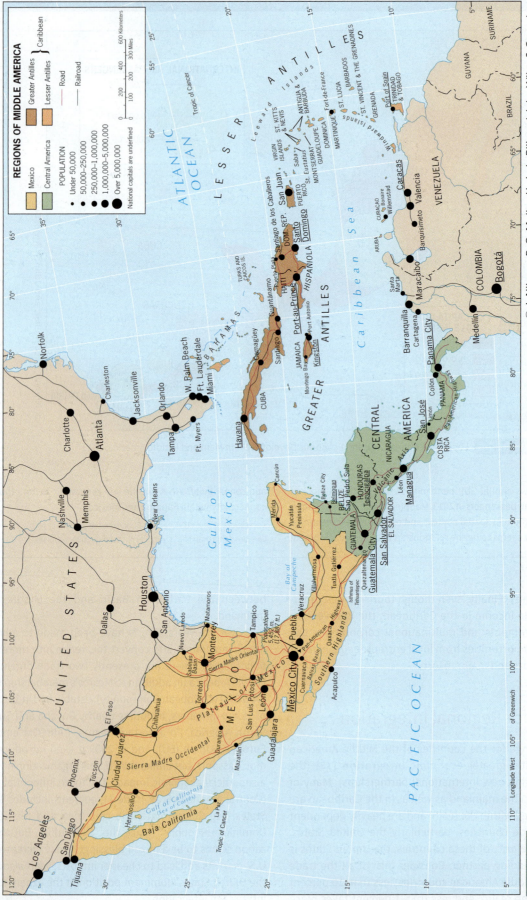

FIGURE 2-7

Box 2-5   Regional Issue: The Tourism Industry in Middle American Economies

Tourism Drives Our Economies

"As the general manager of a small hotel in St. Maarten, on the Dutch side of the island, I can tell you that without tourists, we would be in deep trouble economically. Mass tourism, plain and simple, has come to the rescue in the Caribbean and in other countries of Middle America. Look at the numbers: for many smaller Caribbean countries it's the leading industry and the only one that is growing. Whether it's hotel patrons or cruise-ship passengers, tourists spend money, create jobs, fill airplanes that give us a link to the outside world, require infrastructure that's good not just for them but for locals too. We've got better roads, better telephone service, more items in our stores. All this comes from tourism. I employ 24 people, most of whom would be looking for nonexistent jobs if it weren't for the tourist industry.

"And it isn't just us here in the Caribbean. Look at Belize. I just read that tourism there, based on their coral reefs, Mayan ruins, and inland waterways, brought in nearly U.S. $300 million last year and provides 13 percent of all jobs. That sure beats sugar and bananas. In Jamaica, too, they tell me, one in every four workers has a job in tourism. And the truth is, there's still plenty of room for the tourist industry to expand in Middle America. Those Americans and Europeans can close off their markets against our products, but they can't stop their citizens from getting away from their dreadful weather by coming to this tropical paradise.

"Here's another good thing about tourism. It's a clean industry. It digs no mine shafts, doesn't pollute the atmosphere, doesn't cause diseases, doesn't poison villagers, isn't subject to graft and corruption the way some other industries are.

"Last but not least, tourism is educational. Travel heightens knowledge and awareness. There's always a minority of tourists who just come to lie on the beach or spend all their time in some cruise-ship bar, but most of the travelers we see in my hotel are interested in the place they're visiting. They want to know why this island is divided between the Dutch and the French, they ask about coral reefs and volcanoes, and some even want to practice their French on the other side of the border (don't worry, no formalities, just drive across and start talking). Tourism's the best thing that happened to this part of the world, and other parts too, and I hope we'll never see a slowdown."

The Tourism Industry Is a Mixed Blessing at Best

"You won't get much support for tourism from some of us teaching at this college in Puerto Rico, no matter how important some economists say tourism is for the Caribbean. Yes, tourism is an important source of income for some countries, like Tanzania with its wildlife and Nepal with its mountains, but for many countries that income from tourism does not constitute a real and fundamental benefit to the local economies. Much of it may in fact result from the diversion to tourist consumption of scarce commodities such as food, clean water, and electricity. More of it has to be reinvested in the construction of airport, cruise-port, overland transport, and other tourist-serving amenities. And as for items in demand by tourists, have you noticed that places with many tourists are also places where prices are high?

"Sure, our government people like tourism. Some of them have a stake in those gleaming hotels where they can share the pleasures of the wealthy. But what those glass-enclosed towers represent is globalization, powerful multinational corporations colluding with the government to limit the opportunities of local entrepreneurs. Planeloads and busloads of tourists come through on prearranged (and prepaid) tour promotions that isolate those visitors from local society.

"Picture this: luxury liners sailing past poverty-stricken villages, four-star hotels towering over muddy slums, restaurants serving caviar when, down the street, children suffer from malnutrition. If the tourist industry offered real prospects for economic progress in poorer countries, such circumstances might be viewed as the temporary, unfortunate byproducts of the upward struggle. Unfortunately, the evidence indicates otherwise. Name me a tourism-dependent economy where the gap between the rich and poor has narrowed.

"As for the educational effect of tourism, spare me the argument. Have you sat through any of those 'culture' shows staged by the big hotels? What you see there is the debasing of local culture as it adapts to visitors' tastes. Ask hotel workers how they really feel about their jobs, and you'll hear many say that they find their work dehumanizing because expatriate managers demand displays of friendliness and servitude that locals find insulting to sustain.

"I've heard it said that tourism doesn't pollute. Well, the Alaskans certainly don't agree—they sued a major cruise line on that issue and won. Not very long ago, cruise-ship crews routinely threw garbage-filled plastic bags overboard. That seems to have stopped, but I'm sure you've heard of the trash left by mountain-climbers in Nepal, the damage done by off-road vehicles in the wildlife parks of East Africa, the coral reefs injured by divers off Bonaire or the Virgin Islands. Tourism is here to stay, but it is no panacea."

Regions of Middle America

The Middle American realm can be subdivided into four distinct geographic regions (the first two on the mainland; the other two in the Caribbean Basin): (1) **Mexico**, the giant of the realm in every respect; (2) **Central America**, the string of seven small republics occupying the land bridge from Mexico to South America; (3) the four islands of the **Greater Antilles**—Cuba, Jamaica, Hispaniola (containing Haiti and the Dominican Republic), and Puerto Rico; and (4) the numerous small islands of the **Lesser Antilles** (**Fig. 2-7**).

© Liane Cary/Age Fotostock America, Inc.

Hundreds of thousands of North American retirees, and many affluent purchasers of second homes, have sought the sun and low-cost residential opportunities of Middle America, converting some areas into virtual exclaves of the North. From Mexico to Panama, waterfront real estate is among the attractions for these permanent and seasonal migrants, as here in Cabo San Lucas at the southern tip of Mexico's Baja California peninsula. This is yet another example of international connections that matter.

Region Mexico

Physiography

The physiography of Mexico is reminiscent of that of the conterminous western United States, although its environments are more tropical. **Figure 2-8** shows several prominent features: the elongated peninsula of Baja (Lower) California in the northwest, the far eastern Yucatán Peninsula, and the Isthmus of Tehuantepec in the southeast where the Mexican landmass tapers to its narrowest extent. Here in the southeast, Mexico's physiography most resembles that of Central America: a mountainous backbone forms the isthmus, curves southeastward into Guatemala, and extends northwestward toward Mexico City. Shortly before reaching the capital, this mountain range divides into two chains, the Sierra Madre Occidental to the west and the Sierra Madre Oriental to the east. These diverging ranges frame the funnel-shaped Mexican heartland, the center of which consists of the rugged, extensive Plateau of Mexico (the important Valley of Mexico lies near its southeastern end). Mexico's climates are characterized by dryness, particularly in the broad, mountain-flanked north (also see Fig. G-6). Most of the better-watered areas lie in the southern half of the country where a number of major population concentrations have developed (see Fig. 2-2).

Regional Diversity

Physiographic, demographic, economic, historical, and cultural criteria combine to reveal a regionally diverse Mexico extending from the lengthy ridge of Baja California to the tropical lowlands of the Yucatán Peninsula, and from the economic dynamism of the U.S. borderland to the indigenous traditionalism of far southeastern Chiapas. The country's core area, anchored by Mexico City and extending westward to Guadalajara, lies between the more Hispanic-mestizo north and the dominantly indigenous-mestizo south. East of the core area is the Gulf Coast, once dominated by major irrigation projects and sprawling livestock-raising schemes but now the mainland center of Mexico's petroleum industry. The region south of the core is dominated by the rugged, Pacific-fronting Southern Highlands, where coastal Acapulco's luxurious resorts stand in stark contrast to the indigenous villages and communal-farm settlements of the interior. The dry, far-flung north stands in particularly sharp contrast to these southern regions. To go from comparatively well-off Monterrey in the far north to poverty-mired Chiapas adjacent to southern Guatemala is to observe the entire range of Mexico's regional geography.

Population and Urbanization Patterns

Mexico's population expanded rapidly throughout the closing decades of the twentieth century, doubling in just 28 years; but demographers have recently noted a sharp drop in fertility, and they are predicting that Mexico's population (currently at 129 million) will cease growing altogether by about 2050. That will have important consequences for the country's economy and for the United States as well since it is likely to affect long-term migration patterns.

The distribution of Mexico's population relative to the country's 31 internal States is shown in Figures 2-2 and **2-9**. The largest concentration, containing the core area and more than half of the Mexican people, extends across the densely populated "waist" of the country from Veracruz State on the eastern Gulf Coast to Jalisco State on the Pacific. The center of this corridor is dominated by the most populous State, Mexico (**3** on the map), at whose heart lies the Federal District of Mexico City (**9**). In the dry and rugged terrain to the north of this central corridor lie Mexico's least-populated States. Southern Mexico also exhibits a sparsely peopled periphery in the hot and humid lowlands of the Yucatán, but to the southwest most of the highlands of the continental spine contain sizeable populations.

Another major feature of Mexico's population map is urbanization, driven by the pull of the cities (with their perceived opportunities for upward mobility) in tandem with the push of the economically stagnant countryside. Today, 79 percent of the Mexican people reside in towns and cities, a surprisingly high proportion for a less-developed country. Undoubtedly, these numbers are affected by the recent explosive growth of the region around Mexico City, which has surpassed 20 million and is home to one-sixth of the national population (**Box 2-6**).

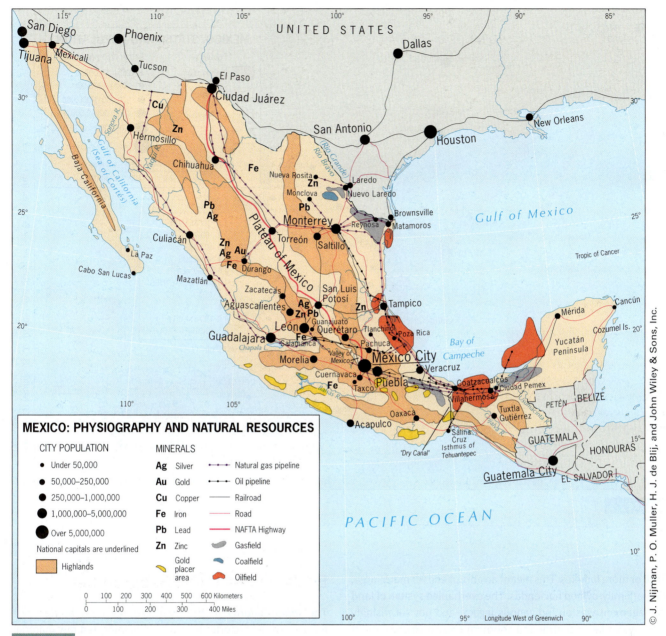

MEXICO: PHYSIOGRAPHY AND NATURAL RESOURCES

CITY POPULATION
- • Under 50,000
- • 50,000–250,000
- • 250,000–1,000,000
- • 1,000,000–5,000,000
- • Over 5,000,000

National capitals are underlined

Highlands

MINERALS
- **Ag** Silver
- **Au** Gold
- **Cu** Copper
- **Fe** Iron
- **Pb** Lead
- **Zn** Zinc
- Gold placer area

- Natural gas pipeline
- Oil pipeline
- Railroad
- Road
- NAFTA Highway
- Gasfield
- Coalfield
- Oilfield

0 100 200 300 400 500 600 Kilometers
0 100 200 300 400 Miles

© J. Nijman, P. O. Muller, H. J. de Blij, and John Wiley & Sons, Inc.

FIGURE 2-8

Urbanization rates are at their lowest in the peripheral southern uplands where indigenous society has been least affected by modernization.

A Mix of Cultures

Nationally, the indigenous imprint on Mexican culture remains prevalent. Today, roughly 60 percent of all Mexicans are mestizos, about 30 percent are predominantly indigenous, (of which 7 percent are full-blooded indigenous); almost all of the remaining 10 percent are Europeans. The Spanish influence in Mexico has been profound, but it has been met with an equally powerful thrust of indigenous culture. It has therefore not been a case of one-way European-dominated **acculturation [19]** but rather **transculturation [20]**—the two-way exchange of

culture traits between societies in close contact. In the southeastern periphery (**Fig. 2-10**), several hundred thousand Mexicans still speak only an indigenous language, and millions more still utilize these languages daily even though they also speak Mexican Spanish. The latter has been strongly shaped by indigenous influences, as have Mexican modes of dress, cuisine, artistic and architectural styles, and folkways. This fusion of heritages, which makes Mexico unique, is the result of an upheaval that began to reshape the country just over a century ago.

Agriculture: Fragmented Modernization

Following the Mexican Revolution of the 1910s and its ensuing land redistribution schemes, around half of the arable land was organized into so-called *ejidos* **[21]**, communally owned farms

FIGURE 2-9

of 20 or more families. This meant an abrupt end for many large, single-family-owned haciendas. The overhauled system of land management was an indigenous legacy, and not surprisingly most *ejidos* lie in central and southern Mexico where nativist social and agricultural traditions are strongest. Today, about half of Mexico's land continues to be held in such "social land-holdings." But some of this land is excessively fragmented and/or managed inefficiently with scarce capital, resulting in both inferior crop yields and widespread, persistent rural poverty. In the 1990s, the Mexican government attempted to privatize *ejidos*, hoping to promote consolidation and increase productivity, but that effort was unsuccessful.

At the same time, larger-scale commercial agriculture has diversified during the past three decades and made major gains in both domestic and export markets. The country's arid northern tier has led the way as major irrigation projects have been built on streams flowing down from the interior highlands. Along the booming northwestern coast of the mainland, which lies within a day's drive of Southern California, large-scale production of fruits, vegetables, cut flowers, and cotton has become the cornerstone of an increasingly profitable export trade to the United States and elsewhere.

Evolving Economic Geographies

The *maquiladoras* in the country's northern borderland—foreign-owned factories that assemble imported raw materials and components and then export finished manufactures back to the United States—account for over one-sixth of Mexico's industrial jobs and a major share of its total exports. *Maquiladoras* presently employ roughly two million Mexican workers. Many companies are headquartered in cities just north of the border, such as San Diego and El Paso, and this is where design, marketing, and other strategic work takes place. But most of the workforce is located south of the border at the assembly plants (which is why the cities on the Mexican side have grown so much faster than their U.S. counterparts). Think electronics, machinery, clothing, construction materials, automobile parts, and much more. Although the impact of NAFTA has been widespread, it should be noted that Mexican employees work long hours for low wages with meager fringe benefits. Moreover, most reside in the overcrowded rudimentary settlements that encircle the burgeoning urban centers of the border zone. In 2015, the minimum daily wage in Mexico was still below six U.S. dollars. It is estimated that 15 percent actually earn less,

Box 2-6 Among the Realm's Great Cities: Mexico City

Middle America has only one great metropolis: Mexico City. With 20.2 million inhabitants, the metro area is home to more than 15 percent of Mexico's population, and it grows by more than 300,000 every year. It ranks among the largest urban areas in the world and is the third-biggest in the Western Hemisphere.

Lakes and canals marked this site when the Aztecs built their city of Tenochitlán here seven centuries ago. The conquering Spaniards made it their headquarters, and following independence the Mexicans made it their capital. Centrally positioned and well connected to the rest of the country, Mexico City, hub of the national core area, became the quintessential primate city.

Vivid social contrasts mark the cityscape. Historic plazas, magnificent palaces, churches, villas, superb museums, ultramodern skyscrapers, and luxury shops fill the city center. Beyond lies a zone of comfortable middle-class and struggling, but stable, working-class neighborhoods. Outside this belt, however, lies a ring containing more than 500 established slum areas and the shacktowns of countless, even poorer *ciudades perdidas*—the "lost cities" where newly arrived peasants live in miserable poverty and squalor. (These squatter settlements contain one-third of the metropolitan area's population.) Mexico City's more affluent residents have also been plagued by problems in recent times as the country's social and political order came close to unraveling. Rampant crime remains a serious concern, much of it associated with corrupt police.

Environmental crises parallel the social problems. Local surface waters have long since dried up, and groundwater supplies are approaching depletion; to meet demand, the metropolis must now import much of its water by pipeline from across the mountains (with almost half of that supply lost through leakages in the city's crumbling water pipe network). Air pollution here is among the world's worst as nearly 5 million motor vehicles and tens of thousands of factories churn out smog that in Mexico City's thin, high-altitude air sometimes reaches 100 times the acceptably safe level. And add to all this a set of geologic hazards: severe land subsidence as underground water reservoirs are overdrawn; the ever-present

threat of earthquakes that can wreak havoc on the city's unstable surface (the last big one occurred in 1985); and even the risk of volcanic activity, as nearby Mount Popocatépetl occasionally shows signs of ending centuries of dormancy.

In spite of it all, the great city continues to beckon, and every year hundreds of thousands of the desperate and the dislocated arrive with hope—and little else.

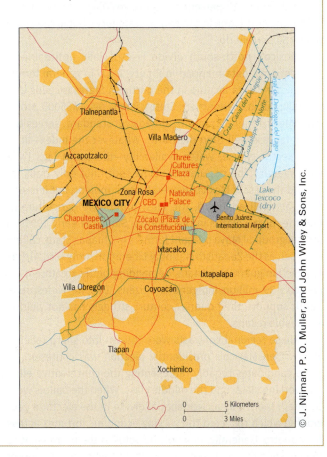

© J. Nijman, P. O. Muller, and John Wiley & Sons, Inc.

© J. Nijman, P. O. Muller, H. J. de Blij, and John Wiley & Sons, Inc.

FIGURE 2-10

whereas more highly skilled workers in the north might make three times the minimum wage. The pay-scale disparity with the adjacent United States, however, remains enormous.

Mexico's government seeks to encourage the growth of higher-paying jobs in the more advanced economic sectors—particularly electronic goods—that are also less likely to be lost to other parts of the world. Expansion of education and training in high-technology and management fields is a critical first step. The northeastern city of Monterrey in the relatively high-income State of Nuevo Léon has become a successful model in this effort, nurturing both an international business community and an ultramodern industrial complex that has attracted leading multinational corporations. Here indeed lies hope for Mexico's future, and one frequently hears reports these days that rising labor costs in China may induce some manufacturers to return to northern Mexico, allowing them to more directly control flexible (so-called "on-time") production at a much closer distance.

States of Contrast

Countries with strong internal regional disparities face serious challenges that can be difficult to overcome and may intensify over time. Mexico's southernmost States—Chiapas, Oaxaca, and Guerrero, all bordering the Pacific Ocean—are by far the poorest. The States bordering the United States in the north, including Nuevo Léon, Chihuahua, and Baja California, have the highest incomes. Using rural poverty as a measure, we find that only about 10 percent of people residing in the countryside in the north rank in the poorest category—but in the south nearly 50 percent do.

Mexico's north-south divide is especially noticeable in the economic data mapped in **Figure 2-11**. In general terms, annual per capita income in the northern States exceeds U.S. $10,000; but in the southern States, it falls below $5000. Moreover, Mexico's infrastructure, already substandard, serves the south far less well than the north. Whatever the index—literacy, electricity use, water availability—the south lags by a wide margin. The population of the south not only trails far behind the rest of Mexico in overall development, but it is also the least well educated, the least productive agriculturally, and the most isolated in the country. Since the early 1990s, a radical group of Maya peasant farmers in Chiapas State, calling themselves the Zapatista National Liberation Army (ZNLA), has engaged in guerrilla warfare, demanding better treatment for Mexico's 39 million indigenous citizens. Despite widespread public support, their quarter-century struggle has not yielded noticeable results.

These regional contrasts were thrown into sharp relief in 2012, when Mexico's presidential election was contested by three candidates broadly representing the conservative elite, the working class, and the middle class. When the ballots were counted, there was a clear spatial divide in the electoral geography, which reflected the varying economic fortunes across the country. Generally, most of the votes in the north went to the conservative and centrist candidates, whereas the poorer southern States supported the leftist candidate (see inset map in Fig. 2-11). It was the centrist candidate, Enrique Peña Nieto, who narrowly won the presidency in 2012—providing the latest evidence of Mexico's sharp politico-geographical cleavages.

The Drug Wars

Over the past two decades, Mexico has been plagued by yet another obstacle in its struggle to achieve sustained economic progress. This problem has received worldwide attention as the cocaine-producing drug cartels centered in Colombia established new bases in U.S.-border cities in northern Mexico and launched a vicious war for supremacy. They responded in part to the success of the antidrug campaign in Colombia, but the cartels also saw new opportunities beckoning in Mexican territory adjacent to their main market in the United States.

Figure 2-12 provides an overview of the geography of this drug war. Cocaine is produced in Colombia, Bolivia, and Peru, and a large share of it enters the United States through Mexico. Bolivian and Peruvian cocaine, constituting approximately half of Mexico's "imports," is shipped by sea and illegally enters the country in Pacific Coast States such as Guerrero and Sinaloa. Most shipments originate in Colombia and find their way into Mexico along various routes: either overland via the Middle American land bridge through Guatemala; by boat across the Caribbean and Gulf of Mexico to be smuggled into the eastern States of Yucatán, Quintana Róo, and Veracruz; or by boat on the Pacific side, along with the cocaine from Bolivia and Peru. The illicit powder is sometimes repackaged (and/or crystallized to form crack) in smaller quantities for "retailing" purposes prior to being smuggled into the United States, mainly through such border cities as Reynosa, Nuevo Laredo, Ciudad Juárez, and Tijuana (Fig. 2-12).

Interestingly, when Mexican criminal organizations were co-opted by the Colombian cartels, they demanded to be paid off in cocaine. This allowed the Mexicans to start up their own distribution networks; it also meant that some of the product "stuck around" and ended up on the streets of Mexican cities, resulting in heightened drug addiction and skyrocketing crime. Several major Mexican cartels sprang up and got involved in this drug-trafficking operation, and competition continues to literally be murderous. Territorial control is key in this "business": each cartel dominates its own turf, with various components being heavily contested from time to time. The cartels themselves are not particularly stable, with countless mergers and splits occurring over time. In 2016, there were less than ten cartels with significant spheres of influence across the country. The most powerful rivals include Los Zetas and the Sinaloa Cartel, which control huge swaths of eastern and western Mexico, respectively.

Much of the worst violence occurs where control is contested, especially in transit areas, along primary transport routes, and near border zones. And it is not only about

ALEX CRUZ/epa/Corbis

At the height of the drug war, students in Mexico City staged this protest against cocaine-trafficking violence. Although a large number of drug-related crimes go unreported, the "official" peak of 22,480 murders was recorded for the year 2011. Since then, such murders have declined by a third as Mexican authorities have clamped down on the narcotics trade, which has in part shifted from Mexico to the Caribbean islands. None theless, drug violence remains a serious threat to Mexican society.

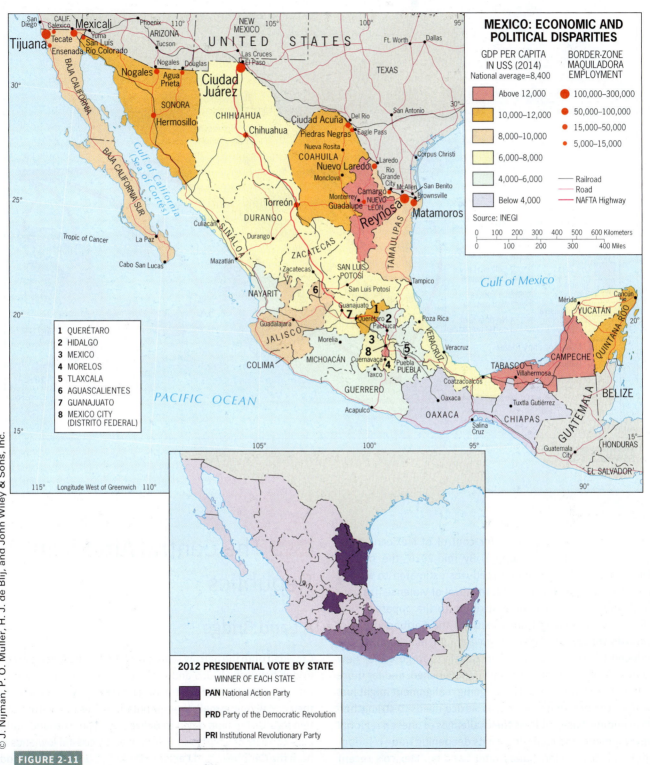

MEXICO: ECONOMIC AND POLITICAL DISPARITIES

GDP PER CAPITA IN US$ (2014)
National average=8,400

Above 12,000
10,000–12,000
8,000–10,000
6,000–8,000
4,000–6,000
Below 4,000

BORDER-ZONE MAQUILADORA EMPLOYMENT

100,000–300,000
50,000–100,000
15,000–50,000
5,000–15,000

Railroad
Road
NAFTA Highway

Source: INEGI

0 100 200 300 400 500 600 Kilometers
0 100 200 300 400 Miles

1 QUERÉTARO
2 HIDALGO
3 MEXICO
4 MORELOS
5 TLAXCALA
6 AGUASCALIENTES
7 GUANAJUATO
8 MEXICO CITY (DISTRITO FEDERAL)

2012 PRESIDENTIAL VOTE BY STATE
WINNER OF EACH STATE

PAN National Action Party
PRD Party of the Democratic Revolution
PRI Institutional Revolutionary Party

© J. Nijman, P. O. Muller, H. J. de Blij, and John Wiley & Sons, Inc.

FIGURE 2-11

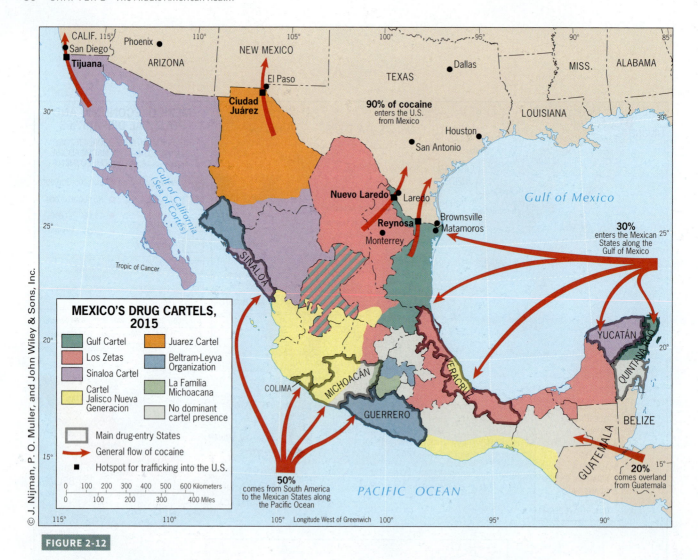

MEXICO'S DRUG CARTELS, 2015

Gulf Cartel
Los Zetas
Sinaloa Cartel
Cartel Jalisco Nueva Generacion
Juarez Cartel
Beltran-Leyva Organization
La Familia Michoacana
No dominant cartel presence

☐ Main drug-entry States
→ General flow of cocaine
■ Hotspot for trafficking into the U.S.

0   100 200 300 400 500 600 Kilometers
0   100   200   300   400 Miles

90% of cocaine enters the U.S. from Mexico

30% enters the Mexican States along the Gulf of Mexico

50% comes from South America to the Mexican States along the Pacific Ocean

20% comes overland from Guatemala

© J. Nijman, P. O. Muller, and John Wiley & Sons, Inc.

FIGURE 2-12

cocaine: the cartels also vie for control of Mexican-grown marijuana and heroin (poppy). By mid-2016, the murder total since the drug wars erupted was estimated to exceed 150,000. After peaking in 2011, the level of violence has generally declined, which some attribute to the apparent shift of cocaine trafficking from Mexico to the Caribbean Basin. Crackdowns by the Mexican government may finally have induced the drug suppliers and cartels to forge easier new routes via the Caribbean's small, impoverished, and far more vulnerable island-states. This ongoing realignment might just provide the breathing space that Mexico needs to strengthen its economic base, address the challenges of uneven regional development, and capitalize on its deepening trade relationship with the United States (and Canada). Mexico's recent upheavals, instabilities, and setbacks notwithstanding, the country's developmental potential remains promising. Yet evidence also shows that a decrease of violence in one place is often balanced by an increase elsewhere—reflecting not only government crackdowns but especially the shifting battlegrounds of the cartels (**Box 2-7** and **Fig. 2-13**).

Region The Central American Republics

A Land Bridge

Crowded onto the narrow segment of the Middle American land bridge between Mexico and the South American continent are the seven countries of Central America (**Fig. 2-14**). Territorially, they are all quite small; their population sizes range from Guatemala's 16.7 million down to Belize's 367,000. The land bridge here consists of a highland belt flanked by coastal lowlands on both the Caribbean and Pacific sides (**Fig. 2-15**). These uplands are studded with volcanoes, and local areas of fertile volcanic soils are scattered throughout them.

The land bridge has a fascinating geologic and evolutionary history, and one famous study referred to it as the Monkey Bridge. For some 50 million years, North and South America were separated; the land bridge was formed only 3 million

Box 2-7 MAP ANALYSIS

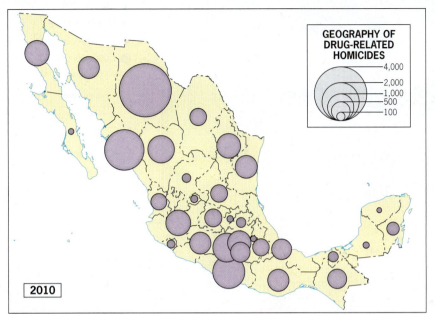

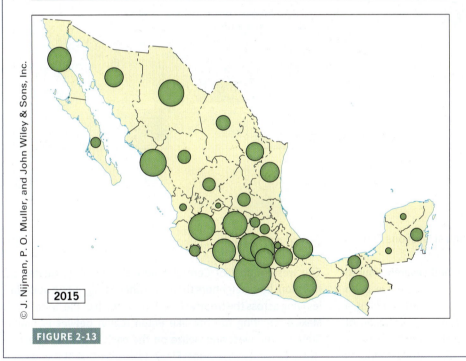

Violence associated with the drug wars has exacted a huge toll across Mexico in this decade. As these maps for 2010 and 2015 show, the geography of violence and death are neither uniform nor stable over space or time. The pattern of drug-war homicides mirrors the ongoing struggles between rival cartels, onshore trafficking in the southern Mexico, as well as the counter-narcotics strategies and tactics of the Mexican government. How would you describe the shifting spatial pattern of violence? Does Figure 2-12 offer clues about possible explanations?

Access your WileyPLUS Learning Space course to interact with a dynamic version of this map and to engage with online map exercises and questions.

GEOGRAPHY OF DRUG-RELATED HOMICIDES
- 4,000
- 2,000
- 1,000
- 500
- 100

2010

2015

© J. Nijman, P. O. Muller, and John Wiley & Sons, Inc.

FIGURE 2-13

years ago, becoming a biologic highway of sorts for evolutionary exchange. The region contains only 1 percent of the Earth's land but 7 percent of all the world's natural species. The southern part (Costa Rica and Panama) is known as a global **biodiversity hot spot [22]**, even though deforestation has been a major problem. Human inhabitants have always been concentrated in the upland zone, where tropical temperatures are moderated by elevation and rainfall is sufficient to support a variety of crops.

Trapped in the Narcotics Trade

Unfortunately for Central America, over the past decade the region has increasingly been forced to accommodate the cocaine trade. Its intermediate location between northern South America and Mexico made the Central American land bridge the only possible route when narco-traffickers decided to expand their overland transit operations. Consequently, the lion's share of the cocaine that now reaches the United

FIGURE 2-14

States through Mexico has passed through at least one Central American country.

Given the region's endemic poverty and severely limited economic opportunities, large numbers of locals have been drawn into the potentially lucrative drug trade, sparking a steep rise in organized criminal activity and an explosion of violence, especially in the countries closest to Mexico—Guatemala, Honduras, El Salvador, and Belize. Since 2011, Honduras has recorded the highest murder rate in the world of around 90 per 100,000 (in the U.S., not one of the safest countries, it is now about 10 per 100,000); at the same time, Belize, El Salvador, and Guatemala were joining Honduras to all rank among the world's top five on this unenviable list. Central American governments, with few resources to call upon, were virtually powerless to respond to the criminal cartels and affiliated vicious youth gangs, while most businesspeople and civilians had learned long ago never to rely on the police or military. Not surprisingly, in the Honduran capital of Tegucigalpa, private security forces (for those able to afford them) today outnumber police personnel four to one.

Guatemala

The westernmost of Central America's republics, Guatemala has more land neighbors than any other. Straight-line boundaries lying across the tropical forest mark much of the border with Mexico, creating the box-like Petén region between Chiapas State on the west and Belize on the east; also to the east lie Honduras and El Salvador (Fig. 2-14). This heart of the ancient Maya Empire, which remains strongly permeated by indigenous culture and traditions, has only a small window on the Caribbean but a longer Pacific coastline. Most populous of the seven republics with 16.7 million inhabitants (mestizos are in the majority with 55 percent, indigenous 45 percent), Guatemala has seen a great deal of conflict, and military regimes have dominated political life. There is a deepening split between the wretchedly poor indigenous populations and the better-off mestizos who continue to control the government, military, and land-tenure system.

Guatemala's economic geography demonstrates considerable potential but has long been shackled by unrelenting

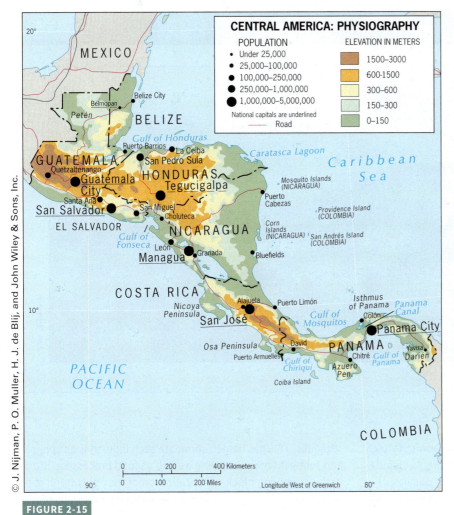

CENTRAL AMERICA: PHYSIOGRAPHY

POPULATION
- Under 25,000
- 25,000–100,000
- 100,000–250,000
- 250,000–1,000,000
- 1,000,000–5,000,000

National capitals are underlined
— Road

ELEVATION IN METERS
- 1500–3000
- 600–1500
- 300–600
- 150–300
- 0–150

FIGURE 2-15

© J. Nijman, P. O. Muller, H. J. de Blij, and John Wiley & Sons, Inc.

demographic complexion of Belize is being reshaped. Thousands of residents of African descent have emigrated (many went to the United States) and were replaced by tens of thousands of Spanish-speaking immigrants. Most of the latter have been land seekers and refugees from strife in nearby Guatemala, El Salvador, and Honduras, and their proportion of the Belizean population has risen from 33 percent in 1980 to well over 50 percent today. With the newcomers now in the majority, Belize's cultural geography is becoming increasingly Hispanicized; in fact, since 2012 Spanish speakers outnumber those who use English or creole as a first language.

The Belizean transformation extends to the economic sphere as well. No longer just an exporter of sugar and bananas, Belize today is producing new commercial crops, and its seafood-processing and clothing industries have become major revenue earners. Also important is tourism, which annually lures more than 200,000 vacationers to the country's Mayan ruins, resorts, and newly legalized casinos; a booming specialty is **eco-tourism**, based on the natural attractions of the country's near-pristine environment, including its magnificent offshore coral reefs. Belize is also known as a center for **offshore banking [23]**, a financial haven for foreign companies and individuals who endeavor to avoid paying taxes in their home countries.

internal conflicts and the widening of one of the hemisphere's biggest gaps between rich and poor. In tandem, they keep the income of well over half the population below the poverty line. The country's mineral wealth includes nickel in the highlands and oil in the lower-lying north. Agriculturally, soils are fertile and moisture is ample over highland areas large enough to produce a wide range of crops, including excellent coffee. The 929-kilometer (577-mi) border with Mexico offers opportunities as well, but progress has been thwarted by out-of-control drug trafficking and violent crime.

Belize

Strictly speaking, Belize is not a Central American republic in the same tradition as the other six. Until 1981, this country, a wedge of land between northern Guatemala, Mexico's Yucatán Peninsula, and the Caribbean Sea (Fig. 2-14), was a dependency of the United Kingdom known as British Honduras. Slightly larger than Massachusetts and containing a minuscule population of only 367,000 (many of African descent), Belize is much more reminiscent of a Caribbean island than of a continental Middle American state. Today, all that is changing as the

Honduras

Honduras, in contrast with Guatemala, has a long Caribbean coastline and a relatively small window on the Pacific (Fig. 2-14). The country also occupies a critical place in the political geography of Central America, flanked as it is by Nicaragua, El Salvador, and Guatemala—all continuing to grapple with the aftermath of years of internal conflict, recent natural disasters, and the ongoing ravages of drug trafficking. Comparable in natural beauty and biodiversity to Costa Rica, Honduras hopes to exploit its potential for ecotourism but is constrained by a defective infrastructure and the lack of funding for new facilities (see **Box 2-8**).

Agriculture, livestock, forestry, and limited mining form the mainstays of the economy, with the now-familiar Central American products—apparel, bananas, coffee, and shellfish—earning most of the external income. With just over 8 million inhabitants, about 90 percent mestizo, Honduras faces huge development challenges that are complicated by escalating crime and social dislocation.

In an effort to move forward, Honduras has unveiled a development strategy aimed at the creation of quasi-independent

Box 2-8 From the Field Notes …

"Invited by the Honduran government to survey the country's ecotourism potential, I toured the beautiful countryside with a National Geographic Society delegation. There is no doubt that Honduras is exceptionally endowed by nature and that there is considerable potential for a vibrant tourist industry. But this is a poor country, and the government has great difficulty making the required investments—in infrastructure, for example. We flew in these small propeller planes to get from the capital, Tegucigalpa, to the Mayan ruins of Copán. There was no airport nearby, and overland travel is arduous and time consuming. Ecotourism by its very nature is small scale and the revenues tend to be relatively modest, so it is a major challenge to raise the substantial funds needed for costly infrastructural improvements."

© Jan Nijman

"special development regions." The idea is to attract foreign investment into these protected zones, which are to be independent in their economic, fiscal, and budgetary policies. This initiative follows the charter-cities model that has proven successful in Hong Kong, but critics fear that it may only clear the way for organized crime to take full control of these new economic enclaves—particularly in and around the besieged city of San Pedro Sula, which in 2015 had the second-highest murder rate in the world.

El Salvador

El Salvador is Central America's smallest country territorially, smaller even than Belize, but with a population 20 times as large (6.1 million) it is the region's most densely peopled. El Salvador adjoins the Pacific in a narrow coastal plain backed by a chain of volcanic mountains, behind which lies the country's heartland (Fig. 2-14). Unlike neighboring Guatemala, El Salvador has a far more homogeneous population (86 percent mestizo and just 1 percent indigenous). Yet ethnic uniformity has not translated into social or economic equality or even opportunity. Whereas other Central American countries were called banana republics, El Salvador was a coffee republic, and the coffee was produced on the huge landholdings of a few landowners and on the backs of a subjugated peasant labor force.

More affluent El Salvadorans who left the country in past decades, and who have succeeded in the United States and elsewhere, send substantial funds back home; these **remittances [24]** now provide the largest single source of foreign revenues. This has helped stimulate such industries as apparel and footwear manufacturing as well as food processing. But a major stumbling block to revitalization of the agricultural sector has again been land reform, and in this regard El Salvador's future still hangs in the balance.

Panama

Panama reflects some of the usual geographic features of the Central American republics. Its population of 4 million is about two-thirds mestizo but also contains substantial indigenous, European, and black minorities. Spanish is the official language, but English is also widely used. Ribbon-like and oriented east-west, much of Panama's topography is mountainous and hilly. Eastern Panama, especially Darien Province adjoining Colombia, is densely covered by tropical rainforest and contains the only remaining gap in the intercontinental Pan American Highway (Fig. 2-14). Most of the rural population lives in the uplands west of the Canal; there, Panama produces bananas, coffee, sugarcane, rice, and, along its narrow coastal lowlands, shrimps and other seafood. Much of the urban population is concentrated in the vicinity of the famous Canal, anchored by the cities at each end of this pivotal waterway.

Panama owes its existence to the idea of an artificial waterway connecting the Atlantic and Pacific oceans to avoid the lengthy circumnavigation of South America. The Panama Canal (see the inset map in Fig. 2-14) was opened in 1914, a symbol of U.S. power and influence in Middle America. It was both a massive feat of engineering and an especially dangerous undertaking in a hostile tropical environment. It is estimated that 25,000

Box 2-9 Panama Canal Expansion

Its 5000 containers piled high, a Panamax-sized cargo carrier squeezes through a lock in one of the Panama Canal's original lanes two years before the 2016 enlargement (left photo). The right photo was taken at the climax of opening day (June 26, 2016) as the first Neo-Panamax ship, carrying 9472 containers, approached the final new lock to complete its interoceanic transit at the Panama City outlet to the Pacific.

NEW FEATURES

1 Retaining basins fill lock chambers and retain 60 percent of the water for re-use.

2 Rolling lock gates slide in recesses, saving space and making maintenance work much easier.

3 Tugboats maneuver vessels into position, replacing costly towing locomotives.

Locomotives

Original locks ▼

New lane ▶

Locks transfer vessels from one elevation to another by filling closed chambers with water, or emptying them. When the level is even with that of the next chamber, the gates open.

3 *Tugboat*

Adapting to larger ships

The most recent container vessels and tankers are known as Neo-Panamax. They can hold up to 13,000 containers, 2.5 times the cargo capacity of ships able to fit in the existing locks.

© Mika Grondahl/NG Image Collection

NEO-PANAMAX SHIP

160 ft
60 ft
180 ft

PANAMAX SHIP
Pre-2016 maximum size

106 ft   42 ft
110 ft

Diagram showing the expansion and upgrading of the Panama Canal's lock system in order to accommodate ships that exceed the waterway's original ("Panamax") capacity. The intended completion date for the new third lane to handle "Neo-Panamax" vessels was August 15, 2014—the Canal's one-hundredth anniversary. Following numerous construction delays, the enlarged Canal finally opened for business in June 2016.

Box 2-10 Technology & Geography

The Shipping Container

Of all the innovations in transportation technology during the past century, the lowly shipping container continues to be one of the most significant to the global economy and trade between (and within) world regions. It is a technological breakthrough that is deceptively simple: how can a metal box measuring 20 (or now 40) feet (6/12 m) in length, 8 feet (2.4 m) wide and 8 feet high—or "twenty-foot equivalent units (TEUs)"—matter so much?

The ingenuity of the shipping container is fourfold. First, its uniformity allows universal, high-volume stacking and storage (increasingly enabling the use of supersized ships). It also facilitates mechanized or automated loading and unloading. Second, containers have driven the growth of highly efficient **intermodal transport systems [25]** involving ships, trains, and trucks. Because all these forms of transport are compatible with containers, they are often referred to as *intermodal freight containers*. Third, transport companies or clients can track individual containers with GPS, allowing for precise logistical planning. Finally, the durable steel frame of the container allows reuse, with the lifespan of the average container exceeding ten years.

An estimated 90 percent of the goods we consume travels by sea. The clothes we wear, the cars we drive, the television sets we watch, and the foods we eat—all mostly travel inside a container on a ship, then onto a railcar or the bed of a semi-trailer before reaching its final destination. Given these advances in shipping technology, logistics, improved port management, and even fuel efficiency, it is frequently more economical to put goods like food into a container to process overseas and then ship it back than it is to process locally.

Trade between the United States and East Asia (especially China and Japan) comprises a huge share of global container

© age fotostock/Alamy Stock Photo

The Mexican intermodal container port of Veracruz on the Gulf of Mexico. This port serves much of eastern and central Mexico via rail and road, and is dominated by trade with the United States, and to a lesser extent Brazil, Venezuela, and Colombia.

shipping, and in 2016 nearly 40 percent of it (equal to about 14,000 ships per year) passed through Middle America's Panama Canal (**Figure 2-16**). This share is likely to increase following the widening of the Canal in 2016 (see **Box 2-9**), and this will likely result in a decrease in the flow of intermodal trade between East Asia and the U.S. West Coast. And if the plans for a second, Nicaragua Canal come to fruition (see **Box 2-11**), the regional reorganization of shipping will be even more significant. It is all part of the dynamic reshaping of global transportation networks propelled by the advent of the shipping container.

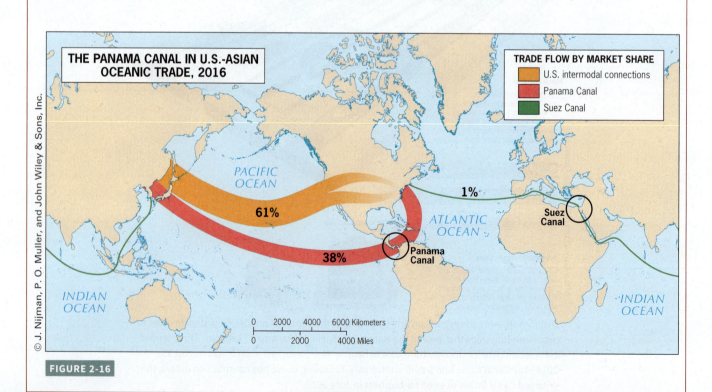

© J. Nijman, P. O. Muller, and John Wiley & Sons, Inc.

THE PANAMA CANAL IN U.S.-ASIAN OCEANIC TRADE, 2016

TRADE FLOW BY MARKET SHARE
- U.S. intermodal connections
- Panama Canal
- Suez Canal

PACIFIC OCEAN
ATLANTIC OCEAN
INDIAN OCEAN
INDIAN OCEAN

61%
38%
1%

Panama Canal
Suez Canal

0   2000   4000   6000 Kilometers
0   2000   4000 Miles

FIGURE 2-16

workers died during the 10 years of excavation and construction. The canal itself was contained within the cross-isthmian **Canal Zone**, an elongated strip of land 16 kilometers (10 mi) wide, with the waterway forming its centerline (Fig. 2-14 inset). This corridor was ceded to the United States in 1904 by newly independent Panama under a treaty granting the United States "all the rights, powers, and authority" in the Zone "as if it were the sovereign of the territory." In the 1970s, Panama sought to terminate U.S. control in the Canal Zone, and delicate negotiations were opened. In 1977, an agreement was reached on a staged withdrawal by the United States from the territory, first from the Canal Zone and then from the Panama Canal itself (a process completed on December 31, 1999).

Near the northern, Caribbean end of the Panama Canal lies the city of Colón, site of the Colón Free Zone, a huge trading hub designed to transfer and distribute goods on their way to South America. Near the southern, Pacific end lies the capital, Panama City, often likened to Miami because of its waterfront location and high-rise-dominated skyline. The capital is the financial center that handles the revenues generated by the Canal, but its skyscrapered profile also reflects the proximity of Colombia's illicit drug industry and associated money-laundering and corruption.

Until 2016, ships sailing through the Panama Canal could not be wider than 32 meters (106 ft). The Canal's much wider third lane, which opened in mid-2016, accommodates ("Neo-Panamax") vessels up to 55 meters (180 ft) in width that can carry two-and-a-half times as many shipping containers (up to 13,000)(see **Box 2-9** and **Box 2-10**). This will boost interoceanic traffic and increase business opportunities in Panama. U.S. ports on the Gulf of Mexico and the Atlantic seaboard will also get busier as they handle this heightened volume of cargoes. The ports of Miami and the New York–New Jersey metropolitan region are expanding their capacities, and leading retailers such as Walmart (a massive importer of Chinese goods) are building additional storage facilities at the Port of Houston and other key coastal distribution points (see also Box 1-3, the Regional Planning Case in Chapter 1).

Nicaragua

Figure 2-14 underscores Nicaragua's regional position, tucked away in the heart of Central America. The Pacific coast follows a southeasterly direction here, but the Caribbean coast is oriented north-south so that Nicaragua forms a triangle of land, with its lakeside capital, Managua, situated on the mountainous, earthquake-prone, Pacific side (the country's core area has always been located here). The Caribbean side, where the uplands yield to a wide coastal plain of tropical rainforest, savanna, and swampland, has for centuries been home to indigenous peoples such as the Miskito, who have lived remote from the focus of national life. Options for this country of more than 6 million are limited: agriculture dominates the economy while manufacturing is weak. Like El Salvador, there is a heavy reliance on remittances from those who have emigrated.

For decades, there has been talk of an interoceanic canal to rival the Pacific-Atlantic waterway across nearby Panama, even though the land bridge here is three times wider. What long seemed like a distant dream took a turn toward reality in 2013 when the Nicaraguan government officially declared that an international consortium of largely Chinese developers would forge ahead with the project. Planning and some preliminary construction began in 2014 even though many observers continued to question its feasibility (Box 2-11).

Costa Rica

If there is one country that underscores Middle America's variety and diversity it is Costa Rica—because it differs significantly from its neighbors and from the norms of Central America as well. Bordered by two volatile countries (Nicaragua to the north and Panama to the east), Costa Rica is a nation with an old democratic tradition and, in this boiling cauldron, no standing army since 1949! Although the country's Hispanic imprint is similar to that found elsewhere on the Middle American mainland, its early independence, its good fortune to lie remote from regional strife, and its leisurely pace of settlement allowed Costa Rica the luxury of concentrating on its economic development. Perhaps most important, internal political stability has predominated over much of the nearly 200 years since its independence from Spain.

Like its neighbors, Costa Rica (with a population of 5 million) is divided into environmental zones that parallel its coastlines. The most densely settled is the central highland zone, lying in the cooler *tierra templada*, whose heartland is the Valle Central (Central Valley), a fertile basin that contains Costa Rica's main coffee-growing area and the leading population cluster focused on the cosmopolitan capital of San José (Fig. 2-14).

The long-term development of Costa Rica's economy has given this country the region's highest standard of living, literacy rate, and life expectancy (though even here, one-fifth of the population is trapped in poverty). Agriculture continues to dominate (with bananas, coffee, tropical fruits, and seafood among the leading exports), and tourism—especially ecotourism—has expanded steadily. Despite the deforestation of more than 80 percent of its original woodland cover, Costa Rica is widely known for its superb scenery and for its (belated) efforts to protect what is left of its diverse tropical flora and fauna.

Box 2-11   Regional Planning Case

The Nicaragua Canal

The Nicaragua Canal is a planned shipping route across Nicaragua to connect the Pacific Ocean with the Caribbean Sea (**Figure 2-17**). If realized, the Canal holds promise to boost the economy of one of the poorest states in the realm by making Nicaragua a center for transport and global logistics between the Americas and Asia. The proposed Nicaragua Canal is a 278-kilometer (173-mi) artificial waterway—wider, deeper, and three-and-a-half times the length of the Panama Canal—and expected to accommodate supertanker vessels that will be too big to pass even through the enlarged Panama Canal.

The main components of the project include two sets of major locks on the eastern and western legs of the Canal; two new ports at Brito and Camilo; major excavation combined with road construction on the eastern land portion of the Canal; several supersized bridges; a dam and a power plant; dredging of a trench that is 170 kilometers (67 mi) long and 280 meters (787 ft) wide across the floor of Lake Nicaragua; and land development to accommodate urban growth at both endpoint ports.

Overall, this is a spectacular project and the biggest in the history of the realm, and it will provide employment for an estimated 50,000 workers during the construction period that would extend through at least the end of this decade. Nevertheless, knowledgable observers stress that two huge challenges must be surmounted. First, from the beginning the project has been criticized by local communities that would be displaced and by environmental groups concerned about its devastating effects, especially on the Lake Nicaragua ecosystem. This is Central America's largest freshwater source on which millions depend for their livelihoods. It is easy to imagine the impacts of major spills and other accidents. In addition, the Canal will cut through part of Indio Maiz Nature Reserve, and it is estimated that 4000 square kilometers (1544 sq mi) of rainforest will be destroyed.

STR/Getty Images, Inc.

Peasant-farmers in the town of Juigalpa, near the northeast shore of Lake Nicaragua, demonstrate against the construction of the proposed canal in mid-2015.

Second, Nicaragua could wind up in a position where it would have little economic and political control of the Canal. The country itself does not have the resources to fund a project of this magnitude (estimated to cost U.S. $50 billion), so it has been handed over to the Hong Kong Nicaragua Development (HKND) Group, a consortium led by a Chinese telecom billionaire. HKND received an exclusive concession to build and manage the Canal for the next 50 years, and this can be renewed for another 50 years. Comparisons with U.S. control of the Panama Canal after 1914 come to mind, but this time it would be China calling the shots. For China, this investment and undertaking is strategically important because the future canal would handle its largest freighters, significantly enhancing Chinese economic and geopolitical influence throughout this central zone of the Americas.

© J. Nijman, P. O. Muller, and John Wiley & Sons, Inc.

FIGURE 2-17

The Caribbean Basin

Fragmentation, Insularity, and Vulnerability

As Figure 2-7 reveals, the Caribbean Basin, Middle America's island arena, is constituted by a broad arc of numerous islands extending from the western tip of Cuba to the southern coast of Trinidad. The four larger islands or *Greater Antilles* (Cuba, Hispaniola, Jamaica, and Puerto Rico) are clustered in the east-west segment of this arc. The smaller islands or *Lesser Antilles* of the eastern segment extend southward within a crescent-shaped chain from the Virgin Islands to Trinidad and Tobago. (Breaking this tectonic-plate-related regularity are the Bahamas as well as the Turks and Caicos, located north of the Greater Antilles, and many other islands too small to appear on a map at the scale of Fig. 2-7.)

On these myriad islands, whose combined land area accounts for only 9 percent of the Middle American realm, lie 33 states and numerous other political entities. (Europe's colonial flags have not totally disappeared from this archipelago, and the U.S. flag flies over both Puerto Rico and the Virgin Islands.) The inhabitants of these states and territories, however, constitute one-fifth of Middle America's total population, making this the most densely peopled part of the Western Hemisphere.

The island-nations of the Caribbean Basin are generally very small, and their territories are separated, often at considerable distances from other islands. As noted earlier in this chapter, these geographic conditions create a set of circumstances that have proven to be quite challenging. Economic opportunities are minimal, most goods and services are relatively expensive, and due to limited interaction with the outside world, the island societies tend to be static.

It is this vulnerability of the small Caribbean islands, perhaps even more so than is the case for the Central American countries, that has triggered the recent sharp intensification of cocaine trafficking across this far-flung island chain. From the main cocaine distribution source in Colombia, a growing quantity of the product is now being routed eastward across the South American continent—via Venezuela, northernmost Brazil, and the three coastal Guianas—thereby entering the Antillean archipelago from the southeast. From there it proceeds to the United States, either directly or, more likely, indirectly through northern Central America and Mexico. The stressful impacts of the expanding drug trade on the fragmented, poverty-mired societies of the Lesser and Greater Antilles are identical to those that plague the central segment of the continental land bridge: widening cocaine use and addiction among local populations; surges of organized criminal activity; and an alarming rise in the overall level of violence.

Social Cleavages

Social stratification [26] in most Caribbean islands is rigid, and upward mobility is decidedly limited. Class structures tend to be closely associated with ethnicity, and island societies still carry imprints of colonial times. The historical geography of Cuba, the Dominican Republic, and Puerto Rico is suffused with Hispanic culture; Haiti and Jamaica carry stronger African legacies. But the reality of this ethnic diversity is that European lineages still hold the advantage. Hispanics tend to be in the best positions in the Greater Antilles; people who have mixed European-African ancestries, and who are often described as **mulatto [27]**, rank next. The great bulk of this social pyramid is constituted by the most underprivileged: the Afro-Caribbean majority. In virtually all Caribbean societies, the advantaged minorities hold disproportionate power and exert overriding influence. In Haiti, the mulatto minority accounts for less than 5 percent of the population but has long held most of the power. In the neighboring Dominican Republic, the pyramid of power puts Hispanics (16 percent) at the top, the mixed sector (73 percent) in the middle, and the Afro-Caribbean minority (11 percent) at the bottom. In the Caribbean social mosaic, historical advantage has a way of perpetuating itself.

The composition of the population of the islands is further complicated by the presence of Asians from both China and India. During the nineteenth century, the emancipation of slaves and subsequent local labor shortages brought some far-reaching solutions. More than 100,000 Chinese emigrated to Cuba as indentured laborers; and Jamaica, Guadeloupe, and particularly Trinidad saw nearly 250,000 South Asians arrive for similar purposes. To the African-modified forms of English and French heard in the Caribbean Basin, therefore, can be added several Asian languages. The ethnic and cultural diversity of the societies of Caribbean Middle America seems endless.

Region The Greater Antilles

The four islands of the Greater Antilles (whose populations constitute 90 percent of the Caribbean Basin's total) contain five political entities: Cuba, Haiti, the Dominican Republic, Jamaica, and Puerto Rico (Fig. 2-7). Haiti and the Dominican Republic share the island of Hispaniola.

Cuba

The largest Caribbean island-state in terms of both territory and population (11.4 million), Cuba lies only 145 kilometers (90 mi) from the southernmost island of the Florida Keys (**Fig. 2-18**). Havana, the now-dilapidated capital, lies almost directly south of outermost Key West on the northwestern coast of the elongated island.

Cuba was a Spanish possession until 1898 when, with U.S. help in the ten-week-long Spanish-American War, it achieved independence. Fifty years later, an American-backed dictator was fully in control, and by the 1950s Havana had become an American playground. The island was ripe for revolution, and in 1959 Fidel Castro's insurgents gained control, thereby

FIGURE 2-18

converting Cuba into a communist dictatorship and a client-state of the Soviet Union. Nearly a million Cubans fled the island for the United States, and Miami swiftly became the second-largest Cuban city after Havana.

In the fall of 1962, the world was on the brink of nuclear war when the United States called on the Soviet Union to remove its newly installed nuclear missiles from Cuba (aimed at the United States) or face reprisals. In the end, the Soviets conceded and in return Washington promised not to attack Cuba. The United States also continued to retain the Guantánamo Bay Naval Base located near the southeastern tip of the island (**Fig. 2-18**), which was perpetually leased to the United States by the Cuban

Tourists strolling through the plaza in front of the Cuban capital's Cathedral of San Cristóbal de la Habana, a Spanish Baroque architectural gem completed in 1777. With the recent lifting of the U.S. embargo, the ranks of these tourists are sure to swell as a new wave of American visitors is expected to descend on Cuba in the second half of this decade.

government in 1903. Since 2002, the Guantánamo base has become notorious as a detention camp for 9/11-related terrorists as well as prisoners of war from both the Iraq and Afghanistan conflicts.

Castro's rule survived the 1991 demise of the Soviet Empire despite the loss of its benefactor's subsidies and sugar markets on which the country had long relied. As the map indicates, sugar was Cuba's economic mainstay for many years; the plantations, once the property of rich landowners, extend all across the island. But sugarcane is losing its position as the top-ranked Cuban foreign exchange earner. Mills have been closing down, and cane fields are being cleared for other crops as well as for pastures. However, Cuba has additional economic opportunities, especially in its highlands.

Cuba has three mountainous zones, of which the southeastern Sierra Maestra is the highest and most extensive. These highlands create considerable environmental diversity as reflected by sizeable timber-producing tropical forests and varied soils on which crops ranging from tobacco to coffee to rice to subtropical and tropical fruits are grown. Rice and beans are the staples, but Cuba is unable to meet its dietary needs and so must import food. The central and western savannas support livestock-raising. Even though Cuba has only limited mineral reserves, its nickel deposits are extensive and have been mined for more than a century.

In 2006, Fidel Castro became too ill to lead the country and turned over the reins to his brother Raúl. Since then, a series of modest liberalization measures have slowly taken effect. These changes were undoubtedly necessitated by the persistently deplorable condition of the Cuban economy. In late 2014, encouraged by signs of reform in Cuba and a growing consensus that the American embargo of the island was now hurting the people more than the regime by depriving the country of the means to develop, the U.S. government announced the lifting of its embargo in a series of stages. In July 2015, diplomatic relations between the two countries were fully restored, and in March 2016, Barack Obama became the first U.S. president to visit the island since 1928.

Jamaica

Across the Caribbean's deep Cayman Trench from southern Cuba lies Jamaica, and a cultural gulf separates these two countries as well. A former British dependency, Jamaica has an almost entirely Afro-Caribbean population. As a member of the British Commonwealth, it still recognizes the British monarch as the head of state, represented by a governor-general. The effective head of government in this democratic country, however, is the prime minister. English remains the official language here, and British traditions still linger.

Smaller in size than Connecticut and with just under 3 million people, Jamaica has experienced a steadily declining national income over the past few decades despite its relatively slow population growth. Tourism has become the largest source of income, but the markets for bauxite (aluminum ore), of which this island was once a leading exporter, have dwindled. And like other Caribbean countries, Jamaica has difficulty making money from its sugar exports. Jamaican farmers also produce crops ranging from bananas to tobacco, but the country faces the disadvantages on world markets common to those in the global periphery. Meanwhile, Jamaica must import all of its oil and much of its food because the densely populated coastal flatlands suffer from overuse and shrinking harvests.

The capital, Kingston, lies on the southeastern coast and reflects Jamaica's economic struggle. Almost none of the hundreds of thousands of tourists who visit the country's beaches explore its Cockpit Country of awesome limestone towers and caverns, or populate the many cruise ships calling at Montego Bay or other points along the north coast get even a glimpse of what life is like for the ordinary Jamaican.

Jamaica has become an investment target of Chinese companies, mainly in infrastructure. A major new expressway running north-south across the island was completed in 2016 at a cost of $600 million, all paid for by the Chinese. In return, the Jamaican government turned over 1200 acres of land along that highway corridor to the company, which will build and manage its own tourist resorts. Among Jamaicans, the overall benefits for the island are disputed.

Haiti

Already the longtime poorest state in the Western Hemisphere, Haiti in the autumn of 2008 was staggered by no less than four tropical cyclones that struck within a few weeks of each other. The storms and the floods they unleashed killed more than 800 people and uprooted some 800,000. Then, barely 15 months later came the monstrous earthquake of January 12, 2010, which produced devastation on a truly unimaginable scale. The capital city, Port-au-Prince, lay in ruins, most of the country's infrastructure collapsed, and the Haitian government became effectively invisible. Schools, hospitals, and railroads as well as the seaport and the airport all came to a standstill. Millions became unemployed overnight. But all of that paled in comparison to the scale of human suffering. In the months following the quake, the estimated death toll reached and then surpassed 300,000; at least another 300,000 victims were injured, many permanently disabled. About 1.5 million homeless people initially fled the capital for the countryside, and, as noted earlier in this chapter, seven years after the disaster these homeless still numbered in the hundreds of thousands (see earlier photo of Port-au-Prince).

It was a tragedy of colossal proportions, its horrendous images broadcast around the world on television and the Internet. Large-scale international emergency aid arrived, but it became clear that, in the long run, much of the affected area would have to be totally rebuilt. In the catastrophe's aftermath, millions of Haitians continue to subsist in the most wretched circumstances imaginable.

The challenge of recovery remains a gargantuan undertaking because nature has not provided this country of just

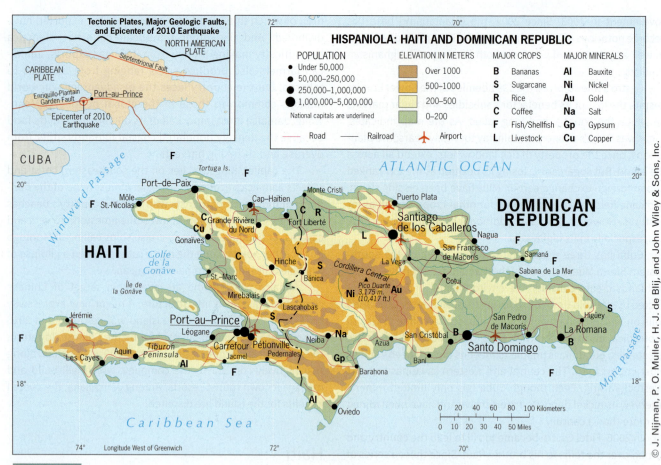

Tectonic Plates, Major Geologic Faults, and Epicenter of 2010 Earthquake

NORTH AMERICAN PLATE

CARIBBEAN PLATE

Septentrional Fault

Enriquillo-Plantain Garden Fault

Port-au-Prince

Epicenter of 2010 Earthquake

HISPANIOLA: HAITI AND DOMINICAN REPUBLIC

POPULATION
- Under 50,000
- 50,000–250,000
- 250,000–1,000,000
- 1,000,000–5,000,000

National capitals are underlined

Road   Railroad   Airport

ELEVATION IN METERS
- Over 1000
- 500–1000
- 200–500
- 0–200

MAJOR CROPS
- B Bananas
- S Sugarcane
- R Rice
- C Coffee
- F Fish/Shellfish
- L Livestock

MAJOR MINERALS
- Al Bauxite
- Ni Nickel
- Au Gold
- Na Salt
- Gp Gypsum
- Cu Copper

© J. Nijman, P. O. Muller, H. J. de Blij, and John Wiley & Sons, Inc.

FIGURE 2-19

Box 2-12 From the Field Notes . . .

"Fly along the political boundary between Haiti and the Dominican Republic, and you see long stretches of the border marked by the starkest possible contrast in vegetation: denudation prevails to the west in Haiti while the forest prospers on the Dominican (eastern) side. On the Haitian side, overpopulation, widespread grinding poverty, and chronic government mismanagement—all exacerbated by the snail's pace of earthquake recovery—combine to create one of the Caribbean's most jarring cultural-landscape divides."

under 11 million with any breaks. Haiti lies along a particularly dangerous tectonic fault zone where the North American and Caribbean plates meet (**Fig. 2-19**, inset map). The quake of 2010, geologists say, had been in the making for centuries, and others could easily follow in the future. This country also lies astride the axis of Hurricane Alley, and even a single year without a major storm is something to be grateful for (fortunately including the six hurricane seasons following the earthquake).

Few countries in the world have a more checkered history than Haiti, where political instability, repression, and material deprivation have been constants since the early nineteenth century. Haiti's GNI per capita is barely one-sixth of Jamaica's, a level below even that of many impoverished African countries. Compared to other countries in Middle and South America, its infant mortality rate is nearly three times as high. A shocking one-third of elementary-school-age girls have never even attended school. The negative consequences of Haiti's irreversible environmental degradation (**Box 2-12**) are staggering. And because of its scant natural resources and almost nothing to offer in the international trading arena, most of Haiti's meager public expenditures depend on direct foreign aid. As for the limited private consumption that takes place, it relies heavily on remittances from the Haitian Diaspora in such cities as Miami, Paris, New York, and Montreal.

Dominican Republic

The mountainous Dominican Republic has a larger share of Hispaniola than Haiti (Fig. 2-19) in terms of territory (64 percent of the island), but at 10.6 million the Dominican Republic's population ranks just behind that of Haiti. "The Dominican," too, has a wide range of natural environments but also a far more robust resource base than its neighbor to the west. Nickel, gold, and silver have long been exported along with sugar, tobacco, coffee, and cocoa, but tourism is the leading industry. Yet the Dominican Republic's economic performance since 2000 has been erratic and today remains below its potential. However, compared to near-destitute, low-income Haiti, the DR still ranks two developmental categories higher as an upper-middle-income country (see Fig. G-11).

Puerto Rico

The largest U.S. domain in Middle America, this easternmost and smallest island of the Greater Antilles region covers 9000 square kilometers (3500 sq mi) and has a population of 3.7 million. Puerto Rico is larger than Delaware and more populous than Iowa. Most Puerto Ricans are concentrated in the urbanized northeastern sector of this rectangular island (**Fig. 2-20**).

Puerto Rico fell to the United States well over a century ago during the Spanish-American War of 1898. Since the Puerto

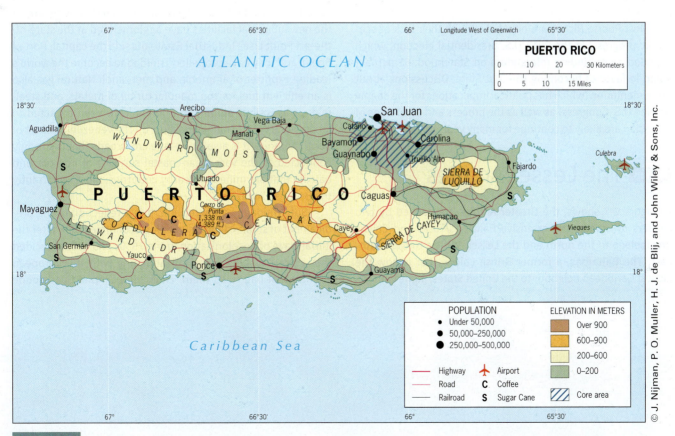

FIGURE 2-20

© J. Nijman, P. O. Muller, H. J. de Blij, and John Wiley & Sons, Inc.

Ricans had been struggling for some time to free themselves from Spanish imperial rule, this transfer of power was, in their view, only a change from one colonial master to another. As a result, the first half-century of U.S. administration was difficult, and it was not until 1948 that Puerto Ricans were permitted to elect their own governor. When the island's voters approved the creation of a Commonwealth in a 1952 referendum, Washington, D.C. and San Juan, the seats of government involved, entered into a complicated arrangement. Puerto Ricans are U.S. citizens but pay no federal taxes on their local incomes. Puerto Rico also receives a sizeable annual subsidy from Washington, averaging approximately U.S. $13 billion in recent years.

Despite these apparent advantages in the poverty-plagued Caribbean, Puerto Rico has not thrived under U.S. administration. Long dependent on a single-crop economy (sugar), the island based its industrial development during the 1950s and 1960s on its comparatively cheap labor, tax breaks for corporations, political stability, and special access to the U.S. market. Consequently, pharmaceuticals, electronic equipment, and apparel top the current list of exports, not sugar or bananas. But this industrialization failed to stem a tide of emigration that carried more than one million Puerto Ricans to the New York City area alone. The same wage advantages that favored corporations kept many Puerto Ricans poor or unemployed. The level of unemployment here in 2015 stood at about 12 percent, more than twice the average in the continental United States. The private sector remains severely underdeveloped, and a major government-debt crisis has emerged in the mid-2010s.

Most Puerto Ricans are now fed up with their lack of economic progress. In the 2012 U.S. presidential election, which included a nonbinding referendum on Statehood, 61 percent voted in favor of becoming the 51st State. Discussions about implementation (which has political implications for the deeply divided U.S. Congress) as well as the process and conditions of accession are likely to continue for many years.

Region The Lesser Antilles

As Figure 2-7 shows, the Greater Antilles are flanked by two clusters of islands: the extensive Bahamas/Turks and Caicos archipelago to the north and the Lesser Antilles to the southeast. **The Bahamas**, a former British colony that is now the closest Caribbean neighbor to the United States, alone consists of nearly 3000 coral islands—most of them rocky, barren, and uninhabited—but approximately 700 carry vegetation, of which roughly 30 are inhabited. Centrally positioned New Providence Island houses most of the country's 400,000 inhabitants and also contains the capital, Nassau, a leading tourist attraction.

The Lesser Antilles are grouped geographically into the Leeward Islands and the Windward Islands, a (climatologically incorrect) reference to the prevailing airflows in this tropical zone. The Leeward Islands extend southward from the U.S. Virgin Islands to the French dependencies of Guadeloupe and Martinique, and the Windward Islands from St. Lucia south and then west to the Dutch-affiliated islands of Aruba and Curaçao off the northwestern Venezuelan coast (Fig. 2-7).

It would be impractical to detail the individual geographic characteristics of each island of the Lesser Antilles, but we should note that nearly all share the insularity and environmental risks of the Caribbean Basin—major earthquakes, volcanic eruptions, and hurricanes; that they confront, to varying degrees, similar social and economic problems in the form of limited domestic resources, overpopulation, soil exhaustion, land fragmentation, and rising levels of crime and violence; and that tourism has become the leading industry for many.

Under these kinds of circumstances one looks for any hopeful sign, and such an indication comes from **Trinidad and Tobago**, the two-island republic at the southern end of the Lesser Antilles (Fig. 2-7). This country (population: 1.4 million) has embarked on a natural-gas-driven industrialization boom that could turn it into a future economic tiger. Trinidad has long been an oil producer and has also become the largest supplier of liquefied natural gas for the United States. Many of the new industrial facilities have agglomerated at the state-of-the-art Point Lisas Industrial Estate outside the capital, Port of Spain, and they have propelled Trinidad to become the world's leading exporter of ammonia and methanol. Natural gas also is an efficient fuel for the manufacturing of metals, and steelmakers as well as aluminum refiners have been attracted to locate here. With Trinidad lying just off the Venezuelan coast of South America, it is also a sea-lane crossroads that is strongly connected to the vast, near-coastal supplies of iron ore and bauxite that are mined in nearby countries, most importantly the Brazilian Amazon.

But Trinidad is an exception. It is a reflection of the predicament of small-island developing economies that, after the initial wave of decolonization in the 1950s and 1960s, decided they were probably better off affiliating with the European country that had ruled them. Thus, for example, Guadeloupe

Boutin/Sipa Press

The capital city of Trinidad and Tobago may have an historic colonial name (Port of Spain), but after nearly three centuries of Spanish rule, the British took control here. English became the *lingua franca* (everyone's second language), democratic government followed independence in 1962, and the development of natural-gas reserves has recently come to propel a thriving economy. As can be seen here, the harbor of Port of Spain is bursting at the seams as a car carrier delivers automobiles from Japan and containers are stacked high on the docks.

remained with France, and the Cayman Islands are still British. The former Netherlands Antilles passed up full independence to continue its affiliation with the Dutch but did so individually and with considerable autonomy: **Curaçao**, **Aruba**, and **St. Maarten** are now 'countries' within the Kingdom of the Netherlands, whereas the other Caribbean islands of the former Dutch Empire have acquired the status of overseas municipalities. Although economic logic seems to dictate continued association with the former colonial powers, local politics and pride demand some minimal degree of autonomy and avoidance of paternalistic European interference. The combined result is often a complicated legal framework that shapes the contemporary political status of the islands.

Points to Ponder

- The United States is the single most important economic partner of just about every country and territory in this realm—is that a plus or a minus for Middle America?

- Is the international "War on Drugs"—spearheaded by the United States but largely fought south of the border—worth continuing?

- Most of the small Caribbean island-nations contain less than half a million people—is that enough to sustain a viable economy?

- Is China's pivotal role in the construction of the Nicaragua Canal a cause for concern for the region and/or the United States?

FIGURE 3-1

CHAPTER 3

Alice Nerr/Shutterstock

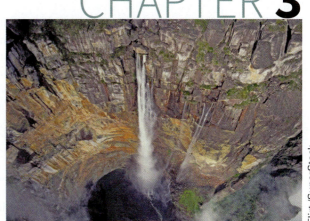

Flirt /SuperStock

Where were these pictures taken?

The South American Realm

IN THIS CHAPTER

- Challenges of regional integration
- South America after the commodity boom
- The strengthening indigenous peoples' movement
- China's economic footprint in South America
- Medellín: From cocaine capital to model city
- Brazil sets an example in controlling deforestation

CONCEPTS, IDEAS, AND TERMS

Unity of place	[1]	Gini index	[18]
Indigenous peoples	[2]	*Dependencia* theory	[19]
Altiplano	[3]	Insurgent state	[20]
Land alienation	[4]	Failed state	[21]
Liberation theology	[5]	Neoliberal policies	[22]
Cultural pluralism	[6]	Landlocked country	[23]
Commercial agriculture	[7]	Human Development Index	[24]
Subsistence agriculture	[8]	Triple Frontier	[25]
Remote sensing	[9]	Primate city	[26]
Uneven development	[10]	Viticulture	[27]
Supranationalism	[11]	Elongation	[28]
Rural-to-urban migration	[12]	Buffer state	[29]
Informal sector	[13]	*Entrepôt*	[30]
Barrio	[14]	Forward capital	[31]
Favela	[15]	*Cerrado*	[32]
Megacity	[16]	Negative externalities	[33]
Central business district	[17]	Growth-pole concept	[34]

Of all the continents, South America has the most familiar shape—a gigantic triangle connected by mainland Middle America's tenuous land bridge to its neighbor in the north. South America also lies not only south but mostly east of its northern counterpart. Lima, the capital of Peru—one of the continent's westernmost cities—lies farther east than Miami, Florida. Thus South America juts out much more prominently into the Atlantic Ocean toward southern Europe and Africa than does North America. But lying so far eastward means that South America's western flank faces a much wider Pacific Ocean, with the distance from Peru to Australia nearly twice that from California to Japan.

Defining the Realm

As if to reaffirm South America's northward and eastward orientation, the western margins of the continent are rimmed by one of the world's longest and highest mountain ranges, the Andes, a giant wall that extends unbroken from Tierra del Fuego near the continent's southern tip in Chile to northeastern Venezuela in the far north (**Fig. 3-1**). The other major physiographic feature of South America dominates its central north—the Amazon Basin; this vast humid-tropical amphitheater is drained by the mighty Amazon River, which is fed by several prominent tributaries (see **Box 3-1**). Much of the remainder of the continent can be classified as plateau, with the most important components being the Brazilian Highlands that cover most of Brazil southeast of the Amazon Basin, the Guiana Highlands located north of the lower Amazon Basin, and the cold Patagonian Plateau that blankets the southern third of Argentina. Figure 3-1 also reveals two other noteworthy river basins beyond Amazonia: the Paraná-Paraguay Basin of south-central South America, and the Orinoco Basin in the far north that drains interior Colombia and Venezuela.

Physiography

Explorers' Continent

It was here in northern South America that the great German explorer and scientist Alexander von Humboldt, one of the founders of the modern discipline of geography, embarked on his legendary expeditions in the early nineteenth century. After landing on the coast of Venezuela and trekking across the continent's northern interior, the 30-year-old Humboldt was struck by the area's biodiversity, its majestic natural beauty, and the adaptive abilities of the human populations. He discovered and named many species of flora and fauna, traversed tropical grasslands and jungles, met with indigenous peoples, crossed dangerous rivers, and reached the summit of the highest mountain in the Americas climbed by Europeans at the time (Ecuador's Chimborazo). He compiled large numbers of maps and was one of the first scientists to note how the coastlines of eastern South America and western Africa fitted together like pieces of a jigsaw puzzle, speculating about the continents' geologic movements. Humboldt was a key figure in the rise of the modern discipline of geography because of his many discoveries as well as his views on the **unity of place [1]**—that in a particular locality or region intricate connections exist among climate, geology, biology, and human cultures. As such, he laid the foundation for modern geography as an *integrative discipline* marked by a spatial perspective.

Almost three centuries earlier, the opposite end of South America was the scene of a crucial stage in the first circumnavigation of the globe and expedition led by Ferdinand Magellan. Once it became clear that Columbus had stumbled upon America, not India, Spanish and Portuguese explorers continued their efforts to discover a westward passage from the Atlantic to the Pacific. This took them to the far south along what is now the Argentinean coast. Magellan's ships set sail for the treacherous waters of what he named the *Estrecho de Todos los Santos*—now called the Strait of Magellan (Fig. 3-1). Of the (remaining) four

---

Box 3-1 Major Geographic Features of South America

1. South America's physiography is dominated by the Andes Mountains in the west and the Amazon Basin in the central north. Much of the remainder is plateau country.

2. Almost half of the realm's area and just under half of its total population are concentrated in one country—Brazil.

3. South America's population remains concentrated along the continent's margins. Most of the interior is sparsely peopled, but portions of it are now undergoing significant development.

4. Interconnections among the states of the realm are improving. Economic integration has become a growing force but is still at an early stage.

5. Regional economic contrasts and disparities strongly persist, both in the realm as a whole and within individual countries.

6. Cultural pluralism prevails in almost all of the realm's countries and is often expressed regionally.

7. Headlong urban growth continues to mark much of the South American realm, and urbanization overall is today on a par with the levels of the United States and western Europe.

8. This realm contains abundant natural resources that, although important for exports, tend to create considerable dependency on economic conditions in world markets.

ships in Magellan's fleet, only three made it through the daring, 600-kilometer (375-mi)-long passage; the fourth turned back and crashed on the rocks in the icy waters of the Southern Ocean.

Myriad Climates and Habitats

On the world map, if the Russia/Central Asia realm is the widest in east-west extent, South America is the longest measured from north to south. (Within South America, no other country is more emblematic of this elongated geography than Chile, averaging only about 150 kilometers [90 mi] in width but 4000 kilometers [2500 mi] in length.) As a consequence of this huge latitudinal span, from about 12°N to 56°S, this realm contains an enormous variety of climates and vegetation. Combine this with substantial variation in elevation from west to east, and it is clear why South America has such an impressive range of natural habitats.

Take another look at the map showing the global distribution of climates (Fig. G-6) and note the variation in climatic types, particularly in the realm's northwest and southern half. Travel northeast from Lima, Peru, for about 600 kilometers (375 mi) and you encounter no less than four different climate zones: arid, highland, and two varieties of humid tropical. A transect of similar length from Santiago, Chile eastward across the continent to Buenos Aires, Argentina will take you through five climate zones: interior highland, arid, and semi-arid environments bracketed by a different humid-temperate climate along each coast. Vegetation in South America varies accordingly, from lush tropical rainforests to rocky and barren snow-covered mountaintops to grasslands, fertile as well as parched. This natural diversity also makes for considerable cultural differences, as we shall soon discover.

States Ancient and Modern

Thousands of years before the first European invaders appeared on the shores of South America, peoples now referred to as **indigenous [2]** had migrated into the continent via North and Middle America and founded societies in coastal valleys, in river basins, on plateaus, and in mountainous locales. These societies achieved different and remarkable adaptations to their diverse natural environments, and by about 1000 years ago, a number of regional cultures thrived in the elongated valleys between mountain ranges of the Andes from present-day Colombia southward to Bolivia and Chile. These high-altitude valleys, called **altiplanos [3]**, provided fertile soils, reliable water supplies, building materials, and natural protection to their inhabitants.

The Inca State

One of these *altiplanos*, surrounding the city of Cuzco in what is now southeastern Peru, became the core area of South

---

Box 3-2 From the Field Notes …

"Climbing up to Machu Picchu, South America's best-known pre-Columbian site, is a thrill. This fifteenth-century "City of the Incas" is located on an Andean ridge (elevation: 2400 meters/ 7900 ft) about 80 kilometers (50 mi) northwest of Cuzco, the long-time capital of the Inca civilization and still a major city today. Machu Picchu is believed to have been built (ca. 1450) as an estate for the emperor Pachacuti but was abandoned less than a hundred years later at the time of the Spanish Conquest. It first became known to the outside world in 1911, and restoration (now nearing the halfway point) has been ongoing ever since. Archeologists have long been fascinated as well, and their intensive on-site fieldwork and research efforts continue. All these buildings were put in place using manual labor at an elevation where most visitors, myself included, have to carefully pace themselves to avoid shortness of breath."

© Jan Nijman

America's greatest indigenous empire, that of the **Inca**. The Inca were expert builders whose stone structures (among which Machu Picchu near Cuzco is the most famous), roads, and bridges helped to unify their vast empire; they also proved themselves to be efficient administrators, successful farmers and herders, and skilled manufacturers; their scholars studied the heavens, and their physicians even experimented with brain surgery. Great military strategists, the Inca integrated the peoples they vanquished into a stable and well-functioning state, an amazing accomplishment given their never-ending need to cope with treacherous, high-relief terrain (**Box 3-2**)

As a minority ruling elite in their far-flung empire, the Inca were at the pinnacle within their rigidly class-structured, highly centralized society. So centralized and authoritarian was their state that a takeover at the top was enough to gain immediate power over all of it—as a small army of Spanish invaders discovered in the 1530s. The European invasion brought a quick end to thousands of years of indigenous cultural development and changed the social order forever (**Box 3-3**).

The Iberian Invaders

The modern map of South America started to take shape when the Iberian colonizers began to understand the location and economies of the indigenous societies. The Inca, like Mexico's Maya and Aztec peoples, had accumulated gold and silver at their headquarters, possessed productive farmlands, and constituted a ready labor force. Not long after the defeat of the Aztecs in 1521, Francisco Pizarro sailed southward along the continent's northwestern coast, learned of the existence of the Inca Empire, and withdrew to Spain to organize its overthrow. He returned to the Peruvian coast in 1531 with 183 men and two dozen horses, and the events that followed are well known. Only two years later, his party rode victorious into Cuzco.

The new order that emerged in western South America placed the indigenous peoples in serfdom to the Spaniards. Great haciendas were formed by **land alienation [4]** (the takeover of indigenously held land by foreigners), taxes were instituted, and a forced-labor system was introduced to maximize the profits of exploitation. Lima, the west-coast headquarters of the Spanish conquerors, soon became one of the richest cities in the world, its wealth based on the exploitation of vast Andean silver deposits. Lima also became the seat of the Viceroyalty of Peru, the political administrative district (later subdivided into a triad of viceroyalties) whereby its South American territories were incorporated into the Spanish Empire.

Meanwhile, another vanguard of the Iberian invasion was penetrating the east-central part of the continent, the coastlands of present-day Brazil. This area had become a Portuguese sphere of influence, and Brazil's boundaries were bent far inland to include almost the entire Amazon Basin.

Box 3-3 From the Field Notes …

"Cuzco is the historic capital of the Inca Empire, located high in the Peruvian Andes at an elevation of 3400 meters (11,200 ft). The city center is a UNESCO-designated World Heritage Site and attracts well over a million tourists each year. I drove out of the valley up the hillside to get a good view of the layout of the center, so typical of Spanish colonial cities. The focal Plaza de Armas with its parks and benches is surrounded by the Cathedral of Santo Domingo and other religious and municipal buildings, and adjacent blocks around them are used for commerce and retailing. Located farther away from the Plaza are newer residential areas. This historic Inca capital was sacked in the early sixteenth century by Pizarro's *conquistadores*; few Inca ruins remain in the city center, but just outside Cuzco lies the impressive Sacsayhuaman fortress, with most of its walls still intact. This is not a very large city—it is too remote for that—but Cuzco's population has tripled over the past quarter-century to nearly 500,000."

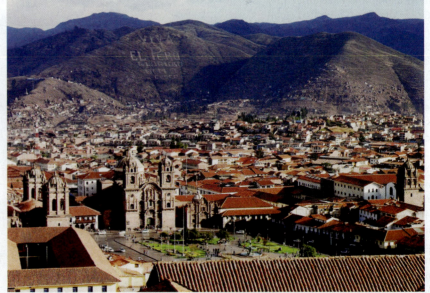

© Jan Nijman

Eventually, the country came to be only slightly smaller in territorial size than all the other South American countries combined. This westward thrust was the result of Portuguese and Brazilian penetration, particularly by the **Paulistas**, the settlers of São Paulo who needed indigenous slave labor to run their plantations.

Independence and Isolation

Despite their adjacent location on the same continent, their common language and cultural heritage, and their shared national problems, the countries that arose out of South America's Spanish viceroyalties (together with Brazil) until quite recently existed in a considerable degree of isolation from one another. Distance as well as physiographic barriers reinforced this separation, and most of the realm's population agglomerations still adjoin the coast, mainly the eastern and north western coasts (**Fig. 3-2**). The viceroyalties existed primarily to extract riches and fill Spanish coffers. In Iberia, there was little interest in developing the American lands for their own sake. Only after those who had a stake in Spanish and Portuguese America had made it their new home and had therefore rebelled against Iberian authority did things begin to change, and then very slowly. Thus South America was saddled with the values, economic outlook, and social attitudes of eighteenth-century Iberia—not the best tradition from which to begin the task of forging modern countries.

Certain isolating factors had their effect even during the wars for independence. Spanish military strength was always concentrated at Lima, and those territories that lay farthest from their center of power—Argentina as well as Chile—were the first to gain independence from Spain (in 1816 and 1818, respectively). In the Caribbean north, Simón Bolívar led the burgeoning independence movement, and in 1824 two decisive military defeats there spelled the end of Spanish control in South America.

This joint struggle, however, did not produce much unity because no fewer than nine countries emerged from the three former viceroyalties. It is not difficult to understand why this fragmentation took place. With the Andes intervening between Argentina and Chile and the Atacama Desert between Chile and Peru, overland distances seemed even greater than they really were, and these obstacles to contact proved quite effective. Hence, from their outset the new countries of South America began to grow apart amid friction that occasionally escalated into local wars. In fact, only within the past few decades have the countries of this realm finally begun to recognize the mutual advantages of increasing cooperation and to make lasting efforts to steer their relationships in this direction.

The Population Map—Then and Now

If we were able to reconstruct a (pre-Columbian) map of South America's population before the arrival of the Europeans, it would vary sharply from the current one displayed in Figure 3-2. Indigenous societies inhabited not only the Andes and adjacent lowlands but also riverbanks in the Amazon Basin, where settlements numbering in the thousands subsisted on fishing and farming. They did not shy away from harsh environments such as those of the island of Tierra del Fuego in the far south, where the fires they kept going against the bitter cold led the Europeans to name the place "land of fire."

Today, the realm's population distribution looks quite different. Many of the indigenous societies succumbed to the European invaders, not just through warfare but also because of the diseases the Iberian conquerors brought with them. It is estimated that 90 percent of native Amazonians died within a few years of contact, and even the peoples of isolated Tierra del Fuego are no longer there to build their fires. From one end of South America to the other, the European intrusion spelled disaster.

Spanish and Portuguese colonists penetrated the interior of South America, but the great majority of the settlers stayed on or near the coast, a pattern still evident today.

© J. Nijman, P. O. Muller, H. J. de Blij, and John Wiley & Sons, Inc.

SOUTH AMERICA: POPULATION DISTRIBUTION, 2017
One dot represents 50,000 persons

0   600   1200 Kilometers
0   300   600 Miles

FIGURE 3-2

Almost all of the realm's major cities have coastal or near-coastal locations, and the current population distribution map gives you the impression of a continent still awaiting penetration and settlement. But another look at Figure 3-2 will reveal a swath of population located well inland from the settlements along the northwestern coast, most clearly in Peru but also extending northward into Ecuador and southward into Bolivia. That is the legacy of the Inca Empire and its incorporated peoples, surviving in their highland redoubt and still numbering in the tens of millions.

The Cultural Mosaic

Indigenous Reawakening

In the early twenty-first century, South America's long-downtrodden indigenous peoples are staging a social, political, and economic reawakening. In several of the realm's countries, where their numbers are large enough to translate into political strength, they have begun to realize their potential. This is especially true in the Andean countries: indigenous peoples, descendants of the societies controlled by the Inca Empire, constitute no less than 45 percent of Peru's national population, and in Bolivia they are in the majority at 55 percent.

Today, newly empowered indigenous political leaders have emerged and are bringing the plight of the realm's aboriginal peoples to both local and international attention. South America's native peoples were conquered, decimated by foreign diseases, robbed of their best lands, subjected to involuntary labor, denied the right to grow their traditional crops, socially discriminated against, and swindled out of their fair share of the revenues from resources in their traditional domains. They may still rank among the poorest of the realm's poor, but they are increasingly asserting themselves. For a number of South American states, the consequences of this movement will be far-reaching.

The indigenous reawakening is in part related to changing religious practices in South America. Officially, just over 80 percent of the population is Roman Catholic, and traditionally South Americans tend to be viewed as devout followers of the Vatican. But recent surveys show that many do not attend church regularly and that more than half of all Catholics describe themselves as believers of the doctrine—rather than adherents of the Church. Since the late nineteenth century, the Catholic Church has often been criticized for its conservative position on social issues and for always siding with the establishment. During the 1950s, a powerful movement known as **liberation theology [5]** emerged in South America and subsequently gained followers around the world. This movement was a blend of Christian religion and humanist philosophy that interpreted the teachings of Christ as a quest to liberate the impoverished masses from oppression. Although the Church has lately been trying to make amends, it could not avoid losing popular support, especially among indigenous peoples.

African Descendants

When the Portuguese began to develop their territory on the coast of what is now Brazil, they turned to the same lucrative activity that their Spanish rivals had pursued in the Caribbean—the plantation cultivation of sugar for the European market. And they, too, found their workforce in the same source region: millions of Africans (nearly half of all who came to the Americas) were brought in bondage to the tropical Brazilian coast north of Rio de Janeiro. Not surprisingly, Brazil now has South America's largest black population, which is still heavily concentrated in the country's poverty-mired northeastern States. With Brazilians of direct or mixed African ancestry today accounting for just over half of the population of 210 million, the Africans decidedly constitute a major component in the immigration of foreign peoples into South America.

Ethnic Landscapes

The cultural landscape of South America, similar to that of Middle America, is a layered one. Indigenous inhabitants had cultivated and crafted diverse landscapes throughout the continent, some producing greater impacts than others. When the Europeans arrived, the cultural transformation that resulted from depopulation severely impacted the environment. Native peoples now became minorities in their own lands, and Europeans introduced crops, animals, and ideas about land ownership and land use that irreversibly changed South America. They also brought in African slaves from various parts of Subsaharan Africa. Europeans from non-Iberian Europe also began emigrating to South America, particularly during the first half of the twentieth century. Japanese settlers arrived in Brazil and Peru during the same era. All of these elements contributed to shaping the present-day ethnic complexion of this realm. **Figure 3-3** shows the distinct concentrations of indigenous and African cultural dominance, as well as areas where these groups are largely absent and people of European ancestry predominate.

Of course, ethnic origins are not always so straightforward, and patterns can change as a result of internal migrations and ethnic mixing. In recent years, research on individual DNA and genetics has taught us a lot about the regional and group origins of people. On the map, Argentina is indicated as having a predominantly European ancestry. More specifically, recent research shows that, for the average Argentine, nearly 80 percent of his or her genetic structure is European, 18 percent aboriginal, and 2 percent African. If nobody had mixed ancestors, these percentages would translate perfectly into European, indigenous, and African shares of the population—but of course that is not the case. Many people do not have precise knowledge of their ethnic ancestry, and although some may be descendants of those with a single origin, many others exhibit at least some degree of mixed ancestry. In the aggregate, however, there is no doubt that the population of cone-shaped southern South America is predominantly of European origin. South America at large is a realm marked by **cultural pluralism [6]**, where

FIGURE 3-3

SOUTH AMERICA:
DOMINANT ETHNIC
GROUPS

- African
- Mestizo
- European
- Indigenous

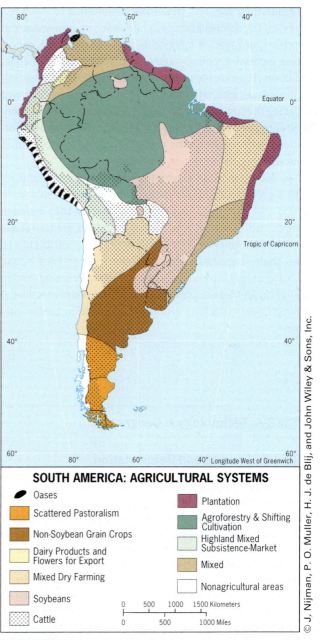

FIGURE 3-4

SOUTH AMERICA: AGRICULTURAL SYSTEMS

- Oases
- Scattered Pastoralism
- Non-Soybean Grain Crops
- Dairy Products and Flowers for Export
- Mixed Dry Farming
- Soybeans
- Cattle
- Plantation
- Agroforestry & Shifting Cultivation
- Highland Mixed Subsistence-Market
- Mixed
- Nonagricultural areas

peoples of various ethnicities and cultures cluster in adjacent areas but generally do not mix. The result is a cultural mosaic of almost endless variety.

Economic Geography

Agricultural Land Use and Deforestation

The internal divisions of this cultural kaleidoscope are further reflected in the realm's economic landscape. In South America, larger-scale **commercial [7]** or market (for-profit) agriculture and smaller-scale **subsistence agriculture [8]** (primarily for own use) exist side by side to a greater degree than anywhere else in the world (**Fig. 3-4**). The geography of plantations and other commercial agricultural systems was initially tied to the distribution of landholders of European background, whereas subsistence farming (such as highland mixed subsistence-market, agroforestry, and shifting cultivation) is historically associated with the spatial patterns of indigenous peoples as well as populations of African and Asian descent.

At a broader level of generalization, agricultural systems and land use in South America also vary in close relationship to

physiography, as one would expect. Forestry and agroforestry prevail in the Amazon Basin; ranching dominates in the grasslands of the Southern Cone; and different forms of agriculture, with or without irrigation, are found in an extensive zone from northeastern Brazil to northern Argentina, as well as scattered among pockets of moderate elevation in the Andean highlands. Land uses throughout South America are changing rapidly today, owing mainly to ongoing deforestation and the introduction or expansion of new crops. The fastest-growing crop is soybeans, the cultivation of which now dominates much of east-central Brazil and spills over into adjacent areas of Paraguay, Uruguay, and Argentina (Fig. 3-4).

Deforestation is a global problem, but it has been a particularly acute problem in Brazil. In the past, deforestation was attributed mainly to small-scale landholders, colonists who had made their way into the Amazon rainforest to eke out a new living. The deforestation crisis intensified as *agro-industrial* operations, engaged in large-scale production for the export market, increasingly penetrated Amazonia and transformed its land cover. The exact loss of rainforest is difficult to measure, but it has been estimated that Brazil lost an area of rainforest exceeding the size of Texas since 1970 (see **Box 3-4**). Since then, the government has succeeded in reducing the rate of deforestation while increasing the area of protected forests—but in absolute terms, deforestation continues.

As South American economies modernize, agriculture remains highly important. Almost one-fifth of this realm's workforce is still employed in the primary sector that includes farming, cattle raising, and fishing—a much greater share than in North America. The South American contribution to global trade in grains, soybeans, coffee, orange juice, sugar, and many other crops is significant and still increasing. And in addition to all these successful agricultural endeavors, illegal farming thrives in the form of narcotics production, particularly cocaine.

The Geography of Cocaine

All of the cocaine that enters the United States comes from South America, mainly from Colombia, Peru, and Bolivia. Within these three countries, cocaine annually brings in billions of (U.S.) dollars and "employs" tens of thousands of workers, thereby constituting a powerful industry. The first of the three stages of cocaine production involves the extraction of coca paste from the coca plant, a raw-material-oriented activity that is located near the areas where the plant is grown. The main zone of coca cultivation is along the eastern slopes of the Andes and adjacent tropical lowlands in Bolivia, Peru, and Colombia (**Fig. 3-5**).

Coca leaves harvested in the source areas of the Andes and nearby interior lowlands make their way to local collection centers, located at the convergence of rivers and trails, where coca paste is extracted and prepared. The second stage of production involves the refining of that coca paste (about 40 percent pure cocaine) into cocaine hydrochloride (more than 90 percent pure), a lethal concentrate that is diluted with substances such as sugar or flour before being sold on the streets to consumers.

Box 3-4 Technology & Geography

Remote Sensing and Deforestation

It is said that "a picture is worth a thousand words," but satellite imagery of the Earth can also leave us speechless. The indirect capture of images by satellites orbiting the Earth is referred to as **remote sensing [9]**. Remote sensing helps us to understand, define, and try to reshape the regions of the world and the complex interrelationships between the human and physical environments. In addition to capturing images illuminated by sunlight, sensors on satellites can also detect portions of the electromagnetic spectrum that are invisible to humans (e.g., infrared) and can also transmit signals like radar to measure the surface of the Earth. Coupled with advances in computer modeling and with the compelling bird's eye perspective of satellite images, new insights and greater understandings of processes and patterns behind phenomena such as urbanization, agricultural development, and climate change can be obtained.

Since 1972, NASA's Landsat program has captured remote sensing images for use in science, government, business, and industry. As the world's longest continuously operating platform for acquiring remotely sensed imagery, Landsat is instrumental in identifying and recording regional-scale changes across the surface of the Earth such as the deforestation of the Amazon rainforest (see photo). Without remote sensing imagery, it was extremely difficult to estimate rates of deforestation, and the information could be easily manipulated by governments or interest groups. In that sense, remote sensing technologies have contributed immensely to our objective knowledge of land-use change, around the world.

National Remote Sensing Centre Ltd/Science Source

This is what the Amazon's equatorial rainforest looks like from an orbiting satellite after the human onslaught in preparation for settlement. The colors on this Landsat image emphasize the destruction of the trees, with the dark green of the natural forest contrasted against the pale green and pinks of the leveled forest. The linear branching pattern of deforestation is the preferred approach here in Rondônia State's Highway BR-364 corridor. But farming is not likely to succeed for very long, and much of the cleared land is likely to be abandoned. Then the onslaught would resume to clear additional land—a pernicious cycle the Brazilian government has committed itself to ending by 2030.

COCA PLANT PRODUCTION IN NORTHWESTERN SOUTH AMERICA

Coca plant growing area
(Colombian territories are not named)

© J. Nijman, P. O. Muller, H. J. de Blij, and John Wiley & Sons, Inc.

FIGURE 3-5

and Bolivia have long struggled to modernize their economies and improve standards of living. Within countries, there is substantial **uneven development [10]** as well. Most manufacturing is concentrated in and around major urban centers, leaving vast empty spaces in the interior of the realm.

Brazil, in particular, increasingly drew attention over the past decade for its momentous growth and the rising sophistication of a number of its economic sectors. In the past, typical Brazilian exports were foods and footwear. Today, the leading exports include oil (thanks to the discovery of huge offshore reserves), steel, and state-of-the-art Embraer aircraft.

Despite recent impressive economic growth across much of the realm, it is clear that the longstanding isolation of its individual countries remains a serious impediment. Most governments agree that tighter integration, political as well as in terms of transportation infrastructure, would greatly enhance economic opportunities.

Economic Integration

The realm continues to be economically fragmented, and only about 25 percent of its exports remain in South America (a very low share compared to Europe, where intra-realm exports account for nearly 70 percent). Nonetheless, things are moving in a positive direction. With mutually advantageous trade the catalyst, a new continent-wide spirit of cooperation is blossoming at every level. Cross-border rail, road, and pipeline projects, stalled for years, are multiplying steadily. And importantly, investments now flow more freely than ever from one country to another, most notably in the agricultural sector.

Recognizing that free trade may well be able to solve many of the realm's economic-geographic problems, governments are now pursuing several avenues of heightened international cooperation. This is a good example of the concept of **supranationalism [11]**, which geographers define as a voluntary association in economic, political, or cultural spheres of three or more independent countries willing to yield some measure of sovereignty for their mutual benefit (a concept elaborated in Chapter 4).

In 2016, South America's republics were affiliating with a number of trading blocs, of which the following pair are the most important:

- **Mercosur/l** (**Mercosur** in Spanish; **Mercosul** in Portuguese): Launched in 1995, this Common Market established a free-trade zone and customs union that now links Brazil, Argentina, Uruguay, Paraguay, and Venezuela (five additional western South American countries are associate members). The organization aspires to be the dominant free-trade association for all of South America, but political agendas tend to prevail over economic needs. As a result, the path toward full economic integration is proving to be both slow-paced and obstacle-ridden.

- **Pacific Alliance (PA):** Inaugurated in 2012, this bloc was founded by Mexico, Colombia, Peru, and Chile. They swiftly formed a free-trade area along the western edge of the continent and announced their commitment to further economic integration "with a clear orientation to Asia." Whereas Mercosur/l countries are often entangled

Cocaine refining requires sophisticated chemicals, carefully controlled processes, and a labor force skilled in their supervision. Most of this processing stage is dominated by Colombia and takes place in the lowland central-south and east, beyond the reach of the Bogotá government.

The final stage of production entails the distribution of cocaine to the marketplace, which depends on an efficient, clandestine transportation network that leads into the United States. Private planes operating out of remote airstrips were the preferred "exporting" method until about 1990, and in the 1970s and 1980s Miami was a major center of cocaine distribution in the United States. But aggressive U.S. measures along the coasts of Florida caused a shift toward Mexico (and also, since the mid-2000s, the countries of northern Central America). Today, most cocaine still reaches the United States through smuggling across the Mexican border (a topic elaborated in Chapter 2). In this decade, another route has opened: small planes fly the product eastward from Colombia to Venezuela and the Guianas; from there it is transshipped northward by boat or plane to various Caribbean islands; subsequently, it is routed westward to Central America or Mexico, sometimes even directly into the United States.

Industrial Development

The geography of industrial production varies markedly in this realm. Brazil and Chile have shown rapid growth; Argentina has been in long-term decline; and countries such as Peru, Ecuador,

in political debates, the PA has advanced steadily, exhibiting a collective economic growth rate almost twice as high. Candidates for new membership, led by Panama and Costa Rica, are lining up; 44 other countries, including Canada, the U.S., Japan, China, and 14 European states, have acquired observer status. The rise of the Pacific Alliance in this decade, and its articulation of a strong liberal economic stance in comparison with Mercosur/l, have spawned notions of an east-west "continental divide" regarding the realm's evolving economic geography.

Dependence on Raw Materials

During the first decade of this century, South America's economy in the aggregate grew robustly by about 5 percent annually. Much of this growth occurred in Brazil, Chile, and Peru, but other parts of the continent also fared well. A leading driver was the steadily rising demand for the realm's abundant raw materials in the global marketplace, in no small part resulting from the explosive growth (and accompanying raw-material appetite) of China. The Asian giant gobbled up Brazilian soybeans, Chilean copper, Peruvian silver, Venezuelan oil, and so much more. Commodity prices skyrocketed, and South America reaped the benefits.

The economic picture in this decade, however, is quite different. Commodity-driven growth has leveled off markedly since 2011, much of it in response to slowing Asian demand and lowered prices on the global market. Although exports of raw materials are still substantial, the heyday hype of the **commodity boom** has faded. In its place has been a resurfacing of the pre-boom realization that heavy reliance on the production of raw materials (slightly more than 50 percent of the realm's exports today) is a risky longer-term proposition—for a variety of reasons.

First, the increasing demand for commodities tends to drive up the value of a producing country's currency, which in turn limits sales of other exports as they become more expensive for foreign trading partners. Second, higher revenues from the export of raw materials can undermine the development of local manufacturing and other processing industries. Some economists now employ the term *premature deindustrialization*, wherein domestic industries erode as government policies overly emphasize the exporting of commodities at a time of rising value in the local currency. Third, resource management, because of complex ownership issues and exploitation rights, invites both corruption and complacency (note, for instance, that major discoveries of oil reserves rarely result in a society's advancement). Finally, and most significantly, the great majority of resources (especially key minerals and energy deposits) are nonrenewable, making it essential that producing countries plan for an economic future without them. Thus in the aftermath of South America's commodity boom in the 2000s, its longer-term benefits may well depend on the ability of national governments to successfully refocus their energies on nurturing home-grown industries.

© NASA image created by Jesse Allen, using EO-1 ALI data provided courtesy of the NASA EO-1 team

An aerial image of one of South America's most spectacular raw-material sites—northern Brazil's Carajás Iron Ore Mine, located where the southeastern rim of the Amazon Basin meets the Brazilian Highlands in the center of Pará State. This may be the world's richest as well as largest proven deposit of iron ore, whose open-pit mines steadily deepen and sprawl outward as minerals are extracted from the surface—one stripped-away, environmentally-degrading layer at a time. Annually, more than 300 million metric tons are hauled away via the dedicated, 850-kilometer (535-mi) railroad to the Atlantic port of São Luis (see Fig. 3-17).

At any rate, by 2016 most South American economies had gone through a half-decade of economic stagnation, with many countries recording near-negative growth. Brazil, having been celebrated for years as the new powerhouse among emerging economies, went through a particularly painful contraction, recording its worst economic performance in two decades. The situation was also particularly dire in Argentina and, most of all, in Venezuela (which is overwhelmingly dependent on world market prices for oil).

Urbanization and Its Regional Expressions

Rural-Urban Migration

As in most other realms, South Americans are leaving the land and migrating to towns and cities. This realm's modern urbanization process got an early start and has been intensifying since 1950 as the growth of urban settlements has averaged 5 percent a year. Meanwhile, rural areas have grown by less than 2 percent annually over those same six-plus decades. As a result, the realm's urban population has climbed to its current level of 83 percent, ranking it with those of western Europe and the United States. These numbers underscore not only the dimensions but also the durability of the **rural-to-urban migration [12]** from the countryside to the cities.

In South America, as in Middle America, Africa, and Asia, people driven by the poverty of rural areas are attracted to the towns and cities (**Box 3-5**). Both *push* and *pull* factors are at work. Rural land reform has been very slow in coming, and for this and other reasons every year tens of thousands of farmers simply give up and leave, seeing little or no possibility of economic advancement. The urban centers lure them because they are perceived to provide opportunity—the chance to earn a regular wage, visions of education for their children, better medical care, upward social mobility, and the excitement of life in a city.

But the move itself can be traumatic. Cities in developing countries are surrounded and often invaded by squalid slums, and this is where the urban immigrant most often finds a first—and sometimes permanent—abode in a makeshift shack without even the most basic amenities and sanitary facilities. But still the people come, hopeful for a better life. Many newcomers earn their first cash income by becoming part of the **informal sector [13]**, in which workers are undocumented and money transactions are beyond the control of government. The entry-level settlements consist mostly of self-help housing, vast shantytowns known as *barrios* **[14]** in Spanish-speaking South America and as *favelas* **[15]** in Brazil. Some of their entrepreneurial inhabitants succeed more than others, transforming parts of these shantytowns into beehives of activity that can propel resourceful workers toward a middle-class existence.

Bambu Productions / Getty Images,Inc.

Brazil's São Paulo is the tenth-largest metropolis in the world and, with more than 20 million people, by far the biggest in this realm. Most of its growth over the past half-century is the result of massive rural in-migration. Severely overcrowded São Paulo has grown so rapidly that certain neighborhoods exhibit some of the highest residential densities on Earth. Moreover, this sprawling megacity has become notorious for its gridlocked automobile traffic and choked transit systems, which produce commuting times that are just about triple those in major U.S. cities. This photo was taken in late 2014 from a taxi-helicopter, an elitist new mode of transportation used routinely by high-level business executives to leapfrog the congestion below.

Regional Patterns

The generalized spatial pattern of South America's urban reconfiguration is displayed in **Figure 3-6**, a *cartogram* (specially transformed map) of the continent's population. Here we see not only the realm's countries in population space relative to one another, but also the proportionate sizes of individual large cities within their total national populations.

Regionally, the Southern Cone (countries colored green) is the most highly urbanized. Today in Argentina, Chile, and Uruguay, almost all of the population resides in cities. Ranking below them in urbanization is Brazil (tan colored). The next highest group of countries (beige) borders the Caribbean Sea in the north. Not surprisingly, the Andean countries (brown) constitute the realm's least urbanized zone. Figure 3-6 tells us a great deal about the relative positions of the largest metropolises within their countries. Four of them—São Paulo, Rio de Janeiro, Buenos Aires, and Lima—rank among the world's **megacities [16]**, in which the metropolitan population exceeds 10 million. The size and density of these megacities pose enormous challenges in terms of housing, mobility, pollution, water provision, and overall environmental sustainability (see **Box 3-6**).

Problems of Inequality

An important indicator of South America's economic growth in this century has been the steady expansion of its middle

---

Box 3-5 Major Cities of the Realm, 2016

City	Metropolitan Area Population (millions)
São Paulo, Brazil	20.6
Buenos Aires, Argentina	14.3
Rio de Janeiro, Brazil	11.8
Lima, Peru	11.0
Bogotá, Colombia	9.5
Santiago, Chile	6.3
Belo Horizonte, Brazil	4.6
Caracas, Venezuela	2.9
Asunción, Paraguay	2.9
Guayaquil, Ecuador	2.8
Brasília, Brazil	2.6
Quito, Ecuador	2.4
Santa Cruz, Bolivia	2.2
Manaus, Brazil	1.9
Montevideo, Uruguay	1.3

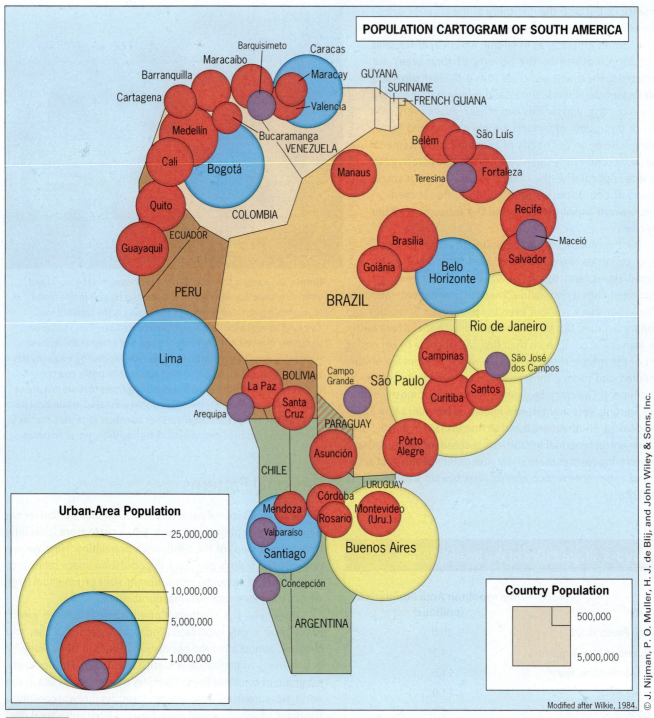

POPULATION CARTOGRAM OF SOUTH AMERICA

Urban-Area Population

25,000,000
10,000,000
5,000,000
1,000,000

Country Population

500,000

5,000,000

Modified after Wilkie, 1984.

© J. Nijman, P. O. Muller, H. J. de Blij, and John Wiley & Sons, Inc.

FIGURE 3-6

class. The World Bank recently reported that this socio-economic stratum had increased its share of the realm's income-earning population from 20 to 29 percent between 2000 and 2010. But that period coincided (as noted above) with strong economic growth that was in part due to the commodity boom. As stagnation set in over the past half decade, it appears that inequality stalled as well; that is, the improvements of the 2000s did not carry over. By most measures, the disparity between rich and poor is still wider in this realm than in any other, with wealth concentrated in

the hands of a tiny minority by an especially lopsided margin (see **Box 3-7** and **Figure 3-7**).

External Relations

The Shadow of the United States

The United States has long played a key role in this realm, beginning in 1823 with the Monroe Doctrine's assertion that European

Box 3-6 Among the Realm's Great Cities: São Paulo

São Paulo, which lies on a plateau 50 kilometers (30 mi) inland from its Atlantic outport of Santos, at first appears to possess no obvious locational advantages. Yet here on this site we find the tenth-largest metropolis on Earth, whose population has multiplied so uncontrollably that São Paulo has more than doubled in size to just under 21 million over the past three decades.

São Paulo was founded in 1554 as a Jesuit mission. The initial choice of location was based on access to the relatively large native population groups of the interior that were targets of conversion to Catholicism—hence the naming after St. Paul. The mission was also situated on the Tietê River, a convenient means of transport that originates near the coast but flows inland toward the northwest interior. In the seventeenth century, São Paulo became the home base of the so-called *bandeirantes* (explorers, gold prospectors, and slave traders), an unruly lot for whom the relatively remote location, far from the rule of law, was an advantage. Important gold mines were soon discovered in neighboring Minas Gerais State, turning São Paulo into a busy gateway that was further enhanced as the fertile lands to the west attracted a growing number of settlers.

Modern São Paulo, however, was built on the nineteenth-century coffee boom. Ever since, it has grown steadily as both an agricultural processing center (soybeans, orange juice concentrate, and sugar besides coffee) and a manufacturing complex (today accounting for about half of all of Brazil's industrial jobs). Along the way, it also evolved into Brazil's primary focus of commercial and financial activity. Twenty-first-century São Paulo's bustling, high-rise **central business district [17]** (see photo) is the very symbol of urban South America and attracts the realm's largest flow of foreign investment as well as the trade-related activities that befit the city's emergence as the business capital of Mercosur/l.

Nonetheless, even for this metropolitan industrial giant of the Southern Hemisphere, the increasingly global tide of postindustrialism is rolling in and São Paulo is learning to adapt. To avoid becoming a Detroit-style rustbelt, the aging automobile-dominated manufacturing zone on the central city's southern fringes is today attracting new industries. Internet companies have flocked here, as well as to the nearby city of Campinas, in such numbers that they now form the country's largest high-tech cluster—increasingly referred to as the "Silicon Valley of Brazil."

Elsewhere in São Paulo's far-flung urban constellation—whose suburbs now sprawl outward up to 100 kilometers (60 mi) from the CBD—many additional opportunities are being exploited. In the outer northeastern sector, new research facilities as well as computer and telecommunications-equipment factories are springing up. And to the west of central São Paulo, lining the ring road that follows the Pinheiros River, is South America's largest suburban office complex replete with a skyline of ultramodern high-rises.

These advances notwithstanding, the colossal recent growth of this megacity has come at a price. Whereas incomes in São Paulo are about 25 percent higher than those in Rio, so is the cost of living. And there are massive problems of overcrowding, pollution, and congestion. Traffic jams here are among the world's worst, with more than twice as many gridlocked motor vehicles on any given day than in Manhattan. The recent expansion of the metro—the last of only four such transit lines—was not completed until 2010; it has somewhat eased commuting, but overcrowding is immense with a staggering 5.2 million passengers riding the trains every work day. The urban region has expanded in all directions and is now relatively well connected by major highways. One of the world's most spectacular expressways leads to Santos, with enormous bridges, tunnels, and crisscrossing north- and southbound lanes—mostly amidst breathtaking scenery.

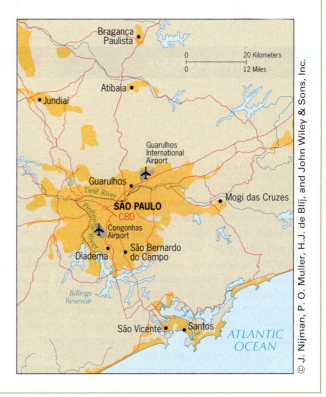

colonial powers no longer had rights to South America. During the Cold War (1945–1990), the United States became politically involved in a number of countries, mainly to keep Soviet influence out of the Western Hemisphere (above all in Chile during the early 1970s). This effort did not always enhance the local standing of the United States. In fact, anti-Americanism never seems far from the surface in South America, based heavily on the U.S. record of supporting right-wing dictatorships as well as continuing suspicions that the Americans still harbor imperialistic intentions.

Relations with the United States have never been smooth as they evolved in a consistently asymmetrical manner. Thus South America today attracts only about 4 percent of all U.S. foreign trade. Yet the U.S. remains the biggest trading partner for the realm as a whole, accounting for about one-fifth of the exports and imports of all South American countries. It was hardly a coincidence that *dependencia* theory [19] originated in South America during the 1960s. It was a new way of thinking about economic development that interpreted the persistent poverty of certain countries as the outcome of their unequal relations with the world's more affluent countries. Whatever the current validity of *dependencia*, the asymmetry remains and U.S. foreign policy, as viewed from South America, is not always credible. Ironically, America's special interests in this realm are now under increased pressure from another major power that would have been furthest from President Monroe's mind back in 1823: China.

China's Growing Economic Clout

China has in recent years displaced the United States as the leading trading partner of Brazil, Chile, Peru, and Ecuador. Other South American countries may soon be next. This remarkable economic rebalancing, and its possible political ramifications, has yet to fully register in the United States. Nearly invisible just twenty years ago, China has made its presence felt all across South America by establishing new embassies and consulates, buying up companies, partnering joint ventures, financing infrastructure projects and development assistance, and sending as well as inviting highest-level trade delegations. Their motives are clear: the Chinese still need large supplies of key raw materials such as oil, copper, and a plethora of minerals to reinforce the ongoing economic development of their country. Just as importantly, they are also seeking to expand their markets for Chinese exports, and the steadily expanding middle-income consumer classes of Brazil and other countries make potentially lucrative targets.

For several South American countries, China in some ways poses the same challenges as the United States did in the past—an asymmetry of relations that can lead to dependency and undercut the development of domestic economic sectors. The economic slowdown of China in recent years, accompanied by downward-trending demand for raw materials, has had immediate and deep repercussions for many South American economies, and it underscores this asymmetry and precarious dependence.

Box 3-7 MAP ANALYSIS

The South American realm has long been considered to be the most unequal in the world. One way to measure inequality is with the **Gini index [18]** or coefficient (named after the Italian statistician Corrado Gini). A Gini value of 0 indicates that income is equally distributed across society, whereas a value of 100 indicates that all income is held by a single recipient. The Gini index for Sweden is a modest 27.3, and the United States comes in at 41.1. Figure 3-7 shows the Gini index for South America's countries and a number of major cities. Examine the geographic patterns of national- and urban-scale income inequality across the realm. How does inequality in South America compare to that in the United States? In which cases is there a discrepancy between the Gini scores of cities and the countries in which these cities are located? How would you explain this variation?

Access your WileyPLUS Learning Space course to interact with a dynamic version of this map and to engage with online map exercises and questions.

INCOME INEQUALITY IN SOUTH AMERICAN STATES AND SELECTED CITIES
GINI INDEX

Above 49.0
46.2 – 49.0
Below 46.2
No data

0   600   1200 Kilometers
0 300   600 Miles

© J. Nijman, P. O. Muller, and John Wiley & Sons, Inc.

FIGURE 3-7

Regions of South America

On the basis of physiographic, cultural, and political criteria, South America can be divided into four clearly defined regions (**Fig. 3-8**):

1. **The Caribbean North**, located almost entirely north of the equator, consists of five entities that display a combination of Caribbean and South American features: Colombia, Venezuela, and those that represent three historic colonial footholds by Britain (Guyana), the Netherlands (Suriname), and France (French Guiana).

2. **The Andean West** is formed by four republics that share a strong indigenous cultural heritage as well as powerful influences resulting from their Andean physiography: Ecuador, Peru, Bolivia, and, transitionally, Paraguay.

3. **The Southern Cone**, for the most part located south of the Tropic of Capricorn (latitude 23½°S), includes three countries: Argentina, Chile, and Uruguay (all with strong European imprints and little remaining indigenous influence) plus aspects of Paraguay.

4. **Brazil** occupies an enormous part of interior and eastern South America. In the Amazon Basin of the north, its own interior overlaps with the continent's "green heart"—often referred to as the world's 'lungs'—the largest tropical rainforest on Earth. Brazil is South America's giant, accounting for just about half the realm's territory as well as population. In Brazil the dominant Iberian influence is Portuguese, not Spanish, and here Africans, not indigenous peoples, form a significant component of demography and culture.

Each of these four regions shares some important commonalities in terms of physical environment, ethnic origins, cultural milieu, and international outlook. But there also are important differences, especially involving economic performance, democratic functioning, and social stability.

Region The Caribbean North

The countries of South America's northern tier have something in common besides their coastal location: each has a coastal tropical-plantation zone based upon the Caribbean colonial model. Especially in the three Guianas, early European plantation development encompassed the forced immigration of African laborers and eventually the absorption of this element into the population matrix. Far fewer Africans were brought to South America's northern shores than to Brazil's Atlantic coasts, and tens of thousands of South Asians also arrived as contract laborers and stayed as settlers, so the overall situation here is not comparable to Brazil's. Moreover, it is also distinctly different from that of the rest of South America.

To this day, Guyana, Suriname, and French Guiana still display the coastal orientation and plantation dependency with which the colonial period endowed them, although the logging of their tropical forests is penetrating and ravaging the interior. In Colombia and Venezuela, however, farming, ranching, and mining drew the population inland, overtaking the coastal-plantation economy and establishing more diversified economies (**Fig. 3-9**).

Colombia

Imagine a country more than twice the size of France but with only three-quarters of the French population, with an environmental geography so varied that it can produce crops ranging from the temperate to the tropical, and possessing world-class oil reserves as well as many other natural resources. This country is situated in the crucial northwestern corner of South America, with 3200 kilometers (2000 mi) of coastline on both Atlantic (Caribbean) and Pacific waters, closer than any of its neighbors to the markets of the north and sharing a border with giant Brazil to the southeast. This nation uses a single language and adheres to one dominant religion. Wouldn't such a country thrive?

The answer, for most of its recent history, is no—and it illustrates that the influence of geography on development is rarely simple or straightforward. Colombia has a history of strife and violence; its politics is unstable, its population divided through extreme inequality, and its economy impaired. Its spectacular, scenic physical geography also divides its population of 48.7 million into clusters not sufficiently interconnected to foster integration; even today, this enormous country has barely 1600 kilometers (1000 mi) of four-lane highways. Its proximity to U.S. markets is a curse as well as a blessing: at the root of Colombia's latest surge of internal conflict lies its role as one of the world's leading producers of cocaine and other illicit narcotics.

People and Resources As Figure 3-9 indicates, Colombia's physiography is mountainous in its Andean west and north and comparatively flat in its interior. Colombia's scattered population tends to cluster in the west and north (Fig. 3-2), where the resources and the agricultural opportunities (including the coffee for which Colombia is famous) lie.

Colombia's population clusters continue to be poorly interconnected. Several lie on the Caribbean coast, centered on Barranquilla, Cartagena, and Santa Marta, old colonial entry points. Others are anchored by major cities such as Medellín and Cali. What is especially interesting about Figure 3-9 in this context is how little development has taken place along the country's lengthy Pacific coast, where the port city of Buenaventura, across the mountains from Cali, is the only settlement of any size.

The map also shows that Colombia and neighboring Venezuela share the oil and gas reserves in and around the "Lake" Maracaibo area ("Lake" because this is really a gulf with a narrow opening to the sea), but Venezuela has the bigger portion. Recent discoveries, however, have boosted Colombia's production along the base of its easternmost Andean cordillera, and our map shows a growing system of pipelines from the interior

FIGURE 3-8

to the coast. Meanwhile, the vast and remote southeastern interior proved fertile ground for that other big Colombian money-maker—illicit drugs.

Cocaine's Curse

The ascendance of the narcotics industry, fueled by outside (especially U.S. but also Brazilian and European) demand and coupled with its legacy of violence, crippled the state for decades and threatened its very survival. Drug cartels based in the cities controlled vast networks of producers and exporters; they infiltrated the political system, corrupted the military and police, and waged wars with each other that cost tens of thousands of lives and destroyed much

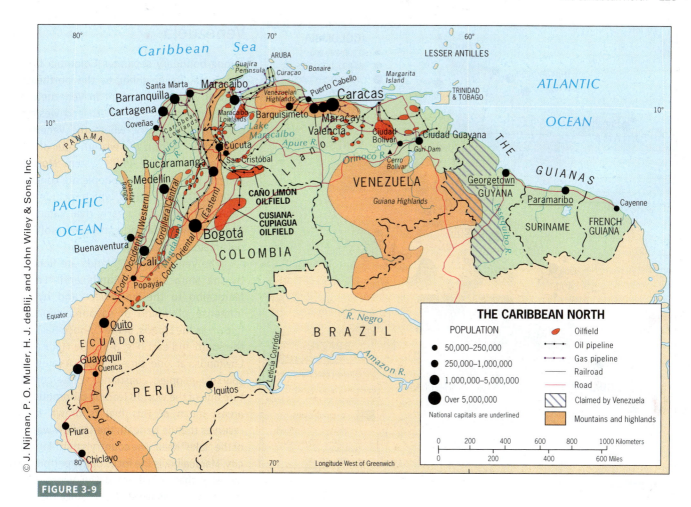

FIGURE 3-9

of Colombia's social order. The drug cartels also organized their own armed forces to combat initiatives by the Colombian government to control the clandestine narcotics economy. Meanwhile, the owners of large haciendas in the countryside hired private security guards to protect their properties, banding together to expand these units into what quickly became, in effect, private armies. Colombia was in chaos as narco-terrorists committed appalling acts of violence in the cities, rebel forces and drug-financed armies of the political "left" fought against paramilitaries of the political "right" in the countryside, and Colombia's legitimate economy, from coffee-growing to tourism, suffered grievously. To further weaken the national government, the rebels even took to blowing up oil pipelines.

Threats of Insurgency   Colombia's problems were compounded by the presence of various rebel groups (most notably, the Revolutionary Armed Forces of Colombia, or FARC) that contested government control of several provinces and that established mutually convenient partnerships with the cocaine cartels. FARC initially pursued a purely political or ideological agenda, but their widespread guerrilla activities soon brought them into contact with the cocaine cartels, which had their own reasons for wanting to exercise territorial control. In essence, this led to close, if murky, ties between the cartels and the rebel armies: the insurgents had found themselves a source

of funding, while the cartels linked up with a convenient source of military support against government forces in times of need. In the course of the 1990s, FARC assumed the role of what political geographers call an **insurgent state [20]**, whereby rebels hold total control over part of the state's territory. The area controlled by FARC was named *Farclandia*, the red-striped zone mapped in **Figure 3-10**. It appeared to be just a matter of time before the national government in Bogotá would lose control, the status quo would slide into disintegration, and Colombia would devolve into a **failed state [21]** (a country whose institutions have collapsed and in which anarchy prevails).

Colombia's Revival   Since 2002, however, Colombia has mounted successful campaigns to defeat the rebels militarily and to persuade them through legal means to give up their arms. Long-running peace talks with FARC finally resulted in a "definitive" ceasefire in June 2016. By then, rebel influence had diminished as the government had restored control over most of its territory—effectively transforming Colombia from an incipient failed state into a more stable, secure, and potentially prosperous country.

The government matched its domestic counteroffensive against the rebels with an international campaign to help revive the economy, promoting market-oriented, business-friendly policies. The country's strong economic performance in this decade

FIGURE 3-10

COLOMBIA

● 50,000–250,000
● 250,000–1,000,000
● 1,000,000–5,000,000
● Over 5,000,000

National capitals are underlined

Rebel controlled areas

Area of abrogated insurgent state, "Farclandia"

Coca-growing area

●——● Pipeline   —— Road

0   100   200 Kilometers

0   50   100 Miles

© J. Nijman, P. O. Muller, and John Wiley & Sons, Inc.

Venezuela

A long boundary separates Colombia from Venezuela, its neighbor to the northeast. Most of what is important in Venezuela is concentrated in the northern and western parts of the country, where the Venezuelan Highlands form the eastern spur of the northern end of the Andes system. Most of Venezuela's 31.5 million people are concentrated in these uplands, which include the capital of Caracas, its early rival Valencia, and the commercial/industrial centers of Barquisimeto and Maracay.

The Venezuelan Highlands are flanked by the Maracaibo Lowlands and "Lake" Maracaibo to the northwest and by an immense plainland of savanna country, known as the *Llanos*, in the Orinoco Basin to the south and east (Fig. 3-9). The Maracaibo Lowlands, at one time a disease-infested, sparsely peopled coastland, today constitute one of the world's leading oil-producing areas; much of this fuel is drawn from reserves that lie beneath the shallow waters of the "lake" itself. The country's third-largest city, Maracaibo, is the focus of the petroleum industry that transformed the Venezuelan economy in the 1970s; however, as we shall see, since then oil has been more of a curse than a blessing.

The *Llanos* on the southern side of the Venezuelan Highlands and the Guiana Highlands in the country's far southeast, like much of Brazil's interior, are in a nascent stage of development. The 300- to 650-kilometer- (180- to 400-mi)-long *Llanos* slope gently from the base of the northernmost Andean spur to the Orinoco's floodplain. Their mixture of savanna grasses and scrub woodland facilitates cattle grazing on higher ground, but widespread wet-season flooding of the more fertile lower-lying areas has thus far inhibited the plainland's commercial farming potential (much of the development of the *Llanos* to date has been limited to the exploitation of its massive oil reserves). Crop-raising conditions are more favorable in the *tierra templada* zones at moderate altitudes in the Guiana Highlands. Economic integration of this even more remote interior zone with the heartland of Venezuela was spearheaded by the discovery of rich iron ores during the 1950s on the northern side of the Guiana Highlands southwest of Ciudad Guayana near

is based on revenues from fossil fuels, metallic ores, tourism, and farm products, particularly exports of coffee and cut flowers (Colombia is the leading U.S. importer). Bogotá, the capital, centers the country's leading economic region and has become one of the realm's primary business complexes. The city of Medellín—former base of the legendary drug kingpin, Pablo Escobar, and designated the world's murder capital in the late 1980s—has made a comeback in recent years and is now internationally regarded as a model for innovative urban design (see photo).

In 2013, Medellín, Colombia was voted the world's most innovative city by the U.S. Urban Land Institute. Striking innovations include cable cars and escalators on the city's steep hillsides that have drastically reduced travel time. Through thriving public-private partnerships, the city has invested in new parks, schools, and libraries, especially in poor neighborhoods. This erstwhile capital of Colombia's cocaine industry may have been plagued by violence in the 1980s and 1990s, but Medellín's turnaround in this decade has attracted attention and acclaim from urban planners around the world.

© Christian Kober/Robert Harding World Imagery/Corbis Images

the mouth of the Orinoco River. Local railroads were built to connect the mines with the Orinoco, and from there ores have long been shipped directly by sea to foreign markets.

A Mounting Crisis

Venezuela is one of those countries in which a rich natural resource base can have destructive consequences for reasons elaborated earlier in this chapter. The country has long been in political turmoil because its governments had the bad habit of living off oil profits while neglecting long-term productive investments in the economy to better the lives of most Venezuelans. This overreliance on oil exports was felt most painfully when prices on the world market fell.

By the mid-1990s, the government was required to sharply devalue the currency, and a political crisis ensued that resulted in a severe recession and widespread social unrest. With more and more Venezuelans enraged at the way their oil-rich country was approaching bankruptcy without making progress toward a more equitable distribution of the national wealth, the way was cleared for the rise to power of the socialist radical, Hugo Chávez, in 1999 (with an apparent mandate from the urban poor and the increasingly penurious middle class).

Chávez's rule, however, quickly turned both autocratic and economically destructive. During his long presidency that ended with his death in 2013, relatively high oil prices brought in consistently high revenues totaling well over (U.S.) $800 billion. But there was very little to show for it: infrastructural improvements were hardly to be seen, bloated public-sector companies chronically underperformed, and the entrepreneurial sector of the population had fled the country soon after Chávez came to power. As for the poor, they intermittently benefited from new social programs, but life did not improve for them. Today, all Venezuelans are worse off, bedeviled by constant shortages of basic foods and vital retail products thanks to the staggering incompetence of the regime of Nicolás Maduro, Chávez's successor. By 2016, the performance of Venezuela's economy was the worst in the realm, shrinking by 7 percent from the year before; reinforced by widening political turmoil, the country seemed to have an even better shot at becoming a failed state than Colombia in the 1990s. On the larger South American stage, its disastrous "Bolivarian Revolution" had isolated Venezuela at a time when most of the realm's countries were embracing **neoliberal policies [22]** aimed at supporting private entrepreneurship and free markets.

The "Three Guianas"

Three small political entities form the eastern flank of the realm's northern region: Guyana, Suriname, and French Guiana. They are good reminders of why the name "Latin" America is so inappropriate: the first is a legacy of British colonialism and employs English as its official language; the second is a remnant of Dutch influence where Dutch is still official among its polyglot of tongues; and the third is still a dependency—of France. None has a population exceeding 800,000, and all three exhibit social characteristics and cultural landscapes far more representative of the Caribbean Basin than South America.

Here British, Dutch, and French colonial powers acquired possessions and established plantations, brought in African and Asian workers, and created economies similar to those that mark a number of the nearby Caribbean islands.

Guyana, with just under 800,000 people, has about as many inhabitants as Suriname and French Guiana combined. When Guyana became independent in 1966, its British rulers left behind an ethnically and culturally divided population in which people of South Asian (Indian) ancestry now make up about 44 percent and those of African background (including Afro-European ancestry) 30 percent. This makes for contentious politics, given the religious mix that is approximately 57 percent Christian, 28 percent Hindu, and 7 percent Muslim. Guyana remains dominantly rural, and plantation crops continue to figure prominently among exports (gold from the interior is the most valuable single product).

Oil may soon become a factor in the economy, however, because a recently discovered (apparently significant) reserve that lies offshore from Suriname extends westward beneath Guyana's waters. Still, Guyana is among the realm's poorest and least urbanized countries, and it is ever more strongly affected by the expanding drug industry of South America's north. Its thinly populated interior, beyond the reach of antidrug campaigns, has become a staging area for cocaine distribution to the United States via Middle America (and increasingly to Brazil and Europe). A recent study suggests that Guyana's take from the drug trade already amounts to at least one-fifth of the country's total economy.

Suriname actually progressed more rapidly than Guyana did after it was granted independence in 1975, but unending political instability swiftly ensued. The Dutch colonists brought South Asians, Indonesians, Africans, and even some Chinese to their colony, making for a notably fractious state. More than 100,000 residents—about one-quarter of the entire population—emigrated to the Netherlands and were it not for support from its former colonial ruler, Suriname's economy would have collapsed. Agriculturally, its rice farms provide Suriname with self-sufficiency, even allowing for some exporting (which is dominated by plantation crops). Suriname's leading income producer continues to be its bauxite (aluminum ore) that is mined in a zone across the middle of the country. To this short list can also be added the promise of those offshore oil finds noted above, which may soon be earning important revenues.

Suriname's cultural geography is seasoned by the many languages spoken by its 550,000 citizens. Dutch is the official tongue, but other than by officials and in schools it is not heard much. A mixture of Dutch and English, *Sranan Tongo*, serves as a common language, but one also hears indigenous, Hindi, Chinese, Indonesian, and even some French Creole on the streets.

French Guiana, the easternmost outpost of the Caribbean North region, is the last remaining dependency on the mainland of South America. This territory is an anomaly in other ways as well. Consider this fact: it is almost as large as South Korea (population: just above 50 million) with a population totaling only 275,000. Its status is that of an Overseas *Département* of France, and its official language is French. Nearly half the population resides in the immediate vicinity of the coastal capital, Cayenne.

In 2016, there still was no prospect of independence for this decidedly underdeveloped relic of the former French Empire. Gold remains the most valuable export, and a small fishing industry sends some exports to France. But what really matters here in French Guiana is the European Space Agency's launch complex at Kourou on the Atlantic coast, which accounts for more than half of the territory's entire economic activity—from plantation farming to spaceport … but with minimal involvement, or benefits, for most ordinary people.

Region The Andean West

The second regional grouping of South American states—the Andean West (**Fig. 3-11**)—is dominated physiographically by the great Andes mountain chain and historically by indigenous peoples. This region encompasses Peru, Ecuador, Bolivia, and transitional Paraguay, the last with one foot in the West and the other in the South (Fig. 3-8). Bolivia and Paraguay also constitute South America's only landlocked countries.

© J. Nijman, P. O. Muller, and John Wiley & Sons, Inc.

FIGURE 3-11

Box 3-8 From the Field Notes…

"It is late January of 2015 and this *altiplano* of south-central Peru offers some panoramic views. Down below, just right of center, is the old Andean town of Chinchero, lying at an elevation of 3600 meters (12,000 ft) and surrounded by mountains up to 5800 meters (19,000 ft) high. The original name of the town has been lost, but it belongs to what the Incas called the 'Sacred Valley' or the 'Land of the Rainbow.' It is easy to get a sense of how the Incas capitalized on the advantages of this highland environment during their pre-Columbian rule. Soils here are fertile, temperatures sufficiently moderate, and the snowpack and glaciers of the enveloping Andean ranges keep local streams and rivers flowing year-round. True to their heritage, the lower slopes all around Chinchero are still being farmed as they have been for centuries, growing potatoes, wheat, and other hardy *tierra fría* crops."

Spanish conquerors overpowered the indigenous nations, but they did not reduce them to small minorities as happened to so many indigenous peoples in other parts of the world. Today, 45 percent of the residents of Peru, the region's most populous country, are indigenous; in Bolivia, they form a majority of 55 percent (although there is considerable mixing of populations). Approximately 25 percent of Ecuador's population can be classified as indigenous, and in Paraguay the ethnic mix, not regionally clustered as in the other three countries, is dominated by those of indigenous ancestry.

This region has long been South America's poorest economically, with lower incomes, higher numbers of subsistence farmers, and fewer opportunities for job-seekers than elsewhere in the realm. For a very long time, the urbane lives of the land-owning elite have been worlds apart from the hard-scrabble existence of the landless peonage (the word **peón** is an old Spanish term for an indebted day laborer). But in recent years, the Andean West, like the realm as a whole, has been stirring, and oil and natural gas are leading the way.

Peru

Peru straddles the Andean spine for more than 1600 kilometers (1000 mi) and is the largest of the region's four republics in both territory and population (32 million). Physiographically and culturally, Peru divides into three parts (Fig. 3-11). Lima and its port, Callao, lie at the center of the **desert coast** subregion, and it is symptomatic of the cultural division prevailing in Peru that for nearly 500 years the capital city has been positioned on the western periphery, not in a central location in a basin of the Andes (moreover, Peru has never elected an indigenous president even though aboriginal people constitute nearly half of the population). From an economic point of view, however, the Spaniards' choice of a headquarters on the Pacific coast proved to be sound, for the coastal strip has become commercially the most productive part of the country. A thriving fishing industry contributes significantly to the export trade; so do the products of irrigated agriculture in some 40 oases dispersed across the arid coastal

plain, which include fruits such as citrus and olives, and vegetables such as asparagus (a big money-maker) and lettuce.

The Andean or **Sierra** subregion occupies just about one-third of the country and is the ancestral home of the largest component in the total population, the speakers of Quechua, the *lingua franca* that emerged during the Inca Empire. Their physical survival during the harsh Spanish colonial regime was made possible by their adaptation to the high-altitude environments they inhabited.* The Andean subregion is not nearly as important economically as the coast, but it does yield copper, zinc, and lead from mining centers, the largest of which is Cerro de Pasco. Peru is also the world's sixth-largest producer of gold, the price of which has been soaring in recent years, contributing to this country's steadily growing economy; most of the mines are located in the "gold belt" of the northernmost Peruvian Andes, but illegal strip mining is widespread in the Amazon rainforest of the southeastern interior. In the high valleys and intermontane basins, the indigenous population is concentrated either in isolated villages, around which many people engage in subsistence agriculture, or in the more favorably located and fertile areas where they are tenants, peons on white- or mestizo-owned haciendas (see **Box 3-8**).

Of Peru's three subregions, the East or **Oriente**—the inland-facing slopes of the Andean ranges that lead down to the Amazon-drained, rainforest-covered *montaña*—is the most isolated. The focus of the eastern subregion, in fact, is Iquitos in the far north, a city that looks east rather than west because it can be reached by oceangoing vessels sailing 3700 kilometers (2300 mi) up the Amazon River across northern Brazil. Iquitos grew rapidly during the Amazon's wild-rubber boom of the early twentieth century but swiftly declined thereafter; now it is finally growing again, reflecting Peruvian plans to open up the eastern interior. Today, this subregion, and perhaps Peru as a whole, appears to

---

* This is a good place to become reacquainted with the concept of **altitudinal zonation** that is elaborated in Figure 2-4 and accompanying text. The subregions of the Andean West represent the best place on Earth to apply this fivefold classification of vertical environments. The five *tierras*, in ascending elevational order, are the *caliente, templada, fría, helada,* and *nevada.*

be on the threshold of a new era owing to major new discoveries of oil and gas reserves in the *Oriente*. Indeed, while still one of the poorer countries in this realm, since the mid-2000s Peru has experienced a higher economic growth rate than any other. A trans-Andean pipeline already transports natural gas from the Camisea reserve (north of Cuzco) to a conversion plant on the coast south of Lima. From there, an underwater connector pipeline extends to an offshore loading platform for tankers taking it to the U.S. market (Fig. 3-11). Additional interior gas reserves are now being tapped, and they only heighten Peruvian aspirations to become a major energy producer.

Peru's economic development is not proceeding without adversity, however, as the government is being challenged on at least two fronts. Environmentalists and supporters of indigenous rights are raising issues throughout the *Oriente* (even as political activists argue against the terms of trade that Peru has accepted with the big oil companies). They complain that the oil- and gas-led exploitation of the unprotected East will harm both the environment and local communities, and that the new energy revenues will only further benefit the already-favored coastal residents of Peru—leaving the ever more disadvantaged inhabitants of the interior even farther behind.

Ecuador

On the map, Ecuador, the smallest of the three Andean West republics, appears to be just a northern corner of Peru. But that would be a misrepresentation because Ecuador also possesses a complete range of regional variations (Fig. 3-11). It has a coastal belt; an Andean zone that may be narrow (less than 250 kilometers [150 mi]) but by no means of lower elevation than elsewhere in the region; and an *Oriente*—an eastern subregion that is as sparsely settled and as economically marginalized as that of Peru. As with Peru, just about half of Ecuador's population (which totals 16.4 million) is concentrated in the Andean intermontane basins and valleys, and the most productive subregion is the coastal strip. There, however, the similarities end.

Ecuador's Pacific coastal zone consists of a belt of hills interrupted by lowlands, of which the most important lies in the south between the hills and the Andes, drained by the Guayas River. Guayaquil—the country's largest city, main port, and leading commercial center—forms the focus of this subregion (**Box 3-9**). Unlike Peru's coastal strip, Ecuador's is not a desert: it consists of fertile tropical plains not afflicted by excessive rainfall. Seafood (especially shrimp) is a leading product, and these lowlands support a thriving commercial agricultural economy built around bananas, cacao, cattle raising, and coffee on the hillsides. Moreover, Ecuador's western subregion is also far less Europeanized because this white component of the national population is only about one-third the size of Peru's.

A greater proportion of whites are engaged in administration and hacienda ownership in the central Andean zone, where most of the Ecuadorians who are indigenous also reside—and, not surprisingly, where land-tenure reform is an explosive issue. The differing interests of the Guayaquil-dominated coastal lowland and the Andean-highland subregion focused on the capital (Quito) have long fostered a deep regional cleavage between the two. This schism has not lessened in recent years, and autonomy and other devolutionary remedies often are openly discussed in the coastlands.

Box 3-9 From the Field Notes ...

© H. J. de Blij

© H. J. de Blij

"I can't remember being hotter anyplace on Earth, not in Kinshasa, not in Singapore ... not only are you near the equator here in steamy Guayaquil, but the city lies in a swampy, riverine lowland too far from the Pacific to benefit from any cooling breezes and too far from the Andean foothills to enjoy the benefits of suburban elevation. Guayaquil is not the disease-ridden backwater it used to be (left photo): its port is modern, its city-center waterfront on the Guayas River has been renovated, its international airport is the hub of a commercial complex. This has grown into a metropolis of 2.8 million, made possible in part by oil revenues. Beating the heat and glare is an everyday priority: entire streets have been covered by makeshift tarpaulins and more permanent awnings to protect shoppers (right). Talk to the locals, though, and you find that there is another daily concern: the people 'up there in the mountains who rule this country always put us in second place.' Take the 45-minute flight from Guayaquil to cool and comfortable Quito, the capital, and you're in another world, and you quickly forget Guayaquil's problems. That's just what the locals here say the politicians do."

In the rainforests of the *Oriente* subregion, oil production is expanding as a result of the discovery of additional reserves. Some analysts predict that eastern Ecuador as well as Peru will prove to contain energy resources comparable to those buried along the inland side of the Andes in Colombia and Bolivia. These developments could well portend the dawning of a new "oil era" accompanied by the economic transformation of the Ecuadorian *Oriente*. Indeed, oil already tops the country's list of exports, and to prepare for the next stage, the energy industry's infrastructure is being modernized and upgraded.

Bolivia

Nowhere are the problems typical of the Andean West more acute than in landlocked [23], volatile Bolivia. Before reading further, take a careful look at Bolivia's regional geography in both Figures 3-11 and 3-12. Bolivia is bounded by remote peripheries of both Brazil and Argentina, mountainous Andean highlands as well as intermontane *altiplanos* (high-elevation basins and valleys) of Peru, and coveted coastal zones of northern Chile. As the maps show, the Andes in this zone broaden into a vast mountainous complex some 700 kilometers (450 mi) wide.

A landlocked location does not always imply insurmountable challenges (think of Switzerland), but in the majority of cases it poses major limitations. There are 45 landlocked

countries in the world and almost all can be considered poor. Indeed, of the 15 lowest-ranking countries on the UN's Human Development Index [24], eight have no coastline. They have no direct access to ports; transport is costly; they are dependent on neighboring countries; the construction of cross-border infrastructures is costly and complicated; and overland trade is easily disrupted by possible conflicts in those adjoining countries. The economic impact is difficult to measure, but a recent research paper estimated that Bolivia's GDP would be 20 percent higher if it had a corridor (through northern Chile) with access to the Pacific Ocean.

On the boundary between Peru and Bolivia, freshwater Lake Titicaca lies at 3700 meters (12,500 ft) above sea level and helps make the surrounding *Altiplano* (see map) livable by ameliorating the coldness in its vicinity, where the snow line lies just above the plateau surface. On the cultivable land that encircles the lake, potatoes and *tierra fría* grains have been raised for centuries dating back to pre-Inca times, and the Titicaca Basin still supports a major cluster of indigenous (Aymara) subsistence farmers. This portion of the *Altiplano* is the heart of modern Bolivia and contains its administrative capital, La Paz.

The Bolivian state is the product of the Hispanic impact, and the country's indigenous peoples (who now comprise over half of the national population of 11 million) no more escaped the loss of their land than did their Peruvian or Ecuadorian counterparts. What made the richest Europeans in Bolivia wealthy, however, was not land but minerals. The town of Potosí in the eastern cordillera became a legend for the immense deposits of silver in its vicinity; tin, zinc, copper, lead, and several ferroalloys were also discovered there. Indigenous workers were coerced into toiling in the mines under the most appalling conditions. Today, natural gas, oil, and zinc are leading sources of foreign revenues.

Landlocked, physiographically bisected, culturally bifurcated, and economically divided, Bolivia is a severely challenged state. Moreover, the country's prospects are worsened by its political geography: look again at Figure 3-12 and you can see that Bolivia's nine provinces are regionally divided between indigenous-majority *departments* (as subnational units are called here) in the upland west and departments with mestizo majorities in the much-lower-lying east. As noted above, La Paz, the national seat of government, is located on the indigenous-majority Andean *Altiplano*, but many mestizo Bolivians do not recognize it as such.

The Santa Cruz Department, like the others in the *Oriente*, stands in sharp contrast to those of the Andes in the west. Here the hacienda system (discussed in Chapter 2) persists almost unchanged from the colonial era, its profitable agriculture supporting a wealthy

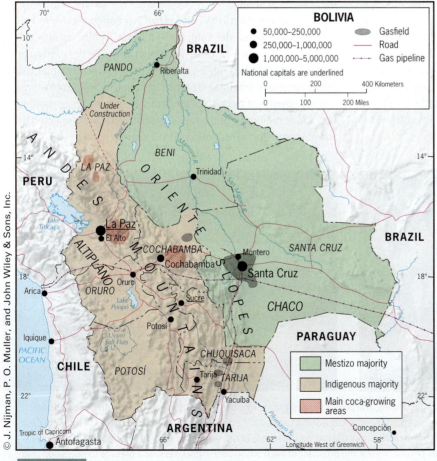

FIGURE 3-12

aristocracy. Recently, this eastern sector of Bolivia has proved to contain abundant energy resources as well, adding to its economic advantages. So there is talk—abetted by frequent public demonstrations—in support of autonomy, even secession here (see photo pair). But it must also be kept in mind that neither the haciendas nor the energy industry could operate without the indigenous labor force.

Paraguay

Paraguay, Bolivia's landlocked neighbor to the southeast, is one of those transitional countries lying between regions and exhibiting properties of each (see Fig. 3-8). Certainly, Paraguay (population: 6.7 million) is not an "Andean" country. It has no highlands of consequence. Its well-watered eastern plains give way to the dry scrub of the Gran Chaco in the northwest. Nor does it have clear, spatially entrenched ethnic divisions between indigenous, mestizo, and other peoples as do Bolivia and Peru. But aboriginal ancestry dominates the ethnic complexion of Paraguay, and continuing protests by landless peasants are similar to those taking place elsewhere in the Andean West. Moreover, the native language, Guaraní, is so widely spoken (by over 90 percent of the people) that this is one of the world's most thoroughly bilingual societies.

Today, Paraguay's longstanding isolation is eroding, most evident in the hot, semiarid Chaco of the northwest. Here, rapid deforestation has opened up a burgeoning new cattle-ranching frontier. For aboriginal peoples like the Ayoreo, however, the impact is disastrous because their traditional livelihoods are threatened with imminent destruction (the word for bulldozer in their language means "attacker of the world"). The contraction of the Chaco has also sounded alarms among biologists and conservationists because so much of it remains unexplored and is certain to contain myriad undiscovered plant and animal species.

Looking southward, there is little in Paraguay's economic geography to compare to Argentina, Uruguay, or Chile. Paraguay's low GDP resembles that of countries of the Andean West, not the far more advantaged Southern Cone region. This is also one of South America's least urbanized states, and poverty dominates the countryside as well as the slums girdling the capital, Asunción, and other towns (Fig. 3-11). Almost a quarter of the population lives at or below the official poverty level. Moreover, research suggests that 1.6 percent of the population owns about 80 percent of the land, which may be a record on this measure of inequality in the South American realm.

In the east, perhaps half a million Brazilians have crossed the border to settle in Paraguay, where they have created a thriving commercial agricultural economy that produces soybeans, livestock, and other farm products exported to or through Brazil. This puts Paraguay in a geographically difficult position, and it often complains that Brazil is not living up to its regional-trade obligations, thereby creating unacceptable difficulties for Paraguayan exporters. Meanwhile, politicians raise fears that Brazilian immigration is creating a steadily expanding foreign enclave within Paraguay, where people speak Brazilian Portuguese (including in the local schools), Brazil's rather than Paraguay's flag flies over public buildings, and a Brazilian cultural landscape is steadily emerging in the eastern border zone.

Another quintessentially geographical problem stemming from Paraguay's long-term weakness lies in the southeast, where the borders of Brazil, Argentina, and Paraguay converge in a chaotic tangle of smuggling, money-laundering, political intrigue, and even terrorist activity. This hub of lawlessness—centered on the town of Ciudad del Este, but with local spillovers into the territory of all three countries—is known as the *Triple Frontera* or **Triple Frontier [25]** (Fig. 3-11). It should also be noted that the Triple Frontier contains a sizeable, tristate Middle Eastern community, identified by the U.S. government as a source of funding for Islamic militant groups—an allegation repeatedly rejected by the governments of Argentina, Brazil, and Paraguay.

Bolivia's political geography is as divided as its physical geography, the power base of the indigenous population lying on the **Altiplano** and in the adjacent mountains, and that of the mestizo minority centered in the interior lowlands. These photos, taken a day apart in 2014 in Santa Cruz (now Bolivia's largest city) where mestizo strength is concentrated, show the crucial difference of opinion: a majority of mestizos want autonomy for their eastern provinces (left); virtually all indigenous people protest the prospect of such autonomy (right), seen as a prelude to secession and independence that would fatally fracture their country.

Aizar Raldes/AFP/Getty Images, Inc.

Rodrigo Buendia/AFP/Getty Images, Inc.

Region # The Southern Cone

Argentina

The largest Southern Cone country by far is Argentina, whose territorial size ranks second only to Brazil in this geographic realm; its population of 44 million ranks third overall after Brazil and Colombia. Although Argentina exhibits a great deal of physical-environmental variety within its boundaries, the overwhelming majority of the Argentines are concentrated in the physiographic subregion known as the **Pampa** (a Spanish word meaning "plain"). **Figure 3-13** underscores the degree of clustering of Argentina's inhabitants on the land and in the cities of the Pampa. It also shows the relative emptiness of the remaining six subregions shown on this map: the scrub-forested **Chaco** in the northwest; the mountainous **Andes** in the west, along whose crestline lies the boundary with Chile; the arid plateaus of **Patagonia** south of the Rio Colorado; and the undulating transitional terrain of intermediate **Cuyo, Entre Rios** (also

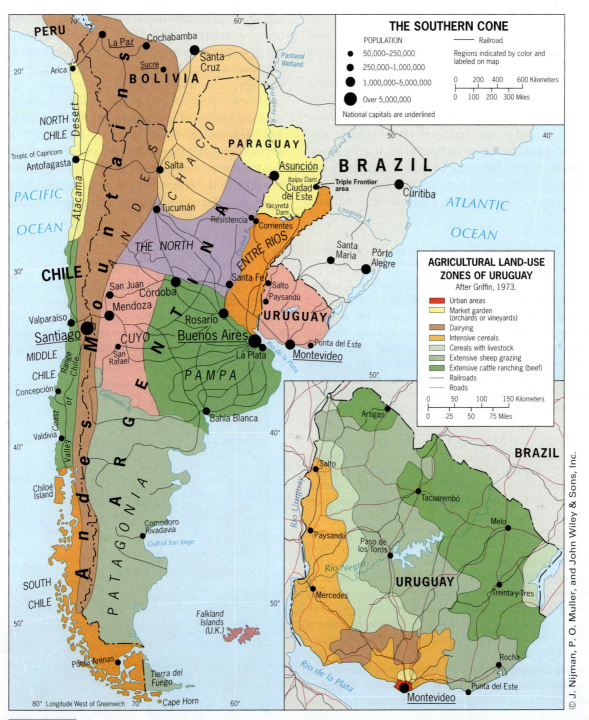

FIGURE 3-13

known as *Mesopotamia* because it lies between the Paraná and Uruguay rivers), and the **North**.

The Argentine Pampa is the product of the past 150 years. During the second half of the nineteenth century, when the great grasslands of the world were being opened up (including those of interior North America, Russia, and Australia), the economy of the long-dormant Pampa began to emerge. The food needs of industrializing Europe grew by leaps and bounds, and the advances of the Industrial Revolution—railroads, more efficient ocean transport, refrigerated ships, and agricultural machinery—helped to make large-scale commercial meat and grain production in the Pampa not only feasible but also highly profitable. Large haciendas were laid out and farmed by tenant workers; railroads were built and radiated ever farther outward from the booming capital of Buenos Aires, soon bringing the entire Pampa into production.

Argentina once was one of the richest countries in the world. Its historic affluence is still reflected in its architecturally splendid cities whose plazas and avenues are flanked by ornate public buildings and private mansions. This is true not only of the capital, Buenos Aires, at the head of the Rio de la Plata estuary—it also applies to interior cities such as Mendoza and Córdoba. The cultural imprint is dominantly Spanish, but the cultural landscape was diversified by a massive influx of Italians and smaller but influential numbers of British, French, German, and Lebanese immigrants. This has also long been one of the realm's most urbanized countries: 92 percent of its population is concentrated in cities and towns, a higher percentage even than western Europe or the United States. Fully one-third of all Argentinians live in metropolitan Buenos Aires, by far the leading industrial complex (**Box 3-10**).

But political infighting and economic mismanagement have combined over the long term to shackle a vibrant and varied economy. Part of the problem, it appears, lies in the country's lopsided geography with an enormous gap between a few highly populated and urbanized provinces that hold almost all the political power and a large number of provinces with very small populations (**Fig. 3-14**). Buenos Aires Province has 17 million people (out of a total 44 million), whereas Tierra del Fuego has barely more than 150,000. A dozen additional provinces contain less than 750,000 inhabitants, so that the larger ones besides dominant Buenos Aires are also disproportionately influential in domestic politics, especially Córdoba and Santa Fe. Corruption and mismanagement reached so high a level in 2012 that the authoritative international weekly, *The Economist*, decided it would no longer include statistics provided by the Argentine government; pointedly, the newsmagazine also announced that it did not want to be part of the

---

Box 3-10 Among the Realm's Great Cities: Buenos Aires

Its name means "fair winds," which first attracted European mariners to the site of Buenos Aires alongside the broad estuary of the muddy Rio de la Plata. The shipping function has remained paramount, and to this day the city's residents are known throughout Argentina as the *porteños* (the "port dwellers"). Modern Buenos Aires was built on the back of the nearby Pampa's grain and beef industry. It is often likened to Chicago and the Corn Belt in the United States because both cities have thrived as interfaces between their immensely productive agricultural hinterlands and the rest of the world.

Buenos Aires (14.3 million) is yet another classic South American **primate city [26]**, housing one-third of all Argentines, serving as the capital since 1880, and functioning as the country's economic core. Moreover, Buenos Aires is a cultural center of global standing, a monument-studded city that contains the world's widest street (*Avenida 9 de Julio*).

During the half-century between 1890 and 1940, the city was known as the "Paris of the South" for its architecture, fashion leadership, book publishing, and performing arts activities (it still has the world's biggest opera house, the newly renovated *Teatro Colón*). With the recent restoration of democracy, Buenos Aires is now trying to recapture its golden years. Besides reviving these cultural functions, the city has added a new one: the leading base of the hemisphere's motion picture and television industry for Spanish-speaking audiences.

During the past few years, the city has been showing signs of economic distress. Inflation in 2016 ran at nearly 33 percent annually and the cost of living skyrocketed. Economic growth has been slow and unemployment on the rise. The homeless became a common sight along the city's elegant boulevards. With a newly elected national government set on economic reforms, backed by the business community, and vowing to end corruption, there was hope in 2016 that the Argentine economy might finally turn the corner. If it does, it will first surface in the Buenos Aires metropolis.

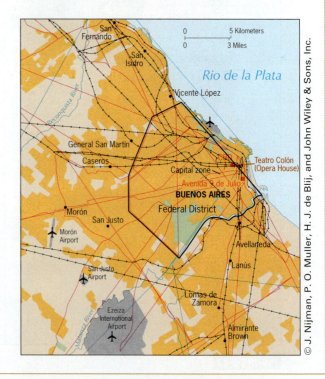

© J. Nijman, P. O. Muller, H. J. de Blij, and John Wiley & Sons, Inc.

government's "deliberate attempt to deceive voters and swindle investors"—a reflection of the country's sorry condition in this decade.

If one compares Argentina to Paraguay, it is striking how much better Argentina's position is in terms of natural endowments and geographic opportunities. But as things now stand, the Argentinean state is more reminiscent of chaotic Venezuela: things could be so much better than they presently are.

Chile

For over 4000 kilometers (2500 mi) between the crestline of the Andes and the coastline of the Pacific lies the narrow strip of land that is the Republic of Chile (**Fig. 3-16**). On average, just 150 kilometers (90 mi) wide (and only rarely over 250 kilometers [150 mi] wide), Chile is the world's quintessential example of what **elongation [28]** means to the functioning of a state. Accentuated by its north-south orientation, this severe territorial attenuation not only results in Chile extending across numerous environmental zones; it has also contributed to the country's external political, internal administrative, and general economic challenges. Nevertheless, throughout most of their modern history, the Chileans have made the best of this potentially disruptive centrifugal force: from the beginning, the sea has constituted an avenue of longitudinal communication; the Andes Mountains continue to form a barrier to prevent encroachment from the east; and when confrontations loomed at the far ends of the country, Chile proved to be quite capable of coping with aggressive neighbors.

As Figures 3-13 and 3-16 indicate, Chile is a three-subregion country. About 90 percent of its 18.1 million people are concentrated in what is called Middle Chile, where Santiago, the capital and largest city, and Valparaíso, the chief port, are located. North of Middle Chile lies the Atacama Desert, wider, drier, and colder than the coastal desert of Peru. To the south of Middle Chile, the coast is punctuated by a plethora of fjords and islands, the topography is mountainous, and the climate—wet and cool near the Pacific—soon turns drier and much colder against the Andean interior.

PROVINCES OF ARGENTINA

— - — International boundary
— - — Provincial boundary
• Location of provincial capital

Provincial population

■ Greater than 5 million
■ 3–5 million
■ 1–3 million
■ 500,000–1 million
■ Less than 500,000

0   200   400   600 Kilometers
0   100   200   300 Miles

FIGURE 3-14

The name Mendoza is synonymous with Argentinean wine. As our maps show, Mendoza is a province as well as a city in the Cuyo subregion of west-central Argentina, and both are massively involved in wine production at some of South America's biggest winemaking facilities. This has always been red-wine country, and its best-known varietals are Malbec and Cabernet Sauvignon. Even though Mendoza Province produces more than 80 percent of Argentina's wine, this still-expanding industry must contend with the area's semiarid environment and increasingly rely on irrigation networks to meet its growing water needs. This view is toward the west, whose horizon is filled entirely by the mighty wall of the Andes; beyond the mountain range lies central Chile, another South American producer of world-class wines.

Not surprisingly, the land in Middle Chile is the most fertile and valuable, and is marked by some important wine-growing regions (see **Box 3-11**). In contrast, there is hardly any agriculture to be found in either the North or South (Fig. 3-4). The three subregions are also apparent on the realm's cultural map, with Europeans dominating the Middle, mestizos prevailing in the North, and indigenous groups forming the ethnic majority in the South (Fig. 3-3).

Nevertheless, prior to the 1990s, the arid Atacama in the North accounted for more than half of Chile's foreign revenues. It contains the world's largest exploitable deposits of nitrates, which was the country's economic mainstay before the discovery of methods of synthetic nitrate production in the twentieth century. Subsequently, copper became the dominant export (Chile again possesses the world's largest reserves, which now accounts for more than half of all export revenues). It is mined in several places, but the main concentration lies on the eastern margin of the Atacama near Chuquicamata, not far from the port of Antofagasta. The Atacama Desert also contains lithium, a rare mineral rapidly gaining strategic value because it is a key component in today's electric batteries (see photo).

Chile has in recent decades embarked on a highly successful program of free-market economic reform that brought stable growth, lowered inflation as well as unemployment, reduced poverty, and attracted foreign investment. The last is of particular significance because these new international connections enabled the export-led Chilean economy to diversify and develop in some badly needed new directions. Copper remains at the top of the export list, but many other mining ventures have been launched. In the agricultural sphere, fruit and vegetable production for export has soared. Industrial expansion is taking place as well, though at a more leisurely pace, and new

Box 3-11　MAP ANALYSIS

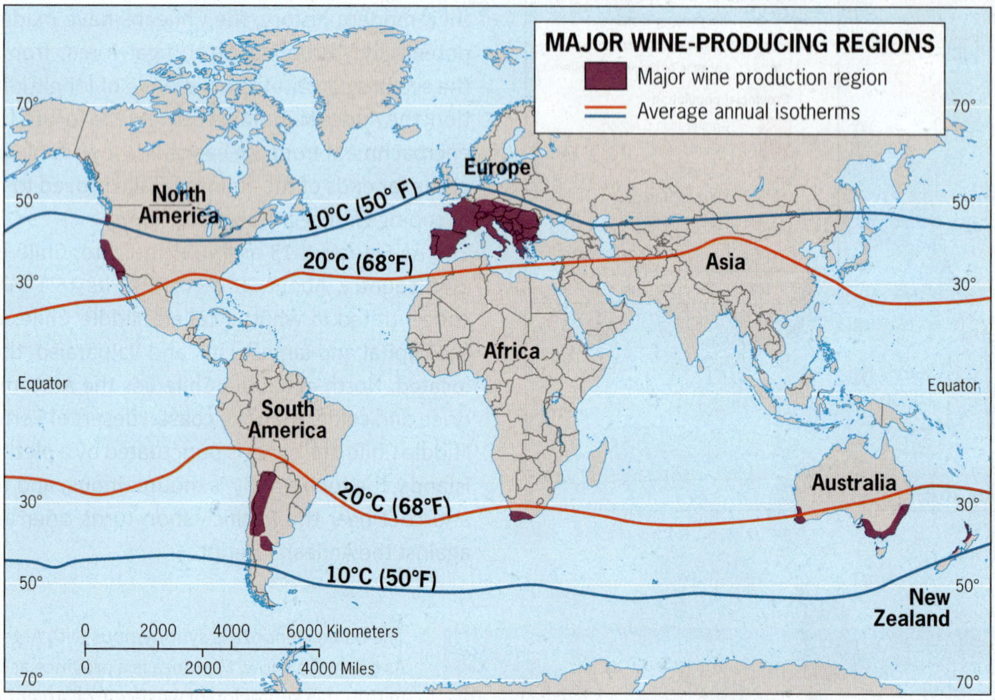

FIGURE 3-15

© J. Nijman, P. O. Muller, and John Wiley & Sons, Inc.

Viticulture [27], or wine growing, in Argentina, Chile, and everywhere else is very much a function of geography and climate (at a finer scale, soil and slopes matter a great deal too). And climatic conditions, especially temperatures, are partially dependent on latitude. **Figure 3-15** shows the location of the world's major wine-producing regions. Note the importance of the upper and lower temperature limits as shown on the map. Which latitudinal zones contain almost all of the world's wine regions? But wine production does not occur everywhere within these latitudinal zones. Why not? Consult Figure G-6 in the Introduction (which displays the global geography of climate types), and explain in greater detail.

Access your WileyPLUS Learning Space course to interact with a dynamic version of this map and to engage with online map exercises and questions.

IVAN ALVARADO/Newscom

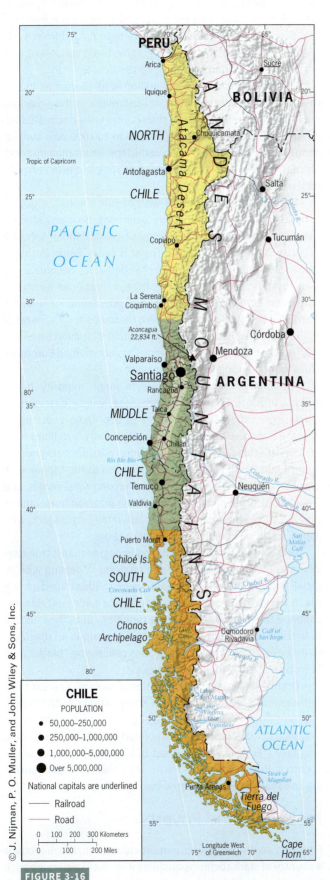

Lithium is a small but critical component of batteries—all batteries. It is a scarce commodity, and the world market price has been going up steadily along with demand. It is quickly turning into what some call a strategic natural resource, even though the overall value (due to the tiny quantities involved) of lithium trade is still quite modest. Bolivia, Chile, and Argentina together contain nearly 60 percent of all known world reserves. This photo was taken in the Atacama Desert of northern Chile, looking east toward the Andes. Lithium is made from the brine (water saturated with salt) obtained from pools in the surrounding salt flats. This brine is pumped into rectangular ponds that, as the water evaporates, turn into different shades of blue, yellow, and green, like a giant artist's palette. Once the brine is sufficiently concentrated, it is collected and transported to factories in which various potassium compounds and the prized lithium powder are extracted.

manufactures have included an array of goods that range from basic chemicals to computer software.

Chile's increasingly globalized economy has propelled the country into a prominent role on the international scene. Its regional commerce has also been growing steadily in conjunction with Chile's new role as a full member of the recently formed Pacific Alliance, which now seems to be outpacing Mercosur/l.

Uruguay

Uruguay, in comparison with Argentina or Chile, is compact, small, and rather densely populated. It is about the size of Florida but with one-sixth the population. This **buffer state [29]** of the early independence era, lying between (then) potentially hostile Argentina and Brazil, has evolved into a fairly prosperous agricultural country—in effect, a smaller-scale Pampa, though possessing less favorable soils and topography. Montevideo, the coastal capital, contains half of the country's population of 3.4 million; from here, railroads and roads radiate outward into the productive agricultural hinterland (Fig. 3-13). In the immediate vicinity of Montevideo lies Uruguay's major farming zone, producing vegetables and fruits for the metropolis as well as wheat and fodder crops; most of the rest of the country is used for grazing cattle and sheep, with beef products, wool and textile manufactures, and hides dominating the export trade (Fig. 3-13, inset map). Tourism is another major economic

FIGURE 3-16

© J. Nijman, P. O. Muller, and John Wiley & Sons, Inc.

activity as Argentines, Brazilians, and other visitors flock to the Atlantic beaches at Punta del Este and other thriving seaside resorts. In terms of per capita income, Uruguay has long been one the realm's better performing countries.

Region Brazil: Giant of South America

By every measure, Brazil is South America's giant. It is so large that it has common boundaries with all the realm's other countries except Ecuador and Chile (Fig. 3-8). Its tropical and subtropical environments range from the equatorial rainforest of the Amazon Basin to the humid-temperate climate of the far south. Territorially as well as in terms of population, Brazil ranks fifth in the world, and in both it accounts for just about half of this entire realm.

The emergence of Brazil as the regional superpower of South America and an economic powerhouse in the world at large was one of the main stories of the early twenty-first century. Known largely for its coffee and other farm products in the past, by 2010 Brazil had become a leading manufacturer of machinery, chemicals, high-tech goods, and aircraft. For about ten years, Brazil's economic performance ranked it among the most productive emerging-market countries on Earth. In the process, it became the world's sixth-largest national economy (before it slipped back to ninth in 2016).

Brazil's rapid post-2000 rise can be attributed to two main factors. First, after a long period of dictatorial rule by a minority elite that used the military to stay in power, Brazil embraced democratic government in 1989 and did not look back. The era of military coups and repeated crushing of civil liberties was over. Second, with its vast storehouse of natural resources, Brazil benefited enormously from increased demands for commodities in the world market, mainly from China and India. In some ways, Brazil's political and economic turnaround ran parallel to that of Chile, but because of its sheer size Brazil matters more to the rest of the realm and, indeed, the world.

However, Brazil also epitomizes the excessive dependence of this realm on raw materials, and its susceptibility to the vagaries of the world market. When, around 2012, the demand for commodities began to slow and prices came down, the Brazilian economy felt the impact almost immediately. By 2013, economic growth had fizzled out, and in 2015 GDP growth was the lowest in 25 years (negative 3.8 percent), with a similar decline projected for 2016. Moreover, for the first time in many years, major political scandals surfaced and had a paralyzing effect on both the government and the recession-plagued economy. Thus the positive hype of the 2000s has given way to talk of continuing crisis.

Population and Culture

Brazil's population grew rapidly during the world's twentieth-century population explosion. But over the past three decades, the rate of natural increase has plunged from nearly 3.0 percent to 0.8 percent today, and the average number of children born to a Brazilian woman has been more than halved from 4.5 in 1975 to 1.8 in 2016. It is a demographic trend consistent with Brazil's overall modernization since 1980.

Brazil's population of 210 million is as diverse as that of the United States. In a pattern quite familiar in the Americas, the indigenous inhabitants of the country were decimated following the European invasion (less than 700,000 now survive, about two-thirds of them deep within the Amazonian interior). Africans came in very large numbers as well, and they currently total more than 15 million.

Brazil's culture is infused with African themes, a quality that has marked it from the very beginning. So many Africans were brought in bondage to the city and hinterland of Salvador in Bahia State (**Fig. 3-17**) from what is today Benin (formerly named Dahomey) in West Africa that Bahia has become a veritable outpost of African culture. Indeed, Brazil can be said to have the second-largest black (African) population in the world, after Nigeria. Significantly, however, there was also much racial mixing, and the 2010 census reported that 97 million Brazilians (50.7 percent of the total population) have combined European, African, and minor indigenous ancestries.

Yet another significant, though small, minority began arriving in Brazil in 1908: the Japanese, whose descendants remain concentrated in the States of São Paulo and Paraná. These Japanese-Brazilians (numbering nearly 3 million) form the largest ethnic Japanese community outside Japan, and in their multicultural environment have risen to the top ranks of Brazilian society as business leaders, urban professionals, and commercial farmers. Committed to their Brazilian homeland as they are, the Japanese community also retains its contacts with Japan, resulting in many a trade connection.

Brazilian society, to a much greater degree than is true elsewhere in the Americas, has made progress in dealing with its racial divisions. Blacks do remain the least advantaged among the country's major population groups, and community leaders continue to complain about discrimination. But ethnic mixing in Brazil is so pervasive that hardly any group is unaffected—so in the end, official census statistics concerning "blacks" and "Europeans" are rather meaningless.

Brazilians have a strong national culture, expressed in a traditional adherence to the Roman Catholic faith (though declining under Protestant-evangelical and secular pressures), universal use of a modified form of Portuguese as the common language ("Brazilian"); and a set of lifestyles in which soccer, "beach culture," distinctive music and dance, and an intensifying national pride are fundamental ingredients.

Inequality and Poverty

For all its accomplishments in multiculturalism, Brazil remains a country of stark social inequalities (**Fig. 3-18**). Although such inequality is hard to measure precisely, South America is frequently cited as the geographic realm exhibiting the world's widest gap between affluence and poverty (Box 3-7). And within

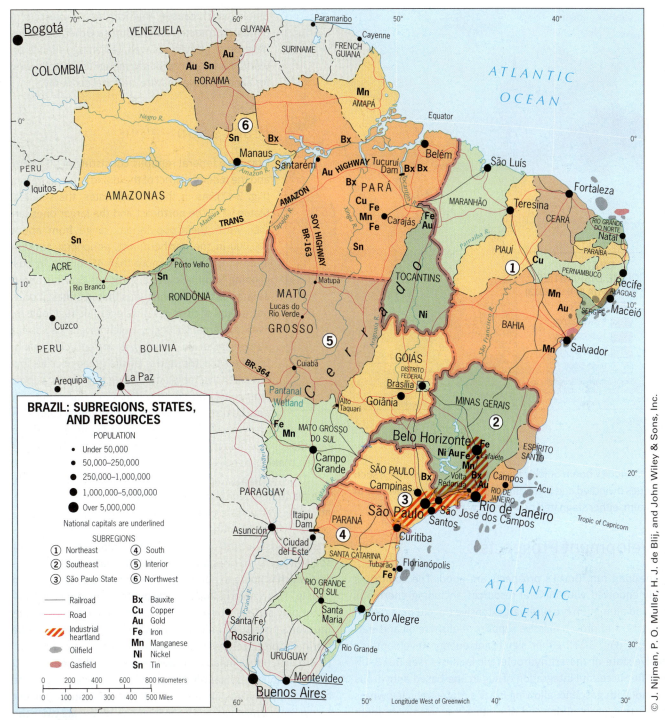

FIGURE 3-17

South America, Brazil is reputed to have one of the most voluminous gaps of all.

At the start of this century, the Brazilian government implemented a set of policies aimed at bringing relief to the poor while maintaining robust economic growth. It enacted land reform and increased access to education for Brazil's masses. It has also launched a subsidy program that has had significant results in the poorest States. This *Bolsa Familia* (Family Fund) plan, instituted under former President Fernando Cardoso in the 1990s, provides families with small

payments of cash to keep their children in school and ensure their vaccinations against diseases that especially afflict the poor. In just a few years this program proved to be so successful that it became a model for antipoverty campaigns the world over.

Regional inequalities were being targeted as well by government policies. Major new legislation was passed in 2012 to allow revenues from oil to be shared by all of Brazil's States (previously this income was mainly channeled to the three oil-producing States of Rio de Janeiro, Espirito Santo, and São Paulo). But, with

States (which caused American steel producers to demand protectionist measures). Other industries have been growing as well, lessening reliance on world market prices for raw materials, and allowing for a more stable, mature, national economy.

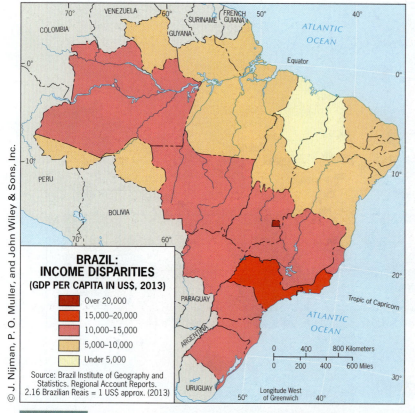

BRAZIL: INCOME DISPARITIES
(GDP PER CAPITA IN US$, 2013)

- Over 20,000
- 15,000–20,000
- 10,000–15,000
- 5,000–10,000
- Under 5,000

Source: Brazil Institute of Geography and Statistics. Regional Account Reports. 2.16 Brazilian Reais = 1 US$ approx. (2013)

FIGURE 3-18

© J. Nijman, P. O. Muller, and John Wiley & Sons, Inc.

the commodity boom quickly becoming a distant memory, such revenues dwindled and inequality in Brazil, as most elsewhere in the realm, either stagnated or resumed widening.

Development Prospects

In the long run, Brazil's prospects are good. The country is richly endowed with mineral resources, including enormous iron and aluminum ore reserves, extensive tin and manganese deposits, and proven massive supplies of offshore oil and natural gas (Fig. 3-17). Other significant energy developments involve state-of-the-art hydroelectric facilities (see Box 3-13) and the successful integration of sugarcane-based anhydrous ethanol with gasoline, allowing Brazil's motor-vehicle owners to use this 'gasohol' instead of costlier imported petroleum (all Brazilian cars today are configured to use this biofuel, whose ethanol content must be at least 18 percent). Besides these endowments, Brazil has more arable land than any other country on Earth. Brazilian soils sustain a bountiful agricultural output that makes the country the world's leading exporter of coffee, orange juice, sugar, tobacco, ethanol, beef, and chicken.

Industrialization has driven Brazil's ascendancy as a global economic power. As a result, annual revenues from industrial exports surpassed those from agriculture more than a decade ago. The relatively low wages of its workers and the mechanized efficiency of its steelmakers enable Brazil to produce that commodity at half the cost of steel made in the United

Brazil's Subregions

Brazil is a federal republic consisting of 26 States and the federal district of the inland capital, Brasília (Fig. 3-17). As in the United States, the smallest States lie in the northeast and the larger ones farther west; their populations range from about 500,000 in the northernmost, peripheral Amazon State of Roraima to more than 45 million in burgeoning São Paulo State. Although Brazil is about as large as the 48 contiguous United States, it does not exhibit a clear-cut physiographic regionalism. Even the Amazon Basin, which covers just about 60 percent of the country, is not entirely a plain: between the tributaries of the great river lie low but extensive tablelands. Given this physiographic ambiguity, the six Brazilian subregions that will now be discussed exhibit no absolute or even generally accepted boundaries. In Figure 3-17, those boundaries have been drawn to coincide with the borders of States, making identifications easier.

The **Northeast** was Brazil's source area, its culture hearth. The plantation economy was establish here at an early date, attracting Portuguese planters, who soon began to import the country's largest contingent of African slaves to work in the sugarcane fields. But the ample rainfall occurring along the coast soon gives way to drier and more variable patterns in the interior, which is home to about half of the region's 57-plus million people. This semiarid inland backcountry—called the *sertão*— is not only seriously overpopulated but also contains some of the worst poverty to be found anywhere in the Americas. The Northeast produces less than one-sixth of Brazil's gross domestic product, but its inhabitants constitute more than one-fourth of the national population. Moreover, this subregion contains half of the country's poor and exhibits a literacy rate 20 percent below Brazil's mean as well as an infant mortality rate twice the national average.

Much of the Northeast's problems are rooted in its unequal system of land tenure. Farms must be at least 100 hectares (250 acres) to be profitable in the hard-scrabble *sertão*, a size that only large landowners are able to afford. Moreover, the Northeast is tormented by a monumental environmental problem: the cyclical recurrence of devastating droughts at least partly attributable to **El Niño** (periodic events of sea-surface warming off the continent's northwestern coast that skew regional weather patterns).

The Northeast is finally receiving greater attention from the central government, largely in the form of federally funded infrastructure projects and incentive-driven investment (think oil money). In cities such as Recife and Salvador, hordes of peasants driven from the land constantly arrive to expand the surrounding shantytowns. As yet, few of the generalizations about emerging Brazil apply here, but there are some bright spots. A petrochemical complex has been built near Salvador, creating thousands of jobs and luring foreign investors. Irrigation projects have nurtured a number of productive new commercial agricultural ventures. Tourism is booming along the entire Northeast coast, whose thriving beachside resorts attract tens of thousands of vacationing Europeans (flight times are under 8 hours). Recife has now spawned a budding software industry and a major medical complex. And Fortaleza is the center of new clothing and shoe industries that have already put the city on the global economic map.

The **Southeast** has been modern Brazil's *core area*, with its major cities and leading population clusters. Gold first drew many thousands of settlers, and other mineral finds also contributed to the influx—with Rio de Janeiro serving as the terminus of the "Gold Trail" and then as the longtime capital of Brazil

until 1960. "Rio" became the cultural capital as well, the country's most internationalized city, *entrepôt* [30], and tourist hub (**Box 3-12**). The third quarter of the twentieth century brought another mineral age to the Southeast, based on the iron ores around Lafaiete carried to the nearby steelmaking complex at Volta Redonda (Fig. 3-17). The surrounding State of Minas Gerais (the name means 'General Mines') formed the base from which industrial diversification in the Southeast has steadily mushroomed. The burgeoning metallurgical center of Belo Horizonte paved the way and is now the endpoint of a rapidly developing, ultramodern manufacturing corridor that stretches 500 kilometers (300 mi) southwest to metropolitan São Paulo (Fig. 3-17, striped zone).

São Paulo State is both the leading industrial producer and the primary focus of ongoing Brazilian development. This economic-geographic powerhouse accounts for nearly half of the country's GDP, with a thriving economy that today matches Argentina's in overall size. Not surprisingly, this subregion is growing phenomenally (it contains more than 20 percent of

---

Box 3-12 Among the Realm's Great Cities: Rio de Janeiro

Nicknamed the "magnificent city" because of its breathtaking natural setting, Rio replaced Salvador as Brazil's capital in 1763 and held that position for almost two centuries until the federal government shifted its headquarters to interior Brasília in 1960. Rio de Janeiro's primacy suffered yet another blow in the late 1950s: São Paulo, its urban rival 400 kilometers (250 mi) to the southwest, surpassed Rio to become Brazil's largest city—a lead that has been widening ever since. Although these events triggered economic decline, Rio (11.8 million) remains a major business center, air-travel and tourist hub, global sports mecca, and leading cultural center with its entertainment industries, universities, museums, and libraries.

On the darker side, this city's reputation is increasingly tarnished by the widening abyss between Rio's affluent and poor populations—symbolizing inequities that rank among the world's most devastating. All great cities experience problems, and Rio de Janeiro has for years been bedeviled by the drug use and crime waves emanating from its most desperate hillside *favelas* (slums) that continue to grow explosively.

Rio's planners recently launched a multifaceted project (known as "Rio-City") to improve urban life for all residents. This ambitious scheme, it is said, aims to reshape nearly two dozen of the aging city's neighborhoods, introduce an ultramodern crosstown expressway to relieve nightmarish traffic congestion, and—most importantly—bring electrical power, paved streets, and a sewage-disposal system to the beleaguered *favelas*.

Much of this was related to Rio's hosting of the soccer World Cup tournament in 2014 followed by the Summer Olympics in 2016, which motivated local and federal authorities to do all they could to showcase their city in the best possible light. When the Olympics were awarded to Brazil in 2009, the country was rapidly moving up in the world and the timing seemed perfect. In mid-2016, however, the economy was in a near state of crisis, the suspended president faced impeachment over corruption charges, and alarms

were being sounded as to whether all the Olympic venues across Rio would be up to the task. On top of all that, Brazil was in the midst of a public health emergency due to the outbreak of the mosquito-borne *Zika virus*, raising concerns about the safety of athletes and spectators at the Olympics—and testing Brazil's and Rio's resilience to their extremes.

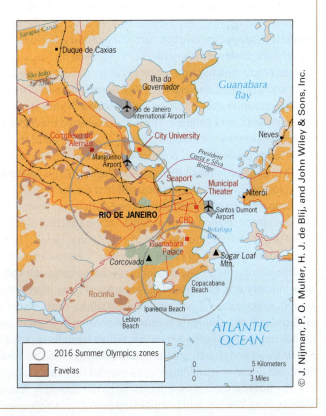

© J. Nijman, P. O. Muller, H. J. de Blij, and John Wiley & Sons, Inc.

Brazil's population) as a magnet for upwardly mobile migrants, especially from the Northeast.

The wealth of São Paulo State was built on its coffee plantations (known as *fazendas*), and Brazil is still the world's leading producer. But coffee today has been eclipsed by other farm commodities. One of them is orange juice concentrate (here, too, Brazil leads the world). São Paulo State now produces more than double the annual output of Florida, thanks to a climate all but devoid of winter freezes, ultramodern processing plants, and a fleet of specially equipped tankers that ship the concentrate to markets around the globe. Another major crop is soybeans, contributing to Brazil's ranking today as the world's number-two producer.

Matching this agricultural prowess is the State's industrial strength. The revenues derived from the coffee plantations provided the necessary investment capital, ores from Minas Gerais supplied the vital raw materials, São Paulo City's outer port of Santos facilitated access to the ocean, and immigration flows from Europe, Japan, and other parts of Brazil contributed the increasingly skilled labor force. As the capacity of the expanding domestic market grew, the advantages of central location and large-scale agglomeration nailed down São Paulo's primacy. This also resulted in metropolitan São Paulo becoming the country's—and South America's—leading industrial complex and megacity. But this rapid progress was accompanied by massive problems of overcrowding, pollution, and congestion, which now severely challenge the sustainability of this gigantic urban dynamo (see Box 3-6).

The **South** consists of three States, whose combined population totals almost 30 million: Paraná, Santa Catarina, and Rio Grande do Sul (Fig. 3-17). Southernmost Brazil's exceptional agricultural potential has long attracted a steady flow of European immigrants. They introduced their advanced farming methods to several areas in this part of the country. Portuguese rice farmers clustered in the valleys of Rio Grande do Sul, where tobacco production has now propelled Brazil to become, by far, the world's number-one exporter. The Germans, specialists in raising grain and cattle, occupied the somewhat higher areas to the north and Santa Catarina. The Italians selected the highest slopes, where they established thriving vineyards. All of these fertile lands proved highly productive, and with growing markets in the mushrooming urban areas to the north, this tri-State subregion has become Brazil's most affluent corner.

Economic development within the South is not limited to the agricultural sector. Coal from Santa Catarina and Rio Grande do Sul is shipped north to the steel plants of Minas Gerais. Local manufacturing is growing as well, especially in Pôrto Alegre and Tubarão. During the 1990s, a major center of the computer software industry was established in Florianópolis, the island-city as well as State capital just off Santa Catarina's coast. Known as *Tecnópolis*, this budding technopole continues to grow by capitalizing on its seaside amenities, skilled labor force, superior air-travel and global communications linkages, and government and private-sector initiatives to support new companies.

The **Interior** subregion—constituted by the sprawling States of Goiás, Mato Grosso, and Mato Grosso do Sul—is also known as the *Central-West*. This is the subregion that Brazil's developers long sought to make a part of the country's productive heartland, and in 1960 the new capital of Brasília was deliberately situated on its margins (Fig. 3-17).

By locating the new capital city in the wilderness 650 kilometers (400 mi) inland from its predecessor, Rio de Janeiro, the nation's leaders dramatically signaled the opening of Brazil's development thrust toward the west. Brasília is noteworthy in another regard because it represents what political geographers call a **forward capital [31]**. A state will sometimes relocate its capital to a sensitive area, perhaps near a peripheral zone under dispute with an unfriendly neighbor, in part to confirm its determination to sustain its position in that contested zone. Brasília does not lie close to a contested area, but Brazil's interior was an internal frontier to be conquered by a growing nation. Spearheading that drive, the newly built capital occupied a decidedly forward position.

Despite the subsequent growth of Brasília to 2.6 million inhabitants today (which includes a sizeable ring of peripheral squatter settlements), it was not until the 1990s that the Interior subregion began its economic integration with the rest of Brazil. The catalyst was the exploitation of the vast **cerrado [32]**—the fertile savannas that blanket the Central-West and make it one of the world's most promising agricultural frontiers (at least two-thirds of its arable land still awaits development). As with the U.S. Great Plains, the flat terrain of the *cerrado* is one of its main advantages because it facilitates the large-scale mechanization of farming with a minimal labor force. Another benefit is rainfall, more prevalent here than in the Great Plains or Argentine Pampa. The increasingly dominant local crop is soybeans (see pink zone in Fig. 3-4), and today only the United States produces more than Brazil.

The **Northwest** is Brazil's territorially largest and most rapidly developing subregion, which consists of the seven States of the Amazon Basin (Fig. 3-17). This was the scene of the great rubber boom of a century ago, when the wild rubber trees in the *selvas* (tropical rainforests) produced huge profits and the central Amazon city of Manaus enjoyed a brief period of wealth and splendor. But the rubber boom had ended by 1920, and for most of the six decades that followed, Amazonia was a stagnant hinterland lying remote from the centers of Brazilian settlement. All that changed quite dramatically during the 1980s as new development began to stir throughout this awakening subregion (**Fig. 3-19**), which continues to be the scene of the world's largest migration into virgin territory as more than 200,000 new settlers arrive each year.

Brazil's Northwest is also widely known for its high rates of deforestation. Between 1995 and 2005, an area of 20,000 square kilometers (7700 sq mi) was cleared away every year. Some believe that the worst might be over because the Brazilian government has pledged to achieve an 80 percent reduction in deforestation by 2020 and to terminate the practice by 2030. At least part of its thinking is that Brazil has enormous potential as a "green economy" with its abundance of land, water, and sunshine. Today, nearly half of the country's energy is already obtained from renewable sources, mostly biofuels

Box 3-13 Regional Planning Case

The Madeira River Dams

State-of-the-art hydroelectric facilities are at the heart of South America's regional economic development planning, most notably in the Amazon Basin. A number of such dams are planned for the Madeira River, the mighty Amazon's largest single tributary (Fig. 3-19). The Madeira River Complex is the single largest project within a massive development plan known as the Integrated Regional Infrastructure for South America (IIRSA), a Brazil-led initiative launched in 2000 with the participation of every South American country except French Guiana.

Two of the four largest hydroelectric dams on the Madeira—the Jirau and the Santo Antônio—lie in Brazil's Rondônia State, not far from the Bolivian border. The third Madeira dam is scheduled to be constructed right on that border, and the fourth will be built further upstream inside Bolivia. Both of the latter two are among the more than 30 large dams being planned for the 2020s as replacements for high-cost, older-generation thermal-electric, gas, and oil-fired plants.

The projects are driven by the urgent need to ease electricity shortages throughout Brazil. Bolivia would also benefit from the power generated at the dams within its territory as well as the improved navigability of the Madeira (whose highly variable water levels will finally be controlled by locks alongside each dam).

Although these dam projects underscore the enormous range of resource-based development opportunities available to Brazil, they have also sparked widespread protests from indigenous peoples, environmental activists, advocacy organizations, and local citizenries. These groups are most concerned about the undesirable impacts or **negative externalities [33]** of dam construction in Amazonia. Undoubtedly, the proliferation of dam-building will further fuel the transformation of the remaining rainforest, exacerbating longstanding crises of rampant deforestation, uncontrolled environmental degradation, and callous displacement of local populations.

Calculating the true price of the Northwest's energy development schemes must not only take into account the financial bottom line, but also their environmental and social costs. This is particularly important in emerging market countries like Brazil, which struggle to keep up with the accelerating demand for electrical power propelled by population growth and economic development. In Brazil's case, that also requires maintaining the ecological integrity of the unsurpassed biodiversity of the Amazon Basin.

© Pulsar Images/Alamy Stock Photo

The Brazilian city of Pôrto Velho, with the Madeira River and its new Santo Antônio Dam nearing completion in the background. The river flows from left to right, and empties into the Amazon near Manaus, about 1000 kilometers (625 mi) to the northeast (Fig. 3-19). Located deep inside the Amazon Basin, Pôrto Velho is now westernmost Brazil's largest city. Its population has doubled since 2000, and today approaches 400,000. Dam construction here began in 2008, and this generating facility is expected to be fully operational by early 2017.

(vs. 13 percent in the United States), and it is becoming easier to envision Brazil as a pacesetting environmental power. Still, it is hard to imagine that the government's ambitious goals will be achieved anywhere near 2020 and 2030, respectively—not least because at the same time it has embarked on major infrastructural projects deep in the Amazon Basin (**Box 3-13**, Fig. 3-19).

Development projects abound in the Amazon Basin today (Fig. 3-19, Box 3-13). One of the most durable is the ***Grande Carajás Project*** in central Pará State, a huge multifaceted scheme centered on the world's largest known deposit of iron ore in the hills around Carajás (Fig. 3-17; see aerial photo in the "Economic Geography" section). Besides its vast iron-mining complex, other

Ian Trower/Getty Images, Inc.

The Ponte Rio Negro, iconic symbol of the ongoing development thrust into South America's interior, with the Manaus riverfront in the foreground. This first major bridge in the Amazon Basin and Brazil's longest, at 3.5 kilometers or about 2 miles, was completed in 2011 and spans the mighty Rio Negro just upstream from its confluence with the even greater Amazon itself. Not surprisingly, this most impressive river junction on the South American continent spawned the central Amazon's largest city, Manaus, well over a century ago. Hemmed in until recently by dense rainforests and these enormous rivers, Manaus's new bridge connects the city to its satellite, Iranduba. The latter is now booming as a result of its much-upgraded accessibility and serves as the jumping-off point on the west side of the Rio Negro, its own suburbs spearheading the outward thrust of metropolitan Manaus into an entirely new development sector.

THE AMAZON BASIN

- ▭ Amazon River basin
- ▨ Rainforest
- ▨ Indigenous groups
- ─── Major road
- ─── Railroad
- ═══ Proposed railroad

HYDROELECTRIC DAMS
- ▪ Planned or inventoried
- ▫ Under construction
- ▪ In operation

OIL PRODUCTION
- ▨ Leased areas
- ▨ Exploration proposed

FIGURE 3-19

new construction in the vicinity includes the Tucuruí Dam on the nearby Tocantins River and an 850-kilometer (535-mi) railroad to the Atlantic port of São Luis. This ambitious development project also emphasizes the exploitation of other minerals, cattle raising, crop farming, and forestry. What is taking place here is a manifestation of the **growth-pole concept [34]**. A growth pole is a location at which a set of activities, given a start, will expand and generate widening ripples of development in the surrounding area. According to this scenario, the stimulated hinterland could one day cover one-sixth of all Amazonia.

Understandably, hordes of settlers have descended on this part of the Amazon Basin. Those seeking business opportunities have been in the vanguard, but they have been followed by masses of lower-income laborers and peasant farmers in search of jobs and land ownership. The initial stage of this

colossal enterprise has boosted the fortunes of many urban centers, most of all Manaus northwest of Carajás. Here, a thriving industrial complex specializing in the production of electronic goods has emerged within the free-trade zone adjoining the city thanks to the outstanding air-freight facilities and operations at Manaus's ultramodern airport. At the same time, in and around Manaus, a new era of urban expansion has been unleashed by the opening of the spectacular Rio Negro bridge at the beginning of this decade (see photo).

But many problems have also surfaced as the tide of pioneers has rolled across central Amazonia. One of the most tragic involved the Yanomami people, whose homeland in Roraima State was overrun by thousands of claim-stakers (in search of newly discovered gold), who triggered violent confrontations that ravaged the fragile aboriginal way of life.

Future Prospects of South America

South American countries have a long history of intermittent political turmoil. Dictatorial regimes ruled from one end of the realm to the other; unstable governments fell with damaging frequency. Widespread poverty, harsh regional disparities, inferior overland connections, limited international contact, and economic stagnation long prevailed.

Since the turn of our century, much of that has been cast aside as most of South America has entered a new era of transition, political stability, and economic progress. Well into the second decade of the twenty-first century, democratic governance have taken hold almost everywhere. Long-isolated countries are becoming better connected through new transportation routes and trade agreements. Dormant settlement frontiers are finally being opened. Energy resources, some long exploited and others newly discovered, are boosting national economies. Representatives of foreign states and corporations have appeared throughout the realm to buy commodities and invest in infrastructure. The pace of globalization has accelerated from Bogotá to Buenos Aires.

But economic growth has slowed in the mid-2010s, in no small part due to the slowing demand for raw materials in the world economy (especially in China). In response, certain national economies, particularly Brazil's, have stagnated. South America's immediate future will continue to partially depend on the global economy, but also on the political choices of its national governments. Venezuela's appalling crisis, for instance, only seems to be reversible through the forging of bold new political directions. Argentina's newly elected government (in 2015) is charged with ending years of mismanagement in order to capitalize on the country's major geographic resources. Brazil needs to weather its political storm over corruption scandals and embark on more balanced strategies of economic development and environmental stewardship. This realm's rich endowments of natural and human resources surely create substantial opportunities for long-term prosperity.

Points to Ponder

- The distance from Bogotá, Colombia to Chile's southernmost Cape Horn (6600 km/4100 mi) is greater than the distance from Paris, France to Delhi, India.

- Even though the Amazon rainforest lies almost entirely within Brazil, it is often referred to as "the lungs of the Earth." Should its preservation be a global responsibility?

- Is it a coincidence that both liberation theology and *dependencia* theory originated in this realm?

- Bolivia, Peru, and Chile together contain nearly 60 percent of the world's known reserves of lithium, a necessary component of electrical batteries of every stripe.

FIGURE 4-1

CHAPTER 4

© Jan Nijman

© Jan Nijman

In which country were these pictures taken?

The European Realm

IN THIS CHAPTER

- Europe's refugee crisis
- The specter of Islamist terrorism in Europe
- Brexit: The United Kingdom votes to leave the EU
- Anti-EU sentiments in Europe's heartland
- The question of Europe's eastern boundary
- Crisis in Ukraine: A conflict in geographic context

CONCEPTS, IDEAS, AND TERMS

Transition zone	**[1]**	Central business district (CBD)	**[14]**	Situation	**[27]**
Geographic information system (GIS)	**[2]**	Centrifugal forces	**[15]**	Estuary	**[28]**
Digital elevation model	**[3]**	Centripetal forces	**[16]**	Conurbation	**[29]**
Land hemisphere	**[4]**	Supranationalism	**[17]**	Landlocked location	**[30]**
City-state	**[5]**	Euro zone	**[18]**	World-city	**[31]**
Local functional specialization	**[6]**	Schengen Area	**[19]**	Metropolis	**[32]**
Industrial Revolution	**[7]**	Four Motors of Europe	**[20]**	Break-of-bulk	**[33]**
Sovereignty	**[8]**	Devolution	**[21]**	*Entrepôt*	**[34]**
Nation-state	**[9]**	Asylum	**[22]**	Shatter belt	**[35]**
Nation	**[10]**	Microstate	**[23]**	Balkanization	**[36]**
Indo-European language family	**[11]**	Urban system	**[24]**	Irredentism	**[37]**
Complementarity	**[12]**	Primate city	**[25]**	Exclave	**[38]**
Transferability	**[13]**	Site	**[26]**		

145

A case can be made that, over the past five centuries, Europeans have influenced and changed the rest of the world more than the people of any other realm. For good or bad, much of the world would look quite different today if it had not been for Europe. Europe's colonial empires spanned the globe and transformed societies far and near. Millions of Europeans migrated from their homelands to the Old World as well as the New, changing (and sometimes nearly obliterating) traditional communities, and creating new societies from Australia to North America. Colonial power and economic incentive combined to propel the movement of millions of imperial subjects from their ancestral homes to distant lands: Africans to the Americas, Indians to Africa, Chinese to Southeast Asia. In agriculture, industry, politics, and other spheres, Europe generated revolutions—and then, through imperialism and colonialism, exported those revolutions across the world.

During the seven decades since World War II (1939–1945), however, and particularly over the past four, Europe has decidedly become a more "ordinary" realm. The days of empire are now long gone. National economies face the same global challenges as their counterparts elsewhere. And Europe no longer plays a leading role in the world economy, even though it continues to rank among the most highly developed geographic realms. If the claim can still be made today that Europe is "special," it is with an eye on the remarkable process of economic and political integration that has shaped the evolution of the *European Union (EU)* since the 1950s. Although the European realm is highly diverse in a cultural-geographical sense, most of it is also tightly integrated, with considerable economic powers devolved from national governments to "Brussels," the EU headquarters city. This integration itself, it should be noted, varies across Europe as well—making for an especially complex geographic realm.

Defining the Realm

Europe's Eastern Boundary

As **Figure 4-1** shows, Europe is a realm of peninsulas and islands on the western margin of the world's largest landmass, Eurasia. It is a realm containing almost 600 million people and 40 countries, but it is territorially quite small. The European realm is bounded on the west, north, and south by Atlantic, Arctic, and Mediterranean waters, respectively (**Box 4-1**). But where is Europe's eastern limit? Each episode in the historical geography of eastern Europe has left its particular legacy in the cultural landscape.

Twenty centuries ago, the Roman Empire ruled much of eastern Europe (Romania is a cartographic reminder of that period); for most of the second half of the twentieth century,

the Soviet Empire controlled nearly all of it. During the intervening two millennia, Christian Orthodox church doctrines spread from the southeast, and Roman Catholicism advanced from the northwest. Turkish (Ottoman) Muslims invaded and created an empire that reached the vicinity of Vienna. By the time the Austro-Hungarian Empire ousted the Turks more than a century ago, millions of eastern Europeans had been converted to Islam (Albania and Kosovo still remain predominantly Muslim countries). Meanwhile, it is often said that western Europe's civilization had its cradle in ancient Greece, but that lies farther still to the southeast, beyond the former Yugoslavia.

As we shall see, the geographic extent of Europe has always been debatable. In this decade, it is a highly contentious issue in terms of European Union expansion and relations

---

Box 4-1 Major Geographic Features of Europe

1. The European geographic realm lies on the western flank of the Eurasian landmass.

2. Though territorially small, Europe is heavily populated and fragmented into 40 states.

3. European natural environments are highly varied, and Europe's resource base is rich and diverse.

4. Europe's geographic diversity, cultural as well as physical, created strong local identities, specializations, and opportunities for trade and commerce.

5. The European Union (EU) is a historic and unique effort to achieve multinational economic integration and, to a lesser degree, political coordination.

6. Europe's relatively prosperous population is highly urbanized and aging rapidly.

7. Local demands for greater autonomy as well as cultural challenges posed by immigration are straining the European social fabric.

8. Despite Europe's momentous unification efforts, east-west contrasts still mark the realm's regional geography.

9. The impacts of the refugee crisis that gathered momentum in the mid-2010s pose a threat to the EU and to political stability across the realm.

with Russia—as underscored by the turmoil that continues to plague eastern Ukraine. Europe's eastern boundary is a dynamic one, and it has changed with history. Some would say that there really is no clear boundary, that Europe's ambience, so to speak, just thins out toward the east. Our present definition places Europe's eastern boundary between Russia and its six European neighbors to the west. That definition is based on several geographic criteria that include European-Russian contrasts in territorial dimensions, geopolitical developments, cultural properties, and history.

Figure 4-1, as you may already have observed, places the Crimean Peninsula inside Russia as a consequence of its *de facto* annexation from Ukraine in 2014. This does not mean that we as geographers endorse that forcible takeover. But it does show that state boundaries frequently materialize through violence and the will of the victors. Thus the global map of states is in many ways the product of power struggles, territorial conflicts, and war—especially in Europe. In fact, it can be argued that much of eastern Europe constitutes a **transition zone [1]** between the European and Russian-dominated realms (which is particularly notable in Belarus and eastern Ukraine). That argument could be made on the basis of language, religion, and/or functional spatial interaction—but it does not translate into simple claims of territorial control.

Climate and Resources

From the balmy shores of the Mediterranean Sea to the icy peaks of the Alps, and from the moist woodlands and moors of the Atlantic fringe to the semiarid prairies north of the Black Sea, Europe presents an almost infinite range of natural environments (**Figs. 4-2** and **4-3**).

Europe's peoples have benefited from a large and varied store of raw materials. Whenever the opportunity or need arose, the realm proved to contain what was required. Early on, these requirements included cultivable soils, rich fishing waters, and wild animals that could be domesticated; in addition, extensive forests provided wood for houses and boats. Later, coal and mineral ores propelled industrialization. More recently, Europe proved to contain sizeable deposits of oil and natural gas.

Landforms and Opportunities

Europe's area may be small, but its physical landscapes are varied and complex. Regionally, we identify four broad units: the Central Uplands, the southern Alpine Mountains, the Western Uplands, and the North European Lowland (Fig. 4-3).

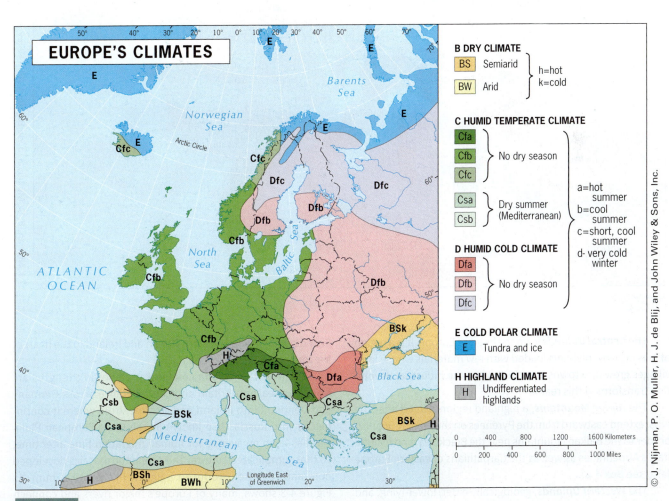

FIGURE 4-2

© J. Nijman, P. O. Muller, H. J. de Blij, and John Wiley & Sons, Inc.

EUROPE'S PHYSICAL LANDSCAPES

- Western Uplands
- North European Lowland
- Central Uplands
- Alpine System
- Canal

0   250   500 Kilometers
0   100   200   300 Miles

© J. Nijman, P. O. Muller, H. J. de Blij, and John Wiley & Sons, Inc.

FIGURE 4-3

The **Central Uplands** form the heart of Europe. It is a region of hills and low plateaus loaded with raw materials whose farm villages grew into towns and cities when the Industrial Revolution transformed this realm.

The **Alpine Mountains**, a highland region named after the Alps, extend eastward from the Pyrenees on the French-Spanish border to the Balkan Mountains near the Black Sea, and include Italy's Appennines as well as the Carpathians of eastern Europe (also see **Box 4-2**).

The **Western Uplands**, geologically older, lower-lying, and more stable than the Alpine Mountains, extend from northern Scandinavia through western Britain and Ireland to the heart of the Iberian Peninsula in Spain.

The **North European Lowland** stretches in a lengthy arc from southeastern Britain and central France across Germany and Denmark into Poland and Ukraine, from where it continues far into Russia. Also known as the Great European Plain, this has been an avenue for human migration time after time, so that complicated cultural and economic mosaics developed here and together produced a jigsaw-like political map. As Figure 4-3 shows, many of Europe's major rivers and connecting waterways serve this populous region, where a number of

---

Box 4-2 Technology & Geography

What is GIS?

Maps are often taken for granted because they are used everywhere, from the Internet to this textbook. The ubiquity of maps and mapping is a relatively new phenomenon and can be attributed to **geographic information systems** or **GIS [2]**. Generally, GIS consists of computer hardware, software, and specialized tools such as algorithms and models that are used to make digital maps. Mapping is also much easier and more efficient with GIS technology. GIS takes the digital data that we collect with satellites, sensors, and even smartphones, and puts them on a map. One feature that distinguishes GIS from other technologies is its ability to integrate different and diverse datasets using geography. This process is called **overlay**. Overlaying one or more data layer on top of another based on location reveals spatial patterns, highlights geographic relationships, and supports analysis and decision-making.

When satellite data are processed and integrated with other data, the power of GIS is revealed. For instance, the image on the left (below) that resembles a black and white X-ray is really a **digital elevation model [3]** (DEM) based on remote sensing imagery of the French Alps near Grenoble. Darker shades (black and dark grays) denote lower elevations, and lighter shades (white and light grays) indicate higher elevations. You can actually "see" valleys and mountain ranges. With a bit of GIS processing, the relief of Earth's surface can be visualized with other data (e.g., railroads, vegetation, snow cover) to get an idea of the accessibility and physiography of this Alpine region (right image).

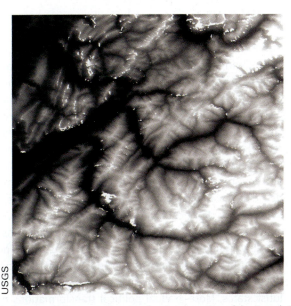

USGS

USGS

Left: a remotely-sensed image, showing a portion of the French Alps near Grenoble, constructed as a digital elevation model; the darker the color, the lower the altitude. Right: a layered GIS of the same terrain that adds vegetation, snow cover, and railway lines.

Europe's leading cities (London, Paris, Amsterdam, Copenhagen, Berlin, Warsaw) are located.

Locational Advantages

Europe's *relative location*, at the crossroads of the **land hemisphere [4]**, creates considerable efficiency for contact with much of the rest of the world (**Fig. 4-4**). A "peninsula of peninsulas," Europe is nowhere far from the ocean and its avenues of seaborne trade and conquest. Hundreds of kilometers of navigable rivers, augmented by an unmatched system of canals, open the interior of Europe to its neighboring seas and to the shipping lanes of the world. The Mediterranean and Baltic seas, in particular, were critical in the development of trade in early modern times and in the emergence of Europe's early **city-states [5]** such as Venice, Italy (see photo) in the south and Lübeck, Germany in the north.

Visions Of Our Land/The ImageBank/Getty Images, Inc.

The Rialto Bridge, one of Venice's famous landmarks, dates from the time that Venice was one of Europe's most powerful and richest city-states. The first bridge over the Grand Canal at this spot dates back to the twelfth century; the present span was completed in 1591.

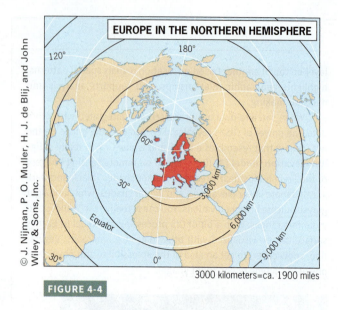

EUROPE IN THE NORTHERN HEMISPHERE

3000 kilometers=ca. 1900 miles

FIGURE 4-4

And do not overlook the scale of the maps in this chapter. Europe is a realm of moderate distances and close proximities. This propinquity, combined with substantial cultural differences, makes for intense interaction, the constant circulation of goods and ideas. That has been the hallmark of Europe's geography since early modern times.

Modern Historical Geography

The Industrial Revolution

From the late eighteenth century onward, Europe experienced a steady process of intensifying **local functional specialization [6]**. The **Industrial Revolution [7]**, the beginning of which is often attributed to the invention of the steam engine by Scotsman James Watt (1736–1819) in 1780, was essential in this process, and it would provide Europe with a critical edge in world affairs for a long time to come. The Industrial Revolution gathered momentum throughout the nineteenth century, and Britain was at its epicenter. The power loom revolutionized the weaving industry. Iron smelters, long dependent on Europe's dwindling forests for fuel, could now be concentrated near coalfields. Engines could move locomotives as well as power looms. Ocean shipping entered a new age. The British had an enormous advantage because they controlled the flow of raw materials and held a monopoly over products that were in demand globally. In Britain, manufacturing regions developed near coalfields in the English Midlands, at Newcastle to the northeast, in southern Wales, and along Scotland's Clyde River around Glasgow.

After 1850, the Industrial Revolution spread eastward from Britain into continental Europe and by 1890 had reached as far as western Russia. In mainland Europe, the axis of industrialization followed a west-to-east belt of major coalfields extending roughly along the southern margins of the North European Lowlands; each gave rise to manufacturing complexes that can still be seen on the map today in northern France and Belgium, Germany's Ruhr Basin, the Czech Republic's Bohemia, southern Poland's Silesia, and eastern Ukraine's Donets Basin (Donbas). Abundant iron ore deposits lie in a similar west-east belt and together with coal provide the key raw materials for manufacturing steel.

As in Britain, this cornerstone heavy-metal industry spawned new concentrations of economic activity, which grew steadily as millions migrated in from the countryside to fill expanding employment opportunities. Densely populated and heavily urbanized, these industrial agglomerations became the backbone of Europe's world-scale population cluster (labeled **D** in Fig. G-7) as well as an enduring major feature on the realm-scale map of Europe's population distribution (**Fig. 4-5**).

Political Transformations

The Industrial Revolution unfolded against the backdrop of the ongoing formation of Europe's national states, a process that had been underway since the 1600s. These political revolutions generally involved the centralization of royal power by monarchies and the usurpation of the former city-states in larger national territories—which frequently produced violent reconfigurations of the realm's evolving political geography. Most historians point to the Peace (Treaty) of Westphalia in 1648 as a key step in the emergence of Europe's state system, ending decades of war and recognizing territories, boundaries, and the **sovereignty [8]** of countries. But this was only the beginning, and these so-called *absolutist states*, in which monarchs held all the power and the people had few if any rights, would not last.

One of the most dramatic episodes to follow was the French Revolution (1789–1795), which soon ended the era of absolutist states. More gradual political transformations occurred in the Netherlands, Britain, and the Scandinavian countries. But a number of other components of Europe would remain much longer under the tight control of authoritarian (dictatorial) regimes headed by monarchs or despots. By the late nineteenth century, Europe became the arena for the competing ideologies of **liberalism, socialism**, and **nationalism** (national spirit, pride, patriotism). However, a generation later the rise of extreme nationalism (**fascism**) in the 1930s plunged the realm into one of the most violent wars the world has ever seen.

A Fractured Map

Europe's political map has always been one of intriguing complexity. As a geographic realm, Europe occupies only about 5 percent of the Earth's land area; but that minuscule territory is fragmented into some 40 countries—more than one-fifth of all the states in the world today. Therefore, when you look at Europe's political map, the question that arises is how did so small a geographic realm come to be divided into so many political entities? Europe's map is a legacy of its feudal and monarchic periods, when powerful kings, barons, dukes, and other rulers, rich enough to fund armies and powerful enough

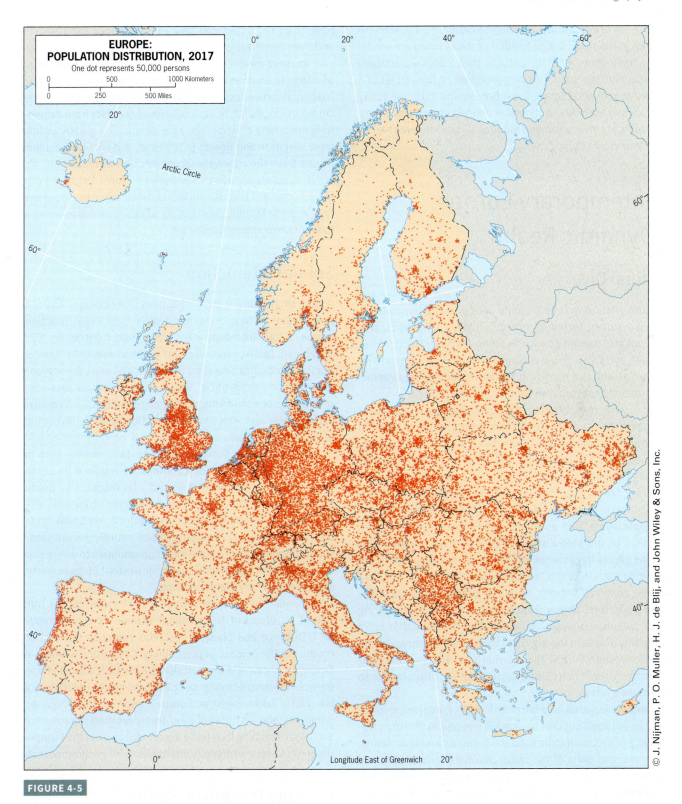

**EUROPE:
POPULATION DISTRIBUTION, 2017**
One dot represents 50,000 persons

0   500   1000 Kilometers
0   250   500 Miles

Arctic Circle

Longitude East of Greenwich

FIGURE 4-5

to forcibly extract taxes and tribute from their domains, created bounded territories in which they reigned supreme. Royal marriages, alliances, and conquests actually simplified Europe's political map. In the early nineteenth century, there still were 39 separate German States; a unified Germany as we know it today would not emerge until the 1870s.

State and Nation

Europe's political transmutations produced a prototype of politico-geographical organization known as the **nation-state [9]**, a territorial state embodied by its culturally distinctive population. The term **nation [10]** refers to a people with a single language, a common history, and a similar ethnic background.

In the sense of **nationality**, it relates to citizenship (legal membership) in the state. Only a handful of states today are so culturally homogeneous that, spatially, their culture practically coincides with the state. Europe's prominent nation-states of a century ago—France, Spain, Great Britain, and Italy—are today multicultural societies, with *nation* defined more by an intangible spirit of nationalism and emotional commitment than by cultural or ethnic homogeneity.

Contemporary Europe: A Dynamic Realm

Cultural Diversity

The European realm is home to peoples of numerous cultural-linguistic stocks, including not only Latins, Germanics, and Slavs but also minorities such as Finns, Magyars (Hungarians), Basques, and Celts. This diversity of ancestries continues to be an asset as well as a liability. It has generated not only interaction and exchange, but also conflict and war.

It is worth remembering that Europe's territory is just over 60 percent the size of the United States, but that the population of Europe's 40 countries is almost twice as large as America's. This population of just under 600 million speaks numerous languages, almost all of which belong to the **Indo-European language family [11]** (see Figure G-8 and **Figure 4-6**). Nonetheless, most of those languages are not mutually understandable; some, such as Finnish and Hungarian, are not even members of the Indo-European family. When Europe's unification efforts began after World War II, one major problem was to determine which languages to recognize as "official." That problem still prevails, although English has become the realm's unofficial **lingua franca** (common language). During a visit to Europe, though, you would find that English is more commonly usable in the big cities than in the countryside and more in western Europe than farther east. Europe's multilingualism remains a rich cultural legacy, but it is also a barrier to further integration.

Another major divisive force confronting Europeans involves religion. Europe's cultural heritage is steeped in Christian traditions, but sectarian strife between Catholics and Protestants, which plunged parts of the realm into bitter and widespread conflict, still divides communities and, as until recently in Northern Ireland, can still provoke violence. Some political parties still carry the name "Christian"—for instance, Germany's Christian Democrats.

More generally, Christianity has gradually lost adherents since the 1960s as secularization gathered momentum, especially in western Europe. The Roman Catholic Church, long powerful in much of Europe, has steadily been losing its grip on society, and many church institutions—schools, universities, unions, political parties, charities, and clubs—have been hollowed out. Moreover, many churches have closed down or have been converted into art galleries, public meeting halls, and even corporate offices.

Today, a new force roils the religious landscape: the rise of Islam. In southeastern Europe, this takes the form of new Islamic assertiveness in an old Muslim bastion: the (Turkish) Ottoman Empire left behind millions of converts from Bosnia to Bulgaria, many of whom are now demanding greater political representation and power. In the West, this Islamic resurgence results from the relatively recent infusion of millions of Muslim immigrants from former colonies in North Africa and other parts of the far-flung Islamic World. Here, while mosques overflow with the faithful, churches stand nearly empty as a witness to secularism among Europeans.

Spatial Interaction

If not a single culture, then what does unify Europe? The European realm is best understood as a large-scale **functional region**, a fully interdependent realm held together by highly developed spatial-economic and political networks. Contemporary Europe has seized on the realm's abundant geographic opportunities to create a huge, intensively used network of spatial interaction linking places, communities, and countries in countless ways. This spatial interaction operates on the basis of two key principles.

First, regional **complementarity [12]** means that one area produces a surplus of a commodity required by another area. The mere existence of a particular resource or product is no guarantee of trade: it must be needed elsewhere. When two areas each require the other's products, we speak of *double complementarity*. Europe exhibits countless examples of such complementarity, from local communities to entire countries. Industrial Italy needs coal from western Europe; western Europe needs Italy's farm products.

Second, the ease with which a commodity can be transported by producer to consumer defines its **transferability [13]**. Distance and physical obstacles can raise the cost of a product to the point of unprofitability. But Europe is small, distances are short, and the Europeans have built a most efficient transport system of roads, railroads, and canals linking navigable rivers. Taken together, Europe's enormously diverse economic regions, its ultra-efficient transportation infrastructure, and its far-reaching free-trade agreements within the European Union make for a supremely interdependent geographic realm.

A Highly Urbanized Realm

About three of every four Europeans live in towns and cities (in western Europe the share of the urban population is much higher still). Large cities are production centers as well as marketplaces, and they also form the crucibles of their national cultures. Europe's major cities tend to have long histories. They are compact, and in general the European cityscape looks quite different from its North American counterpart (**Box 4-3** and **Box 4-4**).

LANGUAGES OF EUROPE

0 200 400 600 Kilometers
0 100 200 300 Miles

MAJOR INDO-EUROPEAN BRANCHES

GERMANIC GROUP

WESTERN GERMANIC
1 Dutch
2 German
3 Frysian
4 English

NORTHERN GERMANIC
5 Danish
6 Swedish
7 Norwegian
8 Icelandic
9 Faeroese

ROMANCE GROUP

10 Portuguese
11 Spanish
12 Catalan
13 Provençal
14 French
15 Italian
16 Rhaeto-Romansch
17 Romanian
18 Corsican-Italian
19 Sardinian-Italian
20 Walloon

SLAVIC GROUP

WEST SLAVONIC
21 Polish
22 Slovak
23 Czech
24 Lusatian

EAST SLAVONIC
25 Russian
26 Ukrainian
27 Belarussian

SOUTH SLAVONIC
28 Slovene
29 Serbo-Croatian
30 Macedonian
31 Bulgarian

OTHER INDO-EUROPEAN BRANCHES

CELTIC GROUP

BRITANNIC
32 Breton
33 Welsh

GAELISH
34 Irish Gaelic
35 Scots Gaelic

BALTIC GROUP

36 Latvian
37 Lithuanian

HELLENIC

38 Greek

THRACIAN/ILLYRIAN GROUP

39 Albanian

INDO-IRANIAN GROUP

40 Romani (dispersed)

URALIC LANGUAGE FAMILY

FINNO-UGRIC GROUP

41 Finnish
42 Karelian
43 Saami
44 Estonian
45 Hungarian
46 Komi

SAMOYEDIC GROUP

47 Samoyedic

ALTAIC LANGUAGE FAMILY

TURKIC GROUP

48 Turkish

OTHER LANGUAGES

BASQUE

49 Basque

Areas with significant concentrations of other languages (usually adjacent national languages)

Boundary between languages

After Murphy, 1998.

© J. Nijman, P. O. Muller, H. J. de Blij, and John Wiley & Sons, Inc.

FIGURE 4-6

Box 4-3 Major Cities of the Realm, 2016

Metropolitan Area	Population* (in millions)
Paris, France	10.9
London, UK	10.4
Madrid, Spain	6.2
Milan, Italy	5.3
Barcelona, Spain	4.7
Berlin, Germany	4.1
Rome, Italy	3.9
Athens, Greece	3.5
Kiev, Ukraine	2.8
Lisbon, Portugal	2.7
Budapest, Hungary	2.5
Warsaw, Poland	2.3
Brussels, Belgium	2.1
Munich, Germany	2.0
Frankfurt, Germany	1.9
Vienna, Austria	1.8
Amsterdam, Netherlands	1.6
Stockholm, Sweden	1.5
Prague, Czech Republic	1.4
Dublin, Ireland	1.2

Seemingly haphazard inner-city street systems impede traffic; central cities may be picturesque, but they are also cramped. European city centers tend to be more vibrant today than cities in the United States. They offer a mix of businesses, government functions, shopping facilities, educational and arts institutions, and entertainment as well as housing for upper-income residents.

Wide residential sectors radiate outward from the **central business district (CBD) [14]** across the rest of the central city, often inhabited by particular income groups. Beyond the central city lies a sizeable suburban ring, but even here residential densities are much higher than those in the United States because the European preference has been to live in apartments rather than detached single-family houses and to set aside recreational spaces in "greenbelts." There also is a greater reliance on public transportation, which further concentrates the realm's urban and suburban development pattern. However, traditions may now be starting to change. One indication is the recent growth of cutting-edge suburban business centers that increasingly compete with the CBD in many parts of urban Europe (see Box 4-4).

European Unification

Realms, regions, and countries can all be subject to dividing and unifying forces that cause them to become more or less cohesive and stable political units. Where the region or country is home to diverse populations with different political goals, it may prove difficult to avoid divergence and territorial fragmentation. Political geographers use the term **centrifugal forces [15]**

Box 4-4 From the Field Notes …

"If you were to be asked what city is shown here, would Paris spring to mind? Most images of the French capital show venerable landmarks such as the Eiffel Tower or Notre Dame Cathedral. I am taking this photo from the top of the famous Arc de Triomphe, looking toward another Paris, the burgeoning northwestern business district just beyond the city line named *La Défense*. There, Paris escapes the height restrictions and architectural limitations of the historic center and displays an ultramodern complexion. Glass-box skyscrapers reflect the vibrant global metropolis this is, and the central landmark is the 'Cube,' a huge open structure admired as well as reviled (as was the Eiffel Tower in its time). *La Défense* is ingeniously incorporated into Paris's urban design, with a straight-line, broad avenue connecting it to the Arc de Triomphe and then connecting, via the Champs Élysées to the Place de la Concorde. In terms of a grand geographic layout, Paris is hard to beat."

© Jan Nijman

to identify and measure the strength of such division, which may result from religious, racial, linguistic, political, economic, or other regional factors.

Centrifugal forces are measured against **centripetal forces [16]**, the binding, unifying glue of the state or region. General satisfaction with the system of government and administration, shared identities, legal institutions, and other functions of the state (notably including its treatment of minorities) can ensure stability at the state level.

Since World War II ended in 1945, Europe has witnessed a steady process of integration and unification at a much broader geographic scale. A majority of European states and their leaders recognized that closer association and regional coordination formed the key to a more stable, prosperous, and secure future. Centripetal forces have thus far prevailed, but, as the EU has expanded to include ever more countries, recent years have witnessed growing dissent and criticism, particularly in the countries where it all began in the late 1950s.

Postwar Motivations

At the end of World War II, much of Europe lay shattered, its cities and towns devastated, its infrastructure wrecked, its economies ravaged. If this was one of the world's most developed realms at the beginning of the twentieth century, its economic prowess had been almost completely destroyed by 1945. One of the primary motives for integration and collaboration among western European countries, therefore, was rapid economic recovery.

The integration process was also driven by political considerations. From the perspective of countries such as France, the Netherlands, and Belgium, one of the key issues was to control defeated Germany in the postwar years, and this could only be done through close political cooperation. Thus European integration was from the start both an economic and a political affair, and concerns about Germany played a major part.

From afar, the United States was a strong supporter of (west) European integration. The ending of World War II had witnessed accelerating tensions with the Soviet Union, which had taken control over the bulk of eastern Europe. Furthermore, communist parties seemed poised to dominate the political life of major western European countries. The United States was intent, therefore, to have a firm hand in that part of Europe and to keep communist influences at bay. Under the so-called *Marshall Plan*, massive U.S. financial support, exceeding $130 billion in today's money, poured in to 16 European countries, including the western half of Germany not controlled by the Soviets as well as Turkey.

As these economies recovered, Germany (West Germany, that is, until the Cold War ended in 1989) became firmly embedded in a pan-European structure, the United States assumed a more distant role, and the motivations driving European unification developed a new emphasis. Economically, the process has increasingly been propelled by European corporate interests—the need to facilitate an ever larger and more efficient open market that could compete globally with the U. S., China, and Japan. Politically, twenty-first-century goals are more about stabilizing a much larger and diverse European community that now approaches the western edge of Russia. Thus the focus has shifted from Germany to an enlarged and far more complicated geopolitical zone. The latter consists of numerous sovereign states that must maintain satisfactory relations with the Russian giant next door.

The Unification Process

European integration has been a complex, gradual, and step-by-step process aimed at the creation of a single territorial economy across Europe. It has involved (1) agreements on free trade across boundarie without tariffs or border controls; (2) the abolition of protectionist economic policies; (3) harmonization of national laws (e.g., on taxation) to establish a level playing field; (4) political cooperation and the creation of pan-European executive, judicial, and legislative bodies; (5) the introduction of a single currency, the *euro*; and (6) agreements to transfer specific powers from national governments to EU institutions.

European integration is a classic example of **supranationalism [17]**, the voluntary association in economic, political, or cultural spheres of independent states willing to yield some measure of sovereignty for their mutual benefit. **Box 4-5** provides a timeline of the key milestones and shows that European integration has in no small part been a matter of steady institution building. **Figure 4-7** maps the territorial expansion of the European Union (the name took hold in 1993) and of the **Euro zone [18]** (not all EU countries have given up their national currency). Croatia became the 28th country to join the EU in 2013; Lithuania, in 2015, became the 19th member-country to adopt the euro; but in a turnaround, the United Kingdom in 2016 voted to bail out.

Consequences of Unification

The European Union is not just a paper organization. It has a profound impact on national economies, on the role of individual states, and on the daily lives of its member-countries' nearly 450 million citizens.

One Market

EU directives are aimed at the creation of a single market for producers and consumers, businesses, and workers. Corporations should be able to produce and sell anywhere within the Union without legal impediments, whereas workers should be able to move anywhere in the EU and find employment without legal restrictions. In order to make this happen and to keep things manageable, member-states have had to harmonize a wide range of national laws from taxation to the protection of the environment to educational standards. The so-called **Schengen Area [19]** is comprised of 26 EU countries that have abolished internal border controls and that have a single visa policy for non-EU visitors.

Box 4-5 Timeline of Supranationalism in Europe

1944 Benelux Agreement signed; this forerunner association aimed to fully integrate the economies of Belgium, the Netherlands, and Luxembourg.

1948 Organization for European Economic Cooperation (OEEC) established.

1949 Council of Europe created.

1957 Treaty of Rome signed, establishing European Economic Community (EEC) (effective 1958); also known as the Common Market and "The Inner Six," the founding member-countries were France, (then) West Germany, Italy, Belgium, the Netherlands, and Luxembourg. European Atomic Energy Community (EURATOM) Treaty signed (effective 1958).

1967 European Community (EC) inaugurated.

1968 All customs duties removed for intra-EC trade; common external tariff established.

1973 United Kingdom, Denmark, and Ireland admitted as members of EC, creating "The Nine."

1979 First general elections for a European Parliament held; new 410-member legislature convenes in Strasbourg, France. European Monetary System established.

1981 Greece admitted as member of EC, creating "The Ten."

1985 The "Schengen Agreement" results in the abolition of border checks between the Benelux countries, Germany, and France. The Schengen Area is gradually expanded to include 26 EU countries by 2016.

1986 Spain and Portugal admitted as members of EC, creating "The Twelve." Single European Act ratified, targeting a functioning European Union in the 1990s.

1993 Single European Market goes into effect. Modified European Union Treaty ratified, transforming the EC into the European Union (EU).

1995 Austria, Finland, and Sweden admitted into EU, creating "The Fifteen."

1999 European Monetary Union (EMU) goes into effect.

2002 The euro is introduced in 12 countries; the Euro zone expands gradually over time (see Fig. 4-7) but not all EU members choose to give up their national currency.

2004 Historic expansion of EU from 15 to 25 countries with the admission of Cyprus, the Czech Republic, Estonia, Hungary, Latvia, Lithuania, Malta, Poland, Slovakia, and Slovenia.

2005 Proposed EU Constitution is rejected by voters in France and the Netherlands.

2007 Romania and Bulgaria are admitted, bringing total EU membership to 27 countries. Slovenia adopts the euro.

2010 Financial crisis strikes heavily indebted Greece, requiring massive EU bailout. Later in the year, Ireland follows and raises fears of similar crises in Portugal, Spain, and Italy. The future of the EMU is clouded; the value of the euro declines after a long rise against the dollar.

2013 Croatia gains entry as the 28th EU member-state.

2014 Latvia adopts the euro.

2015 A popular referendum in Greece results in a majority rejecting another EU bailout and related austerity measures. A "Grexit" was closer than ever before, but the government eventually agrees on a renegotiated package. Lithuania adopts the euro, increasing Euro zone membership to 19 countries.

2016 The intensifying refugee crisis across Europe stresses the EU and produces divisive impacts in member-countries. The United Kingdom holds its 'Brexit' referendum, with a 52-percent majority voting to leave the EU.

One major step toward union came with the introduction of a single central bank (with considerable power over, for example, interest rates) and a single currency, the euro. The euro was also meant to symbolize Europe's strengthening unity and to establish a counterweight to the no-longer-almighty American dollar. In 2002, twelve of the (then) 15 EU member-countries withdrew their own currencies and began using the euro, with only the United Kingdom, Denmark, and Sweden opting to stay out (Fig. 4-7). More recently, Slovakia (2009), Estonia (2011), Latvia (2014), and Lithuania (2015) joined the Euro zone. The single currency, hailed as a major triumph at its inception, has evolved into a major bone of contention because it has significantly reduced the freedom of member-states to formulate fiscal policies. We will return to the ongoing "euro crisis" shortly.

A New Economic Geography

The establishment and continuing development of the EU have generated a new economic landscape that today not only transcends the old but is fundamentally reshaping the realm's regional geography. By investing heavily in new infrastructure and by smoothing the flows of money, labor, and products, European planners have dramatically reduced the divisive effects of their national boundaries. And by acknowledging demands for greater freedom of action by their provinces, States, departments, and other administrative subunits of their countries, European leaders unleashed a wave of economic energy that has transformed some of these subnational units into powerful engines of growth (**Fig. 4-8**).

Four of these burgeoning hubs are especially noteworthy, to the point that many policy analysts refer to them as the **Four Motors of Europe [20]**: (1) France's southeastern **Rhône-Alpes Region**, centered on the country's second-largest city, Lyon; (2) **Lombardy** in north-central Italy, focused on the industrial metropolis of Milan; (3) **Catalonia** in northeastern Spain, anchored by the cultural and manufacturing nucleus of Barcelona; and (4) **Baden-Württemberg** in southwestern Germany, headquartered by the high-tech city of Stuttgart.

Despite economic integration and the harmonization of policies, Figure 4-8 also shows that major inequalities persist

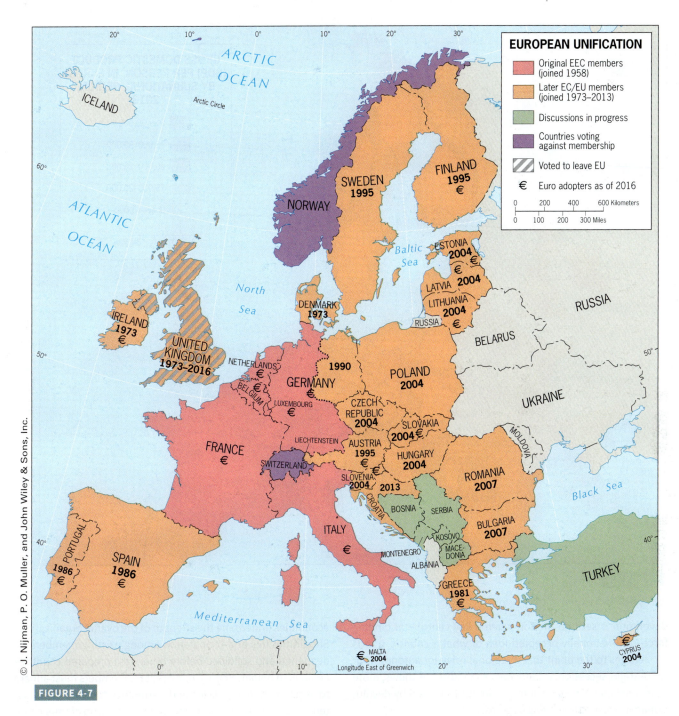

EUROPEAN UNIFICATION

- Original EEC members (joined 1958)
- Later EC/EU members (joined 1973–2013)
- Discussions in progress
- Countries voting against membership
- Voted to leave EU
- € Euro adopters as of 2016

© J. Nijman, P. O. Muller, and John Wiley & Sons, Inc.

FIGURE 4-7

among Europe's regions as well as within EU member-states. This is an important reminder that, as noted in the introductory chapter, uneven development is a pervasive phenomenon that plays out at every geographic scale.

Diminished State Power and New Regionalism

As states relinquish some of their power to Brussels and express agreement with the ideal of realmwide integration, some of the provinces and other subdivisions *within* states have seized the opportunity to assert their own cultural identity as well as

particular economic interests. Often, the local governments in these subregions simply bypass the central government in their national capital, dealing not only with each other but also directly with foreign governments as their business networks expand to the global scale.

Thus, even as Europe's states have been working to join forces in the EU, many of those same states are confronting severe centrifugal stresses. The term **devolution [21]** has come into use to describe the powerful centrifugal forces whereby regions or peoples within a state, through negotiation or active rebellion, demand and gain political strength and sometimes autonomy at the expense of the center. Most states exhibit some level of internal regionalism, but the process of

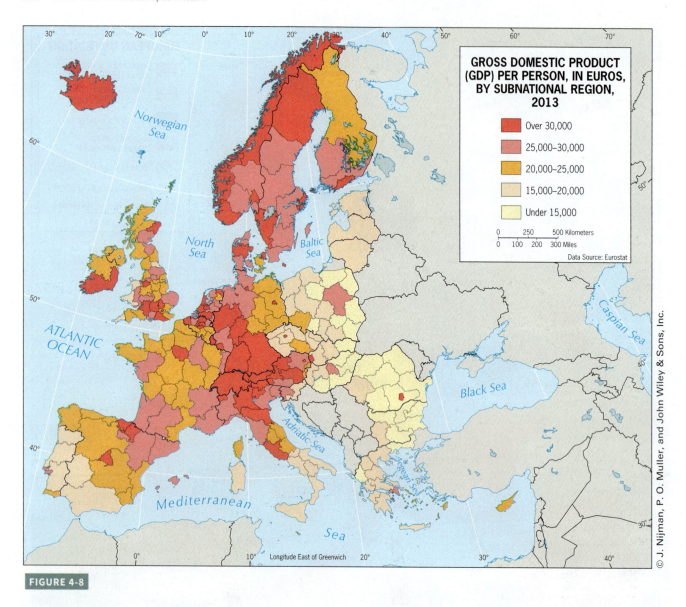

FIGURE 4-8

devolution is set into motion when a key centripetal binding force—the nationally accepted idea of what a country stands for—erodes to the point that a regional drive for autonomy, or for outright secession, is launched. As **Figure 4-9** shows, more than a dozen European countries are now affected by devolutionary pressures to varying degrees.

States respond to such pressures in a number of ways, ranging from accommodation to suppression. One way to deal with these centrifugal forces is to give historic regions (such as Scotland or Catalonia) certain rights and privileges formerly held exclusively by the national government. Another answer is for EU member-states to create new administrative divisions that will allow the state to meet regional demands, often in consultation with, or under pressure from, the EU.

The Changing Risks of Expansion

Expansion was always a EU objective, and the subject never fails to arouse passionate debate. Will the incorporation of weaker economies undermine the strength of the overall organization? Should the ties and cooperation among existing members be deepened and solidified before other, less prepared countries are invited to join? Keep in mind that EU members must adhere to strict economic policies and harmonize their political systems (see **Box 4-6**). This is much easier for prosperous countries with longstanding democratic traditions than for poorer nations with a volatile political past. Despite such misgivings, negotiations to expand the EU have long been in progress, and the gains of the twenty-first century are mapped in Figure 4-7.

Numerous structural implications arise from this expansion, affecting all member-countries. First, some of the pre-2004 EU's less affluent countries, which used to be on the receiving end of EU subsidy programs, now had to pay up to support the much poorer new eastern member-states. Second, disputes also intensified over representation at EU's Brussels headquarters. And, most of all, the incorporation of weaker national economies tends to compromise the ability of the EU to move forward in unison because the weaker members are unable to keep their fiscal deficits within limits, and sometimes

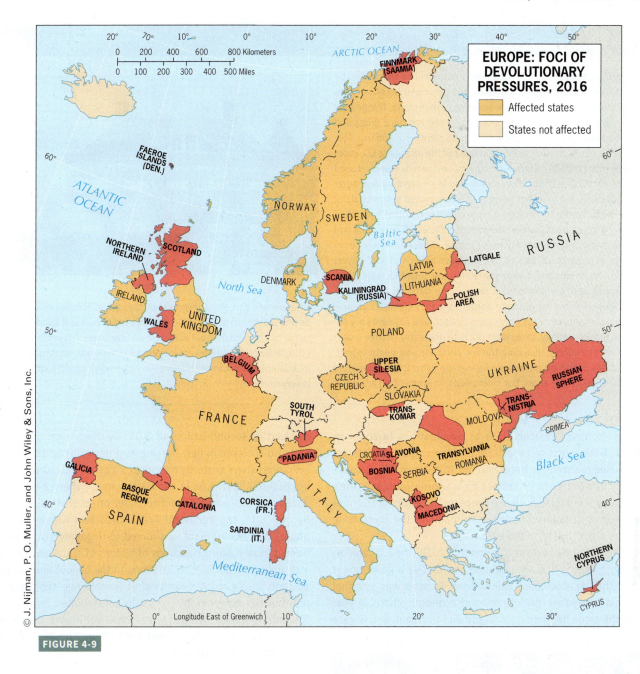

FIGURE 4-9

their industries are hurt by the more developed, competitive industries of the wealthier members. The inclusion of countries like Romania and Bulgaria, especially, has often been criticized in this regard.

The Challenges of a Single Currency

The kind of economic incompatibility among EU countries, described above, has come into sharper focus since the arrival of the single currency. Since the late 2000s, the "euro crisis" has been an almost standard news item throughout the realm as it laid bare the limitations that accompanied the shift to a single currency and a common monetary policy. The evolving problems are not just about the euro itself, but about the inability or

unwillingness of EU members to comply with fiscal standards (e.g., avoiding government deficits) and apply them as solutions within their individual economies. Keep in mind that each EU country still has its own budget as well as its own revenues and expenditures. But even though the EU budget itself remains quite small (about 1 percent of the EU's GDP), the budget rules governing the Union are in crucial ways determined by Brussels.

The economic difficulties confronted by several European national governments in this decade intensified because they had only recently introduced the single European currency—which significantly constrained their policy options. Some of the EU's weaker member-states, most of all Greece, accumulated enormous debts that violated terms in the agreement that established the Euro zone. Moreover, having a single currency implied that countries could no longer independently

Box 4-6   From the Field Notes …

"The Grasshopper, one of Amsterdam's many 'coffee shops.' For the past four decades, the Netherlands has permitted legalized sales and consumption of moderate amounts of marijuana. The general consensus is that marijuana is not as harmful as hard drugs and that legalization helps to decriminalize its commerce and use. This policy is at odds with legislation in most of the rest of the EU, especially with bordering Germany, Belgium, and nearby France. It used to be that coffee shops were found only in Amsterdam and other major cities, but in recent years they have opened in towns along the Dutch border to attract a foreign clientele (another byproduct of heightened cross-border interaction). As a result, once quiet and conservative rural villages (where lifestyles differ enormously from those in Amsterdam) started to experience heavy traffic of buyers and users from abroad—not something the townfolk were likely to appreciate. Largely for this reason, the national government has turned coffee shops in the border zone into membership-only establishments whereby only Dutch citizens and residents are eligible to join. Initially, there were plans to make this change across the entire country, including the big cities. As you might expect, reactions to this proposal varied geographically: it was supported in the rural border areas, but most people in Amsterdam and other

© Jan Nijman

cities viewed it as an infringement on traditional liberties and a bad business decision to boot. Said one Amsterdammer and steady customer of The Grasshopper with a tone of sarcasm: 'I never thought I'd experience this: some provincials trying to decide what happens in Amsterdam. Not in my lifetime!' In Amsterdam's coffee shops, it is very much business as usual."

---

pursue monetary policies such as devaluations (a common approach to promoting exports) or the lowering of their interest rates (used to stimulate the economy). In other words, as countries such as Greece faced deepening economic problems, their hands were tied by their dependence on the European Central Bank—whose primary mission is to maintain the purchasing power (price stability) of the euro.

The euro crisis began in earnest in 2010 when it came to light that Greece had accumulated so massive a government deficit that an emergency bailout from Brussels was essential to avoid immediate bankruptcy and economic chaos. A year later, the same happened to Ireland and Portugal, and Greece

received its second bailout in 2012. By then, fiscal problems had also emerged in Italy (forcing its prime minister to resign), in France (where austerity measures triggered widespread public protests), and even in the Netherlands (where a coalition government fell when it could not agree on severe budget cuts). By mid-2015, the persistent Greek crisis had worsened (see photo). The government was forced to negotiate yet another bailout (now totaling hundreds of billions of euros since 2009) and organize a popular referendum on the bailout and related austerity measures. A majority of 61 percent rejected the bailout, but the government eventually renegotiated terms with Brussels. A "Grexit" loomed larger than ever before, but Greek membership survived—at least until 'Brexit' opened a new era for the European Union.

'Brexit': An Historic Referendum

Perhaps the most critical challenge to the EU came in 2016 from the United Kingdom (UK), so often described as the 'reluctant European'. There had been growing opposition in the UK to continued EU membership and the government decided to hold a popular referendum on June 23, 2016. Those favoring "Leave" argued that EU rules constrained the British economy, that the UK was asked to contribute too much to the spectrum of funds that were distributed among other member-states. More generally, they opposed the idea of 'ever closer union' as a growing threat to the UK's sovereignty. And most importantly, they also claimed that the UK must control its own borders in order to reduce the number of immigrant workers and refugees.

On the other hand, those favoring "Remain" argued that the benefits of membership outweighed the cost. Thanks to the EU,

Robert Geiss/picture-alliance/dpa/AP Images

A Greek pensioner raises his fist as he marches through central Athens as part of a demonstration involving tens of thousands in late 2015. This huge crowd of protesters is shouting slogans against budgetary cuts in pensions and healthcare services—mandated by the EU as part of its continuing bailout of Greece's floundering economy.

they claimed, the UK was able to send more than half of its exports to EU countries, London had become one of the most influential nodes in the global economy, and it enabled the country to play a prominent role in EU decision-making (especially concerning the free movement of people between the European continent and the UK). In their view, severing ties with the EU would not solve the refugee problem and could even make it worse by depriving the British of a voice in realm-wide policy-making on this critical issue. Moreover, the EU supporters maintained, the UK was committed to staying outside the Euro zone (and therefore remain unaffected by its fiscal crises), and that immigrant workers contributed significantly to the UK economy.

The well-known outcome of the 2016 referendum was that approximately 52 percent of the electorate voted for the UK to leave the EU. In every way, this was a hotly contested choice and deeply divided the country. The prime minister of the pro-EU government felt compelled to resign and new elections were called. Throughout the EU and in much of the rest of the world, the result was deplored. Financial markets everywhere were shaken by the news, and the UK economy, in particular, entered a dark period of uncertainty as the long, complicated process of EU severance got underway.

The Remaining Outsiders

'Brexit' notwithstanding, a number of other countries and territories still remain outside the EU (Fig. 4-7). The first group includes five of the seven states that emerged out of former Yugoslavia, plus Albania. Here in the western Balkan Peninsula, most are troubled politically as well as economically. Only Slovenia and Croatia have achieved full EU membership. The rest—Serbia, Bosnia and Herzegovina, Macedonia, Montenegro, and Kosovo—rank among Europe's poorest and most ethnically fractured states, yet they do contain almost 20 million people whose circumstances could worsen outside the "EU club." Better, EU leaders reasoned, to move them toward membership by demanding political, social, and economic reforms.

The second group of outsiders is composed of Ukraine and its neighbors. The government of Ukraine itself has expressed interest, but its electorate is strongly (and regionally) divided—as well as overshadowed by a civil conflict that has put everything on hold (see discussion later in this chapter). Moldova views the EU as a potential supporter in the struggle against devolution along its eastern border, and a country not even on the map—Georgia, clear across the Black Sea from Ukraine—has proclaimed its interest in future membership for a similar reason: to find an ally against Russian intervention in its internal affairs (including a brief 2008 military incursion).

Finally, we should take special note of another long-mentioned candidate for EU membership—Turkey. Some EU leaders have long wanted to include (NATO member) Turkey mainly for geostrategic reasons, but others argued that Turkey was just not "European" enough. The Turkish case for accession, however, has been thrown into a different light today because of the country's pivotal role in the refugee crisis that has engulfed Europe (to be discussed shortly). Because the EU and Turkey need each other to resolve this dilemma, admission may well be accelerated.

NATO and European Security

The emphasis on EU developments tends to overshadow another important pan-European institution—the **North Atlantic Treaty Organization (NATO)**—a far-reaching security alliance that partially overlaps with the EU (Fig. 4-10). NATO, however, is a very different organization. First, it is primarily concerned with politics and security and not with economics: NATO's purpose is to provide military protection for its members. Second, NATO is led by the United States, which provides the bulk of the alliance's military muscle, and also includes Canada. And third, NATO includes non-EU member Turkey, a country that is hugely important because of its strategic location on the southeastern edge of Europe nearest to the Middle East.

NATO was created in 1949 to provide a security umbrella for western Europe under U.S. leadership (in response, the Soviets and their satellites created the Warsaw Pact). When the Cold War ended in 1991 with the disintegration of the Soviet Union, NATO became a more general security organization that would provide stability across Europe. It also became more concerned with emerging, twenty-first-century threats to its member-states: terrorism, attacks by "rogue" states with weapons of mass destruction, and cyber-warfare (a very real and growing danger with critical global implications). However, as NATO opened its doors to membership for eastern European countries (Fig. 4-10), it initiated an eastward expansion—not unlike that of the EU—that is already producing a renewed standoff with Russia.

Europe's Refugee Crisis

The turmoil and conflict that have plagued parts of North Africa and Southwest Asia since 2011 have triggered an ever larger stream of refugees into (relatively nearby) Europe. The contrasts between these two realms are enormous: one is quite rich, politically stable, highly integrated, democratic, progressive, and largely Christian; the other is mostly poor, conflict ridden, often war-torn, fragmented, authoritarian, conservative, and mostly Islamic. The story of this refugee crisis is as much about globalization and the "shrinking" of distance, as it is about stark socioeconomic juxtapositions and cultural encounters.

The statistics are astounding: the total number of verified illegal border-crossings into Europe rose from about 400,000 in 2014 to more than 2 million in 2015. This huge sudden increase coincided with a major shift of refugee flows from the Central Mediterranean to the East Mediterranean and the Balkans (Fig. 4-11). Those coming across the Mediterranean mostly originated in North Africa and (to a lesser extent) Subsaharan Africa, while those seeking entry into Europe through the Balkans come mostly from Syria and other sources in Southwest Asia (see photos).

FIGURE 4-10

Map legend:

NATO MEMBERSHIP IN EUROPE

Joining dates:

- Founders* (1949)
- Joined after 1949
- Potential candidates

*Also includes Iceland, United States, and Canada

0   250   500 Kilometers
0   100   200   300 Miles

© J. Nijman , P. O. Muller, and John Wiley & Sons, Inc.

The EU countries on the frontline are Greece and Italy. From there, refugees or migrants (the distinction has become politically contentious in Europe) typically try to travel to EU countries they believe offer them the best chance of acceptance and a good living. In practice, this means that many are destined for Germany and Sweden. Once they have reached their destination, many apply for **asylum [22]**, legally protected residency status. **Figure 4-12** shows the number of asylum-seekers by country in 2015, with Germany leading, followed by Sweden, Hungary, Austria, and Italy.

These numbers of asylum-seekers lie well below the total of detected illegal border-crossers—indicating that many of those caught at a border are immediately returned, or that a sizeable number of those who do enter are not actually applying for

Refugees trying to reach Europe across the Central Mediterranean (left) and the Balkan routes (right). The lefthand photo was taken on April 17, 2016 off the coast of Libya, from where more than 130 migrants cast off to reach Italy. Within hours, their boat broke in two, and according to news reports, 108 lives were saved as 28 drowned. The righthand photo, taken on March 6, 2016, shows a group of refugees in Greece approaching its northern border with Macedonia. By this time, however, the Macedonian government had erected a border fence, making it virtually impossible to cross.

Patrick Bar / SOS Mediterranee/Starface / Polaris/Newscom

Kay Nietfeld/dpa/picture-alliance/Newscom

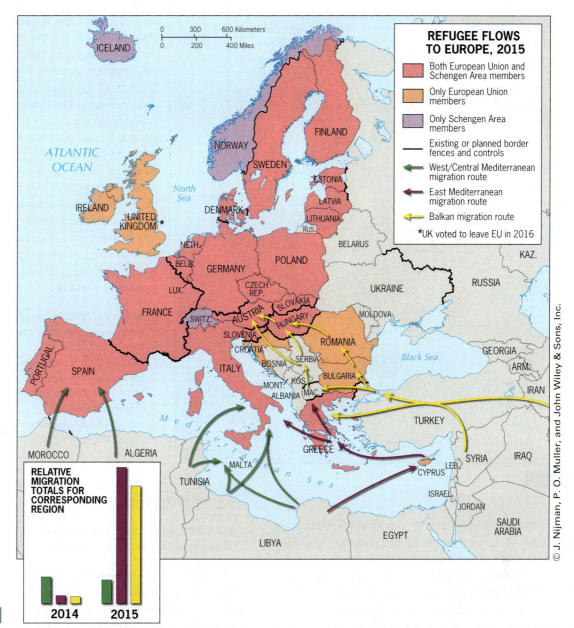

FIGURE 4-11

asylum (presumably out of fear that their application will be rejected). The EU total of asylum applications increased from 300,000 in 2011 to 627,000 in 2014 and 1.3 million in 2015. The main sources of the asylum applicants were Syria (29 percent), Afghanistan (14 percent), and Iraq (10 percent). Overall, 52 percent of the 2015 applications were successful.

Once people have set foot inside EU borders, (e.g., in Italy or Greece), they are free to move around to any of the 26 countries that are part of the **_Schengen Area_** (Fig. 4-11). It has been proposed, especially by Germany, that all EU member-states should accept their share of asylum-seekers, but most countries have been less than forthcoming. Some, particularly those on the Balkan route, have seized the initiative and put up border fences to avoid being overwhelmed by refugees (e.g., Hungary, Macedonia, Austria)(Fig. 4-11). Certain other EU countries have refused to take in any refugees at all, or only minimal numbers (e.g., Poland, Slovenia). As might be expected, most refugees would prefer to settle in Germany or Sweden than in Romania or Lithuania, and the problem of distributing refugees has become both intractable and politically divisive in the EU.

In the meantime, Italy, and especially Greece, became swamped by refugees at the same time they were experiencing difficult economic times. When Macedonia closed its border, thousands of makeshift refugee camps sprang up across northern Greece. It was this increasingly untenable situation in Greece that drove the EU to an agreement with Turkey. More than half the illegal border-crossers into Europe in 2015 (amounting to 1.2 million people) came via Turkey. From Turkey, they either crossed the Aegean Sea to Greece or entered Greece or Bulgaria via overland routes. The deal implied that Turkey would close and control its land borders with Greece and Bulgaria; that it would stop Aegean Sea crossings; and that it would accept returned refugees from Greece. In exchange, the EU pledged U.S. $3.2 billion in financial support to Turkey in order to run its expanded refugee centers, visa-free travel for Turks to the EU, _and_ promises to accelerate Turkey's bid for EU membership.

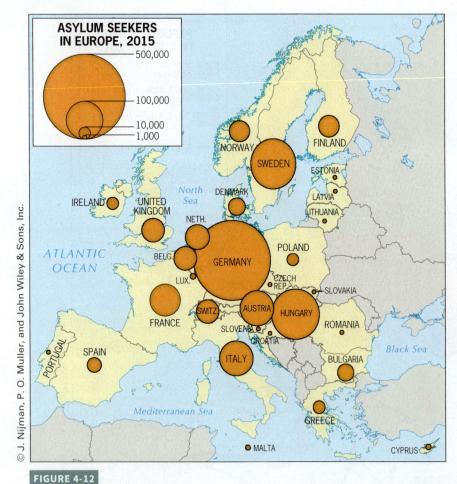

ASYLUM SEEKERS
IN EUROPE, 2015

500,000

100,000

10,000
1,000

FIGURE 4-12

on the rise, and the reasons are not always clear. On one hand, they are grounded in the involvement of the EU and individual European countries (e.g., France) in the conflicts of the Middle East; on the other hand, numerous European cities are home to relatively large numbers of disaffected, radicalized, second-generation Islamic youth—who were born in Europe, but who somehow do not fit in. Although the attackers were linked to Islamic State, it is hard to determine the extent to which their actions were centrally coordinated or locally initiated.

At press time, the Brussels bombings were only the latest in an escalating series of attacks that were claimed by, or attributed to, Islamist terrorists:

- November, 2015: 130 people died and hundreds were wounded in a series of attacks in Paris;
- January, 2015: 12 people were killed in the Paris office of the satirical newspaper *Charlie Hebdo* in retaliation for its articles on Islamic fundamentalism;
- May, 2014: four people were killed at the Jewish Museum in Brussels;
- March, 2012: six people, including Jewish schoolchildren, were killed in Toulouse, France;
- July, 2005: 52 London commuters died when suicide bombers blew themselves up on subway trains and a bus;
- March, 2004: a series of bombs on commuter trains killed 191 people and injured nearly 2000 in metropolitan Madrid.

These events have helped fuel right-wing, anti-immigration, and anti-Muslim sentiments—even if the overwhelming majority of European Muslims are as critical or fearful of Islamist terror as everyone else. The challenge for European governments is to fight terrorism without alienating this Muslim majority, and to prevent radicalization of young Muslims by building a more inclusive society. At the same time, it will require advanced and coordinated intelligence to find and destroy terrorist cells and thwart their murderous planning.

Immigration and the Growing Multicultural Challenge

It seems ironic, amidst all the problems revolving around refugees and Islamist terror, that most European countries are in dire need of immigrants—and that when these immigrants arrive, they largely come from non-Western societies, and are often Muslim.

There was a time when Europe's population was (in the terminology of population geographers) exploding, sending

The refugee problem and its possible solutions have had a divisive effect on both the EU and individual countries (**Box 4-8**). As noted above, it surrounded the UK's 2016 'Brexit' referendum, and it has fueled support for populist right-wing political parties from France to Denmark and from The Netherlands to Poland. The difficulty of distinguishing between "migrants" who seek a better economic future versus "refugees" who want to escape deadly conflict at home has complicated matters, as has the fact that certain aging European countries would clearly benefit from an influx of young immigrants (Box 4-7): of all asylum-seekers in the EU in 2015, 83 percent were under the age of 35. And then there is the challenge of rising Islamist terrorism in Europe during this decade. It may not be directly connected to the refugee crisis, but because myriad asylum-seekers are Muslims, for many Europeans the politics of these issues have become intertwined.

Islamist Terrorism in Europe

The bombings in Brussels, Belgium on March 22, 2016, which killed 32 people and wounded hundreds, were the work of ISIS (Islamic State of Iraq and Syria), the fundamentalist, terrorist organization that controls parts of those countries and Libya (see Chapter 6). Islamist terrorism in Europe has been

Box 4-7 MAP ANALYSIS

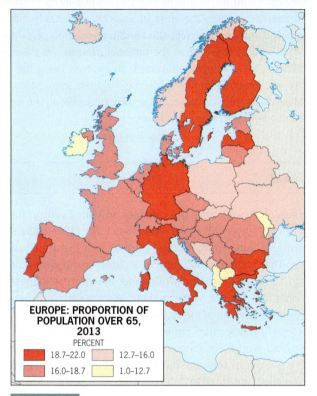

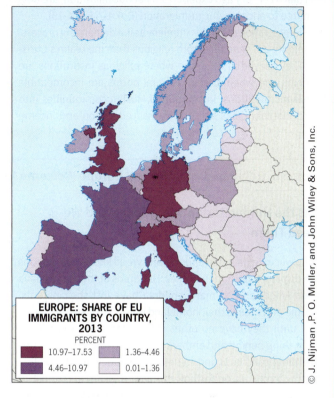

© J. Nijman , P. O. Muller, and John Wiley & Sons, Inc.

FIGURE 4-13

A popular saying proclaims that "Demography is Destiny"—because tomorrow's age-group distribution in any given population depends on today's birth rates and overall growth. Understanding the forces that shape population growth/decline—which also includes migration—is critical to understanding Europe's future. The geographic variation of demographic "destinies" is clearly visible on maps of the European realm. Some countries are recognized to

rank among the oldest and fastest aging in the world, and some are the preferred destinations for migrants from outside of Europe. Take a close look at the patterns shown in **Figure 4-13**. Considering their shares of elderly population, which countries would benefit most from an infusion of immigrants?

Access your WileyPLUS Learning Space course to interact with a dynamic version of this map and to engage with online map exercises and questions.

© epa european pressphoto agency b.v./ Alamy Stock Photo

Five days after the March 2016 terrorist attacks, a crowd gathers in Brussels' Place de la Bourse to memorialize the victims. The death toll was 32 and more than 300 were injured in ISIS-directed attacks on the airport and a downtown metro station.

millions to the New World and other colonies while still growing at home. But these days Europe's native population, unlike most of the rest of the world's, is shrinking. To keep a given population from declining, the (statistically) average woman must bear 2.1 children. For Europe as a whole, that indicator was 1.58 in 2014, and several countries recorded numbers well below that. Such negative population growth poses serious challenges for any nation. When the age-group pyramid becomes top-heavy, the number of workers whose taxes pay for the social services of the elderly goes down, leading to reduced pensions and dwindling funds for healthcare. It is estimated that, for the EU as a whole, the working-age population will shrink by 6 percent between 2015 and 2030. In some countries, that decline will be more severe: in Germany, for example, the working-age population is projected to drop another 13 percent by 2030.

The most obvious way to counter the adverse economic effects of negative population growth is to promote immigration, particularly of youthful working-age people and families with young children. This is why a number of European countries, which half a century ago experienced net out-migration, are now committed to encouraging in-migration (Box 4-7; Fig. 4-13).

The integration of Muslim immigrants is a demanding process. The majority are generally more religious than the realm's Christians. They arrived in a Europe where religious institutions are weakening and where certain cultural norms are incompatible with Islamic traditions. Integration of Muslim communities into the national fabric has been slow; their education and income levels are considerably below those of the host population; and unemployment among Muslims is usually higher. In western European countries, Islamic immigrants are heavily concentrated in metropolitan areas. Therefore, in major cities such as Hamburg, London, Paris, and Brussels, the Muslim population percentage is considerably higher than the national average. Yet there are positive signs of integration as well, an example being the May 2016 election of Sadiq Khan—a son of Pakistani immigrants and a Muslim—as the new mayor of London. Undoubtedly, he would describe himself first as a Londoner.

Box 4-8　Regional Issue: Should Europe Welcome Refugees?

It Is the Humanitarian Thing To Do, and We Need Them

"The war in Syria is horrendous, and we all know that. Millions are displaced from their homes. Hundreds of thousands have been killed, and so many children have lost their parents. People don't leave their homeland unless they are desperate and until they have no other options left. They must leave everything behind and embark on a dangerous journey. They are hungry and poor and vulnerable, in search of a decent life, some dignity, an opportunity to earn a living, and, most of all, to be safe. These are the people who are knocking on our door, here in Germany, and we should let them in—because we can, and because it is the right thing to do.

"When that horrible photo appeared in the news back in 2015, of the 3-year-old Syrian boy, Aylan Kurdi, whose body washed ashore on a Turkish beach, his mother and brother also deceased—that was a moment when many Germans realized the full scope of this tragedy, and that we had to react and open up to these people in need. Because if we don't, then what do we stand for, really?

"Of course we can take in several hundreds of thousands of refugees. It won't add up to even 1 percent of our population in Germany. Yes, it takes a major logistical effort at first, but they will gradually find a place to live and a job. You do realize that Germany needs young immigrants, right? With our aging population, we need young people to join our workforce. So, yes, there is a humanitarian argument and that comes first, but to all those who suggest that the refugees are a recipe for trouble, I say this: remember what these people gave up to come here and what they risked just to be here. And don't you think they will make grateful and productive citizens? They need our help now, and we can sure use their contribution to our country later."

There Is No Way We Can Harbor Them All, and They Don't Fit In

"Looking at this problem from Paris, I cannot believe how naïve some people are. If you put up banners and go on television to welcome the refugees to your country, the incoming surge will never end. Yes, I am talking about Germany. Sure, many refugees are in a desperate situation. That is understood, and in an ideal world we would help them all. But this world and the problems these people face are not of our making. We are very fortunate to be living in France today, and I am all too aware that it's a mess in many other countries, especially in Africa and the Arab World. For its people, war and conflict seem to be the order of the day. But they all get the news, and if our leaders are going to declare that refugees are welcome here, it will be impossible to stop them from coming. And I wouldn't blame them.

"If the EU could set a limit on the number of refugees and they could be equitably allocated among the member-states, I could see a rationale. But what has been going on thus far is sheer foolishness. You know that most of the refugees don't come from countries at war. The majority are opportunity seekers from all over Africa and southwestern Asia. They all received word that they can get in, and now they are on the way. This is why most migrants are men. If they were truly refugees, wouldn't they just be happy to have left Syria and made it to Turkey or Greece, and then stay there? No, the moment they arrive in those countries, they set out for the most desirable places, like Germany, Sweden, or France. Too many are not refugees—they are economic migrants.

"Finally, and I know this may not be politically correct here in Europe today, but a great number of the asylum-seekers are Muslims and it seems to me that this, too, is asking for problems. We already have more than 6 million Muslims in this country and most are not properly integrated. The cultural differences remain significant, and most simply do not want to fit in. We have Islamic militants, too, and the terrorist attacks of 2015 shocked our country to the core. Are we really going to welcome hordes of Muslim men from Syria and Iraq and Afghanistan into this country without knowing their backgrounds, and trusting that they have good intentions? For the sake of our own well-being, security, and future, we must keep these refugees and migrants out."

Regions of Europe

In the past, Europe divided rather easily into Western, Northern (Nordic), Mediterranean (Southern), and Eastern regions, and countries were grouped accordingly. This traditional configuration remains relevant today, but in the era of European unification, a **core-periphery** framework has been superimposed on that regional scheme, defined in terms of how central—or peripheral—countries are to the workings of the EU. In other words, Europe can still be understood as a set of formal, more or less homogeneous, cultural regions; yet at the same time, the entire realm increasingly functions as a tightly interdependent economic and political regional system, a spatial network that revolves around the EU. We distinguish four major regions that constitute the European realm (**Fig. 4-14**): (1) *western Europe*, including Britain and Ireland; (2) *northern Europe*;

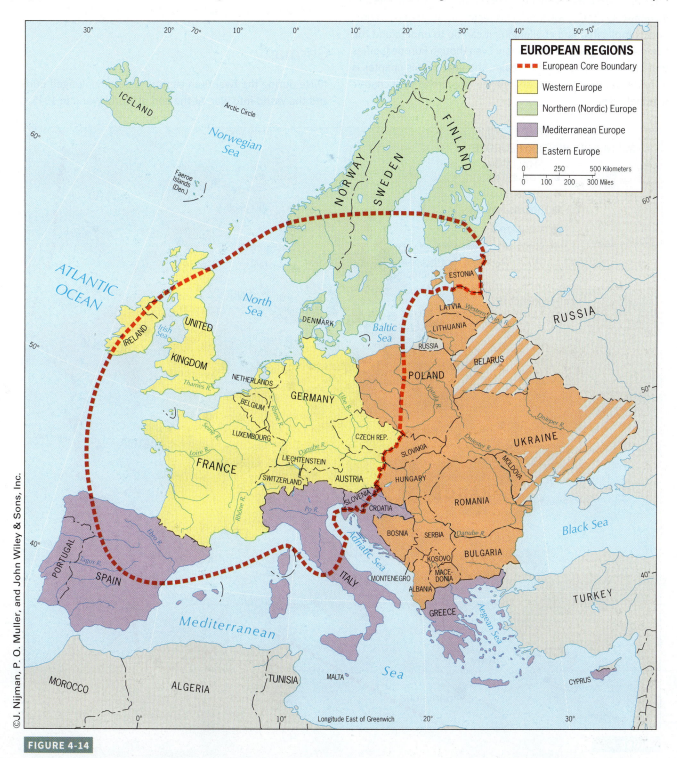

FIGURE 4-14

(3) **Mediterranean Europe**, including Slovenia and Croatia; and (4) **eastern Europe**, the broad eastern tier of the realm that extends to the Russian border.

Take a close look at Figure 4-14 and note that the European Core (bounded by the red dashed line) in many places extends beyond western Europe and rarely coincides with national borders. Thus the European Core is primarily defined on the basis of regional economic performance, not political territorialization. For instance, northern Italy (but not the Italian south) and southern Sweden (but not the Swedish north) are both part of this Core; also note the inclusion of Estonia in the Core, despite its relatively remote location. This regionalization underscores how complicated and challenging it can be to understand the complexity of spatial organization. Even though Europe typifies that challenge like no other realm, the goal of this chapter is to bring clarity to what at first glance appears to be a chaotic regional arena.

Region Western Europe

Western Europe consists of 11 countries including the **micro-state [23]** of Liechtenstein (**Figure 4-15**, **Box 4-9**). This region also lies almost entirely within the European Core that forms the heart of the EU. The large offshore islands of Britain and Ireland (referred to in the past as the British Isles) have been closely involved in western Europe's history as well as its ongoing economic and political development. Let us begin on the mainland with Germany, which is both the realm's most populous country and largest economy.

Germany

Twice during the twentieth century Germany plunged Europe and the world into war. At the end of World War II in 1945, the

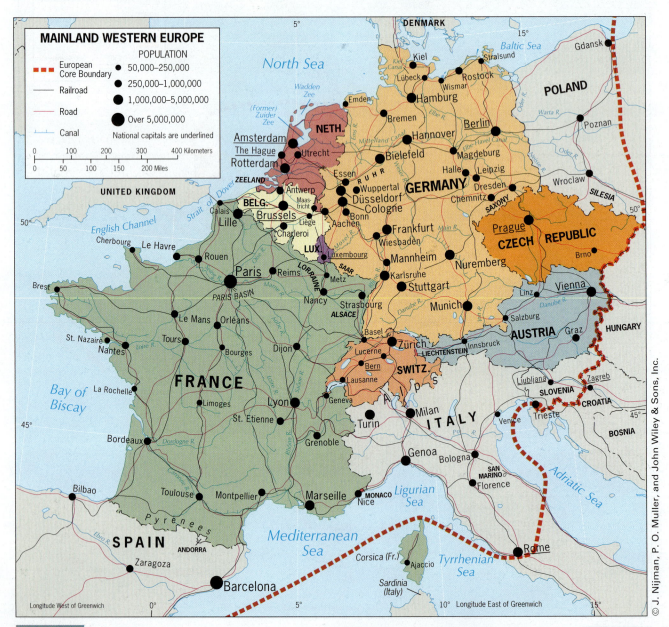

FIGURE 4-15

Box 4-9 From the Field Notes …

"It has been a long day, driving north from Italy through the Swiss Alps. I just crossed the border into the ministate of Liechtenstein to look for a place to spend the night. From my hotel in the town of Triesenberg, I have a panoramic view of an upper stretch of the Rhine River. It flows from left to right in the picture. This is western Europe's leading river and one of its defining geographic features—as a means of transportation, sometimes as a barrier, and frequently as a natural boundary. Here it forms the ministate's border with Switzerland, located on the far side. Even though we are looking at what is a political boundary, the physical and cultural landscapes on both sides of the Rhine have much in common."

© Jan Nijman

defeated and devastated German state was divided into two parts: West Germany and East Germany (their former border is delimited by the red line in **Fig. 4-16**), the latter of which was forced to surrender some important territory to Poland.

West and East Germany then set forth on widely different economic and political trajectories. The Soviet governance of East Germany was established on the USSR's communist model, and as repayment for the extreme hardships the Soviet Union had suffered at German hands during the war, this rule was unremittingly punitive. The U.S.-led authority in West Germany was far less strict and aimed at rehabilitation rather than retribution. West Germany was included when the Marshall Plan was instituted in 1948, and the massive infusion of aid stimulated rapid economic recovery. At the same time, West Germany was reorganized politically into a modern federal state based on democratic foundations.

West Germany's economy soon began to thrive. Between 1949 and 1964, its gross national income (GNI) tripled as industrial output rose by 60 percent. Simultaneously, its political leaders participated enthusiastically in the negotiations that created the six-member Common Market in 1957 (and thereafter evolved to become today's EU). Geography worked in West Germany's favor: it had common borders with all but one of the original member-states. Its swiftly rebuilt transportation infrastructure was second to none in Europe. The reinvigorated spatial economy focused on heavy industrial production in the Ruhr Basin (in the hinterland of the gigantic Dutch port of Rotterdam) in addition to the leading-edge manufacturing complexes centered on Hamburg in the north, Frankfurt (the leading financial and air transport hub as well) in the center, and Stuttgart in the south. West Germany now exported huge

quantities of iron, steel, motor vehicles, machinery, textiles, and farm products. Until just a few years ago, Germany had the largest export economy by value in the world; although it has now been surpassed by China and the United States, Germany's exports per capita are still by far the highest on Earth.

At the time of Germany's reunification in 1990, West Germany had a population of about 62 million and East Germany 17 million. By 2015, former East Germany's population was down to less than 13 million. Decades of communist misrule in the East had yielded outdated factories, crumbling infrastructures, polluted environments, drab cities, inefficient farming, and inadequate legal and other institutions. Thus the 1990 reunification was more of a rescue than a merger, and the cost to the reconstituted German state since then has been gargantuan. Figure 4-16 maps the 16 States (or *Länder*) that constitute today's reunified Federal Republic of Germany and also makes a key point: regional disparity in terms of gross domestic product (GDP) per person remains a serious problem. Note that all of former East Germany's six States (except Berlin) still rank in the lowest income category, whereas nine of the ten former West German States rank in the two higher-income categories.

Another challenge facing Germany, along with other western European countries, is the integration of sizeable immigrant minorities. Germany now has nearly 16 million immigrants, about 20 percent of the total population of 80.7 million. They not only are culturally different but also worse off economically: more than 40 percent of immigrant households live below the official poverty line, largely because the unemployment rate of immigrants far exceeds that of native Germans.

Nevertheless, Germany navigated the recent global recession better than any other European country and retains its

STATES (LÄNDER) OF GERMANY

GDP PER CAPITA,
IN EUROS, 2013

- Over 30,000
- 25,000–30,000
- 20,000–25,000

City population
- • 50,000–250,000
- ● 250,000–1,000,000
- ⬤ 1,000,000–5,000,000
- ⬤ Over 5,000,000

—— Railroads
—— Roads

National capitals are underlined

0   50   100 Kilometers
0   25   50 Miles

Data source: Eurostat, 2013

© J. Nijman, P. O. Muller, and John Wiley & Sons, Inc.

FIGURE 4-16

position as the realm's economic powerhouse. Today, this country is developing closer ties with eastern Europe, and from the Balkans to the Baltic Sea, the Germans are regarded as the region's most desirable trading partner. Germany has also played a leading role in EU bailouts of Greece and other countries, and—in an ironic historical twist, given the initial reasons for European integration—has been a vital force in maintaining a sufficient degree of EU unity.

France

Territorially, France is larger than Germany, and the map suggests that France has a superior relative location, with coastlines

on the Mediterranean, the Atlantic Ocean, and, at Calais, even a window on the North Sea. But France does not have any good natural harbors, and oceangoing ships cannot navigate its rivers and other waterways very far inland.

The map of mainland western Europe (Fig. 4-15) reveals a significant contrast between France and Germany in terms of their **urban systems [24]**, the spatial distribution and interrelationships among cities of various sizes. France has one dominant city, Paris, and no other city comes close either in size or centrality: Paris has 10.9 million residents, whereas its closest rival, Lyon, contains only 1.6 million. Germany has no city to match Paris, but it does have a wider range of medium-sized cities: in the 100,000-plus category, Germany has more than 80, whereas France has less than 40.

Leipzig is eastern Germany's biggest city after Berlin. It has a rich cultural history and has been home to, among others, Johann Sebastian Bach, Johann Wolfgang von Goethe, Richard Wagner, Friedrich Nietzsche, and Felix Mendelssohn. But during communist times Leipzig was dragged down along with the rest of East Germany. Although this city is now finally experiencing something of a revival, there are still many struggling neighborhoods and outlying communities.

Paris is an example of a **primate city [25]**, one that is disproportionately large compared to all others in the national urban system, exceptionally expressive of its country's culture, and (almost) always the capital. Paris grew so large and dominant within the French urban system for reasons of geography

and history. Geographically, Paris has advantages in terms of both **site [26]** (the physical attributes of the place it occupies) and **situation [27]** (its location relative to surrounding areas). The *site* of the original settlement at Paris lay on an island in the Seine River, a defensible spot where the river was often crossed. This island, the Île de la Cité, was a Roman outpost 2000 years ago; for centuries its security ensured continuity. Eventually the island became overcrowded, and the city expanded along the banks of the river (**Fig. 4-17**). Paris's advantageous *situation* lay in its fertile agricultural hinterland and the presence of key waterways. The Seine River is joined near Paris by several navigable tributaries (the Oise, Marne, and Yonne). When canals extended these waterways even farther, Paris was linked to the Loire Valley of central France; the Rhône-Saône Basin of the southeast; Lorraine, a key industrial area in the northeast; and the northern border zone with Belgium. When Napoleon reorganized France and built a radial system of roads (duplicated later by railroads) that focused on Paris from all parts of the country, the city's primacy was assured (Fig. 4-17, **Box 4-10**).

Historically, France early on developed the strong tradition of a highly centralized state in which the central government in Paris maintained tight control over 96 subnational *départements*. It remained so for nearly two centuries, but France decentralized its administrative structure in 1982 with the introduction of a new layer of governance, consisting of 26 larger *regions* (22 in France itself and 4 overseas). These regions were inserted at the level between Paris and the *départements* both to accommodate devolutionary forces and to seize the opportunities for

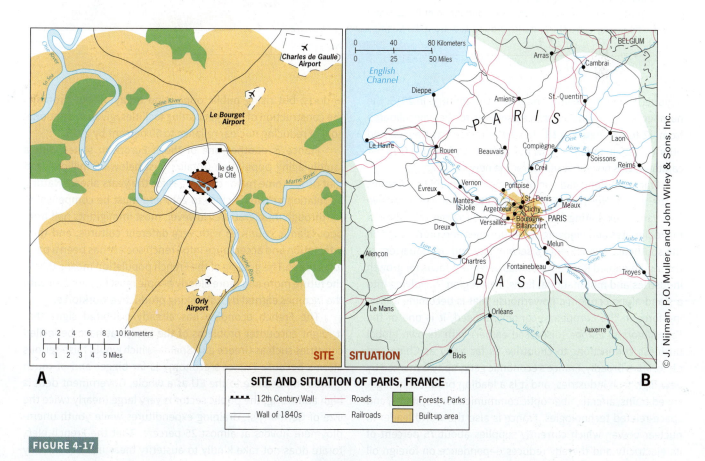

SITE AND SITUATION OF PARIS, FRANCE

FIGURE 4-17

Box 4-10 Among the Realm's Great Cities: Paris

If the greatness of a city were to be measured solely by its number of inhabitants, Paris (10.9 million) would not rank in the world's top 30. But if greatness is measured by a city's historic heritage, cultural content, and international influence, Paris has few peers. Old Paris, near the Île de la Cité that housed the original village where Paris began (Fig. 4-17) and carries the eight-century-old Notre Dame Cathedral, contains an unparalleled assemblage of architectural and artistic landmarks old and new. The Arc de Triomphe, erected by Napoleon in 1806 (though not completed until 1836), commemorates the emperor's victories and stands as a monument to French neoclassical architecture, overlooking one of the world's most famous avenues, the Champs Elysées, which leads to the grandest of city squares, the Place de la Concorde, and continues on to the magnificent palace-turned-museum, the Louvre.

Even the Eiffel Tower, built for the 1889 International Exposition over the objections of Parisians who regarded it as ugly and unsafe, became a treasure. From its beautiful Seine River bridges to its palaces and parks, Paris embodies French culture and tradition. It is perhaps the ultimate primate city in the world.

As the capital of a globe-girdling empire, Paris was the hearth from which radiated the cultural forces of Francophone assimilation, transforming much of North, West, and Equatorial Africa, Madagascar, Indochina, and many smaller colonies into societies on the French model. Distant cities such as Dakar, Abidjan, Brazzaville, and Saigon acquired a Parisian atmosphere. France, meanwhile, spent heavily to keep Paris, especially Old Paris, well maintained—not just as a relic of history, but as a functioning, vibrant center, an example to which other cities can aspire.

Today, Old Paris is ringed by a new and different Paris. Stand atop the Arc de Triomphe and turn around 180° from the Champs Elysées, and the tree-lined avenue gives way to the vista of *La Défense*, an ultramodern high-rise complex that is one of Europe's leading business districts (see Box 4-4). But from the top of the Eiffel Tower you can see as far as 80 kilometers (50 mi) and discern parts of a Paris visitors rarely experience: grimy, aging industrial quarters, and poor crowded neighborhoods where discontent and unemployment fester—and where immigrants cluster in a world apart from the splendor of the old city.

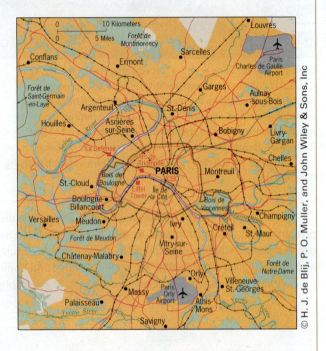

© H. J. de Blij, P. O. Muller, and John Wiley & Sons, Inc

regional and local growth throughout the country. In 2016, the number of regions inside France itself was further consolidated from 22 to 13. **Figure 4-18** shows the regions as they existed prior to 2016 in order to show the regional variation in GDP per capita—it was not yet available for the new regions.

France has long had one of the most diversified economies in the world. Northern French agriculture remains intensive and varied, exploiting the country's wide range of soils and microclimates and enjoying state subsidies and protections. In the southeast, Lyon, France's second city and headquarters of the region named Rhône-Alpes, has become a focus for growth industries and multinational firms. This region is evolving into a stand-alone economic powerhouse that is becoming a driving force in the European economy. Indeed, it is one of the *Four Motors of Europe* (discussed earlier) with its own global business connections to countries as far away as China and Chile. More broadly, France's economic geography is marked by new high-tech industries, and it is a leading producer of high-speed trains, aircraft, fiber-optic communications systems, and space-related technologies. France is also the world leader in nuclear power, which currently supplies about 75 percent of its electricity and thereby reduces dependence on foreign oil

imports; but rising public opposition today all but prohibits the construction of new generating facilities, given the national goal of reducing nuclear reliance to 50 percent by 2025.

France is another one of those European states with a rapidly aging population (totaling just below 65 million), and that poses some major challenges for the years ahead. Natural population growth is close to zero and must be compensated for through immigration. However, most immigrants hail from poor Islamic countries with very different cultures, and their integration and acceptance into French society has been a painful process. The graying of the French population means that in the future more and more elderly people must be cared for with the incomes earned by a shrinking productive workforce.

The French economy has already exhibited signs that it might encounter problems of the kind that have bedeviled countries such as Greece and Spain—which would raise serious alarms because France's economy is far bigger and of much greater importance to the EU as a whole. Government debt is high and rising; the public sector is very large (nearly twice the size of Germany's), draining expenditures while youth unemployment hovers at almost 25 percent. That the French electorate does not take kindly to austerity measures has already

RÉGIONS OF FRANCE

GDP PER CAPITA, IN EUROS, 2013

- Over 30,000
- 25,000–30,000
- 20,000–25,000

City population

- 50,000–250,000
- 250,000–1,000,000
- 1,000,000–5,000,000
- Over 5,000,000

Railroad

National capitals are underlined

Data Source: Eurostat

DÉPARTEMENTS

© J. Nijman, P. O. Muller, and John Wiley & Sons, Inc.

FIGURE 4-18

been demonstrated in their 2012 election of a president from the Socialist Party. And the clearest sign that all is not well came in the response to the horrific 2015 terrorist attack on the Paris office of *Charlie Hebdo* (discussed above). In its wake, France's right-wing *Front National* became the fastest-growing French political party, espousing an extremist, ultranational agenda that is highly critical of Muslim and other immigration as well as EU policies in general.

Benelux

Three small countries are crowded into the northwestern corner of mainland western Europe—Belgium, the Netherlands, and Luxembourg—and are collectively referred to by their first syllables (*Be-Ne-Lux*). Their total population of 29 million reminds us that this is one of the most densely peopled corners of our planet. Not coincidentally, the *Low Countries*, as they are also sometimes called, are situated at the **estuary [28]** (mouth) of the great Rhine and Scheldt rivers and have access to the seas that, throughout modern history, has been the envy of the Germans and the French. They are a highly productive trio: both the Netherlands and Belgium generally rank among the top 20 economies on Earth, and tiny Luxembourg has one of the world's highest per capita gross national incomes.

The Netherlands, one of Europe's oldest democracies and a constitutional monarchy today, has for centuries been expanding its territory—not by warring with its neighbors but by wresting land from the sea. About a quarter of the country lies below sea level. The urban, demographic, and economic center of gravity lies within a **conurbation [29]** (a large urban complex formed by the coalescence of two or more cities) in the west of the country. This so-called *Randstad* ('Ring City'),

roughly triangular in shape, is anchored by Amsterdam, the constitutional capital; Rotterdam at the mouth of the Rhine, Europe's largest port; and The Hague, the national seat of government (**Box 4-11**, **Figure 4-19**).

Belgium also has a thriving economy as well as a leading port in the city of Antwerp. With 11.4 million people, Belgium's

---

Box 4-11 Regional Planning Case

Amsterdam's North-South Line

One of the most complicated and specialized infrastructural urban planning projects in Europe has been under way in Amsterdam since 2003 and is scheduled for completion sometime during 2017. This is the so-called North-South Line, an underground metro (subway) line that extends for about 10 kilometers (6 mi), with its central segment running beneath Amsterdam's seventeenth-century city center. This urban core is the heart of a typical historic European city, but one that is unusually compact and marked by a particularly dense network of buildings and streets. The resulting congestion impedes above-ground trams and buses, and is urgently in need of the much faster public transit that only a modern metro line can supply.

The North-South Line will also significantly improve connectivity among several parts of the city (Fig. 4-19):

- Amsterdam-North is separated from the city by the River IJ and is far less developed than the CBD lying just across this waterway to the south. Improved accessibility to Amsterdam-North is expected to produce considerable development after the new Line goes into operation.
- The Zuidas (meaning "South Axis") is a major high-rise business district and employment hub located just south of the Ring Road. Analogous to *La Défense* at the edge of Paris (see Box 4-4), it specializes in financial services but does not have

good transit linkages to Amsterdam's CBD. There is a nearby railway station, Amsterdam-South, that has been converted into a multimodal, public transportation facility to accommodate commuter trains, the new metro, trams, and buses.
- Schiphol Airport (the world's 14th busiest in 2015) lies to the southwest of the city. It has rail linkages to both Central and South stations, but before 2017 had only poor transit connections to the rest of the metropolis.

What makes the North-South Line such a unique engineering project is that Amsterdam lies entirely below sea level and is a city built entirely on pilings anchored to what is generally soggy soil. Thus every structure rests either on old wooden pilings or on more modern concrete pilings, mostly about 12 meters (40 ft) deep. Under heavier structures, the pilings extend to a greater depths; for example, the builders of the new metro station below the Central Station had to contend with pilings going down as far as 60 meters (200 ft).

Hence, the North-South Line had to be built beneath a city resting on hundreds of thousands of pilings, all of which had to be circumvented and left undisturbed. Thus the Line has to run directly beneath existing main streets (where there are no pilings), and its two tracks were mostly forced to be positioned one above the other. But even Amsterdam's main streets are not that wide, which required the Line to get very close to the pilings of adjacent buildings (see photo). It also explains why progress had at times been so painstakingly slow. Most of all, however, this herculean project showcases the genius of Dutch engineering in water-logged environments. In the tongue-in-cheek words of one of the North-South Line's lead engineers, "Why do we do this? Because we *can*!"

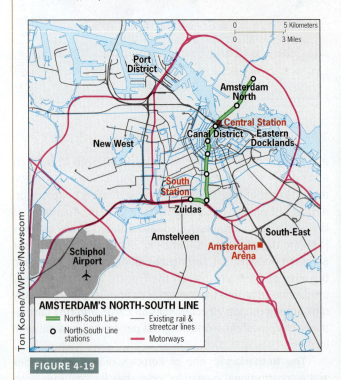

FIGURE 4-19

The North-South Line under construction in the center of Amsterdam during 2011. Because the Line runs entirely below sea level, steel walls were required to keep water out of the construction site. All the buildings adjacent to the site rest on pilings.

regional geography is dominated by a cultural fault line that cuts diagonally across the country, separating a Flemish-speaking majority (59 percent) centered on Flanders in the northwest from a French-speaking minority in southeastern Wallonia (40 percent). Brussels, the predominantly French-speaking capital, forms a linguistic island in the Flemish-speaking sector; but the city also is one of Belgium's greatest assets because it serves as the headquarters, and in many ways as the *de facto* capital, of the European Union (though it is important to note that the EU does not have an official capital city). Nonetheless, devolution is a persistent problem for Belgium, with political parties that agitate for Flemish separatism constantly roiling the social landscape.

Luxembourg, one of the realm's numerous ministates, lies between Germany, Belgium, and France, with a Grand Duke as head of state (its official name is the Grand Duchy of Luxembourg), a territory of only 2600 square kilometers (1000 sq mi), and a population of just 580,000. Luxembourg has translated sovereignty, relative location, and stability into a haven for financial, services, and information-technology industries. There are more than 150 banks in this tiny country and about 14,000 holding companies (corporations that hold controlling stock in other businesses in Europe and worldwide). With its unmatched GDP per capita (more than U.S. $110,000), Luxembourg is in some ways the greatest beneficiary of the advent of the European Union. And in no country in Europe is support for the EU stronger than it is here.

The Alpine States

Switzerland, Austria, and the microstate of Liechtenstein on their border share an absence of coasts and the mountainous terrain of the Alps and not much else (Fig. 4-15). Austria speaks one language; the Swiss speak German in their north, French in the west, Italian in the southeast, and even a bit of Rhaeto-Romansch in the remote central highlands (Fig. 4-6). Austria has a large primate city; multicultural Switzerland does not. Austria has a substantial range of domestic raw materials; Switzerland does not. Austria is twice the size of Switzerland and has a larger population, but far more trade moves across the Swiss Alps between western and Mediterranean Europe than crosses Austria.

Switzerland, not Austria, is in most ways the leading state in the Alpine subregion of western Europe. Mountainous terrain and a **landlocked location [30]** can constitute crucial barriers to economic development but, as is so often the case, geography poses constraints in one way while offering opportunities in another. That is why Switzerland is such an important lesson in the complexities of human geography. Through the skillful maximization of their opportunities (including the transfer needs of their neighbors), the Swiss have transformed their seemingly restrictive environment into a prosperous state. They deftly utilized the waters cascading from their mountains to generate hydroelectric power needed to develop highly specialized industries. Swiss farmers perfected

A view of Lake Lucerne in central Switzerland, with the town of Stansstad in the foreground and the city of Lucerne in the distance. In this country, connectivity has triumphed over the challenging terrain. Near the opposite end of this lake, Switzerland in 2016 opened the Gotthard Rail Tunnel (the world's longest)—a 57-kilometer (35-mi) marvel running below the entire massif of the Alps.

©Jan Nijman

methods to optimize the productivity of mountain pastures and valley soils. Swiss leaders converted their country's isolation into stability, security, and neutrality, making it a world banking behemoth, a global magnet for money. Zürich, in the German-speaking sector, is the financial center; Geneva, in the French-speaking sector, is one of the world's most internationalized cities. The Swiss feel that they do not need to join the EU—and they have no plans to apply in the foreseeable future.

Austria, which joined the EU in 1995, is a remnant of the old Austro-Hungarian Empire and has a historical geography that is far more reminiscent of unstable eastern Europe than that of Switzerland. In fact, even Austria's physical geography seems to demand that the country look eastward: it is at its widest, lowest, and most productive in the east, where the River Danube links it to Hungary, its old ally in the anti-Muslim wars of the past.

Vienna, by far the Alpine subregion's largest city, also lies on the country's eastern perimeter. One of the world's most expressive primate cities with magnificent architecture and monumental art, Vienna today is western Europe's easternmost major city, but its relative location has changed dramatically with EU enlargement since 2000. Even though Vienna found itself in a far more centralized position when the EU border shifted eastward, most Austrians doubted that their neighbors to the east could ever become EU members of potentially equal standing. One result is that Austrian public support for the European Union today has fallen to a record low level.

The Czech Republic

The Czech Republic is focused on the historic Bohemian Basin, the mountain-encircled national core area centered on the capital, Prague. This is another classic primate city, its cultural landscape highly reflective of Czech traditions. Prague is also an important industrial hub and collection center, surrounded by mountains that contain many valleys with small towns that specialize, Swiss-style, in fabricating high-quality manufactures. In the old eastern Europe, even during the communist period, the Czechs always were leaders in technology and engineering; their products could be found on markets in foreign countries near and far. Moreover, Bohemia always was cosmopolitan and Western in its exposure, outlook, development, and linkages; Prague lies in the upper basin of the Elbe River, its traditional outlet through northern Germany to the North Sea. Not long ago, the Czech Republic would have been considered part of "Central Europe," but today it has shifted its gaze entirely to the west, with its solid economic development fully warranting inclusion in the European Core (Fig. 4-14).

The United Kingdom (UK)

It has often been observed that the geographical position of the United Kingdom, just off the European mainland, reflects its traditional hesitation to consider itself part of 'Europe.' It did join the EU in 1973, but was always reluctant to move toward tighter integration. When most member-states in 2002 adopted the new euro in favor of their national currencies, the British kept their pound sterling and delayed their participation in the European Monetary Union (EMU)—a decision they did not regret when the 'euro crisis' unfolded in the late 2000s. Meanwhile, within the UK, opposition to continued EU membership steadily mounted, culminating in the 2016 'Brexit' referendum—whose close vote to break away from the EU deeply divided the country along class lines.

With an area about the size of Oregon and a population of 65 million, the UK is by European standards a fairly large country. Based on a combination of physiographic, historical, cultural, economic, and political criteria, the United Kingdom can be subdivided into four subregions (numbered in **Fig. 4-20**):

1. **England.** So dominant is this subregion of the United Kingdom that the entire country is sometimes referred to by this name. England is anchored by the sprawling London metropolitan area, which by itself contains nearly one-sixth of the UK's total population. Indeed, along with New York, London is regarded as one of the top-tier **world-cities [31]**, with financial services, high-technology, communications, engineering, and associated industries reflecting the momentum of its long-term growth and agglomeration (see **Box 4-12**). Within England, there is a notable difference between the crowded, fast-paced, and globally connected southeast centered on London, and the far less prosperous north and west (see inset map in Fig. 4-20). The latter areas, once the global hearth of the Industrial Revolution, have now completely lost their dynamism as well as segments of their population in our age of deindustrialization. Spatial inequality has also been rising, and today's widening gap between rich and poor within England's residential mosaic shows no sign of abating.

2. **Wales.** This nearly rectangular, rugged territory was a refuge for ancient Celtic peoples, and in its western counties more than half the inhabitants still speak Welsh. Because of the high-quality coal reserves in its southern tier, Wales, too, was engulfed by the Industrial Revolution, and Cardiff, the Welsh capital, was once the world's leading coal exporter. But the fortunes of Wales also declined, and many Welsh emigrated. In several cultural and economic spheres,

Wales is closely tied to England. and independence is not in the cards. In the Welsh Assembly, the nationalist party claims only 11 of the 60 seats. The situation in Scotland, however, is entirely different.

3. **Scotland.** With a population of 5.5 million, Scotland is a major component of the UK—but perhaps for not much longer. It is also well removed from Britain's thriving southeastern core area. Most of the population resides in the Scottish Lowlands, anchored by Edinburgh, the capital, in the east and Glasgow in the west. Attracted there by the labor demands of the Industrial Revolution (coal and iron reserves lay nearby), the Scots developed a world-class shipbuilding industry. Subsequent decline and obsolescence were followed by high-tech development (notably around Glasgow) and Scottish participation, beginning in the 1970s, in the exploitation of oil and gas reserves beneath the adjacent North Sea. Nevertheless, a growing number of Scots feel that they are increasingly disadvantaged within the UK and should play a bigger role in the EU. In 2014, a referendum on the question of full independence won the support of 45 percent of the electorate. Since then, the separatist cause has gained momentum—only fueled further by the 'Brexit' decision (opposed strongly here) and its new opening for independence followed by Scotland rejoining the EU.

4. **Northern Ireland.** Northern Ireland was home not only to Protestants from Britain, but also to a substantial population of Irish Catholics who found themselves on the wrong side of the border when Ireland became independent nearly a century ago. Ever since, conflict has intermittently engulfed Northern Ireland and spilled over into Britain. A declining share of the overall population of 1.8 million is Protestant and traces its ancestry to Scotland or England; meanwhile, the proportion of Roman Catholics is rising, and in the capital (Belfast) they achieved majority status in 2014. Catholics share their religion with the Irish Republic on the southern side of the border. Although Figure 4-20 suggests that there are majority areas of Protestants and Catholics in Northern Ireland, no such clear separation exists; mostly they reside in clusters throughout the territory, including the walled-off neighborhoods that still mark the major cities of Belfast and Londonderry.

Protesters rallying outside the Houses of Parliament on the tumultuous morning after the 'Brexit' referendum. Here in London, the "Remain" vote won decisively but was unable to overcome the large plurality of "Leave" voters in the rest of the UK. These angry thousands mostly blamed the Tory government for mishandling the referendum campaign and stirring up negative populist sentiments; in fact, Prime Minister Cameron was in the process of resigning at that very hour on June 24, 2016.

Jeff J Mitchell/Getty Images

Box 4-12 Among the Realm's Great Cities: London …

Along the banks of the Thames, this megacity of almost 10.5 million inhabitants displays the heritage of state and empire: the Tower and the Tower Bridge, the Victoria Embankment, and the Houses of Parliament (officially known as the Palace of Westminster). But in recent times this historic landscape has become studded with ultramodern office buildings, museums, and entertainment sites such as the EYE, the giant ferris wheel that provides a panoramic view of this giant **metropolis [32]** (central city plus its suburban ring). A bit further downriver, the economic heart of the city (and of the UK) beats 24/7 in what is known as the *City of London*, an incredibly dense concentration of office activity surrounding the Bank of England that houses London's enormous finance sector and related producer services (accounting, advertising, consulting, insurance, etc.). London, together with New York, is one of the most preeminent world-cities, in which business decisions are made daily that have consequences all across the globe.

This is also one of the world's most cosmopolitan cities, with countless nationalities and immigrant ethnic communities. But London has also become hugely expensive, and the gulf between its rich and poor has steadily widened. Lately, a substantial number of lower- and middle-income households have had no choice but to move to the suburbs or even farther away because life in this city has simply become unaffordable. Central London's property has almost become like a "global reserve currency" where the world's rich invest their money. For the year 2012, it was calculated that $120 billion worth of real estate in London was paid for in cash.

The urban layout of the London region reveals much about the internal spatial structure of the European metropolis in general. Such a **metropolitan area** remains focused on the large city at its center, especially the downtown **central business district**, which is the oldest commercial hub of the urban agglomeration. Cities with a preindustrial past are often located on a river (or harbor), expanding from their historical nucleus in a more or less concentric manner—as the case of London illustrates. It should also be noted

that an encircling greenbelt was set aside during the 1930s for recreation, horticultural farming, and other nonresidential, noncommercial uses. Although London's subsequent growth eroded parts of this greenbelt, in places leaving only "green wedges," the design preserved critically needed open space in and around the city, channeling suburbanization outward into the urban fringe more than 40 kilometers (25 mi) from the center.

Compared to large North American cities, London is highly compact. This allows for effective public transportation, but it also requires careful urban planning. Consider this: in 2016, Heathrow was the world's sixth busiest international airport, but it has only two runways. It has long been operating at full capacity, but London's ever-growing air traffic now requires additional landing facilities. There are four other airports in the metropolis—Gatwick, Stansted, Luton, and City—but they are insufficient. So now the government is planning yet another sizeable airport to the east of London near Gravesend on the south bank of the Thames.

From the national government's point of view, London poses a dilemma of sorts: it is absolutely vital to the British economy, but as a highest-echelon global city, it is in certain ways better connected to the rest of the world than it is to, say, the outer reaches of the UK. One result is the still-widening gap in prosperity between southeastern England and the rest of the country (underscored in the 'Brexit' vote result). Yet London's needs are always top-priority: e.g., it receives almost 90 percent of UK government funding for transportation projects in order to keep this premier world-city moving.

© Jan Nijman

In both 2014 and 2015, London was ranked (by the high-end real estate company, Knight Frank) as the best place for the world's wealthiest people to live and/or own a home. Not surprisingly, it is also one of the most expensive cities in the world. To reside here in the Royal Borough of Kensington and Chelsea, a high-cachet neighborhood in central London close to Buckingham Palace, be prepared to pay more than U.S. $2 million for a two-bedroom apartment (houses start at $5 million). Foreign investors, from Russia, China, and the oil-rich Persian Gulf States are most prominent among the buyers. London may be an extreme case, but it is indicative of a trend among a select few "hot" cities (New York, Paris, and San Francisco are others) that are experiencing skyrocketing real estate values due to foreign interest—increasingly to the detriment of established middle-class residents who can no longer afford to remain in their longtime neighborhoods.

© H. J. de Blij, P. O. Muller, J. Nijman, and John Wiley & Sons, Inc.

METROPOLITAN LONDON

Built-up area	Road
Greenbelt	Railroad
Central Business District	

0   10   20   30   40 Kilometers
0   10   20 Miles

Bletchley
Leighton
Hitchin
Luton
Dunstable
Welwyn Garden City
St. Albans
Hemel Hempstead
High Wycombe
Watford
Hertford
Harlow
Chestnut
Chigwell
Brentwood
The City
Silvertown
South Benfleet
Maidenhead
Slough
Thames Barrier
Southend-on-Sea
Windsor
Greenwich
Dartford
Bracknell
Heathrow Airport
LONDON
Gravesend
Wokingham
Chertsey
Chatham
Camberley
Caterham
Leatherhead
Maidstone
Fleet
Guildford
Tunbridge
Aldershot
Godalming
Gatwick Airport
Crawley
Tunbridge Wells

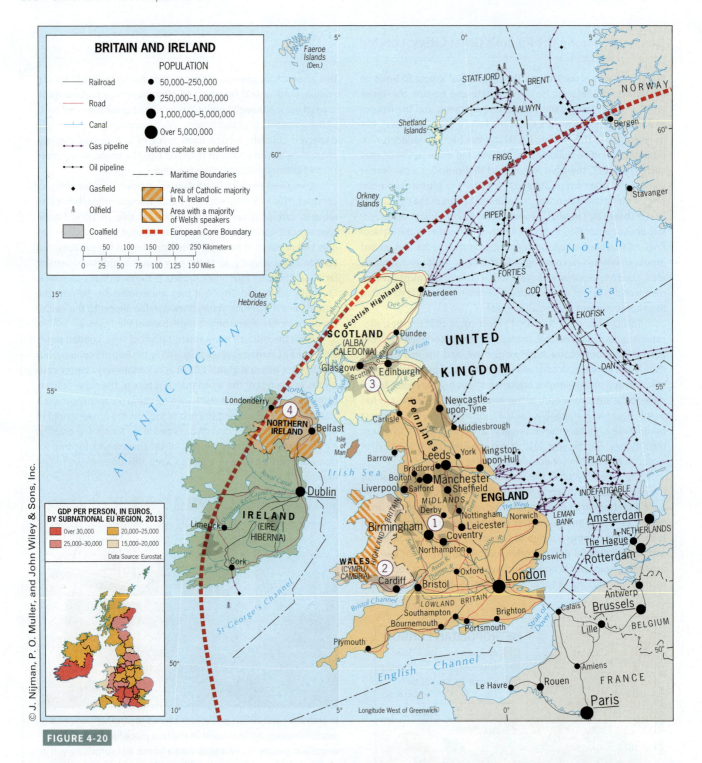

FIGURE 4-20

Republic of Ireland

Hard-won independence from Britain in 1921 did not bring real economic prosperity until the final decades of the twentieth century, when Ireland became known as the **Celtic Tiger** (likening it to the miraculous, Hong Kong-style economic development of the *Asian Tigers* discussed in Chapter 9). Participation in the EU (since 1973), adoption of the euro (in 2002), business-friendly tax policies, comparatively low wages, an English-speaking workforce, and an advantageous relative location combined for a time to produce the highest rate of economic growth in the EU. During the 1990s, Ireland's booming, service-based economy, accompanied by burgeoning towns and cities (especially the capital, Dublin), skyrocketing real estate prices, mushrooming industrial parks, and bustling traffic, transformed a country long known for emigration into a magnet for industrial workers, producing new social challenges for a closely knit, long-isolated society.

In 2007, Ireland's economic boom began to falter as global recession loomed, and a year later everything seemed to fall apart: the real estate market crashed, service industries found more favorable conditions in eastern Europe, and unemployment soared. By 2011, the government had amassed such huge deficits that a EU bailout was required. But if the depth of the crisis had taken the country and foreign observers by surprise, so did its resurgence. From about 2012, a recovery took hold, and by 2016 Ireland, seemingly far removed from the refugee crisis that engulfed continental Europe (and England), once again showed vigorous growth. Ireland today (population: 7.4 million) continues its role as a major EU locus of high-technology investment from the United States.

Region Northern Europe

Northern Europe's remoteness, isolation, and challenging environment, especially at latitudes higher than 60°N, have had positive binding effects for much of this domain and seem to have fostered similar cultures. The Scandinavian Peninsula carrying Sweden and Norway lay removed from most of the wars of mainland Europe and developed in relatively tranquil circumstances (even though Norway was overrun by Nazi Germany during World War II). Three of its major languages (Swedish, Norwegian, and Danish) are mutually intelligible, and in terms of religion there is overwhelming adherence to the same Lutheran church. Democratic and representative governments emerged early on, and individual rights and social welfare have been carefully protected for centuries. Women participate more fully in government and politics here than in any other part of the world.

But consider the implications of Figures 4-14 and **4-21**: the southern, coastal, and more urbanized areas of the region form part of the European Core, but the sparsely populated remainder does not. In the aggregate, economic indicators for most of this region are quite strong, but almost all development is concentrated within that smaller Core zone. Territorially, northern Norway and Sweden as well as most of Finland belong to the European Periphery. The five countries of this higher-latitude domain of Europe have a combined population of only 27 million, yet their core areas make up in prosperity and external linkages what they lack in size. Interestingly, a few of the countries in the region have stopped short of full participation in the EU: Norway is not even a member, and only Finland has joined the Euro zone to date.

Sweden

Sweden is northern Europe's largest country in terms of both population (9.9 million) and territory. Most Swedes live south of latitude 60°N (which passes through Uppsala just north of the capital, Stockholm) in what is climatically the most moderate part of the country (Fig. 4-21). Here lie the primate city, core area, and the main industrial districts; here, too, are the main agricultural areas that benefit from the lower relief, better soils, and somewhat milder climate (**Box 4-13**).

Sweden long exported raw or semifinished materials to industrial countries, but today the Swedes make finished products themselves, including motor vehicles, electronics, stainless steel, furniture, and glassware. Much of this production is based on local resources, including a major iron ore reserve at Kiruna in the far north (there is a steel mill at Luleå). Swedish manufacturing, in contrast to that of several western European countries, is based in dozens of small and medium-sized towns specializing in particular products. Energy-poor Sweden was a pioneer in the development of nuclear power, but a national debate focused on the risks involved has reversed that course. Sweden joined the EU in 1995 but chose to stay out of the Euro zone in 2002.

As Nordic Europe's biggest and most influential country, Sweden has historically dominated both Norway and Denmark. Culturally, Sweden has time and again made its presence known on the international scene, from popular music to films to literary works to the annual awarding of the Nobel Prizes.

Norway

Norway does not need a nuclear power industry to supply its energy needs. It has found its economic opportunities on, in, and beneath the sea. Norway's fishing industry, now augmented by highly efficient fish farms, has long been a cornerstone of the economy, and its merchant marine spans the world. But over the past four decades, Norway's economic life has been transformed by the bounty of oil and natural gas discovered beneath the floor of its sector of the North Sea.

With its limited patches of cultivable soil, high relief, extensive forests, and spectacular fjords, Norway has nothing to compare to Sweden's agricultural or industrial development. Its cities, from the capital Oslo and the North Sea port of Bergen to the historic national focus of Trondheim as well as Arctic Hammerfest, lie on narrow coastal lowlands and have difficult overland connections. The isolated northern province of Finnmark has even become the scene of a devolutionary movement among the reindeer-herding indigenous Saami (Fig. 4-9). The distribution of Norway's population of 5.3 million has been described as a necklace, its beads linked by the thinnest of strands, but this has not constrained national development. Economically, Norway has one of the lowest unemployment rates in Europe, and in terms of income per capita it is one of the world's richest countries.

Norwegians have a strong national consciousness and spirit of independence. In 1995, when Sweden and Finland voted to join the European Union, the Norwegians again said no. They had little desire to exchange their economic independence for the regulations of a broader, even possibly safer, Europe.

Denmark

Territorially small by Scandinavian standards, Denmark has a population of 5.7 million, the North's second-largest country after Sweden. It consists of the Jutland Peninsula and several islands to the east at the gateway to the Baltic Sea; it is on one of these islands, Sjaelland, that the capital of Copenhagen is located. Copenhagen has long been a port that collects, stores, and transships large quantities of goods. This **break-of-bulk [33]** function exists because many oceangoing vessels cannot enter the shallow Baltic Sea, making the city an **entrepôt [34]** where

FIGURE 4-21

transfer facilities and activities prevail. The completion of the Øresund bridge-tunnel link to southern Sweden in 2000 further enhanced Copenhagen's situation (Fig. 4-21).

Denmark remains a kingdom, and in centuries past Danish influence extended far beyond its present confines. Remnants of that period continue to challenge Denmark's governance. Greenland came under Danish rule following union with Norway (1380) and continued as a Danish possession when that union ended in 1814. In 1953, Greenland's status changed from colony to province, and in 1979 its 56,000 inhabitants were granted home rule under the new Inuit name of *Kalaallit Nunaat*.

They promptly exercised their rights by withdrawing from the EU, of which they had become a part when Denmark joined in 1973. Greenland was dependent on financial support from Denmark, but this has changed in recent years following the discovery of oil off Greenland's west coast, north of the Arctic Circle.

Another restive Danish dependency is the Faroe Islands, located between Scotland and Iceland (Fig. 4-14). These 17 minuscule islands and their 50,000 inhabitants were awarded self-governance in 1948, complete with their own flag and currency, but Denmark remains sovereign and debates continue about full independence.

Box 4-13 From the Field Notes ...

"From the water, outer Stockholm's suburban landscape looks very prosperous. I am cruising along Lilla Värtan, the narrow strait that separates the Swedish capital from the adjacent island of Lidingö. Stockholm is home to 1.5 million people, and its long history dates back to the thirteenth century. Its layout is quite unusual in that this metropolis has been built on a cluster of no less than 14 islands. Sweden is a wealthy country, and much of that affluence is concentrated in and around this capital city. Which is not to say that everybody lives like this. The myriad waterways and islands have allowed for considerable segregation, with drab, low-income, highrise residential areas located well away from these obviously prospering suburbs."

© Jan Nijman

Finland

Finland, territorially almost as large as Germany, has only 5.5 million residents, most of them concentrated in the southern triangle formed by the capital, Helsinki, the textile-producing complex of Tampere, and the shipbuilding center of Turku (Fig. 4-21). A land of evergreen forests and glacial lakes, Finland's national income has long been sustained by wood and wood-product exports. But the Finns have now developed a more diversified economy with staple agricultural crops and the manufacture of precision machinery and telecommunications equipment. Finland's environmental challenges and relative location have created cultural landscapes similar to those of northern Sweden and Norway, but the Finns are not a Scandinavian people—their linguistic and historical affinities are with the Estonians across the Gulf of Finland to the south (Fig. 4-6). In fact, other ethnic groups speaking these Finno-Ugric languages are also widely dispersed across what is today western Russia. Finland shares a long border with Russia, and during the Cold War it had to tread very carefully in international affairs. But after the collapse of the Soviet Union it joined the EU in 1995 and was one of the first countries to adopt the euro in 2002. In the twenty-first century, Finland has decidedly positioned itself within Europe.

Iceland

Iceland is a volcanic, glacier-studded island in the frigid waters of the North Atlantic just south of the Arctic Circle. Inhabited by people with Scandinavian ancestries (2016 population: 335,000), Iceland is of special scientific interest because it lies astride the Mid-Atlantic Ridge. Here, the Eurasian and North American tectonic plates of the Earth's crust are diverging (see Fig. G-4), new land can be seen forming, and spectacular volcanic eruptions

are periodically on display (as occurred memorably in 2010 when the Eyjafjallajökull volcano brought trans-Atlantic and domestic air traffic to a standstill for days on end).

Iceland's population is almost completely urban, and the capital, Reykjavik, is home to fully half of the country's inhabitants. The nation's economic geography is traditionally oriented to the frigid surrounding waters, whose seafood harvests have given Iceland a high standard of living. In the 1990s, Iceland embarked on economic liberalization policies, and its financial industries and banks grew rapidly. For a while, Iceland was referred to as the *Nordic Tiger*; but then, not unlike Ireland which experienced a similar boom, the economy floundered with the onset of the global recession in 2008. Iceland's government, closely involved with some of the troubled banks, had to be bailed out by the International Monetary Fund. Subsequently, plans were initiated to join the European Union, but in 2015 the government withdrew its bid, stating that its interests were better served by remaining outside the EU.

Region Mediterranean Europe

As Figure 4-14 reveals, the northern sectors of two major countries of southern Europe—Italy and Spain—form parts of the European Core. Portugal, on the western flank of the Iberian Peninsula, remains outside the Core, far less urbanized, much more agrarian, and not strongly integrated with western Europe. The remaining Mediterranean countries include Greece, the island-states of Cyprus and Malta, Croatia, and Slovenia, of which only the last can be considered part of the European Core in recognition of its rapid economic development. Slovenia's recent upward trajectory is similar to that of the Czech Republic (and, as we shall presently see, Estonia).

Spain and Portugal

At the western end of southern Europe lies the broad Iberian Peninsula, separated from France and western Europe by the rugged mountain wall of the Pyrenees and from North Africa by the narrow Strait of Gibraltar. Spain (population: 46 million) occupies most of this compact Mediterranean landmass, and peripheral Portugal lies in its southwestern corner.

Spain followed the leads of Germany and France and decentralized its administrative structure as a concession to devolutionary pressures. These pressures were especially strong in Catalonia, the Basque Country, and Galicia; in response, the Madrid government created a subnational regional framework of 17 so-called *Autonomous Communities (ACs)* (**Fig. 4-22**). Even so, the new AC mosaic has not defused any secessionist-minded movements. Relations between Madrid and the Basque Country remain tense, and in Catalonia the drive toward independence is strengthening as the Spanish economy struggles while Catalonia continues to forge ahead. In late 2014, the Catalan AC government held a referendum in which, reportedly, about 80 percent of the population voted in favor of outright independence from Spain; however, the Spanish High Court swiftly voided this vote, declaring it to be in violation of Spain's constitution.

Centered on the dynamic, productive coastal city of Barcelona, the Catalonia AC is not only Spain's leading industrial agglomeration but has also become one of Europe's Four Motors. Catalonia is endowed with its own distinctive language and culture that find vivid expression in Barcelona's vibrant urban landscape. In recent years, Catalonia—with 6 percent of Spain's territory and 17 percent of its population—has annually produced at least 25 percent of all Spanish exports and more than one-third of its industrial exports.

As Figure 4-22 shows, Spain's capital and largest city, Madrid, lies near the geographic center of the state. It also lies astride an economic-geographic transition zone. In terms of annual GDP per capita, the Spanish north is far more affluent than the south, a reflection of the country's inequitable distribution of resources, development opportunities, and environmental assets (the south suffers from drought as well as inferior soils). The most prosperous ACs, apart from Madrid, are in the northeastern tier closest to the French border.

Economically, Spain has been going through a difficult stretch. Economic decline has plagued most of the country since 2008 when it was hit hard by recession and a housing market crash. Unemployment in mid-2016 still hovered around

FIGURE 4-22

22 percent (and around 50 percent among youths), and Spain's government debt ranked among the highest in the EU. By the mid-2010s, a modest recovery took hold, and there is hope that this recent economic upturn continues.

Portugal (10.3 million) occupies the southwestern corner of the Iberian Peninsula. Because EU agricultural policies entail transfers from rich to poor areas in the Union and because the EU funds major infrastructural projects in more isolated areas, Portugal at first benefited substantially from its 1986 admission to the EU. Nevertheless, it still remains far behind Europe's leading economies. Unlike Spain, which has major population clusters on its interior plateau as well as its coastal lowlands, the Portuguese are mainly concentrated within their Atlantic coastal zone. The country has several excellent natural harbors, including those of primate-city Lisbon and Sines (Fig. 4-22), with considerable potential to attract a larger share of container shipping (especially for transfer onto freight trains bound for Spain), but these ports need to become more efficient and competitive. The best farmlands lie in the moister western and northern zones of the country; but farms here are small and inefficient, and even though Portugal remains dominantly rural, it must import as much as half of its food. Like Spain, Portugal has suffered a major setback since the onset of global recession, and signs of recovery have so far been minimal.

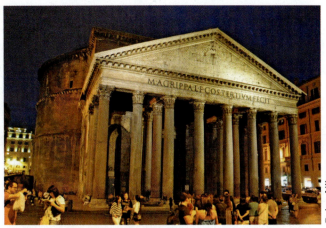

Rome's magnificent Pantheon was built around AD 118–128 by the emperor Hadrian. It is generally thought to have served as a temple for the worship of all Roman (pagan) gods at the time, though this is still debated by historians. From the seventh century on, it has been used as a Catholic Church and two nineteenth-century Italian kings are buried here. The building shows the strong influence of ancient Greece on ancient Rome: the word "pantheon" itself is from the Greek and means "related to all gods," while the Corinthian columns and the tympanum (the triangular pediment above the columns) reflect an essential architectural innovation from ancient Greece. This is one of the best preserved icons of Greco-Roman architecture.

Italy

Centrally located within Mediterranean Europe, most populous of the realm's southern states, best connected to the European Core, and economically most advanced is Italy (population: 59.8 million), a charter member of what is now the EU. The northern half of Italy stands in strong social, economic, and political contrast to such southern regions as Calabria (the "toe" of the Italian "boot") and Italy's two major Mediterranean islands, Sicily and Sardinia. Not surprisingly, Italy is often described as two countries—a progressive north and a stagnant south (known as the **Mezzogiorno**). The urbanized, industrialized north is very much a part of the European Core.

North and south are bound by the ancient (and current) capital, Rome, which lies within the narrow transition zone between Italy's contrasting halves (**Box 4-14**). This clear manifestation of Europe's core-periphery distinction is referred to in Italy as the **Ancona Line**, named after the city on the Adriatic coast where it reaches the eastern side of the peninsula (**Fig. 4-23**, blue line). Whereas Rome remains Italy's governmental and cultural focus, the functional core area of the country has shifted northward into Lombardy in the basin of the Po River. Here lies Mediterranean Europe's leading industrial complex, the Milan–Turin–Genoa triangle. Metropolitan Milan itself personifies the new, modern Italy—not only as its primary manufacturing center (making Lombardy one of Europe's Four Motors) but also as the headquarters of the country's financial and service industries. Although the Milan area contains only

9 percent of Italy's population, it now accounts for more than one-third of the entire country's national income.

The Italian economy also has been slow to recover from the global recession, and here too austerity measures have been a hard sell. Beyond these challenges, Italy is now increasingly burdened by the accelerating flow of undocumented migrants who illegally enter southern Italy from the northern coast of Africa. Since 2014, these boat people, desperately seeking political and/or economic refuge in Europe, have been streaming by the thousands across the Mediterranean in the flimsiest of vessels toward Sicily. Too many of these clandestine voyages are doomed to end tragically, and the 2013 drowning of 360 Africans off the coast of Lampedusa (the small Italian island, located halfway between Libya and Sicily, shown in Fig. 4-23) first brought this crisis to the world's attention. In 2015, propelled by the disintegration of Libya, (and subsequently Syria), an upsurge in this immigration flow sharply intensified the crisis, overwhelming the EU's belated rescue operation (Fig. 4-11) as well as unleashing heightened chaos at intake centers on European shores and larger-scale disasters at sea (in mid-April 2015, as many as 700 people drowned when their overloaded boat capsized just north of Libyan waters).

Slovenia and Croatia

These two countries, both lying in close proximity to Italy (Fig. 4-23), were the most progressive to emerge from the 1990s

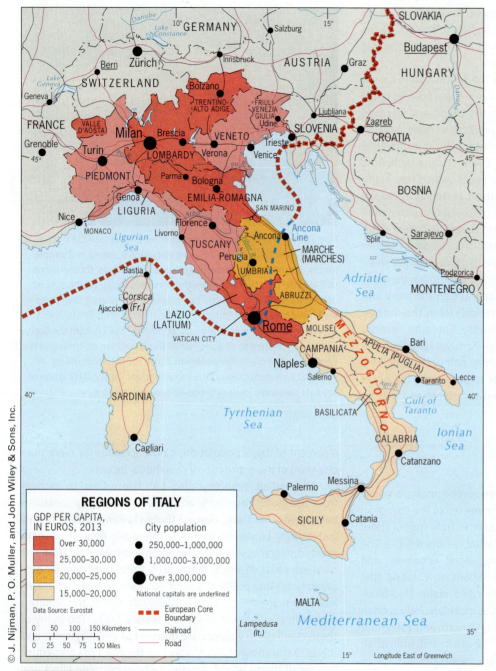

REGIONS OF ITALY

GDP PER CAPITA,
IN EUROS, 2013   City population

- Over 30,000   ● 250,000–1,000,000
- 25,000–30,000   ● 1,000,000–3,000,000
- 20,000–25,000   ● Over 3,000,000
- 15,000–20,000   National capitals are underlined

Data Source: Eurostat

0 50 100 150 Kilometers   ▬▬▬ European Core Boundary
0 25 50 75 100 Miles   ——— Railroad
                                  ——— Road

FIGURE 4-23

Greece, Cyprus, and Malta

Greece is the birthplace of European civilization, a major historic presence in the Mediterranean region, and a longtime member of the EU and NATO. Geographically, it is also relatively far removed from the European Core. Seeming to dangle from the southern end of eastern Europe, Greece was the wellspring of one of the ancient world's greatest civilizations, its scientists and philosophers still cited to this day, its famous tragedies still staged in the amphitheaters built more than 2000 years ago. Its familiar peninsulas and islands are bounded by Turkey to the east and by Bulgaria, Macedonia, and Albania to the north (Fig. 4-14). Note that some of Greece's islands in the Aegean Sea lie on the very doorstep of Turkey; in addition, Greeks represent the majority of the population of Cyprus in the remote northeastern corner of the Mediterranean Sea.

But EU membership had other, far less favorable consequences. When the Union expanded to include poorer states such as Bulgaria and Romania, Greece lost much of its EU subsidy as Brussels had to divert its equalization funds to needier members. Then the global financial crisis hit Greece especially hard in the late 2000s, and it became increasingly difficult for the government to continue to pay for such personal frills as guaranteed lifetime employment in the public sector and generous (early) retirement, and to reward businesses with protection against competition and the loose enforcement of tax regulations. Suddenly, this proud and historic nation of 11 million found itself at the center of an unfolding economic and social calamity as the overmatched Greek government—mired in mountainous debt and facing massive street riots that greeted every effort at urgently needed reform—was forced to conform to the painful bailout terms imposed by the EU agencies providing the indispensable loans.

This southeastern corner of Europe also contains the island country of **Cyprus**, an EU member since 2004, but one whose fractious political geography continues to cause problems for the Union. As Figure 4-14 shows, Cyprus lies closer to Turkey than it does to any part of Europe, and for more than

disintegration of former Yugoslavia. **Slovenia**, wedged against Austria and Italy in the hilly terrain near the head of the Adriatic Sea, has just over 2 million inhabitants, a nearly homogeneous ethnic complexion, and a productive economy. Slovenia joined the European Union in 2004 and entered the Euro zone three years later. **Croatia** (population: 4.2 million), an EU member since 2013, has a long coastline on the Adriatic that has recently become a major European tourist destination. In the long and bitter aftermath of the collapse of Yugoslavia, relations with neighboring Serbia are still strained, and the presence of a Croatian minority of about 600,000 in neighboring Bosnia continues to stir a range of political sensitivities.

Box 4-14 Among the Realm's Great Cities: Rome

From a high vantage point, Rome seems to consist of an endless sea of tiled roofs, above which rise numerous white, ochre, and gray domes of various sizes; in the distance, the urban perimeter is marked by high-rises fading into the urban haze. This historic city lives amid its past as perhaps no other as busy traffic encircles the Colosseum, the Forum, the Pantheon, and other legacies of Europe's greatest empire.

Founded about 3000 years ago at an island crossing point on the Tiber River about 25 kilometers (15 mi) from the sea, Rome had a high, defensible site. A millennium later, with a population some scholars estimate as high as 1 million, it was the capital of a Roman domain that extended from Britain to the head of the Persian Gulf and from the shores of the Black Sea to North Africa. Rome's emperors endowed the city with magnificent, marble-faced, columned public buildings, baths, stadiums, obelisks, arches, and statuary; when Rome became a Christian city, the domes and spires of churches and chapels further added to its luster.

It is almost inconceivable that such a city could collapse, but that is what happened after the center of Roman power shifted eastward to Constantinople (now Istanbul). By the end of the sixth century, Rome probably had fewer than 50,000 inhabitants, and in the thirteenth, a mere 30,000. Papal rule and a Renaissance revival lay ahead, but in 1870, when Rome became the capital of newly united Italy, it still had a population of only 200,000.

Now began a growth cycle that eclipsed all previous records. As Italy's political, religious, and cultural focus (though not an industrial center to match), Rome grew to 1 million by 1930, to 2 million by 1960, and subsequently to 3.9 million where it has leveled off today. The religious enclave of Vatican City, Roman Catholicism's headquarters, makes Rome a twin capital; the Vatican functions as an independent entity and exerts a global influence that Italy cannot equal.

Rome today remains a city whose economy is dominated by service industries; national and local government, finance and banking, insurance, retailing, and tourism employ three-quarters of the labor force. The new city sprawls far beyond the old, walled, traffic-choked center where the Roman past and the Italian future come face to face.

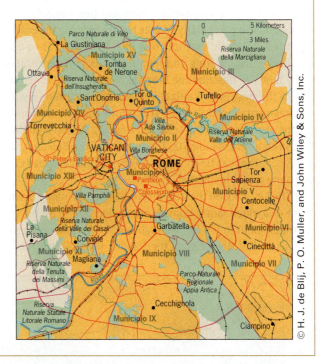

three centuries it was ruled by the Turkish Ottoman Empire. But today's population of 1.2 million is predominantly Greek, the people who have been there the longest. In 1983, the 40 percent of Cyprus under Turkish control, with about 100,000 inhabitants (plus 30,000 Turkish soldiers), declared itself the independent *Turkish Republic of Northern Cyprus*. Only Turkey recognizes this ministate (which today contains a population of less than 300,000), whereas the international community recognizes the government on the Greek side as legitimate. Things got even more complicated when only the Greek side of the island was admitted to the EU in 2004. Resentment was high on both the Turkish side and in Turkey itself, a reminder that Cyprus's "Green Line" separating its Greek and Turkish communities constitutes not just a regional border but a boundary between neighboring geographic realms.

The final component of Mediterranean Europe is the ministate of Malta, a cluster of five islands located just to the south of Sicily (Fig. 4-23). Although Malta contains merely 420,000 inhabitants, its history is that of a crossroads at the very center of the Mediterranean Sea, a cultural nexus that received infusions from Britain (as a former dependency), Italy, North Africa, and Southwest Asia. As an EU member since 2004, Malta enjoys a comparatively high standard of living reinforced by a thriving tourist industry.

Region Eastern Europe

Europe has changed fundamentally in the quarter-century since the Cold War ended. The old Eastern Europe, its western edge sharply defined by the Iron Curtain, no longer exists, and to a certain extent we have seen the revival of an older cultural region known as Central Europe or, in its prevailing German terminology, *Mitteleuropa*. Many of these changes came about as NATO and the European Union expanded eastward. As we saw, the Czech Republic and Slovenia have already been integrated into the European Core. By 2017, several eastern European states had already joined the EU, and more are knocking on the door. But this has been a decidedly uneven process because of the disparate conditions and variable geographies that mark so many countries in the region. Some are unlikely to be admitted because of their very poor economic circumstances (e.g., Albania and Moldova), whereas others—especially Ukraine— would pose serious geopolitical challenges.

The region defined as eastern Europe covers roughly half of the realm's territory and encompasses no fewer than 17 countries. Our coverage is organized within the framework of four subregions: the Balkans, East-Central Europe, the Baltic States, and the Eastern Periphery.

The Balkans

This subregion of eastern Europe was deeply impacted by the violent dismantling of communist Yugoslavia during the final decade of the twentieth century. However, this was but the latest chapter in a particularly volatile history because the Balkans constitute a classic example of a **shatter belt [35]**—a zone of persistent splintering and fracturing. Geographers employ many terms to describe the breakup of established order, and one of the best is **balkanization [36]**—the recurrent division and fragmentation of a region. That term is rooted in this part of Europe, derived from the east-west trending Balkan Mountains that stretch across Bulgaria. Today, none of the Balkan subregion's countries can be considered as part of the European Core, and none are EU members.

The key state on the new map is **Serbia**, the name of what is left of a much larger domain once ruled by the Serbs, who were dominant in the former Yugoslavia. Centered on the historic capital of Belgrade on the Danube River, Serbia (with a population of 8.8 million) is the largest and potentially most important new country in the Balkans subregion. But Serbia faces some tough challenges. First, more than one million Serbs live in neighboring Bosnia and Herzegovina, where they have an uneasy relationship with the local Muslims and Croats (**Fig. 4-24**). Second, the coastal province named **Montenegro** broke away in 2006, when voters there decided to convert it into an independent state. Third, the Muslim-majority, southernmost province of **Kosovo** declared independence in 2008, its sovereignty immediately recognized by the United States and a majority of (but not all) European governments. And fourth, Serbia still incorporates a Hungarian minority of more than 250,000 in its northern province of *Vojvodina*. Clearly, the Balkans contain a virtual kaleidoscope of ethnic identities that severely complicates any kind of orderly territorial arrangement (Fig. 4-24).

When former Yugoslavia collapsed, **Bosnia and Herzegovina** (*Bosnia* in our shorthand form) was the cauldron of calamity. No ethnic group was numerically dominant here, and this multicultural, effectively landlocked triangle of territory, lying between the Serbian stronghold to the east and the Croatian republic to the west and north, fell victim to disastrous conflict among Serbs, Croats, and *Bosniaks* (now the official name for Bosnia's Muslims, who comprise just over half of the population of 3.8 million). At least a quarter-million civilians perished in concentration camps associated with *ethnic*

FIGURE 4-24

cleansing practices; in 1995, a U.S. diplomatic effort resulted in a truce that partitioned the country as shown in Figure 4-24. In 2015, Bosnia reached an initial understanding with the EU that might result in a future candidacy for EU admission.

The southernmost "republic" of the former Yugoslavia was **Macedonia**, which in 1991 emerged from the wreckage as an independent state with a mere 2 million inhabitants, of whom about two-thirds are Macedonian Slavs. As the map shows, Macedonia adjoins Muslim Albania and Kosovo, and its north-western corner is home to the 33 percent of the Macedonians who are nominally Muslims. The remainder of this culturally diverse population are Turks, Serbs, and Roma (Gypsies). Macedonia is one of Europe's poorest countries, landlocked and powerless. Even its very name caused it problems: Greece, Macedonia's neighbor to the south, argued that this name was Greek property and therefore would not recognize it. Macedonia also confronts a long-running autonomy movement among its Albanian citizens, requiring the allocation of scarce resources to hold this fragile state of 2.2 million together.

The only other dominantly Muslim country in Europe is **Albania**, where 59 percent of the population of 2.9 million are adherents of Islam. Albania shares with one other country—Moldova—the status of being Europe's poorest state. It has one of Europe's higher rates of natural population growth, and many Albanians try to emigrate to the EU by crossing the Adriatic Sea to Italy. Most Albanians subsist on livestock herding and farming. They also are eager to join the EU—which is not likely in the foreseeable future.

East-Central Europe

As Figures 4-14 and **4-25** show, the three states of this sub-region—Poland, Hungary, and Slovakia—all adjoin or are

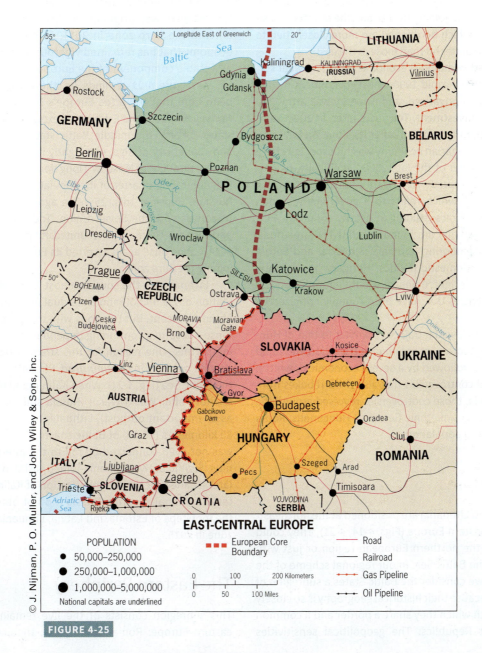

© J. Nijman, P. O. Muller, and John Wiley & Sons, Inc.

EAST-CENTRAL EUROPE

POPULATION
- 50,000–250,000
- 250,000–1,000,000
- 1,000,000–5,000,000

National capitals are underlined

European Core Boundary

Road
Railroad
Gas Pipeline
Oil Pipeline

0   100   200 Kilometers
0   50   100 Miles

FIGURE 4-25

partially integrated into the European Core. All are EU and NATO members, and Slovakia has also joined the Euro zone. Considering that these countries were tightly bound into the Soviet bloc before its collapse in 1991, these changes have been truly remarkable.

The largest and most populous of these countries is **Poland**. With 38.6 million people and lying between two historic enemies, Poland's borders have shifted time and again; but its current, post–1945 reconfiguration appears to be more durable than previous territorial layouts. As the map shows, the historic and once-central capital of Warsaw now lies closer to Russia in the east than to Germany in the west, but the country today has decidedly shifted its gaze westward. After entering the EU in 2004, Poland saw hundreds of thousands of workers depart for jobs in the European Core. But many have since returned as the national economy has grown robustly in this decade (Poland may now be in its best shape ever), and economic relations with Germany are at their closest in history. Overall, Poland has benefited substantially from joining the EU, and the country's westernmost provinces, which border and are more closely tied to Germany, are much better off than those in the east.

Hungary's post-communist economic transition went surprisingly smoothly. By the mid-1990s, most government-owned companies had been privatized, and Hungary was attracting significant foreign investment. Ten years later, in 2004, Hungary joined the EU. But soon thereafter its situation began to deteriorate as the Hungarian economy was hit especially hard by the global recession of the late 2000s. Politically, heightening discord has centered on Hungary's persistent problem of **irredentism [37]**, which refers to a government's open support for fellow ethnics residing in nearby countries. Ethnic Magyars (Hungarians) are found in relatively large numbers in parts of neighboring Romania, Slovakia, and Serbia (**Fig. 4-26**). As for domestic politics, Hungary has recently taken a turn to the far right, unleashing both nationalistic and anti-Semitic tendencies that have been roundly criticized by EU leaders.

Slovakia has followed a trajectory similar to that of Hungary: a relatively smooth post-communist transition and rapid accession to the EU, followed by a severe economic downturn and rising political controversies that also drew rebukes from Brussels. In Slovakia, those controversies revolved around the alleged poor treatment of ethnic Germans, Hungarians, Roma (Gypsies), and LGBTQ populations.

The Baltic States

Estonia, Latvia, and Lithuania are positioned at the northeastern extremity of eastern Europe (Figs. 4-14, 4-21). They could be grouped with the northern European region or just with Finland, around the Baltic Sea. In our regional scheme of the European realm, we consider the Baltic States a subregion of eastern Europe because their historical geography is so closely tied to Russia, with which they share a border and a common history as Soviet Republics. The geopolitical sensitivities

regarding Russia are palpable here, yet the pro-European orientation is crystal clear. All three countries joined the EU and NATO.

Estonia is the northernmost of the three Baltic states, and it has longstanding ethnic and linguistic ties to Finland. During the period of Soviet control from 1940 to 1991, Estonia's demographic structure transformed drastically: today about 23 percent of its 1.3 million inhabitants are Russians, most of whom came here as colonizers. After an initially difficult period of adjustment, Estonia has forged ahead of its Baltic neighbors, Latvia and Lithuania. Busy traffic links Tallinn, the capital, with its nearby counterpart in Finland, Helsinki, and a free-trade zone at Muuga Harbor facilitates commerce with Russia. But more important for this country's future was its entry into the European Union in 2004. Since then, Estonia has gained attention as a result of its economic advances, particularly its creative entrepreneurship in the high-tech and software industries (Skype, for instance, was originally an Estonian company). Having entered the Euro zone in 2011, Estonia has developed an economy that has remained vigorous, and we classify it here as a stable component of the European Core.

As Figure 4-21 reminds us, the boundary of the European Core that traverses Europe's North includes southern Norway, Sweden, and Estonia but excludes the other two Baltic states. **Latvia**, the middle Baltic state centered on the port and capital city of Riga, was tightly integrated into the Soviet system throughout Moscow's long domination. Here, as in Estonia, about one-quarter of its population of 2 million is ethnic Russian. Consider this: 25 years ago, virtually all of Latvia's trade was with the Soviet Union. Today its principal trading partners are Germany, the United Kingdom, and Sweden. In 2014, Latvia once again asserted its European reorientation by joining the Euro zone.

Lithuania is the southernmost Baltic state, with a population of 2.9 million and a residual Russian minority of only about 6 percent. But relations with this gigantic neighbor are tense—despite a continuing dependence on Russia as a trading partner. One reason for this strained relationship has to do with neighboring **Kaliningrad**, Russia's **exclave [38]** facing the Baltic Sea (Fig. 4-21). When Kaliningrad became a Russian territory after World War II (Russia still regards this exclave as an important outpost), Lithuania was left with only about 80 kilometers (50 mi) of Baltic coastline and a small port that was not even connected by rail to the interior capital of Vilnius. Significantly, Lithuania joined NATO in 2004, and a year later it called for the demilitarization of Kaliningrad as a matter of national security. Russia did not oblige. Following in the footsteps of Estonia and Latvia, Lithuania joined the Euro zone in 2015.

The Eastern Periphery

This subregion consists of the five remaining countries of eastern Europe: Romania, Bulgaria, Ukraine, Moldova, and

FIGURE 4-26

Belarus (**Fig. 4-27**). The first two joined NATO in 2004 and were admitted to the EU three years later. For the other three, such membership is not even on the horizon. Ukraine and Belarus find themselves on Russia's doorstep, and separatist-minded eastern Ukraine (with its substantial population of more than 8 million ethnic Russians) descended into a *de facto* state of civil war in 2014.

Just how **Romania** managed to persuade EU leaders to endorse its 2007 accession still remains a mystery to many Europeans. Romania exhibits some of Europe's worst social indicators: its economy is weak, its incomes are low, its governance has not been sufficiently upgraded from communist times, and political infighting and corruption are endemic.

But this is an important country, located in the lower basin of the Danube River and occupying much of the heart of eastern Europe. And like a number of other countries in this region, it was incorporated into NATO before it was allowed to join the EU. With 19.4 million inhabitants and a pivotal location on the Black Sea, Romania is a bridge between Central Europe and the realm's southeastern corner, where EU-member Greece faces its historical adversary, Turkey, across both land and water (Fig. 4-27).

Across the Danube River lies Romania's southern neighbor, **Bulgaria**. The rugged Balkan Mountains form Bulgaria's physiographic backbone, separating the Danube and Maritsa basins. As Figure 4-27 shows, Bulgaria has five neighbors, several of

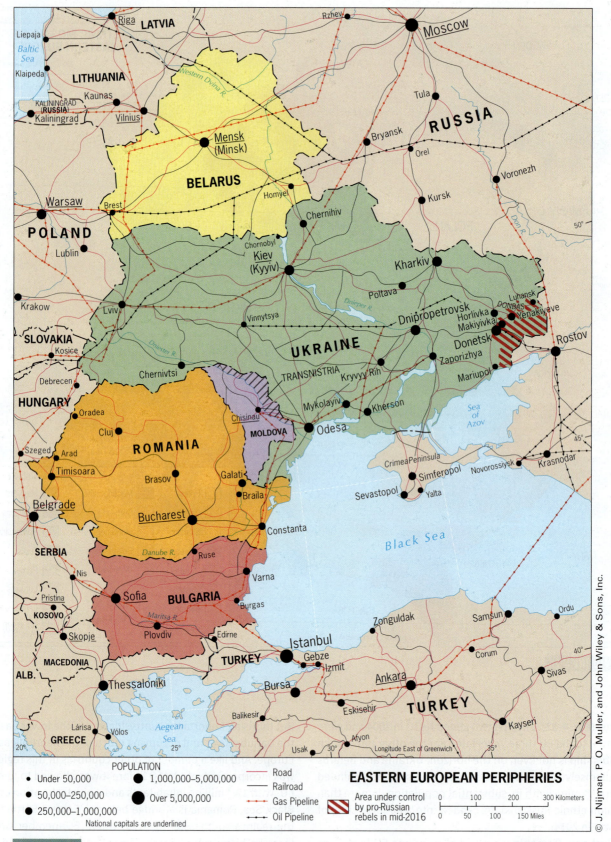

POPULATION

- Under 50,000
- 50,000–250,000
- 250,000–1,000,000
- 1,000,000–5,000,000
- Over 5,000,000

National capitals are underlined

Road
Railroad
Gas Pipeline
Oil Pipeline

Area under control by pro-Russian rebels in mid-2016

0   100   200   300 Kilometers
0   50   100   150 Miles

EASTERN EUROPEAN PERIPHERIES

FIGURE 4-27

which are in political turmoil. In 2007, Bulgaria also became an EU member, once required judicial and other social reforms were finalized. But today its 7.1 million citizens constitute one of the Union's poorest countries, and development prospects are not encouraging. Bulgaria fronts the Black Sea and has a major port at Varna, but the main advantage it derives from its seaside location is the tourism its beaches generate. The capital, Sofia, lies near the opposite, western border with Serbia, and in its core-area hinterland modest foreign investment is slowly modernizing parts of the economic landscape.

Territorially, **Ukraine** is the realm's largest country. Demographically, however, its population of 44.6 million only qualifies Ukraine for second-tier ranking among European states, on a par with Poland and Spain. As Figure 4-27 underscores, Ukraine's relative location is crucial. Not only does it link the core of Russia to the periphery of the European Union: it also forms the northern shore of the Black Sea from Russia to Romania, in the center of which lies the strategically prominent Crimea Peninsula. Perhaps most important of all, oil and gas pipelines connect Russian fields to European markets across Ukrainian territory. Once a land of farmers tilling its famously fertile soils, Ukraine emerged from the Soviet era with the huge (and decaying) Donbas heavy-industrial complex in its east as well as with a substantial (18 percent) Russian minority population. Most of the Russians are concentrated in the zone that stretches eastward from the Dnieper Valley (Fig. 4-26)—which is why that part of the country is mapped as a transition zone in Figure 4-14.

The vital Dnieper River forms a valuable geographic reference for comprehending Ukraine's division that creates the transition zone (Fig. 4-27). To its west lies agrarian, rural, mainly Roman Catholic Ukraine; in the Dnieper's great southern bend and eastward lies industrial, urban, Russified (and Russian Orthodox) Ukraine. For the most part, Ukraine's electorate is divided between a pro-Western (and pro-EU) north and west and a pro-Russian south and east (see **Box 4-15**). Violent protests broke out in Kiev (Kyyiv), the Ukrainian capital, in February 2014 over major political disagreements concerning relations with Russia and possible membership in the European Union. Pro-Russian groups inside Ukraine raised their profile by protesting against EU membership. The Crimea was swiftly annexed by Russia in March 2014 amidst vociferous objections from the Kiev government and most of the international community. Within days, the Russian border zone of eastern Ukraine was engulfed by an escalating conflict that pitted Ukrainian government forces against those of the separatists, now aided directly by Russian military power; effectively, the country had descended into a civil war. As of mid-2016, much of eastern Ukraine was controlled by Russian rebels, with both open and clandestine support from Moscow (Box 4-15, **Figure 4-28**).

Moldova, Ukraine's tiny impoverished southwestern neighbor, was a Romanian province seized by the Soviets in 1940 and converted into a landlocked "Soviet Socialist Republic." A half-century later, along with the USSR's 14 other "republics," Moldova gained independence when the Soviet

Box 4-15 MAP ANALYSIS

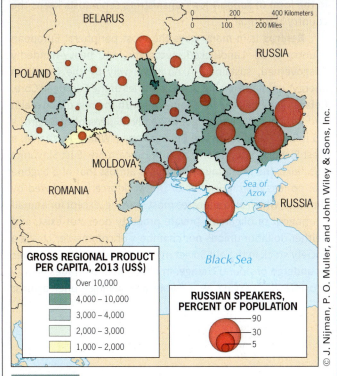

© J. Nijman, P. O. Muller, and John Wiley & Sons, Inc.

FIGURE 4-28

Maps are quite helpful in demarcating and analyzing regional transition zones. Figure 4-28 places the ongoing Ukrainian political crisis in spatial context. For each oblast (province), including Crimea, it shows income per capita as well as the percentage of Russian speakers in 2011. Carefully examine the spatial distribution of both variables, particularly their east-west gradients. Then closely track the Russian-Ukrainian border and consult the relevant maps in this chapter to discover which of Ukraine's neighboring countries are members of the EU and/or NATO. What are the key geographic factors that explain why EU membership is so strongly supported in the *western* half of Ukraine?

Access your WileyPLUS Learning Space course to interact with a dynamic version of this map and to engage with online map exercises and questions.

Union disintegrated. Romanians remain in the majority among its 4.1 million people, but many of the Russians and Ukrainians (each about 6 percent) have migrated across the Dniester River to an industrialized strip of land between that waterway and the Ukrainian border, proclaiming there a "Republic of Transnistria" (see Fig. 4-27). This separatist movement constitutes only one of Moldova's multiple problems. Its economy, dominated by farming, is in decline; an estimated 40 percent of the population works at jobs outside its borders because unemployment rates in Moldova typically hover around 30 percent;

smuggling and illegal arms trafficking are rife. These many misfortunes translate into weakness, and Russia's influence in the form of support for Transnistria's separatists keeps the country in turmoil.

Belarus is in many ways the most peripheral European country of all. Landlocked, autocratic, and heavy-handedly misgoverned, sustained in no small way by the transit of energy supplies from Russia to countries of the European Core, Belarus itself has few functional ties with the rest of the realm and little prospect of progress. Nearly 85 percent of Belarus's 9.5 million inhabitants are Belarussians ("White" Russians), a West Slavonic people; only about 8 percent are (East Slavonic) Russians. The economy of Belarus is almost that of a bygone era: no less than four-fifths of the workforce is employed by the state, which also allocates most housing. Except for a small contingent of oligarchs, private property does not exist, and neither do labor unions nor chambers of commerce. This is a society stuck in the old Soviet mode in a realm that continues to undergo profound change. In effect, Belarus trades energy subsidies for serving as a stable and willing buffer against further European encroachment on Russia's shrinking western doorstep.

Europe's Future Prospects

The future of the European realm, involving both its internal dynamics and external relations, has become ever more tightly intertwined with that of the European Union. The EU today has penetrated deeply into eastern Europe; encompasses 27 members; and has facilitated economic progress as well as political stability as it evolved over the past six decades. This remarkable advance amounts to a tremendous achievement in this historically volatile and fractious part of the world. The EU now has a combined population of almost 450 million that constitutes one of the world's richest markets; its member-states collectively supply about 40 percent of all exports.

But Europe, and the European Union, are under significant stress. First, internally, Europe's political leaders must demonstrate the value and legitimize the costs of integration to their electorates. The problem of the so-called *democratic deficit* is a serious obstacle: from Germany to Greece, citizens often feel that they are not being heard, yet national governments must have their votes to support the Union. Second, the ongoing euro crisis underscores the difficulty of having a single currency across such an economically disparate realm. And third, Europe since 2015 has been faced with a large-scale refugee crisis. Individually, many countries have struggled to cope with the massive influx of refugees. More importantly, the refugee crisis has exposed enormous differences within the EU in terms of economic abilities as well as political cultures.

By 2016, strong political opposition had been marshaled against continued EU membership in a range of member-states. The referendum in the United Kingdom, that resulted in a 52-percent majority vote to leave the EU, was emblematic of this crisis. Nontheless, in a geographic realm as diverse and complex as this one, the European Union amounts to a brilliant achievement—but its continued success is by no means a certainty.

Points to Ponder

- Per capita income in Norway or Luxembourg is nearly twice as high as in the United States.
- In 2016, youth unemployment in Greece, Spain, and Italy exceeded 40 percent.
- Brussels is the main seat of the European Union, yet it is the capital of a country faced with strong divisive forces that at times threaten its national unity.
- In northern European countries, the south is more developed than the north; in southern European countries, the north is more developed than the south.
- In Europe, 49 percent of the population say they do not believe in God, compared to 11 percent in the United States.

FIGURE 5-1

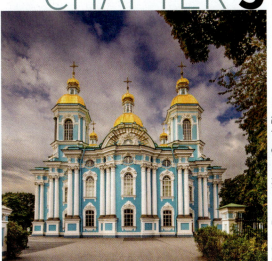

CHAPTER 5

© Igor Saltykov/Alamy Stock Photo

© Sean Pavone/Alamy Stock Photo

Where were these pictures taken?

The Russian/Central Asian Realm

IN THIS CHAPTER

- Military buildup in the Arctic
- Shrinking populations in eastern Russia
- Russia in the Near Abroad
- A New Silk Road across Central Asia
- Astana: Kazakhstan's shining new capital

CONCEPTS, IDEAS, AND TERMS

Every geographic realm has a dominant, distinguishing feature: Europe's jigsaw-puzzle map of 40 countries, South America's familiar triangle, Southeast Asia's thousands of islands and peninsulas. The Russian/Central Asian realm stands out because of its vast areal extent, particularly its enormous, nine-time-zone-wide, east-west dimension. The country of Russia by itself is twice the territorial size of Canada, the world's second-largest country, and three times as big as neighboring Europe. Not surprisingly, Russia dominates this part of the world, making this a classic example of a **monocentric geographic realm [1]**. But not all of this realm is Russian: it also includes the regions of Central Asia and Transcaucasia, which now consist of independent states that were once part of Czarist Russia and its successor Soviet Union.

Defining the Realm

As **Figure 5-1** reveals and as a globe shows even better, Russia's entire northern coast—from its border with Norway in the west to the Bering Strait across from Alaska in the east—faces the Arctic Ocean, with all the environmental consequences you would expect. And for all of its stupendous dimensions, much of this realm lies at higher latitudes, shut off by mountains and deserts from the warmer and moister parts of Eurasia. That has posed a geographic challenge for Russian rulers and governments for as long as Russia has existed as a state. Russia's ports on the Pacific Ocean in the east lie about as far from where most Russians live as they could be. And in the west, where Russia's core area is situated, every maritime outlet is restricted in some way. The Arctic ports operate on seasonal access. The Baltic ports face a shallow sea far from the open ocean. The Black Sea ports (some leased from Ukraine) require maneuvering through narrow straits across Turkish territory—simply to reach the eastern Mediterranean Sea.

The Russia/Central Asia realm may be gigantic in size, but its human numbers (that total almost 230 million) are modest and not much larger than those of such individual countries as Nigeria and Pakistan. Moreover, much of the realm's population has been declining since the 1991 collapse of the Soviet Union in response to changed political, economic, and social conditions that even now continue to produce negative demographic effects.

This realm, as defined by our global framework, consists of nine political entities: the giant Russian state itself; three small nations in Transcaucasia; and five countries in Central Asia. Transcaucasia, the mountainous zone between the Black and Caspian seas, contains Georgia, Armenia, and Azerbaijan. All are small in size, with a combined population of 17 million (7.5 percent of the realm's total); they also lie in a historically turbulent arena of conflict between Russian and numerous non-Russian peoples. The five Central Asian countries are Kazakhstan, Uzbekistan, Turkmenistan, Kyrgyzstan, and Tajikistan—sometimes referred to collectively as the "Stans" (although they do not include Afghanistan, which we assign to the South Asian realm in Chapter 8). In contrast to Transcaucasia, Central Asia covers a

---

Box 5-1 Major Geographic Features of Russia/Central Asia

1. Russia/Central Asia is an enormous realm that is almost entirely lacking in warm-water access to the world ocean.

2. This is by far the "widest" geographic realm on Earth, stretching from the Baltic Sea in the west to the Pacific Ocean in the east for some 10,000 kilometers (6200 mi) and spanning no less than nine time zones.

3. Much of Russia/Central Asia is cold and/or dry. Extensive rugged mountain zones separate the realm from warmer subtropical air, and most of it lies open to Arctic air masses.

4. Russia/Central Asia borders four other realms: Europe, North Africa/Southwest Asia, South Asia, and East Asia. These boundaries have shifted throughout history, and today some of them must be mapped as transition zones.

5. For so large a territory, the population size of Russia/Central Asia (nearly 230 million) is relatively small. In Russia itself, the population has been declining in recent decades, especially in the country's east and in most of the rural areas.

6. This realm's core area is located in western Russia. New development is now mostly concentrated to the west of the Ural Mountains; here are most of the major cities, leading industrial complexes, densest transport networks, and most productive farming areas. Outside Russia, certain parts of Central Asia (most notably Kazakhstan) have experienced rapid development based on their rich natural resources.

7. For a monocentric realm, this one is decidedly multicultural and has a complex political geography. Russian domination has long inflamed tensions within this ethnic mosaic, especially Muslim concentrations along the realm's southern periphery.

8. Central Asia is a region that highly interests China, for both its natural resources and its geopolitical role on China's western frontier. There is now much talk in China about forging a "New Silk Road."

9. The country's powerful position as a leading exporter of oil and natural gas has been a springboard in Russia's quest to reclaim its Soviet-era role as a major player on the international stage.

comparatively large area. Territorially, Kazakhstan is the biggest Central Asian country (about five times the size of France) and also has a lengthy border with Russia. Central Asia's overall population is 68 million, accounting for just about 30 percent of the realm's total.

The political geography of this realm is complex: it contains myriad contested boundaries and territories; **exclaves [2]** from (Russian) Kaliningrad in northwest Europe to (Azerbaijani) Nakhchivan in south Armenia; and a fair number of **irredentist [3]** claims that are at the root of intractable conflicts. The implosion of the Soviet Union in 1991 still resonates across many parts of the realm and beyond, especially near the Russian border. This is a realm, perhaps more than any other, where **geopolitics [4]** is at the forefront (**Box 5-1**).

What Transcaucasia and Central Asia have in common is that both were part of Czarist Russia and the Soviet Union. Today they remain firmly within Russia's sphere of influence (the same could be said for parts of eastern Europe, such as Belarus and eastern Ukraine). It is also worth noting that these southern peripheral components of the realm display characteristics of a **transition zone**: Transcaucasia is positioned in a historically volatile corridor between Russia and Southwest Asia, whereas the "Stans" of Central Asia form a broad arena that has for centuries functioned as a crossroads linking Russia, Southwest Asia, South Asia, and East Asia.

Physical Geography of Russia/Central Asia

Physiographic Regions

One of the first features to observe on the realm's physiographic map is a prominent north-south trending mountain range in western Russia that extends from the Arctic Ocean southward to Kazakhstan (Figs. 5-1 and **5-2**), dividing Russia into two unequal halves: the Russian Plain to the west and the subdivisions of Siberia to the east. This range, the **Ural Mountains** (numbered ① in Fig. 5-2), is sometimes designated as the eastern boundary of Europe. That is a mistake because Russia is not Europe: with both sides of the range equally Russian, the Urals cannot form a cultural boundary. Note, too, that the island of Novaya Zemlya (Fig. 5-1) is the northern extension of the Urals, which at their opposite end also extend southward into Kazakhstan.

The **Russian Plain** ② to the west of the Ural Mountains is the eastward continuation of the North European Lowland, and here lies Russia's *core area*. Travel northward from the centrally located capital of Moscow, and the countryside soon changes to coniferous (needleleaf) forests like those of Canada. The Volga River, Russia's Mississippi, flows southward in a wide arc across this plain from the forested north—but unlike the Mississippi, the Volga drains into a landlocked lake, the Caspian "Sea." Travel southward from Moscow today, and the land is draped in grain fields and pastures. Eastward, the Urals are not high enough to impede surface transportation, but they retain their prominence because this range separates two vast lowlands. At the southern end of the Russian Plain, the **Caucasus Mountains** ③ rise like a wall between Russia and the lands beyond—a multi-tiered barrier over which Russians have fought with their southern neighbors for centuries.

The **West Siberian Plain** ④ to the east of the Urals is often described as the world's largest unbroken lowland, and here all the rivers flow northward. In fact, this plain is so flat that over the last 1600 kilometers (1000 mi) of its course to the Arctic Ocean, the Ob River (the lowland's largest) falls less than 90 meters (300 ft) in enormous meanders flanked by forests until, in the near-Arctic cold, the trees yield to mosses and lichens. Follow an imaginary trail eastward, and in just about the middle of Russia, at the Yenisey River, the West Siberian Plain gives way to the higher relief of the massive **Central Siberian Plateau** ⑤, one of the most sparsely populated areas in the habitable world.

Contrasting landscapes of the Russia/Central Asia realm. On the left is a coniferous forest in early autumn near where the Arctic Circle crosses the West Siberian Plain; to the right are the Tian Shan Mountains that parallel the Chinese border in northern Kyrgyzstan.

Natalya Onishchenko / Alamy Stock Photo

© Ivan Vdovin/JAI

FIGURE 5-2

To the south, beyond the Russia-Kazakhstan border, the West Siberian Plain transitions into the **Caspian-Aral Basin** ⑥. This physiographic region contains almost all of Central Asia and is dominated by windswept plains of moisture-starved steppes and desert. Poorly executed irrigation schemes along the rivers draining into the Aral Sea marked much of the Soviet era, and the price was paid after 1990 as an environmental disaster unfolded that has now resulted in the nearly complete drying up of the Aral Sea and the waterways that supply it. To the southeast, lining the long border with China and Mongolia, lie the **Central Asian Ranges** ⑦ that stretch northeastward from Tajikistan and Kyrgyzstan to Lake Baykal.

Continuing our eastward transect of Russia toward the Pacific, we first encounter the **Yakutsk Basin** ⑧ drained by the Lena River, and then we approach the realm's mountainous eastern perimeter (look again at Figs. 5-1 and 5-2), where the few roads and railroads must wend their way through tunnels and valleys full of hairpin turns. We map this jumble of ranges collectively as the **Eastern Highlands** ⑨ which blanket one of the most spectacular, diverse, and still-remote landscapes on the face of the Earth.

Finally, our survey of the realm's physiographic regions ends at the farthest extremity of Russia, the partially offshore **Pacific Rimland** ⑩ that consists of the huge Kamchatka Peninsula and the long, narrow island of Sakhalin to its southwest across the

Sea of Okhotsk. Here, this immense geographic realm makes contact with the Pacific Ring of Fire (see Fig. G-4). There are no erupting volcanoes in Siberia and earthquakes are rare, but the Kamchatka Peninsula has plenty of both. This is one of the most volatile segments of the Pacific Ring of Fire, with more than 60 volcanoes, many of them active, forming the spine of the peninsula. Nothing here resembles Siberia (the climate is moderated by the Pacific waters offshore; the vegetation is mixed, not coniferous), and Kamchatka is one of the world's most difficult places to inhabit, effectively separated from the rest of Russia by the absence of overland connections. On the island of Sakhalin, earthquakes rather than volcanism are the leading environmental hazard. Long a battleground between Russians and Japanese, Sakhalin has remained in Russian hands since the end of World War II. In subsequent decades, the island and its maritime surroundings proved to contain substantial reserves of oil and natural gas, making it a most valuable asset in Russia's energy-based export economy.

Climate and Vegetation

As the northernmost populous country on Earth, Russia has virtually no natural barriers against the onslaught of Arctic air masses. Moscow lies farther north than Edmonton, Canada, and St. Petersburg lies at the latitude of the southern tip of Greenland. Winters are long, dark, and bitterly cold in most of Russia; summers are short and growing seasons limited. Precipitation, even in western Russia, ranges from modest to minimal because the warm, moist air carried eastward across Europe from the North Atlantic Ocean loses both heat and moisture by the time it reaches Russia. **Figure 5-3** reveals the consequences. Russia's climatic **continentality [5]** (inland climatic environment remote from moderating and moistening maritime influences) is expressed by its prevailing *Dfb* and *Dfc* conditions. Compare the Russian map to that of North America (Fig. G-6), and you will note that, except for a small corner off the Black Sea, Russia's climatic conditions resemble those of the U.S. Upper Midwest and interior Canada. Along its entire northern edge, Russia has a zone of *E* climates, the most frigid on the planet. In these Arctic latitudes originate the polar air masses that dominate its environments.

The Russian realm's harsh northern climates affect people, animals, plants—and even the soil. In Figure 5-2 you can find a direct consequence of what you see in Figure 5-3: a dashed blue line that starts at the shore of the Barents Sea, crosses Siberia eastward, and reaches the coast of the Pacific Ocean north of the neck of the Kamchatka Peninsula. North of this blue line, water in the ground is permanently frozen, creating an even more formidable obstacle to settlement and infrastructure than the severe weather alone. It is referred to as **permafrost [6]**, and it affects other high-latitude environments as well (Alaska, for example). In the soil layer above the permafrost, seasonal

temperature changes cause alternate thawing and freezing, with predictably destructive impacts on buildings, roads, railroad tracks, and pipelines.

Examine Figure 5-3 carefully, and you will see two ecological terms: *tundra* and *taiga*. **Tundra [7]**, as the map suggests, refers to both climate and vegetation. The blue area mapped as *E* marks the coldest ice-affected environmental zone in Russia and elsewhere in the high Arctic: this is frigid, treeless, windswept, low-elevation terrain where bare ground and rock prevail and mosses, lichens, patches of low grass, and a few hardy shrubs are all that grows. The stippled area on the map is the **taiga [8]**—a Russian word meaning "snowforest" (also called boreal forest)—which extends across vast reaches of Eurasia as well as northern North America and is dominated by coniferous (as in pine cone) trees. As the map shows, the taiga extends southward into more moderate environments, and there it becomes a mixed forest of coniferous and deciduous trees. Although the taiga prevails from northern Scandinavia to the Russian Far East, this is the vegetative landscape most often associated with Siberia: endless expanses of rolling countryside draped in dense stands of evergreen pine trees (see left frame of the photo pair).

Climates and Peoples

Climate and weather have always challenged the peoples of this realm. The fierce winds that drive the bitter Arctic cold southward deep into Eurasia, the blizzards of Siberia, the aridity

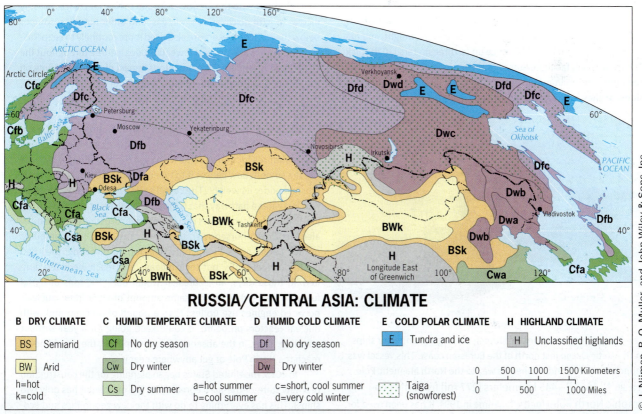

RUSSIA/CENTRAL ASIA: CLIMATE

B DRY CLIMATE	C HUMID TEMPERATE CLIMATE	D HUMID COLD CLIMATE	E COLD POLAR CLIMATE	H HIGHLAND CLIMATE
BS Semiarid	Cf No dry season	Df No dry season	E Tundra and ice	H Unclassified highlands
BW Arid	Cw Dry winter	Dw Dry winter		
h=hot k=cold	Cs Dry summer			

a=hot summer b=cool summer
c=short, cool summer d=very cold winter

Taiga (snowforest)

0   500   1000   1500 Kilometers
0   500   1000 Miles

© J. Nijman, P. O. Muller, and John Wiley & Sons, Inc.

FIGURE 5-3

of Central Asia, the temperature extremes, rainfall variability, and short and unreliable growing seasons even in the more moderate western parts of the realm have always made farming difficult, to the point that grain often had to be imported into Russia.

As pointed out in the Introduction, and will be repeatedly observed throughout this book, humankind's long-term dependence on agriculture remains etched on the population map, even as our planet becomes ever more urbanized and its economy more globalized. By examining the climates of the Russian/Central Asian realm, we can begin to understand what the map of population distribution communicates (**Fig. 5-4**). The overwhelming majority of the realm's inhabitants remain concentrated in western and southwestern Russia, where environmental conditions were least difficult at a time when farming was the mainstay of most of the people. To the east of the Urals, the population is sparser and tends to cluster like beads on a string along the southern margin of the realm, becoming even more thinly distributed east of Lake Baykal. Central Asia's population distribution is very closely tied to local water supplies, which are most readily available in the hilly terrain that lines the rugged mountains of the southeast. And let us also keep in mind that nearly three-quarters of this realm's population now resides in cities and towns, so it is not difficult to imagine just how empty vast stretches of countryside must be—especially in those frigid northerly latitudes.

Climate Change and Arctic Prospects

As every map of this high-latitude realm shows, its northern coast lies entirely poleward of the Arctic Circle (latitude 66.5°N). Such a lengthy coastline on the Arctic Ocean is not exactly an advantage: most of that ocean's surface is frozen much of the year, and only limited warmth from the North Atlantic keeps the western ports of Murmansk and Arkhangelsk open a bit longer (Fig. 5-1). Ports like these (and even St. Petersburg on the Gulf of Finland, an arm of the Baltic Sea) would never have developed the way they did if Russia had better access to the world ocean (see **Box 5-2**).

These days it looks as though nature may be about to give Russia a helping hand. As global warming causes long-term melting of sizeable portions of the Arctic Ocean's ice cover, that body of water may come to play a new and different role in this realm's future. Milder atmospheric conditions will shrink the area of permafrost mapped in Figure 5-2; moister air masses may improve agriculture on the Russian Plain; warmer water may even keep Arctic ports open year-round; and the so-called **Northeast Passage [9]** will open up to directly connect the

---

Russia's Nuclear Icebreakers in the Arctic Ocean

Nuclear-powered icebreaker ships were first built in the 1970s by Russia. They are specialized vessels whose design, weight, and fuel system allow them to crush sea ice up to three meters (10 ft) thick. The icebreakers facilitate access to the frozen seas for commercial, scientific, and military purposes. They can be on dedicated missions, or they may carve a pathway through the ice for other ships to follow (see photo). The introduction of icebreakers has greatly facilitated access to the Arctic Ocean, and their importance has only increased with discoveries of oil, gas, and other natural resources below the ice. Icebreakers are also used more routinely to keep Russia's waterways and ports at lower latitudes ice-free throughout the year. Notwithstanding the momentous impact of global warming on the Arctic (discussed further below and elaborated in Chapter 12), these ships continue to be a vital technology, especially during the winters.

These icebreakers have a double hull of reinforced steel, each about two inches thick at the bow, with a thick synthetic coating that minimizes friction with the ice. When the ship ploughs into the ice, the smooth bow causes it to ride up on the ice, which then gets crushed under its weight. The vessel's relatively wide body ensures that it creates a pathway that is broad enough for other ships to follow in its wake. Icebreakers require massive thrusting power, equivalent to 75,000 horsepower, therefore consuming huge amounts of fuel. Earlier-generation icebreakers ran on diesel oil: they had enormous fuel tanks, but their mileage and reach were very limited. In those days, an icebreaker would burn 100 tons of diesel fuel in a day; today, for the same amount of work, their nuclear-powered engines use up less than a pound of uranium and, with nuclear reactors on board, it could keep going for a year without refueling. Thus, in the absence of nuclear power, they could never reach the North Pole or get anywhere near it.

In 2015, the United States began planning for the construction of a new state-of-the-art icebreaker (at the moment it has only two ships). This has everything to do with the new geopolitics and economics of the Arctic Basin (see Chapter 12).

ИТАР-ТАСС/Newscom

The Russian nuclear icebreaker *Yamal*, carving a channel for ships in the Arctic Ocean just north of the Eurasian coast. This vessel was launched in 1992 and crushed its way to the North Magnetic Pole (ca. 86°N, 166°W—445 kilometers [277 mi] from the geographic North Pole) for the first time in 1994. It has been used for commercial, military, scientific, and even touristic purposes.

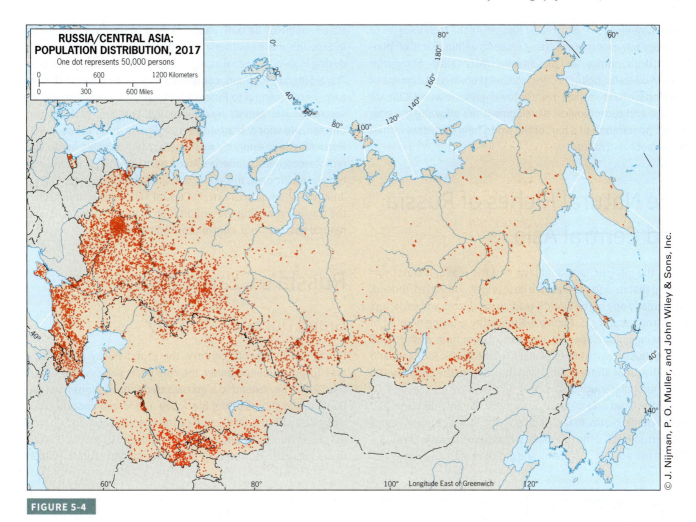

RUSSIA/CENTRAL ASIA:
POPULATION DISTRIBUTION, 2017
One dot represents 50,000 persons

FIGURE 5-4

© J. Nijman, P. O. Muller, and John Wiley & Sons, Inc.

Bering Strait with the North Sea (see Chapter 12), shortening certain international sea routes by thousands of kilometers and ushering in a new era of long-distance trade.

That, at least, is how many Russians like to envision global warming—as a potential advantage nature has long denied them. And indeed, the Arctic provides much evidence for such warming: the melting of parts of the Greenland Ice Sheet, the shrinking of the average extent of permanent Arctic Ocean ice, and the reduced incidence of icebergs. Russian economic planners look at the shallow offshore waters along the Arctic Ocean coastline (see the white- and light-blue areas in Fig. 5-1) and hope that oil and gas reserves beneath the seafloor will come within reach of exploitation. These developments may even have the effect of expanding the realm far into the Arctic. In 2007, the Russian government even placed a metal Russian flag on the seafloor at the North Pole under the permanent ice of the Arctic Ocean to symbolize its intentions (see final photo in Chapter 12).

Ecologies at Risk

If, as most computer models predict, global warming persists and perhaps intensifies in the polar latitudes, these ecologically sensitive environments are sure to be severely affected.

Animals as well as humans have long adapted to prevailing climatic conditions, and such adaptations would be significantly disrupted by rapid environmental change. The intricate web of relationships among species and their environments on the one hand, and among species themselves on the other, can be quickly damaged by temperature shifts. The polar bear is just one prominent example: it depends on ample floating sea ice to hunt as well as raise cubs. When sea ice contracts and wider stretches of open water force polar bears to swim greater distances during the cub-rearing season, fewer infants make it to adulthood. If ice-free Arctic summers are indeed in the offing, the polar bear could become extinct in your lifetime. Rapid ecosystem change would also endanger seal, bird, fish, and other high-latitude wildlife populations.

Such changes affect human populations as well. For example, indigenous communities still pursuing traditional lives in parts of the Arctic domain are similarly adapted to the harsher environments of the past. Their traditions, already under pressure from political and economic forces resulting from their incorporation into modern states, would be further imperiled by environmental changes that are likely to irreversibly modify the ways of life they developed over thousands of years.

And if a new cycle of oil and natural gas exploration and exploitation is indeed about to begin, already-fragile offshore environments face even greater danger. As the technologies of recovery put ever more of these reserves within reach, oil platforms, drills, pumps, and pipelines will materialize—along with risks of oil spills, pollution, and collateral ecological damage of the kind we have seen all over the lower-latitude world. Climate change and technological development are allowing the economic penetration of a part of the world long protected from it by distance and nature.

The Natural Riches of Russia and Central Asia

Given the gigantic territorial layout of the realm, it is not surprising that its natural resource base is vast and varied. Indeed, it is already one of the world's leading producers and exporters of oil and natural gas. **Figure 5-5** maps the growing number of producing fields dispersed across the realm: from western and central Russia eastward to Sakhalin in the Pacific Rimland, and from the Arctic shore of Siberia in the north to the Caspian-Aral Basin in the far south (**Fig. 5-5**).

Petroleum and natural gas are not the only underground riches of this realm. In fact, it is so well endowed with minerals that virtually all of the raw materials required by modern industry are extractable. Nor are oil and gas the only energy reserves.

Major coalfields lie both east and west of the Urals as well as in Siberia's more southerly latitudes (the string of mines producing high-quality coal along the corridor of the all-important Trans-Siberian Railroad, completed in the early twentieth century, were essential to Russia's industrial development and its successful war against Nazi Germany). Large deposits of iron ore, too, are widely scattered across Russia, from the so-called Kursk Magnetic Anomaly near the border with Ukraine to the remotest corners of Siberia's Arctic north. And when it comes to other metallic ores, this realm has everything from gold to lead and from aluminum to zinc. In Central Asia, Kazakhstan has led the way in expanding the exploitation of its rich deposits of oil, natural gas, uranium, chromium, and titanium.

Russia's Czarist Roots

Grand Duchy of Muscovy

During the fourteenth century, the Grand Duchy of Muscovy rose to preeminence under the rule of its princes and dukes (**Fig. 5-6**). They extended Moscow's trade connections from the Baltic to the shores of the Black Sea and forged enduring religious ties with the leadership of the Eastern Orthodox Church in what is now Istanbul, Turkey. Then, beginning around 1450, there began a period of more than three centuries during which

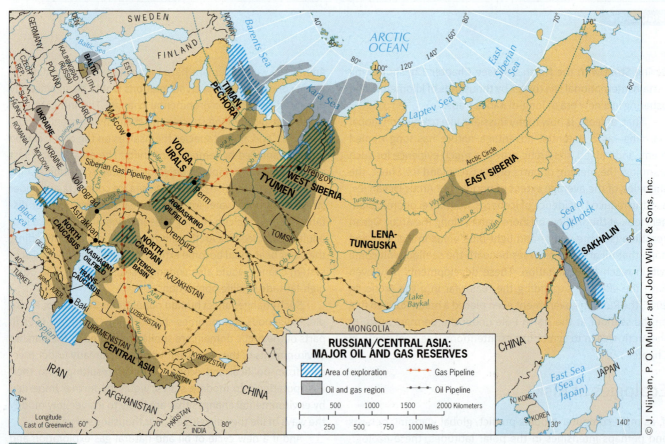

FIGURE 5-5

© J. Nijman, P. O. Muller, and John Wiley & Sons, Inc.

powerful Russian rulers—now called czars—etched Muscovy's imprints ever more widely on the map of Russia, Eurasia, and, by extension, the world.

By the sixteenth century, during the reign of Czar Ivan IV (better known as Ivan the Terrible), the Grand Duchy had been transformed into a military powerhouse and an imperial state. Ivan expanded Moscow's empire first by conquering the Islamic areas of what is now western Russia and then by gaining control over present-day Estonia and Latvia on the Baltic Sea. It was a time of almost constant warfare, and Ivan acquired his reputation from the reign of terror he unleashed in his pursuit of military discipline, centralized administrative control, and retaliation against many members of the nobility.

Eighteenth-Century Czarist Russia

When Peter the Great became czar (he ruled from 1682 to 1725), Moscow already lay at the center of a great empire—great, at least, in terms of the territories it controlled. The Islamic threat had been ended, and the growing influence of the Russian Orthodox Church was represented by its distinctive religious architecture and powerful bishops. Peter consolidated Russia's gains and endeavored to make a modern, European-style state out of his loosely knit country. He built St. Petersburg as a **forward capital [10]** on the doorstep of Swedish-held Finland, fortified it with major military installations, and made it Russia's leading port (Fig. 5-6; **Box 5-3**).

Peter the Great was an extraordinary leader and in many ways the founder of modern Russia. In his desire to remake the country—to pull it from the forests of the interior to the seas of the west, to open it to outside influences, and to relocate its population—he left nothing to chance. Prominent merchant families were forced to move from other cities to St. Petersburg. Ships and wagons entering the city had to bring building stones as an entry toll. The czar himself, aware that to become a leading power Russia had to be strong at sea as well as on land, went to the Netherlands to work as a laborer in the famed Dutch shipyards to learn the most efficient methods for building them. Meanwhile, the czar's forces continued to conquer peoples and territory: Estonia was incorporated in 1721, widening Russia's window to the west; and in Siberia steady expansion soon began in the area south of the city of Tomsk (Fig. 5-6).

Under Czarina Catherine the Great, who ruled from 1760 to 1796, Russia's empire in the Black Sea Basin grew at the expense of the Ottoman Turks. The Crimean Peninsula, the port city of Odessa, and the entire northern coastal zone of the Black Sea fell under Russian control. Also during this period, the Russians made a fateful move: they penetrated the corridor between the Black and Caspian seas, whose spine is the mountainous Caucasus with its dozens of ethnic and cultural homelands, many of which were Islamized. The cities of Tbilisi (now in Georgia), Baku (Azerbaijan), and Yerevan (Armenia) were captured. Eventually, the Russian push toward an Indian Ocean outlet was thwarted by the British, who held sway in Persia (modern Iran), and also by the Turks.

FIGURE 5-6

© J. Nijman, P. O. Muller, and John Wiley & Sons, Inc.

Box 5-3   From the Field Notes … St. Petersburg

"Not only the city of St. Petersburg itself, but also its surrounding suburbs display the architectural and artistic splendor of Czarist Russia. The czars built opulent palaces in these outlying districts (then some distance from the built-up center), among which the Catherine Palace, begun in 1717 and completed in 1723 followed by several expansions, was especially majestic. These photos show the ornate sculptures that decorate the magnificent exterior (left) as well as the interior detail of a set of rooms known as the Golden Suite, of which the ballroom (right) exemplifies eighteenth-century Russian Baroque at its height. The 'glorious past' of Czarist Russia (see Fig. 5-6) seems to be reviving in the hearts and minds of today's most patriotic Russians."

Around the same time, Russian colonists were advancing far to the east, settling along the southeastern frontier, crossing the Bering Strait, and even entering Alaska in 1784. But in less than a century the Russians gave up on their North American outposts (which at their height had reached as far south as central California). The sea-otter pelts that had attracted the early pioneers were running out, European and white American hunters were cutting into the profits, and indigenous American resistance was growing. When the United States offered to purchase Russia's Alaskan holdings for $7.2 million (about $125 million in today's dollars) in 1867, the Russian government quickly agreed, setting Alaska on a new course that culminated in its joining the United States as the forty-ninth State in 1959. From the point of view of the United States, besides the 24 dollars allegedly paid to acquire Manhattan Island, this transaction is often regarded as the most lucrative real estate deal in world history.

Nineteenth-Century Expansion

Although Russia had withdrawn from North America, Russian expansionism during the nineteenth century forged ahead in Eurasia. As they were extending their empire southward, the Russians also took on the Poles, old enemies to the west, and succeeded in conquering most of what is now Poland, including the capital of Warsaw. To the northwest, Russia took over Finland from the Swedes in 1809.

Throughout most of the nineteenth century, however, the Russians were preoccupied with Central Asia, where Tashkent and Samarqand (Samarkand) came under St. Petersburg's control (Fig. 5-6). The Russians here were still bothered by raids of nomadic horsemen, and they sought to establish their authority over the Central Asian drylands as far southeast as the high mountains that marked China's western frontier. As it made headway here, Russia gained many Muslim subjects because it was now penetrating Islamic Asia. Nonetheless, under czarist rule local peoples retained some autonomy.

Thus the Russian state, like the states of Britain, France, and other European powers, expanded through ***imperialism***. Whereas those European mother countries expanded their domains overseas, Russian influence spread overland into Central Asia, Siberia, China's inland peripheries, and the far eastern Pacific coastlands. What emerged in the end was not the greatest empire but the largest territorially contiguous empire in the world.

The czars embarked on their imperial conquests in no small part because of Russia's relative location—most importantly, the lack of warm-water ports. Had the early-twentieth-century revolution not intervened, their southward push might have reached the Indian Ocean or even the Mediterranean Sea. Czar Peter the Great envisaged a Russia open to trading with the entire world, and he made sure his new capital of St. Petersburg on the Baltic Sea developed into Russia's leading port. But in truth, throughout the long arc of its history, Russia's geography has mostly been one of remoteness from international mainstreams of change and progress, repeatedly punctuated by episodes of self-imposed isolation.

A Multinational Empire

Centuries of Russian expansionism did not confine themselves to empty land or unclaimed frontiers. The Russian state became an imperial power that annexed and incorporated numerous nationalities and cultures. This was done by employing force of arms, by overthrowing uncooperative rulers, by seizing and annexing territory, and by stoking the fires of ethnic conflict. By the time the ruthless Russian regime had begun to confront revolution among its own citizens, Czarist Russia controlled peoples representing more than 100 nationalities. The winners in the revolutionary struggle that ensued—the communists who forged the Soviet Union—did not liberate these subjugated peoples. Rather, they changed the empire's framework, binding the peoples colonized by the czars into a new system that would in theory give them autonomy as well as identity. In practice, however, it doomed those peoples to bondage and, in certain cases, extinction. The Soviet Union, which arose in the wake of the communist revolution of 1917, expeditiously undertook another effort to integrate all these peoples and regions into this vast multinational state—notwithstanding the mantra-like socialist condemnations of imperialism.

The Soviet Union (1922–1991)

Communism found fertile ground in the Russia of the 1910s and 1920s. In those days, Russia was infamous for the wretched serfdom of its peasants, the cruel exploitation of its workers, the excesses of its nobility, and the ostentatious palaces and riches of the czars. Ripples from Europe's Industrial Revolution during the 1890s had introduced a new age of misery for those laboring in factories. There were workers' strikes and ugly retributions, but when the czars finally tried to better the lot of the poor, it was too little too late.

There was no democracy, and the people had no way to express or channel their grievances. Europe's democratic revolution had bypassed Russia, and its economic revolution touched the czars' domain only slightly. Most Russians, as well as tens of millions of non-Russians under czarist control, faced exploitation, corruption, starvation, and harsh subjugation. When the people finally began to rebel in 1905, there was no hint of what lay in store. Over the next two decades, it would result in nothing less than a full-scale revolution that transformed Czarist Russia into a behemoth communist state. In the process, the lives of nearly 150 million now-Soviet citizens were fundamentally reorganized, and the seeds sown for what would eventually become one of the twentieth century's superpowers on the global political stage.

The Soviet Territorial Framework

Russia's great expansion had brought myriad nationalities under czarist control; now the revolutionary government set out to restructure this heterogeneous ethnic mosaic into a smoothly functioning state, the **Union of Soviet Socialist Republics (USSR)**—or **Soviet Union** in short form. The czars had conquered, but they had done little to bring Russian culture to the peoples they ruled. When the old order was overthrown in 1917, ethnic Russians themselves constituted only about half of the population of their empire.

The political framework for the new Soviet Union was based on the ethnic identities of its numerous incorporated peoples. Given the size and cultural complexity of the now Soviet Empire, it was impossible to allocate territory of equal political standing to each of the nationalities (keep in mind that the communists now controlled the destinies of well over 100 peoples, ranging from large nations to small isolated ethnic pockets). It was decided to subdivide this gargantuan realm into 15 *Soviet Socialist Republics (SSRs)*, each of which was delimited to correspond broadly to one of the major nationalities. (As Figure 5-6 shows, much later these SSR territories morphed into the framework of the national state borders that emerged after the 1991 dissolution of the USSR.) Although Russians constituted barely 50 percent of the nascent Soviet Union's population, they were (and still are) the most widely dispersed ethnic group in the realm (see **Fig. 5-7**). Thus the Russian Republic was by far the largest designated SSR, encompassing just over three-quarters of the Soviet Empire's territory.

As designed by Vladimir Lenin, the Soviet Union's founder, the new political system claimed to be a **federation [11]**, which suggests power sharing among the various republics—but the reality was quite different. The Russian leadership in Moscow maintained absolute control over the SSRs, and they made it Soviet policy to relocate entire peoples from their homelands in order to better fit the grand design as well as rewarding or punishing localities—in many instances capriciously. The overall effect, however, was to move minority peoples eastward and replace them with Russians. In time, this **Russification [12]** of the Soviet Empire produced substantial ethnic Russian minorities in each of the 14 other republics.

The centerpiece of the tightly controlled Soviet "federation" was the Russian Soviet Federative Socialist Republic. Containing half the USSR's population, the capital city, the realm's core area, and 76 percent of all Soviet territory, Russia was unmistakably the Empire's nucleus. Outside the Russian Republic, "Soviet" almost always meant "Russian"—the reality with which the lesser republics lived. Russians came to these 14 republics to teach (Russian was taught in the colonial schools), to organize (and frequently dominate) the local Communist Party, and to implement Moscow's economic decisions. Clearly, this was colonialism, but somehow the communist disguise—how could socialists, as the communists called themselves, be colonialists?—and the contiguous spatial nature of the Empire made it appear to the rest of the world as something else. Indeed, on the international stage the Soviet Union became a champion of oppressed peoples, a force in the post–1950 decolonization process. It was an astonishing contradiction that would, over time, be fully exposed.

The Soviet Economic Framework

The geopolitical changes that resulted from the formation of the Soviet Union were accompanied by a gigantic economic experiment: the conversion of the Empire from a czarist autocracy with a capitalist veneer to communism. From the early

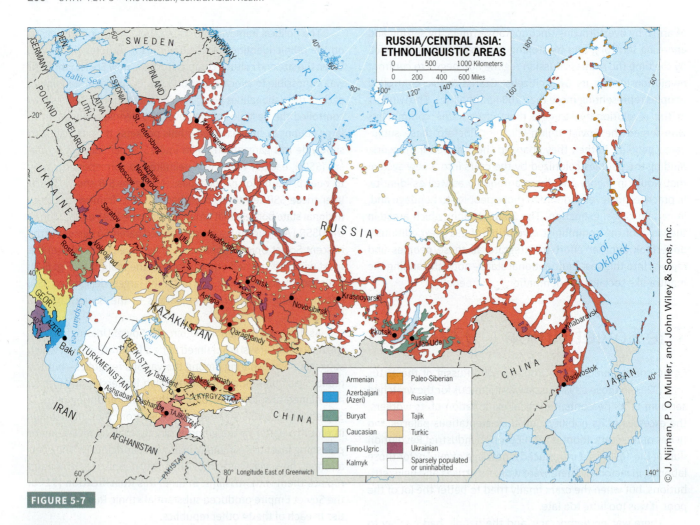

RUSSIA/CENTRAL ASIA: ETHNOLINGUISTIC AREAS

Legend:
- Armenian
- Azerbaijani (Azeri)
- Buryat
- Caucasian
- Finno-Ugric
- Kalmyk
- Paleo-Siberian
- Russian
- Tajik
- Turkic
- Ukrainian
- Sparsely populated or uninhabited

© J. Nijman, P. O. Muller, and John Wiley & Sons, Inc.

FIGURE 5-7

1920s onward, the country's economy would now be *centrally planned*—the communist leadership in Moscow would make all decisions regarding economic planning and development. Soviet planners had two principal objectives: (1) to accelerate industrialization and (2) to collectivize agriculture. For the first time ever on such a scale, an entire country was organized to work toward national goals prescribed by a central government.

But Soviet agriculture never came close to achieving its intended level of productivity because farmers were inadequately motivated. Most opposed collectivization, but any who obstructed the communists' grand design suffered a dreadful fate. It has been estimated that between 30 and 60 million people lost their lives through imposed starvation, constant political purges, Siberian exile, and/or forced relocation. Even though this historic Soviet "experiment" produced an incalculable human tragedy, the secretive nature of Soviet officialdom made it possible to hide this disaster from the world.

The USSR operated a **command economy [13]**, in which state planners assigned the production of particular manufactures to particular places, often disregarding the basic rules of economic geography. For instance, the manufacture of railroad cars might be assigned (as indeed it was) to a factory in Latvia. No other factory anywhere else would be permitted to produce this equipment—even if supplies of raw materials would make

it cheaper to build them near, say, Volgograd 2000 kilometers (1250 mi) away. Yet, despite an expanded and improved transportation network (see Fig. 5-11), such policies made manufacturing in the USSR extremely expensive, and the absence of competition made managers complacent and workers far less productive than they might have been.

The Downfall of the Soviet Union

On December 25, 1991, the inevitable came to pass: the Soviet Union expired. The centrally planned economy went into terminal structural failure; the arms race with the United States during the Cold War had drained too many resources; and the long effort at Russification had finally backfired, fueling the drive for independence among most of the peoples who had been forcibly incorporated into the USSR. The last Soviet president, Mikhail Gorbachev, resigned, and the Soviet hammer-and-sickle flag flying atop the Kremlin in Moscow was lowered for the final time and was immediately replaced by the white, red, and blue Russian tricolor. Yet even though a new and turbulent era had opened literally overnight, deeply entrenched Soviet institutions and systems will long continue to cast their shadows over transforming Russia.

Robert Harding Productions / Getty Images

The spirit of the communist revolution found expression in "social-ist-realist" art and political propaganda. This famous landmark at the entrance to Moscow's Mosfilm Studios dates from 1937 and glorifies the working class in typical fashion: a heroic statue of individual male and female workers hoisting the iconic hammer and sickle above their heads. Russians call this sculpture *Rabochiy i Kolkhoznitsa* (Worker and Kolkhoz Woman—a kolkhoz being a collectivized farm). Even today, statues like this stud urban landscapes all across Russia.

When the Soviet system collapsed and the 15 Soviet Socialist Republics became independent states in 1991, Russia was left without the empire that had taken centuries to build and consolidate—and that contained crucial agricultural as well as industrial resources. No longer did Moscow control the farms of Ukraine or the oil and natural gas reserves of the Caspian Basin and Central Asia. The former Russian Soviet Federative Socialist Republic had suddenly become the newborn country of Russia, still a state of enormous territorial size but no longer able to claim sovereignty over the valuable lands beyond its new borders.

Post-Soviet Russia and Central Asia

The Soviet Union's implosion signaled a major reorientation of the Russian Republic as well as the other 14 former SSRs that

became independent countries. Central planning came to an end, and free markets had to be introduced; national industries and corporations had to be privatized; and the communists lost their political monopoly to a new multiparty system that (at the outset) offered real elections. Although individual freedoms increased, state support in a variety of forms, from pensions to rent subsidies, mostly came to a halt. As all the components of the former Soviet Union went through an initial phase of shock, the overriding hope was that all this would be transitional, and that temporary dislocation and occasional chaos would soon be followed by a better life for most citizens. Before we examine the new political and economic forces unleashed at the regional level, we need to establish their broader context by reviewing two important developments reshaping the realm's human geography in the twenty-first century: (1) the continuing evolution of its complex cultural mosaic, and (2) Russia's geopolitical repositioning in its "Near Abroad" (the former non-Russian SSRs) as well as the world beyond.

A Complex Cultural Mosaic

For centuries, this realm has been a culturally and ethnically diverse part of the world whose traditions and customs spill over into neighboring realms even as neighbors have come to live here. Ethnic Russians still form the overall majority, but sizeable parts of this realm are home to a growing number of non-Russians—and not just in Transcaucasia and Central Asia. As Figure 5-7 shows, the realm contains clusters of Finnish, Turkic, Armenian, Kazakh, and dozens of other minorities, but the scale of this map is too small to reflect the complexity of the total ethnic mosaic (especially within Russia itself). Here is an interesting example: in the southwest, facing the Caspian Sea, lies a small "republic" named Dagestan that is part of modern Russia. It is half the size of Maine, has a population of 3 million, and contains nearly 30 ethnic communities that speak their own languages, most of which are variants of languages spoken in the nearby Caucasus, Turkey, and Iran.

As Figure 5-7 shows, the Slavic people collectively known as the Russians not only form the majority of the realm's population but also are the most widely dispersed across it. Although the large lowland west of the Urals is the historic hearth and modern core area of the Russian state, the distribution of Russians today still extends from the shores of the Arctic Sea southward to the Black Sea coast and from St. Petersburg on the Gulf of Finland all the way east to Vladivostok on the Pacific's Sea of Japan. Discontinuous nuclei and ribbons of Russian settlement are even scattered across parts of Siberia. Nonetheless, another glance at Figure 5-4 reminds us that the population as a whole decidedly concentrates in the realm's southwestern sector.

Traces of the penetration of the Islamic Ottomans hundreds of years ago can clearly be discerned on today's cultural map (Fig. 5-7): the large swath of Turkic peoples (shaded tan) extending from near Nizhniy Novgorod southeastward across Kazakhstan to the fringes of Central Asia is a solid piece of evidence. Contemporary Russia has a relatively higher percentage of Muslims in its population (roughly 13 percent) than western European countries do. As for Central Asia—whose total population is 48 percent of Russia's—the combined percentage of Muslims living in its five countries today is more than 75 percent.

Long before Islam appeared on the scene, Slavic peoples from one end of their domain to the other had accepted the teachings of the Eastern Orthodox Church. It stayed that way until the communists overthrew the czar and promptly began the disestablishment of the Church. For the next seven decades, atheism was official policy; but following the demise of the USSR in the early 1990s, the Russian Orthodox Church has made a powerful comeback, today increasingly accompanied by nationalistic propaganda.

The Near Abroad

The ending of the Cold War upon the breakup of the USSR implied an abrupt termination of Soviet geopolitical strategies that had been aimed at confronting U.S.-led alliances in Europe, the Middle East, and South Asia as well as Southeast Asia. Soviet foreign policy and power had been projected well beyond Soviet borders to other geographic realms from Africa to Middle America. The downfall of the Soviet Union meant that the new Russia, shorn of its **satellite states [14]** in eastern Europe and abandoned by all the former SSRs, had to completely reinvent its position in the world, and most immediately within Eurasia. Moreover, it had to meet this challenge in the knowledge that a total makeover of Russia's economy was the foremost priority.

In 1992, the new Russian government introduced the term **Near Abroad [15]** to refer to the newly formed countries that surround Russia today. For the most part, these encompass the 14 former Soviet republics, stretching from Estonia to Tajikistan. Those states may now be located "abroad," but seen from Moscow they are so "near"—and collectively constitute such a pronounced encirclement of Russia—that the Russian government feels entitled to take a special interest in them (**Box 5-4**). What this really means is that Russia will not permit any of these countries to mount a threat of any kind and that it reserves the right to intervene if it deems that is necessary. It is not difficult to envision this sense of territorial encirclement in Russia, particularly with the eastward expansion of the EU and **NATO [16]** (now almost at Russia's western doorstep), heightened Islamic militancy in the southern periphery, and the swift rise and growing assertiveness of China in the Far East as well as Central Asia.

And then there is the issue of ethnic Russian minorities who now find themselves living "abroad." Although many Russian nationals have returned from the former SSRs since 1991, millions more stayed on. A large proportion of these expatriates reside in areas where Russians were in the majority (for example, in parts of eastern Ukraine, northern Kazakhstan, and eastern Moldova) and where post-Soviet life was generally unchanged from what it was during Moscow's rule. But elsewhere, Russians who remained in the erstwhile Soviet republics were instantly reduced to minority status; many were mistreated in various ways, and they often appealed to Moscow for assistance. Thus the concept of a Russian "Near Abroad" found its way into Russia's national discourse. Beyond Russia's borders, but of growing concern to its leadership and citizens, the "Near Abroad" became geographic shorthand for an enduring Russian sphere of influence, whereby Moscow stood ready to help its fellow ethnics in case of trouble. So far, the most acute and dangerous confrontation continues to unfold in Ukraine following Russia's unexpected, forcible annexation of the Crimean Peninsula in 2014. Within weeks, parts of eastern Ukraine had descended into *de facto* warfare as separatist Russian rebels, strongly supported by Moscow (both openly and surreptitiously), battled the Kiev-based Ukrainian government in their struggle to secede from Ukraine and achieve annexation by Russia (see Chapter 4, especially Box 4-15).

© Geodakyan Artyom/ITAR-TASS Photo

On March 18, 2015, Russia celebrated the first anniversary of its "reunification" with Crimea. This pivotal peninsula on the northern coast of the Black Sea, which contains Russia's major naval base at Sevastopol, is of the utmost geostrategic importance. It was forcibly annexed from Ukraine in 2014 despite the strenuous objections of the Ukrainian government and most of the international community led by the United States. These Russian Black Sea Fleet officers in Sevastopol's Nakhimov Square are gearing up for a high-profile parade through the city's streets. The Crimean Peninsula was "given" to the Ukrainian Soviet Socialist Republic in 1954 by the USSR's leader, Nikita Khrushchev, as a gesture of good will and friendship—at a time when the Russians were firmly in control. It was never imagined that an independent Ukraine might one day turn its back on Russia. But that is precisely what has been happening over the past several years, culminating in President Vladimir Putin's 2014 decision to invade, occupy, and forcibly reattach Crimea to the Russian Federation.

Box 5-4 Regional Issue: Does Russia Need to Build Up Its Military?

We Are Threatened and Must Show Our Strength!

"It always amazes me that the West does not seem to understand our position. It has been like this since World War II. Russia suffered more deaths than any other country during that war, in which it was attacked by Germany. My own father was a soldier in that war and he never came back…. After the war, we had to make sure that we controlled part of Eastern Europe so that it could not happen again. What did the West do? They created NATO and encircled us—while we had given so much to defeat the Nazis in an alliance with the United States, France, and England…. We did not organize our own military alliance, the Warsaw Pact (1955), until after they created NATO (1949). They were aggressive, and we had to respond.

"And the story repeats itself. After the Soviet Union tragically fell apart and we voluntarily stepped back from former Soviet Republics such as Kazakhstan, the Baltic States, or Ukraine, what does the West do? It expands the European Union *and* NATO to our doorstep! How would the United States feel if we set up a military alliance with Mexico? Well, that is how it is for us. We must show our strength these days, as we have done in Ukraine and Chechnya.

"We also face a serious threat from Islamic militants inside our borders and just to our south, in the Caucasus. It requires an iron fist and investments in military intelligence and high-tech warfare. Again, I would ask you: if you had neighboring countries that were known to harbor terrorists, would you not act in the strongest way to secure your safety? And, mark my words: the world will be grateful for what we are doing in Syria—because it is us, not the United States or those opportunistic free-riding Europeans, who are destroying ISIS.

"It is a simple fact that Russia is encircled with challenges to its security, also including the rise and growing assertiveness of China in the Far East, and all the increased activity in the Arctic, of various countries that shouldn't even be there, like China. Maybe, if you actually *wanted* to understand our situation, you would. We need a strong military more than ever."

Russia Needs Democracy and Economic Progress, not a Stronger Military

"What is it with this Russian obsession with security and feeling threatened? Do you really think the Europeans or Americans, or the Chinese for that matter, are interested in invading our territory? Such geopolitical notions belong to the past. Nobody is interested anymore in actually invading other countries. Today, it is all about economic and political stability and international collaboration. And about being clever in international trade and finance. And most of all, for us Russians, it should first be about our economic well-being. I am finishing up my business degree in college in Moscow, but it will be awfully hard to find a decent job.

"The Soviet Union failed, that is a simple truth we must accept, and we should be trying to join the global economy and a world that favors freedom, economic progress, and democracy. From 1917 to 1991, we have been on a dead-end course, and what we should have done over the last 25 years is rebuild this great country. Yes, ours is a great country of poets, mathematicians, writers, and engineers; and we are blessed with immense natural resources and millions of people who want to work hard and get ahead. But we have to get over this paranoia that is, unfortunately, perpetuated by our own leaders.

"While many of us live in poverty, lack proper healthcare, vote in elections that are meaningless, and have no real economic prospects, our government keeps distracting the people with talk about external threats and keeps playing on feelings of national pride. It still works for the older crowd, but the young people I know here in the city are not impressed. All that money spent on new military bases in the Arctic, or on sending our troops to Syria, would be of much better use if it were invested in the economy, or better healthcare. Maybe, then, in the end we could compete economically with the West and with China."

Playing Russia's Energy Card

During Soviet times, the non-Russian republics all became dependent to some extent on Russia's prodigious oil, natural gas, and mineral supplies. Today, this dependency on Russian energy reserves in particular plays a critical role in the evolving geopolitics of the realm. In parts of Central Asia, however, Russia has actually been losing ground because a number of countries contain their own abundant fuel supplies. Most of all, Kazakhstan has begun to exploit its huge reserves of oil and gas—attracting growing interest as well as substantial investments from around the world. Between 2000 and 2012, Kazakhstan's exports to Russia were almost halved from 17 to 9 percent, while exports to China practically tripled from 7 to 20 percent. Oil constitutes the bulk of these exports, and the Chinese are now working with the Kazakhs to complete a pipeline clear across western China and Kazakhstan in order to tap the latter's high-yielding northern Caspian Sea oilfields. Similarly, oil- and gas-rich Turkmenistan has in recent years intensified relations with China at Russia's expense: 81 percent of Turkmen exports (mostly natural gas) now flow to China. The remaining (energy-poorer) Central Asian countries—Uzbekistan, Kyrgyzstan, and Tajikistan—continue to be tightly tethered to the Russian economy, and Moscow's political influence in all three remains considerable.

As for the European countries to Russia's west, dependence on Russian energy supplies is enormous, implying significant leverage for Moscow in its political dealings with that realm. Consider these current percentages of oil and gas imports from Russia: Belarus, 100; Slovakia, 98; Lithuania, 92; Bulgaria, 90; Hungary, 86; Finland, 76; Latvia, 72; and Estonia, 69. Farther west, Russian imports are also substantial: Sweden, 40; Germany, 30; the Netherlands, 30; and Italy, 28. Russia also is the single biggest fuel supplier to the European Union as a whole, providing 35 percent of its oil and 30 percent of its natural gas. In the mid-2010s, as criticism of Russia accelerated in Europe during the deepening crisis in Ukraine, there has been much talk about the need for Europe to reduce its energy reliance on Russia. But this cannot happen overnight because, quite apart from the economic and political issues involved, energy infrastructures are complex, very expensive, and can take years to build. **Figure 5-8** maps both existing and proposed gas pipelines connecting Russia and Europe. Note that a great deal of Russian gas being pumped to Europe passes through Ukraine—yet another potentially critical factor in the evolving crisis there. Ukraine itself is hurriedly trying to decrease its reliance on Russia by seeking new suppliers in the West. In 2013, 95 percent of its gas came from Russia; but in 2015, the Ukrainian government claims, that percentage was reduced to 33.

Militarizing the Arctic

Unlike the Antarctic, the Arctic Ocean Basin is not officially divided up among countries claiming sovereignty (compare Figs. 12-5 and 12-6). Yet claims are being made by the United States, Canada, and Denmark (all three members of NATO) as well as Russia. Clearly, Russia has a longstanding interest in the Arctic because its northern border fronts nearly half of it. With increasing evidence of major mineral deposits below the seafloor and with global warming improving Arctic accessibility, Russia has in recent years stepped up its military presence there. **Figure 5-9** shows Russia's existing and new/upgraded military bases, along with those of Canada, Denmark, and the United States (in Alaska). Seen from Moscow, a strong military presence in the region is paramount, especially if the **Northeast Passage** increasingly opens up to international shipping in the coming years (see map of shipping routes in Fig. 12-6).

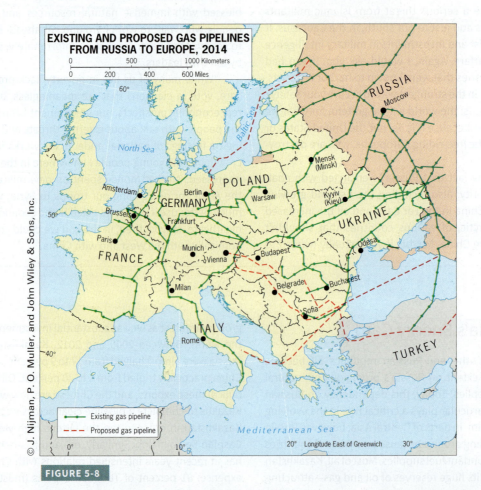

© J. Nijman, P. O. Muller, and John Wiley & Sons, Inc.

FIGURE 5-8

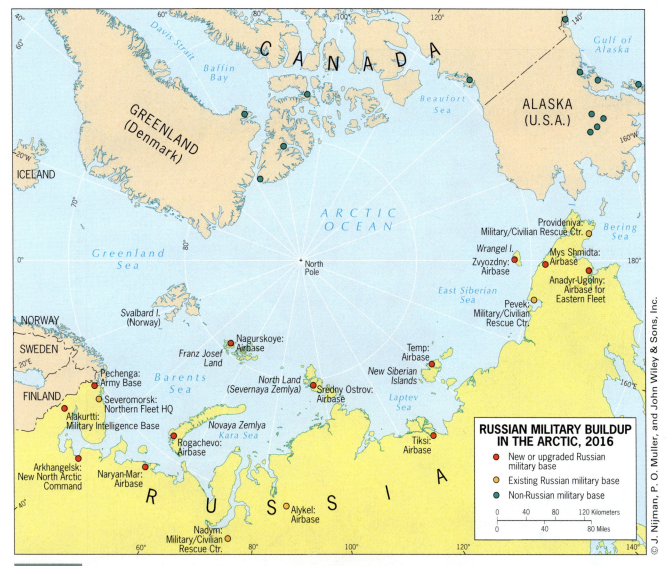

FIGURE 5-9

Regions of Russia and Central Asia

Few other single states dominate their geographic realm to the extent that Russia does, but we also need to keep in mind all those recent changes that have occurred within and beyond the realm's peripheries. After the USSR imploded nearly three decades ago and all the former SSRs had gone the way of independent statehood, Russia was left by itself—still of unmatched proportions territorially, but a nation that had lost control of bordering regions in Europe and Southwest Asia, from the Baltic Sea to the Tian Shan Mountains lining China's westernmost boundary. Today, we refer to the Russian/Central Asian realm, which consists of Russia, the three comparatively tiny countries of Transcaucasia, and, to the east of the Caspian Sea, the five Central Asian countries that were all once part of the Soviet Union.

As we are aware, Russia remains a gigantic country, one so diverse in terms of its human and physical geography that

it could be regionally subdivided in a variety of ways and at different scales. We will keep things simple by delimiting what can be called six first-order regions, recognizing that each of them contains its own notable subdivisions. Four of these six regions to be discussed, which are mapped in **Figure 5-11**, lie entirely inside Russia, and the other two are located in separate sectors of Russia's southern frontier zone that is almost transcontinental in east-west length:

1. The Russian Core
2. The Southeastern Frontier
3. Siberia
4. The Russian Far East
5. Transcaucasia
6. Central Asia

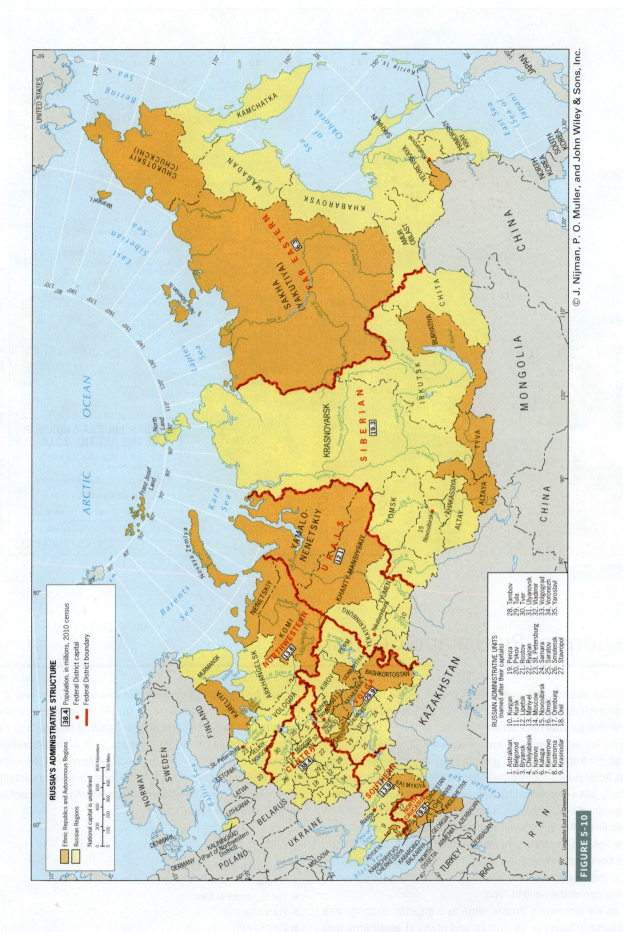

RUSSIA'S ADMINISTRATIVE STRUCTURE

38.4 Population, in millions, 2010 census
● Federal District capital
▬ Federal District boundary

National capital is underlined

Ethnic Republics and Autonomous Regions
Russian Regions

RUSSIAN ADMINISTRATIVE UNITS
(named after their capitals)

1. Astrakhan
2. Belgorod
3. Bryansk
4. Chelyabinsk
5. Ivanovo
6. Kaluga
7. Kemerovo
8. Kostroma
9. Krasnodar
10. Kurgan
11. Kursk
12. Lipetsk
13. Mariyel
14. Moscow
15. Novosibirsk
16. Omsk
17. Orenburg
18. Orel
19. Penza
20. Pskov
21. Rostov
22. Ryazan
23. St. Petersburg
24. Samara
25. Saratov
26. Smolensk
27. Stavropol
28. Tambov
29. Tula
30. Tver
31. Ulyanovsk
32. Vladimir
33. Volgograd
34. Voronezh
35. Yaroslavl

© J. Nijman, P. O. Muller, and John Wiley & Sons, Inc.

FIGURE 5-10

REGIONS OF THE RUSSIAN/CENTRAL ASIAN REALM

Transition Zone

© J. Nijman, P. O. Muller, and John Wiley & Sons, Inc.

FIGURE 5-11

The New Russia

Russia's Political-Administrative Structure

When Russia's new government was installed after the disintegration of the USSR, it immediately faced an old Soviet problem in the form of an exceptional geographical challenge—the sheer size of the country, its vast distances, and the remoteness of many of its far-flung spatial components. Geographers refer to the principle of **distance decay [17]** to explain how increasing distances between places tend to reduce interactions among them. Because Russia is the world's largest country, distance is a significant factor in the relationships between the capital and outlying areas. Moreover, Moscow and St. Petersburg, decidedly the two biggest cities of the realm (**Box 5-5**), lie in the far west of this behemoth country, literally half a world away from Russia's Pacific coast. Still another problem is the astonishing size variation among the country's myriad administrative units, as can be seen in **Figure 5-10**. Whereas the (territorially) smallest divisions are concentrated in the Russian core area, the largest lie far to the east, where the Republic of Sakha is nearly 1000 times the size of the Republic of Ingushetiya. On the other hand, the populations of the enormous eastern divisions are minuscule compared to those of the smaller ones in the Russian Core. Such diversity spells administrative difficulty.

Figure 5-10 reflects this especially complex Russian administrative system, which in large part is a legacy of the Soviet Union. There are eight Federal Districts (newly created in 2000) that are subdivided into 21 Ethnic Republics and Autonomous Regions, and 35 Russian Regions. These administrative units function with varying degrees of power and autonomy from the federal government. The 21 Republics, established to recognize and accommodate the largest ethnic minorities, include a cluster in the far south, another lying to the east of Moscow, and a third that parallels the far-off Mongolian border.

In 1992, almost all of Russia's administrative units signed a document known as the Russian Federation Treaty, which committed them to cooperate within the new **federal system**. At first, a few holdouts refused to sign, including a Republic in the remote Caucasus periphery, Chechnya-Ingushetiya, where Muslim rebels had launched a campaign for independence (Fig. 5-10). Chechnya-Ingushetiya subsequently split into two separate Republics. Eventually, only Chechnya refused to sign the Russian Federation Treaty; Russian military intervention soon followed and triggered a prolonged, violent conflict, producing disastrous consequences for Chechnya's people and infrastructure (the capital, Groznyy, was completely destroyed).

FIGURE 5-12

	Metropolitan Area Population (in millions)
Moscow, Russia	16.6
St. Petersburg, Russia	5.1
Tashkent, Uzbekistan	2.8
Baki (Baku), Azerbaijan	2.7
Almaty, Kazakhstan	1.5
Novosibirsk, Russia	1.5
Yekaterinburg, Russia	1.4
Yerevan, Armenia	1.3
Chelyabinsk, Russia	1.2
Tbilisi, Georgia	1.1
Kazan, Russia	1.1
Volgograd, Russia	1.0
Vladivostok, Russia	0.6

In 2000, the newly elected administration of President Vladimir Putin moved to diminish the influence of the Regions by creating a new geographic framework that combined the 83 Regions, Republics, and other entities into a set of eight new Federal Districts—not to enhance their clout in Moscow, but to increase Moscow's authority over them. As Figure 5-10 shows, each of these new Federal Districts has a capital, elevating cities such as Rostov and Novosibirsk to a status secondary only to Moscow. In a related move supported by Russia's parliament, Putin decreed that Regional governors would be appointed rather than elected—thereby concentrating even more power in Moscow. Thus Russia's federal system began its steady shift toward a **unitary state system [18]**—one dominated by a centralized government that exercises power equally throughout the country. This administrative reorganization not only reinforced the position of western Russia as the unrivaled political focus of the state: this heartland is also the demographic center of gravity (Fig. 5-4) and is by far Russia's most accessible as well as urbanized region (**Fig. 5-12**).

Russia's Shrinking Population

Some of these administrative problems are exacerbated by an ominous trend: the Russian population has been shrinking steadily since 1990. In fact, certain parts of the country have witnessed a veritable **population implosion [19]**. When the Soviet Union disintegrated in 1991, Russia's population was about 149 million. Today, that total has dropped by 4 percent and is down to 143.4 million—despite the in-migration of several million ethnic Russians from the old non-Russian SSRs.

Undoubtedly, the transition away from Soviet rule is one reason: uncertainty is known to cause families to have fewer children; moreover, abortion is widespread in Russia. Although the birth rate has stabilized at 13 per thousand, it is Russia's death rate that has been rising and now exceeds 13 per thousand. According to the latest data, this produces an annual net change of population that hovers around zero.

Russian males are the ones most affected. Male life expectancy dropped from 71 in 1991 to 65 in 2015 (over the same period, the rate for females dipped only slightly to 76—a life expectancy fully 17 percent higher). Males are more likely to be afflicted by alcoholism and related diseases, by AIDS (chronically underreported in Russia), by heavy smoking, and by suicide, accidents, and murder. On average, a Russian male is nine times more likely to die a violent or an accidental death than his European counterpart. Shockingly, only about half of today's male Russian teenagers will survive to age 60.

In 2012 and 2013, according to its own demographic statistics, Russia seemed to be bucking the trend, for the first time in 20 years the overall population expanded, albeit minimally. It is too early to tell if this turnaround will last, and it may well prove to be but a momentary reprieve. Whatever the outcome, the situation is grimmest at the regional level, where a few Federal Districts have been hit hard by very substantial population losses. Out-migration unquestionably plays a role because people are no longer forced to stay in place by a totalitarian regime. Villages in particular continue to empty out all across Russia, and at least 20,000 have turned into "ghost villages"—more than 15 percent of all such settlements in the country two decades ago. **Box 5-6** and **Figure 5-13** provide further insights into the spatial dimension of Russia's ongoing population decline.

A Volatile Economy

At the beginning of this century, Russia's economy appeared to have turned completely around from what it was under its communist predecessor. Private property, start-up companies, trade, foreign investment, and functioning stock exchanges—all the things we regard as vital components of a free market economy—seemed to have found their way into post-Soviet Russia. The city of Moscow changed beyond recognition with a new high-rise skyline and unprecedented linkages to the global economy. After some major (and for many people painful) adjustments were made, Russia's GDP grew robustly as the price of shares on the Moscow stock exchange skyrocketed. In common parlance, this performance catapulted Russia into membership in the exclusive club of the so-called **BRICs [20]**, the world's four biggest emerging market economies (Brazil, Russia, India, China—though in subsequent years the BRICs lost much of their shine as momentum slowed in China and Brazil, and Russia itself stagnated). Russia's *nouveaux riche*, nonetheless, became well known for their extravagant consumption on shopping trips to Paris, Milan, and New York; for sending their children to the highest-cachet Swiss and British boarding schools; and even for purchasing premier soccer-team franchises in Europe.

Box 5-6 MAP ANALYSIS

RUSSIA: GDP AND POPULATION DECLINE
GDP in $ billions, 2010

- Over 715
- 250 – 350
- 200 – 250
- 100 – 150
- Under 50

38.4 2010 population in millions

-6% Percent decline in population since 1989

—— Federal District boundary

0   500   1000 Kilometers
0   300   600 Miles

80° Longitude East of Greenwich

© J. Nijman, P. O. Muller, and John Wiley & Sons, Inc.

FIGURE 5-13

Between the dismantling of the Soviet Union in 1991 and 2016, Russia's total population dropped from 149 million to just above 143 million. Because this population decline has varied markedly from region to region, its geographic impact needs to be more closely examined. Figure 5-13 shows the percentage of population loss from 1989 to 2010 in Russia's eight Federal Districts. Can you identify the spatial pattern of population shrinkage? This map also shows the size of the economy (GDP) in each District for 2010. Is it possible to determine which is the better predictor of population decline in a District, geographic remoteness or economic robustness?

Access your WileyPLUS Learning Space course to interact with a dynamic version of this map and to engage with online map exercises and questions.

By 2016, however, it became quite apparent that Russia's advancement had been precarious at best. Too much of it was based primarily on exports of oil and gas; incoming foreign direct investment was anemic in comparison with the other emerging markets; the get-rich-quick schemes of a handful of oligarchs (highly successful businessmen—but almost never women) did not benefit the masses; and the gargantuan state apparatus left by the Soviets was responsible for widespread corruption as well as close connections to organized crime syndicates. The value of the "bribes market" was consistently estimated to be an alarming 20 percent of the national economy. Thus in 2014 when world oil prices came crashing down and stayed there, the nose-diving Russian ruble ignited a severe economic downturn from which it had not recovered by 2017.

Even though statistics on the Russian economy since 2000 suggested abundant growth, ample evidence has also accumulated to indicate that most citizens were unable to better their standard of living. Frustrations with the economic situation have boiled over into the political arena. The rigged parliamentary elections of 2011 that handed victory to Putin's United Russia Party triggered massive demonstrations on Moscow's streets. Nonetheless, a few months later Putin himself easily won election to serve as president until 2018—when his previously engineered constitutional change kicks in to allow him to run for reelection for yet another six-year term (his fourth). Prospects for true democracy in Russia, therefore, seem likely to be deferred until at least the mid-2020s.

Region The Russian Core

The first of Russia's regions—the Russian Core—extends from the western border of the realm to the Ural Mountains in the east (Fig. 5-11). This is the Russia of Moscow and St. Petersburg, of the Volga River and its industrial cities. This is the most developed component of this enormous realm and the focus of its transportation network (Fig. 5-12).

Moscow is the megacity hub of a wider area encompassing some 50 million inhabitants (more than one-fifth of the realm's entire population), most of them concentrated in cities that, during the communist period, specialized in assigned industries such as automaking in Nizhniy Novgorod (the "Soviet Detroit") and textile fabrication in Ivanovo. But today, Russia is less a diversified manufacturing country than a commodity producer and exporter (dominated by oil and natural gas). The tall steel-and-glass towers you see rising above the CBDs of the biggest cities mainly belong to energy companies such as Gazprom. And the cars being built are more likely to display foreign brand names such as Fiat or Volkswagen.

The Moscow metropolitan area itself has grown rapidly in recent years and today has a population of over 16 million, of which an estimated 2 million are undocumented migrants (Box 5-7). It is the economic and business opportunities that continue to draw people in, but steadily worsening congestion and an increasingly brutal social environment mean that many dream of leaving the city as soon as they can afford to. The Moscow metro system (now the world's fifth largest) carries more than nine million passengers daily, and new route extensions and metro stations are built nearly every year. To handle all this overcrowding, the government has announced plans for a massive expansion of the capital to the southwest over the next two decades (Fig. 5-14).

St. Petersburg (formerly Leningrad) remains Russia's second city, with a population of 5.1 million. In the late seventeenth and early eighteenth centuries, Czar Peter the Great and his architects transformed this city at the head of the Baltic Sea's Gulf of Finland into the "Venice of the North." St. Petersburg's palaces, churches, waterfront façades, bridges, and monuments still give the city a decidedly European look unlike that of any other city in Russia. It was proclaimed the Russian capital in 1712, with Moscow serving as the distant second city. The Imperial Winter Palace and the adjoining Hermitage (one of the biggest art museums on Earth) at the heart of the city are among a host of surviving architectural treasures (see Box 5-3). Today, St. Petersburg is again important to Russia's trade with the rest of the world, and the city has become a major destination for many cruise ships. But it possesses none of Moscow's locational advantages, at least not with respect to the domestic market. With Moscow these days increasingly defined in terms of global finance and producer services, St. Petersburg's niche involves trade, culture, and tourism.

Central Industrial Region

At the heart of the Russian Core lies the Central Industrial Region (Fig. 5-15). The precise definition of this subregion varies because all regional definitions are subject to debate. Certain geographers prefer to call this the Moscow Region,

thereby emphasizing that for over 400 kilometers (250 mi) in all directions from the capital, everything is oriented toward this historic focus of the state. As Figure 5-12 underscores, Moscow has maintained its decisive centrality [21]: roads and railroads converge in all directions—from Ukraine in the south; from Minsk (Belarus) and the rest of eastern Europe in the west; from St. Petersburg and the Baltic coast in the northwest; from Nizhniy Novgorod and the Urals in the east; from the cities and waterways of the Volga Basin in the southeast (a canal links Moscow to the Volga, the country's single most important navigable river); and even to the subarctic northern periphery that faces the Barents Sea, where the strategic naval port of Murmansk and the lumber-exporting outpost of Arkhangelsk lie.

Povolzhye: The Volga Region

Another major subregion within the Russian Core is the **Povolzhye**, the Russian name for an area that extends along the middle and lower valley of the Volga River as well as its tributary, the Kama, above the city of Kazan. It would be appropriate to call this the Volga Region, for this greatest of Russia's rivers is its lifeline and most of the Povolzhye's cities lie on its banks (Fig. 5-15). The city of Volgograd is centrally located within this subregion, on the west bank of the Volga. It was named Stalingrad from 1925 to 1961, and that name will always be associated with the decisive battle here that became the turning point of World War II on Nazi Germany's Eastern Front.

The elongated Povolzhye (the *vol* refers to the Volga) forms the eastern flank of the Central Industrial Region, its cities and farms sustained by this Mississippi River of Russia. In the 1950s, the crucial Volga-Don Canal was opened, creating new linkages and opportunities. Next, the full dimensions of the oil- and gasfields in and near the Povolzhye became clear (see Fig. 5-5), and for some time these reserves were the largest in the entire Soviet Union. And finally, the Volga Region's northwestward linkages, via the Moscow and Mariinsk canals to the Baltic Sea, were improved and augmented by upgraded rail and road connections. Today, the industrial composition of the Volga's riverside cities continues to evolve—but if the Central Industrial Region is Russia's heart, then the Povolzhye is its aorta.

The Urals Region

The Ural Mountains form the eastern edge of the Russian Core. They are not particularly high; in the north they consist of a single range, but southward they broaden into a hilly zone. Nowhere are they an obstacle to east-west surface transportation. An enormous storehouse of metallic mineral resources located in and near the Urals has made this subregion a natural place for industrial development. Today, the Urals Region, well connected to the nearby Volga and Central Industrial Regions, extends from Serov in the north to Orsk in the south (Fig. 5-15).

Box 5-7 Among the Realm's Great Cities: Moscow

In the vastness of Russia's territorial expanse, Moscow, capital of the Federation, seems to lie far from the center, close to its western margin. But in terms of Russia's population distribution, Moscow's centrality is second to none among the country's cities (see Fig. 5-4). The Russian population as a whole has been shrinking, but Moscow is bursting at the seams. The metropolis lies at the heart of Russia's core area and at the focus of its circulatory systems.

On the banks of the meandering Moscow River, Moscow's skyline of onion-shaped church domes and modern buildings rises from the forested flat Russian Plain like a giant oasis in a verdant setting. Archeological evidence points to ancient settlement of the site, but Moscow enters recorded history only in the middle of the twelfth century. Forest and river provided defenses against Tatar raids, and when a Muscovy force defeated a Tatar army in the late fourteenth century, Moscow's primacy was assured. A huge brick Kremlin (citadel; fortress), with walls 2 kilometers (1.2 mi) in length and with 18 towers, was built to ensure the city's security. From this base Ivan the Terrible expanded Muscovy's domain and laid the foundations for Russia's vast empire.

The triangular Kremlin and the enormous square in front of it (Red Square of Soviet times), flanked by St. Basil's Cathedral and overlooking the Moscow River, is the center of old Moscow and is still the heart of this metropolitan complex of 16.6 million (see photo). From here, avenues and streets radiate in all directions to the Garden Ring and beyond. Until the 1970s, the Moscow Circular Motorway enclosed most of the built-up area, but today the metropolis sprawls outward far beyond this beltway.

In the postcommunist era, the center of Moscow has undergone a substanial makeover as Soviet-dictated urbanism made way for the emergence of a fast-paced, international business complex—one that increasingly resembles those of other global cities. Once the free market, however imperfectly, took root,

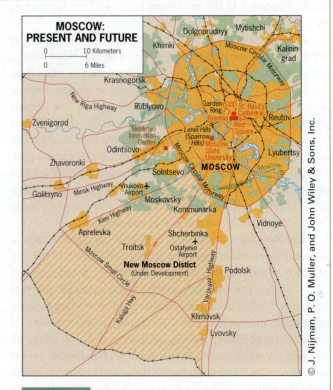

FIGURE 5-14

© J. Nijman, P. O. Muller, and John Wiley & Sons, Inc.

Moscow's CBD became the preferred location for foreign investors, multinational corporations, upscale shopping, and elite housing. It quickly became one of the most expensive residential areas in the world, its exclusiveness even more striking in the heart of an urban agglomeration that is still mostly dominated by drab, austere Soviet architecture.

Moscow has grown so fast and congestion has been so overwhelming that the government has announced an enormous urban expansion plan. It began by incorporating a huge swath of land southwest of the capital into Moscow City proper that will double the size of the metropolitan area, in the process swallowing up extensive tracts of forest as well as numerous municipalities. A consortium of design and planning companies is now believed to be working on the plan for *New Moscow*—a project unprecedented in scale and ambition that is to include a new federal government complex, a highest-order financial district, a world-class high-technology cluster, and a broad range of housing and amenities for nearly two million new residents. But on the ground, not much has happened so far, and it is said that full implementation of the plan will not occur until 2030 at the earliest (Fig. 5-14).

Travelwide/Alamy

The heart of central Moscow, with the triangular Kremlin (right of center) fronting on adjacent Red Square.

FIGURE 5-15

© J. Nijman, P. O. Muller, H. J. de Blij, and John Wiley & Sons, Inc.

Region The Southeastern Frontier

From the southeastern flanks of the Ural Mountains to the headwaters of the Amur River, and from the latitude of Tyumen to the northern zone of neighboring Kazakhstan, lies Russia's vast Southeastern Frontier region, the product of a gigantic Soviet experiment in the eastward extension of the Russian Core along the axial Trans-Siberian Railroad (Fig. 5-11). As the maps of Russia's cities and transportation (Fig. 5-12) as well as population distribution (Fig. 5-4) suggest, this region is more densely peopled and more fully developed in the west than in the east. To the east of the Yenisey River, linear settlement prevails, marked by ribbons and clusters along the Trans-Siberian Railroad and the newer Baykal-Amur Mainline Railroad that parallels it to the north. Two subregions dominate the human geography: the Kuznetsk Basin in the west and the Lake Baykal area in the east.

The Kuznetsk Basin (Kuzbas)

Some 1500 kilometers (950 mi) east of the Urals lies another of Russia's primary regions of heavy manufacturing resulting from the communist period's national planning: the Kuznetsk Basin, or **Kuzbas** (Fig. 5-15). The leading city, located just outside the Kuzbas, is Novosibirsk, which lies at the intersection of the Trans-Siberian Railroad and the Ob River as the symbol of

Russian enterprise in the vast eastern interior. To the northeast lies Tomsk, one of the oldest Russian towns east of the Urals, founded in the seventeenth century and now caught up in the modern development of the Kuzbas. Southeast of Novosibirsk lies Novokuznetsk, a city that produces steel for this subregion's machine and metal-working plants as well as aluminum products from Urals bauxite.

Wolfgang Kaehler/SuperStock

The river port of Krasnoyarsk, where the south-to-north-flowing Yenisey River meets the Trans-Siberian Railroad, Russia's east-west transport axis, near the center of the Southeastern Frontier region (see Fig. 5-11). Supplies brought by train from the west are shipped northward to settlements in the Siberian interior; raw materials from Siberia arrive by boat and barge for dispatch to Russia's factories and markets. Large storage facilities stand across the river.

The Lake Baykal Area (Baykaliya)

To the east of the Kuzbas, development becomes more insular, and distance becomes an even stronger adversary. North of the central segment of the Mongolian border zone and eastward around Lake Baykal, larger and smaller settlements cluster along the two railroads to the Pacific coast (Fig. 5-15). West of the lake, these rail corridors lie in the headwater zone of the Yenisey River and its tributaries. A number of dams and hydroelectric projects serve the Angara Valley, particularly the city of Bratsk. Mining, lumbering, and some farming sustain life here, but isolation dominates it. The city of Irkutsk, near the southern end of Lake Baykal, is the principal service center for the enormous Siberian region to the north as well as for a lengthy east-west stretch of southeastern Russia.

Beyond Lake Baykal, the Southeastern Frontier really lives up to its name: this is southern Russia's most rugged, remote, and forbidding country. Settlements are rare, with many being mere camps. The Republic of Buryatiya (see Fig. 5-10) is part of this zone; the territory bordering it to the east was taken from China by the czars and may yet become an issue in the future. Where the Russian-Chinese boundary turns southeastward along the Amur River, the region labeled the Southeastern Frontier peters out and Russia's Far East begins.

Region Siberia

Before we assess the potential of Russia's Pacific Rim, we should remember that the ribbons of settlement just discussed hug the southern perimeter of this gargantuan country, avoiding the huge Siberian region to the north (Fig. 5-11). Siberia extends from the Ural Mountains to the Kamchatka Peninsula—a vast, frigid, forbidding land. Larger than the conterminous United States but inhabited by fewer than 20 million people, Siberia quintessentially symbolizes the Russian environmental plight: immense distances, cold temperatures aggravated by ferocious Arctic winds, difficult terrain, unproductive soils, and limited options for survival.

But Siberia does possess valuable resources. From the days of the first Russian explorers, Siberia's riches have beckoned. Gold, diamonds, and many other precious minerals were found. Later, metallic ores including iron and bauxite were discovered. Still more recently, the Siberian interior proved to contain massive deposits of oil and natural gas (Fig. 5-5) that now contribute significantly to Russia's energy supply and exports.

As the physiographic map (Fig. 5-2) reveals, major rivers—the Ob, Yenisey, and Lena—flow gently northward, traversing Siberia and the Arctic Lowland before emptying into the Arctic Ocean. Hydroelectric power development in the basins of these rivers has generated electricity used to extract and refine local ores, and to run the lumber mills that have been built to exploit the vast Siberian forests.

Understandably, the population geography of Siberia is extremely fragmented. Most of the region is devoid of inhabitants, with hundreds of square kilometers of empty territory separating the relatively few islands of settlement (Fig. 5-4). Nonetheless, some small clusters of ethnic Russians have developed. Accordingly, the Yenisey River can be traced on the map of the realm's peoples (Fig. 5-7) as a thin ribbon of Russian settlements north of Krasnoyarsk (see photo); farther east, the upper Lena Valley can similarly be recognized.

Region The Russian Far East

Officially, the huge Far Eastern Federal District incorporates six Russian administrative regions and four residual ethnic republics (Fig. 5-10). But many Russian citizens have a different view of what constitutes the geographic region they call the Far East. Justifiably, they see most of the northern zone of the official Far East as a continuation of Siberia. To them, the real Far East is constituted by the territory beyond the Southeastern Frontier to the Pacific coast, the island of Sakhalin, the Kamchatka Peninsula, and a narrow strip of land along the northern shore of the Sea of Okhotsk (Figs. 5-11 and **5-16**).

In Soviet times, citizens who agreed to move to this most distant region to work in factories or for the administration were rewarded with special privileges to compensate for their hardship. Make no mistake—even though the Pacific Ocean's proximity tends to moderate natural environments here, life anywhere in the Russian Far East is tough. Moreover, Moscow's favors were withdrawn early in the postcommunist era, so under the new democracy hundreds of thousands of inhabitants voted with their feet and simply departed. Petropavlovsk, the largest urban center on the Kamchatka Peninsula, exemplifies the devastating consequences of this exodus (see **Box 5-8**).

Dean Conger/

Oymyakon is known as the coldest town on Earth. It is located north of the city of Yakutsk in eastern Siberia. The average daily temperature during winter hovers around −46° Celsius (−51° Fahrenheit), and the record low recorded here in 1933 was −68°C (−90°F). A few hundred people live here, and there is even a school—which does not open if the temperature drops below −52°C (−62°F). Cell phones would not work here, even if there was satellite coverage, and visitors are advised not to wear glasses—they freeze to your face.

FIGURE 5-16

The Far East, too, has significant reserves of what underpins today's Russian economy—oil and natural gas, most importantly, the deposits located on and around the island of Sakhalin. As Figure 5-16 also reminds us, the southern sector of this region lies between additional reserves and potential markets: China next door and Japan across a narrow sea. We are

A 2014 naval exercise of the Russian Pacific Fleet near its home base of Vladivostok, the port city that serves as the eastern terminus of the Trans-Siberian Railroad. Russia's growing visibility and military strength in the Far East over the past few years suggests a widening geostrategic posture in the East Asian arena—an impression reinforced in 2015 by the rapid warming of relations with China.

Box 5-8 From the Field Notes … Far East

"The Russian Far East is losing people at the fastest rate among all Russian Federal Districts, and my 2009 field trip to Petropav-lovsk-Kamchatsky on the Kamchatka Peninsula only reinforced what I had learned in Vladivostok some years earlier. During Soviet times, Russians willing to move to the Far East were given special privileges (an older man told me that 'being as far away from Moscow as possible was reward enough'), but today locals here feel abandoned, even repressed. In 1989, Petropavlovsk had about 270,000 residents; today that total is estimated to be more than 100,000 fewer. It is not just the bleak environment—the ubiquitous ash and soot from dozens of active volcanoes on

Kamchatka, covering recent snowfalls and creating a pervasive black mud—it is also a loss of purpose that people feel here. There was a time when the Soviet windows on the Pacific were valuable and worthy; now the people scramble to make a living, and when they find a way, for example, by selling used cars bought cheaply in Japan and brought to the city by boat, Moscow sends law enforcement to halt the illegal trade. Petropavlovsk has no surface links to the rest of the world, and this springtime view of the city at the foot of the volcano Koryakskaya (left) and a street scene near the World War II museum (right) underscore why the Far East is depopulating."

not used to thinking about Russia as an Asia-Pacific power, but it certainly qualifies. The distance to Moscow notwithstanding, the Far East provides Russia with highly valuable geopolitical as well as economic assets. In 2012, Russia hosted the Asia-Pacific Economic Cooperation (APEC) summit in Vladivostok (the home base of the Russian Pacific Fleet), spotlighting its strengthening military presence in this part of the world (see photo). And by 2016, the Putin regime was implementing new policies in line with its latest mantra that developing the Far East "must be a 21st century national priority." Its centerpiece is a free land program offering one-hectare (2.5-acre) properties to new rural settlers.

Region Transcaucasia

Now let us focus on the smallest but most complicated region of all—the area where Russia's leaders, communists as well as czars, pushed the expanding state's influence across and beyond the Caucasus Mountains. On regional maps (Figs. 5-10 and 5-17), it looks as though the Russian Core extends to the international boundary (which coincides, more or less, with the crestline of this towering range). However, things are not that simple. When you travel southward from Volgograd or Rostov, the landscape—physical as well as cultural— does not change; the Russian Plain's gently rolling countryside with its villages and farms is very much like that south of Moscow. But taking the train south from Astrakhan in the Volga River Delta, you will

soon see the Caucasus wall looming in the distance, but you may not realize that you have entered a different world until you get off near its base at Groznyy, the capital of Chechnya, one of Russia's 21 Ethnic Republics (Fig. 5-17). Just observing the mix of peoples here makes you aware that you have left the heartland and reached Russia's periphery.

Figure 5-17 is important enough to spend some serious time studying it. Here the communist leaders who succeeded the czars confronted multiple challenges. First, inside the Russian state itself, there were many minority peoples (such as the Chechens) who had fought the czars' armies and who now faced new rulers. Second, there were peoples who had sided with the czars and who wanted to cooperate with the Soviet regime, such as the Ossetians. To recognize the aspirations of these peoples, some (but not all) of whom were Muslim, the communist administration designated ethnic "republics" for them. On our map, these are colored pink to distinguish them from the rest of Russia, which is shown in tan.

Russia's Internal Periphery and the Case of Chechnya

From Buddhist-infused Kalmykiya on the perimeter of the Caspian Sea westward to mainly Orthodox Christian Adygeya near the Black Sea, the *Internal Periphery* of southern Russia is defined by eight ethnic Republics, each with its own identity (Fig. 5-17).

Box 5-9 MAP ANALYSIS

In geopolitical terms, Southern Russia and Transcaucasia comprise an especially complex corner of this realm. These complexities apply not only to areas inside Russia but also to the spatial configuration of the countries of Georgia, Armenia, and Azerbaijan. Take a close look at the layout of their national borders and the location of various ethnic groups and exclaves south of the Russian border. Which of these three countries do you think is the hardest to govern from a geopolitical point of view? Why?

Access your WileyPLUS Learning Space course to interact with a dynamic version of this map and to engage with online map exercises and questions.

FIGURE 5-17

Physiographically and culturally divided Chechnya lies astride a transition zone between plains and mountains, with the mountains serving as a refuge for those opposed to Moscow's rule. When the post-Soviet Russian government asked all of the country's Regions and Republics to sign its proposed Russian Federation Treaty in 1992, Chechnya's leaders refused, believing that they now had the opportunity to terminate Russian control. Its western neighbor, Ingushetiya, got caught up in the anti-Russian activism, and the ensuing drawn-out conflict also took a terrible toll in adjacent North Ossetia where, in 2004, more than 350 students, teachers, and parents died in a terrorist attack on a school. Still farther west, things remained calmer, and no problems at all arose in generally pro-Moscow Adygeya.

This tier of ethnic Republics along Russia's southernmost border displays all the properties of a disadvantaged periphery: not only do they share a history of subjugation by the powerful Russian Core to the north, but they also lag far behind the rest of the Federation by almost any measure of social progress. In terms of health, education, income, opportunity, and other social indicators, the peoples of the Internal Periphery, whether they are pro-Russian or not, have a long way to go to achieve current Russian living standards. And, as noted above for Chechnya, there is much simmering resentment of Russian rule, past and present.

In Chechnya's case, such anger has its most recent roots in the Second World War, when Soviet leaders accused the Chechens of sympathizing, and even collaborating, with the Nazi invaders. Soviet dictator Josef Stalin then ordered the entire Chechen population loaded onto trains and exiled to the desert of Kazakhstan across the Caspian Sea. Tens of thousands died along the way or after arriving there, and even though Stalin's successor, Nikita Khrushchev, in 1957 allowed the survivors to return to their homeland, the Chechens never forgave their tormentors. Then in 1992, just after the Soviet Union disbanded, they saw their opportunity: they refused to sign the Russian Federation Treaty that would make their republic an integral component of the new Russian state, and they mounted a campaign for independence that led to a bitter and costly war that destabilized a number of neighboring republics as well. Chechen terrorists eventually carried this campaign to the very heart of Moscow, killing hundreds of civilians there and forever altering Russia's political landscape. When presidential candidate Vladimir Putin promised victory, the Russians elected him overwhelmingly in 2000. Subsequent military intervention did overpower the Chechens, and the war finally came to an end in 2007; Chechnya now has a pro-Russian regime that receives substantial economic support from Moscow in exchange for fierce military loyalty. Nonetheless, the Internal Periphery as a whole remains restive and barely stable.

Russia's External Periphery

Russian (and later Soviet) manipulation of the ethnic map did not terminate at the international boundary we see in Figure 5-17, but it did take a different turn. Beyond the Russian border, the Soviet Empire extended its control over the three adjoining Transcaucasian entities where the Russians had for centuries exerted influence and waged war (**Box 5-9**). These three political units were **Georgia** (mapped in yellow in Fig. 5-17), facing the Black Sea; **Azerbaijan** (green), bordering on the Caspian Sea; and **Armenia** (lavender), landlocked in the middle. During Soviet times, Georgia was a loyal partner (Stalin was a Georgian by birth). Christian Armenia appreciated membership in the Soviet Union because of the security it provided next door to hostile Islamic Turkey. And Islamic Azerbaijan, whose ethnic majority is constituted by a people known as the *Azeris* (with close kinship ties to Iranian Azeris directly across their southern border with Iran), was a valued member of the USSR because of its rich oil reserves.

But note in Figure 5-17 that Azerbaijan has a large exclave on the Iranian border (Nakhchivan), separated from the main part of the country by Armenian territory—and that the Armenians hold another exclave named Nagorno-Karabakh inside what would appear to be Azerbaijan's territory (which is home to more than 150,000 Christian Armenian citizens who were placed under Islamic Azerbaijan's jurisdiction by Soviet mandate).

Soviet rule kept the lid on potential conflict in these three Transcaucasian countries. After the collapse of the USSR, Russia sought to retain its influence and maintain stability in this vital sector of the Near Abroad. It wanted to keep Azerbaijan's oil flowing to and through Russia (see the pipeline connecting Baki [Baku] to the Black Sea terminal at Novorossiysk in Fig. 5-17); Russia also wanted to keep Georgia in the fold, and it looked for ways to mitigate the long-running clash between Azerbaijan and Armenia over Nagorno-Karabakh.

But things have not gone well for Moscow in post-Soviet Transcaucasia. Driven by Western investments, Azerbaijan in 2006 began exporting oil via a new pipeline across Georgia and Turkey to the latter's Mediterranean port of Ceyhan, providing it with a more favorable economic as well as political alternative to Russian transit. Meanwhile, Armenia and Azerbaijan have not been able to resolve their sometimes violent quarrel over Nagorno-Karabakh. And things really fell apart in Georgia, where the government pursued pro-European policies and allegedly mistreated its pro-Russian Ossetian minority, and where Russia openly supported devolutionary initiatives in the northwestern corner of the country, a small ethnic entity known as Abkhazia (Fig. 5-17; see **Box 5-10**).

Soon after the USSR fell apart in 1991, the disagreement between Russia and Georgia escalated into serious discord. The Russians closed their border with Georgia, shutting Georgian farmers out of their crucial market. Then the Russians started issuing Russian passports to Abkhazian citizens of Georgia. Next, the Russians signaled even stronger support for their Ossetian allies, and in a still-disputed sequence of events during 2008 Russian forces entered South Ossetia and engaged the Georgian military in a brief but costly conflict. After a ceasefire was brokered by international mediators, Moscow took the extraordinary step of recognizing the two pro-Russian autonomous regions of Georgia—Abkhazia and South Ossetia—as independent states! Russian military intervention here sent shudders across Europe and much of the rest of the world—and

Box 5-10 From the Field Notes ...

"Upon first entering Sukhum, the capital of the *de facto* state of Abkhazia on the Black Sea coast (see Fig. 5-17), I visited the former Government House—now an eerie and unintended war memorial. It was the scene of an intense battle in 1993 when the Georgian government unsuccessfully tried to suppress Abkhazian separatists. The building bears the marks of bullets, bombs, and fires, and is deliberately maintained in this condition as a reminder of the war. The Abkhazians repelled the Georgians and insist on their own spelling of the city's name (Georgians use *Sokhumi*); also note that it is the Abkhazian 'national' flag flying on the top of the burnt-out building. The legacies of war in the Caucasus are dramatically evident in the landscape, with the omnipresence of ruined buildings, war memorials, empty houses of displaced persons, refugee camps, military checkpoints, minefields, abandoned farms, ruined roads, and destroyed statues such as the one of Lenin that used to stand on the now empty base in front of the building. Like many other structures (over half by some estimates) in Sukhum, this one remains as it was when the fighting ended, but its considerable size and high-rise prominence in a city of small, low buildings make it visible from everywhere in town and a daily reminder of the war's brutality."

John O'Loughlin

turned out to be the prelude to similar Russian aggression in 2014, which saw the forcible annexation of Crimea followed by the outbreak of hostilities in nearby eastern Ukraine.

Unlike Azerbaijan, predominantly agricultural Georgia has no energy card to play—although it did agree to permit the new pipeline from Azerbaijan to Turkey to traverse its territory, thereby angering the Russians even further. In addition, the Georgian government has courted the European Union and even NATO, seeking the security Georgians have lacked for centuries. The Azeris of Azerbaijan, whose prevailing religion is **Shi'ite Islam [22]** like that of neighboring Iran, are caught in the global web (and political geography) of energy demand and supply, but most ordinary people have benefited little from their country's petroleum wealth. The landlocked 3 million Armenians live in a world of unsettled scores—with the Turks over their decimation by Turkish Ottoman Muslims during World War I; with the Azeris over their exclave of Nagorno-Karabakh; and with the international community over the proper recognition of their plight. Transcaucasia may be the smallest region of the Russian/Central Asian realm, but it is densely populated by nearly 17 million people who inhabit one of the most volatile peripheries in this turmoil-plagued part of the world.

Region Central Asia

The sixth and final region of the realm is Central Asia, which consists of five states that simultaneously became independent upon the disintegration of the USSR in 1991: Kazakhstan, Uzbekistan, Turkmenistan, Kyrgyzstan, and Tajikistan. Until quite recently, Central Asia was regarded as both isolated and remote; today Central Asia's relative location in the very heart of Eurasia is being translated into an increasingly strategic regional position. Geopolitically, this region lies directly between Russia and China as well as bordering Transcaucasia, Iran, and portions of South Asia. In terms of economic geography, Central Asia contains huge deposits of oil and natural gas, and key parts of it are being developed as a bridge between China and western Eurasia in the form of major new overland transport corridors that are being likened to a twenty-first-century version of the historic Silk Road.

In Central Asia today, Turkic cultural influences blend with Islam and Soviet legacies. During its hegemony over this region, the USSR tried to suppress Islam and install secular regimes. Nevertheless, Islam reasserted itself strongly after the Soviet withdrawal and has now become one of the defining qualities of Central Asia. From Almaty to Samarqand (the region's most densely populated area as shown in Fig. 5-4), mosques are being restored and reactivated, and Islamic dress has again become part of the cultural landscape. National leaders make high-profile visits to Mecca, and most are sworn into office with a hand on the holy Quran.

Central Asia is marked by enormous ethnic complexity and diversity. Figure 5-7, however, is but a generalization of the highly intricate mosaic of peoples and cultures, because every cultural territory on this map contains clusters of minorities that cannot be captured at this scale. Thus people of different faiths, languages, and lifestyles rub shoulders everywhere, and this often produces friction that can escalate into ethnic conflict. Therefore it should not come as a surprise, in this post-Soviet geographic context, that democracy is hard to come by: each of these five countries is either autocratic or has a flawed representative government (Tajikistan, in fact, is already being labeled by some as a failing state).

The Five States of Central Asia

Kazakhstan is the region's giant but lies between two even greater territorial titans: Russia and China (**Fig. 5-18**). During the Soviet era, the northern tier of Kazakhstan was heavily Russified, and in the 2010s ethnic Russians still account for about 22 percent of the total national population of 17.9 million. To stay on good terms with post-Soviet Russia, the Kazakhs carefully maintain the railway and highway links that connect their north to Russia—and they even moved their capital to Astana in the heart of Russified Kazakhstan (**Box 5-11**). Today, this country is in the forefront of Central Asia's energy boom: it already ranks among the world's leading producers of uranium, and in 2014 it ranked 15th among the oil-producing countries. Figure 5-18 further reveals Kazakhstan's intermediate position between the huge oil reserves of the northern Caspian Basin and China. New oil and gas pipelines operating across Kazakhstan are serving to reduce China's dependence on oil imports carried by long-haul tankers along distant international sea lanes. Kazakhstan's economic growth has averaged 8 percent annually over the past decade, a development thrust that is certain to continue as the country becomes even more of an oil power as well as a leading beneficiary of China's New Silk Road project (which is described in the next section).

Uzbekistan occupies the heart of the region and borders every state within it. Uzbeks not only make up more than 80 percent of the population of 30.3 million, but also form substantial minorities in a number of neighboring countries. The capital, Tashkent, lies in the eastern core area of the country, where most of the people live in towns and farm villages, and where the densely-settled Fergana Valley is the focus. In the desert northwest lies what is left of the still-shrinking Aral Sea, whose feeder rivers were diverted into cotton fields and croplands during the Soviet occupation (see photo pair). Heavy use of pesticides contaminated the groundwater of the Aral Basin, and countless thousands of local people continue to suffer from cancers and other catastrophic medical problems as a result.

Turkmenistan, the autocratic desert republic that extends all the way from the Caspian Sea to the borders of Afghanistan, has a population of 5.4 million, of which nearly 90 percent are ethnic Turkmen. During the days of Soviet control, communist planners initiated work on a massive project—the Garagum (Kara Kum) Canal designed to transfer water from Turkestan's eastern mountains into the heart of the desert. Today the canal is 1100 kilometers (700 mi) long, and it has enabled the cultivation of more than one million hectares (2.5 million acres) of cotton, vegetables, and fruits. The plan is to extend the canal all the way to the Caspian Sea. Meanwhile, Turkmenistan has greatly expanded its oil and gas extraction from Caspian coastal and offshore reserves, most of which is transported to China via a new pipeline that runs through Kazakhstan to northwestern China.

Kyrgyzstan's topography and political geography are reminiscent of the Caucasus. The agricultural economy is generally weak, consisting of pastoralism in the highlands and farming in the valleys. Mapped in yellow in Figure 5-18, Kyrgyzstan lies intertwined with Uzbekistan and Tajikistan to the point of having exclaves and enclaves along their common borders. Ethnic Kyrgyz constitute about 75 percent of the population of 6 million. More than three-fourths of the people profess allegiance to Islam, and **Wahhabism**, an orthodox form of **Sunni Islam [23]**, has achieved a strong foothold here as well; the town of Osh is often referred to as the Central Asian headquarters of this movement. In 2010, ethnic rioting erupted in Osh between the Kyrgyz majority and the Uzbek minority, killing hundreds and putting the country in danger of breaking apart. Tens of thousands of Uzbeks, who make up roughly 14 percent of the total population and form a prosperous merchant class, fled their homes.

Tajikistan's mountainous scenery is even more spectacular than Kyrgyzstan's, and here too topography is a barrier to the integration of a multicultural society. The Tajiks, who constitute around 80 percent of the population of 8.7 million, are ethnically Persian (Iranian), not Turkic, and speak an Indo-European language related to Persian. Most Tajiks, despite these cultural affinities, are Sunni Muslims, not Shi'ites. Territorially minute though it is, regionalism to the point of state failure plagues Tajikistan: the government in Dushanbe is repeatedly at odds with the barely connected northern part of the country (Fig. 5-18), a boiling cauldron not only of Islamic fundamentalism but also of anti-Tajik, Uzbek activism.

A New Silk Road?

Central Asia is still very much a region in evolution—redefining its relations with Russia, searching for new ways to tame its ever-turbulent ethnic mosaic, and working to lure the interest of investors worldwide in the quest to more fully develop its bountiful resource base. As noted above, once-dormant Central Asia has now become a key crossroads linking the different realms and powerful countries that surround it. Of the utmost importance today is China's growing challenge to Russia's longstanding regional influence here. The centerpiece of that effort is an enormous project the Chinese announced in 2013: the **New Silk Road [24]**. The historic Silk Road, of course, was the ancient linear trade network that connected Mediterranean Europe to Asia, stretching eastward from Constantinople (modern Istanbul) to Xian and beyond in eastern China—and named after the prized commodity that had been perfected by the Chinese: silk.

FIGURE 5-18

The geographic idea that forms the cornerstone of the New Silk Road is the Chinese initiative to construct an ultramodern overland routeway extending from East Asia to Europe via Central Asia—a new "Eurasian land bridge." The Chinese claim that it will benefit everyone, particularly the Central Asian countries hungry for investment and cutting-edge industrial technology. China's funding of the project will mainly go into building infrastructure—railroads, highways, and freight terminals—but Chinese companies would also become a driving force in the industrialization of Central Asia. Because of its relative location, Kazakhstan is of supreme importance, and in 2015 the Kazakh and Chinese governments completed their first set of deals said to be worth U.S. $23 billion.

China's motives are fourfold. The first entails the insatiable Chinese appetite for raw materials and energy supplies, and Central Asia is richly endowed with both. Second, as industrial production in China shifts westward (discussed in Chapter 9), Europe-bound goods become costlier to ship if they must be first sent east to the Pacific coast and then transshipped via the far longer, circuitous oceanic route around the entire southern rim of Eurasia. Besides shipping costs, significant time would be saved as well: an existing railroad that moves freight from China to western Kazakhstan and connects to a rail line that crosses Russia and Belarus on its way to Germany is said to require only 14 days versus the 60 days needed to sail the much slower and indirect oceanic route.

Box 5-11 Regional Planning Case

Astana, Kazakhstan: Planned City and Venue for World Expo 2017

In 1998, the capital of Kazakhstan was moved from Almaty—the biggest city as well as commercial and cultural center, located in the southeast corner of the country—to the provincial city of Akmola (known as Tselinograd during Soviet days) in the central north (Fig. 5-18). Akmola was renamed Astana (which means 'capital' in Kazakh), and quickly received a complete makeover.

It was in some ways the creation of a typical **forward capital**, not too different from Brasília (see Chapter 3). Astana was rebuilt from scratch, as a symbol of national identity and unity, at a more centralized location that underscored the territorial integrity of *all* of Kazakhstan (something Almaty did not). Tellingly, Astana also lies much closer to the Russian border, and the government of this former Soviet republic was especially keen to maintain cordial relations with Moscow while deepening economic ties with China. Moreover, 22 percent of Kazakhstan's population are ethnic Russians, most of them concentrated in the northern tier of the country (see Fig. 5-7). Thus having the capital in their midst helps the government to cultivate good relations with this sizeable Russian minority.

Astana is also the urban showcase of a country awash in money—mostly from oil—and even though oil and gas revenues levelled off during the mid-2010s, in the longer run Kazakhstan is well positioned to capitalize on its rich endowment of natural resources. Vibrant Astana is now home to 730,000 (mostly younger) inhabitants, and its promising fortunes rest upon its governmental functions, high-order economic activities centered on the energy industry, educational resources including a university, an increasingly sophisticated modern infrastructure, futuristic skyscrapers, and a fantastical urban landscape studded with innovative architectural designs (see left photo).

To signal both Astana's arrival and Kazakhstan's meteoric rise on the global scene, the city is hosting the World Expo in the summer of 2017—the first such international fair ever held in Central Asia. The Expo's overriding theme is "energy of the future," focusing on innovative and practical energy solutions, alternative energy sources, green technologies, and global sustainability. More than a hundred countries plan to participate and at least three million visitors are expected to attend.

World Expo 2017 takes place on 113 hectares (280 acres) of land in the southeastern sector of Astana, with easy access to the city center, the international airport, and the main railway terminal. After an international architectural competition, the lead design contract was awarded to a Chicago-based firm. The fairground's plan is organized around a new 'National Pavilion of Kazakhstan,' a spherical structure at the center of the site (right photo). There will be hotels, trade and entertainment complexes, new transit lines, and an urban design that optimizes pedestrian mobility and access. In keeping with the energy theme, all the buildings will use solar or wind energy networked on a smart grid. Following the Expo, these buildings are expected to be converted into office and research facilities set amidst new residential neighborhoods marked by an abundance of green space.

Astana aspires to become the "world-city" for all of Central Asia. This is the Bayterek Tower, which depicts a mythical giant poplar tree containing a golden egg in its branches.

Ongoing construction for Astana's World Expo. The spherical building to the left of the center of the photo will be the "National Pavilion of Kazakhstan." The fair is scheduled to open in June 2017.

The third reason for the New Silk Road involves a geostrategic imperative—the need to secure open overland access in case of maritime blockades (especially in the hotly disputed South China Sea [discussed in Chapter 10]). And fourth, the New Silk Road will propel the economic development of China's far western periphery, particularly the restive Autonomous Region of Xinjiang, through which all routes to Kazakhstan are channeled.

As new freight-rail lines open (see photo of one inaugural run), along with oil and gas pipelines that connect Central Asia to China, trade between the two—at the expense of Russia—is steadily increasing. If China and Russia are seen to be competing for influence in Kazakhstan, then China has taken a major step forward. **Figure 5-19** is a map redrawn from a Chinese government publication that first appeared in 2014, and reveals a most interesting perspective on this contest: the proposed main route of the New Silk Road veers southward from Kazakhstan to Uzbekistan before turning westward towards Europe. That not only avoids Russian territory in the layout of the main east-west corridor, but also puts Moscow at the end of a branch line that hives off from eastern Europe.

Russia's latest response is a half-hearted attempt to strengthen the so-called **Eurasian Customs Union [25]**, a supranational organization it created in 2010 with Belarus and Kazakhstan that was subsequently joined by Armenia and Kyrgyzstan in 2015. Even so, Russia may already have lost out to China's initiative in Central Asia—perhaps providing another reason why Russia is strengthening its presence in the Far East as it keeps an ever watchful eye on China all around their common, 4200-kilometer (2600-mi)-long border.

A Realm in Flux

A quarter-century ago, one of the most remarkable politico-geographical events in history took place: the sudden, peaceful expiration and dismemberment of the Soviet Union, global superpower and anchor of the world's next-to-last colonial empire. What remains today is the centerpiece of the Soviet Empire, recast as the Russian Federation—still the world's biggest territorial state, but fringed with a string of countries along many of its edges that were once part of the old USSR. To the south, most of these states are still considered to be part of this geographic realm, while others, to the west, have gravitated into Europe's orbit. At the same time, Russia's relations with this so-called Near Abroad are far from stable.

The 1991 breakup of the USSR may seem like a long time ago, yet the current situation still bears many imprints of the Soviet era and ensures that this realm will remain in flux. Soviet times were repressive and economically stagnant, whereas the new Russia has offered somewhat greater freedom and business opportunities. Geopolitically, however, the Putin years (he has served as president since 2000, interrupted by a four-year, behind-the-scenes term as prime minister) bear all the hallmarks of **geopolitical revanchism [26]**—retaliatory policies aimed at recovering lost territory. Unmistakably, the eastward advance of the European Union as well as NATO, along with the anti-Russian disposition of former SSRs such as Ukraine and the Baltic states, have fueled reactionary Russian policies.

It is hard to separate Russia's foreign relations from its internal political developments. Inside Russia, the promise of democracy has faded away for now as the benefits of the post-Soviet era have almost exclusively been conferred on the favored few while income inequality steadily widened. Although many long for a better future, others harbor a persistent nostalgia for the past. Not surprisingly, half the Russians surveyed in a recent poll agreed that it was "a great misfortune that the Soviet Union no longer exists."

FIGURE 5-19

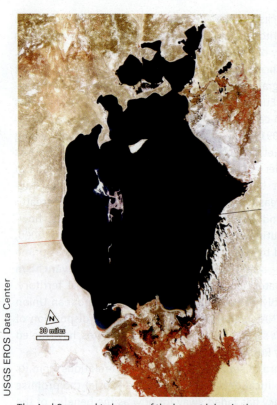

USGS EROS Data Center

http://earthobservatory.nasa.gov/Features/WorldOfChange/aral_sea.php

The Aral Sea used to be one of the largest lakes in the world, straddling the border between Kazakhstan and Uzbekistan (Fig. 5-18). It centered a major drainage basin originally fed by the Syr Darya and Amu Darya rivers that brought water from the mountains to the southeast. Large-scale diversion and a canal system, which served to irrigate agriculture in this area, deprived the lake of its inflow, which in this desert environment soon caused it to all but completely dry up. These images show the areal extent of the Aral Sea in 1977 (left) and 2014 (right). The parched eastern side of the former Sea is now known as the Aralkum Desert, whose very existence is widely considered to be an environmental disaster of the highest magnitude.

The first train (on February 25, 2015) pulls out of the new portside container terminal in Lianyungang on China's Yellow Sea coast to begin its 3750-kilometer (2300-mile) westbound run to Almaty, Kazakhstan's largest city. The freight-rail corridor it will traverse is the pioneering component of the New Silk Road, the Chinese high-profile Eurasian transportation initiative that may fundamentally change relations between China, Russia, and Central Asia.

Points to Ponder

- Russia has land borders with 14 countries and three other geographic realms. Compare that to the conterminous United States, which borders only two countries and only one other realm.

- High-speed rail travel, so urgently needed in this far-flung realm, has been extremely slow to develop. The first such route opened in 2010, linking the capital, Moscow, with the leading port and second city, St. Petersburg; but since then, not a kilometer of new track has been laid.

- One out of seven Russian villages has been abandoned, with many turning into ghost towns.

- According to a poll marking the twentieth anniversary of the Soviet Union's demise, 70 percent of Russians agreed that the changes since 1991 had not benefited ordinary people.

- Landlocked, once-remote Central Asia could soon become a first-order global geopolitical hotspot involving Russia, China, India, and a host of lesser regional powers.

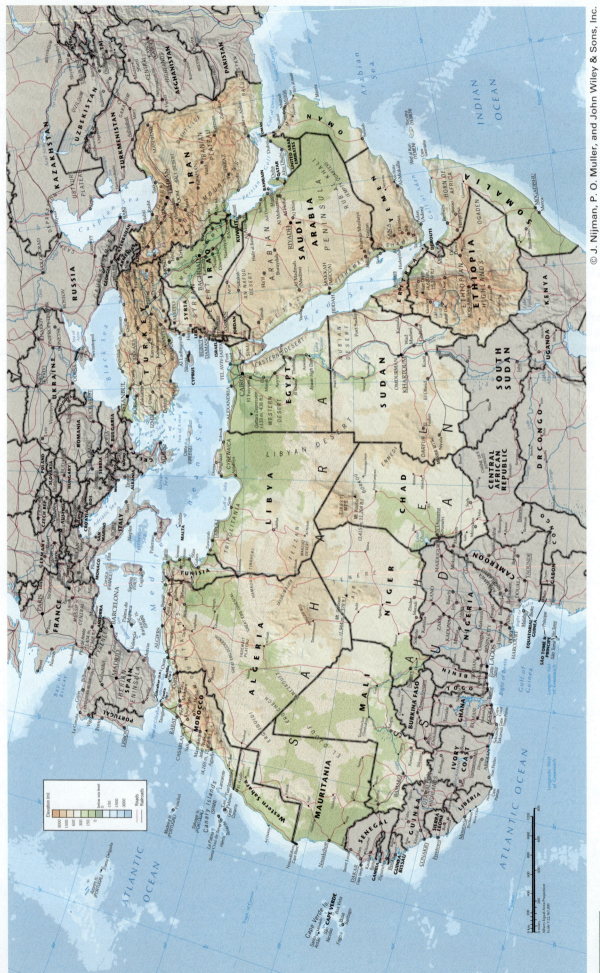

FIGURE 6-1

CHAPTER 6

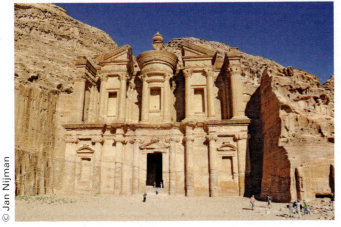

© Jan Nijman

© Jan Nijman

Where were these pictures taken?

The North African/ Southwest Asian Realm

IN THIS CHAPTER

- Escalating terrorism unleashed by ISIS
- Syria's civil war refugees
- Saudi Arabia plans for life after oil
- The steady rise of the Gulf States
- Libya's disintegration
- An Iranian Spring?

CONCEPTS, IDEAS, AND TERMS

From Morocco on the shores of the Atlantic to the Iranian Plateau, and from the Horn of Africa to the Turkish Straits, lies a vast geographic realm of enormous cultural complexity. It stands at the crossroads where Europe, Asia, and Africa meet, and it is part of all three (**Fig. 6-1**; **Box 6-1**). Throughout history, its influences have radiated across Eurasia and Africa and reached most other parts of the world as well. This is one of humankind's primary source areas. On the Mesopotamian Plain between the Tigris and Euphrates rivers (in modern-day Iraq) and on the banks of the Egyptian Nile arose several of the world's earliest civilizations. In its soils, plants were domesticated that are now grown from the Americas to Australia. Along its paths walked prophets whose religious teachings are still followed by hundreds of millions. And in this second decade of the twenty-first century, the heart of this realm has been engulfed by political turbulence, religious strife, and some of the most dangerous conflicts on Earth.

Defining the Realm

As you may conclude from its long and somewhat cumbersome name—North Africa/Southwest Asia—this is a sprawling, geographically complex realm. In our era of high-speed communication you will sometimes see it referred to as *NASWA*, the first letters of its regional deployment. And it is tempting to refer to it in other kinds of geographic shorthand, based on some of its outstanding features. But such generalizations can be misleading, and some examples follow.

A "Dry World"?

This realm is, for instance, sometimes called the Dry World, containing as it does northern Africa's immense Sahara as well as the Arabian Desert. But most of the realm's people live where there is water—along the Nile River, along the hilly Mediterranean coastal strip or *tell* (meaning mound in Arabic) of northwesternmost Africa, along the Asian eastern and northeastern shores of the Mediterranean Sea, in most of the Tigris-Euphrates Basin, in far-flung desert oases, along the lower mountain slopes of Iran south of the Caspian Sea, and in the Ethiopian Highlands. **Figure 6-2** reflects this co-varying relationship between population distribution and available water. Quite a few of this realm's prominent physiographic features—river valleys, basins, and deltas; moist coastlines; adequately watered mountain basins and plateaus—are virtually defined by the clusters you see on this map.

So we know this geographic realm as a place where water is almost always at a premium; where consumers (such as Israelis and their Arab neighbors) compete for limited supplies; where peasants often struggle to make soil and moisture yield a small harvest; where nomadic peoples and their livestock still migrate across dust-blown flatlands; where oases are green islands in the desert that sustain local farming, supply weary travelers, and support trade across vast expanses of aridity.

Box 6-1 Major Geographic Features of North Africa/Southwest Asia

1. North Africa and Southwest Asia were the scene of several of the world's great ancient civilizations, based in its river valleys and basins.

2. The boundaries of the North African/Southwest Asian realm consist of volatile transition zones in several sectors in both Africa and Asia. Afghanistan forms a transition zone vis-à-vis South Asia (i.e., Pakistan), and the African Transition Zone connects to Subsaharan Africa.

3. Islam dominates the cultural geography of this realm, and religion plays a leading role in political developments and conflicts.

4. The conflict-ridden Middle East, one of six constituent regions, lies at the heart of this realm. It includes Israel, but also contains Syria and Iraq—where ISIS has triggered the realm's most dangerous hostilities.

5. Drought and unreliable precipitation dominate natural environments, and population concentrations occur where water supply is adequate to marginal. Conflict over water sources and supplies looms as a steadily rising threat all across this realm, where population growth rates are high by current world standards.

6. Certain countries of the realm have enormous reserves of oil and natural gas, creating great wealth but also regional disparities.

7. Democracy and individual freedoms are rare in this realm, and the high hopes that accompanied the short-lived Arab Spring of 2011 have vanished almost everywhere.

8. Several countries in the realm are ravaged by warfare, especially Syria, Iraq, Yemen, and Libya. Terrorist organizations such as ISIS and al-Qaeda have proven to be powerful forces. External powers, including the United States, are repeatedly drawn into regional conflicts.

9. The severe dysfunction that afflicts parts of this realm is expressed in a massive and expanding refugee crisis. Most pressing are the plight of Syrian war refugees and the flows of desperate migrants aimed at Europe, with tens of thousands risking their lives at the hands of smugglers in the hazardous sea journey across the Mediterranean.

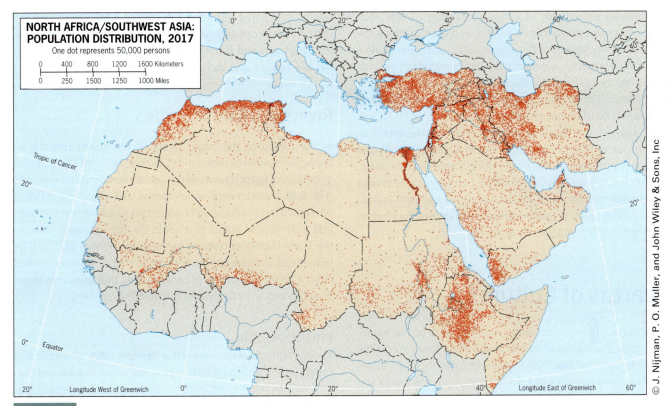

NORTH AFRICA/SOUTHWEST ASIA:
POPULATION DISTRIBUTION, 2017
One dot represents 50,000 persons

FIGURE 6-2

© J. Nijman, P. O. Muller, and John Wiley & Sons, Inc

But this is also the land of the Nile Valley and Delta as well as the Tigris-Euphrates Basin—respectively, the lifelines of Egypt and Iraq; the crop-covered *tell* of coastal Algeria; and the verdant shores of western Turkey. Compare Figure 6-1 to Figure G-6, and the dominance of **B** climates becomes evident, underscoring just how water-dependent this realm's highly clustered population is.

Is *This* the "Middle East"?

This realm is commonly referred to as the Middle East. That must sound odd to someone in, say, India, who might think of a Middle West rather than a Middle East! The name, of course, reflects the biases of its source—the Western world—which saw a Near East in Turkey; a Middle East in Egypt, Arabia, and Iraq; and a Far East in China and Japan. Still, the term has taken hold, and it can be seen and heard in everyday usage by scholars, journalists, and members of the United Nations. Even so, we feel that it should only be applied to one of the regions of this far-flung realm, not to NASWA as a whole.

An "Arab World"?

Another shorthand label often used for North Africa/Southwest Asia is the Arab World. This term implies a uniformity that does not actually exist. First, the name *Arab* is applied loosely to the

peoples of this area who speak Arabic and related languages, but ethnologists normally restrict it to certain occupants of the Arabian Peninsula—the Arab "source." In any case, Turks are not Arabs, and neither are most Iranians or Israelis. Moreover, although the Arabic language prevails from Mauritania in the west across all of North Africa to the Arabian Peninsula, Syria, and Iraq in the east, it is not spoken in certain parts of this realm. In Turkey, for example, Turkish is the major language, and it has Ural-Altaic rather than Arabic's Semitic or Hamitic roots; the Iranian language is a member of the Indo-European linguistic family (see Fig. G-8). Other "Arab World" languages that have separate ethnological identities are spoken by the Jews of Israel, the Tuareg people of the western Sahara, the Berbers of northwestern Africa, and the myriad peoples of the wide transition zone between North Africa and Subsaharan Africa to the south.

An "Islamic World"?

Finally, and perhaps most significantly, this realm is routinely referred to as the Islamic World. Indeed, the Prophet Muhammad was born on the Arabian Peninsula in AD 571, and Islam, the religion to which he gave rise, spread across this realm and beyond after his death in 632. This great saga of Arab conquest saw Islamic armies, caravans, and fleets carry Muhammad's teachings across deserts, mountains, and oceans, penetrating Europe, converting the kings of West African states, threatening the Christian strongholds of Ethiopia, invading Central Asia, conquering most of what is now India, and even reaching the peninsulas and islands of Southeast Asia. Today, the world's largest Muslim state is Indonesia, and the Islamic faith extends far outside the realm under discussion (Fig. G-10). In this context, the term *Islamic World* is also misleading in that it suggests that there is no Islam beyond NASWA's borders.

In any case, this "World of Islam" is not entirely Muslim either. Christian minorities continue to survive in all of the regions and many of the countries of the North Africa/Southwest Asia realm. Judaism has its base in NASWA's Middle East region; and smaller religious communities, such as Lebanon's Druze and Iran's Bahais, many of them under pressure from Islamic regimes, continue to diversify the realm's religious mosaic.

Hearths of Cultures

This geographic realm occupies a pivotal part of the world: here Africa, the source of humankind, meets Eurasia, crucible of human cultures. Two million years ago, the ancestors of our species walked from East Africa into North Africa and Arabia and then spread all across Asia. Less than 100,000 years ago, *Homo sapiens* crossed these lands on their way to Europe, Australia, and, eventually, the Americas. Ten thousand years ago, human communities in what we now call the Middle East began to domesticate plants and animals, learned to irrigate their fields, enlarged some of their settlements into towns, and formed the earliest states. The world's dominant monotheistic religions originated here: first came Judaism, followed by the rise of Christianity, and most recently the heart of the realm was stirred and mobilized by the teachings of Muhammad and the **Quran** (Koran), and Islam arose.

Dimensions of Culture

In the introductory chapter, we discussed the concept of culture and its regional expression in the cultural landscape. **Cultural geography [1]**, we observed, is a wide-ranging and comprehensive field that studies spatial aspects of human cultures, focusing not only on cultural landscapes but also on **culture hearths [2]**—the crucibles of civilization, the sources of ideas, innovations, and ideologies that transformed regions and realms. Those ideas and innovations spread far and wide through a set of processes that we study under the rubric of **cultural diffusion [3]**. Because we understand these processes better today, we can reconstruct many of the ancient routes by which the knowledge and achievements of culture hearths spread—that is, **diffused**—to other areas.

Another aspect of cultural geography, particularly noteworthy in the context of the North Africa/Southwest Asia realm, is the study of **cultural landscapes [4]** that a dominant culture creates. Human cultures exist in long-term accommodation with (and adaptation to) their natural environments, exploiting

opportunities that these habitats present and coping with the extremes they can impose. In the process, they fuse their physical landscape and cultural landscape into an interacting unity between nature and society.

Rivers and Communities

In the basins of the major rivers of this realm (the Tigris-Euphrates system of modern-day Turkey, Syria, and Iraq as well as the Nile of Egypt) lay two of the world's earliest culture hearths (**Fig. 6-3**). Mesopotamia (which means land between the rivers) possessed fertile alluvial soils, abundant sunshine, ample water, and animals and plants that could be domesticated. Here, in the Tigris-Euphrates lowland between the Persian Gulf and the uplands of present-day Turkey, arose one of humanity's first culture hearths, a cluster of communities that grew into larger societies and, eventually, into the world's first states.

Mesopotamia

Mesopotamians were innovative farmers who domesticated cereal grains such as wheat and knew when to sow and harvest various other crops, water their fields, and store their surpluses. Their knowledge spread to villages near and far, and a **Fertile Crescent**, a region of significant agricultural productivity, evolved, curving westward from Mesopotamia across Syria, the eastern Mediterranean coastlands, and terminating at the lower Nile Valley (**Box 6-2**; **Fig. 6-4**).

Irrigation was the key to prosperity and power in Mesopotamia, and urbanization was its major reward. Among many settlements in the Fertile Crescent, some thrived, grew, enlarged their hinterlands, and diversified socially as well as occupationally; others failed. What determined success? One theory, the **hydraulic civilization theory [5]**, holds that cities that could control irrigated farming across large hinterlands held power over others, used food as a weapon, and prospered. One of these cities, Babylon on the Euphrates River, endured for nearly 4000 years (from 4100 BC). A busy port, its walled and fortified center endowed with temples, towers, and palaces, Babylon for a time was the world's largest city.

Egypt and the Nile

Egypt's cultural evolution may have started even earlier than Mesopotamia's, and its focus lay upstream from (south of) the Nile Delta and downstream from (north of) the first of the Nile's series of rapids, or cataracts. This part of the Nile Valley is surrounded by inhospitable desert, and unlike Mesopotamia (which lay open to all comers), the Nile provided a natural fortress here. The ancient Egyptians converted their security into progress. The Nile was their leading avenue of trade and interaction; it also supported agriculture through irrigation (**Box 6-3**). The Nile's cyclical ebb and flow was much more predictable than that of the Tigris-Euphrates river system. By the time Egypt fell victim to outside invaders (around 1700 BC), a full-scale urban civilization had emerged. Ancient Egypt's artist-engineers left a

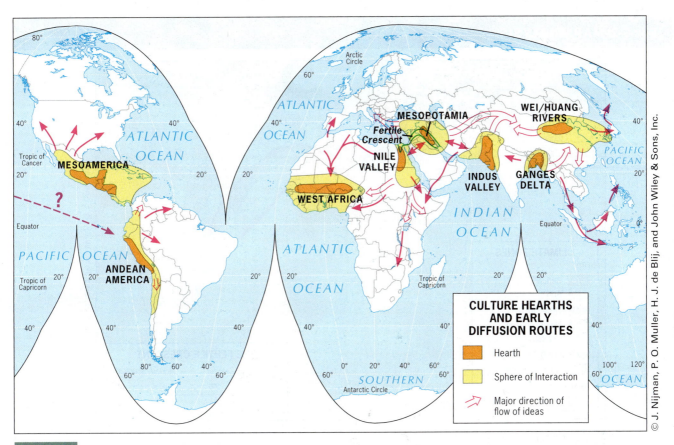

FIGURE 6-3

magnificent legacy of huge stone monuments, some of them containing treasure-filled crypts of god-kings called Pharaohs. These tombs have enabled archeologists to reconstruct the ancient history of this culture hearth.

Today, the world continues to benefit from the accomplishments of the ancient Mesopotamians and Egyptians. They domesticated cereals (wheat, rye, barley), vegetables (peas, beans), fruits (grapes, apples, peaches), and several animals (horses, pigs, sheep). They also advanced the study of the calendar, mathematics, astronomy, government, engineering, metallurgy, and a host of other skills and technologies. In time, many of their innovations were adopted and then modified by other cultures in the Old World and eventually in the New World as well. Europe was the greatest beneficiary of these legacies of Mesopotamia and ancient Egypt, whose achievements constituted the foundations of Western civilization.

Stage for Islam

In a remote place on the Arabian Peninsula, where the foreign invasions of the Middle East had had little effect on the Arab communities, an event occurred early in the seventh century that was to change history and affect the destinies of people in many parts of the world. In a town called Mecca (Makkah), about 70 kilometers (45 mi) from the Red Sea coast in the Jabal

Mountains, a man named Muhammad in the year AD 611 began to receive revelations from Allah (God). Muhammad (571–632) was then in his early forties. Arab society was in social and cultural disarray, but Muhammad forcefully taught Allah's lessons and began to transform his culture. His personal power soon attracted enemies, and in 622 he fled from Mecca to the safer haven of Medina (Al Madinah), where he continued his work until his death. This moment, the *hejira* (meaning migration), marks the starting date of the Muslim era, Year 1 on Islam's calendar. Mecca, of course, later became Islam's holiest place.

The Faith

The precepts of Islam in many ways constituted a revision and embellishment of Judaic and Christian beliefs and traditions. All of these faiths have but one god, who occasionally communicates with humankind through prophets; Islam acknowledges that Moses and Jesus were such prophets but considers Muhammad to be the final and greatest prophet.

Islam brought to Arabian culture not only the unifying religious faith it had lacked but also a new set of values, a new way of life, a new individual and collective dignity. Islam dictated observance of the Five Pillars: (1) repeated expressions of the basic creed, (2) daily prayer, (3) a month each year of daytime fasting (Ramadan), (4) the giving of alms, and (5) at least one pilgrimage to Mecca in each Muslim's lifetime (the *hajj*). Islam

Box 6-2 MAP ANALYSIS

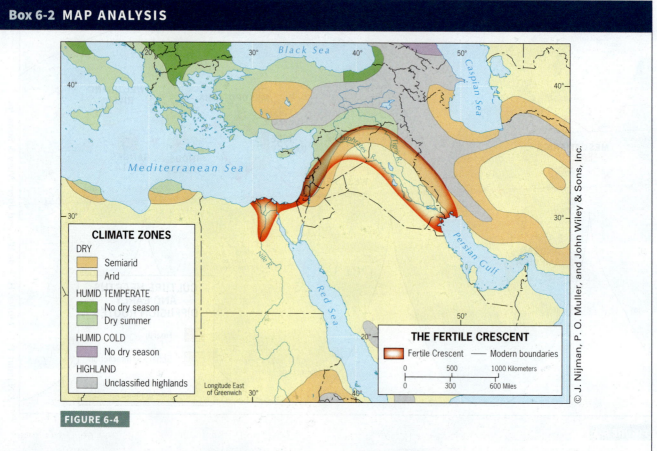

CLIMATE ZONES

DRY
- Semiarid
- Arid

HUMID TEMPERATE
- No dry season
- Dry summer

HUMID COLD
- No dry season

HIGHLAND
- Unclassified highlands

THE FERTILE CRESCENT

Fertile Crescent — Modern boundaries

0   500   1000 Kilometers
0   300   600 Miles

© J. Nijman, P. O. Muller, and John Wiley & Sons, Inc.

FIGURE 6-4

The Fertile Crescent was so named because of the major civilizations that blossomed there in ancient times and that owed much to its advanced agricultural traditions. Not coincidentally, this was also where some of the world's earliest cities arose. Take a close look at Figure 6-4, and think of two commonalities within the area spanned by the Fertile Crescent. Can you think of two physiographical explanations as to why this environment was conducive to the rise of these pioneering civilizations?

Access your WileyPLUS Learning Space course to interact with a dynamic version of this map and to engage with online map exercises and questions.

---

prescribed and proscribed in other spheres of life as well. It forbade the use of alcohol, smoking, and gambling. It tolerated polygamy, even though it acknowledged the virtues of monogamy. Mosques appeared in Arab settlements, not only for the (Friday) sabbath prayer, but also as social gathering places to knit communities closer together. Mecca became the spiritual center for a divided, widely dispersed people for whom a collective focus was something new.

The Arab-Islamic Empire

Muhammad provided such a powerful stimulus that Arab society was mobilized almost overnight. The prophet died in 632, but his faith and fame spread like wildfire. Arab armies carrying the banner of Islam formed, invaded, conquered, and converted wherever they went. As **Figure 6-5** shows, by AD 700 Islam had reached far into North Africa, into Transcaucasia, and into most of Southwest Asia. In the centuries that followed, it penetrated both southern and eastern Europe, Central Asia, West Africa, East Africa, South Asia, and Southeast Asia, even reaching into China by AD 1000.

The spread of Islam provides a good illustration of a set of processes known as **spatial diffusion [6]** that focus on the way ideas, inventions, and cultural practices propagate through a population in space and time. Diffusion takes place in two forms: **expansion diffusion [7]**, when propagation waves originate in a strong and durable source area and spread outward, affecting an ever larger region and population; and **relocation diffusion [8]**, in which migrants carry an innovation or an idea (or belief). Islam spread through expansion as well as relocation diffusion, reaching such faraway places as the Ganges Delta in South Asia, what is now Indonesia in offshore Southeast Asia, and East Africa.

The heart of Islam, however, remained in Southwest Asia. There, Islam became the cornerstone of an Arab Empire with Medina as its first capital. As the Empire expanded, its headquarters shifted from Medina to Damascus (in what is now Syria) and later to Baghdad on the Tigris River (in modern-day Iraq). And it prospered. In architecture, mathematics, and science, the Arabs overshadowed their European contemporaries (now mired in their medieval standstill). The Arabs established institutions of higher learning in many cities, including Cairo,

Box 6-3 From the Field Notes...

"Flying northward over central Egypt from Luxor's ruins toward Cairo along the Nile's ribbon of green, I pondered the famous remark by the ancient Greek historian Herodotus that 'Egypt is the gift of the River Nile.' The Nile brings water from the interior of Africa that crosses the vast northern deserts and is crucially important to a country with as little rainfall as Egypt. About 95 percent of the Egyptian population lives within 20 kilometers (12 mi) of the river's banks. Almost all of Egypt's agriculture has always depended on irrigation, which is how this narrow strip of desert turned fertile. Notice the razor-sharp boundaries between irrigated and nonirrigated land. Undoubtedly, the Nile was the lifeline that supported one of the ancient world's leading civilizations."

© Jan Nijman

Baghdad, and Toledo (Spain), and their distinctive cultural landscapes united their far-flung domain (Box 6-4). Non-Arab societies in the path of the Muslim drive were not only Islamized, but also Arabized, adopting other Arab traditions as well. Islam had spawned a culture; it still continues to function at the heart of that culture today.

Islam and Other Religions

Two other major faiths also originated in the **Levant** (the lands bordering the Mediterranean's eastern coast, extending inland for about 300 kilometers [200 mi])—Judaism and Christianity—with both predating the founding of Islam. Islam's rise submerged many smaller Jewish communities, but the Christians, not the Jews, waged centuries of holy war against the Muslims, seeking, through the Crusades, not only to drive Islam back but to reestablish Christian communities where they had dominated before the Islamic conquest. The aftermath of that lengthy campaign still marks the realm's cultural landscape today. A substantial Christian community remains in Lebanon, and Christian minorities also survive in Israel, Egypt, Syria, and Jordan. Since World War II, the most intense conflict to date has pitted Israel, the Jewish state formed in 1948, against its Islamic neighbors near and far. Jerusalem—holy city for Judaism, Christianity, and Islam—lies squarely in the middle of this confrontation.

Islam Divided

For all its vigor and success, Islam is still fragmented into sects. The earliest and most consequential division arose after

Muhammad's death between those who believed that only a blood relative should follow the prophet as leader of Islam (***Shi'ites***) and others who felt that any devout follower of Muhammad was qualified to take over (***Sunnis***). Overall, the Sunnis prevailed, and, in the centuries that followed, the great expansion of Islam was mainly propelled by Sunnis; the Shi'ites survived as minorities scattered throughout the realm. Today, even though roughly 90 percent of all Muslims are Sunnis, the Shia–Sunni schism has strongly resurfaced to become the root of a number of the realm's ongoing conflicts.

The minority Shi'ites have always vigorously promoted their version of the faith. In the early sixteenth century, their work finally paid off: the royal house of Persia (present-day Iran) made Shi'ism the only legal religion throughout its vast empire. That domain extended from Persia into lower Mesopotamia, into Azerbaijan, and into western Afghanistan and Pakistan. As the world map of religions (Fig. G-10) shows, this created for Shi'ism a large culture region and gave the faith unprecedented strength. Iran remains the bastion of Shi'ism in the realm today, and its appeal continues to radiate into neighboring countries and even farther afield.

The Ottoman Empire and Its Aftermath

If we want to understand this sprawling realm today, we must appreciate the historical role of the Ottoman Empire, its enormous expansion, and the ways it was undone by other great powers. It is ironic that Islam's last great advance into Europe

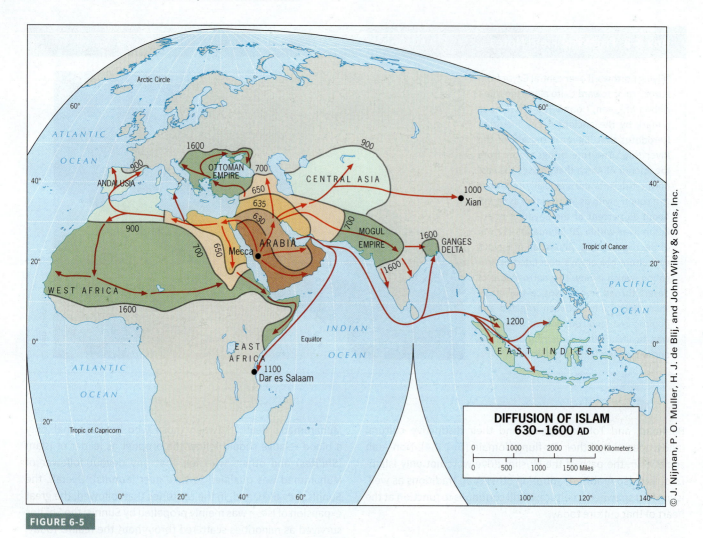

DIFFUSION OF ISLAM
630–1600 AD

FIGURE 6-5

eventually led to European occupation of Islam's very heart-land. The Ottomans (named after their leader, Osman I), based in what is today Turkey, conquered Constantinople (now Istan-bul) in 1453 and then pushed into southeastern Europe. Soon Ottoman forces were on the doorstep of Vienna; they also invaded Persia, Mesopotamia, and North Africa. At its height, the Ottoman Empire under Suleyman the Magnificent (ruled 1522–1560) was the most powerful state in western Eurasia.

Although the Ottoman Empire survived for more than four centuries until its dissolution in 1923, it lost territory as time went on. First came losses to the Hungarians, then to the Rus-sians, and later to the Greeks and Serbs until, after World War I, the European powers took over its provinces and turned them into colonies—colonies we now know by the names of Syria, Iraq, Lebanon, and Yemen (**Fig. 6-6**). As the map shows, the French and the British took several possessions; even the Ital-ians annexed the part of the Ottoman domain that is now Libya.

The boundary framework that the colonial powers cre-ated after World War I to delimit their new holdings was unsat-isfactory. As Figure 6-2 illustrates, this realm's population is clustered, fragmented, and strung out in river valleys, coastal zones, and crowded oases. The colonial powers laid out long stretches of boundary as ruler-straight lines across uninhabited

territory; they saw no need to make those boundaries conform to cultural or physical features in the landscape. Other borders, even some in desert zones, were poorly defined and never marked on the ground. Later, when the colonies had become independent states, such boundaries led to quarrels, even armed conflicts, among neighboring Muslim states. In some instances, they left entire nations, such as the Kurds and the Palestinians, without any territory of their own, thereby render-ing them **stateless nations [9]**.

Population and the New Political Framework

Although Islam and its cultural expressions overspread this geo-graphic realm, its political and social geographies are divided and often fractious. We encounter states with strong internal divisions (Iraq, Lebanon), stateless nations (Palestinians, Kurds, Berbers), and territories still in the process of integration (West-ern Sahara, Palestine). Of its two dozen states and territories, the three largest and in many ways the most important are Egypt in North Africa, Turkey on the threshold of Europe, and Iran at the margins of Central Asia. All three have populations

Box 6-4 From the Field Notes...

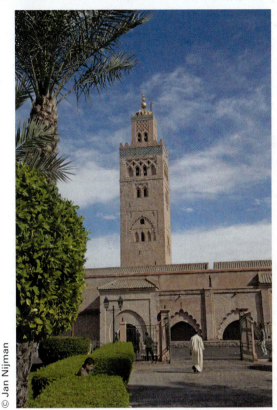

"As I walked through the medina (old city) of Marrakech (Morocco) in early 2015, I came upon the famous Koutoubia Mosque. Its minaret is the city's signature landmark, standing 77 meters (250 ft) tall. It was built in the twelfth century and reminds me of the masterful architecture and cosmopolitanism that flourished in parts of the Arab-Islamic Empire while European culture lay dormant during the depths of the so-called Dark Ages. In fact, this Islamic cultural imprint also extended into southern Europe, and comparable architectural wonders still stand in the Spanish cities of Seville and Granada. When, in the late fifteenth century, Spain expelled its Jewish population, some found refuge across the Strait of Gibraltar in Morocco. I tried to take a look inside the Mosque, but non-Muslims were refused entry (has that always been the case?). Today, Morocco is a relatively stable outpost on the western edge of a decidedly turbulent and deeply challenged realm."

between 80 and 93 million, modest by world standards. But at the other end of this continuum are ministates such as Bahrain, Qatar, and Djibouti, each with fewer than 2.3 million inhabitants. It is quite important to study Figure 6-1 attentively because the boundary framework—much of it inherited from the colonial era—now lies at the root of so many of the realm's most difficult problems.

It is also useful to take another careful look at Figure 6-2 because it shows how differently populations are distributed within the countries we will be examining. Compare the three largest countries just mentioned: Egypt's population forms a

river-valley ribbon and delta cluster; Turkey's, between the Black and Mediterranean seas, is far more evenly dispersed, although its west is more populous than its east; and in Iran you can draw a line down its center from the Caspian Sea to the Persian Gulf that separates the comparatively crowded west, where the capital is located, from the drier and far more sparsely peopled east. In North Africa, almost everyone seems to live in the vicinity of the Mediterranean coast, but no such coastal concentrations mark the Arabian Peninsula except in its far south, where Yemen's inhabitants cluster near the Red Sea.

The Power and Peril of Oil

About 30 of the world's countries have significant oil reserves, with six of the nine leading countries located in this realm: Saudi Arabia, Iran, Iraq, Kuwait, the United Arab Emirates, and Libya. In general terms, NASWA's deposits of oil (and associated natural gas) are concentrated within two discontinuous zones (**Fig. 6-7**). The most productive of these zones extends from the southern and southeastern part of the Arabian Peninsula north-westward around the rim of the Persian Gulf, reaching into Iran and continuing northward into Iraq, Syria, and southeastern Turkey, where it peters out. The second zone lies across North Africa and extends from north-central Algeria eastward across northern Libya to Egypt's Sinai Peninsula.

Winners and Losers

Saudi Arabia is the world's largest oil exporter, and the combined oil production of the NASWA countries far exceeds that from all other sources. Throughout recent times, the United States has been the world's leading importer, followed closely by China. Today, the United States is both the biggest consumer of oil as well as the top producer, making it less dependent on imports than it would otherwise be.

As Figure G-11 indicates, the production and export of oil and gas has vaulted a number of this realm's countries into the higher-income categories. But petroleum wealth also has entangled these Islamic societies and their governments in global strategic affairs. When regional conflicts create instability in producing countries that have the potential to disrupt international supply lines, powerful consumers are tempted to intervene and have done so.

When the colonial powers laid down the boundaries that partitioned this realm among themselves, no one knew about the riches that lay beneath the ground. A few wells had been drilled, and production in Iran had begun as early as 1908 and in Egypt's Sinai Peninsula in 1913. But the major discoveries came later, in certain cases after the colonial powers had already withdrawn. Some of the newly independent countries, such as Libya, Iraq, and Kuwait, found themselves endowed with wealth undreamed of when the Ottoman Empire collapsed. As Figure 6-7 shows, however, others were less fortunate. A few countries had

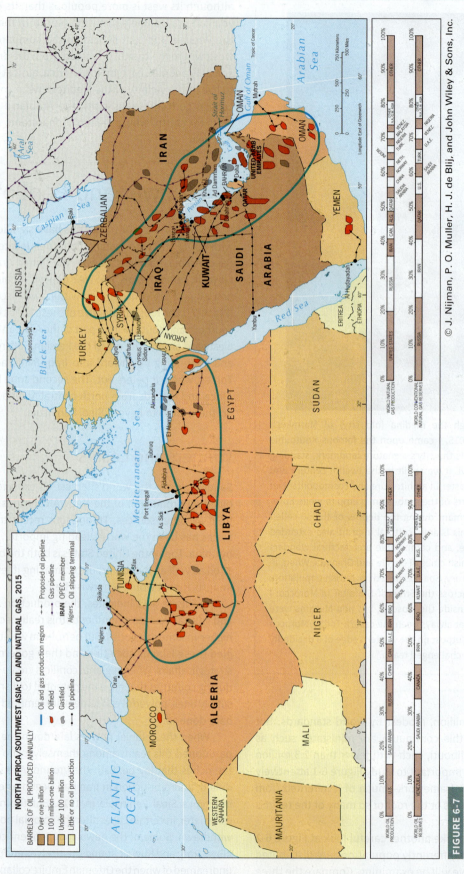

NORTH AFRICA/SOUTHWEST ASIA: OIL AND NATURAL GAS, 2015

BARRELS OF OIL PRODUCED ANNUALLY

- Over one billion
- 100 million-one billion
- Under 100 million
- Little or no oil production

— Oil and gas production region
— Oil pipeline
— Gas pipeline
— Proposed oil pipeline
IRAN OPEC member

Oilfield
Gasfield
Oil pipeline
Algiers⚓ Oil shipping terminal

© J. Nijman, P. O. Muller, H. J. de Blij, and John Wiley & Sons, Inc.

FIGURE 6-7

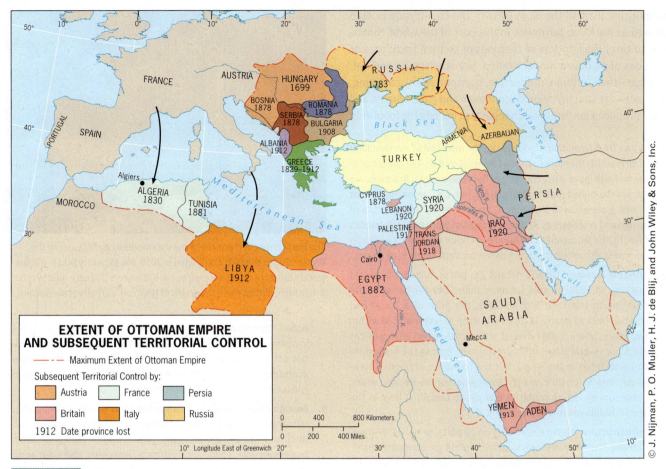

© J. Nijman, P. O. Muller, H. J. de Blij, and John Wiley & Sons, Inc.

FIGURE 6-6

(and still have) potential. The smaller, weaker emirates and sheikdoms on the Arabian Peninsula always feared that powerful neighbors would try to annex them. The unevenly distributed oil wealth, therefore, created yet another source of division and distrust among Islamic neighbors.

A Foreign Invasion

The oil-rich countries of the realm found themselves with a coveted energy source but lacking in the skills, capital, or equipment to exploit it. Those resources had to come from the Western world and entailed what many tradition-bound Muslims feared most: a strong foreign presence on Islamic soil, outside intervention in political as well as economic affairs, and penetration of Islamic societies by the vulgarities of Western ways. The transport of oil from these countries to destinations in other geographic realms also heightened the importance of strategic arteries that connect NASWA to the rest of the world. Most significant are **choke points [10]**, the narrowing of international waterways that increase the risk of ship collisions and vulnerability to piracy and other forms of attack. This realm contains no less than five of the world's most critical choke points: the Suez Canal and Bab-el-Mandeb Strait at opposite ends of the Red Sea; the Strait of Hormuz at the mouth of the Persian Gulf; the Turkish Straits that connect the Black and Mediterranean seas; and the Strait of Gibraltar at

the Mediterranean's outlet to the Atlantic Ocean (Fig. 6-1). Just before the end of this chapter, we discuss the virtual "invasion" of a number of foreign powers in Djibouti, which occupies the crucially important west shore of the Bab-el-Mandeb Strait. In the case of Djibouti, it is not about oil but all about geostrategic location.

Foreign intervention has often aggravated clashes between the traditional and the modern. Oil revenues created cultural landscapes in which gleaming skyscrapers towered over ornate mosques and centuries-old communities (see photo of Dubai). The social chasms between the rich, well-connected, Westernized elites and for less fortunate local citizenries bred resentment. In Iran, these issues played a role in the alienation of the ruling shah (king) from his own people, fueling a religious revival that ultimately led to the monarch's ouster in 1979. Thus oil, foreign involvement, economics, and religion are inextricably intertwined in the realm's petroleum-producing countries.

The Geography of Oil's Impact

As often occurs with exploited natural resources, oil has completely transformed cultural landscapes in certain parts of the North African/Southwest Asian realm and left others virtually unchanged. For hundreds of millions of this realm's inhabitants, a patch of tillable soil and a source of water still mean more to daily life than all the oil in OPEC.* The countryside in Arab as well as non-Arab territories in this part of the world continues to carry the imprints of centuries of cultural tradition, not decades of oil-driven modernization. Nonetheless, oil and natural gas—their location, production, transportation, marketing, and sale—are vital to national economies and governments and have produced fundamental changes that include the following:

1. **Urban Transformation.** Undoubtedly, the most visible manifestation of oil wealth is the modernization of cities (**Box 6-5**). The tallest building on Earth (as of 2016) rises above Dubai (see photo) in a ministate named the United Arab Emirates (UAE), but it is only one in a forest of new glass-encased skyscrapers, many of which test the limits of design and engineering. Besides the capitals (such as Saudi Arabia's Riyadh), the landscapes of a number of other cities on the Arabian Peninsula also reflect the riches oil has brought—most of all Mecca.

2. **Variable Incomes.** Oil and natural gas prices fluctuate on world markets. In 2008, oil sold for more than $140 per barrel before plunging to less than $40; by 2013, the price of a barrel had climbed back to nearly $120, only to drop to less than $50 within two years, where it still hovered in mid-2016. When energy prices are high, several states in this realm rank among the highest-income societies in the world, but the countries without oil have to pay dearly for their energy. Even when prices decline, many petroleum-exporting countries manage to remain at least in the upper-middle-income category (Fig. G-11).

3. **Infrastructure.** Massive spending on airports, seaports, bridges, tunnels, four-lane highways, public buildings, shopping malls, recreational facilities, and other components of national infrastructure creates an image of comfort and affluence quite unlike that prevailing in countries without significant oil or gas revenues. Saudi Arabia, for instance, is now engaged in a gigantic modernization project extending from coast to coast.

4. **Industrialization.** A number of far-sighted governments among those with oil wealth, realizing that reserves will not last forever, are investing some of their income in industries that will survive the oil-exporting era. Petrochemical manufacturing using domestic supplies, aluminum, steel, and fertilizers are among those industries, although others potentially more promising, especially in high-technology fields, are only slowly becoming a part of this important initiative.

5. **Regional Disparities.** Oil wealth, like other high-value resources, tends to create strong regional contrasts both within and among countries. Saudi Arabia's ultramodern east coast is a world apart from most of its interior, which remains a land of barren deserts, widely scattered oases, enormous distances, and isolated settlements. Among countries, the contrasts can also be striking, for example, between Qatar and Yemen.

6. **Foreign Investment.** Governments and private entrepreneurs have invested enormous amounts of oil-generated wealth in foreign countries, buying financial securities as well as acquiring prestigious

The Burj Khalifa, centerpiece of the ultramodern city of Dubai on the Persian Gulf in the UAE. Completed in 2010, the 163-story Burj is still the world's tallest building, topping out at 828 meters (2716 ft)—just over half a mile high, almost twice the height of New York's Empire State Building.

hotels, upscale retailers, and other high-cachet properties. These investments have created a network of international linkages that not only connects NASWA states and individuals to the national economies of other realms but also links them to the growing Islamic communities in those countries.

7. **Foreign Involvement.** To many inhabitants of this realm, particularly those with strong Islamic-revivalist convictions (a topic to be discussed shortly), the inevitable presence of foreigners (including businesspeople, politicians, architects, engineers, and even armed forces) on Islamic soil is an unwelcome byproduct of the energy era. In fact, public opinion in Saudi Arabia forced its ruling regime to negotiate the 2003 departure of U.S. troops from the country's territory.

8. **Intra-Realm Migration.** Petroleum wealth enables governments, industrialists, and private individuals to hire workers from less-favored parts of the realm to labor in the oilfields, ports, and numerous menial service occupations. This has brought many

---

*OPEC, the Organization of Petroleum Exporting Countries, is the international oil cartel (syndicate) formed by 14 producing countries to promote their common economic interests through the setting of joint pricing policies and the limitation of market options for consumers. Its eight NASWA members are Algeria, Iran, Iraq, Kuwait, Libya, Qatar, Saudi Arabia, and the United Arab Emirates.

Box 6-5 Major Cities of the Realm, 2016

Metropolitan Area	Population (in millions)
Cairo, Egypt	15.9
Tehran, Iran	13.7
Istanbul, Turkey	13.5
Baghdad, Iraq	6.8
Riyadh, Saudi Arabia	5.8
Khartoum, Sudan	5.2
Algiers, Algeria	3.7
Casablanca, Morocco	3.2
Tel Aviv-Jaffa, Israel	3.0
Damascus, Syria	2.6
Beirut, Lebanon	2.2
Tripoli, Libya	1.1
Jerusalem, Israel	0.9

Shi'ites to the countries of the eastern Arabian Peninsula, where they now form significant sectors of national populations; hundreds of thousands of Palestinian Arabs also sought temporary employment in local industries here. In 2016, it was estimated that Saudi Arabia, with a permanent population of 32 million, hosted more than 8 million foreign workers, about 5 million of whom come from elsewhere in the NASWA realm.

9. *Migration from Other Realms.* The labor market in such burgeoning places as the UAE's Dubai and Abu Dhabi has also attracted hordes of workers from well beyond the realm's borders during recent periods of rapid growth. Wages in South Asian and African countries are considerably lower than those paid by the building industry or private employers in oil-boom-driven NASWA states. Besides low-level construction jobs, these workers serve as domestics, gardeners, trash collectors, and the like. In 2016, almost 90 percent of the total population of the United Arab Emirates and Qatar consisted of foreign employees! Over the past few years, substandard working conditions and wage issues have also led to protests that have exposed the harsh conditions faced by many guest workers throughout this realm.

10. *Diffusion of Revivalism.* Islamist regimes utilize oil and gas revenues to support Muslim communities as well as their mosques and cultural centers throughout the world. No NASWA country has spent more money on such causes than Saudi Arabia, and thousands of mosques from England to Indonesia prosper as a result. This example of relocation diffusion creates myriad nodes of recruitment for the faith—and ensures the dissemination of revivalist principles.

Fragmented Modernization

More than any other realm, this one is marked by a human spatial mosaic of **fragmented modernization [11]**. To a certain degree NASWA is highly modern and prosperous, but in many more respects it remains traditional, stagnant, and quite poor. These disparities that characterize so many of the realm's societies are best explained by the uneven impact of oil, the lack of democracy, and the mercurial role of religion.

Oil-Generated Inequalities

Oil brought this realm into contact with the outside world in ways unforeseen just a century ago. Oil has strengthened and empowered some of its peoples but has dislocated and imperiled others. It has truly been a double-edged sword.

We should remind ourselves that the great majority of NASWA's inhabitants are not, in their daily lives, directly affected by the changes the energy era has brought. Most Moroccans, Tunisians, Egyptians, Jordanians, Yemenis, Lebanese, and countless millions of others—Kurds, Palestinians, Berbers, Tuaregs—make ends meet by trading or farming or working at jobs their parents and grandparents performed. Take the case of Iran, which as a country earns about two-thirds of its income from oil and natural gas. Only one-half of 1 percent of that country's workforce (just over 100,000 out of more than 20 million workers) is employed in the energy-related sector. By far, the largest number engaged in any single occupation are Iran's five-million-plus farmers. And for all their oil, Iranians in 2015 earned an average yearly income of $5,000, less than half the earnings in neighboring, energy-poor Turkey. As we shall discover, cultural-geographic rather than economic-geographic forces mainly shape the regionalization of this realm.

The two dozen states that constitute NASWA often display stunning degrees of variety and diversity. This difference is accentuated by the severe disparities that result from the uneven distribution of the realm's oil wealth. And even within those countries possessing large energy reserves, glaring inequalities have arisen between their have and have-not citizens. Ultramodern urban skylines on the Arabian Peninsula stand in sharp contrast to Nile Valley villages virtually unchanged for millennia. Modern expressways cross deserts still traversed by nomadic traders riding camels. Social geographies vary just as greatly—for instance, the female literacy rate in Lebanon is 86 percent but only 48 percent in Yemen. As noted above, having or not having oil explains most of these disparities. This even applies to access to water, a commodity that, in the long run, is even more vital than oil (**Box 6-6**, **Fig. 6-8**).

The Absence of Democratic Traditions

Democracy has been a rarity in this realm. Although specific reasons vary, the majority of states share a history of autocratic, conservative regimes, usually backed strongly by their military forces. Most countries were part of the Ottoman Empire before World War I (see Fig. 6-6) and only became independent after a period of (attempted) European rule that was, in some cases, sanctioned by the League of Nations (the predecessor of the United Nations that functioned from 1920 to 1946). Accordingly, Egypt, Iraq, Palestine, Transjordan, and parts of Yemen were administered by Britain; Lebanon, Syria, Tunisia, Morocco, and Algeria were ruled by France; and Libya was controlled by Italy.

In most cases, the Europeans faced persistently tenacious opposition from local populations but were reluctant to let go. Independence was usually preceded by intense conflict or even war; at any rate, none of these countries can be said to have been prepared to govern themselves as European-style democracies. Indeed, virtually everywhere the newly independent countries fell into the hands of the military (such as Egypt) or restored monarchs with strong military support (such as Yemen).

After World War II, these autocratic regimes were often cemented into place through foreign involvement. Mostly, this was about oil: foreign powers needed oil, showed little interest in domestic politics, and provided incumbent regimes with abundant revenues to build up their military forces, secure their political position, and maintain order. Saudi Arabia, the world's richest petro-state, is the quintessential example. In other cases, it was Cold War politics that effectively supported the realm's nondemocratic governments.

Political regimes in this realm differ, but at least until the Arab Spring movement emerged in late 2010, most heads of

Box 6-6 Technology & Geography

Global Water Needs and Desalination

No single natural resource defines a world geographic realm as oil does for North Africa/Southwest Asia. Yet it is access to fresh water that permits oil-rich economies, some of which are located in the extremely arid climates of the Sahara and Arabian deserts, to endure and even thrive. The fate of the ancient civilizations, cities, and communities of this realm were once solely tied to river systems like the Nile and Tigris-Euphrates. Today, the sustainability and future of developed economies and societies of this realm depend on desalination technology. **Desalination [12]** refers to the process of obtaining fresh water from saline (salt) water. There are two primary methods of desalination. The first, *multistage flash distillation,* uses heat to evaporate water, leaving the salt behind. The second, *reverse osmosis,* entails the forcing of salt water through specialized filters at extremely high pressures to purify the liquid. Although simple in theory, desalination is very expensive because it is such an energy-intensive process.

Desalination technology has advanced in this century, thereby cutting the cost, but it remains expensive. For that reason it is not uniformly available across this particularly uneven realm. Geographic variations in access to desalination, and more generally to potable water, are related to the standards of living, economic development, and levels of health found across the realm. Figure 6-8 shows the global variation in water stress levels (measured as available renewable fresh water per person) and the ten leading countries in desalination capacity (also on a per capita basis). Note the exceptional position of the rich oil states in the NASWA realm. For Saudi Arabia, one of the world's wealthiest countries and leading producers of oil, it is estimated that 15 percent of annual oil income is allocated to desalination operations. The combination of volatile oil prices, regional political instability, and rapidly increasing demand for both electrical power and water is pushing the Kingdom to construct nuclear power plants to supply the energy for desalination and to wean itself from oil dependency. It remains to be seen whether or not, and how, the development of nuclear power will contribute to greater access to fresh water, economic growth and sustainability, and realm-wide security.

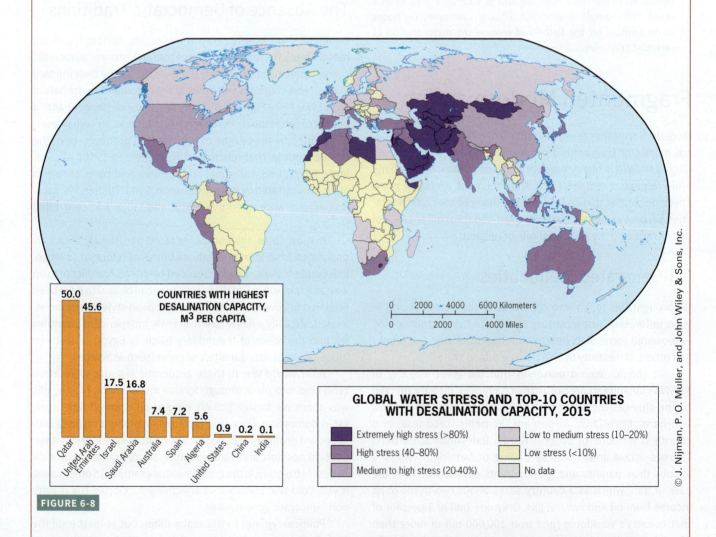

COUNTRIES WITH HIGHEST DESALINATION CAPACITY, M³ PER CAPITA

Qatar 50.0
United Arab Emirates 45.6
Israel 17.5
Saudi Arabia 16.8
Australia 7.4
Spain 7.2
Algeria 5.6
United States 0.9
China 0.2
India 0.1

GLOBAL WATER STRESS AND TOP-10 COUNTRIES WITH DESALINATION CAPACITY, 2015

- Extremely high stress (>80%)
- High stress (40–80%)
- Medium to high stress (20-40%)
- Low to medium stress (10–20%)
- Low stress (<10%)
- No data

© J. Nijman, P. O. Muller, and John Wiley & Sons, Inc.

FIGURE 6-8

Box 6-7 MAP ANALYSIS

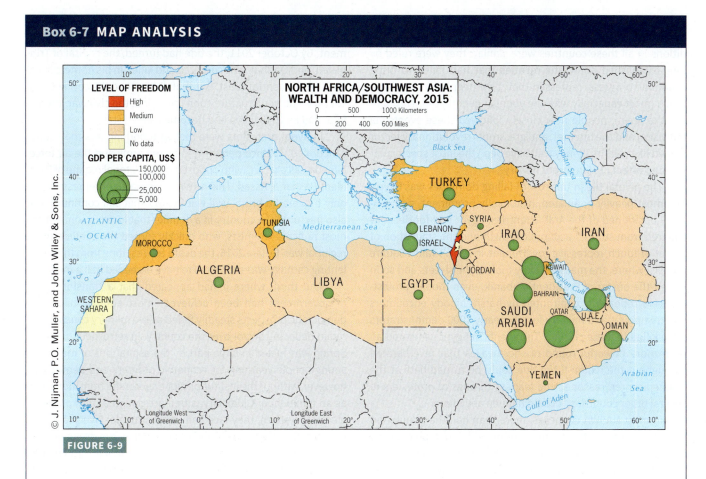

FIGURE 6-9

Worldwide, high levels of national development tend to be associated with democratic political systems that offer a full range of personal freedoms. Conversely, it is thought that the poorer the country, the less likely it is to be a democracy (for example, think of North America versus Subsaharan Africa). But all this can range considerably from one realm to another or within realms. Figure 6-9 maps income per capita against level of democracy for selected NASWA countries in 2015. Can a clear relationship between wealth and democracy be observed in this realm? Can you discern any regional patterns? Comparing this map with the one in Figure 6-7, do you think oil and gas production is associated with the spatial variations you observe?

Access your WileyPLUS Learning Space course to interact with a dynamic version of this map and to engage with online map exercises and questions.

government had been in power for a substantial number of years, and their top priority was to preserve their domination and the status quo. Overall, their rule had been generally conservative and based on top-down control, with minimal concern for the "people on the street" (**Box 6-7**, **Fig. 6-9**).

The Arab Spring of 2011 and Its Aftermath

On December 17, 2010, in the Tunisian capital of Tunis, a 26-year-old fruit-and-vegetable vendor named Mohammed Bouazizi became so distraught when police, following months of constant harassment, confiscated his unlicensed produce stand, that he set himself on fire. His act of desperation resonated with countless Tunisians whose frustration with their government

had reached the breaking point. Bouazizi died soon afterward, but his self-immolation quickly came to symbolize Tunisia's "Jasmine Revolution." Within a month, popular opposition to the regime had grown so fierce that Tunisia's president was forced to resign and flee to Saudi Arabia after 23 years in power.

In Arab countries, according to conventional wisdom, not only democratic governance but even the desire for such a political system was nonexistent. That myth exploded as Tunisians of all backgrounds, including legions of women as well as youths, took to the streets to demand an end to official corruption, repression, cronyism, and economic mismanagement. The Tunisian people were insisting that they be heard—and their voices swiftly and loudly echoed across much of the realm.

The Diffusion of Popular Unrest

What originated in Tunisia, as 2010 turned into 2011, rapidly diffused eastward across North Africa and into adjacent parts of Southwest Asia. Within a few months, popular uprisings had spread into Egypt, Libya, Syria, Yemen, and Bahrain. The conditions in all these countries were similar: they were ruled by long-established autocratic regimes that had failed to bring economic progress, repressed their own people, and had lost touch (most especially with the younger generation). Throughout the realm, people now closely followed these events via television and the Internet, and grassroots revolutionary fervor spread like wildfire. It may well have been the most spectacular **domino effect [13]** in recent history.

The pivotal tile in this row of falling dominoes was Egypt. The most heavily populated, historically prominent, and centrally positioned of the Arab countries, Egypt had for decades suffered from stunted economic growth and the stagnation of an out-of-touch, repressive government. In Cairo's Tahrir Square and elsewhere, the masses boldly took to the streets and soon forced the ouster of President Mubarak, even though protesters were repeatedly confronted by harsh, bloody encounters with the police and military forces; nearly 1000 demonstrators were killed in these clashes. The events in Egypt generated an even more powerful round of shockwaves that reverberated throughout the realm. If the tide could be turned here at the Arab epicenter, reasoned the fearless throngs of protesters in several other countries, it could happen anywhere.

In most NASWA countries, more than half the population is under 25 years of age. This enormous demographic cohort has grown up in a world in which information flows almost uncontrollably, so these young people know only too well that their home countries deny them many opportunities that exist elsewhere. Not surprisingly, the 2011 uprisings were predominantly (though not exclusively) led by the younger generation, ranging from teenagers to thirty-somethings—and including many women. In Tunisia, Egypt, and Syria, students took the lead, using the Internet's social media to organize demonstrations. This emphasized another critical dimension of the Arab Spring: the disconnect between the realm's aging, entrenched autocrats and their young, globally aware subjects.

Unfulfilled Arab-Spring Promises

If the very term *Arab Spring* expressed the hopefulness of a new political dawn, such romantic notions were rapidly dispelled. First of all, even though the spate of uprisings resonated and jangled nerves across the Arabian Peninsula, they did not seriously challenge any of that region's governments (except in peripheral Yemen). It was here—as well as in Jordan, Morocco, and Iran—that the established monarchies and/or theocracies prevailed, with only modest responses to popular demands for reform.

Second, the more overtly tyrannical regimes in Libya and Syria responded with stubborn resistance and iron fists. Libya quickly degenerated into civil war as President Muammar Qadhafi unleashed his military forces against his unarmed people. That attack was so vicious that a makeshift alliance of France, the United Kingdom, and the United States (later subsumed by NATO) intervened with aerial bombing raids to help protect the Libyans from annihilation by their own government. By October 2011, the tide had turned, the capital of Tripoli fell, and Qadhafi was cornered and killed in his hometown of Sirte on the central coast. But that did not end the conflict. The country's power structure soon imploded, the civil war intensified, and by 2015 Libya had become a failed state.

In Syria, protesters were met with nothing less than state-sanctioned violence perpetrated by the army and the air force, aimed at all civilians who opposed the dictatorial regime of President Bashar al Assad. This ruthless campaign of repression not only triggered a brutal civil war in 2011 that still continues, but also fractured the country in a manner that lured in some of the most radical extremists, who quickly turned the conflict into a regional war. By late 2016, the death toll resulting from this unrelenting carnage was estimated to have exceeded half a million.

And third, where it had appeared that a transformation of government had been achieved—most notably in Egypt—democratic prospects steadily faded as the military conspired to retain *de facto* control over the country. In retrospect, the Arab Spring was fueled in large part by the aspirations of masses of young people yearning for economic progress and democracy; however, without the institutions for a successful transition (such as an independent judiciary, a free press, and a multiple-party electoral system), they faced an uphill struggle. Five years after it began, the short-lived Arab Spring had fully run its course—its initial promise and excitement gradually eroding into crushed hopes, with the inexorable restoration of the business-as-usual autocratic regimes.

Politics and Religion

The Arab Spring was also driven by populist grievances concerning the suppression of religious freedoms. Keep in mind that almost all of the governments in this realm were (and are) secular militaristic regimes that view Islamic parties and organizations with suspicion and as threats to political stability. Religious leaders and activists saw their opportunity as part of the wider revolt that was primarily aimed at the removal of repressive regimes. Thus the protest movement, overall, was both complex and potentially conflicted. The notions of democracy that stirred so many of the young protesters ran counter to the fundamentalist views of reactionary Islamists, a divide that added yet one more obstacle to much awaited social change in this realm.

Religious Revivalism

To the extent that the Arab Spring exposed the vulnerability of some long-entrenched secular regimes, it may have unintentionally fueled Islamic fundamentalism—or, more broadly **religious revivalism [14]**. That resurgence had already been in the making for decades. During the 1970s, the *imams* (mosque worship leaders) in Shi'ite Iran sought to reverse the shah's moves toward liberalization and secularization: they wanted to (and soon did) recast society in a traditional, revivalist Islamic mold. An *ayatollah* (leader under Allah) replaced the deposed shah in 1979; Islamic rules and punishments were instituted. Urban women, many of whom had been considerably liberated and educated during the shah's regime, were forced to return to more traditional roles.

The rise of Islamic revivalism was neither limited to Iran nor confined to Shi'ite communities. Many Muslims—Sunnis as well as Shi'ites—in various parts of the realm disapproved of the erosion of traditional Islamic values, the corruption of society by European colonialists and later by Western modernizers (which brought economic benefits only to the elites), and the declining power of the faith within the secular state. As noted earlier, most governments in the NASWA realm continue to be both secular and repressive, and they do not provide the economic progress their citizens yearn for. Put in those terms, it is not that difficult to understand why many people seek solace in religion as a way to regain hope and dignity.

Some have taken their return to the faith much farther and imbued it with a fanaticism that is alien to most Muslims—and this has set Muslim against Muslim in various parts of the realm. Some revivalists fired the faith with a new militancy, aggressively challenging the status quo from Iran to Algeria. These militants forced their governments to ban "blasphemous" books, to re-segregate schools according to gender, to enforce traditional dress codes, to legitimize religion-based political parties, and to heed the wishes of the *mullahs* (teachers of Islamic ways). Militant Muslims also proclaimed that Western ideas about modernization, inherited from colonialists and adopted by Arab nationalists, were incompatible with the dictates of the Quran.

Terrorism in the Name of Islam

The great majority of Muslims are not fundamentalists; not all fundamentalists are militants; and not all militants are terrorists. Most religious scholars point out that the Islamic faith is no more predisposed toward militancy than other religions. And we should not forget that terrorism is a tool of war that has been used around the world, and indeed throughout modern history, by people of several different beliefs and ethnic backgrounds.

But nowadays the religion of Islam has proven to be a significant vehicle for political mobilization and a powerful source of inspiration for militants with terroristic agendas. Certainly, many Muslim adherents contest this extreme interpretation of the Quran; yet, on the other hand, there is little doubt that militants justify their actions as expressions of their particular fundamentalist beliefs. Theirs is a "holy war"—a *jihad* [15]—and as such it is a deeply reactionary movement that looks to the past rather than to the future.

Broadly speaking, two extremist Islamist organizations are dominant today in spreading terrorism within this realm. The first is *al-Qaeda* (the name means "base" or "foundation"), an

internationally networked organization whose goal is to establish Islamic rule across NASWA. They originated in Afghanistan in the late 1980s and have since masterminded terrorist acts in many other countries, from Kenya to Indonesia to the United States (where al-Qaeda was responsible for the 9/11 attacks). After their leader, Osama Bin Laden, was killed by U.S. Special Forces in Pakistan in 2011, the organization was weakened and forced to decentralize. Currently, al-Qaeda's main "branch" is constituted by *AQAP*—shorthand for *Al-Qaeda in the Arabian Peninsula*. It is especially active in Yemen, where it controls key territory and worked to destabilize that country to the point of its 2015 collapse into chaotic civil war.

The other leading Islamist terrorist organization is ISIS, which has rapidly gained importance since 2013. The acronym *ISIS* stands for *Islamic State of Iraq and Syria*; the organization is also known by the acronyms *IS* (*Islamic State*) and *ISIL* (*Islamic State of Iraq and the Levant*). ISIS is notoriously violent and employs unprecedented tactics of intimidation, its videos of decapitations of "infidels" sending shockwaves around the globe. This highly militarized organization emerged in Iraq after the U.S. invasion of 2003 and initially declared allegiance to al-Qaeda. Thereafter, it steadily assumed a more independent role in Iraq and then expanded into unguarded, sparsely populated eastern Syria following the outbreak of civil war in 2011. Exploiting the chaos of that war and Sunni–Shi'ite schisms in both countries, ISIS ever more effectively rallied Sunnis against secular governments that were dominated by Shi'ites.

By 2015, ISIS had become an established force that controlled sizeable swaths of territory in eastern Syria as well as in northern and western Iraq (**Fig. 6-10**). Arguably, ISIS evolved over a short period of time from a dispersed terrorist organization into a territorial quasi-state (perhaps even a formative **insurgent state [16]**, with the Syrian city of Raqqa serving as its capital). Its armed forces are estimated by the U.S. government to total somewhere between 20,000 and 30,000 fighters (ISIS claims it has 200,000). Strict Islamic (*sharia*) law prevails in its territories; common people and shopkeepers are taxed; and control of oilfields and infrastructure provides sufficient revenues to fund its military operations. As ISIS tries to hold and consolidate its territorial gains, its ambitions in this realm are clearly not limited to parts of Iraq and Syria. Most notably, in March 2015 the organization claimed responsibility for a terrorist attack at a museum in Tunisia's capital that left 21 people dead, 18 of them foreign tourists (or in ISIS parlance, "citizens of Crusader countries"). And far more ominously, the designs of ISIS reach far beyond NASWA: it has declared itself to be nothing less than a **caliphate [17]**—an imperial-scale Islamic government led by a direct successor to the Prophet Muhammad who rules and exerts moral authority over Muslims worldwide. History's most recent caliphate was the Ottoman Empire, whose caliphs ruled from 1299 until abolition in 1922. Were the ISIS caliphate to succeed, it would first be governed by "Caliph Ibrahim"—better known today as the terrorist leader, Abu Bakr al-Baghdadi. In 2015 and through mid-2016, ISIS also claimed responsibility for attacks outside of the realm, most notably in the western European countries of France and Belgium (see Chapter 4).

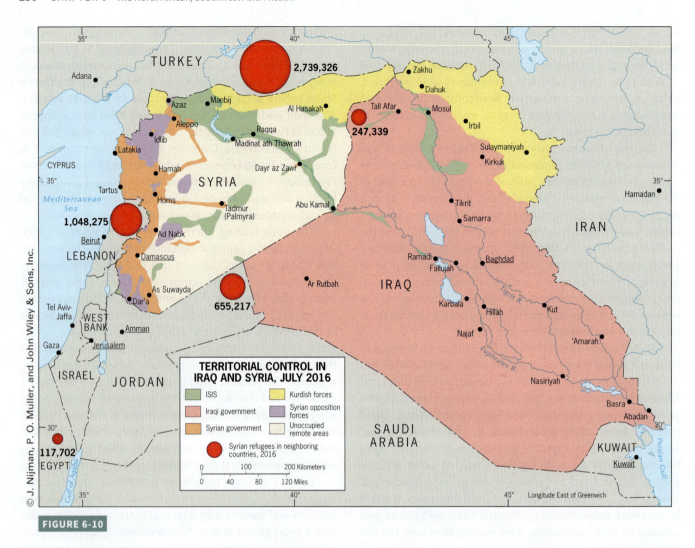

FIGURE 6-10

The threat posed by ISIS in Syria intensified so rapidly that it essentially dissipated much of the international opposition to the Assad regime and its war crimes against hundreds of thousands of Syrian civilians. By the end of 2013, ISIS came to be regarded as the greater danger, and the U.S.-led allied coalition (which included Saudi Arabia, the UAE, the UK, and France) started a campaign of air strikes on ISIS strongholds. Russia, which had emphatically declined to join the U.S. coalition and which had continued its close relationship with the Syrian regime, initiated its own air strikes in 2015 but in a way that clearly supported the Syrian government forces. The Russians bombed both ISIS strongholds as well as other Syrian rebel forces, including Kurdish positions.

Russia's involvement complicated things further on the ground in Syria, but these air strikes together with those of the U.S.-coalition forces, did weaken ISIS. By the summer of 2016, ISIS had lost an estimated one-third of its controlled territories (Fig. 6-10). Unavoidably, a parallel result was the strengthening of the Syrian government's position. For the Syrian people little had changed, and in late 2016 there was no end in sight to this barbaric conflict.

Meanwhile, in Iraq the battle continued between Shi'ite-dominated government forces and ISIS. Here, too, ISIS has lost territorial control during 2015 and 2016. Government forces, with support from the U.S. in the form of air strikes and military training, are also aligned with Kurdish forces in the northeast. In April 2016, Iraqi forces retook the city of Ramadi from ISIS and were gearing up for sieges of Fallujah and Mosul (see Fig. 6-10; photo). Nevertheless, ISIS was continuing to destabilize Iraq through deadly serial bombings of civilians in the capital, Baghdad. We continue the discussion of Syria and Iraq in our coverage of the Middle East region later in this chapter.

Iraqi security forces escort civilians to safe areas in the city of Ramadi (see Fig. 6-10) during days of heavy fighting with ISIS in March 2016. Several weeks later, the city was recaptured from ISIS.

A Realm in Turmoil

The Arab Spring movement failed to achieve the diverse goals of the protesting masses in the streets of Cairo and other cities, but it did destabilize those parts of the NASWA realm in and around the countries that experienced the greatest violence. A good example is Libya, whose political breakdown unleashed a wide spectrum of radical Islamist groups across northern Africa, especially in Mali. And what began as an internal conflict in Syria, soon escalated into a dangerous regional war when it spread into neighboring Iraq—drawing in a wide range of combatants that now include ISIS, Turkey, Iraqi Kurdistan, Iran (traditional supporter of the Shi'ite cause in Syria and Iraq), bordering states swamped by refugee flows (such as Lebanon and Jordan), and the outside powers of the U.S.-led, anti-ISIS coalition. NASWA, and its Middle East region in particular, have a long history of conflict, and the present turmoil is as intractable as ever. As one gloomy Arab author recently observed, history in this part of the world does not move forward: it slips sideways. That view is deeply etched on the realm's ever-changing political map.

NASWA's political instabilities, raging conflicts, social dislocations, and associated economic problems have generated a rising tide of refugee flows in recent years. No longer limited

May 25, 2016: Migrants in a capsizing overcrowded boat in the Mediterranean Sea between Libya and Italy. The Italian Navy reported saving about 500 migrants and recovering seven dead bodies, with the total number of deaths probably considerably higher.

to internal, short-distance, cross-border population shifts, these migratory movements today are increasingly directed toward Europe (also see Chapter 4). Most refugees from the Syrian war are housed in camps in neighboring countries that have been overwhelmed, especially Lebanon where refugees now account for one out of every six inhabitants (Fig. 6-10). To date,

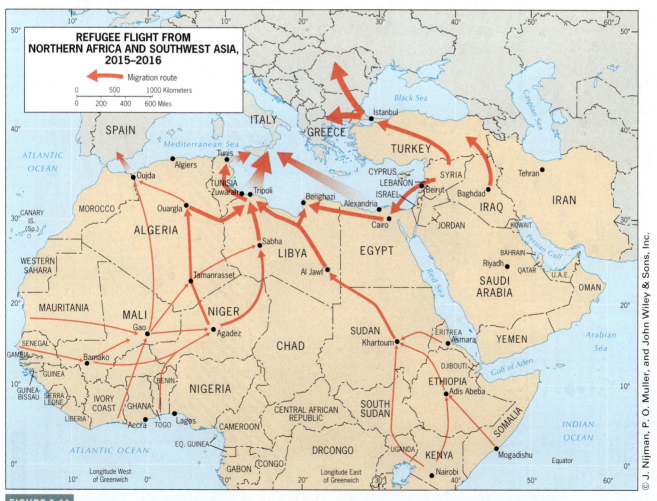

FIGURE 6-11

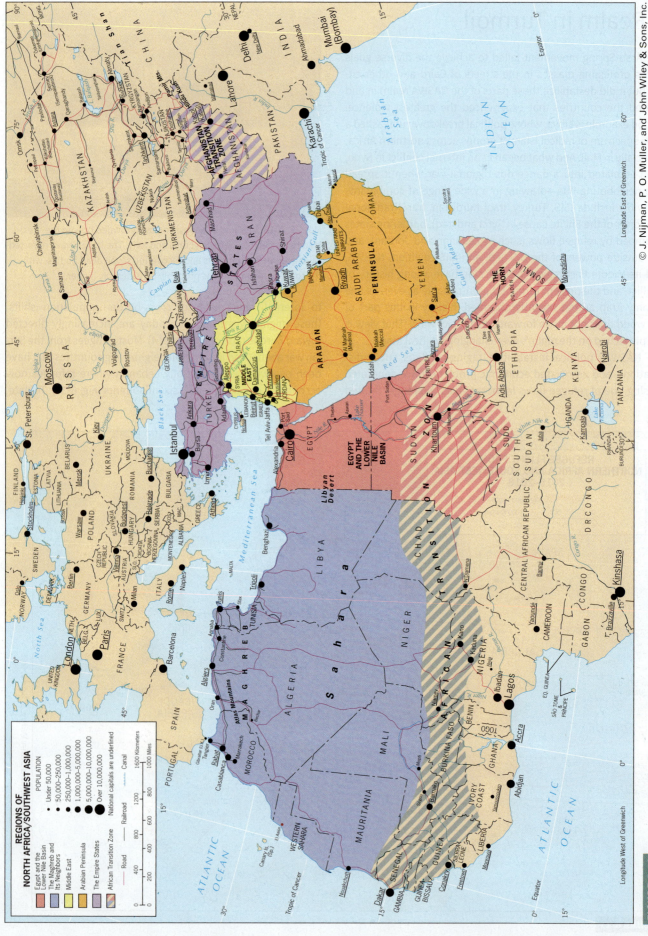

**REGIONS OF
NORTH AFRICA/SOUTHWEST ASIA**

POPULATION
• Under 50,000
• 50,000–250,000
• 250,000–1,000,000
• 1,000,000–5,000,000
• 5,000,000–10,000,000
• Over 10,000,000
National capitals are underlined

Egypt and the
Lower Nile Basin
The Maghreb and
Its Neighbors
Middle East
Arabian Peninsula
The Empire States
African Transition Zone

—— Railroad
—— Road
----- Canal

FIGURE 6-12

tens of thousands of these refugees in adjoining countries have taken flight and attempted to gain entry to Europe. But they are hardly alone, and join broad streams of Europe-bound migrants that originate in a wide range of troubled areas in Southwest Asia, North Africa, and Subsaharan Africa (**Fig. 6-11**).

The number of detected illegal border-crossings into Europe rose from around 400,000 in 2014 to more than 2 million in 2015. The two main refugee routes are either to cross the waters of the central and eastern Mediterranean Sea or to proceed overland through Turkey into Greece or Bulgaria. Most attempts to cross the Mediterranean are made from the coasts of Libya and, to a lesser extent, Tunisia and Egypt; the distance to Lampedusa, Italy's southernmost island south of Sicily, is just over 100 kilometers (65 mi) from Tunis and about

250 kilometers (150 mi) from Tripoli (Fig. 6-11). These desperate illegal migrants, however, are largely in the hands of smugglers, who charge exorbitant fees to cram as many as possible into their rickety boats for the dangerous voyage north.

For most of these refugees, whatever their source, the greatest challenge of their journey begins at the Mediterranean shore. Among those fortunate to have completed the treacherous maritime passage, thousands are refused asylum and are forced to return to their country of origin. Nobody knows how many undocumented migrants perish at sea: in 2015, about one million are known to have crossed the Mediterranean, and the most realistic estimates tell us that many tens of thousands drowned. From every indication, this colossal human tragedy has continued unabated into 2016 (see photo of capsized boat).

Regions of North Africa/Southwest Asia

Identifying and delimiting regions in the sprawling North Africa/Southwest Asia realm is quite a challenge. Population clusters are widely scattered in some areas but highly concentrated in others. Cultural transitions and landscapes—internal as well as peripheral—make it difficult to discern a regional framework. Furthermore, this is a highly changeable realm and always has been. Nonetheless, we can distinguish the following six regions that make up this far-flung realm today (**Fig. 6-12**):

1. ***Egypt and the Lower Nile Basin.*** This region in many ways constitutes the heart of the realm as a whole. Egypt (together with Iran and Turkey) is one of the realm's three most populous countries. It is the historic focus of this part of the world as well as a major political and cultural force. Also included is Sudan, the truncated, northern portion of the much larger former Sudan that split apart in 2011.

2. ***The Middle East.*** This pivotal region includes Israel, Jordan, Lebanon, Syria, and Iraq. In effect, it is the crescent-shaped zone of countries that extends from the eastern coast of the Mediterranean to the head of the Persian Gulf.

3. ***The Arabian Peninsula.*** Dominated by the enormous territory of Saudi Arabia, the Arabian Peninsula also encompasses the United Arab Emirates (UAE), Kuwait, Bahrain, Qatar, Oman, and Yemen. Here lies the source and focus of Islam, the holy city of Mecca; here, too, lie many of the world's greatest oil deposits.

4. ***The Empire States.*** Consisting of two of the realm's giants—Turkey and Iran—we refer to this region as the Empire States because each functioned as the core of a major historic empire.

5. ***The Maghreb and Its Neighbors.*** Western North Africa (known as the *Maghreb*) and the lands that border it also form a region, consisting of Algeria, Tunisia, and Morocco at the center, and Libya, Chad, Niger, Mali, Burkina Faso, and Mauritania along the volatile African Transition Zone where the Arab-Islamic realm of North Africa merges into Subsaharan Africa.

6. ***The African Transition Zone.*** This area is the complicating factor on the regional map of Africa. In Figure 6-12, note that this striped zone of increasing Islamic influence to the north fully incorporates certain countries (e.g., Somalia in the east and Senegal in the

west) while cutting across others, in the process creating Islamized northern sectors and non-Islamic southern ones (e.g., Nigeria, Chad, Ivory Coast). Note, too, that an elongated strip of Islamic dominance extends southward from the Horn of Africa, lining the Indian Ocean coastlands of both Kenya and Tanzania.

Region Egypt and the Lower Nile Basin

Egypt occupies a pivotal location in the heart of a realm that extends more than 9600 kilometers (6000 mi) longitudinally and some 6400 kilometers (4000 mi) latitudinally. At the northern end of both the Nile and the Red Sea, at the eastern end of the Mediterranean Sea, in the northeastern corner of Africa across from Turkey to the north and Saudi Arabia to the east, adjacent to Israel, to Sudan,* and to Libya, Egypt unmistakably lies in the crucible of the NASWA realm. Because it also extends into the Sinai Peninsula, Egypt, alone among countries on the African continent, has a foothold in Asia. This foothold gives it a coast overlooking the strategic Gulf of Aqaba (the northeastern arm of the Red Sea). Egypt also controls the Suez Canal, the vital artificial waterway that links the Indian and Atlantic oceans and forms one of Europe's lifelines. The capital, Cairo, is the realm's largest city and a leading center of Islamic civilization (**Box 6-8**).

Egypt has six subregions, mapped in **Figure 6-13**. Most Egyptians live and work in Lower (i.e., northern) and Middle Egypt, subregions ① and ②. This is the country's core area, anchored by Cairo and flanked by the leading port and second-largest manufacturing center, Alexandria. The other four

---

* *Sudan* refers to the truncated, northern portion of the former country of Sudan that remained after South Sudan (discussed in Chapter 7) achieved independence and split away in 2011. The much larger country that existed before this breakup is referred to in this book as *former Sudan*.

Box 6-8 Among the Realm's Great Cities: Cairo

Stand on the roof of one of the high-rise hotels in the heart of Cairo, and in the distance, to the west, you can see the great pyramids, monumental proof of the longevity of human settlement in this area. The main settlement in ancient times was Memphis, to the south on the map, on the west bank. The present city was not founded until Muslim Arabs chose the site as the center of their new empire in AD 969. Cairo (al Qahira) became and remains Egypt's primate city, situated where the Nile River opens onto its delta, home today to more than one-sixth of the entire country's population.

Cairo is not only the dominant city of Egypt but also serves as the cultural capital of the entire Arab World, with high-quality universities, splendid museums, world-class theater and music, as well as magnificent mosques and Islamic learning centers. Although Cairo has always primarily been a focus of government, administration, and religion, it also is a river port and an industrial complex, a commercial center, and, as it sometimes seems, one giant bazaar. Cairo is truly the heart and soul of the Arab World, a creation of its geography and a repository of its history.

Cairo's population of 15.9 million ranks it among the world's 20 largest urban agglomerations, and it shares with other cities of the less-advantaged realms the staggering problems of crowding, inadequate sanitation, crumbling infrastructure, and substandard housing. The social contrasts can be stunning. Along the Nile waterfront, elegant skyscrapers rise above carefully manicured, Parisian-looking surroundings. Exclusive gated communities are built in new towns (e.g., "Dreamland") that are well connected to the central city but stand apart from it. But look eastward from the center and the metropolitan landscape extends gray, dusty, almost featureless as far as the eye can see. Not far away, more than a million squatters live in the sprawling cemetery known as the City of the Dead. On the urban outskirts, millions more survive in overcrowded shantytowns of mud huts and hovels, while exclusive new residential developments for the middle and upper classes sprawl well into the desert beyond.

The growing population and rising demand from the middle and upper classes for better living environments has resulted in the creation of numerous new settlements within the metropolitan area at a distance from the old city (Box 6-9). New Cairo City to the southeast is the most important of these settlements, because it is the planned new center of the entire metropolis. However, most of the construction of this new heart of **megacity [20]** Cairo has yet to begin.

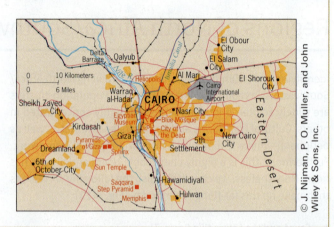

© J. Nijman, P. O. Muller, and John Wiley & Sons, Inc.

Gift of the Nile

subregions are remote and sparsely populated, especially Upper Egypt ③, the Western Desert ④, and the Sinai ⑥. The northern stretches of the Eastern Desert ⑤ are now in the path of Cairo's encroaching urban sprawl (see photo in Box 6-9).

Egypt's Nile is the aggregate of the two great branches upstream: the White Nile, which originates in the streams that feed Lake Victoria in East Africa, and the Blue Nile, whose source lies in Lake Tana in Ethiopia's highlands. The two Niles converge at Khartoum, capital of present-day Sudan. About 95 percent of Egypt's 93.4 million people live within 20 kilometers (12 mi) of the great river's banks or in its delta (Figs. 6-13, 6-2).

Before dams were constructed on the Nile, the ancient Egyptians used **basin irrigation [18]**, building fields with earthen ridges and trapping the annual floodwaters with their fertile silt, to grow their crops. That practice continued for thousands of years until, during the nineteenth century, the construction of permanent dams made it possible to irrigate Egypt's farmlands year-round. These dams, with locks for navigation, controlled the Nile's annual flood, expanded the country's cultivable area, and allowed farmers to harvest more than one crop per year on the same field. The largest of these dams, the Aswan High Dam completed in 1970, created Lake Nasser (the reservoir that extends southward 150 kilometers [100 mi] into northern Sudan) and increased Egypt's irrigable land by nearly 50 percent. Today it still provides a sizeable proportion of the country's electricity.

Much farther upstream, in the western highlands of Ethiopia where the Blue Nile originates, its government is constructing the Grand Ethiopian Renaissance Dam, with more than one-third of the funding coming from China (Fig. 6-13). Note that the location of this dam is near the Sudanese border. It is a common strategy that minimizes possible negative consequences for Ethiopia itself, such as reduced water flow and entrapment of fertile sediments. Instead, such possible consequences, also called **externality effects [19]**, will be felt downstream in Sudan and Egypt. After completion in 2017, the dam is expected to boost electrical power production to meet all domestic demands as well as earn new revenues from sales to Ethiopia's energy-deficient neighbors. But the project has also attracted strong opposition (and raised the potential for conflict) in Sudan and Egypt.

Egypt's elongated oasis along the Nile, just 5 to 25 kilometers (3 to 15 mi) wide, broadens north of Cairo across a delta

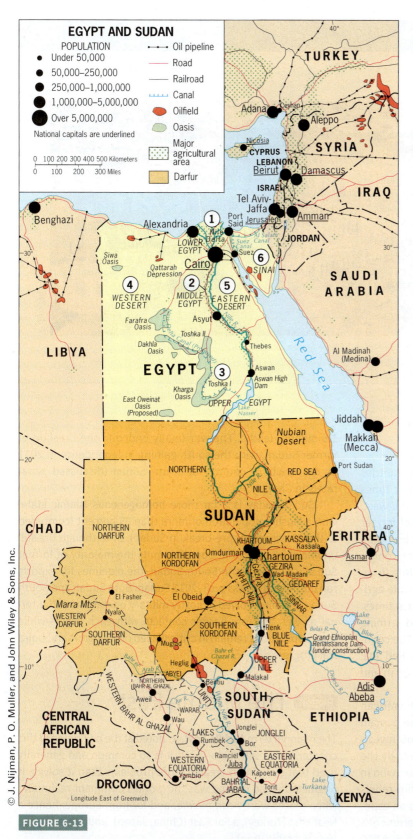

FIGURE 6-13

low-lying delta is also at risk from geological subsidence and sea-level rise, increasing the danger of salt-water intrusion from the Mediterranean that can severely damage soils.

Political Upheaval and Economic Challenges

Before 2011, Egypt had experienced steady economic growth for almost a decade at an annual rate of around 5 percent. At the same time, its annual rate of population growth declined while literacy and per capita consumption increased. Egypt became an exporter of fruits, vegetables, rice, and textiles in addition to its longtime staple, cotton. The country also nurtured a thriving tourist industry that accounted for about one-tenth of GDP. And the exploitation of (modest) oilfields in the Sinai and adjacent Red Sea (Fig. 6-13) resulted in reduced expenditures on fuel imports.

But then came the tumultuous events of the Arab Spring and its aftermath, which devastated the tourist industry and damaged much of the rest of the economy. Despite the efforts of the military-controlled Egyptian government to facilitate a return to stability, annual economic growth rates have hovered around 2 percent since 2011. Politically, it was hoped that Egypt might be moving toward what is sometimes called the **Turkish Model [21]**—an espoused multiparty democracy (though marked by autocratic tendencies) that has a place for Islamic parties, yet is not dominated by them. But, if anything, Egypt has returned to the old ways of repression of both Islamists and more Western-oriented democratic movements. Indeed, some dissidents have indicated to the international news media that the level of repression in 2015 and 2016 was the worst in decades.

Meanwhile, to regain economic momentum, the government has initiated major infrastructural projects to spur employment and reduce popular discontent. In the mid-2010s, the Suez Canal was enlarged to accommodate two-way traffic, double the Canal's capacity (unlike the Panama Canal, there is minimal relief and locks are unnecessary), as well as increase traffic and revenues. Newly constructed suburbs now dominate Cairo's outer metropolitan ring (see **Box 6-9**), and work is beginning on the centerpiece, New Cairo City, just southeast of old Cairo (see **Box 6-8**). Here, with the backing of foreign investors, this showcase urban center is being designed and built in accordance with "principles of sustainability, emphasizing state-of-the-art healthcare, educational, and recreational facilities for its planned five million residents."

anchored in the west by the historic city of Alexandria and in the east by Port Said, gateway to the Suez Canal. The delta contains extensive farmlands, but it is a troubled area today. The ever more intensive use of the Nile's water and silt upstream is depriving the delta of desperately needed replenishment. The

Box 6-9 From The Field Notes…

"Taking off from Cairo International Airport at the northeastern edge of this megacity of nearly 16 million, we flew over one of Cairo's newest housing developments at the edge of the northern sector of the Eastern Desert. Look at this residential density—in a desert! With a birth rate more than twice that of the United States, Egypt's critical problem is the provision of a sufficient water supply—which right now is steadily decreasing. Earlier on this day in November 2009, we took a field trip through this area led by a local geographer. These are middle- and upper-middle-class residential complexes. All those new apartment towers will require water, and so will anything green planted here. Where will all these people work, and how will they get there? Is this sustainable urban development?"

© Jan Nijman

Sudan

As Figure 6-13 indicates, Egypt is bordered by two countries that have recently gone through even more turbulent times than Egypt itself: Sudan to the south, and Libya to the west. Sudan, the northern remnant of the 2011 breakup of former Sudan, is almost twice the size of Egypt (but with less than 45 percent of the population) and lies centered on the confluence of the White Nile (from Uganda) and the Blue Nile (from Ethiopia). Here the capital, Khartoum, anchors a sizeable agricultural area where cotton was planted during colonial times. Almost all of Sudan is desert, with irrigation-based farming along the banks of the While Nile and the Blue Nile. The country also has a 500-kilometer (300-mi) coastline on the Red Sea, where Port Sudan lies almost directly across from Jiddah and Makkah (Mecca) in Saudi Arabia.

Note that a large part of Sudan is situated in the African Transition Zone (Figs. 6-12 and 6-21), which helps to explain some long-running regional conflicts. The triad of southwestern provinces named **Darfur** (Fig. 6-13) are inhabited by the *Fur*, a mix of Arabized pastoralists and settled farmers who in past years apparently sympathized with anti-Muslim rebels in the far south of former Sudan. This pretext led to an especially gruesome conflict with the Khartoum government's military forces in the early 2000s, and as many as 400,000 (mostly civilians) may have died. The southern part of former Sudan, in contrast to the north, was in many ways a typical Equatorial African entity, dominated culturally by Christianity and animist African religions. Ongoing conflicts between north and south were exacerbated by the discovery of oil on contested land. A long, bloody war ensued, with an estimated

1.5 million deaths. The war finally ended in the breakup of former Sudan, with the south gaining independence in 2011 as the newborn country of South Sudan (discussed separately in Chapter 7).

Sudan today is a far more homogeneous Islamic state, with a Muslim population of more than 95 percent (compared to the 68 percent of pre-breakup Sudan). Even though the still unofficial international boundary puts the majority of former Sudan's oil reserves in South Sudan, the pipelines for exports run northward across post–2011 Sudan. Both Sudan and South Sudan remain involved in costly military operations and internal as well as cross-border conflicts with each other—while their populations remain in critical need of economic and political progress.

Region The Middle East

The regional term **Middle East**, as noted at the beginning of this chapter, is not satisfactory, but it is so common and generally used that avoiding it creates more problems than it solves. It originated when Europe was the world's dominant realm and when places were *near, middle,* and *far* from Europe: hence a Near East (Turkey), a Far East (China, Japan), and a Middle East (Egypt, Arabia, Iraq).

Five countries constitute the Middle East (**Fig. 6-14**): Iraq, largest in population and territorial size, facing the Persian Gulf; war-torn Syria, next in both categories and fronting the Mediterranean; Jordan, linked by the narrow Gulf of Aqaba to the Red Sea; Lebanon, whose survival as a unified state is a perennial question; and Israel, Jewish nation in the crucible of the

Islamic domain. The complexity of this region today is heightened by the presence of two non-state entities: Kurdistan and ISIS. Because of the extraordinary importance of this region in world affairs, we focus on its cultural, economic, and political geography.

Iraq's Enduring Importance

Even a glance at the regional map (Fig. 6-14) underscores why Iraq is crucial among the states of the Middle East. With about 60 percent of the region's total area, 48 percent of its predominantly Arab population, and most of its valuable energy and agricultural resources, Iraq is key to the Middle East's fortunes.

Having no less than six neighbors, it is no surprise that Iraq is practically landlocked, and this map shows just how narrow and congested its single outlet to the Persian Gulf is. As a result, a network of pipelines across Iraq's neighbors must carry much of Iraq's oil export volume to coastal terminals in other countries.

Iraq is divided into three cultural domains in which religion, ethnicity, and tradition form the basis of its political geography (Fig. 6-14). The largest population sector is composed of about 20 million Shi'ites (ca. 63 percent of the national total of 37.5 million) in the southeast, whose religious affinities are with neighboring Iran. The next largest is the Sunni minority of some 12 million (about 32 percent) in the north and west. Perhaps the most striking feature on this map is the striped

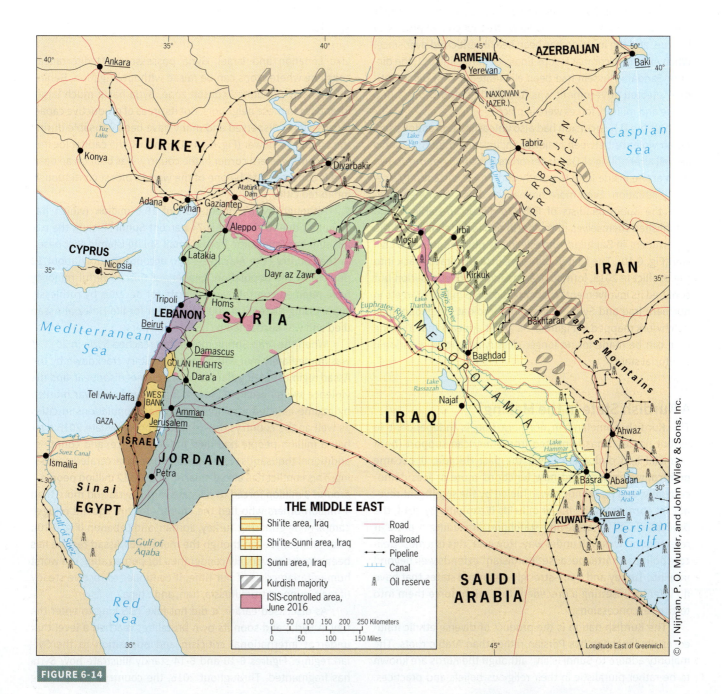

FIGURE 6-14

© J. Nijman, P. O. Muller, and John Wiley & Sons, Inc.

zone that blankets northeastern Iraq, large parts of Turkey and Iran, and a few pockets of northern Syria. In this territory, the majority of inhabitants are not Arabs but Kurds—regarded by many as the world's biggest stateless nation. More specifically, about 6 million Kurds (16 percent of the Iraqi population) live in the northern highlands of Iraq; adjacent southeastern Turkey is the traditional home of at least 15 million more; another 6 million or so inhabit northwestern Iran; and 2 million congregate in the scattered Kurdish clusters of northern Syria. The large majority of Kurds, by the way, are also largely Sunni Muslims.

The U.S.-Led Invasion and Its Aftermath

The United States invaded Iraq in 2003, claiming that it possessed nuclear and other weapons of mass destruction and that al-Qaeda had important bases there. Neither charge proved to be true, and the invasion has been heavily criticized ever since. When American combat troops finally departed in 2011, the dictator Saddam Hussein had been eliminated, a new democratically elected government was functioning, and U.S.-trained police and military forces were in place.

But the U.S. incursion had also unleashed horrendous sectarian violence between majority Shi'ites and minority Sunnis as well as enabling al-Qaeda terrorists (who had not previously been active inside Iraq) to greatly undermine the country's security situation. Even more ominously, these events empowered the Sunni jihadists of ISIS (as discussed earlier), who established themselves in the north and west, challenged the government in Baghdad, and effectively plunged Iraq into civil war (Fig. 6-10). The United States, reluctantly drawn back into the conflict in 2015, formed a large multinational coalition but confined its involvement to air strikes on ISIS military targets in northern Iraq and Syria. Take another look at the geopolitical patchwork mapped in Figure 6-14, and it appears that a strong case can be made for no longer classifying Iraq as a national territorial state.

A Kurdish State in the Making?

The Kurdish homeland in northern Iraq is independent in all but name. It has benefited from American support (both financial and diplomatic) over the past decade; it became politically stabilized and now functions more or less democratically; and its economy was fueled by the significant revenues it earns from the major oilfields it exploits. By 2014, the process aimed at achieving statehood (including a popular referendum) was well under way. However, this is a most delicate political matter because "Kurdistan" extends well beyond Iraq into Turkey and Iran—strongly organized states that have no interest in setting a precedent that might force them into territorial concessions.

The Kurdish nation is the product of diverse ethnic influences. Its language has Persian rather than Arabic roots. The majority adhere to Sunni Islam, although the Kurds are known to be rather pluralistic in their religious beliefs and practices.

Like many other Sunni communities in Iraq and Syria, the Kurds have no sympathy for ISIS, whose violent jihadist offensives continue to be met with fierce resistance from Kurdish armed forces supported by the air power of the U.S.-led coalition. Notwithstanding, in 2015 and 2016 the Kurdish cause has suffered a setback for three reasons. First, the war with ISIS and the latter's territorial control of the city of Mosul and vital nearby oil infrastructure (Fig. 6-14) have undermined Kurdish stability and economic progress. Second, the substantial drop in oil prices since 2014 has severely decreased revenues. And third, the necessary level of political and financial support from the Iraqi government has been lacking because the government itself is dysfunctional. Bottom line: the Kurdish dream has been put on hold.

Syria

Like Lebanon and Israel, Syria possesses a Mediterranean coastline where crops can be raised without irrigation. Behind this densely populated coastal zone, Syria has a much larger interior than these neighbors, but its areas of productive capacity are widely dispersed and many have been unusable during the ongoing civil war (Fig. 6-14). Damascus, the capital located in the southwestern corner of the country, was built on an oasis and is considered to be one of the world's oldest continuously inhabited cities.

Although Syria's population of 18.6 million (which has dwindled due to the war) is 74 percent Sunni Muslim, the ruling elite has long come from a small Shi'ite Islamic sect based in the country—the Alawites—who account for only about 12 percent of the population. Leaders of this powerful minority have retained control over the state for decades by ruthlessly suppressing dissent. In 2000, president-for-life Hafez al-Assad died and was succeeded by his son, Bashar.

When the Arab Spring movement reached Syria in 2011, it was immediately met with a savage military crackdown by the Assad regime. But popular resistance did not disappear, and the violence swiftly escalated into a full-fledged civil war marked by regime-inflicted atrocities on civilian populations in cities as well as the countryside. As noted above, by late 2016 over half a million people had died in the violence, and countless additional thousands were reported to be incarcerated or missing. This conflict has also displaced at least 12 million people—around 8 million forced to relocate within Syria and more than 4 million others who fled the country for the squalid refugee camps of neighboring Turkey, Jordan, and Lebanon (Fig. 6-10). Internationally, even though the murderous Assad regime has been almost universally condemned for perpetrating the worst humanitarian crisis of our time, it continues to receive steady (if tacit) support from Russia, Iran, and China.

As discussed earlier, it did not take ISIS long to enter the fray in Syria, and soon its own brutalities reached a level that undercut international criticism and opposition to the Syrian regime. Figures 6-10 and 6-14 starkly illustrate how Syria has fragmented. Throughout 2016, the country was spiraling deeper into what seemed an unsolvable crisis. Even more than in Iraq, it is extremely difficult to imagine a future restoration of Syria as an integrated national territorial state.

Jordan

With a poverty-stricken capital (Amman), a lack of oil reserves, and possession of only a small and remote outlet to the Gulf of Aqaba, Jordan (population 7.7 million) has survived with U.S., British, and other aid. It lost its West Bank territory in a 1967 war with Israel, including its sector of Jerusalem (then the kingdom's second-largest city). No third country has a greater stake in a settlement between Israel and the Palestinians than does Jordan, especially because Palestinians form the majority here.

Jordan did not experience the kind of upheaval that shook many other Arab countries in 2011 because the monarchy had enjoyed widespread support and government policies lacked the level of repression found in Syria or Tunisia. Yet there were some demonstrations to air grievances, and King Abdullah II had to dismiss his cabinet to quell the protests. Jordan's long-term challenges are primarily economic. The wider regional turmoil has not only deterred tourism (which fell by a third between 2011 and 2015) but brought in more than 600,000 refugees across Jordan's northern border with war-ravaged Syria. It is also important to recognize the remarkable long-term stability of this country within this most volatile of political environments—perhaps comparable to Morocco in this regard, and suggesting that responsible NASWA monarchies have the ability to sense and respond to the pulse of the people. Jordan has exhibited considerable generosity toward Syrian refugees, but the country has suffered a severe decline in tourism (e.g., to the historic site of Petra—see photos at the beginning of the chapter) due to its close proximity to the region's wars.

Lebanon

The map suggests that Lebanon has significant geographic advantages in this region: a lengthy coastline on the Mediterranean Sea; a well-situated capital, Beirut, on its shoreline; oil terminals along its coast; and a major capital (Syria's Damascus) in its hinterland. The scale of Figure 6-14 cannot reveal yet another asset: the fertile, agriculturally productive Bekaa Valley in the eastern interior. Nevertheless, if Lebanon is well endowed from a physical-geographic perspective, its chaotic cultural geography and fragile geopolitical circumstances are problematic. As a result, the country has gone back and forth between good times and bad. Lebanon (population 6 million) is sometimes aptly described as a "garden without a fence" because of its vulnerability to outside interference (especially from neighboring Syria), its latest burden being the more than one million refugees it has taken in from Syria's civil war. This means that in mid-2016 no less than one in five persons in Lebanon was a refugee. With its economy facing severe challenges and so many external powers and influences involved in its affairs, Lebanon is unlikely to achieve stability and significant progress in the foreseeable future.

Israel and the Palestinian Territories

Israel, the Jewish state, lies in the center of the Arab domain (Fig. 6-12). Since 1948, when Israel was created as a homeland for the Jewish people on the recommendation of a United Nations commission, the Arab-Israeli conflict has cast a deep shadow across this entire realm. Figure 6-15 helps us to understand the complex issues involved here. In 1946, the British, who had administered this area in post-Ottoman times, granted independence to what was then called Transjordan, the kingdom to the east of the Jordan River. In 1948, the orange-colored area became the UN-sponsored state of Israel—including, of course, land that had long belonged to Arabs in this territory named Palestine.

As soon as Israel proclaimed its independence, it was attacked by neighboring Arab states. Israel, however, not only held its own but pushed the Arab forces back beyond its borders, gaining the green-colored areas in 1949 (Fig. 6-15). Meanwhile, Transjordanian armies crossed the Jordan River and annexed the yellow-colored area named the West Bank (of the Jordan River), including part of the city of Jerusalem. The king called his newly enlarged country Jordan.

Further conflict was to follow. In 1967, a six-day-long war produced a major Israeli victory: Israel took the Golan Heights from Syria, the West Bank from Jordan, and the Gaza Strip from Egypt; it also conquered the entire Sinai Peninsula all the way to Egypt's Suez Canal (Fig. 6-15). In later peace agreements, Israel returned the Sinai to Egypt but not the Gaza Strip.

All this strife produced a huge outflow of Palestinian Arab refugees and displaced persons. The Palestinian Arabs, therefore, constitute another of this realm's stateless nations; about 1.5 million continue to live as Israeli citizens within the borders of Israel, but more than 2.6 million now reside in the West Bank and another 1.7 million in the Gaza Strip. Many Palestinians also live in neighboring and nearby countries, including Jordan (3.2 million), Syria (630,000), Lebanon (400,000), and Saudi Arabia (280,000); another 400,000 or so reside in Iraq, Egypt, and Kuwait; and some live in other countries around the world, including the United States (ca. 250,000). The global total of the stateless Palestinian population in 2016 was estimated to be 11.8 million.

Israel is approximately the size of Massachusetts and has a population of 8.2 million (including its 1.5 million Arab citizens), but because of its location, its powerful military, and its strong international links, these data do not reflect Israel's importance. Israel has been a recipient of massive U.S. financial aid, and American foreign policy has been to seek an accommodation between Jews and Palestinians, as well as between Israel and its Arab neighbors. Washington continues to strongly push in the direction of a *two-state solution*, but it has been, in the words of U.S. Vice President Biden in 2016, an "overwhelmingly frustrating" endeavor.

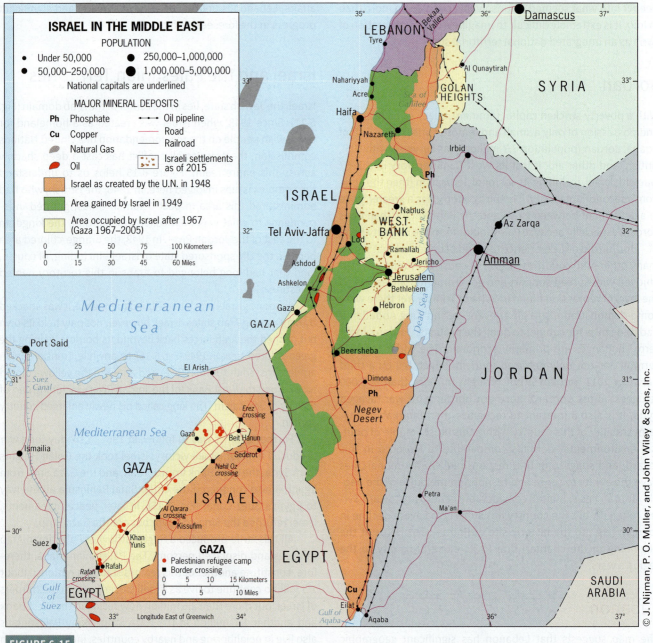

ISRAEL IN THE MIDDLE EAST

POPULATION

- Under 50,000
- 50,000–250,000
- 250,000–1,000,000
- 1,000,000–5,000,000

National capitals are underlined

MAJOR MINERAL DEPOSITS

Ph Phosphate
Cu Copper
Natural Gas
Oil

Oil pipeline
Road
Railroad
Israeli settlements as of 2015

Israel as created by the U.N. in 1948
Area gained by Israel in 1949
Area occupied by Israel after 1967 (Gaza 1967–2005)

0 25 50 75 100 Kilometers
0 15 30 45 60 Miles

Mediterranean Sea

GAZA
Palestinian refugee camp
Border crossing
0 5 10 15 Kilometers
0 5 10 Miles

Longitude East of Greenwich

© J. Nijman, P. O. Muller, and John Wiley & Sons, Inc.

FIGURE 6-15

Geographic Obstacles to a Two-State Solution

1. **The West Bank.** This would be the centerpiece of a future Palestinian state, but its 2.6 million Palestinians are being joined by growing numbers of Israeli Jewish settlers—who in mid-2015 numbered around 370,000, dispersed across a discontinuous network of 237 settlements fortified by walls and/or security fences (**Fig. 6-16**). Moreover, the West Bank is almost entirely bounded by the Israeli-built Security Barrier.

2. **The Security Barrier.** In response to the infiltration of suicide terrorists, Israel is completing the walling-off of the West Bank along the Security Barrier border (see photo) mapped in Figure 6-16. This reinforcement, which in places penetrates deeply into former Palestinian areas, imposes much inconvenience for Palestinians as well as outright hardship for many where it runs

between their homes and farm fields. Palestinians demand that the fences and walls be taken down; Israelis reply that the Palestinian (National) Authority (the interim self-government body) must control the terrorists.

3. **Jerusalem.** The United Nations intended Tel Aviv to be Israel's capital, and Jerusalem an international city (see photo). But the 1948–1949 war ended with Israel holding the western part of the city and Arab forces the eastern sector. Then, in the 1967 war, Israel conquered all of the West Bank, including East Jerusalem; in 1980, the Jewish state reaffirmed Jerusalem's status as its capital, calling on all countries to relocate their embassies from Tel Aviv (some did for a while, but none are there today). Meanwhile, the government redrew the map of the ancient city, building Jewish settlements in a ring around East Jerusalem that would obliterate the old distinction between a Jewish west and an Arab east. This enraged Palestinian

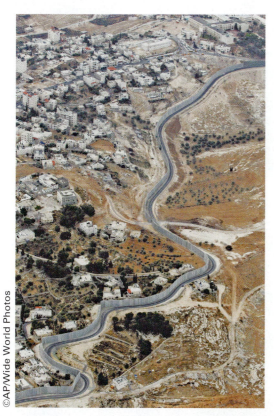

©AP/Wide World Photos

Israel's Security Barrier takes several forms—a concrete wall, a spiked metal fence—and its effect also has multiple dimensions. This section of the Barrier separates the village of Abu Dis on the outskirts of Jerusalem (left) from the West Bank (right). Israel defines this demarcation as a security issue; but the Arabs, citing its location inside Palestinian territory, argue that it is motivated primarily by politics.

leaders, who still view Jerusalem as the eventual headquarters of a future Palestinian state. Renewed U.S. efforts to persuade Israel to halt the construction of new settlements in East Jerusalem have so far been futile. Also note, in Figure 6-16, that the Security Barrier runs well to the east of Jerusalem today.

© Harm deBlij

Jerusalem's urban landscape is studded with cultural contrasts—and is also home to the key holy sites that mean so much to Jews, Christians, and Muslims.

4. **The Gaza Strip.** Israel in 2005 decided to withdraw from the Gaza Strip, removing all Jewish settlements there and yielding control of this territory to the Palestinian Authority (Fig. 6-15, inset map). Gaza has few natural resources, and its near-encirclement by Israel puts it in a most precarious position. Relations between Israel and Hamas (the extremist Palestinian faction that rules Gaza) are severely strained and repeatedly lapse into open conflict—usually in the form of rocket attacks from Gaza into Israel, followed by punishing Israeli retaliation.

Israel lies in the path of an impetuous geopolitical storm. The four obstacles just discussed above are only part of the overall predicament: others involve water rights, compensation for expropriated land, Arabs' "right of return" to pre-refugee abodes, and the exact form a Palestinian state would take.

Region The Arabian Peninsula

The regional identity of the Arabian Peninsula is clear: south of Jordan and Iraq, the entire peninsula is encircled by water. This is a region of (mainly) old-style sheikdoms made fabulously wealthy by oil, modern-looking emirates, and the locale where Islam originated. But it also includes, at its southern periphery on the Gulf of Adan, the economically deprived country of Yemen, which in 2014 collapsed into political chaos, warfare, and state failure.

As a region, the Arabian Peninsula is environmentally dominated by a harsh desert habitat and politically dominated by the Kingdom of Saudi Arabia (**Fig. 6-17**). More than twice the territorial size of Egypt, Saudi Arabia is the realm's second-biggest state after Algeria. Saudi Arabia's neighbors (proceeding clockwise from the head of the Persian Gulf) are Kuwait, Bahrain, Qatar, the United Arab Emirates (UAE), the Sultanate of Oman, and the Republic of Yemen. Together, these countries on the eastern and southern fringes of the peninsula contain 49 million inhabitants; the largest by far is Yemen, with more than 27 million.

Saudi Arabia

Saudi Arabia itself has only 32.2 million residents (plus about 8 million expatriate workers, at least one-fourth of whom are undocumented) in its vast territory, but we are already aware of the Kingdom's importance on the region's oil map (Fig. 6-7): the Arabian Peninsula contains the world's largest concentration of known petroleum reserves. As Figure 6-17 shows, the Saudi reserves lie in the eastern part of the country, particularly in the vicinity of the Persian Gulf, and extend southward into the Rub al Khali (Empty Quarter).

Figure 6-17 also reveals that most of the economic activities in Saudi Arabia are concentrated in a wide belt across the "waist" of the peninsula, from the oil boomtown of Dhahran on the Persian Gulf through the national capital of Riyadh in the interior to the Mecca–Medina anchors of the far west near the Red Sea. A modern transportation and communications network

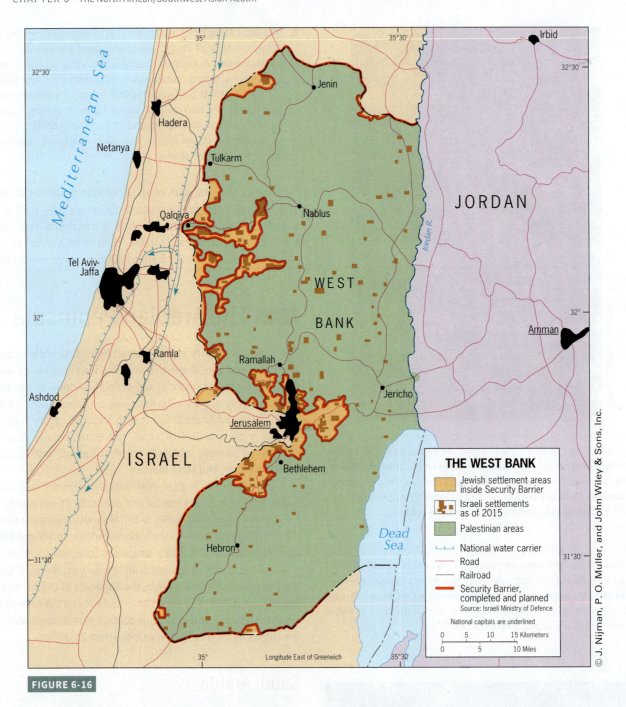

THE WEST BANK

⬜ Jewish settlement areas inside Security Barrier

⬛ Israeli settlements as of 2015

🟩 Palestinian areas

National water carrier

Road

Railroad

Security Barrier, completed and planned
Source: Israeli Ministry of Defence

National capitals are underlined

0   5   10   15 Kilometers
0   5   10 Miles

© J. Nijman, P. O. Muller, and John Wiley & Sons, Inc.

FIGURE 6-16

has recently been completed, but in the more remote zones of the interior, Bedouin nomads still ply their ancient caravan routes across the vast trackless deserts.

Efforts to reduce Saudi Arabia's regional economic disparities have been impeded by the enormous cost of bringing water from deep underground sources to the desert surface to stimulate agriculture, by the sheer size of the country, and by a population growth rate that is still 30 percent above the global average. Nonetheless, housing, healthcare, and education have experienced major improvements, and industrialization also has been stimulated in such newer cities as Jubail on the Persian Gulf and Yanbu on the Red Sea. The latter is part of the historic Mecca–Medina corridor, which is now modernizing rapidly.

Saudi Arabia's archconservative monarchic rule, official friendship with the West, and enormous social contrasts stemming from ongoing economic growth have raised political opposition, for which no adequate channels exist. Consider this: women are still not allowed to drive and were allotted the right to vote for the first time in 2015 (in municipal elections); women cannot even travel today without a male guardian. Even though things remained relatively quiet here, the Saudi regime watched nervously as the Arab-Spring uprisings reverberated across the realm in 2011. By no coincidence, the government was quick to announce a new jobs program involving billions of dollars while beefing up police presence on the streets. But Saudi Arabia's planners have also learned to think much bigger than this, and their grandiose plans now

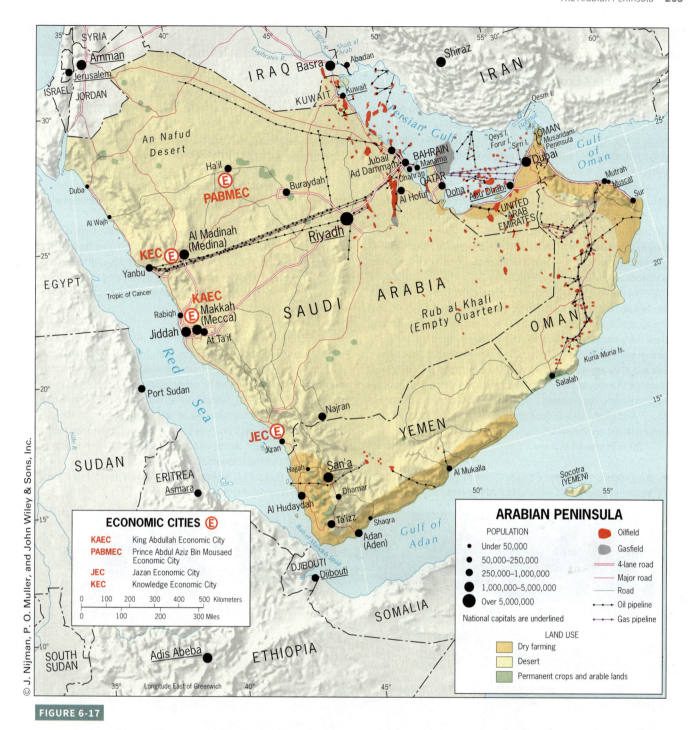

ECONOMIC CITIES Ⓔ

KAEC	King Abdullah Economic City
PABMEC	Prince Abdul Aziz Bin Mousaed Economic City
JEC	Jazan Economic City
KEC	Knowledge Economic City

ARABIAN PENINSULA

POPULATION
- Under 50,000
- 50,000–250,000
- 250,000–1,000,000
- 1,000,000–5,000,000
- Over 5,000,000

National capitals are underlined

- Oilfield
- Gasfield
- 4-lane road
- Major road
- Road
- Oil pipeline
- Gas pipeline

LAND USE
- Dry farming
- Desert
- Permanent crops and arable lands

© J. Nijman, P. O. Muller, and John Wiley & Sons, Inc.

FIGURE 6-17

call for nothing less than a complete economic transformation of the country.

The Saudis have good reason to invest in their economic future. Today, 75 percent of the population is under the age of 30, and unemployment among these young people may run as high as 40 percent. In order to accelerate the generation of the millions of jobs that will be needed, the Saudi royal rulers are spending their country's oil money on four new *economic cities* that by 2020 should create 1.3 million new jobs (see Fig. 6-17; **Box 6-10**). Overall, in anticipation of the end of the oil era, the ultimate goal is to transform Saudi Arabia from a petrodollar-based economy into a world-class industrial power. One

other key ingredient of this plan is the intended sale of all major state-owned oil assets that are now part of government controlled Aramco, considered the most highly valued company in the world. The assets would be sold to multinational oil companies and other investors to create a **sovereign wealth fund [22]** that would amount to $2 trillion, by far the world's largest.

Saudi Arabia is an important player in the geopolitics of this realm and a longtime American ally. Frequently, this partnership involves the writing of checks by Saudi Arabia and the provision of military power by the U.S.—as in Syria, or simply taking the form of Saudi purchases of American weapons. But the United States and Saudi Arabia make a strange pair at

Box 6-10   Regional Planning Case

Saudi Arabia's Economic Cities

Saudi Arabia, the world's richest petro-state, is seeking to diversify its economy for two main reasons. First, its excessive reliance on oil has resulted in a highly skewed labor force and relatively high unemployment, especially among the young. Second, in the longer term, the government is aware that it is just a matter of decades before the oil era will come to an end, either because of depleted reserves or because alternative energy sources will phase out oil revenues.

One of the key components in this strategy of economic diversification is regional planning. Specifically, the government is planning four new *Economic Cities*, each to be developed around at least one globally competitive cluster or industry—from trade to mining and from agribusiness to biotechnology.

The economic geography of the four cities has been carefully worked out to meet three different criteria (see Fig. 6-17). First, they are located in or near existing urban areas in order to benefit from their infrastructure, link up with established economic activities, and provide new opportunities for a growing urban labor force. Second, they are dispersed across the country at a relatively large distance from Riyadh—by far the largest city with more than five million people—to promote a more evenly developed national territory. And third, each Economic City (EC) is strategically located, economically and/or politically.

The four new cities are:

- *King Abdullah Economic City (KAEC)*, the biggest EC, covers 168 square kilometers (65 sq mi); located near the major cities of Mecca and Jiddah; planned to be a hub for global trade as well as a center for industry and investment.
- *Prince Abdulaziz bin Mousaed Economic City (PABMEC)* covers 156 square kilometers (60 sq mi); located near the northernmost city of Ha'il; will become a center of logistics, agribusiness, minerals, and construction materials.
- *Jazan Economic City (JEC)* covers 113 square kilometers (44 sq mi); located in the southwestern corner of Saudi Arabia, near the town of Jizan and close to the Yemeni border; will serve as a manufacturing and raw materials complex.
- *Knowledge Economic City (KEC)*, the smallest of the four, is more like a "science park," focusing on "knowledge-based" industries such as biotechnology and higher education; covers only 4.8 square kilometers (2 sq mi) and is located outside the holy city of Medina and near the modern Red Sea port and industrial city of Yanbu.

As Saudi Arabia seeks to secure its future prosperity, the ambitious economic diversification plan has the potential to be transformative at a number of other levels. First are the physical landscape changes marked by ultramodern real estate and infrastructure development. Second is the badly needed liberalization of business regulations required to attract local and foreign private-sector developers and workers. Third and most interesting are the anticipated cultural changes in view of the relaxation of social rules, and the unavoidable blending of Western and Saudi behavior in these ECs, which some Saudi planners refer to as "controlled islands of change."

© dpa picture alliance archive/AlamyStock Photo

A foreign business delegation, visiting in nearby Rabiqh, Saudi Arabia, takes a look at a model of the planned King Abdullah Economic City.

times, given the latter's highly imperfect democratic record and their clashing interests—for instance, regarding Iran. The Saudis are mostly orthodox Sunnis, and they eye the rising regional influence of Shi'ite Iran with great trepidation. Since 2015, the Saudis have intervened in the escalating conflict in neighboring Yemen (see below) with air strikes to push back Houthi rebels, a Shi'ite militia that was advancing against the Sunni forces of the beleaguered Yemeni government.

The Gulf States

Five of Saudi Arabia's six neighbors on the Arabian Peninsula face the Persian and Oman gulfs: Kuwait, Bahrain, the United Arab Emirates, Qatar, and Oman (Fig. 6-17). All are known for their oil reserves, from which they earn a consequential foreign income. On a GDP per capita basis, these five are all high-income countries; they are also highly urbanized and quite modern in certain ways but traditional in others; and all are monarchies in the (Sunni) Islamic tradition. Their response to popular demands during the short-lived Arab Spring was comparable to that in Saudi Arabia: overly modest political concessions combined with increased public spending to appease the masses. Another similarity is that they each contain a substantial population of foreign workers.

The Gulf States, overall, maintain cooperative relations with the United States. They are generally anti-jihadist, and their economies are becoming ever more globalized. The rise of these states over the past decade or so is the result of a planned diversification away from a singular reliance on oil revenues—similar to what Saudi Arabia is doing. They seek various niches in the global economy, such as Bahrain's emergence as a center of international finance. Most spectacular is the UAE's burgeoning Dubai. This ultramodern city has turned itself into a world-scale air transportation hub (its airport traffic surpassed London's Heathrow in 2014 to become the busiest on Earth) as well as tourist destination—to be crowned by *Dubailand*, the world's largest entertainment complex that will open in 2019. Moreover, these countries are particularly adept at promoting themselves through global public relations campaigns—think

Foreign laborers at work on the construction of Qatar's al-Wakrah Football Stadium in 2015. This is one of nine new stadiums being built for the 2022 World Cup tournament. The decision to award this highest-profile sports event to Qatar has been widely criticized, not least because summer temperatures here are among the hottest on Earth (an average July day in Doha, the capital, records a high of 42°C [108°F] and a low of 31°C [88°F]). The Qataris have responded by promising to build covered, air-conditioned stadiums—yet another extravagance in a Gulf State that seems to have unlimited money to spend. A far more serious criticism, however, has also been leveled at Qatar: the consistently poor treatment of its (overwhelmingly foreign) workforce.

of the high visibility of the UAE's state-supported airlines (Emirates and Etihad) airlines, or the 2022 FIFA World Cup tournament being awarded to Qatar (see photo), home of the global media giant, Al Jazeera. All of the Gulf States, it should be noted, have suffered from the downturn in oil prices over the past few years, experiencing a drop in GDP per capita between 5 percent (Saudi Arabia) and 15 percent (Qatar) between 2009 and 2016. Oil is still critical to these economies.

These five states have learned how to take advantage of their location on the energy-rich Persian Gulf, whose busy oil-tanker traffic must navigate the treacherous Strait of Hormuz outlet, one of the most strategic choke points in the world ocean (Fig. 6-17). At the regional scale, the leading geographic asset of the Gulf States is their central location in Southwest Asia, providing a stable political and economic haven in an otherwise turbulent realm; moreover, none of these small countries even has a hostile neighbor today. Not a single one of these advantages, however, applies to Yemen, the country that occupies the southern corner of the Arabian Peninsula.

Yemen

With a population of 27.5 million—approximately 56 percent Sunni and 43 percent Shi'ite—Yemen is plagued by grinding poverty, centrifugal political forces, and incessant terrorist activity. In the chaotic mid-2010s, an ongoing Shi'ite rebellion against the central government has destabilized the north even as a secessionist movement continues to gain ground in the south, where oilfields encourage such ambitions.

Figure 6-17 highlights the geographic appeal of Yemen for al-Qaeda and other Islamic terrorist groups. Not only does Yemen lie at the back door of Saudi Arabia: its southern tip overlooks another of the world's most crucial and busiest choke points at the Indian Ocean entrance to the Red Sea—the Bab-el-Mandeb ("Gate of Grief") Strait—where ships tightly converge and run the risk of capture by pirates. Across the Gulf of Adan to the south lies another terrorist haven (and pirate stronghold), the failed state of Somalia, and just across the Red Sea to the west lies Eritrea, a likely target for extremists of the al-Qaeda stripe.

The mass demonstrations of Yemen's Arab Spring in 2011 against the 32-year-old regime of President Ali Abdullah Saleh led to his resignation a few months later. But here, too, the removal of an autocratic government was not followed by a democratic turn. On the contrary, Yemen experienced a political transition that mirrored Libya's—a brief spasm of destabilization, an ensuing power vacuum, and the swift emergence of opposing, armed factions.

By 2014, Yemen's government had been rendered powerless and the country descended into anarchy as three opposing forces began their battle for supremacy. One of the combatants is the **Houthi** tribe, based in the country's far west, that is dominated by Shi'ites; another contender is the powerful jihadist branch of **al-Qaeda of the Arabian Peninsula (AQAP)**, which is expanding its territory westward from the central Gulf of Adan coast; the third belligerent force consists of a Sunni-led secessionist movement in Yemen's southwest. The conflict is further exacerbated by foreign involvement: for example, Iran actively supports the Shi'ite Houthis, whereas Saudi Arabia has conducted air strikes against them to shore up what is left of the country's Sunni-dominated government. By 2015, Yemen had joined Syria, Iraq, and Libya in the gloomy NASWA league of failed states.

Region The Empire States

Two major states, both with long imperial histories, dominate the region that lies immediately to the north of the Middle East and Persian Gulf (Fig. 6-12), where Arab ethnicity gives way but Islamic culture endures—Turkey and Iran. Although they share a short border and are both Muslim countries, they display significant differences as well. Even their versions of Islam are different: Turkey is an officially secular but dominantly Sunni state, whereas Iran is the heartland of Shi'ism. Turkey's leaders have worked to establish satisfactory relations with Israel; on several occasions, Iran's leadership promised to "wipe Israel off the map." Turkey has been involved in negotiations aimed at entry into the European Union. Iran's goals have been quite different as that country has defied Western efforts to constrain its nuclear ambitions—though a recent agreement may alter that direction.

These parts of Southwest Asia are a long way from North Africa, and given their location, they are geographically linked to regions outside the realm, particularly Central Asia and westernmost South Asia. For centuries, Turkey and Iran have been involved in what are now, respectively, Turkmenistan and Afghanistan. Iran's southeastern Baluchi nation extends well into southwestern Pakistan, whereas its far northwestern Azeri minority spills over the border into Azerbaijan. Most importantly, Afghanistan constitutes a transition zone between NASWA and South Asia because that war-ravaged country has been tightly enmeshed in the geopolitics and cultures of both realms (see Fig. 6-12). Today, however, it can be argued that Afghanistan is more firmly connected to South Asia, and as such it will be discussed separately in Chapter 8.

Turkey

As **Figure 6-18** indicates, Turkey is a mountainous country with generally moderate relief; it also exhibits considerable environmental diversity, ranging from steppe to highland (see Fig. G-6). On central Turkey's dry Anatolian Plateau, villages tend to be small, and subsistence farmers grow cereals and raise livestock. Coastal plains are not large, but they are productive as well as densely populated. Textiles (from home-grown cotton) and farm products dominate the export economy, but Turkey also has substantial mineral reserves, some oil in the southeast as well as major dam-building projects on the Tigris and Euphrates rivers, and a modest steel industry based on domestic raw materials.

The ancient capital of Turkey was Constantinople (long renamed Istanbul), located on the Bosphorus, a narrow waterway component of the strategic Turkish Straits that connect the Black and Mediterranean seas (**Box 6-11**). But the struggle for Turkey's survival had been waged from the heart of the country, the Anatolian Plateau, and it was here that Kemal Atatürk (modern Turkey's founding leader who ruled from 1923 to 1938) decided to place his seat of government (Fig. 6-18). Ankara, the new capital, possessed some unique advantages: it would remind the Turks that they were (as Atatürk always said) Anatolians, it lay much closer to the center of the country than Istanbul, and it could therefore act as a stronger unifier.

Even though Atatürk moved the capital eastward and inward, his orientation was westward and outward. To implement his plans for Turkey's modernization, he initiated reforms in almost every sphere of life within the country. Islam, formerly the state religion, lost its official status, and Turkey became a secular state whose armed forces ensured that the Islamists would not take over again. The state took over most of the religious schools that had controlled education. The Roman alphabet replaced the Arabic. A modified Western code supplemented Islamic law. Symbols of old—growing beards, wearing the fez—were prohibited. Monogamy was made law, and the emancipation of women was initiated. The new regime emphasized Turkey's separateness from the Arab World, and in many ways it has remained aloof from the affairs that engage other Islamic states.

In present-day Turkey, the divide between traditional Muslims and secular Westerners is quite active. This can be seen in Figure 6-18, which maps support for the country's three leading political parties in the general election of 2015. The AKP (Justice and Development Party) holds a narrow parliamentary majority, has been the governing party since 2002, and is led by President Recep Tayyip Erdoğan; it is a conservative Muslim party, with its power base located in the rural provinces of central Turkey. The CHP is the secular Republican People's Party that was founded by Atatürk in 1923; its support is based in several pockets of western Turkey that lie close to Europe. The HDP is the People's Democratic Party, dominated by Kurds in the southeastern margin of the country, and emphasizes equal rights for minorities. These internal divisions dramatically surfaced in July 2016, when a faction of the military attempted (and failed) to ignite a *coup d'état* to overthrow the government. This rebellion was largely aimed at President Erdoğan, whom they (and others) criticize for his increasingly autocratic behavior and appeasement of conservative Islamic groups.

Turkey and Its Neighbors

From before Atatürk's time, Turkey had a history of intolerance of minorities. Soon after the outbreak of the First World War, the (pre-Atatürk) regime decided to expel the Armenians, concentrated in the country's northeast. Nearly 2 million Turkish Armenians were uprooted and brutally forced out; well over a million died in an eight-year genocidal campaign, the centennial of whose beginning was widely marked in 2015.

Box 6-11 Among the Realm's Great Cities: Istanbul

From both shores of the Sea of Marmara, northward along the narrow Bosphorus waterway toward the Black Sea, sprawls the fabulous city of Istanbul, known for centuries as Constantinople, headquarters of the Byzantine Empire, capital of the Ottoman Empire, and, until Atatürk moved the seat of government to Ankara in 1923, capital of the modern Republic of Turkey as well.

Istanbul's site and situation are exceptional. Situated where Europe meets Asia and where the Black Sea joins the Mediterranean, the city was built on the requisite seven hills (as Rome's successor), rising above a deep inlet on the western side of the Bosphorus, the famous Golden Horn. A sequence of empires and religions endowed the city with a host of architectural marvels that give it, when approached from the water, an almost surreal appearance.

Turkey's political capital may have moved to Ankara, but Istanbul remains its cultural and commercial headquarters. It also is the country's leading urban magnet, luring millions from the countryside. Istanbul's population is doubling every 15 years at current rates, having reached 13.5 million in 2016 to become one of the realm's three largest cities (the other two are Cairo and Tehran). But its infrastructure is crumbling under this massive influx, threatening the legacy of two millennia of cultural landscapes. The city has been expanding swiftly and now covers an enormous, elongated metropolitan region, from the Esenyurt district in the west to Gebze in the southeast, the length of this crescent exceeding 70 kilometers (45 mi).

The heart of the city is known as Stamboul (inset map). Here population densities peak, and this is where you find most of the historic Byzantine and Ottoman architecture (**Box 6-12**). In the "foreign" area north of the Golden Horn called Beyoğlu, modern buildings vie for space and harbor views. Greater Istanbul accounts for about one-third of Turkey's economy. Incomes are much higher here than in most of the rest of the country, but so is the cost of living. Istanbul's youth tends to be cosmopolitan (contrasts with more traditional inhabitants of Turkey's rural areas can be stark), and their opposition to the autocratic tendencies of Turkey's government is expressed regularly in demonstrations in the city's main public spaces, especially Taksim Square.

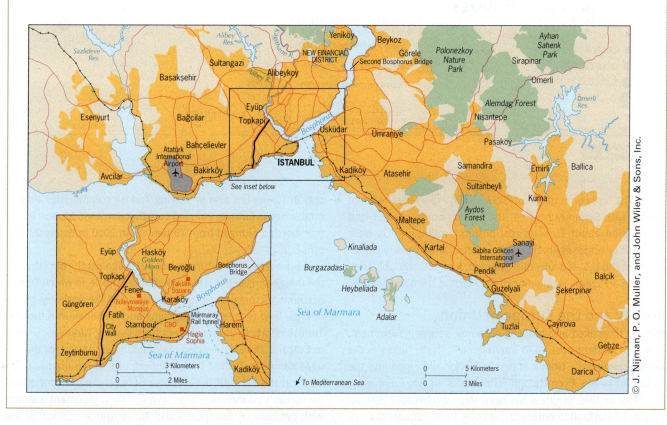

© J. Nijman, P. O. Muller, and John Wiley & Sons, Inc.

In modern times, Turkey also has been widely criticized for its harsh treatment of its large and regionally concentrated Kurdish population. About one-fourth of Turkey's population of 80 million is Kurdish, and successive Turkish governments have mishandled relationships with this minority nation. The historic Kurdish homeland lies in the southeast of Turkey, centered on Diyarbakir (Fig. 6-18), but millions of Kurds have moved to the outskirts of Istanbul—and to jobs in EU countries. With Kurdish nationalism rising just across the border with embattled Iraq, Turkey has responded by suppressing Kurdish insurgent movements while awarding more rights and freedom to those Kurds not involved in armed resistance.

Turkey's overriding foreign policy concern today is Syria. By late 2016, nearly three million Syrian refugees had settled on Turkish soil, and the alleged cross-border movements of Syrian rebel forces had drawn fire from Assad-regime troops inside

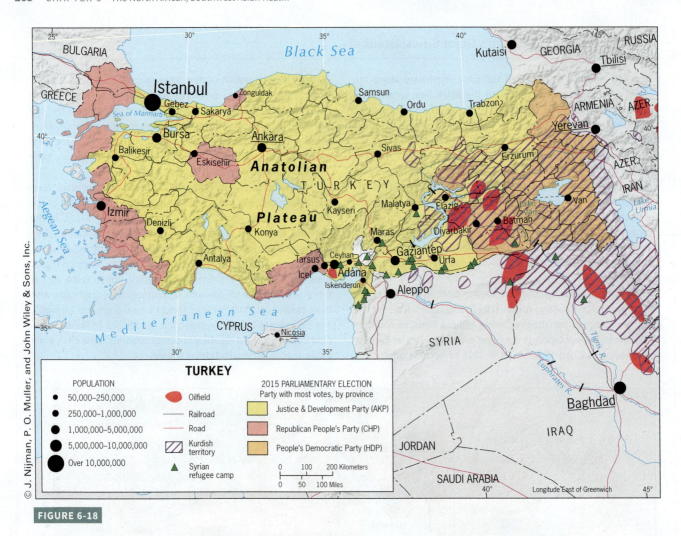

FIGURE 6-18

Syria. The rapid emergence of ISIS in Syria, especially the fierce battles with Syrian Kurds near the Turkish border, has invited speculation about Turkey's real position: although it is clear that Turkey opposes jihadist Islamic forces and formally participates in the U.S.-led anti-ISIS coalition, it would arguably benefit from an overall weakening of the Kurds at the hands of ISIS.

Turkey and the European Union

With its diversified and growing economy, secular government, and independent posture in international affairs, Turkey is the key predominantly Islamic state eligible for admission into the European Union. Even though formal talks on admission were launched in 2005, progress has been very slow because Europeans have been fundamentally divided over Turkey's qualifications. Whereas some focus on the country's hugely important strategic value (Turkey is also a member of NATO—see Chapter 4), others consider it "not European enough," particularly when it comes to so-called European democratic values. Among many Turks, enthusiasm for joining the EU has cooled over the past decade. For one thing, Europe has been forced to deal with severe economic problems, while Turkey's economy has performed comparatively well. For another, many Turks take offense at the unrestrained public debate in Europe concerning their candidacy, often including criticisms of Turkish customs and traditions, not

only in Turkey but also in existing Turkish communities already established in Europe. Polls indicate that domestic support for EU membership plunged from 75 percent in 2000 to barely 27 percent in 2014.

But events took a turn in late 2015, against the backdrop of the refugee crisis that is overwhelming Europe, in which Turkey occupies a crucial geographic position in the pathway of the refugee flow (especially of Syrians and Afghans). In return for Turkish efforts to block refugees from leaving Turkey for Greece or Bulgaria, the EU provided major financial support for the refugee camps in Turkey, the scrapping of visa requirements for Turks in the EU, and the promise of accelerated negotiations for Turkey's EU membership. However, this agreement was met with cynicism by Europeans and Turks who are critical of the undemocratic rule of Turkey's government.

Turkey is a crucial state in a unique geographical position that is trying to steer an independent and singular course within one highly volatile realm while accommodating the demands of another. Its vital importance was dramatically confirmed in 2011 when Arab Spring protesters across NASWA repeatedly espoused the *Turkish Model* as the ideal future scenario for their countries—a "mildly" Islamist government that does not alienate the religiously devout, that walks a fine line on democratic reform, and that prioritizes economic growth.

"Having spent the night in Istanbul's Beyoğlu District on the European side, I walked across the Atatürk (Unkapani) Bridge to the Eminönü District on the south side of the Golden Horn, intending to make it to the upscale Fatih District later. But getting a clear view of the Hagia Sophia 'Museum' was not easy from this angle, standing amid the congested buildup in this hilly area. In a line of taxis at a taxi stand one had a sign saying 'Speak English,' so I asked the driver about getting to a good vantage point, and he showed me on my map how to find a lookout point on the terrace of a local restaurant. From there, I could see the bridge I had crossed to my left, the Hagia Sophia Church with its later minarets in the distance, the Sultan Haseki Hürrem Baths to the right, and in the foreground a simple, local mosque. 'You should know the rules about mosques,' he said. 'The number of minarets reveals the importance of any mosque. One minaret, and it's local, neighborhood. Two, and it's likely to be more prosperous, perhaps the work of a very successful citizen or community. Any more than two, and the builders are required to have not only religious sanction but also state permission to build. And if they get that, they go for four.' Thousands of mosques, many of them architectural treasures, grace the urban landscape of Istanbul."

© H. J. deBlij

Iran

Iran has its own history of conquest and empire, collapse and revival. Oil-rich and vulnerable, Iran already was just a remnant of a once-vast regional empire that waxed and waned, from Europe's Danube Basin to the Indian subcontinent's Indus Valley, over more than two millennia. Long known to the outside world as *Persia* (although the people had called themselves Iranians for centuries), the state was officially renamed *Iran* by the reigning shah (king) in 1935. His intention was to stress his country's Indo-European cultural heritage as opposed to its Arab and Mongol infusions. But for all the rhetoric about modernization, the shahs failed to advance the lot of the common people, and by 1979 a revivalist revolution had engulfed the country. The monarchy was replaced by an Islamic republic ruled by an ayatollah (leader under Allah), and Islamists continue to tightly control this Shi'ite bastion today.

A Crucial Location and Dangerous Terrain

As Figure 6-12 shows, Iran occupies a critical position in this tempestuous realm. It controls the entire corridor between the Caspian Sea and the Persian Gulf. To the west it adjoins Turkey and Iraq, both historic adversaries. To the north (west of the Caspian Sea) Iran borders both Azerbaijan and Armenia, where Islam meets Christianity. To the east Iran borders Afghanistan and Pakistan, and to the northeast lies similarly volatile Turkmenistan.

As **Figure 6-19** reveals, Iran is a country of mountains and deserts. The heart of it is an extensive upland, the Iranian Plateau, that lies surrounded by even higher mountains, including the Zagros in the west, the Elburz in the north along the Caspian Sea coast, and the mountains of Khorasane to the northeast. This mountainous terrain also signals geologic danger: here the Eurasian and Arabian tectonic plates actively converge, causing major and often devastating earthquakes (Fig. G-4). The Iranian Plateau is actually a huge highland basin marked by salt flats and wide expanses of sand and rock. The highlands wrest some moisture from the air, but elsewhere only oases interrupt the arid monotony—oases that for countless centuries have been stops on the area's caravan routes.

The Ethnic-Cultural Map

Today, Iran's population of 80 million is 74 percent urban, and the capital, Tehran (population 13.7 million), lies far to the north on the southern slopes of the Elburz Mountains, at the heart of modern Iran's core area.

Iranians dominate Iran's affairs, but the country is no more ethnically unified than Turkey. Figure 6-19 shows that three corners of Iran are inhabited by non-Iranians, two in the west and one in the southeast. Iran's northwestern corner is part of greater Kurdistan, the multistate territory in which Kurds are in the majority and of which southeastern Turkey also is a part (Fig. 6-12). Between the Kurdish area and the Caspian Sea the majority population is Azeri, who number more than 16 million in Iran and also form the ethnic majority (of about 9 million) in the neighboring state of Azerbaijan to the north. In the southwest, concentrated in Khuzestan Province at the head of the Persian Gulf, an Arab cluster predominates. A significant part of Iran's huge oil reserve is located here (Abadan is the main export terminal, shown in Fig. 6-19), but this Arab community is among Iran's poorest and most restive—in a province whose stability could not be more essential to the country's economy. The third ethnically non-Iranian corner lies in the southeast, where a coastal Arab minority focused on the key port of Bandar-e-Abbas on the Strait of Hormuz gives way eastward to

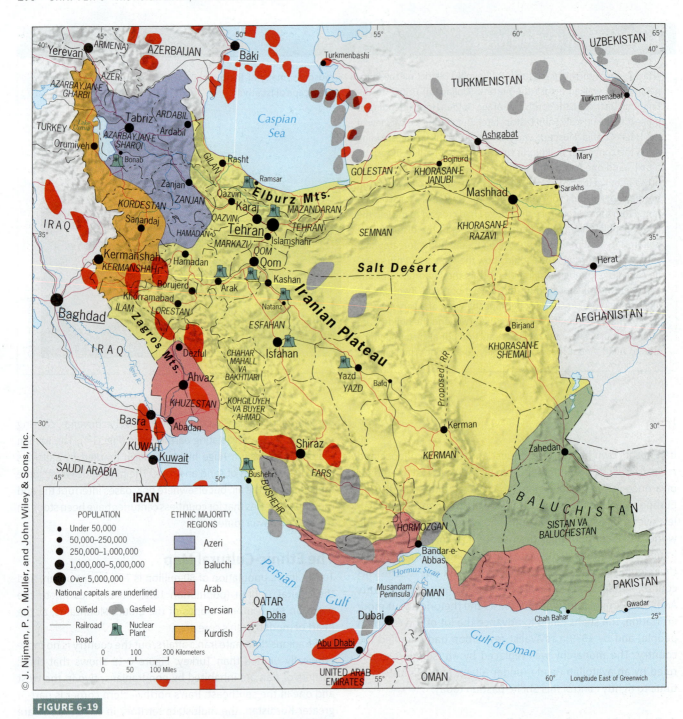

© J. Nijman, P. O. Muller, and John Wiley & Sons, Inc.

IRAN

POPULATION

• Under 50,000
• 50,000–250,000
● 250,000–1,000,000
⬤ 1,000,000–5,000,000
⬤ Over 5,000,000

National capitals are underlined

🔴 Oilfield   ⬤ Gasfield
— Railroad   ⚛ Nuclear
— Road   Plant

ETHNIC MAJORITY
REGIONS

■ Azeri
■ Baluchi
■ Arab
■ Persian
■ Kurdish

0   100   200 Kilometers
0   50   100 Miles

FIGURE 6-19

a scattered and still-nomadic population of Baluchis, who have strong ties to fellow ethnics in Balochistan Province just across the border of neighboring Pakistan. The Iranian component of the Baluchi homeland is a remote and inconsequential area, but in Pakistani Balochistan that country's government faces a persistent rebellion by tribes who feel slighted by official neglect.

Iran's Choices

Iran is a crucial player in the geopolitics of Southwest Asia. Its regional ambitions are hardly surprising. Tehran has a major stake in developments in Iraq and Afghanistan, both neighbors; it has an already-nuclear and unstable Pakistan on its southeastern border; it has a strong interest in the fate of Shi'ite majorities and minorities elsewhere in the realm; it has long been an avowed enemy of Israel and a strong supporter of Palestinian causes; and it has in the past funded organizations labeled as terrorist whose actions have even reached across the Atlantic. In 2016, Iran played an important role in Iraq, providing ground military support to (Shi'ite-dominated) government forces in their battles with ISIS.

But Iran has also suffered for decades from the effects of Western economic sanctions because of its efforts to acquire nuclear capabilities. It possesses the realm's second-largest concentration of oil reserves, which amount to fully half of those in Saudi Arabia. Iran is overly dependent on oil and lacks a diversified economy. Slow economic development and the traditional disposition of the Islamic state (even under the ayatollahs) have kept most Iranians in a state of poverty and backwardness. And yet, young urban Iranians can be surprisingly modern (and Western) in their outlook; there is a palpable yearning among them for openness, democratization, and economic progress.

Iran's worsening economic plight, accentuated by those international embargoes as well as rising domestic discontent, has made the regime ever more receptive to voluntarily downsizing its nuclear ambitions in exchange for an end to sanctions. On July 14, 2015, just such a pact was agreed upon by Iran and the United States (partnered by China, Russia, Germany, France, and the UK).

Region The Maghreb and Its Neighbors

The countries of northwestern Africa are collectively called the **Maghreb**, but the Arab name for them is more elaborate than that: *Jezira-al-Maghreb*, or "Isle of the West," in recognition of the great Atlas Mountain range rising like an enormous island from the Mediterranean Sea to the north and the sandy flatlands of the immense Sahara to the south (see Fig. 6-1).

Whereas Egypt is the gift of the Nile, the Atlas Mountains facilitate the settled Maghreb. These high ranges wrest from the rising air enough orographic rainfall to sustain life in the intervening valleys, where good soils support productive farming.

From the vicinity of Algiers eastward along the coast into Tunisia, annual rainfall averages greater than 75 centimeters (30 in), a total more than three times as high as that recorded for Alexandria in Egypt's delta. Even 240 kilometers (150 mi) inland, the slopes of the Atlas still receive over 25 centimeters (10 in) of rainfall. The effect of this topography can even be read on the world climate map (Fig. G-6): where the highlands of the Atlas terminate, aridity immediately takes over.

The Atlas Mountains are aligned southwest-northeast and begin inside Morocco as the High Atlas, with elevations over 4000 meters (13,000 ft)(**Box 6-13**). Eastward, two major ranges dominate the landscapes of Algeria proper: the Tell Atlas to the north, facing the Mediterranean, and the Saharan Atlas to the south, overlooking the great desert. Between this pair of mountain chains, with each consisting of several parallel ranges and foothills, lies a series of intermontane basins markedly drier than the northward-facing slopes of the Tell Atlas. Within these valleys, the **rain shadow effect [23]** of the Tell Atlas is reflected not only in the semiarid, steppe-like natural vegetation but also in land-use patterns: pastoralism replaces cultivation, and stands of short grass and shrubs blanket the countryside.

The Maghreb Countries

The countries of the Maghreb include Morocco, the last remaining North African kingdom; Algeria, a secular republic besieged by religious-political problems; and Tunisia, smallest and most Westernized of the three (**Fig. 6-20**). **Morocco**, a relatively

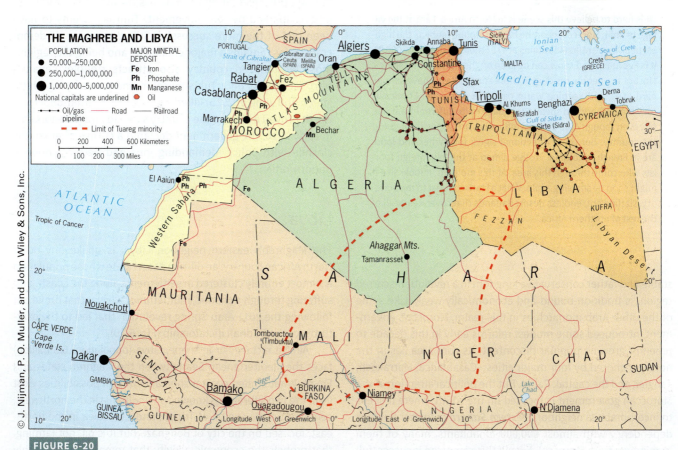

FIGURE 6-20

© J. Nijman, P. O. Muller, and John Wiley & Sons, Inc.

Box 6-13 From the Field Notes...

© Jan Nijman

"It is hard to believe this scene is in North Africa. I am at the village of Imlil, Morocco, near the base of Mount Toubkal, the highest peak in the Atlas Mountains at 4167 meters (almost 14,000 ft). It is February, and snow can be seen on the north-facing slopes. Most of the people residing in these small villages have simple, hard lives, and keeping warm can be a challenge. Ethnically, they are Berbers and speak a language quite different from Arabic, even though Arab cultural influences are strong throughout the country. But these directional signs, for shelter and food, are in French, a colonial vestige aimed at the few foreign tourists who come through this area. At this moment, it looks like it might soon be snowing again. What a stark contrast to the bone-dry, sizzling hot, endless desert flatlands that cover nearly all of the rest of northern Africa."

stable and rather conservative kingdom in a revolution-scarred region, is tradition-bound and economically weak. Like some of the other Arab monarchies in this realm, Morocco's government introduced some modest reforms early in this decade to preempt growing popular discontent. Its core area lies in the north, anchored by four major cities—Casablanca, Marrakech, Tangier, and the capital, Rabat. But the political attention of the Moroccan government is focused on the south, where it seeks to incorporate its neighbor, **Western Sahara** (a former Spanish dependency with almost 600,000 inhabitants, many of them immigrants from Morocco). Even if this campaign is successful,

it will do little to improve the lives of most of Morocco's 35 million people, of whom hundreds of thousands have migrated to Europe. Of greater value to Morocco would be the free access of their (mostly agricultural) products to European markets.

Algeria, France's one-time colony whose major agricultural potential drew more than one million European settlers to its arable coastal strip known as the **Tell**, now has an economy based primarily on its substantial oil and natural gas reserves. The country's primate city, Algiers, is centrally situated within the densely settled corridor lining the Mediterranean that contains most of Algeria's 40 million inhabitants (Fig. 6-2). But a sizeable contingent of Algerians—officially estimated at about 2 million—has emigrated to France, where they form one of Europe's largest Muslim population sectors. Hard-nosed military control combined with generous social spending kept the Arab Spring movement at bay. But jihadists affiliated with ISIS have become active in the remote Sahara of southeastern Algeria, where they took control of some gasfields and beheaded a French hostage, arousing much indignation in Algiers, France, and beyond.

The smallest of the Maghreb states, **Tunisia**, lies at the eastern end of the region. Tunisia in many ways outranks surrounding countries: it has a higher urbanization level, more favorable social indicators, and a much lower growth rate (among its population of 11.4 million) than elsewhere in the Maghreb. Most of the country's productive capacity lies in the north in the hinterland of its historic capital, Tunis. As noted previously, this country took center stage as the source of the Arab Spring uprisings that diffused across NASWA. Of all the revolutions sparked by this movement, Tunisia's seems to have been one of the more successful. Still, preserving its democratic gains remains a major challenge, and being the region's most Westernized country also attracts trouble: separate terrorist attacks on a Tunis museum and a nearby beach (in March and June of 2015) killed a total of 59 international tourists. It proved a blow to the country's already fragile economy. In early 2016, the government declared a curfew after youths clashed with police, demanding jobs and better economic opportunities.

Libya

The Maghreb's eastern neighbor, Libya, is unlike any other North African country: an oil-rich desert state whose population is almost entirely clustered in settlements along the coast—and suffering through a chaotic, multiethnic civil war that broke out following the 2011 Arab Spring revolution that led to the overthrow of the Qadhafi dictatorship (Fig. 6-20).

Almost rectangular in shape, Libya (population: 6.3 million) faces the Mediterranean Sea between the Maghreb states and Egypt. What limited coastal-zone agricultural possibilities exist lie in the two subregions known as Tripolitania in the northwest, centered on the capital of Tripoli, and Cyrenaica in the northeast, focused on the city of Benghazi. But it is oil, not farming, that propels the economic activity that remains in this highly urbanized country. The oilfields are located well inland from the central Gulf of Sidra, linked by pipelines to coastal terminals. Libya's two interior corners, the desert Fezzan district in the southwest and the Kufra oasis in the southeast, are connected to the coast by two-lane roads subject to frequent sandstorms.

Muammar Qadhafi had been one of the realm's most violent despots, clinging to power for more than four decades. His brutal ground and air attacks on antiregime demonstrations in several Libyan cities in early 2011 were hardly unexpected. Within a few months, however, advancing rebel forces captured a number of cities including Benghazi, which became their main base of operations. Assisted by NATO air support, which enforced a no-fly zone that grounded the government's military aircraft, the rebels prevailed; they soon tracked down and killed Qadhafi, who had fled Tripoli to take refuge in his hometown of Sirte.

But even though the Libyan people had rid themselves of a despised tyrant, their country collapsed into political chaos in late 2011 because a new democratic government could not be formed. Local tribes, ethnic minorities, and jihadists of various stripes, long suppressed under Qadhafi, swiftly reemerged to contest each other's turf, but only succeeded in fragmenting the country (Fig. 6-20). Even ISIS gained a territorial foothold, around the coastal city of Sirte, in proximity to oil installations and refineries. The beheading in 2015 of 21 Egyptian Coptic Christians in a jihadist-dominated eastern border town met with immediate reprisals by Egypt's air force, but in the summer of 2016 that area was still ISIS-controlled. Unmistakably, Libya today is another of NASWA's failed states.

Adjoining Saharan Africa

Between the chain of North Africa's coastal states and the southern margins of the Sahara lies an east-west tier of five states, all but one of them landlocked, dominated by the world's greatest desert, sustained by modest ribbons of water, and thoroughly under the sway of Islam: Mauritania, Mali, Niger, Burkina Faso, and Chad (Fig. 6-12). Although their huge overall territory is sparsely populated (Fig. 6-2), the populations of these countries total just over 76 million, not insignificant by African standards. Furthermore, all five are at least partially located within the *African Transition Zone*—a region bedeviled by awesome political and environmental challenges (as we shall shortly see).

The destabilizing effects of Libya's breakdown are especially felt in **Mali**, territorially the size of Texas and California combined, with a population of 18 million that mostly inhabits the country's southern half. Until recently, Mali was cited by many as an African democratic model; but that ended in 2012 when the country descended into chaos after a military coup ousted its president. The military was said to be frustrated with the government's handling of the Tuareg insurgency in the sparsely populated far north (Fig. 6-20). But the coup sparked a rebel surge, and shortly thereafter Mali was effectively split in two. A Tuareg-Jihadist alliance took control of the fabled city of Timbuktu—located near the center of this hourglass-shaped country—and declared that the north was now an independent state named *Azawad*. In 2013, France sent thousands of its troops to help its former colony recapture Timbuktu, did so, and then turned over the effort to restore central-government rule over northern Mali to a UN-mandated, all-African military force. That approach proved ineffective, and by 2015 sporadic fighting had resumed between government and rebel forces in the country's pivotal central waist. Throughout 2016, UN peacekeeping troops were involved in repeated violent clashes with Tuareg-Jihadist forces.

Region The African Transition Zone

Mali's well-populated southern half, which contains its core area, lies in the final region to be discussed in this chapter—the African Transition Zone (Fig. 6-12). From the Atlantic coast of Mauritania and Senegal in the far west to the easternmost Horn of Africa that protrudes into the Indian Ocean, a semiarid belt of land, about 500 kilometers (300 mi) wide, lies between the bone-dry Sahara to the north and the more humid savanna zone to the south. This is the *Sahel* (meaning "shore"—with the bordering Sahara analogous to an ocean), where steppe (*BS*) climates separate desert (*BW*) from tropical savanna (*Aw*) climates (see Fig. G-6).

Life in this environmentally stressed belt of steppelands is quite unpredictable. During the nine-month dry season, there is virtually no rain and temperatures can rise to 50°C (122°F). When the erratic midsummer rains finally come, some areas receive a modest amount while others are flooded. In wet years the harvest can be good, but at other times it is almost nonexistent, particularly during the Sahel's recurrent droughts. Thus human–environment relationships here are always fragile, and it is not surprising that, according to UNICEF, malnutrition kills about a quarter-million children annually in the five westernmost Sahelian countries. But the problems of the Sahel go far beyond climate and food provision.

As both Figures 6-12 and **6-21** indicate, the African Transition Zone is unlike the five other regions of this realm. This is where Subsaharan Africa's cultures to the south intersect with those of the Muslim domain to the north. This Islamic penetration into Subsaharan Africa engulfs some countries in their entirety (such as Senegal and Burkina Faso) and divides others into Muslim and non-Muslim sectors (Nigeria, Ivory Coast, and Chad). As elsewhere in the world where certain geographic realms meet and overlap, complications abound within the African Transition Zone. In some countries, the geographic transition from Muslim to non-Muslim society is gradual, as in Sierra Leone and Ivory Coast. In other areas, the cultural divide is far more abrupt, as in eastern Ethiopia and the two halves of former Sudan, where the border between Christian/animist African cultures and Islamic African communities is practically razor-sharp. The southern border of the African Transition Zone,

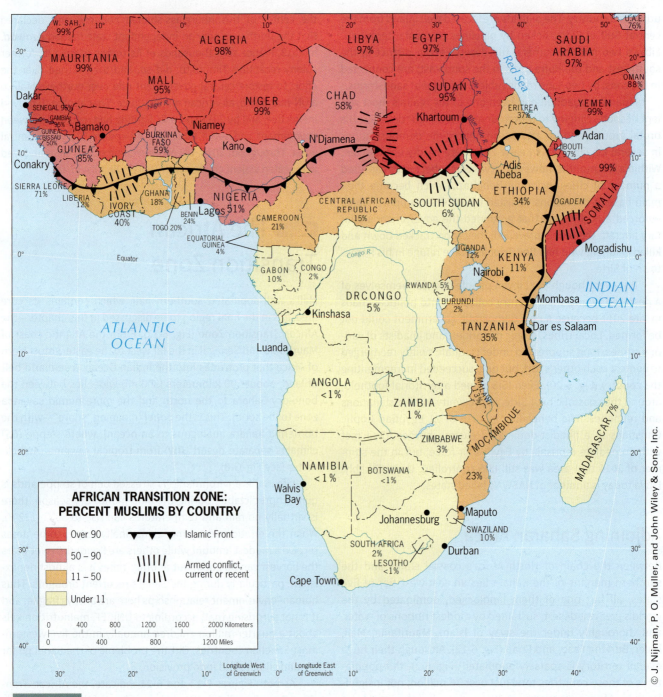

AFRICAN TRANSITION ZONE: PERCENT MUSLIMS BY COUNTRY

Over 90	Islamic Front
50 – 90	Armed conflict, current or recent
11 – 50	
Under 11	

FIGURE 6-21

© J. Nijman, P. O. Muller, and John Wiley & Sons, Inc.

the religious frontier we refer to as Africa's **Islamic Front [24]** (mapped in Fig. 6-21), is therefore neither static nor advancing uniformly.

Conflict marks most of the African Islamic Front: in former Sudan, where a decades-long war for independence by non-Islamic Africans in the south against the Arabized, Muslim north—which finally resulted in the 2011 creation of South Sudan—cost millions of lives; in Nigeria, where violent Islamic revivalism continues to severely threaten that country's prospects; more recently in Ivory Coast, where political rivalries reawakened religious strife in a once stable country; and in the

Horn of Africa—nowhere more than in Somalia, the fifth in the category of failed states in this realm (also see the discussion of certain countries in the African Transition Zone in Chapter 7, which covers the Subsaharan African realm).

The Horn of Africa

The most potentially explosive subregion of the African Transition Zone is the Horn of Africa, which consists of Somalia, Djibouti, Ethiopia, and Eritrea (**Fig. 6-22**). Once again, you will find some

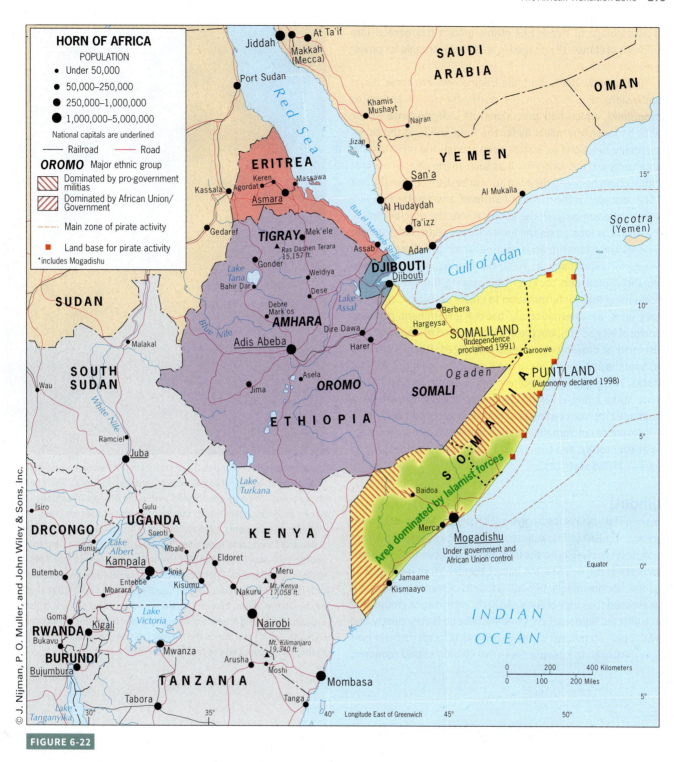

HORN OF AFRICA

POPULATION
- • Under 50,000
- • 50,000–250,000
- ● 250,000–1,000,000
- ● 1,000,000–5,000,000

National capitals are underlined
— Railroad   — Road
OROMO Major ethnic group

⬛ Dominated by pro-government militias

⬛ Dominated by African Union/ Government

–·– Main zone of pirate activity

🟥 Land base for pirate activity

*includes Mogadishu

FIGURE 6-22

discussion of the Horn that is more pertinent to Subsaharan Africa in Chapter 7. We will focus here on the first three countries. Regarding the fourth, the small state of **Eritrea** (about the size of Ohio), suffice it to say that it came into existence in 1993 when it separated from Ethiopia, with whom it still engages in boundary disputes and clashes that have damaged the economies of both countries. Eritrea's early hopes for representative government and economic progress were dashed by war, dictatorial rule, and, most recently, its involvement in Islamic sectarian strife.

Somalia

The key component in the eastern sector of the African Transition Zone is Somalia, where 11 million people, almost all of them Muslim, live at the mercy of a desert-dominated environment that requires cross-border migration into Ethiopia's easternmost Ogaden zone in pursuit of seasonal pastures. As many as 5 million Somalis reside permanently on the Ethiopian side of the border, but this is hardly the only such division the Somali "nation" faces. Somalia's cultural geography comprises an assemblage of five major ethnic groups fragmented into hundreds of clans—all engaged in an endless struggle for power as well as survival.

In the early 2000s, Somalia fragmented into three major subdivisions (Fig. 6-22). In the north, the sector known as **Somaliland**, which had proclaimed its independence in the 1990s and which remains by far the most stable of the three, essentially functions as a state even though the international community will not recognize it as such. In the east, a conclave of local chiefs declared their territory to be distinct from the rest of Somalia, gave it the name of **Puntland**, and asserted varying degrees of autonomy. In the south, where the official capital, Mogadishu, is located on the Indian Ocean coast, local secular warlords and Islamic militias continue their grim, long-running contest for supremacy. In 2006, the Islamic militias stormed into the capital and took control, ousting the warlords and proclaiming their determination to create an Islamic state. Six years later, an internationally backed government regained control of Mogadishu, but the al-Qaeda-linked **al-Shabaab** Islamist terrorist movement it ousted reemerged strongly in southern Somalia and continues to use it as a base to launch its bold attacks. The government receives American support, and a U.S. drone strike killed al-Shahaab's leader in 2014. Nonetheless, the jihadist mosaic continues to blanket and control large territorial expanses of Somalia, political stability is as hard to imagine as ever today, and the country continues to be regarded as a classic **failed state [25]**.

Djibouti

This ministate counts less than a million people and covers a territory about the size of New Jersey, but it is increasingly significant because it lies directly across from Yemen as it overlooks the narrow entry to the Red Sea, the Bab-el-Mandeb Strait (see Fig. 6-22 and the satellite image). Djibouti is fully 97 percent Muslim, but it has escaped the kinds of entanglements with jihadist groups that have afflicted Somalia. Economically, Djibouti is very closely tied to Ethiopia—90 percent of the shipments through its main port come from or go to Ethiopia. Given that country's rapid economic

Jacques Descloitres, MODIS LandRapid Response Team, NASA/GSFC

The narrow Bab-el-Mandeb Strait connects the Red Sea (upper left) to the Gulf of Aden arm of the Indian Ocean (lower right). This classic choke point—the only one located within the African Transition Zone—accommodates one of the world's most crucial sea lanes, especially for oil shipments. With the unstable African Horn on the western shore and failed-state Yemen on the eastern side, it is not hard to envision the Strait's surrounding turmoil escalating into attempts to disrupt traffic here. That would trigger outside intervention, not only by such NASWA countries as Saudi Arabia and Egypt, but also by the United States, Europe's powers, and very likely China. It also explains the strong interest of foreign powers in gaining a foothold in Djibouti—which surrounds the narrow inlet on the western shore just to the southwest of where the Strait meets the Gulf.

growth, Djibouti has benefited as well. But it is Djibouti's geostrategic position that has attracted a significant international presence. This country now contains military bases operated by the United States (4500 troops are permanently stationed there), France, Japan, the EU, and China, and Saudi Arabia and India are also believed to be interested. These land leases also bring in considerable revenues: "We don't have anything else but location" said a government official recently—but it has been turned into a very significant economic asset. The broader challenge for Djibouti is to diversify its economy: nearly three-quarters of its GDP is directly associated with the port, and unemployment is estimated at more than 50 percent.

Ethiopia

Ethiopia, the last country covered in this chapter, allows us to end on a fairly positive note. The highlands of Ethiopia, where the country's core area, capital (Adis Abeba), and its historic Christian heartland lie, are sandwiched between dominantly Muslim societies (Fig. 6-21). And Figure 6-22 reminds us that, Ethiopia's eastern Ogaden is traditional Somali country and almost entirely Islamic. Today, one-third of Ethiopia's population of 102 million is Muslim, and here the Islamic Front is sharply defined. With the continent's second-largest population, Ethiopia is also an important African cornerstone state, and is discussed more fully in Chapter 7 as part of the region of East Africa. It not only faces some of the usual challenges of the African Transition Zone but has been landlocked as well since it lost Eritrea in 1993.

Yet Ethiopia's highland climate and soils, quite conducive for productive agriculture, strongly contrast with the less encouraging conditions that prevail in the stressed semiarid environments across so much of the African Transition Zone. Indeed, Ethiopia exports fruits, vegetables, flowers, coffee, and even tea. Moreover, this country has in recent years attracted considerable investment from China to finance badly needed infrastructure, construction, and agribusiness ventures (see Chapter 7); and in terms of economic growth, Ethiopia has ranked among the world's top 25 countries for most of this decade. Politically, Ethiopia remains stable despite some questionable recent elections managed by a hard-nosed, oppressive regime. But economically, at least, Ethiopia's recent performance far exceeds that of other countries in the African Transition Zone. This not only reminds us of the importance of geography but also of the pitfalls of **environmental determinism [26]**: geography explains a great deal because it poses constraints and offers opportunities—but geography never dictates.

Points to Ponder

- The Arab Spring movement of the early 2010s did not produce major advances toward democracy, but it did contribute to dangerous political destabilization in the realm's regions and a number of individual countries.

- In Saudi Arabia, one of America's foremost trading partners and military allies, women are forbidden to drive; and only in 2015 were they given limited voting privileges for the first time.

- In 2015, Dubai International Airport (DXB) in the United Arab Emirates was the world's busiest airport measured in terms of international passengers (London Heathrow came in second).

- In 2016, the North Africa/Southwest Asia realm counted no less than six failed states (Iraq, Syria, Yemen, Libya, Mali, and Somalia)—more than in any other realm.

- In Lebanon today, a small country the size of Connecticut, one in five people are refugees, mostly from Syria.

- Turkey has long been considered to be a highly stable Western ally in the heart of the Islamic World. The attempted military coup of 2016 detracted from that image even though the government quickly regained control.

FIGURE 7-1

© J. Nijman, P. O. Muller, H. J. de Blij, and John Wiley & Sons, Inc.

CHAPTER 7

Where were these pictures taken?

The Subsaharan African Realm

IN THIS CHAPTER

- Africa's growing economies
- China's foothold in Africa
- Impacts of the Ebola crisis
- Boko Haram in Nigeria's north
- The paradox of export manufacturing in Ethiopia
- Xenophobia in South Africa

CONCEPTS, IDEAS, AND TERMS

The African continent occupies a special place in the physical as well as the human world. This is where **human evolution [1]** began. In Africa we formed our first communities, spoke our first words, and created our first art. From Africa our ancestor hominins spread outward into Eurasia more than 2 million years ago. From Africa our species emigrated, beginning perhaps 95,000 years ago, northward into present-day Europe and eastward via southern Asia into Australia and, much later, farther afield into the Americas. Disperse our forebears did, but we should remember that, at the source, we are all Africans.

Defining the Realm

For millions of years, Africa served as the cradle for the emergence of humankind. For tens of thousands of years, Africa served as the source of human cultures. Yet Africa has not experienced economic development on a par with other world regions. Its underdevelopment is difficult to separate from foreign involvements: from imperialism to colonialism to the geopolitical impact of the **Cold War [2]**. Today, there are hopeful signs that this catastrophic era is ending as a number of states are being propelled forward by the new global economic relationships they are building.

The focus in this chapter will be on Africa south of the Sahara, also known as **Subsaharan Africa**, an unsatisfactory but popular name that applies to the extensive landmass south of the great desert (**Fig. 7-1**; **Box 7-1**). The African continent contains two geographic realms: the African, extending from the southern margins of the Sahara to the Cape of Good Hope, and the western flank of the North African/Southwest Asian realm. Although the great desert constitutes a formidable barrier between the two, the powerful influences of Islam crossed the Sahara from North Africa centuries before the first Europeans set foot in West Africa. By that time, the African kingdoms lining the southern margins of the Sahara had been converted, creating an Islamic foothold in the African Transition Zone (the final region discussed in Chapter 6) that now marks the northern periphery of the realm we are about to investigate.

Africa's Physiography

Africa accounts for about one-fifth of the Earth's entire land surface. Territorially, it is as big as China, India, the United States, Mexico, and Europe combined. The north coast of Tunisia lies 7700 kilometers (4800 mi) from the southernmost coast of South Africa. Coastal Senegal, in West Africa, lies 7200 kilometers (4500 mi) from the tip of Africa's Horn in easternmost Somalia. These distances have critical environmental implications. Much of Africa lies far from maritime sources of moisture. In addition, as Figure G-6 shows, large parts of the landmass lie in latitudes where global atmospheric circulation systems produce arid conditions. As a result, water supply is one of Africa's foremost problems.

As the physiographic map (**Fig. 7-2**) reveals, Africa's topography is marked by several properties that are not replicated on other landmasses. Alone among the continents, Africa does not possess a mountain backbone; neither the northwestern Atlas nor the far southern Cape Ranges are in the same league as the Andes or Himalayas. Where Africa does have mountainous terrain, as in Ethiopia and South Africa, it takes the form of deeply eroded plateaus or, as in East Africa, high snowcapped volcanoes. Furthermore, Africa is one of only two continents containing a cluster of Great

---

Box 7-1 Major Geographic Features of Subsaharan Africa

1. Physiographically, Africa is a massive plateau continent without a major "spinal" mountain range but with a set of Great Lakes, several major river basins, variable rainfall, soils of generally inferior fertility, and dominantly savanna and steppe vegetation.

2. Hundreds of distinct ethnic groups make up Subsaharan Africa's culturally rich and varied population mosaic. They far outnumber the states in this realm, and rarely do state and ethnic boundaries coincide.

3. Approximately three-fifths of Subsaharan Africa's peoples still depend on farming for their livelihood.

4. The realm is rich in raw materials vital to industrialized countries, but many economies continue to rely on primary activities—the extraction of resources—and not the greater income-generating activities of manufacturing and assembly.

5. The realm is famous for its wildlife, but many species are threatened. Combining bioconservation together with sustainable development is a big challenge when the outside world covets elephant tusks and rhinoceros horns.

6. Severe dislocation still affects a number of Subsaharan African countries, from South Sudan to Zimbabwe.

7. Foreign interest in the realm's natural resources, ranging from commodities to agricultural land, has expanded steadily in this century.

8. Subsaharan Africa experienced rapid economic growth in the 2000s, but growth rates after 2009 became subdued as commodity prices declined and China's economic slowdown took hold. Nevertheless, certain countries still record strong growth.

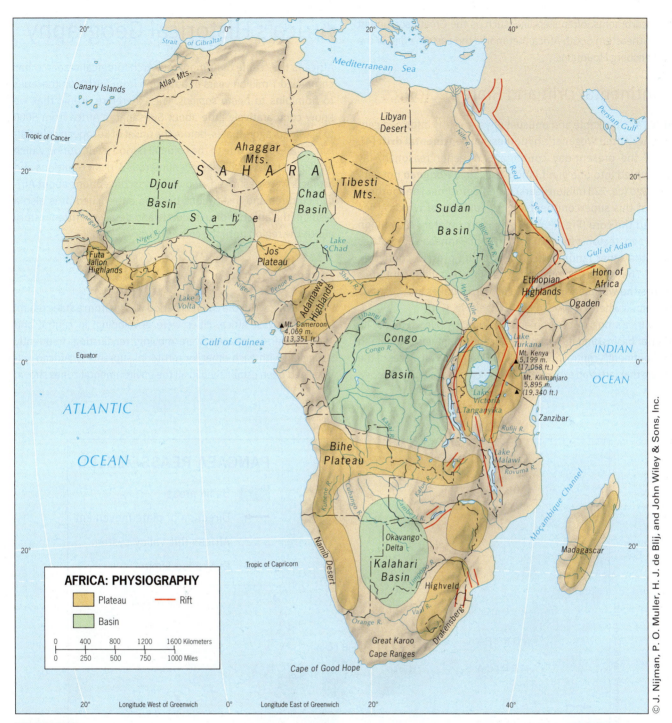

FIGURE 7-2

Lakes. These lakes (with the exception of Lake Victoria) lie in deep trenches called **rift valleys [3]**, which form when huge parallel fractures appear in the Earth's crust and the strips of crust between them sink to form great, steep-sided, linear valleys. These rift valleys, stretching more than 9600 kilometers (6000 mi) from the Red Sea to Swaziland, are mapped by the red lines in Figure 7-2.

Africa's rivers, too, are unusual: their upper courses frequently bear landward, seemingly unrelated to the coast toward which they eventually flow. Several rivers, such as the

Nile and the Niger, have inland as well as coastal deltas. Major waterfalls, notably Victoria Falls on the Zambezi, or lengthy systems of cataracts separate the upper from the lower river courses.

Finally, Africa may be described as the plateau continent. Except for some comparatively limited coastal plains, almost the entire continent lies above 300 meters (1000 ft) in elevation, and half of it lies over 800 meters (2500 ft) high. As Figure 7-2 indicates, the plateau surface has sagged under the weight of accumulating sediments into six major basins (three of them in the Sahara), but the margins of Africa's plateau are marked by escarpments, often step-like, such as the Great Escarpment of southeastern South Africa that marks the eastern edge of the Drakensberg Mountains.

Continental Drift and Plate Tectonics

Africa's remarkable and unusual physiography informed geographer Alfred Wegener's hypothesis of **continental drift [4]**. All of the present-day continents, Wegener reasoned, lay assembled into one giant landmass called **Pangaea** not very long ago (ca. 220 million years) in geologic time. The southern part of this supercontinent was *Gondwana*, of which Africa formed the core (**Fig. 7-3**). When, approximately 200 million years ago, tectonic forces began to split Pangaea apart, Africa (and the other landmasses) acquired its present configurations. That process, now known as *plate tectonics*, continues across the world and is marked by earthquakes and volcanic eruptions where it is most intense (see Fig. G-4).

Africa's ring of escarpments, its rifts, its river systems, its interior basins, and its lack of significant mountains all relate to the continent's central position within Pangaea—all pieces of the puzzle that led to an explanation based on plate tectonics theory.

Africa's Historical Geography

Africa is the cradle of humankind. Archeological research has chronicled 7 million years of transition from Australopithecenes to hominins to *homo sapiens*. It is, therefore, ironic that we know comparatively little about Subsaharan Africa from 5000 to 500 years ago—that is, before the onset of European colonialism. This is partly due to the colonial period itself, during which African history was neglected, numerous African traditions and artifacts were destroyed, and many misconceptions about African cultures and institutions became entrenched. It is also a result of the absence of an accurate and detailed written history about the region until the contemporary period.

African Genesis

Africa on the eve of the colonial period was far from a static realm. In interior West Africa, cities were developing; in central and southern Africa, peoples were moving, readjusting, frequently struggling with each other for territorial supremacy. African cultures had been established in all the environmental zones shown

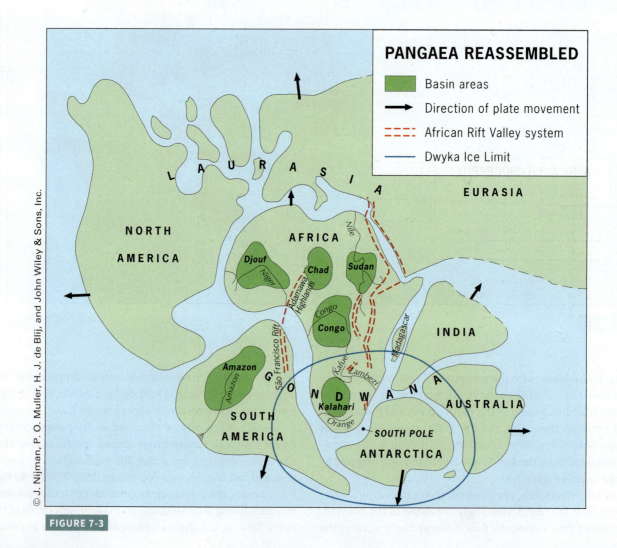

© J. Nijman, P. O. Muller, H. J. de Blij, and John Wiley & Sons, Inc.

FIGURE 7-3

in Figure G-6 for thousands of years and thus long before the intrusions of Islamic and/or European influences. One of these, the Nok culture, endured for over eight centuries (north of the Niger-Benue confluence in what is now central Nigeria) from about 500 BC to the third century AD (**Fig. 7-4**). The Nok people made stone as well as iron tools, and they left behind a treasure of art in the form of clay figurines representing humans and animals.

Early Trade

West Africa, over a north-south span of a few hundred kilometers, displayed a wide range of contrasting environments, economic opportunities, modes of life, and products. The peoples of the tropical forest in the south produced and needed goods that were different from the products and requirements of the peoples of the dry, distant north. For example, salt was a prized commodity in the forests, where humidity precluded its formation, but it was plentiful in the semiarid steppe and the desert beyond. This enabled the desert-margin peoples to provide salt to the forest peoples in exchange for ivory, spices, and dried foods. Accordingly, there evolved a degree of *regional complementarity* between the peoples of the forest and those of the drylands. And the savanna peoples—those located in between—found themselves in a position to channel and handle this profitable trade.

The markets in which these goods were exchanged prospered and grew, some into urban centers. One of these old cities, now an epitome of isolation, was once a thriving center of commerce and learning and one of the leading crossroads in the world—Timbuktu (now located in Mali). In fact, its university is one of the oldest anywhere, with a library that holds irreplaceable documents, thankfully now being preserved. Others, predecessors as well as successors of Timbuktu, declined, a few of them into oblivion. Still other savanna cities, such as Kano in northern Nigeria, remain important today.

Early States

Strong and durable premodern states arose in the interior of West Africa (Fig. 7-4). The oldest known state was ancient Ghana, located to the northwest of present-day Ghana and centered on parts of what are now Mali, Burkina Faso, and Ivory Coast. Ancient Ghana managed to weld various groups of people into a stable territory between the ninth and twelfth centuries AD. Taxes were collected from its citizens, and tribute was extracted from subjugated peoples along Ghana's periphery; tolls were levied on goods entering the state, and an army maintained control. Muslims from the drylands to the north invaded Ghana in 1067, when it was in decline, hastening the state's demise as it shattered into smaller units.

In the centuries that followed, the focus of politico-territorial organization in this West African *culture hearth* (see Fig. 6-3) shifted almost continuously to the east—first to ancient Ghana's successor state of Mali, which was centered on Timbuktu and the middle Niger River Basin, and then to the state of Songhai, whose focus was Gao, a city farther down the Niger that still exists (Fig. 7-4). This eastward movement may have been the result of the growing influence and power of Islam.

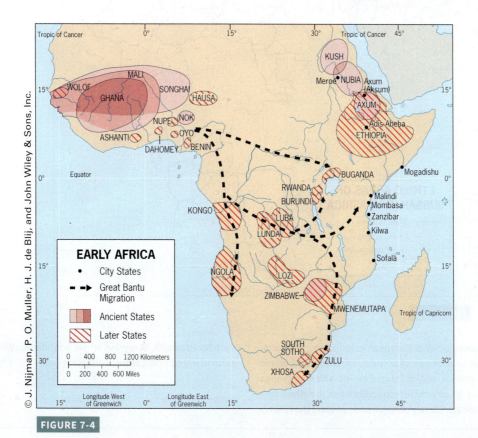

© J. Nijman, P. O. Muller, H. J. de Blij, and John Wiley & Sons, Inc.

FIGURE 7-4

Traditional animist religions prevailed in ancient Ghana, but Mali and its successor states sent massive, gold-laden pilgrimages to Mecca along the savanna corridor south of the Sahara, then turning northward to follow a Nile route that passed through present-day Khartoum and Cairo.

Beyond West Africa

West Africa's savanna zone undoubtedly experienced momentous cultural, technological, and economic developments, but other parts of ancient Africa made progress as well. Early states emerged in present-day Sudan, Eritrea, and Ethiopia. Influenced by key innovations from the Egyptian culture hearth to the north, these kingdoms were stable and durable: the oldest, Kush, lasted 23 centuries (Fig. 7-4). The Kushites built elaborate irrigation systems, forged iron tools, and created impressive structures as the ruins of their long-term capital and industrial center, Meroe, reveal. Nubia, to the southeast of Kush, became Christianized until the Muslim tide overtook it in the eighth century. And Axum was the richest market in northeastern Africa, a powerful kingdom that controlled Red Sea trade and endured for more than six centuries.

The process of **state formation [5]** diffused throughout Africa and was still in progress when the first European contacts occurred in the late fifteenth century. Large and effectively organized states developed on the equatorial west coast (notably Kongo) and on the southern plateau from the southern margin of the Congo River Basin southeastward to Zimbabwe. East Africa, as well, gave rise to a number of city-states, including Mogadishu, Mombasa, and Sofala.

At a broader scale, a crucial series of events commencing about 5000 years ago affected virtually all of West, Equatorial, and Southern Africa: the Great Bantu Migration from present-day Nigeria and Cameroon southward and eastward across the continent. This epic expansion appears to have occurred in waves that populated the Great Lakes area and ultimately penetrated South Africa, where it culminated in the formation of the powerful Zulu Empire in the nineteenth century (Fig. 7-4).

All this reminds us that, prior to European colonization, Africa was a realm of rich and varied cultures, diverse lifestyles, technological advances, mobility, and trade. But it was also a highly fragmented realm with a cultural mosaic (**Box 7-2**; **Fig. 7-5**) constituted by myriad **ethnicities [6]** sharing a common ancestry, culture, language, and/or religion. After the Europeans intervened and redrew the social and

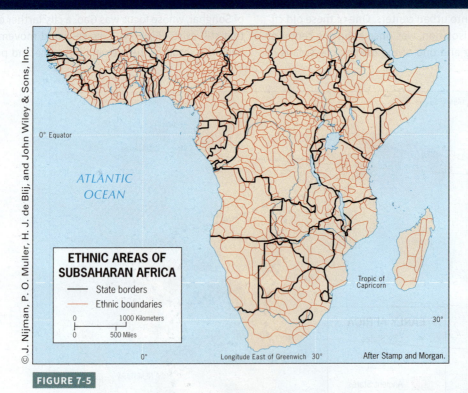

Box 7-2 MAP ANALYSIS

ETHNIC AREAS OF SUBSAHARAN AFRICA

— State borders
— Ethnic boundaries

© J. Nijman, P. O. Muller, H. J. de Blij, and John Wiley & Sons, Inc.

After Stamp and Morgan.

FIGURE 7-5

Despite the common legacies of European colonization, Subsaharan Africa is marked by a mosaic of cultural, linguistic, and ethnic diversity. The coexistence and/or integration of different groups can be a major challenge anywhere, especially when the ethnic patterns become so intricate. Closely observe and compare the realm's

state borders with its myriad ethnic boundaries. Which states have particularly complex ethnic configurations?

Access your WileyPLUS Learning Space course to interact with a dynamic version of this map and to engage with online map exercises and questions.

political map, the modern African state emerged as an amalgamation of multiple ethnic groups that almost entirely disregarded the preexisting cultural-territorial layout. Thus the European idea of the **nation-state [7]**, where state borders coincide with ethnic boundaries, is a rarity within the patchwork of pluralistic countries that marks this realm (Fig. 7-5).

The Colonial Transformation

European involvement in Subsaharan Africa began in the fifteenth century. It disrupted the path of indigenous African development and irreversibly altered the entire cultural, economic, political, and social complexion of the continent. Starting quietly in the late 1400s, Portuguese ships groped their way all along the west coast and by 1488 had rounded the Cape of Good Hope. Their goal was to find a sea route to the spices and riches of India and points east. Soon other European countries were dispatching their vessels to African waters, and a string of coastal stations and forts sprang up. In West Africa, the part of the continent closest to the European spheres in Middle and South America, the initial impact was strongest. At their coastal control points, the Europeans bartered with African intermediaries for the slaves who were destined to work the New World plantations, for the gold that had been flowing northward across the Sahara, and for highly prized ivory and spices.

Suddenly, the centers of activity lay not with the inland cities of the savanna belt but with the foreign stations on the Atlantic coast. As the interior declined, the coastal peoples thrived. Small forest states gained unprecedented wealth, transferring and selling slaves captured in the interior to the European traders on the coast. Dahomey (now called Benin) and Benin (now part of neighboring Nigeria) were two modern states built on the slave trade.

Impact of the Slave Trade

Millions of Africans were forced to migrate from their homelands to the Americas, especially Brazil, the Caribbean, and the United States. The slave trade was a catastrophe that was facilitated, in part, by the peril of proximity. The northeastern tip of Brazil, by far the largest single destination for the millions of Africans forced from their homes in bondage (**Fig. 7-6**), lies about as far from the nearest West African coast as South Carolina lies from Venezuela. This was a much shorter maritime intercontinental journey because the distance from West Africa to South Carolina is more than twice as far.

Slavery was not new to West Africa. In the interior of Africa and within city-states, kings, chiefs, and prominent families traditionally took a few slaves, but the status of these slaves was relatively benign. Indeed, large-scale slave trading was first introduced to East Africa long before the Europeans brought it to western Africa. African intermediaries from the coast raided the interior for able-bodied men and women of various ethnic

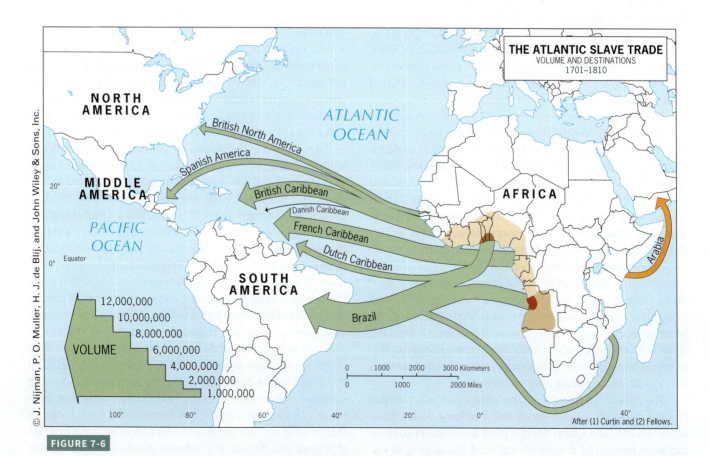

© J. Nijman, P. O. Muller, H. J. de Blij, and John Wiley & Sons, Inc.

FIGURE 7-6

backgrounds and marched them in chains to the Arab markets on the coast (the island of Zanzibar was one such slave trading hub). There, packed into specially built sailboats, they were carried off to Arabia, Persia, and India.

When the European slave trade took hold in West Africa, however, its volume and impact were far greater than anything seen before, and so were the horrors. Europeans, Arabs, and collaborating Africans now ravaged the continent, forcing perhaps as many as 30 million persons away from their homelands in captivity (Fig. 7-6). Families, entire communities, and even cultures were destroyed; those surviving in exile suffered unfathomable misery. The European presence also completely reoriented West Africa's trade routes. Even though the Europeans' insatiable demand for slaves ravaged the inland population, it did not lead to any major European thrust toward the interior or produce colonies overnight. African intermediaries were well organized and strong, and they held off their European competitors for centuries.

Colonization

In the second half of the nineteenth century, the Europeans finally laid claim to virtually all of Africa. In 1884, a conference was convened in Berlin to carve up the African map among the leading European powers. On maps spread across a large table, representatives from these powers drew boundaries, swapped territory, and forged a new map that would become a liability in Africa decades later. Amazingly, at no point in the proceedings were African voices heard. As **Figure 7-7** indicates, while the three-month conference was in progress, most of Africa remained under traditional African rule. Then, in the early 1900s, the European colonial powers moved to take control of all the areas they had marked off and acquired on their maps.

It is important to examine Figure 7-7 carefully because the colonial powers governed their new dependencies in very different ways, and their contrasting legacies are still in evidence today. Some colonial powers were democracies at home (Great Britain and France); others were dictatorships (Portugal and Spain). The British established a system of **indirect rule [8]** over much of their domain, leaving indigenous power structures in place and making local rulers representatives of the British Crown. This was unthinkable in the Portuguese colonies, where harsh direct control predominated. The French sought to establish culturally assimilated elites that would disseminate French ideals in the colonies. But King Leopold II of Belgium claimed Congo Free State as his own personal fiefdom. Subsequently, he unleashed a savagely ruthless campaign of exploitation and genocide. Leopold's reign of terror has become known as Africa's most brutal demographic disaster (after the slave trade): as many as 10 million people were exterminated.

The post-Berlin colonial map entailed a profound political reorientation—yet it lasted barely half a century. Ghana (then known as the Gold Coast) offers a good example. Its colonial ties were formalized despite the constant challenging of the

mother country by indigenous ethnic groups. But in 1957, Ghana became the first African country to gain independence—triggering a movement that swept across the realm (compare the 1958 and 1970 maps in Fig. 7-7). These days, most of Subsaharan Africa has already marked 50 years of independence, and in retrospect the colonial period should now be viewed as only an interlude rather than a paramount chapter in African history. It did, however, leave a deep and enduring geographic imprint, especially on the realm's cultural landscapes (**Box 7-3**).

Human–Environment Relations

Colonial Legacy

Africa's tropical forests and extensive savannas constitute the world's last refuges for many species of wildlife, from primates and zebras to giraffes to wildebeests. But since colonial times, these natural environments have been under constantly rising threat. Early on, the Europeans introduced hunting as a "sport" (a practice that did not exist within African cultural traditions), thereby triggering the mass destruction of animals. Later, a number of colonial rulers and postcolonial national governments laid out game reserves and other conservation areas, but these were neither sufficiently large nor adequately connected to enable herd animals to adhere to their seasonal migration routes. As we shall see, the same climatic variability that plagues farmers also affects wildlife. When the rangelands wither, the animals seek better pastures; but if the fences of game reserves wall them off, they cannot survive. Where there are no fences, wildlife invasions of farmland destroy crops, and the farmers retaliate. Clearly, after thousands of years of equilibrium, the competition between humans and animals in Africa has entered a new era of intensifying stress.

It is not solely population pressure that threatens African wildlife. Today, international criminal syndicates are heavily engaged in the illegal exporting of wild animals and animal parts, and this activity has become the world's fourth most profitable illicit trade (after narcotics, human trafficking, and arms smuggling). The horn of the rhinoceros became a valued property for Arabs who fashion them into dagger handles and, in powdered form, an alleged aphrodisiac and cancer treatment for wealthy East Asians. As a result, the northern white rhinoceros is nearing extinction, and the number of all species of rhino has fallen from half a million a century ago to barely 25,000 today. A terrible tragedy is also befalling the African elephant as the value of its ivory tusks has tripled since 2010 (in 2015, a kilogram of ivory in Beijing peaked at U.S. $2100); if this illicit Chinese market is not quickly terminated, this entire species could disappear in the next decade. Note that even though Africans are complicit in international smuggling and that small quantities of ivory are for sale in informal markets in Nigeria and Angola, it is external demand that drives this

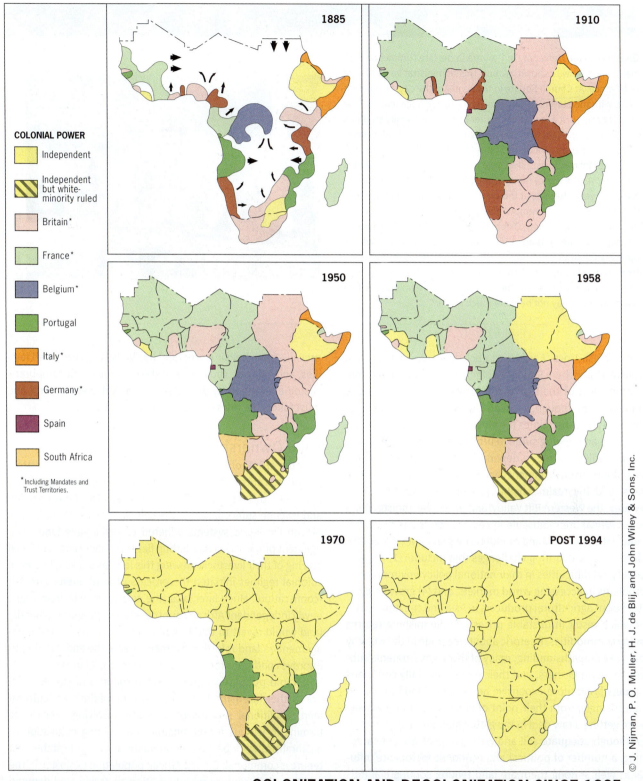

COLONIAL POWER

- Independent
- Independent but white-minority ruled
- Britain*
- France*
- Belgium*
- Portugal
- Italy*
- Germany*
- Spain
- South Africa

*Including Mandates and Trust Territories.

1885

1910

1950

1958

1970

POST 1994

© J. Nijman, P. O. Muller, H. J. de Blij, and John Wiley & Sons, Inc.

COLONIZATION AND DECOLONIZATION SINCE 1885

FIGURE 7-7

clandestine trade. Meanwhile, competition for land has more to do with the decline of Africa's lion population (from around 250,000 in 1975 to less than 20,000 today), whereas illegal logging and mining threaten a host of other species including the great ape.

The Geography of Farming

It would seem that space should be plentiful for both wildlife and humans. **Figure 7-8** does not present a picture of a densely peopled realm—although there are major clusters of population

Box 7-3 From the Field Notes...

"On a visit to the University of Pretoria in the late summer of 2015 (their winter), I was shown around this administrative capital of South Africa. At Church Square in the center of the city, I came face to face with this statue of Paul Kruger, which had been fenced in by the authorities just a few weeks earlier because university students and allied groups wanted to tear it down. Kruger was a leading colonial figure and president of the South African Republic from 1883 to 1900. Colonial memorials like these are reminders of an era that is long gone, yet the legacy remains and offends political sensitivities. That same year, students at the University of Cape Town (the legislative capital) voted to remove the statue of another colonialist—Cecil Rhodes (interestingly, Oxford University in England declared it will keep *its* statue). It is an ironic sight: once, South Africa's whites appropriated all of the lands of this country; now, a former president's statue is caged and confined to a minimal space. Kruger's face seems to express a mixture of disdain and resignation, as if he knows his effigy may not be there much longer."

© Jan Nijman

in coastal West Africa, East Africa encircling Lake Victoria, and the African Horn where Ethiopia's highlands sustain a large concentration. Most of the rest of the realm is relatively sparsely populated, even though its overall annual rate of growth is almost two-and-a-half times the global average of 1.1 percent.

Despite an ongoing surge in urbanization, a large majority of Africans still continue to depend on farming for their livelihood. But African environments are difficult for millions of small farmers, and the realm's highly productive areas (the Great Lakes area, the Western Rift Valley, and the cooler, moister parts of South Africa) are minuscule in comparison to the huge, fertile, alluvial basins of China or India. The great majority of African farmers face daunting challenges that include: (1) climatic variability, (2) difficulties in their national policy environments, (3) problems in accessing world markets, and (4) growing competition from agricultural producers in Asia and the Americas—all of which result in unstable prices and diminishing returns for many commodities. Rhetoric about free trade in the wealthy Global Core notwithstanding, many of those governments subsidize their own farmers, and their environmentally conscious consumers favor more expensive home-grown food products. Not surprisingly, from their point of view, African farmers say they are getting a raw deal in the global marketplace.

Although adequate soil and water supplies are critical for farming, a number of political and economic factors are influential as well. Among them are land tenure; commercial (for-profit) versus subsistence (household-level) agriculture; type of farming system (rotational, shifting cultivation); crop prices; government farm policies; and degree of mechanization.

With roughly 60 percent of the realm's livelihoods dependent on farming, the issue of land tenure may be the most important of all. **Land tenure [9]** refers to the way people own, occupy, and use land. African traditions of land tenure are different from those found in Europe or the Americas. In most of Subsaharan Africa, communal, not individual, rights prevail—often held in perpetuity by chiefs, large extended families, and villages. Thus users of the land have only temporary, custodial rights to it; they cannot sell that piece of land and must follow the rules of their community leaders.

Land Ownership

After the arrival of the Europeans, their colonial administrators prioritized control of the most fertile areas of each colony. In certain cases, this required military conquest; elsewhere, coercion and eviction sufficed. Before the colonial era, many traditional African livelihood systems adhered to sustainable land management practices such as letting fields lie fallow and rotational grazing of their livestock. Viewing this land as unused, however, colonial regimes initiated the process of **land alienation [10]** (confiscation). Much later, after the colonizers withdrew, many newly independent African countries started programs whereby land would revert to traditional forms of ownership and management—a land reform movement that in the end did little to improve conditions for still-marginalized small farmers.

Rapid population growth since the end of the colonial era further intensified pressures on the realm's agricultural land. Traditional systems of land use, involving subsistence farming in various forms ranging from shifting cultivation to pastoralism, work best when populations are fairly stable and tenure is communal. But the African population explosion of the mid-twentieth century set into motion an accelerated demand cycle in which food producers could no longer rest soils and pasturelands—thereby unleashing new rounds of farmland degradation that steadily drove down crop yields.

Subsistence Farming

Although commercial farming prevails in parts of Africa, most farmers are engaged in subsistence agriculture that entails grain crops (corn, millet, sorghum) in drier areas and root

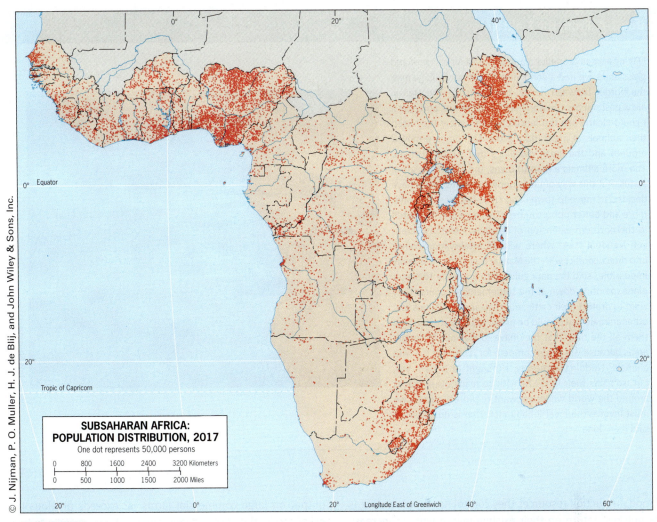

© J. Nijman, P. O. Muller, H. J. de Blij, and John Wiley & Sons, Inc.

**SUBSAHARAN AFRICA:
POPULATION DISTRIBUTION, 2017**
One dot represents 50,000 persons

FIGURE 7-8

crops (yams, cassava, sweet potatoes) where moisture is adequate. Others herd livestock, mostly cattle and goats, as they deal with the constant challenges of environmental variability (**Box 7-4**). But too many farmers and pastoralists have been unable to secure regular access to markets as well as stable prices for their products because of government policies that often promote a particular export-oriented crop (e.g., peanuts in Gambia; tea and coffee in Kenya).

The decline of African agriculture has been disproportionally hard on women who, according to prevailing estimates, produce about two-thirds of all local food in Subsaharan Africa. Development policies frequently pay too little attention to the role of gender in agricultural production (women account for at least 15 percent of all farmland holders), so that farms operated by women are smaller, more isolated, and endowed with poorer-quality soils than those cultivated by men. Moreover, women always have a harder time than men in obtaining credit.

The **Green Revolution [11]**—the development of more productive, drought-tolerant, pest-resistant, higher-yielding

varieties of grain—had less impact in Africa than elsewhere during the latter decades of the twentieth century. Belatedly, a renewed effort has now been launched to jump-start Africa's New Green Revolution. Its twin emphases involve the adoption of more effective farming methods and the raising of higher-value crops in order to boost farmer income as well as enhance the realm's food production output. However, the introduction of genetically modified crops is controversial and encounters substantial resistance today. Many African governments, influential media, and individual farmers regard biotechnological solutions with increasing hostility.

Neocolonial Land Grabs

During the postcolonial era, agricultural land in Africa was of little interest to outside investors, but over the past decade an enormous change has taken place. Many African countries now offer large tracts of land for sale or long-term leases to major investors, a number of whom are associated with multinational **agribusiness [12]** corporations and private equity firms. Governments and other large landowners view the sale and leasing

Box 7-4 From the Field Notes...

"Flying a small propeller plane over East Africa's semiarid grasslands, we had a good view of a Maasai village near the border between Kenya and Tanzania. The Maasai are a pastoral people who herd cattle, sheep, and goats. I could clearly see the fenced 'kraals' where they keep their animals at night (during the day they are moved around) and the small dwellings that surround them. The more animals a man (!) owns, the higher his social status. But this way of life is under pressure: many have decided to move to towns and cities because they offer more and better opportunities. This area is also located in the northern extension of Tanzania's famous Serengeti National Park, where wildlife preservation comes into direct conflict with the Maasai way of life. 'It is really impossible,' said the park guide with us on the 10-seater plane, pointing down at the settlement. 'Of course they want to protect their cattle, but they are killing lions and other animals that should be protected. And we must tell them all the time that they have to stay out of certain areas because the cattle compete for food with the wild animals.' Nodding agreeably, I thought to myself: yes, but isn't this their land? As with so many other places around the world where we want to protect nature, we must involve the local inhabitants. That is much easier said than done."

©Jan Nijman

of food production resources they control (land, soil, water) as the quickest return on their existing investments to increase agricultural output and maximize foreign revenues. As for the investors, they are motivated by the steady rise in global food prices, brightening prospects for biofuels, and growing shortages of arable land in their own countries. They also see African smallholder agriculture as achieving only a tiny fraction of its potential productivity—which the outsiders believe can be significantly upgraded through the widespread application of contemporary agricultural strategies and technology. Although investors from other geographic realms form the majority of the participants in this land rush, certain African elites (many based in cities) have joined in as well—either directly or in partnership with public and private institutions (see **Box 7-5**).

By the mid-2010s, more than 4 million hectares (9.8 million acres) of African land were under negotiation or already managed by foreign investors—an area greater than the combined agricultural land of France, Germany, Belgium, the Netherlands, Denmark, and Switzerland. Most of this land was concentrated in Ethiopia, Congo,* Moçambique, and Sudan. So far, the impact of these land grabs has been mixed. In some cases,

output on large commercial farms far exceeds previous levels of production. But in many others, investors seem to have had speculative intentions and are content to let farmland lie idle. Either way, substantial numbers of small farmers are now dispossessed, with adverse economic effects that reverberate throughout affected localities. For countless thousands of such displaced workers, without any say of their own, traditional pathways to access land and livelihoods have been upended by these institutional changes in land use. Understandably, mistrust is rising rapidly among local populations, and the success of these land deals is increasingly variable: for example, Sudan has performed very poorly, whereas Moçambique's situation is said to have improved notably.

Population and Health

Medical geography [13] is concerned with the tracking of disease outbreaks to identify their sources, detect their carrier agents, trace their spatial diffusion, and prevent their recurrence. Alliances between medical personnel, epidemiological researchers, and geographers yield significant results. Medical specialists know how a disease ravages the body; geographers know how environmental conditions (such as variations in wind direction and river flow) can affect the spread of disease carriers (*vectors*), and their application of GIS techniques to analyze

---

*Two countries in Africa have the same short-form name, *Congo*. In this book, we use ***DRCongo*** for the much larger Democratic Republic of the Congo, and ***Congo*** for the smaller Republic of Congo just to its northwest.

Box 7-5 Regional Issue: **Neocolonial Land Grabs?**

These Land Deals Are a Solution to Our Problems.

"When the dust of all this drama settles around so-called land grabs, it should be clear to anybody that our government has come up with a brilliant solution to the country's problems. As an accountant here in the capital of Adis Abeba, I know a few things about costs and benefits, and in these land deals Ethiopia comes out as a winner.

"We mustn't fool ourselves: decade after decade this country has been hit by famines and food shortages. It was not just people going hungry; hundreds of thousands have died as a result. Surely one reason for this has been that we have failed to produce enough food. The landholdings are too small to be efficient, and most farmers don't have the means or the knowledge to improve yields.

"The way forward for us and several other countries in Africa is to use our abundant land and natural resources to our advantage, to allow foreign investors to come in and exploit these resources for us. It's not like we are giving it away. Not at all! In most cases it is not a permanent sale but a lease, and the agreements contain provisions that they must sell a minimum share of the harvest in Ethiopia so all of it cannot be exported. And they must hire Ethiopian workers.

"Food in this country will become cheaper, you will see, simply because there will be more of it. Foreign agribusinesses know their stuff. Government income will rise from these deals, and these revenues can be invested in education and infrastructure. That, in turn, will create more employment opportunities. I can understand the fears of people about these land deals, but they are misplaced and old-fashioned. In the global economy today, you must be willing to deal with the outside world, use the resources you've got to your advantage. That is what we are doing."

Africa Gets Robbed, Once Again.

"It is hard to believe this is really happening, but it's true and I've seen it with my own eyes here in Ethiopia. Families who have been farming for many years are forced off the land so the government can lease it to foreign investors. It is a disaster because we Ethiopians no longer control our own land and what we grow on it. I own a shop here in a small town in the western part of the country, and I have seen prices go up. Many of my customers are rural people who used to grow their own food, but now they must pay in cash. The government promised them all kinds of things when they terminated the leases on these small farms, yet little has come of it.

"Where is a country going when it sells its land to outsiders? That is my question to you. How can you ever believe that big foreign companies will keep the interests of the Ethiopian people in mind? Of course they don't! Why do you think they come here in the first place? They grow just one or two crops to export back to their own country. Quite a few are said to be rich Arabs from the Gulf States, where fertile land is scarce. Many just grow rice, a food we Ethiopians don't eat much of. I even heard a story of one company growing sorghum for their camels in Arabia!

"And don't believe the argument that these lands were not being cultivated. Do you really think we Africans would let good land sit idle? We have millions of small farmers looking for every bit of fertile soil they can find on which to earn a living. Worst of all is when the outsiders buy up the land and don't do anything with it. Foreign investors get such good deals from our government that they just sit on the land as a speculative investment. They believe Africa is rising and that the land will increase in value, so they wait until they can sell it for a profit in the future. In the meantime, we can't touch our own natural resources. It is a scandalous situation."

---

spatial dimensions of diseases is making myriad contributions to medical research these days.

In Subsaharan Africa today, hundreds of millions of people carry one or more maladies, often without knowing exactly what ails them. A disease that infects many people (the *hosts*) in a kind of equilibrium, without causing rapid and widespread deaths (such as hepatitis or hookworm), is said to be **endemic [14]** to that population. People affected may not die suddenly or dramatically, but their quality of life and productive capacity are hindered as their overall health is weakened and can deteriorate rapidly when a more acute illness strikes.

Epidemics and Pandemics

When a disease outbreak occurs at the local or regional level, it is called an **epidemic [15]**. It may claim thousands, even tens of thousands, of lives, but it remains geographically confined, perhaps limited by the spatial range of its vector. Trypanosomiasis, the disease known as sleeping sickness and vectored by the tsetse fly, is a good example of a regional-scale epidemic

(**Fig. 7-9**). The extensive herds of savanna wildlife form the *reservoir* of this disease, and the tsetse fly transmits it both to livestock and people. It is endemic to wildlife, but it also kills cattle, so Africa's herders try to keep their animals in tsetse-free zones. African sleeping sickness appears to have originated in a West African source area during the fifteenth century, and from there it spread throughout much of tropical Africa. Its epidemic range is limited by that of the tsetse fly: where there are no tsetse flies, there is no sleeping sickness.

When a disease spreads worldwide, it is known as a **pandemic [16]**. Subsaharan Africa's and the world's most deadly vectored disease is malaria, transmitted by a mosquito and the killer of at least 500,000 people each year. Because people living in the world's poorest (lower-latitude) countries are most vulnerable to malaria, that burden falls overwhelmingly on Africa—which not only accounted for 89 percent of all cases worldwide in 2015, but also 91 percent of the deaths from this disease (mostly children under 5 years of age). Ongoing eradication campaigns against this deadly mosquito vector are scoring some successes, and the UN's World Health Organization reports that better preventive and control measures have

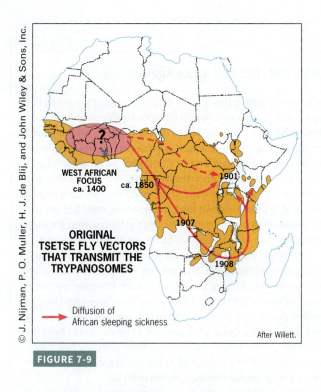

WEST AFRICAN
FOCUS
ca. 1400

ca. 1850

1901

1907

1908

ORIGINAL
TSETSE FLY VECTORS
THAT TRANSMIT THE
TRYPANOSOMES

Diffusion of
African sleeping sickness

After Willett.

FIGURE 7-9

reduced malaria mortality in Africa by over 60 percent since 2000. Yet despite these gains, the realm's long war against this scourge is far from being won and is certain to continue well into the future.

HIV/AIDS: The Continuing Struggle

AIDS stands for Acquired Immune Deficiency Syndrome, the body's failure to protect itself against the slow-acting, human immunodeficiency virus (HIV). If unchecked, this virus multiplies in the body, causing damage to the immune system that produces AIDS. Worldwide, more than 78 million people have contracted the disease, which killed nearly half of them. Globally, as many as 20 million people do not know they have the virus and need to be reached by HIV-testing services. Once again, Africans are most disproportionately affected: AIDS first erupted in this realm during the 1930s, and a subsequent large-scale outbreak here in the 1970s swiftly became the springboard for a pandemic that soon reached the United States in 1981. There is mounting evidence that AIDS in Subsaharan Africa peaked around 1997 and that the rate of new infections has been slightly reduced since then. Uganda, in particular, has won wide acclaim in this effort, and its success has inspired a new wave of public health strategies and assistance programs in several AIDS-affected countries.

The geography of HIV/AIDS is complex, but the Subsaharan Africa realm stands out prominently (with more than 75 percent of the world's cases), as do its remarkable internal patterns. The lack of testing in rural Africa has meant that HIV-positive mothers have unknowingly infected their children during pregnancy, childbirth, and breastfeeding. Recent World Health Organization reports also note that in 2015 Subsaharan Africa

was the realm with the greatest number of children (<15 years of age) infected with HIV in this manner, and today they total nearly 3 million.

With regard to internal regional patterns, most notable are the 10 countries of the Southern Africa region that bear the brunt of the global AIDS burden; collectively, they alone account for *just over 40 percent* of the world total of AIDS cases. When standardized for population size, the health crises of individual countries become clear: no less than 29 percent of the adult population of Swaziland and 22 percent of both Lesotho's and Botswana's are HIV-positive, and close behind at 19 percent is neighboring South Africa. Moreover, Swaziland now has the highest occurrence of this disease in the world; yet even though the number of new cases appeared to be levelling off in 2015, the country's life expectancy of 51 years ranks as the fourth lowest on Earth. (In absolute numbers, Southern Africa's national leader is South Africa, whose HIV-infected population now totals about 7 million.) The HIV prevalence rate is significantly lower in West and Equatorial Africa, ranging from 2 to 5 percent in most countries. That is also generally the case in the region of East Africa, where the highest incidence of HIV/AIDS occurs in Uganda (7 percent) and Kenya (6 percent).

Recent studies have shown that HIV/AIDs in Africa also has important social and economic dimensions. More women than men are infected, underscoring greater female vulnerability mainly associated with their inferior social and economic status. Children by the millions have been orphaned, sharply driving up the cost of caring for them. Societies have been deprived of productive workers and are heavily burdened by related social costs, to the extent that HIV/AIDS can siphon off as much as 1.5 percent from GDP in the most seriously afflicted countries.

Why must Africa suffer so disproportionately? First, HIV/AIDS originated in tropical African forest margins and spread rapidly through all segments of society. Second, the social stigma associated with this sexually transmitted disease makes acknowledging and treating it especially problematic. Third, life-extending medications are both expensive and difficult to provide in remote rural areas. Fourth, government leadership during the AIDS crisis has varied from highly aggressive and effective (as in Uganda) to catastrophically negligent (South Africa stands out here), and there is always the risk that complacency can set in.

Although Africa's HIV/AIDS situation remains serious today, a major battle is being waged against the disease and progress is being made. Public health campaigns throughout the realm—promoting better education, condom usage, and voluntary testing—are having beneficial effects. Generic anti-HIV drugs are becoming cheaper and more widely available through international assistance. At present, more than half of Africa's HIV victims are receiving antiretroviral (ART) therapy, and universal coverage is the goal until a vaccine can be developed (which experts believe could take at least another decade). But even though the number of new infections is declining across the realm, several high-risk populations remain virtually unaffected by the latest advances—particularly refugees, sex workers, prison inmates, and those living in extreme poverty.

The Ebola Crisis of 2014–2016

Another very dangerous malady that can ravage parts of Subsaharan Africa is Ebola Virus Disease (EVD)—as the world discovered in 2014–2015. That is when the largest EVD outbreak on record occurred in West Africa, resulting in thousands of deaths, the devastation of inadequate healthcare systems, and the damaging of the economies of countries still recovering from years of ruinous civil war. The hyped-up overreporting of the Western media notwithstanding, this epidemic was almost entirely limited to three countries: Sierra Leone (where the outbreak originated) and its two eastern neighbors, Liberia and Guinea. As of August 2016, the WHO had reported a total of 28,616 Ebola cases, with all but 36 confined to these countries; 11,310 of those cases were fatalities (40 percent), but only 15 occurred outside the three countries. Of the few EVD cases beyond Sierra Leone, Liberia, and Guinea, almost all were confined to West Africa's Nigeria and Mali, as were 14 of the 15 deaths. In fact, the remaining fatality (a misdiagnosed traveler from Liberia) occurred in Dallas, Texas; the three other EVD patients in the United States all fully recovered. On the positive side there are more than 10,000 survivors of EVD, and a vaccine is in the advanced stages of development with an anticipated release date of 2017.

By mid-2016, the West African EVD epidemic (the worst Ebola outbreak in history) had almost ended (and will officially in September 2016 if no new case occurs). The WHO had already declared Sierra Leone virus-free, but flare-ups occurred in Liberia and Guinea after their virus-free declarations; however, the last patients were discharged in May 2016. Initial assessments of the epidemic were also emerging and show that its impacts went far beyond the healthcare arena. Ebola had caused serious disruptions to travel, transportation, and cross-border trade in this part of West Africa, damaging its ties to the outside world; agricultural production had diminished, exacerbating an already stressed food security situation; and, more ominously, many major international investors, fearing the disease threat, had (at least temporarily) withdrawn from the afflicted countries as did international tourists. Unfortunately, Sierra Leone, Liberia, and Guinea will have to contend with these and other negative economic consequences long after their last new Ebola case has been recorded.

Cultural Patterns

Today, just about 60 percent of Subsaharan Africa's population still inhabit—and work

in or near—Africa's hundreds of thousands of villages. They speak one of the 2000-plus languages in use in the realm. Africa's (numerically) largest groups of people constitute major nations, such as the Yoruba of Nigeria and the Zulu of South Africa. Africa's smallest cultural groups number just a few thousand. Overall, as Figure 7-5 amply illustrates, this geographic realm has the most complex and fragmented cultural mosaic on Earth.

African Languages

Africa's linguistic geography (**Fig. 7-10**) is a key component of that cultural-territorial intricacy. It is clear that the Subsaharan Africa realm begins approximately where northern Africa's Afro-Asiatic language family (mapped in yellow) gives way, a changeover at its sharpest in West Africa.

The realm's dominant language family is the Niger-Congo family (mapped in purple), of which the Kordofanian subfamily is a tiny, historic, northeastern outlier (Fig. 7-10), and the Niger-Congo languages extend across the realm from West to East and Southern Africa. The Bantu language

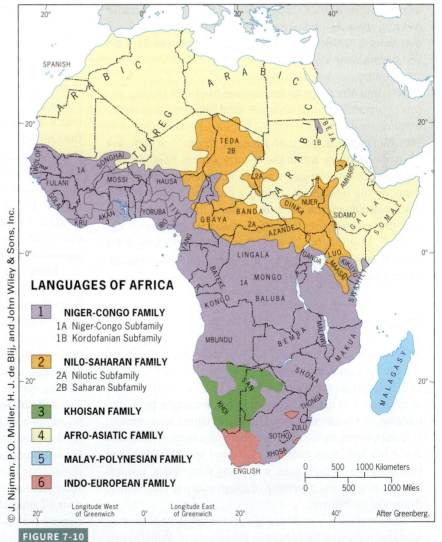

© J. Nijman, P.O. Muller, H. J. de Blij, and John Wiley & Sons, Inc.

LANGUAGES OF AFRICA

1 **NIGER-CONGO FAMILY**
  1A Niger-Congo Subfamily
  1B Kordofanian Subfamily

2 **NILO-SAHARAN FAMILY**
  2A Nilotic Subfamily
  2B Saharan Subfamily

3 **KHOISAN FAMILY**

4 **AFRO-ASIATIC FAMILY**

5 **MALAY-POLYNESIAN FAMILY**

6 **INDO-EUROPEAN FAMILY**

After Greenberg.

FIGURE 7-10

forms the largest branch of this subfamily, but Niger-Congo languages in West Africa, such as Yoruba and Akan, also have millions of speakers. Another important language family is the Nilo-Saharan family (mapped here in orange), extending from Maasai in Kenya northwest to Teda in Chad. No other language families are of similar extent or importance: the Khoisan family, of ancient origins, still survives among the dwindling Khoi and San peoples of the Kalahari in the southwest; the small white minority in South Africa speak Indo-European languages; and Malay-Polynesian languages prevail in Madagascar, which was peopled from Southeast Asia long before Africans reached it.

The Most Widely Used Languages

About 40 African languages are spoken by 1 million people or more, and a half-dozen by 10 million or more: Hausa (34 million), Yoruba (28 million), Ibo, Swahili, Lingala, and Zulu. Although English and French have become important *linguae francae* in countries such as Nigeria and Ivory Coast (officially *Côte d'Ivoire* in the Francophone manner), African languages also serve this purpose. Hausa is a common language across the West African savanna; Swahili is widely used in East Africa. And pidgin languages—mixtures of African and European tongues—are found along West Africa's coast; millions of Pidgin English (called *Wes Kos*) speakers use this medium in Nigeria and Ghana. However, at least 60 African languages are on the verge of extinction, and Nigeria alone has 17 endangered, near-extinct languages.

Multilingualism [17] can be a powerful centrifugal force in society, and African governments have tried with varying success to promote national alongside local languages. Nigeria, for instance, made English its official language because none of its 500-odd languages, not even Hausa or Yoruba, had sufficient interregional usage within the country. On one hand, using a European, colonial language as an official medium invites criticism; on the other hand, making a dominant local language official would unleash negative reactions from ethnic minorities. One creative country today is Rwanda, which promotes its Bantu language (Kinyarwanda) in tandem with English for instruction in schools to hasten the separation from its Francophone past as a Belgian dependency.

Religion

Africans had their own belief systems long before Christians and Muslims arrived to convert them. Despite their diversity, all of Subsaharan Africa's cultural groups held a consistent view of their place in nature. Spiritual forces, according to African tradition, are manifest everywhere in the natural environment: in forests, rivers, mountains, and communal lands—a religious worldview known as **animism**. Gods and spirits are believed to affect people's daily lives, witnessing every move, rewarding the virtuous and punishing (through injury or crop failure) those who misbehave.

As with land tenure, the religious views of Africans clashed fundamentally with those held by the colonizers. Monotheistic

The Muslim faithful kneel during Friday prayers outside a mosque in the city of Kano in northern Nigeria. Whereas most northerners want to practice Islam within the multireligious, multiethnic society of Nigeria, areas governed by strict Islamic (*Sharia*) law are being expanded, bringing official and *Sharia* law into confrontation throughout the north. Nigeria's viability as a unified state is now being threatened by the rise of Boko Haram, the brutally violent jihadist organization that is seeking to expand from its base in the far northeast after declaring itself a caliphate in 2014. Can this country avoid the fate of Syria and thwart the formation of another ISIS-type "Islamic State"? And how will Nigeria's moderate Muslims respond to the terrorist tactics of the extremist Boko Haram cult?

Christianity first penetrated Africa in the northeast when Nubia and Axum were converted, and Ethiopia has been a Coptic Christian stronghold since the fourth century AD. But the Christian churches' real invasion did not commence until the onset of colonialism after the turn of the sixteenth century. Christianity's various denominations made inroads in different locales: Roman Catholicism in much of Equatorial Africa, mainly at the behest of the Belgians; the Anglican Church in British colonies; and Presbyterians and others elsewhere. And these days, as in South America, Evangelical churches steadily gain adherents.

Islam had a rather different arrival and impact. Today, about one-third of the realm's population is Muslim (see Fig. 6-21). By the time of the colonial invasion, Islam had advanced out of Arabia, across the Sahara, and partway down the Indian Ocean coast of Africa. Muslim clerics converted the rulers of African states and commanded them to convert their subjects. They Islamized the savanna states and penetrated into present-day northern Nigeria, Ghana, and Ivory Coast. They encircled and isolated Ethiopia's Coptic Christians and Islamized the Somali people in Africa's Horn. They established beachheads on the coast of Kenya and took over offshore Zanzibar. Muslim and European-Christian proselytizers competed for African minds, but Islam proved to be a far more pervasive force. Today, tension between Islam and Christianity is especially evident all along the **Islamic Front**, one of the defining regional characteristics—and the dynamic southern border—of the African Transition Zone (discussed at the end of Chapter 6).

Urbanization and Africa's Dual Economies

Subsaharan Africa may still be one of the least urbanized geographic realms, but people everywhere are moving to towns and cities at an accelerating pace (**Box 7-6**). Today, just about 40 percent of the realm's inhabitants reside in urban settlements (up from 28 percent as recently as 2000), and in absolute numbers they total more than the entire population of the United States. Even though the growing tide of migrants is overwhelming the realm's already inadequate urban infrastructure, rapid urbanization is projected to continue far into the future—a current crisis that will soon be transformed into a social catastrophe because half of all babies born into the world between now and 2050 will be African.

The biggest African cities operate as the anchors of incipient national core areas, housing government offices and the headquarters of the most influential corporations and extractive industries. The **formal economy [18]**, under governmental control that regulates the civil service, business, and industry as well as their workforces, produces a formal urban landscape featuring a modern government district, clusters of newer highrise buildings, upscale hotels, elite gated residential communities, and even branded fast-food outlets. From a distance, such cityscapes resemble those of urban centers in the Global Core, but on closer inspection it becomes evident that the formal sector is far from dominant here. In the streets, on sidewalks next to shop windows, in downtown-area slums, and on the rough-hewn urban periphery, hustling street hawkers, garment makers, jewelry sellers, and unlicensed taxis signal the functioning of a parallel economy that lies almost entirely outside of government control. This is the **informal economy [19]**, which looms large across every African country (as well as those countries throughout the Global Periphery), and entails an alternative way of living, working, and operating outside of the mainstream national economy. An overwhelming majority of Africans participate in the informal sector by residing in unofficial housing (usually shacks—see below), transacting business by avoiding banks and the payment of taxes, and by relying on networks of family, friends, and acquaintances for their survival.

Urban Housing Innovations: The i-Shack

Shacks are ubiquitous in this realm's poverty-suffused urban landscapes. Substandard dwellings are always unacceptable, but with some carefully planned and practical design improvements shacks can be creatively enhanced to yield a better quality of life for their disadvantaged inhabitants. The Sustainability Institute of South Africa's Stellenbosch University has produced a new prototype shack design called the *i-shack* ("improved shack"), and a pilot project to develop and apply this concept is underway at Enkanini, an informal settlement in the town of Stellenbosch that lies at the eastern edge of metropolitan Cape Town. I-shacks are rudimentary solar-powered dwellings with the capacity to support two indoor lights, a motion-sensitive outdoor light, and a mobile phone charger. Its windows are strategically positioned for optimal air circulation and sunlight heating, while the roof is sloped to harvest rainwater. Recycled cardboard boxes and "Tetra Pak" containers are used for insulation between the exterior zinc surface and the interior, and flame-retardant paint is employed to minimize the risk of fire. I-shacks can be constructed and fitted for U.S. $660, and represent a significant and affordable upgrade on the everyday self-built shack made out of corrugated iron and discarded wood—which are too damp in winter and especially fire-prone in summer (**Box 7-7**).

Africa Rising?

Despite burdensome colonial legacies, formidable environmental challenges, and a long history of serious adversity, twenty-first century Africa is steadily emerging on the global scene. Although conditions remain harsh in many places, the achievements of the past decade are quite encouraging, even stunning in certain cases (such as the lengthening of life expectancy in the realm by a remarkable eight years since 2000). Gathering economic momentum is the new watchword, and the development outlook for a growing number of countries appears resilient even in the face of challenges such as the slowdown of

Box 7-6 Major Cities of the Realm, 2016	
Metropolitan Area	**Population (in millions)**
Lagos, Nigeria	12.8
Kinshasa, DRCongo	11.4
Johannesburg, South Africa	8.7
Nairobi, Kenya	4.9
Abidjan, Ivory Coast	4.8
Dar es Salaam, Tanzania	4.5
Accra, Ghana	4.1
Cape Town, South Africa	3.9
Durban, South Africa	3.5
Adis Abeba, Ethiopia	3.5
Dakar, Senegal	3.2
Ibadan, Nigeria	2.9
Lusaka, Zambia	2.7
Harare, Zimbabwe	2.2
Mombasa, Kenya	1.2

A typical urban landscape in Subsaharan Africa. This aerial view of the city of Arusha in northern Tanzania (population: ca. 440,000) shows several features of the settlement layout in urban Africa. The irregular pattern of low-rise, low-density development reflects a general lack of planning, complex land ownership issues, and piecemeal as well as patchwork investment in construction.

the Chinese economy and commodity price declines, both key determinants of the realm's rapid growth of the past decade. However, the commodity declines and China's slowdown have negatively impacted the economic outlook for South Africa,

Nigeria, Angola, and Zambia, all economies that rely heavily on minerals and oil (**Box 7-8**). Yet commodity price declines are no longer the kiss of death to the region, and this time around many countries are holding up despite the challenging international economic arena.

Still, Subsaharan Africa has averaged annual economic growth of more than 5 percent since 2002, outperforming all other realms (**Fig. 7-11**). Several African countries—led by DRCongo, Ethiopia, Ivory Coast, and six more—are among the world's top 20 in terms of GDP growth rates in 2015. More than half of the top African performers are authoritarian regimes, suggesting that dictatorial rule is related to strong economic performance. At the same time, total foreign investment in the realm rose dramatically from U.S. $11 billion in 2001 to $80 billion in 2014, much of it focused on the burgeoning resource-extraction sector but with ever more industrial and services projects added to the mix. Nowadays, that financial influx is increasingly channeled into telecommunications, banking, infrastructure, and even manufacturing.

What underlies these far more positive economic assessments of Subsaharan Africa's development potential? First, the realm is demonstrably closing the gap between its economic performance and that of the Global Core. Second, the management of Africa's national economies is gradually improving, and the resolution of political conflicts together with the expansion of democratic governance has led to more favorable

Box 7-7 From the Field Notes...

"Shacks are a standard form of housing in the cities of Subsaharan Africa, and have proliferated across the realm as the population continues to rapidly urbanize. Because national and local governments have only limited funds to rehouse slum dwellers, shacks are going to remain a signature landscape of the African city well into the future. Given this status quo, various prototypes have been developed to upgrade the experience of living in shacks. The i-shack is a good example of just such an improved dwelling unit, and also demonstrates how a slum environment can be incrementally upgraded in a meaningful way. A pioneering effort is being spearheaded by Stellenbosch's Enkanini i-shack pilot project, which I visited during my 2015 stay in that city. Since then, Enkanini has progressed to the point that now more than 1000 of its shacks are powered directly by solar energy. This has attracted the attention of the Bill and Melinda Gates Foundation, which wants to facilitate the diffusion of such innovative housing to cities throughout South Africa and nearby countries."

Box 7-8 Fastest-Growing Subsaharan African Economies and Leading Exports, 2015

	% Annual GDP Growth	1st and 2nd Biggest Export Products
DRCongo	9.1	diamonds, copper
Ethiopia	8.6	coffee, hides
Ivory Coast	7.7	oil and natural gas, cocoa
Chad	7.5	cotton, oil
Tanzania	7.2	gold, sisal
Rwanda	7.0	coffee, tea
Kenya	6.9	horticulture, tea
Zambia	6.7	copper, minerals
Djibouti	6.5	live animals, oil

business environments. Third, a new generation of policymakers, activists, and business leaders is emerging across most of Subsaharan Africa—they are Africans to the core, but with an increasingly global mind-set derived from their participation in the Internet age, easier international travel and communications, and greater exposure to current worldwide trends and thinking. Fourth, the African debt crisis is winding down, and economic relations with Western donor countries are improving. And fifth, most important of all, China has exploded onto Africa's economic scene, influencing trade, investment, foreign aid, and diplomatic relations.

China's Role in Emerging Africa

Through much of modern history, Westerners have perceived Africa to be both remote and peripheral to international affairs. Although the subordinate, asymmetrical relationship that long tied this realm to Europe's imperial powers is still visible (in terms of foreign investment, trade, and external oversight), evidence is mounting that things have begun to change. The recent rise of emerging-market countries such as Brazil, Russia, India, and China—the so-called **BRICs**—as well as the retreat of various Western development models as the only game in town, mean that African leaders now have a wider range of options in their international relations.

Many African states are now forging political and economic partnerships of their own choosing. They are casting aside the

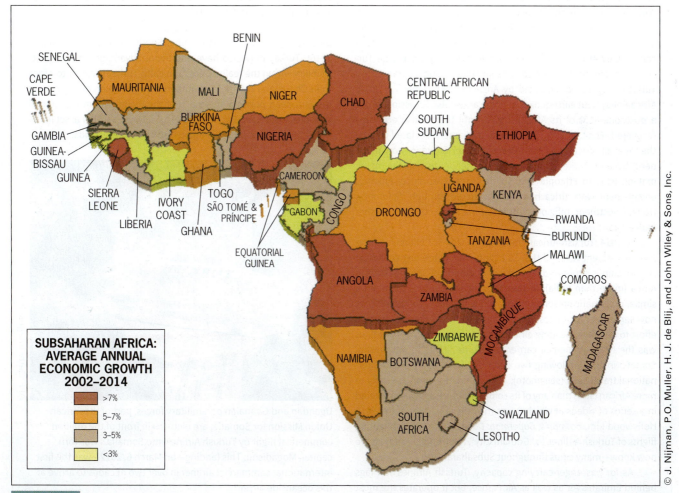

SUBSAHARAN AFRICA: AVERAGE ANNUAL ECONOMIC GROWTH 2002–2014

- >7%
- 5–7%
- 3–5%
- <3%

© J. Nijman, P.O. Muller, H. J. de Blij, and John Wiley & Sons, Inc.

FIGURE 7-11

shackles of colonial thinking and its postcolonial legacies and taking more control of their own international economic policies. Such enhanced agency enables African states to become full-fledged players in international relations rather than the objects of other countries' foreign policies—the dawning of a new era that some are calling a "second independence" for Africa.

Subsaharan Africa is now engaging a host of foreign investors (see **Box 7-9**) but nothing marks the elevation of this realm's stature more than China's continuing surge of investments in infrastructure, energy-producing facilities, and civic construction. For the period 2005-2015, Nigeria tops the list with U.S. $32.2 billion of Chinese investments followed by Ethiopia with U.S. $17 billion of investments. Several other resource-rich African countries (e.g., South Africa, Moçambique, Guinea, Chad, Sudan) have been awarded about U.S. $10 billion worth of Chinese-funded projects now under development (**Fig. 7-12**). Whereas investment in energy and mining dominated over the past decade as priority sectors for investment, the mid-2010s have seen a shift toward investments in infrastructure (particularly transport) and real estate (mostly in South Africa, Kenya, and offshore Mauritius).

China's **New Silk Road** project (discussed at the end of Chapter 5) includes a Subsaharan Africa component (Fig. 7-12). Beijing's "One Belt, One Road" strategy (see Fig. 5-19) relies heavily on upgrading infrastructure in several African ports—especially Mombasa (Kenya), Massawa (Eritrea), and Djibouti (Djibouti). There also are Chinese plans to build a new U.S. $10 billion port at Bagamoyo, Tanzania, located 60 kilometers (37 mi) north of the country's capital, Dar es Salaam. The Bagamoyo plan, if ever fully implemented, is projected to become the largest port in East Africa. On land, China is investing heavily in building new and/or upgraded railroads which include: (1) the "New East African Railway" to connect the Kenyan cities of Mombasa and Nairobi (the capital) with lines that serve Kampala (Uganda), Kigali (Rwanda), and Juba (South Sudan); (2) expansion of the Tanzanian rail network that will include long-distance connections to Djibouti on the Red Sea as well as Angola's Atlantic ports clear across the continent; and

---

Box 7-9 Turkish Airlines Becomes Africa's Leading Air Carrier

The landscape of Africa's external engagement is always shifting. With the world's attention focused on China's expanding ties with Africa, Turkey has been steadily increasing its presence in the realm as well. Of course, Ottoman Turks had ties to North Africa and coastal East Africa in earlier times, but over the past decade Turkey has dramatically intensified its linkages to Subsaharan Africa. Turkish interest was initiated by Ankara's 1998 "Opening Up to Africa Policy," but earthquakes at home and global recession forced a postponement of its African thrust until the Turkish economy recovered. Then, in 2005, the Turkish parliament declared it to be the "Year of Africa," and then-Prime Minister Erdoğan (now president) became Turkey's first government leader to visit the realm, making trips to Ethiopia and South Africa. Ever since, Ankara's engagement with Africa has been furthered through diplomacy, trade, investment, religion, culture, airline connectivity, and other endeavors.

In 2014, Turkish-African trade rose to U.S. $23 billion, and foreign investment reached U.S. $6 billion. This tightening economic relationship is supported by Turkey's operation of 39 embassies in Africa (more than India); in turn, 32 African countries maintain a similar diplomatic presence in Ankara. Significantly, Turkish Airlines now serves 44 African destinations and has made a major strategic effort to be Africa's leading air carrier. For instance, Turkish Airlines was the first commercial carrier to resume flights to Mogadishu, Somalia in 2012, following two decades of civil conflict and international travel bans (see photo). Turkish Airlines' ambition to fly to more African cities than any of its competitors was heavily marketed in a series of videos in which it was the official partner of the 2016 Hollywood film, *Batman v. Superman: Dawn of Justice*, highlighting flights of Turkish Airlines to "Gotham City", "Metropolis", and—as we now know—many cities throughout Subsaharan Africa.

As for passenger-carrying capacity, Turkish Airlines still lags behind Emirates Air as well as Air France, but it operates the most spatially extensive air-travel network in terms of total Subsaharan African destinations. No doubt, some routes are struggling to make a profit, and at times political instability requires that certain routes be suspended (e.g., flights to Juba, South Sudan in the spring of 2016). Turkey may also have additional advantages in raising its visibility within the realm. Geographically, Turkey is closer to Africa than China, and it is not encumbered by the colonial baggage of the European powers. And many of the realm's countries contain sizeable Muslim populations (see Fig. 6-21), yet another asset in the strengthening of cultural, religious, and economic bonds.

Ugandan and Ghanaian paramilitary forces, part of the African Union Mission for Somalia, are pictured in front of the maiden commercial flight by Turkish Airlines into Somalia's war-torn capital, Mogadishu. This landing—on March 6, 2012—was the first international commercial airliner in over two decades to arrive at this oceanside airport.

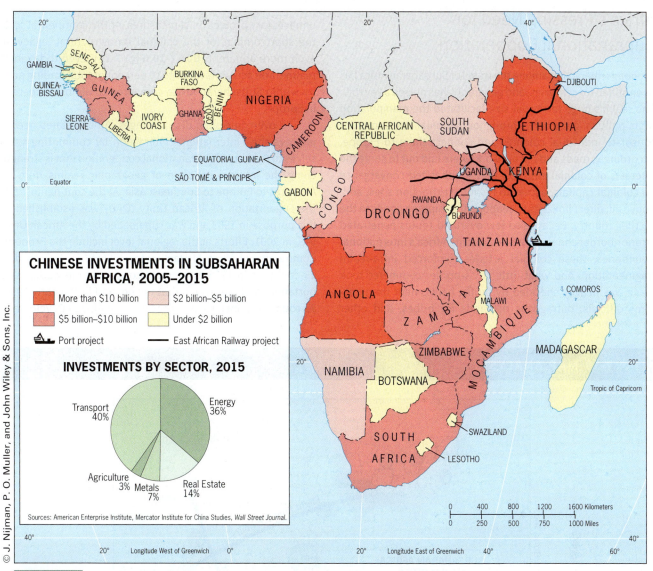

CHINESE INVESTMENTS IN SUBSAHARAN AFRICA, 2005–2015

- More than $10 billion
- $5 billion–$10 billion
- $2 billion–$5 billion
- Under $2 billion
- Port project
- East African Railway project

INVESTMENTS BY SECTOR, 2015

Transport 40%
Energy 36%
Agriculture 3%
Metals 7%
Real Estate 14%

Sources: American Enterprise Institute, Mercator Institute for China Studies, *Wall Street Journal*.

© J. Nijman, P. O. Muller, and John Wiley & Sons, Inc.

FIGURE 7-12

(3) an upgraded rail connection between the port of Djibouti and Ethiopia's capital, Adis Abeba (Fig. 7-12). China has further bolstered its African presence with its first-ever overseas military base in Djibouti—alongside that small country's U.S. naval base, the only place on Earth where warships of these two countries' navies are moored practically next to each other. As Figure 6-22 and its surrounding discussion emphasize, Djibouti occupies a highly strategic geographic position at the choke-point southern outlet of the Red Sea, through which passes at least one-fourth of the world's oil shipped by supertanker.

Underscoring this deepening engagement, China has now become the realm's largest trading partner as well as foreign-aid provider, and is also a leading source of immigrants (elaborated by Howard French in his monumental 2014 book, *China's Second Continent: How a Million Migrants Are Building a New Empire in Africa*). Moreover, China holds valuable assets in the form of oil leases, joint mining ventures, and timber concessions. Perhaps most significant is the difference between the

Chinese and Western engagement with contemporary Subsaharan Africa: Beijing does not require African states to emulate its values. Therefore, this new kind of engagement establishes an alternative enabling environment, wider options for maneuverability, and the opportunity for Africans to chart untried development pathways—ideally of their own choosing.

Africa's ongoing rise as a strategic component of the global economy is based on the realm's natural resources. This has brought major new opportunities to develop infrastructure and real estate (e.g., satellite cities, industrial parks, modern CBDs, shopping complexes, and the like), provide commercial and producer services for Africa's mushrooming urban population, and accelerate the emergence of a middle class. Furthermore, many of the realm's countries are now being courted as diplomatic allies in global politics. In a world where votes matter more than ever, those of the 54 Subsaharan African states increasingly resonate in international forums that shape trade, environmental, and security-related policies.

Africa's Pressing Need for Supranational Cooperation

Economic interaction among the states of Subsaharan Africa has been stifled for centuries, and geography has played a key role. The realm's huge territorial size facilitated the manipulation of its political geography throughout colonial times in order to isolate individual colonies and connect them to Europe. Accordingly, roads and rail networks were laid out to serve the interests of colonial empires: they emphasized convergence on colonial capitals, access to resource extraction sites, and, above all, direct linkages to seaports for shipping goods to the mother country. Because so few overland routes penetrated interior border zones, let alone crossing Africa's international boundaries, these colonial arteries reinforced intra-realm fragmentation. And what little cross-border commerce took place was heavily shackled by tariffs, harassment from customs officials, outrageous demands for bribes, and other

unnecessary protectionist measures. At the continental scale, the shortcomings of the colonial infrastructural legacy are even more apparent (**Box 7-10**; **Fig. 7-13**). Even today intra-realm trade accounts for only 16 percent of total trading activity within Subsaharan Africa compared to almost 70 percent in Europe.

To overcome such disadvantages, African states need to improve their international interaction by committing themselves to the kind of supranational cooperation that is strengthening the internal cohesion of geographic realms from the Americas to Europe to Southeast Asia. At the continental scale, the Organization of African Unity (OAU) was established for this purpose in 1963 and was superseded by the African Union (AU) in 2001. Efforts at the regional level include the Economic Community of West African States (ECOWAS), launched by 15 countries in 1975 to promote cooperation in trade, transportation, and industry in that region; and in the early 1990s, 12 countries joined to form the Southern African Development

Box 7-10 MAP ANALYSIS

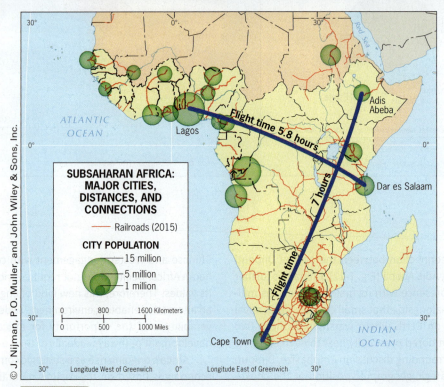

FIGURE 7-13

With one-fifth of the Earth's land surface, continental Africa is larger than China, India, the United States, Mexico, and Europe combined. This map shows the major cities of Subsaharan Africa (greater than 1 million) along with existing railroad networks. With the recent surge of natural-resource-based foreign investments in this realm, the need for transporting raw materials and other commodities to ports and markets has risen exponentially. In Figure 7-13, observe

where most major cities are located, the distances among them, where rail networks are concentrated, and typical long-distance flight times. How would you define the realm's challenge in terms of overland transportation? What do you think should be the primary transportation development strategy?

Access your WileyPLUS Learning Space course to interact with a dynamic version of this map and to engage with online map exercises and questions.

Middle-class youths at a recent midsummer cultural festival in Kampala, Uganda. Demands for education, cultural goods, computer and Internet use, and access to leading entertainment media are skyrocketing, and upscale consumption is steadily on the rise in many of Subsaharan Africa's cities.

Community (SADC) and were soon followed by the 20-odd states that founded the Common Market for Eastern and Southern Africa (COMESA).

Although institutional structures for economic integration already exist, the various initiatives need streamlining. A promising step in that direction was taken in 2015 with the signing of the Tripartite Free Trade Agreement (TFTA) that will form a "Cape-to-Cairo" free-trade zone to be known as the Continental Free Trade Area. Coordinated by the AU, this pact involves the 26 countries participating in the amalgamation of SADC, COMESA, and the EAC (East African Community), and will create Africa's largest common market (with a total population of 630 million) when it begins operating in 2017. However, such commendable intentions aside, supranational organizations have yet to prove they can be mechanisms for meaningful economic progress. So now more than ever, the countries of Subsaharan Africa must summon considerable political will to build lasting regional cooperation and realm-wide interdependence.

Regions of Subsaharan Africa

On the face of it, Africa is so massive, compact, and continuous that any attempt to justify a contemporary regional breakdown is fraught with difficulties. Nevertheless, on the basis of environmental distributions, ethnic patterns, cultural landscapes, distance and proximity, historic culture hearths, and other spatial data, we identify a four-region structure (**Fig. 7-14**). By focusing on the contents of that framework, however, we should not overlook a fifth region that straddles the broad divide between Subsaharan Africa and North Africa—the **African Transition Zone**, discussed separately in Chapter 6 as the final region of the North Africa/Southwest Asia realm.

Beginning in the north at the southern margin of the African Transition Zone, we work our way east and then south through the following regions:

1. **West Africa**, which includes the countries of the western coast north of the equator and those on the margins of the Sahara in the interior. This populous region of fifteen countries is anchored in the southeast by the realm's demographic giant, Nigeria.

2. **East Africa**, where equatorial natural environments are moderated by elevation and where plateaus, lakes, and mountains, some carrying permanent snow, exemplify the countryside. Six countries, including the highland part of Ethiopia, comprise this region. The island-state of Madagascar, marked by Southeast Asian influences, is neither East nor Southern African but is included here because of its relative location.

3. **Equatorial Africa**, much of it defined by the basin of the Congo River, where elevations are lower than in East Africa, temperatures are higher and moisture more abundant, and most of Africa's surviving tropical rainforests stand. Among the nine countries that make up this region, which now includes recently formed South Sudan, DRCongo dominates territorially and demographically.

4. **Southern Africa**, extending from the southern tip of the continent to the northern borders of Angola, Zambia, Malawi, and Moçambique. Ten countries constitute this region, which extends southward from the tropics and whose giant is South Africa.

Region West Africa

West Africa occupies most of Africa's northwestern Bulge, extending southward from the margins of the Sahara to the Gulf of Guinea coast and westward from Lake Chad to Senegal (**Fig. 7-15**). A total of 17 states (also counting Chad and offshore Cape Verde) constitute this region. In addition to once-Portuguese Guinea-Bissau and long-independent Liberia, West Africa encompasses four former British and nine former French dependencies. Four countries, comprising an enormous territory dominated by dry environments but sparsely populated, are landlocked: Burkina Faso, Mali, Niger, and Chad.

As Figure 7-15 shows, political boundaries extend perpendicularly from the coast into the interior, so that from Mauritania to Nigeria, the West African habitat is parceled out among parallel, coast-oriented states. The southern half of the region is home to most of the people (Fig. 7-8). Mauritania, Mali, and Niger include too much of the *Sahel*—the unproductive belt of Sahara-bordering, semiarid environment (mapped in Figure 7-15)—to sustain populations as large as those of the most populous countries of Nigeria, Ghana, and Ivory Coast. Another key feature of the region's human geography is West Africa's bifurcation by a religious divide in upheaval—the turmoil-ridden *Islamic Front* (see Fig. 6-21) that forms the southern edge of the African Transition Zone (the black dashed line in Fig. 7-14).

© J. Nijman, P.O. Muller, H. J. de Blij, and John Wiley & Sons, Inc.

REGIONS OF SUBSAHARAN AFRICA
- Southern Africa
- East Africa
- Equatorial Africa
- West Africa
- African Transition Zone
- Approximate southern boundary of Islam

FIGURE 7-14

Nigeria

Nigeria, this region's cornerstone, is home to 187 million people. This is by far the largest population of any African country, and within the region more people live in Nigeria than in all the other countries of West Africa combined. Moreover, in 2014 Nigeria surpassed South Africa to become the largest economy in the realm. But this country is also bisected by one of the most contentious segments of the Islamic Front, which in this decade has unleashed the vicious Boko Haram insurgency based in the northeast. And another current security crisis affects the exporting of Nigeria's most valuable resource, oil; pirate attacks are on the rise along the Gulf of Guinea coast, aimed at the main sea lane used by tankers to transport oil to Global Core countries from the extensive fields that lie beneath the Niger River Delta in the southeastern corner of the country.

Nigeria achieved independence from Britain in 1960, but its new government faced the herculean task of managing a European political creation containing three major nations and nearly 250 other peoples. For reasons obvious from the map (**Fig. 7-16**), Britain's colonial imprint was always stronger in the two southern subregions than in the north. Christianity became the dominant faith in the south, and southerners, especially the Yoruba, took a lead role in the transition from colony to independent state. The choice of Lagos (**Box 7-11**), the primary port of the Yoruba-dominated southwest, as the capital of a federal Nigeria reflected British aspirations for the country's future. A three-region federation, two of which lay in the south, would ensure the primacy of the non-Islamic part of the state. But this framework did not last long and required frequent modification. Nigeria today has 36 States, and the capital was moved from Lagos to more centrally located Abuja more than a quarter-century ago (Fig. 7-16).

Oil Booms and Busts

Large oilfields were discovered beneath the Niger Delta in the 1950s, and this transformed the agriculture-dominated

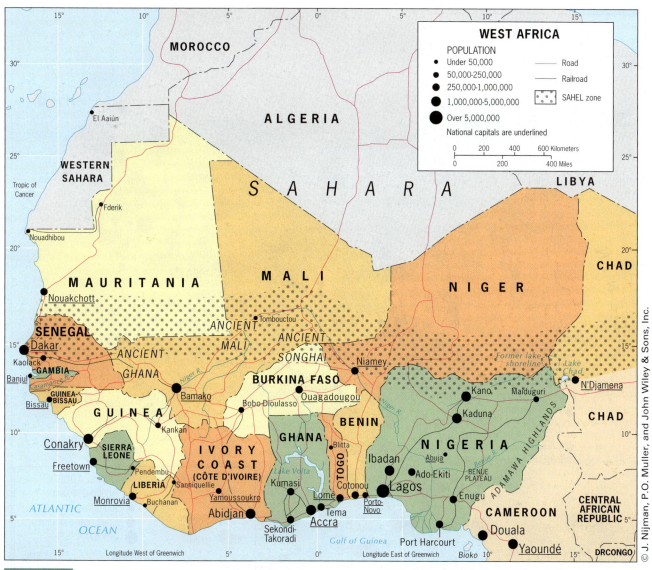

FIGURE 7-15

economy into one based on petroleum exports. But before long, Nigeria's energy riches brought more bust than boom. New misguided development plans focused on ill-considered industrial schemes and the like, while farming, still the mainstay of most Nigerians, fell into neglect. Worse, poor management, corruption, outright theft of oil revenues during military misrule, and excessive borrowing against future oil income led to economic disaster. Much of the country's infrastructure collapsed; in cities, basic services broke down; and in rural areas, clinics, schools, water supplies, and roads to markets crumbled.

Nonetheless, Nigeria has now become the realm's largest economy. Most of that success results from holding its position as the world's twelfth-largest oil producer, with the United States one of its main customers. But these economic benefits have yet to reach the great majority of the fast-growing Nigerian population (birth rates here are double the global average), 80 percent of which still subsists on the equivalent of less than one U.S. dollar a day.

Islamic Revivalism and Boko Haram

In 1999, Nigeria's hopes were heightened as a democratically elected president was sworn into office. But shortly thereafter its challenges magnified when northern States, beginning with Zamfara, began to promulgate strict-Islamic **Sharia law [20]**. When Kaduna State followed suit, violence between Christians and Muslims was unleashed and devastated that State's venerable capital city, Kaduna. There, and in the 11 other northern States mapped in green in Figure 7-16, the imposition of Sharia law led to the departure of thousands of Christians, intensifying the cultural fault line that we now know as the Islamic Front. That schism continues to deepen, and the kind of Islamic revivalism now taking hold in the north has reached a level that is imperiling the stability—perhaps even the unity— of West Africa's cornerstone state.

In 2002, a jihadist Islamic organization calling itself **Boko Haram [21]** (meaning "Western education is sinful"), emerged in the northeasternmost State of Borno (Fig. 7-16), an impoverished, underrepresented corner of Nigeria where repressive

FIGURE 7-16

central government policies had triggered growing militancy. Support for Islamic militants from the State's heavily Muslim population is mainly drawn from young, unemployed, northern Nigerian men, who have lost faith in the politicians and in their own future. Boko Haram's "religious cleansing" campaign was initially confined to Borno, but since 2008 its barbaric activities have expanded across the entire northeastern quadrant of Nigeria, including high-profile terrorist attacks in Kano and even the federal capital, Abuja, well to the south. Targeting civilians, Boko Haram attacks during 2015 killed more than 6000 and forced tens of thousands to flee their home villages (see photo). On April 14, 2014, the world's attention was seized by the dramatic kidnapping of 276 female adolescents from a school in the village of Chibok, to be forcibly married to Boko Haram fighters—igniting

the global *#Bringbackourgirls* social media campaign. By mid-2016 the girls had been missing for over two years.

Boko Haram's aggressive terror campaign against Western education has resulted in the death of 611 teachers and the flight of 19,000 educators since 2009. Geography teachers, in particular, have emerged as a top terrorist target because of their knowledge of the political geography of nations and states and their global perspectives. For example, Boko Haram followers hold bizarre worldviews, contending that the Earth is flat and that rainfall patterns are attributable to Allah's divine will and not to climate-physical processes and **anthropogenic [22]** forces. Anthropogenic causes of drought emphasize adverse human influences on the environment such as land degradation (deforestation, continuous cropping, overgrazing) and

Box 7-11 Among the Realm's Great Cities: Lagos

In a realm that is barely 40 percent urbanized, Lagos, former capital of federal Nigeria, stands out: a teeming megacity of 12.8 million (though some estimates are considerably higher), bigger than Paris or Rio de Janeiro. Lagos boasts the busiest port in West Africa, the country's financial center, the main telecommunications complex, and leading service-sector firms including Nigeria's booming film industry, "Nollywood".

The city evolved over the past three centuries from a Yoruba fishing village, Portuguese slaving center, and British colonial head-quarters into Nigeria's (and Africa's) largest city, principal port, fore-most industrial center, and first capital (1960–1991). Situated on the country's southwestern coast, it consists of a group of low-lying barrier islands and sand spits between the swampy shoreline and Lagos Lagoon. The center of the city still lies on Lagos Island, where the high-rises adjoining the Marina overlook Lagos Harbor and, across the water, Apapa Wharf and the Apapa industrial district. The city expanded southeastward onto Ikoyi Island and Victoria Island, but after the 1970s most urban sprawl took place to the north, on the western side of Lagos Lagoon.

Lagos's cityscape is a teeming mixture of modern high-rises, dilapidated residential areas, and squalid slums. From the top of a high-rise one sees a seemingly endless vista of rusting corrugated roofs, the houses built of cement or mud in irregular blocks sepa-rated by narrow alleys. On the outskirts lie the shantytowns of the least fortunate, where shelters are made of plywood and cardboard and lack even the most basic facilities. Slum demolitions are a reg-ular occurrence, often to make way for new construction. The gov-ernment has proclaimed its ambition to make Lagos into Africa's premier business center.

Traffic in Lagos is as bad as it gets anywhere, with daily com-mutes of three or more hours the rule rather than the exception. During rush hour (which is almost always) cars move at about 5 kilometers (3 mi) an hour. The only beneficiary of this chaos is the *okada*, the motor-bike taxi that in recent years has taken the city by storm. These small vehicles sometimes carry entire families, weav-ing swiftly through traffic and violating every rule in the book—and proliferated so rapidly that they became a problem and are now banned from certain roads and local areas.

By world standards, Lagos ranks among the most severely polluted, congested, and disorderly cities. Mismanagement and official corruption are endemic. However, the building of the new Eko-Atlantic City CBD is a major effort to upgrade the international reputation of Lagos (see **Box 7-12**).

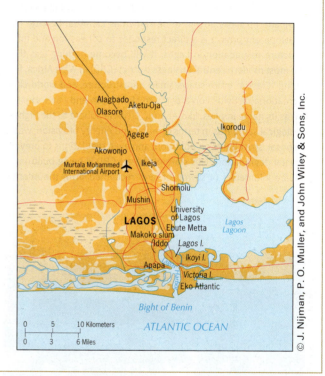

overpopulation, which lead to more extreme weather, climate change, and devastating impacts on Sahelian society. Children across the Sahel, particularly in northern Nigeria, face severe malnourishment and food insecurity. That is one of the driv-ers of displacement, political instability, and conflict. Thus an understanding of state security and its relationship to complex human-climate processes is essential to combat the vicious cycle of conflict and food insecurity.

In early 2015, following the lead of ISIS, Boko Haram declared itself an independent caliphate consisting of the Nigerian territory it controlled, centered on Borno's capi-tal, Maiduguri, a city of nearly 700,000 (Fig. 7-16). Operating under the banner of its own black flag, Boko Haram at its height dominated an area larger than West Virginia. However, since Nigeria's newly elected regime entered office in April 2015, Boko Haram has been pushed back, with the govern-ment claiming that about 40,000 Boko Haram fighters have surrendered and laid down their arms. Some 1.5 million displaced people have now returned to northern Nigeria, and investors are scrambling to acquire land in Borno State. Even Maiduguri is experiencing a real estate boom following the retreat of the insurgents. In mid-2016, Boko Haram is being decimated, yet has still not been defeated; but its retreat is enhancing stability in the Lake Chad area that encompasses Nigeria's far northeast as well as neighboring lands in Chad, Cameroon, and Niger.

Coast and Interior

West Africa's other countries also exhibit distinct regional geo-graphies. Typical of the Gulf of Guinea coastal zone west of Nige-ria is neighboring **Benin**, with a population of 11 million and growing cultural and economic ties with northeastern Brazil's Bahía State, where many of its people were taken in bondage and where elements of West African culture survive. Benin's internal

Box 7-12   Regional Planning Case

The New Lagos: Eko-Atlantic City

Explosive urban growth in Subsaharan African cities has far outpaced the provision of basic infrastructure such as electricity, potable water supply, waste disposal, and public transportation. It seems that nowhere are these infrastructural deficits more visible than in the megacity of Lagos. Some 2000 newcomers arrive in Lagos every day. The housing shortage is so severe that somewhere between 85,000 to 250,000 slum dwellers are crammed into the world's largest floating slum, Makoko (see photo), forced to live in self-built shacks on stilts in the bay as well as on reclaimed land (see map in Box 7-11).

At the same time, the city is constructing a world-class urban node and business district to be known as Eko-Atlantic City. This development project is a direct response to congestion and infrastructure degradation in Lagos. It will become nothing less than the most modern central business district (CBD) in West Africa, featuring "Class A" office space, smart infrastructure, and world-class recreational activities on a par with Manhattan and Paris. Most of all, as its name implies, Eko-Atlantic City will adhere to the highest ecological standards in terms of water and energy provision, waste disposal, strict limits on pollution, and even regulations to prevent coastal degradation. This state-of-the-art CBD is being built on 10 square kilometers (4 sq mi) of land reclaimed from the Atlantic

Ocean's Gulf of Guinea at the southern tip of Victoria Island (map, Box 7-11).

The master plan consists of various components including marinas, street-lined boulevards, luxurious residences with oceanfront views, ultramodern office parks, and upscale retail facilities, hotels, and schools. Nigeria's oil-rich and political elites, wealthy returnees, affluent expatriate professionals, and upwardly-mobile urbanites are being targeted as future residents and business tourists. If all goes as planned, Eko-Atlantic City will be home to 250,000 permanent residents and 100,000 daily commuters. The project, estimated to cost U.S. $6 billion, is being financed by private developers and donors.

Time will tell whether Eko-Atlantic City can become the model of a 21st century eco-friendly city. Not surprisingly, this project elicits both praise and criticism. On the positive side, it is viewed by some as an affirmation of Nigeria's position as this region's economic powerhouse as well as its global ambitions. However, the stark reality is that 80 percent of Nigerians continue to live on the equivalent of less than one U.S. dollar per day. Persistent poverty, economic and spatial polarization, and widening income inequality have led critics to contend that Eko-Atlantic City is but a contemporary version of the colonial city—the main difference being that wealth rather than race is shaping the spatial segregation that results.

The floating slum of Makoko, near downtown Lagos, on the west side of the outlet of Lagos Lagoon.

Ongoing construction of Eko-Atlantic City on Lagos' Victoria Island in early 2016 (see map in Box 7-11).

geography is comparable to that of neighboring **Togo** (population: 7.5 million): both are markedly elongated, narrow political units with savannas in the north and humid tropical lowlands near the coast, but without usable rivers. In contrast, **Burkina Faso** is representative of West Africa's much drier interior—impoverished and landlocked, but containing undeveloped reserves of gold and a commercial economy that relies on exporting cotton. Importantly, with Muslims constituting 59 percent of Burkina Faso's 18.6 million inhabitants, this country—together with Mali (93 percent), Niger (99 percent), and transitional Chad (58 percent)—lies firmly within the Islamic orbit of northern West Africa.

Among the coastal states, **Ghana** (population: 28 million), known in colonial times as the Gold Coast, was the first modern West African state to achieve independence, with a democratic government and a sound commercial economy based on cocoa exports. Both democracy and the economy soon failed. In the 1990s, a military regime was replaced by a stable representative government, and recovery began. By 2007, Ghana celebrated 50 years of independence, its democracy maturing, its economy forging ahead, corruption persistent but declining, and its international stature rising. The same year also brought a U.S. $500-million-plus aid package from America's Millennium

epa european pressphoto agency b.v. / Alamy Stock Photo

The aftermath of a 2013 Boko Haram attack on Kawuri in northeastern Nigeria, which resulted in the murder of 70 villagers and the destruction of every home. About a year later, the jihadist killers struck again in a nearby town as 50 gunmen stormed a school at dawn, stabbing 43 students to death to avoid attracting attention by using their firearms and homemade bombs.

Challenge Corporation to expand commercial agriculture, further improve infrastructure, and combat poverty—based in no small part on Ghana's solid reputation as an African trailblazer with a growing middle class (**Box 7-13**).

Proceeding westward from Ghana along the southern coast of the African Bulge, **Ivory Coast** (officially *Côte d'Ivoire*) has had a turbulent history. Following independence from France in 1960, it translated the next three decades of autocratic but stable rule into economic progress; however, by the late 1990s the political succession had become badly entangled in the north-south, Muslim-Christian schism that degenerated into two rounds of civil war between 2002 and 2011. Conflict has now largely subsided, but this country of 23 million remains deeply troubled as it strives to rebuild its shattered society and cocoa-based commercial economy. Farther up the coast is another key African Bulge country, **Senegal**. This democratic state demonstrates what stability can facilitate: without oil and other major income-generating resources, and with an overwhelmingly subsistence-farming population of 15.6 million, Senegal nevertheless managed to achieve some of the region's highest GNI levels.

Between Senegal and Ivory Coast lie three other coastal countries that are just emerging from a quarter-century of civil war and horror. **Liberia**, founded in 1822 by freed American slaves who returned to Africa with the help of U.S. colonization societies, was governed by Americo-Liberians for six generations and sold rubber and iron ore abroad. A military coup in 1980 ended that era, and by 1989 full-scale civil war had embroiled virtually every ethnic group in the country. This brutal protracted conflict claimed the lives of about one-tenth of the population (4.6 million today), and hundreds of thousands of others fled the country. Liberia finally turned the corner in 2006 with the election of Ellen Johnson Sirleaf

as Africa's first female president, who succeeded in stabilizing the now-democratic political system. **Sierra Leone** followed a similar path from British dependency to republic to one-party state and military dictatorship. In the 1990s, a rebel movement, financed by diamond sales, inflicted dreadful punishment on the local population, but a remarkable turnaround resulted from British intervention during the late 2000s. Diamonds are still the leading export, and now-stabilized Sierra Leone (population: 6.6 million) ranked among Africa's fastest growing economies until Ebola struck in 2014. The same cannot yet be said of neighboring **Guinea**, where dictatorial mismanagement and a violent power struggle combined to ruin economic opportunities in both agriculture and mining. As Figure 7-15 shows, Guinea borders the troubled countries just discussed plus Ivory Coast, and it has involved itself in the affairs of all of them—receiving in return a stream of refugees in its border zones. Guinea, with a population of 13 million, has long been under dictatorial rule and remains one of Africa's lowest-income states. This need not be: the country possesses major deposits of bauxite (aluminum ore) and gold, can produce far more coffee and cotton than it does, and controls productive offshore fishing grounds.

The numerous conflicts noted in this chapter were and are geographically concentrated in relatively few areas and are not representative of vast, populous West Africa as a whole. Tens of millions of farmers and herders who manage to cope with fast-changing environments in the demanding Sahel live remote from the newsmaking conflicts along the well-watered coast. Overall, to take fuller advantage of the development opportunities sweeping across the realm, West Africa's biggest challenge is to achieve both economic survival and sustain nation-building through political stability.

Region East Africa

To the east of the chain of Great Lakes that marks the eastern border of DRCongo, the land rises from the Congo Basin to the East African Plateau. Hills and valleys, fertile soils, and copious rains mark this transition in Rwanda and Burundi. Eastward the rainforest disappears and the open savanna cloaks the countryside. Great volcanoes tower above a rift-valley-dissected upland. At the heart of the region lies Lake Victoria. Farther north the surface rises above 3300 meters (10,000 ft), and so deep are the trenches cut by geologic faults and rivers there that the land was appropriately called Abyssinia (now Ethiopia). Five countries, in addition to the highland component of Ethiopia, constitute this East African region: Kenya, Tanzania, Uganda, Rwanda, and Burundi (**Fig. 7-17**). Kenya is the most important country from a Western perspective, but Tanzania and Ethiopia have larger populations. And the landlocked countries of Rwanda and Burundi may be small in size, but their fierce ethnic conflicts have spilled over into neighboring countries and contributed to the high regional level of geopolitical instability.

Box 7-13 From the Field Notes...

"Even in the poorer countries of the world, you see something that has become a phenomenon of globalization: gated communities. Widening wealth differences, security concerns, and real estate markets in societies formerly characterized by traditional forms of land ownership combined to produce this new element in the cultural landscape. This is the entrance to Golden Gate, the first private gated-community development in Accra, Ghana, begun in 1993 as a joint venture between a Texas-based construction company and a Ghanaian industrial partner. Upscale gated communities are now found in major urban areas throughout the realm and are most widespread in the upper and upper-middle income neighborhoods of both suburbs and cities."

© Jan Nijman/ Richard Grant

Tanzania

Territorially, Tanzania (its name being a hybrid derived from the 1964 union of **Tan**ganyika and offshore **Zan**zibar) is the largest East African state. Its 55.2 million people are drawn from more than 100 ethnic groups. No single ethnic group is large enough to dominate the state; moreover, Muslims constitute 35 percent of the population, highly concentrated on the coast within the southern, tail end of the African Transition Zone (Fig. 7-17). This is also a country without a core because its clusters of population and zones of productive capacity lie dispersed—mostly on

its margins along the Indian Ocean coast, near the shores of Lake Victoria in the northwest, adjoining Lake Tanganyika in the far west, and facing Lake Malawi in the interior south (Fig. 7-8). Experimenting with socialism and farm collectivization in the post-independence period scared off international investors and tourists, and Tanzania soon bottomed out as one of the world's poorest countries. Since 1990, the government has pursued a different course, and prospects have improved. Tanzania today is a leading gold exporter, an emerging oil and natural gas producer, an increasingly popular safari destination (led by well-managed Serengeti National Park in the central north), and its economy has grown at a rate of around 7 percent a year for the past decade.

Ethiopia

Ethiopia, at 102 million, is Subsaharan Africa's most populous country after first-place Nigeria. The highland zone of Ethiopia forms the northernmost component of East Africa, but its extreme north and northeast lie within the African Transition Zone behind the Islamic Front (Fig. 7-17, red-dashed line). Culturally, as well as physiographically, Ethiopia is part of East Africa: the two peoples that together dominate its ethnic complexion (the Oromo and the Amhara) are neither Arabized nor Muslim—they are Africans. Uniquely, deep colonial penetration was avoided here. Adis Abeba, the historic capital and headquarters of a Coptic Christian, Amharic empire, held its own against Italy's attempted colonial takeover (except for a brief period in the late 1930s). Indeed, the Ethiopians, based in their mountain fortress (Adis Abeba lies more than 2400 meters/8000 ft above sea level), became colonizers themselves, at times taking control of much of the Islamic part of the African Horn to the east.

© Jan Nijman

An aerial view, near the Tanzanian-Kenyan border, of an escarpment (right) that lines the edge of East Africa's great Eastern Rift Valley. The lush landscape of this hillside extends clear across the wide valley's well-watered floor, a savanna environment of alternating trees and grasslands, teeming with wildlife.

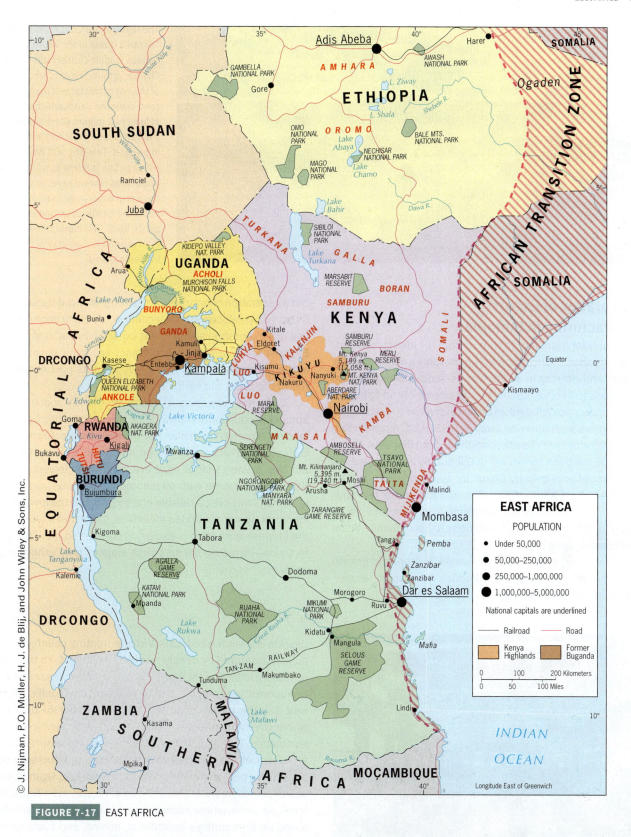

FIGURE 7-17 EAST AFRICA

During the past decade, Ethiopia has made considerable economic progress and ranks as one of the world's fastest-growing economies. Much of this growth has been attributed to government initiatives, such as selling enormous tracts of publicly-owned arable land to foreign agribusiness corporations and welcoming a wide range of Chinese foreign investment. These gains notwithstanding, the country's economy remains focused on such agricultural commodities as coffee, tea, spices, and livestock: no less than 85 percent of the Ethiopian workforce is still composed of subsistence farmers.

The Ngorongoro Crater in northern Tanzania is one of the world's most famous wildlife conservation areas. The background ridge is actually part of the rim of this largest inactive volcano crater on Earth, which formed at least two million years ago when a huge volcano exploded and caved in on itself. This UNESCO World Heritage Site is home to wildebeests and zebras (shown here) as well as dozens of other large mammals, including giraffes, elephants, leopards, lions, and hippos. Ecotourism is an economic mainstay of this immediate area.

The Paradox of Export Manufacturing in Ethiopia

Remarkably for a landlocked country with the lowest road density in Subsaharan Africa, Ethiopia has become a major exporter of shoes and clothing in addition to its traditional export base of livestock and coffee. Ethiopia relies extensively on Djibouti's port complex since it fought an unsuccessful war to keep neighboring Eritrea from breaking away in 1993 (it has longstanding political tensions with its other neighbor, Somalia). Adis Abeba since 2105 has been connected to Djibouti by an upgraded rail link, financed and built by China. By rail, goods can now be transported from Adis to Djibouti in 10 hours versus two days via road. Both countries are benefiting from increased trade, and Djibouti has geopolitical ambitions to become the African Horn's main trading hub.

Ethiopia has experienced phenomenal economic growth (annualized near 10 percent for a decade). The economy is overseen by an authoritarian regime that restricts press freedom, but a generation of returnees and strong financial backing from China has produced greater economic diversification and the rise of a major shoe and garment export manufacturing sector. Ethiopia now is a state that combines economic development and authoritarianism as it emerges from its past communist ties of the 1960s and early 1970s.

With the relocation of much of China's footwear and apparel production to Adis Abeba, the country has become a leading exporter, particularly of women's shoes. Granted that Ethiopian workers are not as productive as Chinese workers, but this country's labor rates are among the lowest globally (U.S. $60 per month with no minimum wage laws versus $629 in China). Moreover, a reliable energy grid and the low price of electrical power are added incentives for manufacturers. And the Chinese work ethic is also being transferred into Ethiopian factories—including banners with slogans that exhort maximum labor output (see photo).

Uganda

Uganda is a landlocked, Lake Victoria-fronting country of 40.2 million, bordered by deeply troubled neighbors. To the north lies newly independent South Sudan, by many measures Africa's poorest state and plagued by incessant civil conflict. To the west lies the heavily contested eastern frontier of DRCongo, where government forces and insurgents fight seemingly endless battles. To the south lies Rwanda, scene of recurrent mass genocides. All this violence has not left Uganda untouched, a destination for persistent refugee flows. In addition, Uganda itself had to overcome the legacies of one of Africa's most brutal dictators, Idi Amin, whose economic nationalization drive wreaked havoc on the country's agricultural, mining, and trade sectors throughout most of the 1970s. Another calamity followed in the 1980s at the height of the late-twentieth-century HIV/AIDS pandemic: more than a million Ugandans succumbed, orphaning 1.2 million children. Responding massively in the aftermath, the government launched such an effective public-health campaign that the country became a global poster-child in the struggle against this disease.

Workers at the assembly line at the Huajian shoe factory in Dukem, Ethiopia, one of 36 in the Adis Abeba area that collectively export about 5 million pairs of shoes annually. (In the process, Ethiopia has also become a global leader in leather production.) As China's own economy matures, companies such as Huajian are increasingly moving to Africa to find more affordable labor.

Rwanda and Burundi

Rwanda and Burundi, each with a population of just under 12 million, are Africa's most densely inhabited countries. Here, Tutsi pastoralists from the north subjugated Hutu farmers, who had themselves made serfs of the local Twa people. Belgian colonial rulers promoted the minority Tutsi population at the expense of the Hutu majority, and nothing changed following independence in the early 1960s. Not until three decades later did the oppressed majority finally erupt in the horrific genocide of 1994, when almost one million Rwandans were slaughtered in 100 days. Although Rwanda has been stabilized and is one of the fastest-growing African economies (7 percent annually over the past decade), it still relies on a narrow export base of coffee and tea—while two-thirds of the population live in grinding poverty. Unfortunately, the longstanding conflict continues to simmer in this corner of East Africa, having spilled over into the DRCongo in the form of armed combat to control strategic minerals, particularly diamonds and coltan (a valuable mineral used in mobile phones and other electronic devices).

Kenya

Over the past half-century, Kenya has played a leading role in East Africa. This country of 47.3 million contains both the region's largest city, its capital of Nairobi, and its principal port, Mombasa. After gaining independence from Britain in 1963, Kenya chose a capitalist path of development, aligning itself with Western interests. Even though it lacked major mineral deposits, Kenya built a relatively prosperous economy based on exports of coffee, tea, and other food products as well as a thriving tourist industry keyed to its magnificent landscapes and wildlife preserves.

But the good times lasted barely a generation as increasingly serious problems arose. Kenya during the 1980s had the highest rate of population increase in the world, and pressure on farmlands and fringes of the wildlife preserves steadily mounted—leading to widespread poaching and a concomitant decline in tourism. Beginning in the mid-1990s, excessive rains triggered massive flooding that was followed by the onset of a long and disastrous drought, resulting in famine across most of the interior. Meanwhile, government corruption siphoned off public funds that should have been invested, and ethnic tensions were allowed to explode into outbreaks of violence, especially after the 2008 fraudulent presidential election. Moreover, the HIV/AIDS pandemic dealt the country a devastating blow, particularly in its western provinces. Kenya's latest misfortune is the growing danger of terrorist attacks in this decade. The Somali-based jihadist group, **al-Shabaab**, has been in the forefront, killing dozens of shoppers in its high-profile 2013 assault on Nairobi's upscale Westgate Mall—sending shockwaves of insecurity through Kenyan society. By 2015, the jihadists had transformed Kenya's coastal-strip portion of the African Transition Zone (Fig. 7-17)—where most of Kenya's 5 million Muslims reside—into a territorial stronghold overrun by a Boko Haram-style reign of terror, directed at all non-Muslims.

Kenya's greatest challenge is its continuing failure to build a unified nation out of its 41 constituent ethnic groups. Geography, history, and politics have placed the Kikuyu—who account for only 17 percent of the population—in a position of dominance. That position is hotly contested by the other groups, especially the Luhya, Kalenjin, Luo, and Kamba, who collectively constitute nearly 50 percent of the population; and in Kenya's periphery, there are additional major peoples such as the Maasai, Turkana, Boran, and Galla (Fig. 7-17). Adding yet another dimension to this ethnic cauldron, large numbers of Somali refugees from that country's never-ending conflicts have concentrated in Nairobi's Eastleigh neighborhood—triggering worsening ethnic discrimination as well as police harassment, increasingly in response to the threatening stance of the Islamic Front rooted along Kenya's Indian Ocean seaboard.

Nairobi, like all major African cities, struggles daily with these and many other growing pains of uncontrolled urbanization (**Box 7-14**). But there is also a brighter side to the continuing development of this busy skyscrapered capital that lies at the heart of a metropolis of five million. Nairobi today is widely regarded throughout the realm as an emerging

The mobile phone revolution has made a dramatic impact on farmers in many parts of the developing world, especially in Subsaharan Africa where distances are lengthy, land lines are absent, and market information is scarce. Mobile phone users in the realm totaled no less than 600 million in 2014, becoming not only the leading communication medium but also a primary platform for a variety of tasks normally performed on computers. Farmers and market women, such as this entrepreneurial Kenyan, now employ smartphones to access real-time commodity market prices, transfer money to suppliers and transport companies, and send funds to relatives back in their home villages. Remarkably, all of this can be accomplished without maintaining a bank account (see **Box 7-15** and **Fig. 7-18**).

Box 7-14 Among the Realm's Great Cities: Nairobi

Nairobi is the quintessential colonial legacy: there was no African settlement on this site when, in 1899, the railroad the British were building from the port of Mombasa to the shores of Lake Victoria reached it. However, it possessed something even more important: water. The fresh stream that crossed the railway line was known to local Maasai cattle herders as Enkare Nairobi (Cold Water). The railroad was extended farther into the interior, but Nairobi grew, and Indian traders set up shop. The British established their administrative headquarters here. Not surprisingly, when Kenya became independent in 1963, Nairobi was chosen to be the national capital.

Nairobi owes its primacy to its governmental functions, which ensured its priority through the colonial and independence periods, and to its favorable situation. To the north and northwest lie the Kenya Highlands, the country's leading agricultural area and the historic base of the largest nation in Kenya, the Kikuyu. Beyond the rift valley to the west lie the productive lands of the Luo in the Lake Victoria Basin. To the east, elevations descend rapidly from Nairobi's 1600 meters (5000 ft), so that highland environments make a swift transition to tropical savanna; to the north, increasing aridity produces semiarid steppe.

A moderate climate, a modern city center, several major visitor attractions (including Nairobi National Park, on the city's doorstep), and an ultramodern airport have boosted Nairobi's fortunes as a

major tourist destination, though wildlife destruction, security concerns, and political conditions have damaged the industry in recent years.

Home to almost 5 million, Nairobi is Kenya's principal commercial, industrial, and educational center. But its growth has come at a price: its modern central business district stands in stark contrast to the squalor in the shantytowns that house the countless migrants its perceived opportunities attract.

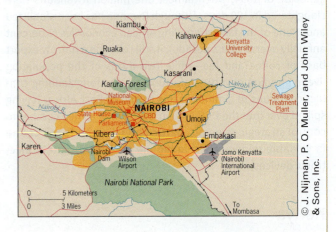

© J. Nijman, P. O. Muller, and John Wiley & Sons, Inc.

---

information technology hub. The centerpiece is Konza Techno City, the new planned satellite city to be built on the suburban frontier south of Nairobi. This future technopole is being billed as Africa's **Silicon Savannah**. The Kenyan government envisions the creation of an ultramodern sustainable city, driven by a burgeoning IT sector capable of generating as many as 100,000 jobs by 2030.

Madagascar

Only 400 kilometers (250 mi) off Africa's central east coast lies the world's fourth-largest island, Madagascar (Fig. 7-14). But Madagascar is part of neither East Africa nor Southern Africa. The first human settlers arrived here about 2000 years ago—not from nearby Africa but from distant Southeast Asia. Subsequently, the powerful Malay kingdom of the Merina came to flourish in the highlands, and its language, Malagasy, became the indigenous tongue of the entire island (see Fig. 7-10). Today the Merina remain the largest of nearly 20 ethnic groups in the population of 25 million. Like the African mainland, Madagascar experienced colonial invasion and competition, but the Merina successfully resisted conquest for centuries. Madagascar finally was forced to become part of France's empire in 1897; as soon as it could, it split from France in 1960 at the height of Africa's independence movement.

Because of its deep Southeast Asian imprint, Madagascar's staple food is rice, and many of its people also depend on fish

(**Box 7-16**). It has some minerals, but the economy is weak and constantly bedeviled by political turmoil. The country has also been burdened for decades by its high population growth rate (it is still almost double the global mean), and its colonial-built infrastructure has crumbled. Madagascar's geographic isolation allowed for the evolution of unique tropical flora and fauna—but the disastrous human onslaught has destroyed almost three-quarters of the island's rainforest home of this biological treasure.

Region Equatorial Africa

The term *equatorial* has both locational and environmental connotations. The equator bisects Africa, but only the western two-thirds of central Africa displays the conditions associated with the low-elevation tropics: intense heat, extreme humidity, copious rainfall, little seasonal variation, rainforest vegetation, and enormous biodiversity. To the east, where East Africa begins beyond the Western Rift Valley, elevations rise, and cooler, more seasonal climatic regimes prevail. Equatorial Africa, lying west of this regional boundary, is physiographically dominated by the gigantic bowl-shaped Congo Basin. This region consists of nine states, of which the Democratic Republic of Congo—formerly Zaïre, and for convenience labeled **DRCongo**—is by far the largest and most important country (**Fig. 7-19**). Five of the remaining eight states—Gabon, Cameroon, Congo (Republic), Equatorial Guinea, and São Tomé and Príncipe—have Atlantic

Box 7-15 Technology & Geography

Mobile Money in Subsaharan Africa

Did you know that Subsaharan Africa leads the world in mobile banking? For instance, it is much easier to send and receive money via a text message in Nairobi than it is in Paris. In Rwanda, Ghana, and Uganda, customers can send a text message to a money transfer service to receive a one-time code to use to withdraw money from an ATM—for example, money "sent" to them from a relative or from a small-business partner. For a small fee, mobile-to-mobile payments, withdrawals, and transfers can swiftly be executed via a text message—no bank card or bank account needed! Such services also provide international transfers, payments, and small loans.

According to the World Bank, 15 of the top 20 countries that lead the world in mobile monetary transactions are located in Africa, with Kenya claiming 80 percent of all such activity globally. This system of **mobile money [23]** challenges the more traditional retail banking networks in terms of accessibility and use. The recent rise of mobile money and cashless transactions across Subsharan Africa can be attributed to a general lack of banking infrastructure, the high costs associated with long-distance financial transactions, as well as the risks associated with all-cash transactions—and to the explosive growth of cellphone usage across this realm. In 2014, it was estimated that more than 70 percent of adults in countries such as Ghana, Kenya, and Tanzania owned cellphones, ownership rates expected to soon match those of more developed economies.

In certain African countries, mobile phone penetration and the innovative use of text messaging are replacing local **cash economies [24]** with cashless transactions. In East Africa, other local factors such as the high cost of financial transactions, permissive regulations, and network effects—plus the fact that the more people who use such services, the likelier they are to attract even more users—help explain the success of mobile money. The lack of a banking infrastructure has also facilitated the development of mobile money networks that rely on local gas stations, bars, and convenience stores—bypassing banks altogether. This is not to say that all of Subsaharan Africa is going cashless or that financial inclusion and banking are now universal across Africa. The point here is that the future of mobile money is being shaped in this realm.

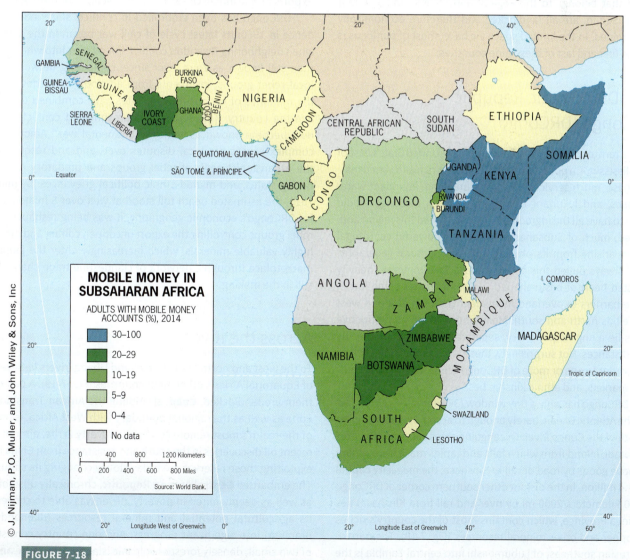

MOBILE MONEY IN SUBSAHARAN AFRICA

ADULTS WITH MOBILE MONEY ACCOUNTS (%), 2014

- 30–100
- 20–29
- 10–19
- 5–9
- 0–4
- No data

Source: World Bank.

© J. Nijman, P.O. Muller, and John Wiley & Sons, Inc

FIGURE 7-18

Box 7-16 From the Field Notes...

"Taking an early morning stroll along the beach near the town of Maroantsetra on Madagascar's northeastern coast, I stopped to observe a group of fishing people at work. It was a very basic way of fishing, with minimal equipment. Using a couple of small rowboats, they dropped their net about 20 meters (65 ft) from shore and then pulled it in toward the beach. It was a meager haul. Most of this catch, they said, was intended for consumption by their families, and the rest would be sold informally. More than 70 percent of the people in this country live below the national poverty level, and in rural and coastal areas the percentage is even higher. These economic problems stand in sharp contrast to the country's natural beauty—or what remains of it after years of severe environmental degradation."

©Jan Nijman

coastlines. The Central African Republic, the southern part of Chad that belongs to this region, and recently independent South Sudan, are all landlocked and therefore economically constrained. In the limited space we have, most of our focus is on the first and last of these nine countries.

The Democratic Republic of Congo (DRCongo)

With a territory not much smaller than the United States east of the Mississippi, a population of 80 million, a rich and varied mineral base, much serviceable agricultural land, abundant water supplies, and luxuriant tropical rainforests, DRCongo would seem to have all the ingredients needed to lead this region and, indeed, much of Subsaharan Africa. But powerful centrifugal forces, arising from its physiography and cultural geography, have always pulled this country apart. The immense, heavily forested heart of basin-shaped DRCongo creates some formidable barriers to overland movement between east and west as well as north and south. Many of the country's productive areas lie along its periphery, separated from one another by vast distances. Not surprisingly, these areas tend to look across the border to one or more of DRCongo's nine neighbors for outlets, markets, and ethnic kinship ties.

DRCongo has only a tiny window (37 kilometers/23 mi) facing the Atlantic Ocean, barely enough to accommodate the wide mouth of the Congo River. Oceangoing ships can reach the port of Matadi, inland from which falls and rapids make it necessary to move goods by road or rail to Kinshasa, the megacity capital of 11.4 million. In the distant other southern corner of DRCongo (3200 kilometers/2000 mi by river and rail from Kinshasa) lies Katanga Province, which contains most of the country's riches of copper and other mineral resources (Fig. 7-19). The narrow protrusion southeast of Lubumbashi into central Zambia is the heart of the Copperbelt, from which these minerals have flowed

onto world markets for over a century via rail lines that connect to ports on the Indian Ocean coast of East and Southern Africa.

DRCongo has been embroiled in conflicts since independence in 1960. Its latest cycle of civil war began in the 1990s when neighboring Rwanda's conflict spilled over into what was then still known as Zaïre. Ever since, competing networks of local rebel groups, insurgents from adjoining countries, and the militaries of various African states have combined to ravage the country. This series of interconnected conflicts has now become known as "Africa's continental war." Among its complex causes are local disputes over land and resources, the expansionist goals of rebel groups and predatory neighboring states, and myriad ethnic political grievances. By mid-2016, the estimated death toll stood at well over 5 million. As for DRCongo's economic geography, it was being reshaped by rebel groups controlling the export of copper, coltan, and other highly valuable minerals, which increasingly enter the global marketplace through a network of clandestine trade channels still in the making.

Across the River

To the west and north of the Congo and Ubangi rivers lie seven of Equatorial Africa's other eight countries (Fig. 7-19). A pair of them are landlocked. **Chad**, straddling the African Transition Zone as well as the regional boundary with West Africa, is one of the realm's most remote countries; poverty is rife, although recent oil discoveries in the south and assistance from China in exploiting those reserves are today slowly changing its status. The embattled **Central African Republic**, chronically unstable as well as deeply mired in poverty, never was able to convert its agricultural potential and mineral resources (diamonds, uranium) into significant progress. And one country consists of two small, densely forested volcanic islands: **São Tomé and Príncipe**, a ministate containing a population of just 195,000

EQUATORIAL AFRICA

POPULATION
- Under 50,000
- 50,000–250,000
- 250,000–1,000,000
- 1,000,000–5,000,000
- Over 5,000,000

National capitals are underlined

Area most affected by rebel activity, 2017
Oilfields
Potential oilfields
Railroad
Road
Pipeline
Projected oil pipeline

0   200   400 Kilometers
0   100   200 Miles

Longitude East of Greenwich

© J. Nijman, P. O. Muller, H. J. de Blij, and John Wiley & Sons, Inc.

FIGURE 7-19

whose economy, like several others in the realm, is being transformed by recent oil discoveries.

The four coastal states present a different picture. All possess oil reserves and share the Congo Basin's equatorial forests (Fig. 7-19); thus petroleum and timber figure prominently among their exports. In **Gabon**, which rejoined OPEC in 2016, this combination has produced an upper-middle-income economy. Of the four, coastal Gabon also has the largest proven mineral resources, including manganese, uranium, and iron

ore. Its capital, Libreville (the only coastal capital in the region), reflects all this in its high-rise downtown, bustling port, and mushrooming shantytowns.

Cameroon (population: 24 million), not as well endowed with oil or other raw materials, possesses the region's strongest agricultural sector by virtue of its higher-latitude location and higher-relief topography. Western Cameroon is one of the more developed parts of Equatorial Africa and includes the capital, Yaoundé, and the principal port of Douala.

With five neighbors, **Congo** (the "other" [former French] Congo) could be a major transit hub for this region, especially for DRCongo, if it ever recovers from civil war. Its capital, Brazzaville, lies just across the Congo River from Kinshasa and is linked to the port of Pointe Noire by road and railway. But devastating, unceasing power struggles have thus far neutralized Congo's geographic advantages.

As Figure 7-19 shows, **Equatorial Guinea** consists of a rectangle of mainland territory and the island of Bioko, where the capital of Malabo is located. A former Spanish colony that remained one of the realm's least-developed territories, Equatorial Guinea was dramatically affected by increased oil production and exporting, and in the 2000s became Africa's only high-income economy (Fig. G-11) with GDP growth rates exceeding 10 percent. However, as in so many other oil-rich countries, this bounty had minimal impact on the well-being of most of the people—while ostentatious displays of wealth by the ruling elite elicited international condemnation. Overreliance on these energy resources has since resulted in a heavy toll on Equatorial Guinea's economic performance, and by 2015 plummeting world oil prices had produced a GDP growth rate of negative 10 percent.

One other coastal territory would also seem to be a part of Equatorial Africa: *Cabinda*, wedged between the two Congos just to the north of the Congo River's mouth. But Cabinda is one of those odd colonial legacies on the African map—it belonged to the Portuguese and was administered as part of Angola. Today it is an exclave of independent Angola, and a most valuable one because it contains sizeable oil reserves.

South Sudan

This newest state to appear on the world political map became independent in 2011, the outcome of a near-unanimous referendum in the southern provinces of former Sudan (the now-truncated country still called *Sudan*, discussed in Chapter 6). South Sudan's birth came in the aftermath of six disastrous decades of postcolonial strife within former Sudan, a brutal conflict in which more than 1.5 million died, magnified by the bifurcation of the country along the religious divide between Islam and Christianity-animism that we know as the Islamic Front (see Fig. 6-21).

The British effectively ruled former Sudan from the 1890s until independence in 1956. One of the roots of the country's long-running internal conflict, which peaked in the resumption of its post-independence civil war between 1983 and 2005, lay in the decision of the British colonial administration to combine northern Sudan, which was heavily Arabized and Islamized, with a sizeable African/Christian-dominated area to the south. As soon as the British departed in the mid-1950s, the Khartoum-based regime in the north sought to impose its Islamic rule on southern Sudan, immediately triggering the first civil war that lasted until 1972. By the time the bitter second civil war erupted a decade later, former Sudan's strife had impoverished the country to the point that its per capita income ranking was among the lowest on Earth. That slowly

began to change in the 1990s when oil was discovered, including major deposits in the embattled south (Fig. 7-19). Eventually, leaders in the south saw oil as a ticket to self-sufficiency, and the north felt that it too would walk away with control of several oilfields.

Animosity between the two halves of former Sudan persists. Indeed, a low-grade, oil-based conflict broke out soon after the political split of 2011—and may finally be ending following a truce in 2015. Nonetheless, South Sudan continues to be plagued by grinding poverty, violent internal ethnic conflict, and ineffective governance (the country's population size can only be estimated—at around 13 million). Meanwhile, the overwhelming majority of the people are either subsistence farmers or cattle herders—a deeply traditional livelihood in which individual and family status depends on the number of livestock owned.

Region Southern Africa

As a geographic region, Southern Africa consists of all the countries lying south of Equatorial Africa's DRCongo and East Africa's Tanzania (**Fig. 7-20**). Thus defined, this region extends southward from Angola and Moçambique (on the Atlantic and Indian Ocean coasts, respectively) to South Africa, and includes half a dozen landlocked states. Also marking the northern limit of the region are Zambia and Malawi. As noted above, Zambia is nearly cut in two by the elongated, mineral-rich land extension from southeastern DRCongo; similarly, Malawi penetrates deeply into Moçambique.

Africa's Resource-Rich Region

Southern Africa constitutes a geographic region in both physiographic and human terms. Its northern periphery marks the southern limit of the Congo Basin in a broad upland that stretches across Angola and into Zambia (the narrow tan lobe extending eastward from the Bihe Plateau in Fig. 7-2). Most of this region is plateau country, and much of the interior upland is framed by the prominent, near-coastal Great Escarpment. There are two consequential river systems: the Zambezi (which forms the border between Zambia and Zimbabwe) and the Orange-Vaal (South African rivers that combine to demarcate southern Namibia from South Africa).

Southern Africa has long been the continent's richest region materially. A great north-south corridor of major mineral deposits extends through the eastern heart of the region from the Zambia/DRCongo Copperbelt through Zimbabwe's Great Dyke and South Africa's Bushveld Basin and Witwatersrand to the goldfields and diamond mines scattered around Bloemfontein in the geographic center of South Africa (Fig. 7-20). Ever since these minerals began to be exploited in colonial times, many migrant laborers have come to work in the mines of these countries. And in Southern Africa's far northwest, huge onshore and offshore oil reserves are propelling the economic

FIGURE 7-20

development of Angola. The region's agricultural diversity matches its mineral wealth, especially so in South Africa: vineyards drape the slopes of South Africa's Cape Ranges, and the country's relatively high latitudes and its range of altitudes create environments for apple orchards, citrus groves, banana plantations, pineapple farms, and myriad other crops.

Despite this considerable wealth and potential, only a few of the ten countries of Southern Africa have prospered, and the gap between them is steadily widening (see Fig. 7-11). As Figure G-11

indicates, four countries—South Africa, Angola, Botswana, and Namibia—now rank in the World Bank's upper-middle-income category; in contrast, three remain mired in the low-income category (Malawi, Moçambique, and Zimbabwe). The three remaining states are classified as lower-middle-income countries: Zambia plus the two ministates of Lesotho and Swaziland. Among the four most prosperous countries, Botswana and Namibia are both desert-dominated, sparsely settled states, with a combined population of less than 5 million. Booming

STRINGER/REUTERS/Newscom

By the time you read this, Angola will probably be Subsaharan Africa's leading oil-producing country. This 2015 panorama of the center of its burgeoning capital city, Luanda, is reminiscent of the Persian Gulf's petrostates during their initial building booms. But oil is a double-edged sword: even though Luanda is the world's most expensive city to live in, the rest of the Angolan population lives in a world of extreme poverty. With 95 percent of the country's foreign income dependent on oil sales, the recent drop in the price of oil has forced government budget cutbacks that only intensify the hardships of life in one of the world's most unequal societies.

Angola has become one of the realm's oil powers, a fast-growing country on the move. South Africa, however, stands apart—the region's giant by every measure.

Since we are working our way through the Subsaharan African realm from north to south, we will also approach the region of Southern Africa from that perspective. We begin with the Northern Tier of countries, proceed southward into the Middle Tier, and then put preeminent South Africa in the geographic spotlight.

The Northern Tier

In the four countries that extend across the Northern Tier of Southern Africa—Angola, Moçambique, Zambia, and Malawi—change is occurring everywhere. **Angola**, Southern Africa's most resource-dependent country, has been the region's star economic performer in this decade. This country of 26 million—which also includes its tiny **exclave [25]** of *Cabinda* located north of the Congo River's mouth—had a thriving (colonial) economy based on a wide range of mineral and agricultural exports at the time of independence from Portugal in 1975. But the Portuguese did not prepare the country for self-rule, and soon Angola fell victim to the Cold War, with its northern peoples opting to follow a communist course and the southerners falling under the sway of a rebel movement backed by South Africa and the United States. The results included devastated infrastructure, abandoned farms, looting of diamonds, hundreds of thousands of casualties, and millions of landmines that continue to kill and maim.

Since the mid-2000s, Angola's economic performance has been nothing short of spectacular, and statistically it became the world's fastest-growing national economy. Today, its capital, Luanda, has achieved the dubious distinction of being the most expensive city on Earth (see photo). Almost all of this growth comes from oil, much of which is exported to China, and Angola is on track to surpass Nigeria in 2017 as Subsaharan Africa's top petroleum producer. Yet this bonanza is decidedly a two-edged sword because the country's reliance on foreign oil income is too great: this fuel now accounts for half of Angola's GDP, more than 90 percent of its exports, and at least 75 percent of government revenues. Those impressive statistical indicators notwithstanding, they mask both Angola's vulnerability to fluctuations in global oil prices and catastrophic and intensifying domestic inequalities that force a huge majority of the population to survive on no more than two U.S. dollars a day.

On the opposite coast, the other major former Portuguese dependency, **Moçambique** (29 million), experienced a comparable trajectory. Upon independence, Moçambique also chose a Marxist course with unfortunate economic and political consequences. Here, too, a rebel movement supported by South Africa caused civil conflict, created famines, and generated a huge stream of refugees toward Malawi.

Moçambique possesses neither oil nor as much agricultural land as Angola. But it does have considerable bauxite and coal deposits, and its long Indian Ocean coastline provides an advantageous relative location (as well as access to promising offshore natural gas reserves). In recent years, port traffic has been revived, and the country is working with South Africa on building a joint Maputo Development Corridor (Fig. 7-20). But despite the rapid growth of Moçambique's economy over the past decade, here again we have an African country that depends too heavily on a single commodity—bauxite (aluminum ore), which accounts for about one-third of all exports. And, like Angola, this country has performed poorly in distributing the benefits of economic development, barely affecting the severe poverty that still reigns among Moçambique's masses.

Landlocked **Zambia** (16.7 million), another product of British colonialism, shares the mineral riches of the Copperbelt with Katanga Province in its northern neighbor, DRCongo. However, the stable commodity prices on which the country depends have fluctuated wildly since independence, and the railways leading to Zambia's outlet ports in Angola and Moçambique were rendered inoperative by Cold War conflicts. More recently, the Chinese have taken an interest in Zambia's minerals, and they have invested in railroad repairs as well as the expansion of mining operations—resulting in an annual economic growth rate averaging about 7 percent over the past decade.

Neighboring **Malawi** (17.7 million), like Zambia, has been able to sustain democracy for more than two decades. Malawi has an almost totally agricultural economic base and is also unremittingly challenged by environmental degradation. This country's dependence on corn as its food staple, its variable climate, and its severely fragmented land-use system have inhibited economic growth—which nevertheless has risen steadily at nearly 6 percent a year since the turn of this century.

The Middle Tier

Five states lie between the countries of the Northern Tier and South Africa. They form Southern Africa's Middle Tier subregion, and all border the Republic of South Africa: Botswana, Namibia, Zimbabwe, and the two ministates of Lesotho and Swaziland (Fig. 7-20). As the map reveals, four of these five are landlocked. Diamond-exporting (and upper-middle-income) **Botswana** occupies the heart of the Kalahari Desert and surrounding steppe. Despite its lucrative diamonds, the great majority of its 2.3 million inhabitants are subsistence farmers, and in recent years Botswana has been the most severely AIDS-afflicted country in all of tropical Africa. **Lesotho** and **Swaziland** (2.2 and 1.3 million, respectively), both traditional kingdoms, depend very heavily on remittances from their workers in South African mines, fields, and factories.

Southern Africa's youngest independent state, **Namibia** (2.5 million), is the former German colony of South West Africa that was administered by South Africa from 1919 to 1990. The country is named after one of the world's driest deserts (the Namib, which lines its coast). Territorially about as large as Texas and Oklahoma combined, only its far north receives enough moisture to enable subsistence farming, which is why most of the people live close to the Angolan border. Mining and ranching constitute the leading commercial activities, but even though orderly land reform is underway, much of the population still lives in poverty.

The Tragedy of Zimbabwe

Zimbabwe (population: 16 million) lies at the heart of Southern Africa, between the Zambezi River in the north and the Limpopo River in the south and between the Great Escarpment to the east and the desert to the west. Landlocked but endowed with good farmlands, cool uplands, and a wide range of mineral resources, Zimbabwe at independence had one of Southern Africa's most vibrant economies and seemed to have a bright future. Its core area straddles the mineral-rich Great Dyke that extends across the heart of the country, which contains copper, gold, asbestos, chromium, and platinum. Its farms are capable of producing tobacco, tea, sugar, and cotton in addition to staples for the local market.

But going back almost four decades, these have not been good times for Zimbabwe. During the colonial period, the tiny minority of whites that controlled what was then called Southern Rhodesia (after Cecil Rhodes, the British capitalist of diamond fame and scholarship honors), took the most productive farmlands and organized and ran the agricultural economy. Following independence, their descendants continued to hold huge estates, only parts of which were being farmed. In the absence of comprehensive land reform, let alone progress

toward democratic rule, the persistence of this monstrous inequality impeded broad-based development.

In the wake of their successful joint campaign to end white-minority rule in 1980, the two peoples that constitute most of Zimbabwe's population—the Shona (71 percent) and the Ndebele (19 percent)— engaged in acrimonious ethnic conflict. Proper legal reforms did not take effect, and the rule of newly elected Robert Mugabe turned from ineffectiveness to disaster to extreme paranoia. As he encouraged squatters to invade white farms (where a number of owners were killed and many others fled), the agricultural economy began to collapse. Corruption in the Mugabe government resulted in the transfer of land, not to needy squatters but to cronies of top officials. Foreign investment in other Zimbabwean enterprises, including the mining industry, dried up. By the mid-1990s, Zimbabwe was in economic free fall.

Mugabe also turned against the vital informal sector of the economy, which was all most Zimbabweans had left when jobs on farms and in factories and mines disappeared. He ordered his henchmen to destroy the dwellings of some 700,000 "informal" urban slum dwellers, who had little choice other than to join a growing army of persecuted homeless refugees who soon began to stream out of Zimbabwe, many headed for South Africa.

Heroic Zimbabweans and persistent outsiders have put pressure on 92-year-old Mugabe to end his catastrophic dictatorial rule, but in 2016 he was plotting for his spouse, Grace, to take the reins. But the country remains isolated internationally—except for the embrace of China and South Africa for very different reasons. Beijing is pursuing numerous economic opportunities in Zimbabwe and is almost entirely responsible for the country's period of recent (unsteady) GDP growth. South Africa's government, however, remains silent in gratitude for Zimbabwean support of South African revolutionaries during the struggle against their own white-minority-government oppressors before 1994. Through it all, Mugabe has remained in power, at press time the oldest and longest serving leader in Africa, and still trying to lay the groundwork for his succession. Today more than ever, Zimbabwe stands in the sharpest contrast to what is being achieved in the other countries of Southern Africa.

South Africa

The Republic of South Africa (RSA) is the giant of the region of Southern Africa, with a population of 55 million and a modern economy second in size only to Nigeria's within this realm. Its historical geography differs somewhat from much of the rest of Subsaharan Africa. The country's lands were fought over by various African nations before the Europeans arrived and the colonial "scramble for Africa" took place. Indigenous peoples had migrated southward—first the Khoisan-speakers and then the Bantu peoples—into the South African *cul-de-sac*. The Zulu and Xhosa nations were fighting over these lands at about the time the first European settlers arrived.

South Africa is one of the most strategic places on Earth, the gateway from the Atlantic to the Indian Ocean, a key source of provisions on the route to Asia's riches. The Dutch East India Company founded Cape Town as early as 1652, and the Hollanders and their descendants, known as **Boers**, have been a part of the South African cultural mosaic ever since. The British took over about 150 years later, and both colonial powers vied for control. The British came to dominate the Cape, while the Boers trekked into the South African interior and, on the high plateau they called the *highveld*, founded their own republics. A war ensued, but by 1910 the Boers and the British had negotiated a power-sharing arrangement, although the Boers eventually achieved dominance that lasted from 1948 to 1994. Having long since shed their European ties, they called themselves **Afrikaners**, their word for Africans.

For more than 40 years, between about 1950 and the mid-1990s, multicultural South Africa was in the grip of the world's most notorious racist policy, **Apartheid [26]**. The word itself means "apartness," but in practice it promotes strict racial segregation and severe discrimination. Out of the concept of Apartheid grew a notion, promulgated by the white (European) minority then in control of the state, known as **separate development**. This would apply Apartheid to the entire country, carving it up into racially-based territories whose inhabitants were citizens of those ethnic domains ("homelands")—but not of South Africa as a whole. Predictably, such racist social engineering aroused strong opposition within South Africa and beyond, leading to worldwide condemnation and international sanctions.

By the late 1980s, South Africa seemed headed for a violent revolution, but disaster was averted by a most unexpected turn of events. A leader of the white-minority government that for decades had ruthlessly pursued its Apartheid policies, and a revered leader of the multicultural majority who had for 28 years languished in an island prison not far from Cape Town, struck a momentous accord, ushering in a new South Africa virtually overnight. Nelson Mandela walked out of prison a free man on February 11, 1990, and in the RSA's first democratic election in 1994 he became the country's president—alongside a new, fully representative parliament that was now indeed the country's first "rainbow assembly." At the same time, the country's internal political geography was reorganized: before 1994, the RSA had been divided into four provinces, but this was now replaced with a fairer federal arrangement consisting of nine provinces (see **Fig. 7-21**, including the inset map).

Ethnicity and Xenophobia

South Africa has long been the realm's most pluralistic and heterogeneous society (see **Box 7-17** table). In addition to the various indigenous African nations and the Europeans who settled in South Africa came peoples from Asia. The Dutch brought thousands of Southeast Asians to the Cape to serve as domestics and laborers, as the British did later from their South Asian colonies on the Indian subcontinent. Moreover, a substantial population of mixed ancestry clustered at the Cape, today known as the *Coloured* component of the country's citizenry.

Although heterogeneity marks the urban spatial demography of South Africa, regionalism pervades the national-scale human mosaic. The Zulu nation, more than 11 million strong, remains largely concentrated in today's Kwazulu-Natal Province (Fig. 7-21). The Xhosa (8.4 million) still cluster in the Eastern Cape below the Great Escarpment, and the Tswana (3.5 million) continues to occupy ancestral lands along the border with Botswana. Metropolitan Cape Town remains the heartland of the Coloured population (4.8 million); the city of Durban still has the strongest South Asian imprint.

Since the end of the Apartheid era more than two decades ago, international migrations from all over the realm have streamed into South Africa. These influxes have intensified the RSA's multiculturalism and are most visible in major urban areas, especially in their takeover of much of inner-city Johannesburg (**Box 7-18**). But peaceful coexistence does not prevail in this purportedly "rainbow nation": tensions persist at an explosive level, and violent attacks directed at poor foreign immigrants (mainly those with a high profile in street trade) periodically erupt. Such expressions of **xenophobia [27]** are motivated by the perception that these immigrants are stealing jobs from South Africans—and are further fueled by the dominance of low-end commerce by

© Jan Nijman/ Richard Grant

The newest immigrants from Somalia to the RSA often start out by running spaza shops—such as this one in Thokoza Township south of Johannesburg. (*Spaza* is a South African word for a small convenience store in the informal economy.) These Somalis work long hours, live on the premises, and constantly face the hostility of South African spaza competitors as well as unemployed youths. Sometimes they are shot at, and a few have even had their shops burned down (as in nearby Soweto Township in early 2015). For many desperate immigrants, working in the spazas of dangerous "locations" (their term for the townships) is a rite of passage and the only job that a freshly arrived immigrant can find. Tellingly, the Somali shop-owners who employ them remain in central Johannesburg, choosing to send out only new Somali arrivals to handle the challenges of working the dangerous suburban "locations."

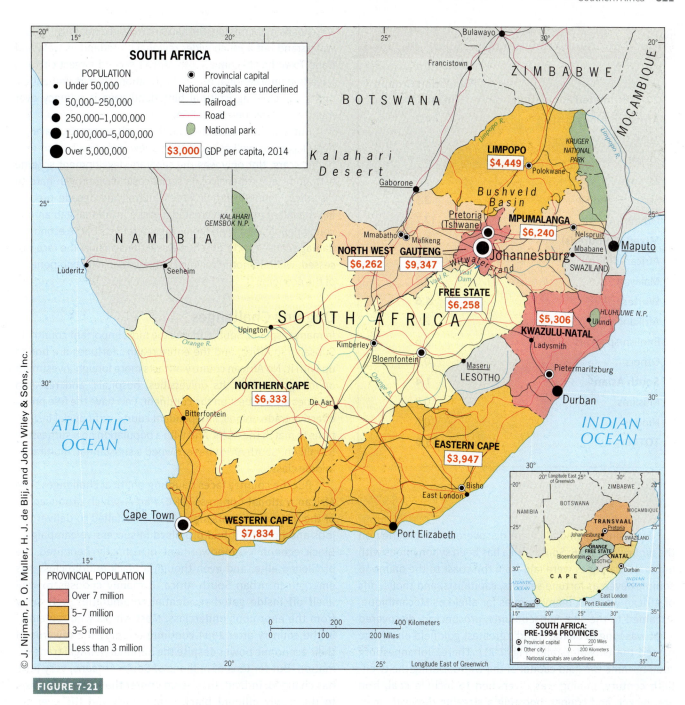

FIGURE 7-21

foreign entrepreneurs whose business skills exceed those of natives (see photo of spaza shop). Efforts to defuse xenophobic sentiments continue: South African schoolchildren learning the alphabet are now taught that the letter "X" stands for xenophobia.

South Africa's Economic Geography

South Africa's economy, the realm's most modernized, has extensive regional influence. On the global scene, the RSA is regarded as an emerging-market economy as well as a gateway to Subsaharan Africa, particularly concerning corporate command and control. With a land area in excess of 1.2 million square kilometers (470,000 sq mi), South Africa contains the bulk of the region's minerals, most of its good farmlands, its biggest cities, best harbors, most productive factories, and the best developed transport network. Mineral exports from the Copperbelt and the Great Dyke move through South African ports. Workers from as far away as Malawi and as close as Lesotho toil in South Africa's mines, industrial plants, and fields. And entrepreneurs from Somalia, Ethiopia, Bangladesh, and China are increasingly entrenched in the commercial transactions of the country's informal economy.

Box 7-17 Demographic Data for South Africa

2016 Estimated Population	Population Groups (in millions)
African nations	**44.2**
Zulu	11.1
Xhosa	8.4
Tswana	3.5
Sotho (N and S)	3.5
Others (6)	17.7
Mixed (Coloureds)	**4.8**
African/White	4.6
Malayan	0.2
Whites	**4.6**
Afrikaners	2.9
English-speakers	1.6
Others	0.1
South Asian	**1.4**
Muslims	0.8
Hindus	0.6
TOTAL	**55.0**

Ever since diamonds were discovered at Kimberley on the *highveld* in 1866, South Africa has been synonymous with minerals. Rail lines were laid from the coast to the diamond complex even as fortune seekers, capitalists, and thousands of African workers, many from as far afield as Moçambique, streamed to the site. Subsequently, prospectors discovered what was long to be the world's greatest goldfield on a ridge named the Witwatersrand (Fig. 7-21). There, Johannesburg soon became the gold capital of the world. During the twentieth century, mining was diversified to include coal, iron ore, nickel, and copper, boosting a sizeable domestic metallurgical and manufacturing sector as well as yielding large revenues on world markets. Before the imposition of severe international sanctions in the 1980s to protest the growing outrages of Apartheid, capital flowed into the country, white immigration expanded, farms and ranches were laid out, and overseas markets multiplied.

Along the way, South Africa's cities grew rapidly. Johannesburg became an industrial agglomeration, major financial center, and immigration hub. To its north, Pretoria became the country's administrative capital, and the two cities have now coalesced into the Gauteng conurbation that contains more than 13 million people who produce over one-third of the national economy's output. In Orange Free State (simply Free State since 1994), substantial industrial development matched the expansion of mining. Durban's port served not only the Witwatersrand but a much wider regional hinterland as well. And Cape Town has become South Africa's second-largest city and a leading international tourist destination; its port, industries, and productive agricultural hinterland gave it primacy over much of the southernmost RSA.

But Apartheid frustrated South Africa's prospects, leaving a deep legacy as well as more than a few major social and economic scars. Not only was the separate development scheme astronomically expensive: its social costs have proven hard to dislodge. Mounting ethnic unrest during the decade preceding the fall of Apartheid created a vast educational gap among young people. And even today, the aftereffects of the economic damage caused by international sanctions against the race-obsessed, white-minority regime continue to plague components of the economy.

Ongoing Challenges

In many respects, South Africa is the most important country in Subsaharan Africa, and the entire realm's fortunes are bound up with it. No African country attracts more foreign investment or foreign workers. Its universities, hospitals, and research facilities are the best on the continent. Few have the free press, effective trade unions, independent courts, or financial institutions to match the RSA's. And with a population now surpassing 55 million, South Africa has spawned a substantial, multiracial middle class.

But the country faces daunting political challenges. The African National Congress (ANC) Party—which produced all of the RSA's presidents since 1994—has failed to deliver the jobs and housing it had promised in successive campaigns. Black economic empowerment has mostly benefited the privileged elite; and even though a black middle class has emerged in urban South Africa, it mainly chooses to wall itself off inside gated residential communities well away from the enormous underclass. Most whites who remained in the country after 1994 continue to enjoy the rewards of their economic power despite the national political transformation. But for tens of millions of South Africans, very little has changed: indeed, they seem poorer than ever compared to the newly affluent black populations and the ever-entrenched whites.

It should therefore come as no surprise that in 2015 South Africa recorded the greatest inequality within any of the world's countries. Predictably, there are massive gaps in education, personal income, and service provision. South Africa today ranks 132nd out of a total of 144 countries in elementary school performance. Urgently needed land reform is stalemated, supposedly by a lack of funding. Although millions of houses have been built and were connected to utilities during the post-Apartheid era, this total falls drastically short of the number of such dwelling units actually needed. The government reports an unemployment figure of about 25 percent, but that proportion is easily doubled in the destitute slums of the inner city and the outlying townships. In the rural areas, farm

Box 7-18 Among the Realm's Great Cities: Johannesburg

Subsaharan Africa may be undergoing a rapid urban transformation, but it still has only one true conurbation, and South Africa's Johannesburg lies at the heart of it. Little more than a century ago, Johannesburg was a small (though rapidly growing) mining town based on the newly discovered gold deposits of the Witwatersrand. In its early history, this future national core area was dominated by the ancestors of Dutch settlers, and the still-remaining names on the map are a reminder of that era.

Today, Johannesburg forms the focus of a conurbation (that anchors a province named *Gauteng*) of more than 13 million people, extending from South Africa's administrative capital, Pretoria, in the north to Vereeniging in the south, and from Springs in the east to Krugersdorp in the west. The population of metropolitan Johannesburg, now 8.7 million, has grown so swiftly since 1994 that by the early 2000s it had overtaken Cape Town to become South Africa's largest urban agglomeration.

Johannesburg's skyline is the most impressive in all of Subsaharan Africa, a forest of skyscrapers reflecting the wealth generated here over the past century. Look southward from a high vantage point, and you see the huge mounds of yellowish-white slag from the mines of the "Rand," the so-called mine dumps, partly overgrown today, interspersed with suburbs and townships. In a general way, Johannesburg developed as a white city in the north and a black city in the south. Well-known Soweto (South Western Townships) lies to the southwest. Houghton and other spacious, upper-class suburbs, formerly exclusively white residential areas, lie to the north.

Johannesburg has neither the scenery of Cape Town nor the climate and beaches of Durban. The city lies at an elevation of 1750 meters above sea level (5750 ft—more than a mile high), and its thin air often is polluted from smog created by motor vehicles, factories, mine-dump dust, and countless cooking fires in the townships and shantytowns that ring much of the metropolis.

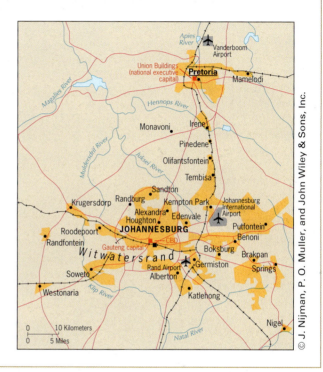

© J. Nijman, P. O. Muller, and John Wiley & Sons, Inc.

workers and miners organize protests and strikes to demand fairer wages, and many white landowners have been intimidated by homeless invaders. It goes without saying that these developments neither inspire confidence nor help the cause of attracting new investment.

South Africa has many of the necessary geographic endowments for a strong economy, but it will take time, stability, and competent management to achieve its full potential. Joining the exclusive BRICs—the world's four biggest emerging-market economies (Brazil, Russia, India, and China) in 2011—raised South Africa's global standing, even though Pretoria has always been the minor power of the group. However, the strengthened ties with China brought new challenges in 2015 and 2016 as China's appetite for South Africa's precious metals, coal, and steel declined sharply; at the same time, the RSA currency (the rand) lost one-quarter of its value, pushing the economy toward recession. Continued heavy reliance on commodity sales in fickle world markets is enhancing volatility, and the RSA also appears to be losing its regional advantage in education and skills. Most of all, South Africa lacks capable leadership that can do for the economy what Mandela did for the society.

As this chapter has made clear, we can no longer accept the "conventional wisdom" conclusion of a Subsaharan Africa described in terms of negative stereotypical judgments. All too often, this realm is still portrayed as a single, monolithic, economic basket-case, where dysfunctional government goes hand in hand with economic underperformance, frequent famines, frightening disease outbreaks, and aimless ethnic violence. On the other hand, neither can the Africa Rising narrative be the broad brush to help us interpret this realm. In the late-2010s, Subsaharan Africa harbors the world's newest emerging-market economy; in many countries, economic forecasts tell of another decade lying ahead in which growth and development may exceed that of all other geographic realms; and a number of African societies are demonstrating incredible resilience in fighting such monumental challenges as HIV/AIDS and Ebola while continuing to adjust to global economic change.

These positives and potentials notwithstanding, we have also seen that Subsaharan Africa still faces a formidable set of obstacles. Democracy is taking hold in an increasing number of countries, but elsewhere war and dislocation remain untouched as authoritarian regimes continue to be deeply entrenched. Nigeria may have overtaken the RSA to become the realm's largest economy, but the Boko Haram insurgency is not defeated. As for South Africa, its foremost internal threat may well be its appalling, widening inequalities of opportunities, lifestyles, and personal well-being. And as far as the bigger geographic picture is concerned, energy and mineral resources are the cornerstones of this realm's progress in the 2010s—but plunging commodity prices to their lowest level this century, combined with the slowdown in China's economy since 2015, confirm that greater diversification in both economic activity and foreign partnering are essential pursuits.

Points to Ponder

- The realm contains more than four-dozen countries; the number of languages spoken within Subsaharan Africa is estimated to exceed 2000.

- China's impact on Subsaharan Africa has expanded steadily through activities ranging from trade to investment to education. Turkey and India are also increasing their ties with this realm.

- Shipping a new car from China to a Tanzanian port costs less than hauling it inland from Tanzania to neighboring Uganda.

- At the end of the first two decades of this century, Subsaharan Africa is likely to have recorded the fastest economic growth among all world geographic realms.

- Too many Subsaharan African countries remain resource-dependent economies. Will this realm have a future in manufacturing?

- South Africa was freed from life under racist Apartheid over two decades ago, but it now ranks as one of the most unequal societies on Earth.

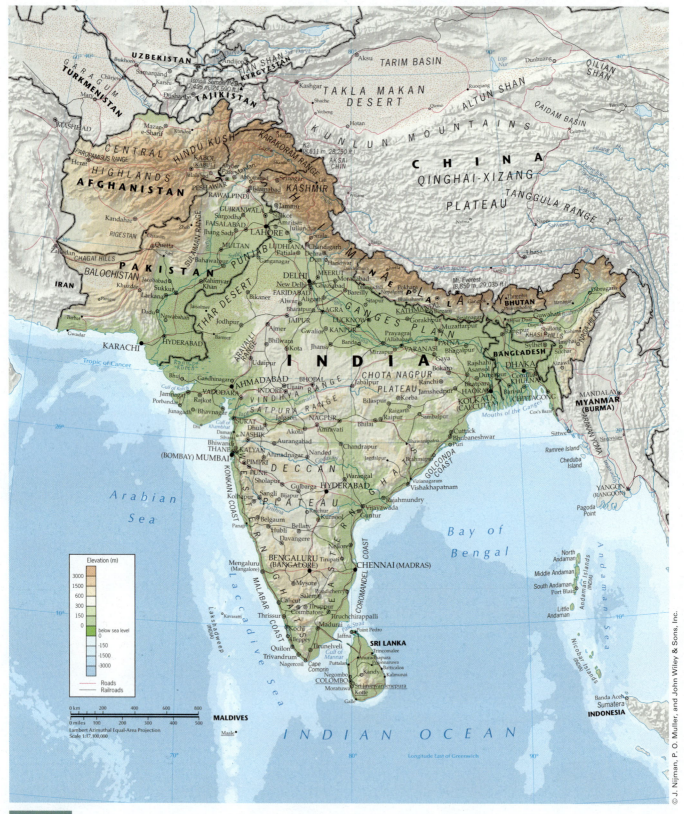

FIGURE 8-1

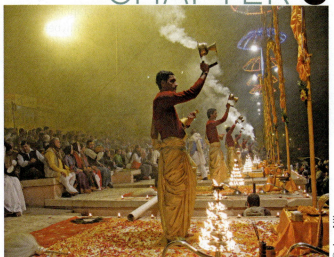

CHAPTER 8

Where were these pictures taken?

The South Asian Realm

IN THIS CHAPTER

- Afghanistan-Pakistan-India: A dangerous triad
- South Asia's demographic gender imbalance
- Global climate change and extreme weather events
- Caste in twenty-first-century India
- Cutting-edge IT, backward agriculture
- The China-Pakistan Economic Corridor
- Drone warfare in western South Asia

CONCEPTS, IDEAS, AND TERMS

Transition zone [1]

Monsoon [2]

Caste system [3]

Indo-European languages [4]

Dravidian languages [5]

Partition [6]

Forward capital [7]

Neoliberalism [8]

Megacity [9]

Extreme weather events [10]

Population geography [11]

Population density [12]

Physiologic density [13]

Demographic transition [14]

Fertility rate [15]

Demographic burden [16]

Population pyramid [17]

Sex ratio [18]

Buffer state [19]

Urban primacy [20]

Taliban [21]

al-Qaeda [22]

China-Pakistan Economic Corridor [23]

Drone warfare [24]

Communal tension [25]

Hindutva [26]

Informal sector [27]

Double delta [28]

Non-governmental organization (NGO) [29]

Micro-credit [30]

Rising sea level [31]

South Asia is a realm of almost magical geographic names: Mount Everest, the Ganges River, Kashmir, the Khyber Pass. There was a time when this realm was legendary and prized. Remember that it was "India" and its fabled wealth that the European explorers were after, from Vasco da Gama to Columbus to Magellan. Before them, the fourteenth-century North African geographer Ibn Battuta had traveled overland to South Asia, and his writings about its riches were met with astonishment and even disbelief. From the sixteenth century onward, European trading companies derived enormous profits from commerce in this realm.

By the late nineteenth century, however, South Asia had become remote from the affairs of the world—hungry, weak, exploited, the prototype realm of the global periphery. Even after independence in 1947, India as well as the other countries of South Asia long remained among the world's poorest. For decades, population growth outstripped economic expansion.

Today, for a number of reasons, South Asia commands the world's attention once again. It became the most populous geographic realm on Earth in 2011. Two of its states, India and Pakistan, often find themselves in conflict, and both are nuclear powers. In Afghanistan and the remote mountain hideaways of western Pakistan, a terrorist organization's leaders planned attacks that changed the skyline of New York and presaged the battleground of Iraq. In the ports of India, a growing navy reflects the emergence of the Indian Ocean as a new global geopolitical arena in which China, too, is asserting itself. Meanwhile, outsourcing by U.S. companies to India became a hot topic, and India's spectacular rise in information technology has changed that industry. Our daily lives are increasingly affected by what happens in this fascinating part of the world.

Defining the Realm

The Eurasian landmass incorporates all or part of 6 of the world's 12 geographic realms, and of these half-dozen very few are more clearly defined by nature than the one we call South Asia. **Figure 8-1** shows us why: the huge triangular Indian subcontinent that divides the northern Indian Ocean between the Arabian Sea and the Bay of Bengal is so sharply demarcated by mountain walls and desert wastes that you could take a pen and mark its boundary, from the Naga Hills in the far east through the Great Himalaya and Karakoram in the north to the Hindu Kush and the Iran-bordering wastelands of Balochistan and Afghanistan in the west (**Box 8-1**). Note how short the distances are over which the green of habitable lowlands turns to the dark brown of massive, snowcapped mountain ranges.

This realm extends from Afghanistan in the west to Bangladesh in the east. India, the realm's giant, is flanked by six countries (in clockwise order, Pakistan, Nepal, Bhutan, Bangladesh, Sri Lanka, and the Maldives) as well as Kashmir, a remaining disputed territory in the far north. South Asia's kaleidoscope of cultures may be the most diverse in the world, proving that neither formidable mountains nor forbidding deserts could prevent foreign influences from further diversifying an already variegated realm. We will encounter many of these influences in this chapter, but most of South Asia also possessed one unifying force of sorts: the British Empire, which from its late-nineteenth-century heyday through the late 1940s came to hold sway over almost all of it.

Since Islam prevails in Afghanistan and Pakistan, why are they included with South Asia and not with the North African/Southwest Asian realm? The answer lies in historical geography. First, Pakistan's history is closely tied to that of India—going back to the ancient civilization of the Indus Valley and, more recently, to the role of Islam and of British hegemony in the subcontinent. The tight integration of Pakistan with the rest

---

Box 8-1 Major Geographic Features of South Asia

1. Most of South Asia is clearly defined physiographically, and much of the realm's boundary is marked by mountains, deserts, and the Indian Ocean.

2. South Asia's great rivers, especially the Ganges, have for tens of thousands of years supported enormous population clusters.

3. South Asia, and especially northern India, was the birthplace of major religions including Hinduism and Buddhism.

4. Because of the realm's natural boundaries, foreign influences in premodern South Asia arrived mainly via a narrow passage in the northwest, the Khyber Pass.

5. Afghanistan is the realm's westernmost country and constitutes a cultural and political transition zone with the North Africa/Southwest Asia (NASWA) realm.

6. The South Asian realm covers less than 4 percent of the Earth's land area but contains almost one-quarter of the world's human population.

7. South Asia's annual monsoon continues to dominate life for hundreds of millions of people. This realm is also particularly vulnerable to the effects of global climate change.

8. Certain disputed territories in South Asia's remote northern mountain perimeter are potentially a source of dangerous friction between India and both Pakistan and China.

9. This realm is still predominantly rural and studded with hundreds of thousands of small villages; but it also contains some of the biggest cities in the world.

of South Asia will not surprise you after you study the realm's physiography in Figure 8-1: the natural boundary in this part of the realm lies west of the Indus River, not east of it. Pakistan today remains part of a realm that begins to change not in the Punjab, but at the Khyber Pass, the highland gateway to Afghanistan. Second, Afghanistan was a staging ground for Islam in the Indian subcontinent, and the cultural boundary between Afghanistan and Pakistan has remained fluid ever since. In political terms, too, there continue to be intricate connections between the two countries that, in turn, impact relations between Pakistan and India. At the same time, Afghanistan today is caught up in developments across Southwest Asia (as well as Central Asia), which is why that country is considered to be a **transition zone [1]** between South Asia and NASWA.

South Asia's Physiography

From snowcapped peaks to tropical forests and from bone-dry deserts to lush farmlands, this part of the world presents a virtually endless array of environments and ecologies, a diversity that is matched by its cultural mosaic. The broad outlines of this realm's physiography are best understood against the backdrop of its fascinating geologic past.

A Tectonic Encounter

As **Figure 8-2** shows, the spectacular relief in the north of this realm is the product of a collision between two of the planet's great **tectonic plates** (see Fig. G-4). About 10 million years ago, after a lengthy geologic journey following the breakup of the supercontinent Pangaea (diagrammed in Fig. 7-3), the Indian Plate encountered Eurasia. In this huge, slow-motion, accordion-like collision, the Earth's crust surrounding the contact zone was pushed upward, thereby creating the mighty Himalaya mountain range. That process is still going on—at the rate of 5 millimeters (0.2 in) per year—and the massive 2015 earthquake in Nepal proved once again that this is one of the most quake-prone areas in the world. One major outcome of this tectonic collision was that the northern margins of the realm were upwardly thrusted to elevations where permanent snow and ice make the landscape appear polar. The march of the seasons melts enough of this snowpack in spring and summer to sustain the great rivers below, providing water for farmlands that support hundreds of millions of people. The Ganges, Indus, and Brahmaputra all have their origins in the Himalaya. Only south of the Ganges Basin does the enormous plateau begin that marks the much older geologic core of the Indian Plate as it drifted northeastward toward Eurasia.

The Monsoon

Physical geography, therefore, is crucial here in South Asia—but not just on and below the surface. What happens in the overlying atmosphere is critical as well. The name "South Asia"

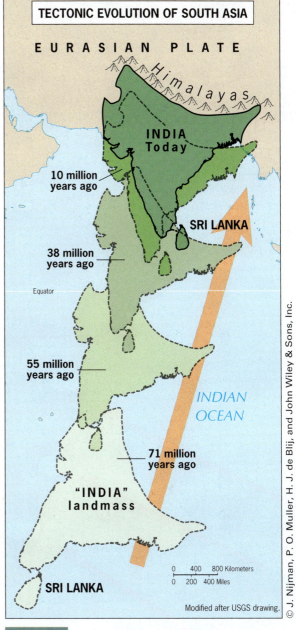

TECTONIC EVOLUTION OF SOUTH ASIA

EURASIAN PLATE

Himalayas

INDIA Today

10 million years ago

SRI LANKA

38 million years ago

Equator

55 million years ago

INDIAN OCEAN

71 million years ago

"INDIA" landmass

0   400   800 Kilometers
0   200   400 Miles

SRI LANKA

Modified after USGS drawing.

© J. Nijman, P. O. Muller, H. J. de Blij, and John Wiley & Sons, Inc.

FIGURE 8-2

is almost synonymous with the term **monsoon [2]** because the annual rains that accompany its onset, usually in June, are indispensable to all forms of agriculture in the realm's key country, India.

Figure 8-3 shows how the monsoon works. As the subcontinental landmass heats up during the spring, a huge low-pressure system is formed above it. This low-pressure system begins to draw in vast volumes of air from over the ocean onto the subcontinent. When the inflow of moist oceanic air reaches critical mass in early June, the **wet monsoon** has arrived. It may rain for 60 days or more. The countryside turns green, the paddies fill, and another dry season's dust and dirt are washed away—the region is reborn (see photo pair). The moisture-laden air flowing onshore from the Arabian Sea is forced upward against the Western Ghats, cooling as it rises

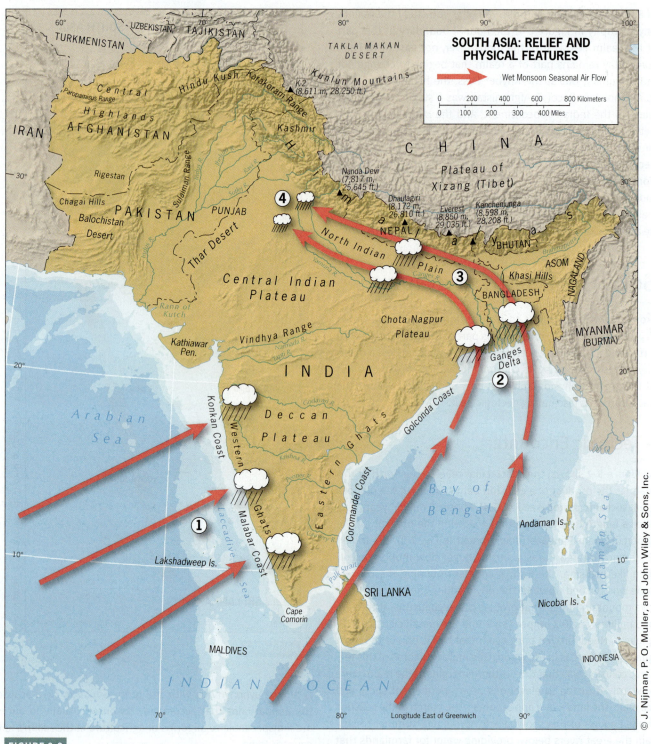

SOUTH ASIA: RELIEF AND PHYSICAL FEATURES

→ Wet Monsoon Seasonal Air Flow

FIGURE 8-3

© J. Nijman, P. O. Muller, and John Wiley & Sons, Inc.

and condensing large amounts of rainfall. The other branch of the wet monsoon originates in the Bay of Bengal and gets caught up in the convection (rising hot air) over northeastern India and Bangladesh. Seemingly endless rain now inundates a much larger area, including the entire North Indian Plain. The Himalaya mountain wall blocks the onshore-flowing airstream from spreading into the Asian interior and the rain from dissipating. Thus the moist airflow is steered westward, drying out

as it advances toward Pakistan. After persisting for weeks, this pattern finally breaks down, and the wet monsoon gives way to periodic rains and, eventually, another dry season. Then the anxious wait begins for the next year's wet monsoon because without it India would face disaster. It is estimated that two-thirds of India's agricultural areas do not have irrigation and depend fully on rainfall. For many people, life can hang by a meteorological thread.

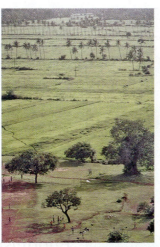

The arrival of the annual rains of the wet monsoon transforms the Indian countryside. By the end of May, the paddies lie parched and brown, dust chokes the air, and it seems that nothing will revive the land. Then the rains begin, and blankets of dust turn into layers of mud. Soon the first patches of green appear on the soil, and by the time the monsoon ends all is green. The photo on the left, taken just before the onset of the wet monsoon in the west-coast State of Goa, shows the paddies before the rains begin; three months later when the rice crop is ready to harvest, this same countryside looks as on the right.

Physiographic Regions

Figure 8-3 underscores South Asia's overall subdivision into three physiographic zones: the northern mountains, the southern plateaus, and, in between, a wide crescent of river lowlands.

The **northern mountains** extend from the Hindu Kush and Karakoram ranges in the northwest through the Himalaya in the center (Everest, the world's tallest peak, lies on the crestline that forms the Nepal–China border) to the ranges of Bhutan and the Indian State of Arunachal Pradesh in the east. Dry and barren in Afghanistan and western Pakistan, the ranges become green and tree-studded in Kashmir, forested in the lower-lying elevations of Nepal, and even more densely vegetated in Arunachal Pradesh. Transitional foothills, with many deeply eroded valleys cut by rushing meltwater, lead to the river basins below.

The belt of **river lowlands** extends eastward from Pakistan's lower Indus Valley (the area known as Sindh) through India's wide Gangetic Plain and then on across the great double delta of the Ganges and Brahmaputra in Bangladesh (Fig. 8-3). In the east, this physiographic region is often called the North Indian Plain. To the west lies the lowland of the Indus River, which rises in Tibet, crosses Kashmir, and then bends southward to receive its major tributaries from the area known as **Punjab** ("Land of Five Rivers") to the east.

The **southern plateaus** constitute peninsular India, dominated by the **Deccan**, a massive tableland built of lava sheets that poured out when India separated from Africa during the sundering of Pangaea. The Deccan (meaning "South") tilts

toward the east, so that its highest areas are in the west and most of the rivers flow into the Bay of Bengal. North of the Deccan lie two other plateaus, the Central Indian Plateau to the west and the Chota Nagpur Plateau to the east (Fig. 8-3). On the map, also note the Eastern and Western Ghats: "ghat" means step, and it connotes the descent from Deccan Plateau elevations to the narrow coastal plains below. The onshore winds of the annual wet monsoon bring ample precipitation to the Western Ghats; as a result, here lies one of India's most productive farming areas and one of southern India's largest population concentrations.

Birthplace of Civilizations and Religions

Indus Valley Civilization

A complex and technologically advanced civilization emerged in the Indus Valley by about 2500 BC, simultaneous with other Bronze Age "urban revolutions" in Egypt and Mesopotamia (see Fig. 6-3 for the map of world culture hearths). The Indus Valley civilization was centered on a pair of major cities, Harappa and Mohenjo-Daro, which may have been capitals during different periods of its history (**Fig. 8-4**); in addition, there were more than 100 smaller urban settlements. The locals apparently called their state **Sindhu**, and both **Indus** (for the river) and **India** (for the later state) very likely derived from this name. Although the influence of this civilization extended as far east as present-day Delhi, it did not last because of environmental change and, perhaps, because the political center of gravity shifted southeastward into the Ganges Basin.

Aryans and the Origins of Hinduism

Around 1500 BC, northern India was invaded by the **Aryans** (peoples speaking Indo-European languages based in what is now Iran). As the Iron Age dawned in India, acculturated Aryans began the process of integrating the Ganges Basin's isolated tribes and villages into a newly-created cultural system, and urbanization made a comeback. The Aryans brought their language, **Sanskrit** (related to Old Persian), as well as a new social order to the vast riverine flatlands of northern India. Their settlement here was also accompanied by the emergence of a religious belief system, **Vedism**. Out of the texts of Vedism and local creeds there arose a new religion—**Hinduism**—and with it a new way of life.

It is thought that the arrival and accommodation of the Aryans in this reorganized society forged a layered system of social stratification that would solidify the powerful position of the Aryans and be legitimized through religion. Starting about 3500 years ago, a combination of regional integration, the consolidation of villages into controlled networks, and the emergence of numerous small city-states produced a hierarchy of

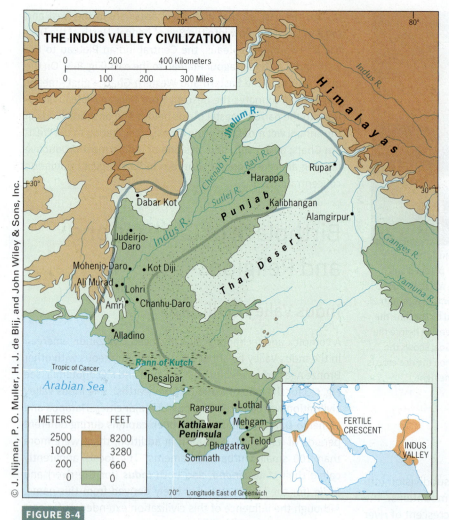

THE INDUS VALLEY CIVILIZATION

FIGURE 8-4

© J. Nijman, P. O. Muller, H. J. de Blij, and John Wiley & Sons, Inc.

histories. Today, Telugu, Tamil, Kanarese (Kannada), and Malayalam are spoken by nearly 300 million people. In India's northern and northeastern fringes, Sino-Tibetan languages predominate, and smaller pockets of Austro-Asiatic speakers can be found in eastern India and neighboring Bangladesh. To the west, Afghanistan consists of a mosaic of Altaic and Indo-European languages, and Pashto (spoken by two-thirds of the Afghan population) spills across the border into neighboring Pakistan.

Buddhism and Other Indigenous Religions

Hinduism is not the only religion that emerged in this realm. Around 500 BC, **Buddhism** arose in the eastern Ganges Basin in what today is the Indian State of Bihar. The famous story of the "enlightenment" of the Prince Siddhartha (the Buddha) took place in the town of Bodh Gaya, and his adherents soon fanned out in every direction. The appeal of Buddhism was (and is) especially strong among lower-caste Hindus, and large numbers of them have converted through the ages. Interestingly, Buddhism evolved inside what is now India, but its most powerful influences were felt beyond the realm in Southeast and East Asia. Today, only 0.7 percent of the population of India adheres to Buddhism (80 percent are Hindu), but it is the state religion in Bhutan and the great majority (around 70 percent) of Sri Lankans are Buddhist as well.

Another (much smaller) indigenous religion that has evolved alongside Hinduism since ancient times is **Jainism**, often described as a more purist, principled, and deeply spiritual form of Hinduism. It is well known for its uncompromising stand on nonviolence and vegetarianism. Jains today constitute less than one-half of 1 percent of the population in India. Finally, we should take note of **Sikhism** as another of the realm's indigenous religions, a blend of sorts of Islamic and Hindu beliefs. This religion, practiced by 1.7 percent of the population, is of course much younger; it emerged around AD 1500, a few centuries after Islam became a dominant force in much of the South Asian realm.

Foreign Invaders

The Reach of Islam

In the late tenth century, Islam came rolling like a giant tide across South Asia, spreading across Persia and Afghanistan,

power among the people, a ranking from the especially powerful (Brahmins—the highest-order priests) to the weakest. This class-based **caste system [3]** of Hinduism is highly controversial in the West (and among many Indians too) because of its rigidity and the ways in which it justifies structural inequality. Those in the lowest castes, deemed to be there because they earned that assignment in their past lives, are worst off, without hope of advancement and at the mercy of those ranking higher on the social ladder. In recent years, the caste system appears to be eroding from the combined effects of globalization, economic growth, and urbanization; this is true for India's bigger cities, but much less so for the majority of people who still reside in rural areas.

Although Hinduism spread across South Asia and even reached the Southeast Asian realm (especially Cambodia and Indonesia), Indo-European languages never took hold in the southern portion of the subcontinent. As **Figure 8-5** indicates, **Indo-European languages [4]** (several of which are rooted in Sanskrit) predominate in the western and northern parts of South Asia, whereas the southern languages belong to the **Dravidian [5]** family—languages that were indigenous to the realm even before the arrival of the Aryans. But these are not fossil languages: they remain vibrant and have long literary

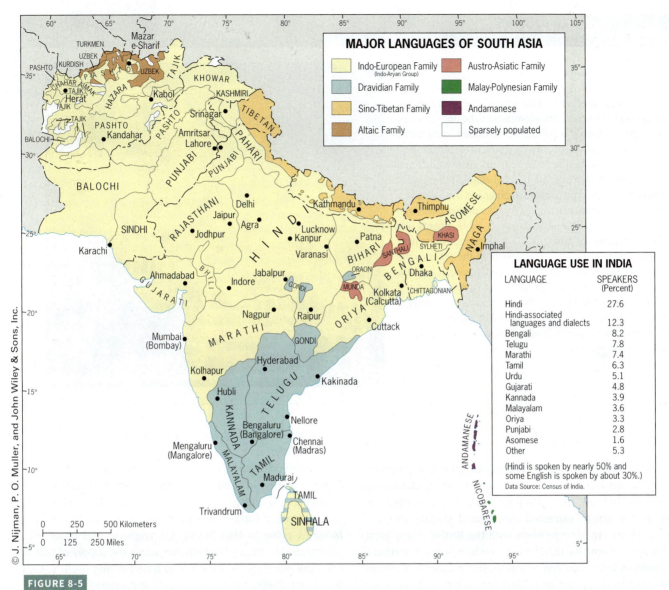

MAJOR LANGUAGES OF SOUTH ASIA

- Indo-European Family (Indo-Aryan Group)
- Dravidian Family
- Sino-Tibetan Family
- Altaic Family
- Austro-Asiatic Family
- Malay-Polynesian Family
- Andamanese
- Sparsely populated

LANGUAGE USE IN INDIA

LANGUAGE	SPEAKERS (Percent)
Hindi	27.6
Hindi-associated languages and dialects	12.3
Bengali	8.2
Telugu	7.8
Marathi	7.4
Tamil	6.3
Urdu	5.1
Gujarati	4.8
Kannada	3.9
Malayalam	3.6
Oriya	3.3
Punjabi	2.8
Asomese	1.6
Other	5.3

(Hindi is spoken by nearly 50% and some English is spoken by about 30%.)
Data Source: Census of India.

© J. Nijman, P. O. Muller, and John Wiley & Sons, Inc.

FIGURE 8-5

through the Khyber Pass into the Indus Basin, and across Punjab into the Ganges Basin—converting virtually everybody in the Indus Valley and foreshadowing the emergence, several centuries later, of the Islamic Republic of Pakistan. By the early thirteenth century, the Muslims had established the long-surviving and powerful **Delhi Sultanate** that controlled much of the northern tier of peninsular India.

In the early 1500s, a descendant of Genghis Khan named Babur conquered Kabol (Kabul) in Afghanistan, and from that stronghold he penetrated the Indus Basin and Punjab and directly challenged the Delhi Sultanate. In the 1520s, his Islamicized Mongol forces ousted the Delhi rulers and established the **Mughal Empire**. By most accounts, Mughal rule was at times remarkably enlightened, especially under the leadership of Babur's grandson Akbar, who aggressively expanded the Empire but adopted tolerant policies toward the Hindus under his sway; Akbar's grandson, Shah Jahan, made his enduring mark on India's cultural landscape by

commissioning magnificent architectural creations, particularly the Taj Mahal in the city of Agra (see photo).

Nonetheless, by the early eighteenth century, the Mughal Empire was in decline. Maratha, a Hindu state in the west, expanded not only into the peninsular south but also northward toward Delhi, capturing the allegiance of local rulers and weakening Islam's hold. Fractured India now lay open to still another foreign intrusion, this time from Europe.

Despite more than seven centuries of Islamic rule in South Asia, it is remarkable that Islam never achieved proportional dominance over the realm as a whole. Whereas today Pakistan is 96 percent Muslim and Bangladesh 90 percent, India—where the Delhi Sultanate and the Mughal Empire were centered—remains only about 15 percent Muslim today. Islam may have arrived like a giant tide, but Hinduism stayed afloat and eventually outlasted the incursion. It also withstood the European onslaught that culminated in the incorporation of the entire realm into the British Empire.

Box 8-2 From the Field Notes...

"More than a half-century after the end of British rule, the centers of India's great cities continued to be dominated by the Victorian-Gothic buildings the colonizers constructed here. This also is evidence of a previous era of globalization, when European imprints transformed urban landscapes. Walking the streets of some parts of Mumbai (the British called it Bombay), you can turn a corner and be forgiven for mistaking the scene for London, double-decker buses and all. One of the British planners' major achievements was the construction of a nationwide rail system, and railway stations were given great prominence in the urban architecture. I had walked up Naoroji Road, having learned to dodge the wild traffic around the circles in the Fort area, and watched the throngs passing through Victoria (now Chhatrapati Shivaji) Station. Inside, the facility is badly worn, but the trains continue to run, bulging with passengers hanging out of doors and windows."

© H.J. de Blij

The European Intrusion

The British took advantage of the weakened and fragmented power of the Mughals and followed a strategy commonly known as "indirect rule." They left local rulers in place as long as the British extracted the desired trading arrangements. Thanks to arrangements with the British, many local maharajas became wealthier than ever before: from northern to southern India, you can find beautiful palaces (now often converted to museums or hotels) that were built by these

The Taj Mahal, India's supreme iconic seventeenth-century legacy of the Mughals, is one of the world's most admired architectural accomplishments. This city of Agra is also a popular destination for growing numbers of domestic tourists, such as this visiting group from a village in the northern, Himalayan State of Uttarakhand.

rulers, often as recently as the late nineteenth or early twentieth century.

British colonialism in South Asia coincided with the Industrial Revolution in Europe, and the impact of Britain on the realm must be understood in that context. South Asia became, in large part, a supplier of raw materials needed to keep the factories going in Manchester, Birmingham, and other British industrial centers. For instance, when the supply of cotton from the American South came to a halt during the U.S. Civil War in the 1860s, the British quickly encouraged (and indeed enforced) cotton production in what is today western India.

When the British took full power in South Asia during the mid-nineteenth century, this was a realm with already considerable industrial development (notably in metal goods and textiles) and an active trade with both Southeast and Southwest Asia. The colonialists saw this as competition, and soon India was exporting raw materials as well as importing manufactured products—from Europe, of course. Local industries declined, and Indian merchants lost their markets.

The British colonial legacy included one of the most extensive transport networks of the colonial era, particularly the railway system—even though the network focused on interior-to-seaport linkages rather than fully interconnecting the various parts of the country. British engineers laid out irrigation canals through which millions of hectares of land were brought into cultivation. Coastal settlements that had been founded by Britain developed into major cities and bustling ports, led by Bombay (now renamed Mumbai), Calcutta (now Kolkata), and Madras (now Chennai). These three cities still rank among India's largest urban centers, and their cityscapes bear the unmistakable imprint of colonialism (Box 8-2).

The British *raj* (the term for Britain's direct rule between 1857 and 1947) also produced a new elite within the native Indian population. They had access to education and schools that combined English and Indian traditions, and their Westernization was reinforced through university education in Britain. This elite drew from Hindu and Muslim communities, and it was to play a major part in the rising demands for self-rule and independence. Those demands started to gather momentum in the early twentieth century and could no longer be denied after World War II came to an end in 1945.

The Geopolitics of Modern South Asia

Partition and Independence

Even before the British government decided to yield to demands for independence, it was clear that British India would not survive the coming of self-rule as a single political entity. As early as the 1930s, Muslim activists were promoting the idea of a separate state. As the colony moved toward independence, a major political crisis developed that eventually resulted in the separation of India and Pakistan. But **partition [6]** was no simple matter. True, Muslims were in the majority in the western and eastern sectors of British India, but smaller Islamic clusters were scattered throughout the realm. Furthermore, the new boundaries between Hindu and Muslim communities had to be drawn right through areas where both sides coexisted—thereby displacing millions (see photo).

The consequences of this migration for the social geography of India were especially far reaching. Comparing the country's 1931 and 1951 distributions of Muslims in **Figure 8-6**, you can see the impact on the Indian Punjab and in what is today the State of Rajasthan. (Since Kashmir was mapped as three entities before partition and as one afterward, the change there represents an administrative, not a major numerical, alteration.) A Muslim exodus occurred even in the east, as is reflected on the map by the State of West Bengal, adjacent to East Pakistan (now Bangladesh), where the Islamic component of the local population declined substantially.

The world has seen many refugee migrations but none involving so many people in so short a time as the one resulting from British India's partition (which occurred on Independence Day, August 15, 1947). Scholars who study the refugee phenomenon differentiate between "forced" and "voluntary" migrations, but as this case underscores, it is not always possible to distinguish between the two. Many Muslims, believing they had

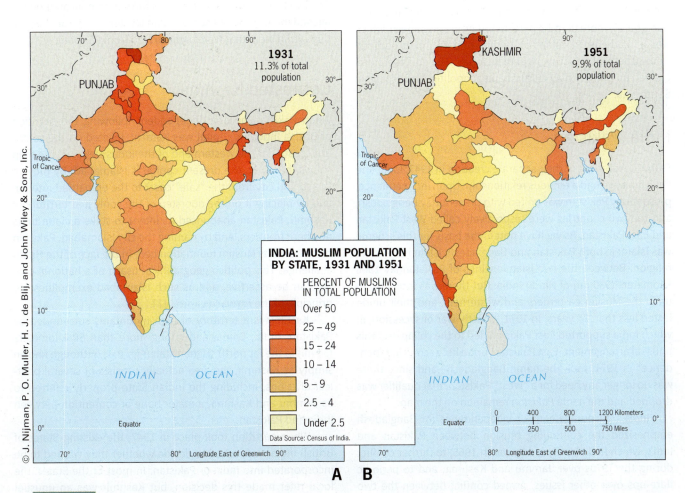

INDIA: MUSLIM POPULATION BY STATE, 1931 AND 1951

PERCENT OF MUSLIMS IN TOTAL POPULATION

- Over 50
- 25 – 49
- 15 – 24
- 10 – 14
- 5 – 9
- 2.5 – 4
- Under 2.5

Data Source: Census of India.

© J. Nijman, P. O. Muller, H. J. de Blij, and John Wiley & Sons, Inc.

FIGURE 8-6

© Bettmann/CORBIS

Flight was one response to the 1947 partition of what had been British India, resulting in the greatest mass population transfer in human history (involving at least 14 million migrants). Here, two trainloads of eastbound Hindu refugees, fleeing then West Pakistan, arrive at the station in Amritsar, the first city inside India.

no choice, feared for their future in the new India and joined the stampede. Others had the means and the ability to make a decision to stay or leave, but even these better-off migrants undoubtedly sensed a threat.

The great majority of Hindus who lived on the "wrong" side of the border moved as well. The Hindu component of present-day Pakistan may have been as high as 16 percent in 1947 but is barely more than 1.4 percent today; in Bangladesh, which was named East Pakistan at the time of partition, it declined from 30 percent to just over 8 percent today. Partition, therefore, created an entirely new cultural and geopolitical landscape in South Asia.

India–Pakistan

From the moment of their separate creation in 1947, India and Pakistan have had a tenuous relationship. Upon independence, present-day Pakistan was united with present-day Bangladesh, and the two countries were respectively called West Pakistan and East Pakistan. As we have noted, the basis for this scheme was Islam: in both Pakistan and Bangladesh, Islam is the state religion. Between the two Islamic wings of Pakistan lay 1500 kilometers (930 mi) of Hindu India. But there was little else to unify the Muslim easterners and westerners, and their union lasted less than 25 years. In 1971, a costly war of secession, in which India supported East Pakistan, led to the collapse of this unusual arrangement. East Pakistan, upon its "second independence" in 1971, took the name Bangladesh—and since there was no longer any need for a "West" Pakistan, that qualifier was dropped and the name *Pakistan* remained on the map.

India's encouragement of independence for Bangladesh emphasized the continuing tension between Pakistan and India, which had already led to war in 1965, to further conflict during the 1970s over Jammu and Kashmir, and to periodic flare-ups over other issues. Armed conflict between the two South Asian countries seemed to be a regional matter—until

the early 1990s, when their arms race took on ominous nuclear proportions. Since then, the specter of nuclear war has hung over the conflicts that continue to embroil Pakistan and India, a concern not just for the South Asian realm but for the world as a whole. No longer merely a decolonized, divided, and disadvantaged country trying to survive, Pakistan has taken a crucial place in the political geography of a world realm in turbulent transition.

The relationship between India and Pakistan is especially sensitive because so many Muslims still live in India. Massive as the 1947 refugee movement was, it left far more Muslims in India than those who had departed. The number of Muslims in India declined sharply, but it remained a huge minority, one that was growing rapidly to boot. Today, it is approaching 200 million (14.2 percent of India's population)—the largest cultural minority in the world and just about equal in size to Pakistan's entire population.

What this means is that a substantial proportion of India's population tends to have more or less "natural" sympathies toward Pakistan, even though India's Muslims are generally known to be "moderate." Their presence in many instances works as a brake on hawkish Indian policies toward the Islamabad government. At the same time, conflict with Pakistan can have detrimental effects on Hindu-Muslim relations inside India, and over the years it has led to communal violence and deadly clashes. Today that issue is further complicated by the alleged involvement of some Indian Muslims in terrorist activities within India that were orchestrated from Pakistan.

Contested Kashmir

When Pakistan became independent following the partition of British India, its capital was Karachi, located on the southern coast near the western end of the Indus Delta. As the map shows, however, the present capital is Islamabad. By moving the capital city from the "safe" coast to the embattled interior, and by placing it on the doorstep of the contested territory of Kashmir, Pakistan announced its intent to stake a claim to its northern frontiers. And by naming the city Islamabad, Pakistan proclaimed its Muslim foundation here in the face of the Hindu challenge. This politico-geographical usage of a national capital can be assertive, and as such Islamabad exemplifies the principle of the **forward capital [7]**.

Kashmir is a territory of high mountains surrounded by Pakistan, India, China, and, along more than 50 kilometers (30 mi) in the far north, Afghanistan (**Fig. 8-7**). Although known simply as Kashmir, the area actually consists of several political divisions, including the Indian State properly referred to as Jammu and Kashmir, a major bone of contention between India and Pakistan.

When partition took place in 1947, the existing States of British India were asked to decide whether they wanted to be incorporated into India or Pakistan. In most of the States, the local ruler made this decision, but Kashmir was an unusual case. It had about 5 million inhabitants at the time, nearly

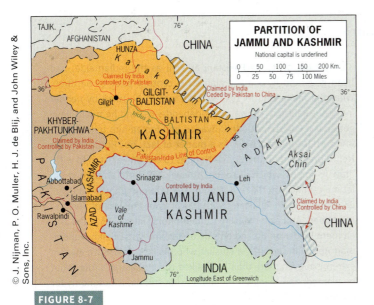

© J. Nijman, P. O. Muller, H. J. de Blij, and John Wiley & Sons, Inc.

FIGURE 8-7

a Pakistan-based organization that among other things aims to return Kashmir to Islamic rule.

In the meantime, Pakistan's northwestern frontier is effectively controlled by the Taliban, the Afghan Islamic extremists who also have a history of collaboration with al-Qaeda. This entire border zone with Afghanistan seems to lie well beyond the control of the Pakistani government. U.S. efforts to defeat the Taliban in Afghanistan continued in 2016 but were in part thwarted by the Taliban's ability to move back and forth across this border—despite the pressure the United States exerts on Islamabad to confront the Taliban on the Pakistani side of the border in this treacherous mountain refuge.

This remains an especially delicate geopolitical chess match for the Pakistani regime. It must be very careful not to alienate its Islamic base, even though it despises the border-zone extremists, and it also fears an economically stronger India with growing ties to the United States. India is deeply concerned about

three-quarters of them Muslims, but the maharajah of Kashmir himself was a Hindu. When he decided not to join Pakistan and instead aimed to retain autonomous status, this was answered with a Muslim uprising supported by Pakistan. The maharajah, in turn, called for help from India. After more than a year's fighting and through the intervention of the United Nations, a cease-fire line left most of Jammu and Kashmir (including nearly four-fifths of the territory's population) in Indian hands. Eventually, this line—now known as the **Line of Control**—began to appear on maps as the final boundary settlement, and Indian governments have proposed that it be so recognized.

With about two-thirds of the Kashmiris adhering to Islam, Pakistan has for decades demanded a referendum in Jammu and Kashmir in which the people can decide for themselves to remain with India or become part of Pakistan. India has refused, arguing that there is a place for Muslims in secular India but not for Hindus in the Islamic Republic of Pakistan. Given the specter of terrorism in India and the dangerous precedent of a concession in light of India's prodigious ethnic and regional diversity, the Kashmir conflict has been left to smolder and is unlikely to be resolved for some time to come (**Box 8-3**).

The Specter of Terrorism

Comparatively successful as the integration of India's Muslim communities into the fabric of the Indian state has been, the risk of Islamic violence, directed against Indian society in general, has been rising. The highest-profile attack to date was in 2008, when terrorists targeted Mumbai's two most upscale, Westernized hotels. Nearly 200 people perished and hundreds were wounded. Live pictures of smoke billowing from the famous Taj Mahal Hotel in southern Mumbai were broadcast around the world (see photo). The group responsible for these attacks was *Lashkar-e-Taiba* (the Party of the Righteous),

AP/Wide World Photos

Smoke billows from the landmark Taj Mahal Hotel in Mumbai on November 29, 2008. The Taj was one of several sites in southern Mumbai that were simultaneously targeted for attack by Islamic militants. The siege lasted four days, and almost 200 people died in the violence. Indian commandos killed all of the terrorists but one, who was captured, tried, sentenced to death, and executed in 2012. It was not the first time that such terrorism had struck India, and several smaller-scale attacks have occurred in this decade.

Box 8-3 Regional Issue: Who Should Govern Kashmir?

Kashmir Should be Part of Pakistan!

"I don't know why we're even debating this. Kashmir should and would have been made part of Pakistan in 1947 if that colonial commission hadn't stopped mapping the Pakistan–India boundary before they got to the Chinese border. And the reason they stopped was clear to everybody then and there: instead of carrying on according to their own rules, separating Muslims from Hindus, they reverted to that old colonial habit of recognizing "traditional" States. And what was more traditional than some Hindu potentate and his minority clique ruling over a powerless majority of Muslims? It happened all over India, and when they saw it here in the mountains they couldn't bring themselves to do the right thing. So India gets Jammu and Kashmir and its several million Muslims, and Pakistan loses again. The whole boundary scheme was rigged in favor of the Hindus anyway, so what do you expect?

"Here's the key question the Indians won't answer. Why not have a referendum to test the will of all the people in Kashmir? India claims to be such a democratic example to the world. Doesn't that mean that the will of the majority prevails? But India has never allowed the will of the majority even to be expressed in Kashmir. We all know why. About two-thirds of the voters would favor union with Pakistan. Muslims want to live under an Islamic government. So people like me, a Muslim carpenter here in Srinagar, can vote for a Muslim collaborator in the Kashmir government, but we can't vote against the whole idea of Indian occupation.

"Life isn't easy here in Srinagar. It used to be a peaceful place with boats full of tourists floating on beautiful lakes. But now the place is heavily militarized and there's tension and occasional violence. Of course we Muslims get the blame, but what do you expect when the wishes of a religious majority are ignored? So don't be surprised at the support our cause gets from Pakistan across the border. The Indians call them terrorists…. In the long run, we will prevail, simply because this is our land and we are in the majority."

Kashmir Belongs to India!

"Let's get something straight. This stuff about that British boundary commission giving up and yielding to a maharajah is nonsense.

Kashmir (all of it, the Pakistani as well as the Indian side) had been governed by a maharajah for a century prior to partition. What the maharajah in 1947 wanted was to be ruled by neither India nor Pakistan. He wanted independence, and he might have gotten it if Pakistanis hadn't invaded and forced him to join India in return for military help. As a matter of fact, our Prime Minister Nehru prevailed on the United Nations to call on Pakistan to withdraw its forces, which of course it never did. As to a referendum, let me remind you that a Kashmir-wide referendum was (and still is) contingent on Pakistan's withdrawal from the area of Kashmir it grabbed. And as for Muslim 'collaborators,' in the 1950s the preeminent leader on the Indian side of Kashmir was Sheikh Muhammad Abdullah (get it?), a Muslim who disliked Pakistan's Muslim extremism even more than he disliked the maharajah's rule. What he wanted, and many on the Indian side still do, is autonomy for Kashmir, not incorporation.

"In any case, Muslim states do not do well by their minorities, and we in India generally do. As far as I am concerned, Pakistan is disqualified from ruling Kashmir by the failure of its democracy and the extremism of its Islamic ideology. Let me remind you that Indian Kashmir is not just a population of Hindus and Muslims. There are other minorities—for example, the Ladakh Buddhists—who are very satisfied with India's administration but who are terrified at the prospect of incorporation into Islamic Pakistan. You already know what Sunnis do to Shi'ites in Pakistan. You are aware of what happened to ancient Buddhist monuments in Taliban Afghanistan (and let's not forget where the Taliban came from). Can you imagine the takeover of multicultural Kashmir by Islamabad?

"To the Muslim citizens of Indian Kashmir, I, as a civil servant in the Srinagar government, say this: look around you, look at the country of which you are a subject. Muslims in India are more free, have more opportunities to participate in all spheres of life, are better educated, have more political power and influence than Muslims do in Islamic states. Traditional law in India accommodates Muslim needs. Women in Muslim-Indian society are far better off than they are in many Islamic states. Kashmir belongs to India. All inhabitants of Kashmir benefit from Indian governance. What is good for all of India is good for Kashmir."

Pakistan's role in terrorism on Indian soil and, even worse, the possibility of jihadists taking control of Pakistan's nuclear arsenal; at the same time, it must guard against heightening tensions between Hindus and Muslims inside India. India is also frustrated by American reluctance to choose sides in the Kashmir conflict. The United States, in turn, is sympathetic to the world's largest democracy but needs Pakistan as an ally in the global counterterrorism campaign. Each of these parties is walking a tightrope, where the slightest mistake could have deadly consequences.

Chinese Border Claims

Any overview of this realm's geopolitical framework today must also include China's increasingly powerful and sometimes

invasive presence. As Figure 8-7 shows, China claims the northeastern extension of India's Jammu and Kashmir State. This issue has been stalemated in recent years, but officially neither China nor India shows any sign of conceding.

China's control over Tibet (called Xizang by the Chinese) and its efforts to influence the lives of Tibetans both within and outside Tibet create additional issues. To the dismay of Beijing, Tibet's exiled Dalai Lama calls India his second home. In recent years, China has been pressuring the government of Nepal, wedged between India and Tibet, to discourage Tibetan immigration and to constrain the activities of Tibetans already in the country. Farther to the east, China claims the bulk of the Indian State of Arunachal Pradesh ("land of the dawn-lit mountains") based on its assertion that the boundary, established in 1914, was never ratified by Beijing— even though it was approved by the then-independent Tibetans themselves. And China even claims rights to the small area of northern India lying between Tibet-adjoining Sikkim and Bhutan, again on the basis that its inhabitants are Tibetans and therefore belong under Chinese jurisdiction.

China's power in South Asia is felt in other ways as well. A few years ago, the Chinese announced plans to construct dams on the Tibetan headwaters of the upper Brahmaputra River, potentially jeopardizing the main water supplies of both north-eastern India and Bangladesh. Thus by no means is Kashmir the only place in the realm's lengthy northern frontier zone where trouble can erupt at any time—and within which India directly borders China for 3500 kilometers (2200 mi).

Indian Ocean Geopolitics

China needs access to markets for its products and supplies of raw materials to sustain its steadily growing industrial production, and a major part of that access involves the Indian Ocean. In order to protect these interests, China is projecting political and military power by expanding its naval presence in this maritime arena and building bases in Pakistan, Bangladesh, Myanmar (Burma), and Djibouti. Meanwhile, increasingly concerned about Chinese intentions around the subcontinent's territorial waters, India is responding by strengthening its own navy and forging new alliances with such Southeast Asian states as Indonesia and Vietnam.

From a broader pan-Asian perspective (South and East), geopolitical developments are increasingly a function of the U.S.–China–India relationship. These developments are propelled by China's assertions in the Indian Ocean Basin and along its land borders with India; India's economic rise, which gives it an ever higher profile in the global political arena; and the ideological propensity of the United States to side with India. The near-term future will likely see a continued informal alliance between the United States and India that seeks to limit the expansion of China's boldness in South Asia. American support for admitting India to membership in the United Nations Security Council is but one manifestation of the strengthening connection between the two countries.

These underlying issues aside, there have also been recent indications of more conciliatory exchanges between the two Asian giants, including mutual state visits of their presidents. In a notable turnaround, China in 2015 avowed that it would not oppose Indian membership in the Security Council. More importantly, growing economic interdependence is proving to be advantageous for both countries. China wants to penetrate India's enormous, expanding consumer markets. China–India trade has increased rapidly in recent years, with about two-thirds going from China to India and one-third in the other direction. So far, China has been running a significant surplus, reflecting the fact that the terms of trade are more favorable to China: India mainly supplies raw materials, whereas China largely sells finished, higher-value-added products.

Emerging Markets and Fragmented Modernization

Over the past 15 years or so, optimistic reports in the media have been proclaiming a new era for South Asia, marked by rising growth rates for the realm's national economies, rewards from globalization and modernization, and tightening integration into the world economy. India, obviously the key to this realm, has even been described as "India Shining" during this enthusiastic wave of optimism.

Nevertheless, well over half of India's 1.3 billion people continue to live in poverty-stressed rural areas, their villages and lives virtually untouched by what is happening in the cities (where tens of millions of urban dwellers inhabit some of the world's poorest slums). Afghanistan and Nepal rank among the poorest countries on Earth; fully one-third of Pakistan's population lives in miserable poverty, and female literacy still lingers below 50 percent. It is estimated that about 40 percent of the children in South Asia are malnourished and underweight, a majority of them girls—this at a time when the world is able to provide adequate calories for all its inhabitants, if not sufficiently balanced daily meals. It still remains to be seen whether the benefits of newfound economic growth can be spread around widely enough to improve the living conditions of South Asia's masses.

Economic Liberalization

Most countries in this realm have liberalized their economies since the 1980s as part of a worldwide turn toward **neoliberalism [8]**. This entails privatization of state-run companies, lowering of international trade tariffs, reduction of government subsidies, cutting of corporate taxes, and overall deregulation to stimulate business activity.

The results of these reforms can be observed across South Asia in terms of increased economic growth and higher per capita incomes. Generally, this progress has continued into this decade, even though variations between countries are quite pronounced (**Fig. 8-8**). Most of the growth is in manufacturing, services, and, in India, information technology (**Box 8-4**). More open economies have attracted increased foreign investment, and during the past two decades a new, urban middle class has emerged. This steadily expanding social stratum may account for only about 25 percent of the realm's population, but in South Asia that translates into some 400 million people—a huge new consumer market for an array of products ranging from smartphones to refrigerators to automobiles. At the same time, it leaves well over a billion South Asians unaffected by this upward social mobility. For them almost nothing has changed; they remain overwhelmingly dependent on (stagnant) agriculture and they are not likely to log on to the new information economy anytime soon. As in so many other parts of the world, here too increased overall growth has mostly been accompanied by ever greater economic and social inequality.

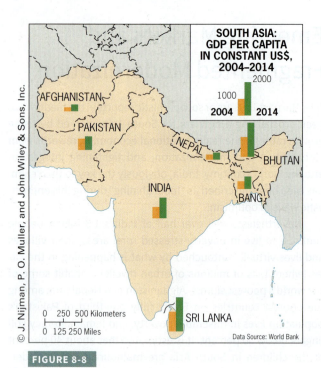

FIGURE 8-8

One striking feature of this realm is that even though the majority of its people live traditional lives in rural villages, there also are eight **megacities [9]** in which social and economic change is the order of the day. The metropolitan areas centered on Delhi–New Delhi, Mumbai, Karachi, Dhaka, Kolkata, Lahore, Bengaluru, and Chennai all contain populations that exceed 10 million and have grown rapidly over the past two decades (see Box 8-8). Their densities are often overwhelming, environmental quality is poor and worsening, and the widening income gap between rich and poor is staggering. But almost always it is better to be poor in the city than in the countryside, simply because cities offer opportunities that do not exist in the villages—which helps to explain the surging flow of rural-to-urban migrants that drives the growth of towns and cities all across South Asia. As **Figure 8-9** demonstrates, there is substantial variation in the internal densities and layouts of megacities; those differences also extend to levels of prosperity, and the contrasts between megacities in the Global Core and the Global Periphery are stunning (**Box 8-5**).

Box 8-4 From the Field Notes...

"Strolling the grounds of one of India's leading high-tech companies, Wipro, in the southern city of Bengaluru (formerly Bangalore), I was struck by the thought that their use of the word 'campus' was right on target: the great majority of the 33,000 employees at this facility (120,000 in total) are in their mid-20s, and almost all have a college degree; a large number of them are involved in research and product development; and the layout of the premises could easily be mistaken for a (well-funded) university. This Silicon Valley-like campus is the heart of the company's booming outsourcing and consulting business. Wipro is active in most major Indian cities and has a sizeable global presence. I was shown around by Rohit, a 26-year-old with a bachelor's degree in engineering from an Indian college and an MBA from Singapore National University. He reveled in the opportunity: 'I joined the company just four months ago in their marketing division and feel extremely fortunate. Everybody wants this job! It's a great professional opportunity and the salary is very good.' And Bengaluru is a pleasant place to live: at 900 meters (3000 ft) above sea level, even the summer weather is tolerable. Bengaluru, however, is growing so rapidly that its infrastructure cannot keep up: thus congestion, traffic jams, and commuting times are all growing as well."

© Jan Nijman

Box 8-5 MAP ANALYSIS

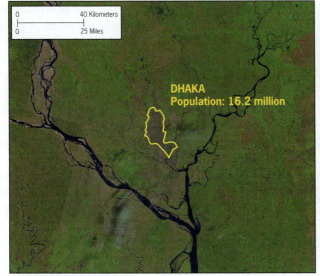

DHAKA
Population: 16.2 million

LOS ANGELES
Population: 15.1 million

© J. Nijman, P. Muller, and John Wiley & Sons, Inc.

FIGURE 8-9

South Asia is the most heavily populated geographic realm on Earth. But its overall level of urbanization is quite low at 32 percent because the great majority of people inhabit rural areas. In absolute terms, however, the bigger South Asian cities are growing rapidly, and the realm now contains eight megacities. In comparison, the North American realm has only two: New York and Los Angeles. It is important to realize that significant contrasts mark the world's megacities, as shown in the pair of maps above that differentiate the metropolitan areas of Dhaka and Los Angeles. Analyze the two maps and provide a concise description of the fundamental differences. Can you think of some reasons that explain these differences?

Access your WileyPLUS Learning Space course to interact with a dynamic version of this map and to engage with online map exercises and questions.

The Significance of Agriculture

More than half of South Asia's workforce is employed in agriculture, ranging from about 40 percent in Pakistan to nearly 70 percent in Nepal. But overall productivity is low, and the contribution of agriculture to the national economy averages only around 20 percent. Incomes in rural areas are much lower than in the towns and cities, and the same is true for the standard of living. With two-thirds of the realm's population living outside urban areas, most people either work in agriculture or rely on it indirectly.

Millions of lives every year depend on a bountiful harvest. As Figure 8-3 shows, the wet monsoon brings life-giving rains to the southwestern (Malabar) coast, and a second branch from the Bay of Bengal sweeps across north-central India toward Pakistan, losing strength (and moisture) as it marches westward. This means that in amply watered eastern India, States like Asom (Assam) (**Box 8-6**) that surround Bangladesh as well as the southwestern coastal strip, rice can be grown as the staple crop. But drier northwestern India and Pakistan can only support wheat.

Farmers' fortunes tend to vary geographically, as can be illustrated by comparing the two sides of India's Western Ghats upland (Fig. 8-3). The monsoon rains are generally plentiful along the Arabian Sea-facing western slopes of this linear highland, from the southern tip of the subcontinent as far north as the vicinity of Mumbai. Here you can see the hillside vegetation assume all shades of fresh green after the rains begin in June, a promising sign that the harvest will be abundant. But on the eastern (rain shadow) side of the Western Ghats and deeper into the interior of the Deccan Plateau, the rains are sporadic and do not last as long; farming becomes a gamble with nature, and life becomes precarious. Many of the farmers here belong to the lowest-ranking castes, landless and indebted, and have the most difficulty in making ends meet. Almost every year, Maharashtra's inland districts report several thousand farmer suicides as desperately poor peasants end their lives because they can no longer support their families.

It is clear that the majority of people in this realm depend on agriculture and that governments must target economic policies at improving farm productivity to raise rural standards of living. But they have a long way to go. Demands on governments are voluminous, and they often seem distracted by economic sectors that can make a faster and greater contribution to the tax base—such as manufacturing, financial services, and information technology—economic activities that almost always are centered on the bigger cities far from the impoverished countryside.

Box 8-6 From the Field Notes...

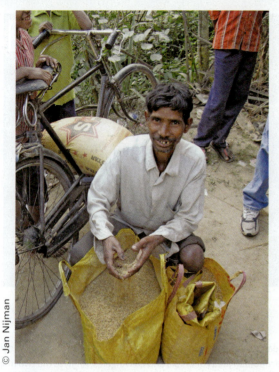

"During my travels in northeastern India in early 2011, I visited some agricultural areas in the Brahmaputra Valley that benefit from plentiful irrigation. Tea and rice are two major crops. Because big companies run the large tea plantations, many of the rice farmers have to make do with small plots of land. Here a farmer in Asom shows off part of his rice harvest. I encountered him and several others at a small, machine-operated mill shop. Note that the rice is brown—milling it removes the bran layer and makes the rice white. 'We have the best rice in all of India,' he proudly said, as he slowly poured it from his hand back into the bag. Who was I to disagree…."

Climate Change and Food Production

It is the direct dependence of well over half the population of South Asia on agriculture that makes it particularly vulnerable to climate change. Exposure to global climate change in this realm is part of its physical-geographic package: the effects of global warming on the Himalaya Mountains, whose glaciers and powerful rivers are the main water source for billions of people; the low-lying terrain of areas such as the Ganges and Brahmaputra deltas, and archipelagos such as the Maldives, that make them so susceptible to inundation from rising sea level; and the high ocean-surface temperatures, moisture conditions, and air circulation patterns that are conducive to the formation of tropical cyclones (hurricanes).

Climate change vulnerability, however, is not simply about atmospheric warming trends. According to the latest meteorological research in India, the South Asian wet monsoon is becoming increasingly unpredictable, with the number of **extreme weather events [10]** having tripled over the past 35 years (see photo pair). These events take the form of:

1. Severe heat waves, such as northern India's record-breaking summer temperatures in 2015 and again in 2016.

2. Excessive torrential downpours outside the summer monsoon season, unleashing major flooding and/or landslides. Examples include the massive flooding of Pakistan's lower Indus Valley in 2010; the deadly floods and landslides in northern India's Uttarakhand State in 2013; and the record-breaking unseasonal rains in the southern Indian State of Tamil Nadu in 2015.

3. Intensified droughts, such as the one that accompanied the heat wave of 2016 across central India, from Maharashtra State in the west to Orissa in the east. Across much of this belt, the monsoons of 2014 and 2015 had been inadequate to fulfill the needs of farmers.

In this densely populated realm, it is the exposure of so many people to the impacts of extreme weather events that necessitates more comprehensive and coordinated climate change mitigation. That involves infrastructural and disaster planning to

Recent extreme weather events in India may be related to global climate change. On the left, people wade along a water-logged roadway in the southern megacity of Chennai during December 2015 (well beyond the monsoon season). Torrential rains caused the city's airport to close and thousands of residents were temporarily stranded in their homes or in traffic. On the right, villagers in the northern State of Uttar Pradesh (India's most populous), near the city of Prayagraj (Allahabad), walk on parched soil in the bed of the shrunken Mansaita River in May 2016. A third of a billion people in northern India were suffering from drought at this time, desperately awaiting the arrival of the tardy monsoon (it finally did about four weeks later, and by early August copious "normal" rains were pelting this crucial part of the Ganges Basin).

adapt to these growing threats (e.g., better water storage, levies, dikes). But myriad institutional and bureaucratic hurdles stand in the way, and financial resources in South Asia are limited.

South Asia's Population Geography

Given its enormous human population, South Asia's areal size is relatively small and totals less than 40 percent of the territory of similarly populous East Asia. Comparing the world's two giants shows that China's territorial extent is almost three times as large as India's. The total population of Subsaharan Africa is less than half of South Asia's, in an area almost five times as large. Adjectives such as "teeming," "overcrowded," and "crammed" are often used to describe the realm's habitable living space, and with good reason. South Asia's intricate cultural mosaic is tightly packed, with only the deserts in the west and the mountain fringe in the north displaying extensive empty spaces. The outlines of the densely populated river basins are clearly visible in the clustering patterns on the population distribution map (**Fig. 8-10**).

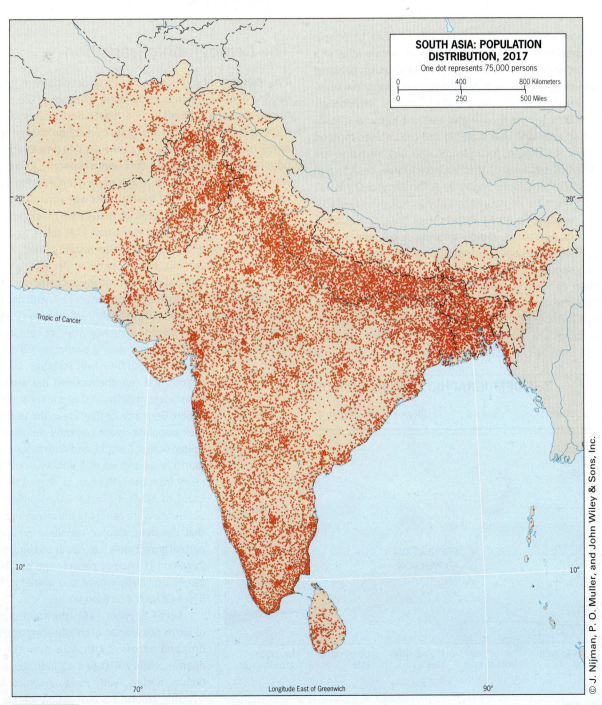

FIGURE 8-10

The field of **population geography [11]** focuses on the characteristics, distribution, growth, and other aspects of spatial demography in a country, region, or realm as they relate to soils, climates, land ownership, social conditions, economic development, and other factors. In the South Asian context, it is useful to concentrate on four demographic dimensions: the role of density; the demographic transition; demographic burdens; and the gender bias in birth rates. As we shall see, population issues are often more complicated than they first appear.

Population Density and the Question of Overpopulation

Population density [12] measures the number of people per unit area (such as a square kilometer or square mile) in a country, province, or an entire realm. We distinguish between two types of measures. *Arithmetic density* is simply the number of people per areal unit, usually a country. **Physiologic density [13]** is a more meaningful measure because it takes into account only land that is arable and can be used for food production. In Pakistan, for example, the two measures are quite different because of that country's large deserts and inhospitable mountainous zones: its arithmetic density is 242 per square kilometer compared to a physiologic density of 834.

Until recently, South Asia's persistent poverty was often related to its enormous, rapidly growing population and its high population densities. The idea was that there were simply "too many mouths to feed"—the realm was "overpopulated." The notion of *overpopulation* can be compelling and seems to make sense at an intuitive level because every country or region can be thought of as having a limited "carrying capacity."

But things are more complex than that. Some countries with high densities, such as the Netherlands or Japan, are highly developed, and not necessarily because they possess an impressive natural resource base (neither does). The point is that high density in itself is not always a problem and that, in certain circumstances, population can be considered as a human resource. If productivity is high, there does not appear to be a problem; but if productivity is low, then large populations can be a drain on the economy. Countries with high education levels, institutional effectiveness, and technological know-how are able to use their resources more efficiently. Thus in South Asia, with its substantial number of undereducated and illiterate inhabitants, population functions as an impediment rather than a resource—the problem being that too many people are not sufficiently productive.

The Demographic Transition

The relevance of population issues to development goes far beyond density, which is really just a snapshot of the population pattern at a moment in time. It gets more interesting—and more complicated—when we relate population change to economic trends. For instance, for a considerable time South Asia's population grew faster than the realm's economy. Clearly, that was a problem because more and more people had to survive on less. Today, fortunately, it is the other way around: in most of the realm, the economy is growing faster than the population.

The term **demographic transition [14]** refers to a structural change in birth and death rates resulting, first, in rapid population increase and, subsequently, in decreasing growth rates and then a stagnant or even a slowly declining population (**Fig. 8-11**). Note that Stage 2 and part of Stage 3, with high birth rates and low death rates (due to healthcare advances), entail a sizeable population expansion. The key issue, of course, is for birth rates to come down so that overall growth rates will diminish and the population will stabilize. Different regions and countries around the world find themselves at different stages in the transition. Japan, Germany, or Italy are in the fifth stage; their populations are currently shrinking. The United States, Canada, and Australia are in the fourth (stagnant) stage. South Asian countries range from the latter part of Stage 2 (Afghanistan) to Stage 4 (Sri Lanka), meaning their populations are generally still increasing. Note that the demographic transition pertains to natural growth only (i.e., death and birth rates); migration is another piece of the puzzle and can be especially critical to countries in Stage 5 (see Chapter 4 on Europe).

Fertility rates [15] (the total number of births per woman of childbearing age) have dropped across South Asia over the past quarter-century. **Figure 8-12** indicates a continuing decline, with most countries today hovering around 2.5 births. Considering that a fertility rate of about 2.3 reflects a stable

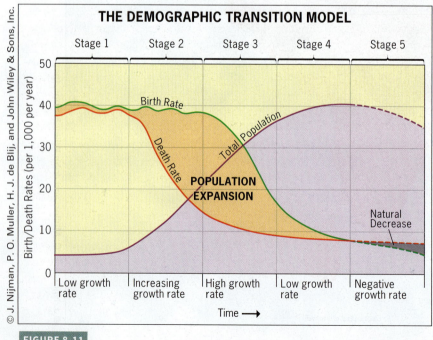

THE DEMOGRAPHIC TRANSITION MODEL

Stage 1   Stage 2   Stage 3   Stage 4   Stage 5

Birth Rate

Death Rate

Total Population

POPULATION EXPANSION

Natural Decrease

Birth/Death Rates (per 1,000 per year)

Low growth rate | Increasing growth rate | High growth rate | Low growth rate | Negative growth rate

Time →

© J. Nijman, P. O. Muller, H. J. de Blij, and John Wiley & Sons, Inc.

FIGURE 8-11

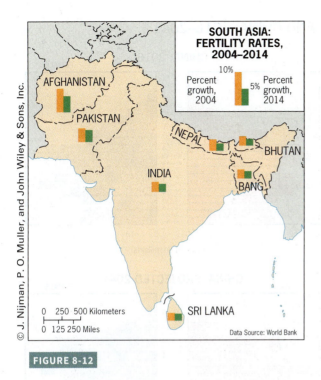

SOUTH ASIA: FERTILITY RATES, 2004–2014

Percent growth, 2004   10%   5%   Percent growth, 2014

0   250   500 Kilometers
0   125 250 Miles

Data Source: World Bank

FIGURE 8-12

population for a realm at this level of development, it is clear that South Asia is still growing. But Afghanistan's extremely high fertility rate is an outlier: recording 4.9 in 2015, double the realm's average.

Demographic Burdens

The immediate significance of demography to economics lies in what is called the **demographic burden [16]**. This term refers to the proportion of the population that is either too old or too young to be productive and that must be cared for by the productive population. Typically, the most productive population in developing countries is represented in the age cohorts between 20 and 50 years. A country with low death rates and high birth rates will have a relatively large share of old and young people, and thus a large demographic burden. Obviously, the way to reduce this burden is to lower birth rates.

Let us now examine **Figure 8-13** and compare today's **population pyramids [17]** (diagrams showing the age–sex structure) for India and China. China has been more successful in curtailing births since 1980, so it now faces a lower demographic burden than India. But, interestingly, what is advantageous today can turn into a disadvantage down the road. Figure 8-13 also displays the population profiles that are projected to exist 25 years from now. Assuming that India will be able to further reduce its birth rate in the coming years, its demographic burden will be less than China's one generation from now. In China, today's productive cohort will have moved on to old age, thereby adding to the national demographic burden. This is yet another reason for India boosters to be optimistic about the future. However, other major challenges remain, and certain forms of birth control are as morally reprehensible as they are economically counterproductive.

Gender Bias: The Missing Girls

Issues of family planning and birth control shed some fascinating light on the "fragmented modernization" of South Asia. As we have seen, the realm finds itself in an advanced stage of the demographic transition wherein birth rates have begun to decline. But take a good look at India's population pyramid for 2016 and note that among young children boys far outnumber girls. In fact, males outnumber females well into middle age.

Traditionally, boys are valued more than girls because they are thought to be more productive income earners, because they are entitled to land and inheritance, and because they do not require a dowry at the time of marriage. When a couple gets married (often arranged and at a young age), the bride comes into the care of the groom's family, where she also contributes her work in and around the house. For this, the bride's family must provide a dowry, which may impose a major expense for her parents. For these reasons, the birth of a boy is a greater cause for celebration than that of a girl. "Raising a daughter," as one saying goes, "is like watering your neighbor's garden."

One reason for the high fertility rates of the past was that families would continue to have children until there were enough sons to take care of the parents in their old age (the girls, after all, would be taking care of their future husband's parents). Hence, this gender tilt is in itself a major factor in South Asia's population growth—but that is not the only problem. When a poor couple repeatedly produces girls and not boys, in some instances the family decides to end the life of the newborn

A Rajasthani infant peering over his mother's shoulder at the market in the northwestern Indian town of Pushkar. Across the subcontinent, women tend to marry and bear children at an early age. Yet fertility rates have been declining, and in India the average number of children per woman has dropped from 3.9 in 1990 to 2.4 today—a substantial decrease but still generating a national population growth rate well above the global average.

POPULATION PYRAMIDS: INDIA AND CHINA, 2016–2041

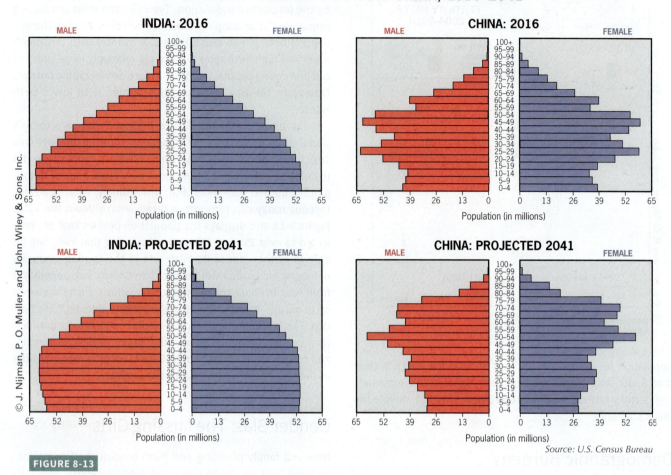

Source: U.S. Census Bureau

FIGURE 8-13

daughter. It is this **female infanticide** or *gendercide* that causes the unnatural gender bias in South Asia's population profiles (the same applies to parts of China as well).

A group of girls in a park in Delhi showing off their henna tattoos. "Mehndi" is a centuries-old festive tradition involving decorative painting on the hands and wrists. These designs usually wear off within a few weeks.

But why—as the economic situation has improved, as birth rates have receded, and as modernization has begun to set in—do we still observe this skewed **sex ratio [18]**? The answer is that with fewer children, the importance of having at least one boy has for many families become even more pressing. And here is where "modernization" throws another curve ball: newly available technologies of ultrasound scanning plus rising incomes (i.e., the growing affordability of a scan) have induced many families to determine the gender of the unborn child and decide on abortion if the child is female. Thus in recent years the sex ratio has become more, not less, skewed, and the most extreme ratios are now found in some of the most developed parts of South Asia, such as the Indian States of Punjab and Haryana (**Box 8-7**).

In the long run, of course, this leads to a shortage of females, which becomes particularly apparent at marriage age. In some areas, families now face a problem in finding brides for their sons, and this "bachelor angst," in turn, is leading to a change in attitudes. Look at India's population pyramid for 2041 and note that over the next quarter-century the sex ratio is projected to become less skewed.

More broadly, discrimination against women and girls in South Asia is expressed in a variety of ways in daily life. One matter that is now receiving more attention is the lack of

Box 8-7 MAP ANALYSIS

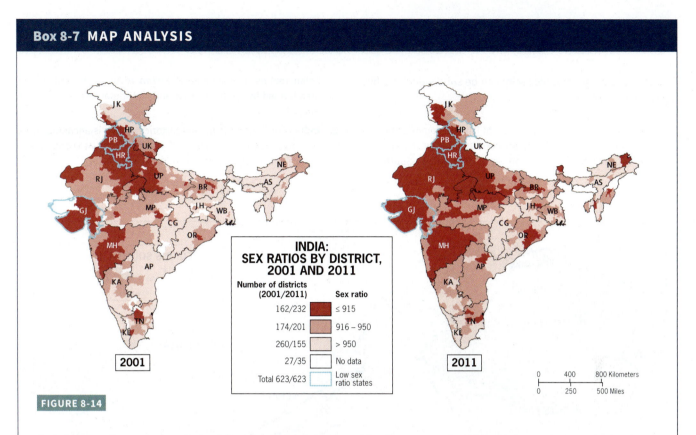

INDIA: SEX RATIOS BY DISTRICT, 2001 AND 2011

Number of districts (2001/2011)	Sex ratio
162/232	≤ 915
174/201	916 – 950
260/155	> 950
27/35	No data
Total 623/623	Low sex ratio states

2001

2011

FIGURE 8-14

The imbalance of sex ratios in India is attributed to female infanticide. Over time, with modernization, improving economic circumstances, and better medical care, ratios are expected to become more balanced. **Figure 8-14** displays child sex ratios (females per 1000 males) across India's States in 2001 and 2011. A lower number indicates a greater imbalance. How have things changed over that decade? Are changes more or less uniform spatially? Can you identify a State with a high imbalance where conditions deteriorated further? Can you identify a State with a low imbalance where the situation became even better? What are some possible explanations for these changes?

Access your WileyPLUS Learning Space course to interact with a dynamic version of this map and to engage with online map exercises and questions.

Adapted from The *Lancet*, Vol. 377, Prabhat Jha, Maya A. Kesler, Rajesh Kumar, Faujdar Ram, Usha Ram, Lukasz Aleksandrowicz, Diego G. Bassani, Shailaja Chandra, Jayant K. Banthia, "Trends in selective abortions of girls in India: Analysis of nationally representative birth histories from 1990 to 2005 and census data from 1991 to 2011," page 8, Copyright (2011), with permission from Elsevier.

sanitation facilities for females. In many workplaces, schools, slum neighborhoods, and public spaces, women still have no (or insufficient) access to toilets. For safety and personal reasons women need more privacy, but in India at least 300 million females lack proper access to toilets, with major implications for health and social functioning. Imagine a small rural school with one toilet (if indeed there is one): it will be used by boys at the exclusion of girls. Girls will have no available access all day. This simple fact leads to girls missing school and even dropping out. Recent efforts by development organizations to install toilets for girls have so far only produced significant reductions in female-student dropout rates.

The recent increase of highly publicized incidents of rape in India has expanded the debate surrounding the country's gender relations. It is an alarming trend, especially where it signals a reaction by men against liberalized behavior among young urban women. But it also reflects the increasingly open disposition of India's mass media as well as a new assertiveness among women (though it is hard to know to what extent this increase is due to a higher rate of reporting and not of rape itself). And, relatively speaking, the incidence of (reported) rape in India is still far below the level of most Western countries, including the United States.

It is not easy to make generalizations about gender relations across this most populous of all geographic realms, in part because of religious and regional diversity as well as rural-urban differences. To be sure, these are in many ways male-dominated societies, especially in the younger age cohorts, and more so in Afghanistan and Pakistan than in India, Sri Lanka, and Bhutan. Nonetheless, keep in mind that Pakistan, India, Sri Lanka, and Bangladesh have all had female prime ministers who held their countries' most powerful political office. It was not until 2016 that such a possibility first arose, with respect to the presidency, in the United States.

Regions of South Asia

The South Asian geographic realm can be subdivided into five regions (Fig. 8-14):

1. **The West** region consists of a pair of Islamic-dominated countries. War-ravaged **Afghanistan** currently forms a transition zone between the South Asian and North African/Southwest Asian realms. Beleaguered **Pakistan** revolves around a core area formed by the Indus River and its tributaries in east-central Punjab.

2. **India** is the realm's giant and historic heart. It is anchored by its core area focused on the broad Ganges Basin. It also contains key subregions based on cultural and other criteria.

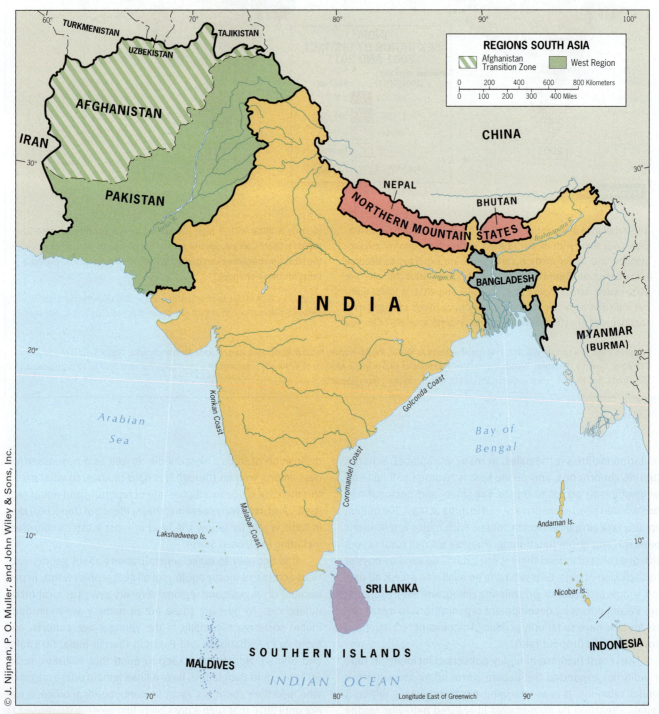

FIGURE 8-15

© J. Nijman, P. O. Muller, and John Wiley & Sons, Inc.

© Jan Nijman

Camel traders gearing up at sunrise in northwestern India's Thar Desert, not far from the border with Pakistan.

3. **Bangladesh** centers the densely-populated, low-lying double delta formed by the Ganges and Brahmaputra rivers, which creates a unique and challenging environment.

4. **The Northern Mountain States** region comprises two noncontiguous Himalayan states: *Nepal* and *Bhutan*. Nepal is plagued by a continuing political struggle following the failure of longtime royal rule; Bhutan remains a peaceful traditional kingdom.

5. **The Southern Islands** region consists of two very different countries. Dominantly Buddhist, ethnically divided Sri Lanka is located a short distance to the southeast of India's southern tip. Farther offshore, on the southwestern side of the subcontinent's tip, lies the island chain known as The Maldives, an Islamic ministate imperiled by rising sea levels.

Region The West: Afghanistan's Unstable Transition Zone

Geography and history appear to have conspired to divide Afghanistan. As Figure 8-1 reveals, the towering Hindu Kush range dominates the center of the country, creating three broad environmental zones: the comparatively well-watered, fertile northern plains and basins; the rugged, earthquake-prone central highlands; and the desert-dominated southern plateaus. Kabol (Kabul), the capital, lies on the southeastern slopes of the Hindu Kush, linked by narrow passes to the northern plains and by the Khyber Pass to Pakistan and the rest of South Asia (**Fig. 8-16**).

Across this rugged, variegated landscape moved countless peoples: Greeks, Turks, Arabs, Mongols, and others. Some settled here, their descendants today speaking Persian, Turkic, and other languages (see Fig. 8-5). Others left archeological remains or no trace at all. The present-day population of Afghanistan (33.4 million) contains no ethnic majority. This is a country of minorities in which the Pushtuns of the east are the most numerous but make up only 42 percent of the total. The second-largest minority are the

Tajiks (27 percent), a world away across the formidable Hindu Kush, concentrated in the far north near Afghanistan's border with Tajikistan. Other groups include Hazaras, Uzbeks, Turkmen, and Balochs (Fig. 8-5).

Afghanistan has a long history of conflict, complicated by the persistent involvement of outside powers vying for influence in this strategically located territory. The protracted nineteenth-century competition between the Russian and British empires over control of Afghanistan—with the wider purpose of controlling adjacent Central Asia—went down in history as "the Great Game." In fact, modern Afghanistan owes its existence to this contest: the Russians and British agreed to tolerate it as a geopolitically neutral cushion—or **buffer state [19]**—between them. During the second half of the twentieth century, Afghanistan was pulled into the Cold War, suffered a Soviet invasion and ten years of occupation, the emergence of violent Islamic militants (particularly al-Qaeda), the presence of U.S. troops since 2002, and most recently a resurgence of the vicious Taliban jihadist forces. And yet, despite this turmoil, Afghanistan remains much as it was before the first inning of "the Great Game"—a feudally balkanized country with a weak and ineffectual government in Kabol. It has also now become a *transition zone*, an arena of spatial change in which the peripheries of adjacent realms overlap.

As Afghanistan has found itself time and again drawn into geopolitical contests between foreign powers and disturbed by conflict and war, its economic development has been compromised throughout. Even by South Asian standards, Afghanistan is desperately poor, its per capita income at U.S. $590 per year barely one-third of India's, while fertility rates are the highest in the realm. Afghanistan has been urbanizing very slowly: just over a quarter of the population lives in cities. It is also marked by **urban primacy [20]**: Kabol is by far the biggest city with 3.7 million people, but the second biggest, Kandahar, contains only about half a million (**Box 8-8**).

The Emergence of the Taliban and al-Qaeda

In 1994, what at first seemed to be just another one of fractious Afghanistan's warring factions appeared on the scene: the **Taliban [21]** (seekers of religion), who originated at religious schools (*madrassas*) in Pakistan. Their avowed aim was to end Afghanistan's chronic factionalism and endemic corruption by imposing *Sharia* (strict Islamic) law. Popular support across the war-weary country, especially among the Pushtuns, led to a series of successes, and by 1996 the Taliban had taken Kabol. The harsh Taliban version of *Sharia*, however, was uncompromising: restrictions on the activities of women ended their professional education, employment, and freedom of movement; the lives of children were even more devastatingly impacted; public stonings and amputations rigidly enforced the Taliban's code.

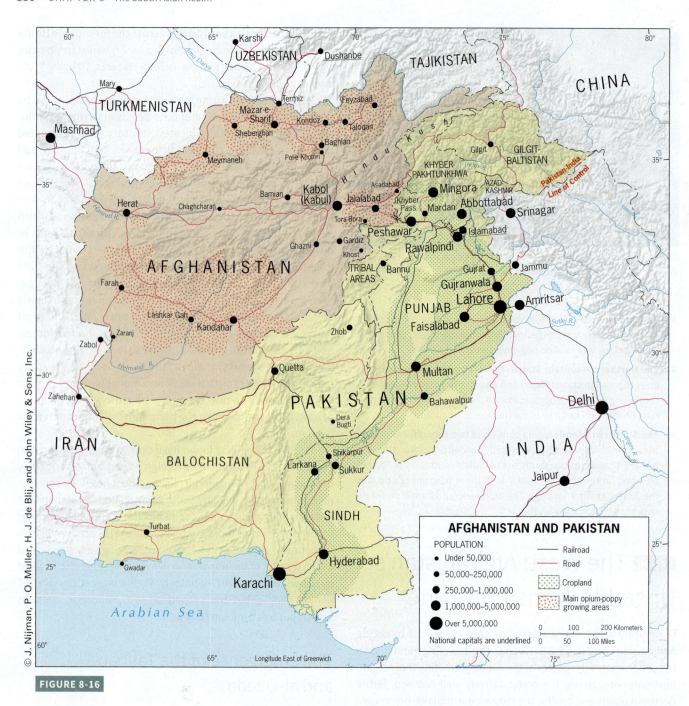

FIGURE 8-16

Afghanistan became a haven for Islamic revolutionary groups whose agendas and goals went far beyond those of the Taliban: they plotted attacks on Western interests throughout southwestern Asia and threatened regimes they deemed to be compliant with Western priorities. Taliban-ruled, cave-riddled, remote and isolated Afghanistan turned out to be an ideal refuge for these outlaws. Already in possession of arms and ammunition (Soviet as well as American) left over from the Cold War, they also took advantage of Afghanistan's huge, thriving, illicit opium trade. Afghanistan has long served as the world's foremost producer of opium, accounting for more than three-quarters of the global heroin supply, and much of this drug revenue was quickly diverted into the coffers of the conspirators.

In 1996, a terrorist organization named **al-Qaeda [22]** took root in the country, an expanding global network that worked to further the aims of the once loosely allied revolutionaries. Afghanistan now became al-Qaeda's headquarters, and Osama bin Laden, a notorious Saudi renegade, its director. Al-Qaeda's most fateful operation came with the suicide attacks of September 11, 2001, on New York and Washington, D.C., which destroyed the World Trade Center's twin towers, damaged the Pentagon, killed over three thousand people, and inflicted massive psychological damage. Several weeks later, U.S. and British forces, with the acquiescence of Pakistan, attacked both the Taliban regime and al-Qaeda in Afghanistan.

Box 8-8 Major Cities of the Realm, 2016	
Metropolitan Area	**Population 2016 (in millions)**
Delhi–New Delhi, India	25.7
Mumbai (Bombay), India	22.9
Karachi, Pakistan	22.8
Dhaka, Bangladesh	16.2
Kolkata (Calcutta), India	14.8
Lahore, Pakistan	10.4
Bengaluru (Bangalore), India	10.2
Chennai (Madras), India	10.0
Hyderabad, India	7.8
Ahmadabad, India	7.4
Kabol, Afghanistan	3.7
Kathmandu, Nepal	2.8
Colombo, Sri Lanka	2.2
Varanasi, India	1.6

The U.S.-Led Invasion of Afghanistan and Its Aftermath

The American-led invasion following 9/11 initially appeared to be successful. The Taliban were swiftly driven off, a more representative government was taking shape, local warlords were co-opted or sidelined, Pushtun and other refugees were streaming back into Afghanistan from Pakistan and elsewhere, and life in Kabol and other urban centers returned to a semblance of normal. By 2004, Afghanistan had even held an election that produced

Local farm workers harvesting opium sap from a poppy field in southern Afghanistan's Kandahar Province in the spring of 2015. This country is by far the world's largest producer of opium (the raw material of heroin). Cultivating these poppies is considerably more profitable than raising legal crops and is believed to involve various networks of organized crime, including al-Qaeda.

a representative government. It took nearly a decade of U.S. intelligence work, but in 2011 Bin Laden was finally discovered and killed in an American military raid—not in Afghanistan but in the city of Abbottabad in neighboring northern Pakistan. With al-Qaeda markedly weakened and having allowed for a transitional period in which to fully train Afghan security forces, the United States at the end of 2014 withdrew the last of its combat troops from Afghanistan (a residual U.S. advisory force of 8400 was planned to remain into 2017).

During that multiyear transition, American and NATO advisors found it extremely challenging to foster a workable consensus among various Afghan factions in order to build a more democratic state. One serious problem was the country's economy, which had been all but shattered by decades of unremitting conflict and war, forcing the government to depend heavily on foreign aid. Gauging the new status quo, by mid-2015 Taliban fighting units had regrouped into smaller militias and were again attacking security forces, government facilities, and Western compounds. Much of this activity first took place in the far north adjacent to Tajikistan (Fig. 8-16), but by mid-2016 was spreading across the Hindu Kush to resume hostilities in and around Kabol. Although al-Qaeda in Afghanistan had been dealt a stunning blow at the core, the organization has since proven resilient. As discussed in Chapter 6, by 2013 it had branched out beyond Afghanistan to become a major player in unstable countries, such as Pakistan, as well as in the now-failed states of Somalia, Yemen, and Libya.

Region The West: Pakistan

In the Indus Valley of Pakistan lay South Asia's earliest urban civilizations, whose innovations diffused southeastward into the massive triangular peninsula. Today, this elongated river basin contains South Asia's Muslim frontier, contiguous to the great Islamic realm to the west and irrevocably linked to India's enormous Muslim minority to its east. Pakistan's cultural landscapes bear witness to its transitional location. Teeming, disorderly Karachi is the typical South Asian megacity. Historic, architecturally Islamic Lahore is reminiscent of the scholarly centers of Muslim Southwest Asia. In Pakistan's east, the boundary of the 1947 partition divides a Punjab subregion that continues well into India—a land of villages, wheatfields, and irrigation canals. In the northwest, Pakistan closely resembles the Afghanistan Transition Zone in its huge migrant populations and mountainous borderland.

The Indus River and its principal tributary, the Sutlej, nourish the ribbons of life that form the heart of this populous country (Fig. 8-16). Territorially, Pakistan is not large by Asian standards; its area is about the same as that of Texas plus Louisiana. But Pakistan's population of 193 million makes it one of the world's ten most populous states. Among Muslim countries (its official name is the Islamic Republic of Pakistan), only Southeast Asia's Indonesia is larger.

A Hard Place to Govern

At independence (West) Pakistan had a bounded national territory, a capital, a cultural core, and a population—but few centripetal forces to bind state and nation. The disparate subregions of Pakistan shared the Islamic faith but little else. Karachi and the coastal south, the southwestern desert of Balochistan, the city of Lahore and surrounding Punjab, the rugged northwest along Afghanistan's border, and the mountainous far north remain worlds apart. Urdu is the official language, yet English remains the *lingua franca* of the elite. Several other major languages, however, prevail in different areas (see Fig. 8-5), and ways of life vary enormously as well.

Successive Pakistani governments, civilian as well as military, turned to Islam to provide the common bond that history and geography had denied this state. In the process, Pakistan became one of the world's most theocratic countries. But even Islam itself is not unified in restive Pakistan. Roughly 80 percent of the people are Sunni Muslims, and the Shia minority accounts for most of the remaining 20 percent. Sunni fanatics intermittently attack Shi'ites, leading to retaliation and establishing grounds for subsequent counterattacks.

To govern so diverse and fractious a country would challenge any political system, and so far Pakistan has failed the test. Democratically elected governments have time and again squandered their opportunities, only to be overthrown by military coups. Pakistan's recent economic boom has not filtered down to the poor; literacy rates are not rising; health conditions are not improving significantly; national institutions are weak (for instance, there are only about 2.5 million registered taxpayers); and one consequence of the global antiterrorism campaign is that Pakistanis, who used to be overwhelmingly secular in their political choices, are now increasingly joining Islamic parties.

Meanwhile, too little is being done to confront a growing water-supply crisis, an insurgency festers in Balochistan, the army is incapable of establishing control over mountainous Waziristan (where al-Qaeda and Taliban militants reign), the issue of Kashmir costs Pakistan dearly, and relations with neighboring India (which if satisfactory would bring enormous benefits) remain conflicted.

The Four Subregions of Pakistan

Punjab constitutes Pakistan's core area, the Muslim heartland across which the post-independence boundary between Pakistan and India was superimposed (**Fig. 8-17**). (As a result, India also has a State named Punjab on the other side of the border.) Pakistan's Punjab is home to just over half of the country's population; in the triangle formed by the Indus and Sutlej rivers live more than 100 million people. Punjabi is the language here, and wheat farming is the mainstay.

Three cities anchor this core-area subregion: Lahore, the foremost center of Islamic culture in the realm; Faisalabad; and Multan. Lahore, now home to more than 10 million

people, lies close to the border with India. Founded about 2000 years ago, Lahore was situated favorably to become a great Muslim center during the Mughal period when Punjab was the main corridor into India. Punjab's relationship with Pakistan's other three provinces is one of the country's weaknesses. Both the governments and residents of those provinces feel uneasy about the dominance of Punjab, the populous, powerful core of the country from which most of the military is drawn.

The lower Indus River is the key to life in **Sindh**, but Punjab controls the waters upstream, which is one of the issues that divide Pakistan (Fig. 8-17). When the Punjab-dominated regime proposes to build dams across the Indus and its major tributaries, Sindhis (who make up almost one-fifth of the national population) are reminded of their underrepresentation in government and talk of greater autonomy. The ribbon of fertile, irrigated, alluvial land along the lower Indus, where the British laid out irrigation systems, makes Sindh a Pakistani breadbasket for wheat and rice. Commercially, cotton is king here, supplying textile factories in the cities and towns (textiles account for more than half of Pakistan's exports by value).

But the dominant presence in southern Sindh is the chaotic megacity of Karachi, a cauldron of searing contrasts under a broiling sun (**Box 8-9**). With little effective law enforcement, Karachi has become a hotbed of terrorist activity, but somehow the city still functions as Pakistan's (and Afghanistan's) major maritime outlet and the seat of Sindh's provincial capital.

As Figure 8-17 indicates, the province of **Khyber Pakhtunkhwa** lies wedged between powerful Punjab to the east and troubled Afghanistan to the west, with the territory long known as the *Tribal Areas* intervening in the south. The name "Pakhtunkhwa" translates as "belonging to the Pushtuns," the Afghan-associated tribes that inhabit this subregion.

The Pakistani government's reach into these parts has always been quite limited. The province's mountainous physical geography reflects its remoteness and the isolation of many of its people. Mountain passes lead to Afghanistan; the Khyber Pass, already noted as the historic route of invaders, is legendary (see photo). Coming from Afghanistan, the Khyber's road leads directly to the provincial capital, Peshawar, which lies in a broad fertile valley where wheat and corn drape the countryside. Khyber Pakhtunkhwa is a conservative, deeply religious, militant province, where Islamic political parties and movements are stronger than in any other part of the country.

The fourth subregion mapped in Figure 8-17 is **Balochistan**. By far the biggest of Pakistan's four provinces, it accounts for nearly half of the national territory (not including Kashmir) but is inhabited by only about 14 million people, barely 5 percent of the country's population. For a sense of its terrain, take a look at Figure 8-1: much of this vast territory is desert, with mostly barren mountains that wrest some moisture from the air only in the northeast. Sheep-raising is the leading livelihood here, and wool is the primary export. In its northern extremity, Balochistan abuts the Tribal Areas and Afghanistan. This

Box 8-9 Among the Realm's Great Cities: Karachi

Located next to the Indus Delta and on the Arabian Sea, Karachi is Pakistan's biggest city as well as its economic and financial center. In former days, it was the capital too, but that function was shifted far inland to Islamabad in the 1960s. Until independence in 1947, Karachi was a typical colonial port city, designed to serve British interests. That function can still be seen in the urban landscape: the city was organized around the port, and this is where you can still find the main railroad terminals. Raw materials and agricultural goods were brought in from the hinterland to be loaded onto ships headed for England's industrial complexes. The old city center lies just north of the port, and today's CBD is located just to the east of it.

Karachi has grown enormously over the years, from around 100,000 at the beginning of the twentieth century to about half a million in 1950 to nearly 23 million today. Most recently, the city has expanded rapidly to the west and to the north (note "New Karachi" on the map) to form a sprawling metropolis. This is Pakistan's "world-city" because it dominates the country's linkages to the global economy (even more so than Mumbai does for India). Karachi is home to most of the foreign multinational firms, accounts for about three-quarters of international trade, and produces about 20 percent of Pakistan's GDP. It has the busiest airport and two seaports that together handle more than 90 percent of all transoceanic trade (the newer port of Qasim was built east of the metropolis in the 1970s). The city's economic importance is manifest in the imposing skyline of its central business district.

But this is a city besieged with serious problems. Rampant and unplanned growth has produced congestion and pollution. Millions live in grinding poverty, and the contrast with the small but very wealthy urban elite is stark. Compared to most other cities in this realm, street crime is rampant and Karachi ranks near the bottom of the list of the world's "most liveable" cities. The great majority of its residents are first- or second-generation immigrants from a wide range of ethnic backgrounds (many originally hailing from northern India prior to the 1947 partition), and recent years have witnessed both growing communal tension and violent conflict. In addition, the city is widely described as a nurturing ground for Islamic fundamentalists and increasingly serves as a base for Taliban leaders and followers. Karachi's fraying image as Pakistan's most cosmopolitan center is under enormous strain.

© J. Nijman, P. O. Muller, and John Wiley & Sons, Inc.

Courtesy Barbara A. Weightman

is where Quetta lies, capital of a province that could easily be called the "South West Frontier Province."

Yet Balochistan is of considerable economic-geographic importance. Beneath its parched surface lie possibly substantial reserves of oil as well as coal, and already this province produces most of Pakistan's natural gas. The seaport of Gwadar on the southwestern coast not only handles raw materials from Balochistan but is also a transshipment point for oil and gas from Iran and the Caspian Basin, destined for markets in East Asia. China has a key interest in Gwadar and in a much larger infrastructural project to connect it all way to the China-Pakistan border in the north (**Box 8-10** and Fig. 8-17). But the Baloch people are treated poorly by Pakistan's central government—fully 90 percent of them have no energy supply, and eight out of ten do not even have access to clean water. Short-lived local rebellions have flared up here since Pakistan became independent, and during the 2000s they escalated into

The pivotal Khyber Pass that links Afghanistan and Pakistan across the Hindu Kush (see Fig. 8-16). One of the most strategic passes in the world, it was no easy passage for invading armies in times past. Nowadays, the road and its tunnels facilitate the movement of refugees, drugs, and weapons, which are used by militant separatists on both sides of the border. Pakistan's Khyber Pakhtunkhwa, especially its provincial capital of Peshawar, is a hotbed of such activities.

Box 8-10 Regional Planning Case

The China-Pakistan Economic Corridor

As Figure 8-17 shows, Gwadar is a remote fishing town in Pakistan's Balochistan Province that is located west of Karachi on the shore of the Arabian Sea—and even closer to the outlet of the Persian Gulf that begins at the nearby Iranian border. Gwadar is also in the process of becoming the coastal terminus of the so-called **China-Pakistan Economic Corridor [23]**, yet another international mega-project planned, financed, and executed mostly by China. The Pakistani government had constructed a deepwater port here in 2006, but it was never completed. In 2013, China took over the port's operation and planned expansion, setting into motion a much bigger (U.S. $46 billion) scheme that will connect Gwadar, via a 2000-kilometer (1250-mi)-long corridor of roads and railways, to western China'a booming Xinjiang Uyghur Autonomous Region. In Gwadar itself, as part of a 43-year lease, the Chinese have acquired the rights to more than 800 hectares (2000 acres) of land to house the China Overseas Port Holding Company.

The new Economic Corridor will provide China with greater access to oil-rich Southwest Asia, Africa, and Europe through Pakistan—a much shorter and faster connection than the main sea route through the South China Sea, the Strait of Malacca, and the Indian Occan. In addition to the new rail and road infrastructure, a network of pipelines is being built to transport oil and gas from the Persian Gulf across the length of western Pakistan, and then over the towering Hindu Kush and Karakoram ranges into Xinjiang. Most of this mega-project's funding is being used to construct a network of power stations, both to run the new infrastructure and to alleviate Pakistan's chronic energy shortages.

Pakistani officials portray the project as a huge economic opportunity, with Gwadar positioned to be developed into a full-fledged regional hub and transshipment *entrepôt*. They anticipate substantial revenue from port cargoes and freight handling fees that will boost the country's sagging economy. And there is another potential benefit for Pakistan, one not openly discussed: economic progress in surrounding, restless Balochistan will help to counter the province's separatist and insurgent tendencies.

As for China, this project has great economic and political value. Closer collaboration with Pakistan in China's southwest may help in the controlling of dissident Islamic groups in that region, and the new Corridor also aligns nicely with China's highly-touted ***New Silk Road*** (covered in Chapters 5 and 9). And geostrategically, Gwadar provides a key window on the Indian Ocean from which China can maintain a low-profile military presence. Finally, none of this distracts from Pakistan's own perceived benefits, which might also include some new leverage in its relations with India.

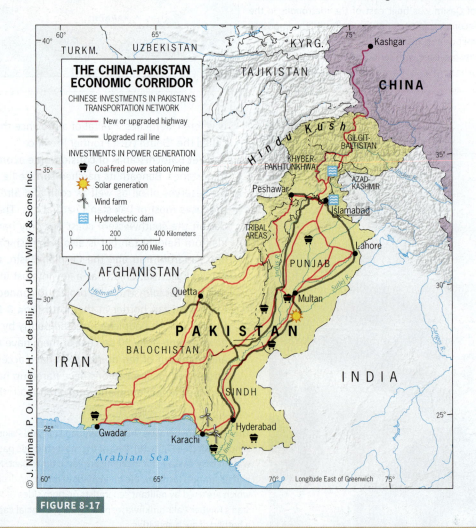

FIGURE 8-17

a serious insurgency led by the Balochistan Liberation Army (BLA), whose attacks on security forces continue.

Prospects for Western South Asia

This western flank of South Asia is the realm's most critical region, today more so than ever. Afghanistan is a fractured, partially functional state that has proven impossible to control by outside powers for any prolonged period of time. Pakistan's coherence and stability have now become crucial at a time when the global struggle against terrorism continues to entangle its leaders in Western priorities and foreign policies. Thus the United States' fight against the Taliban is placed mainly in Afghanistan, but at times it uncomfortably spills over into Pakistan, often aided by the tactics of drone warfare (**Box 8-11**). The role of Pakistan's government in this struggle is disputed and resented by many of its own citizens, who express their distaste by voting for activist Islamic parties and even voicing support for the Taliban resurgence in neighboring Afghanistan. Moreover, militancy and instability are no longer confined to Pakistan's northwest and far north: key cities such as Lahore and Karachi are now increasingly targeted by terrorists (the government blames the Taliban). Meanwhile, relations between Pakistan and India remain tense and at times perilously close to open conflict, with India's regime accusing the Pakistani government of failing to crack down on the militants in Kashmir and other northern flashpoints. The geopolitical situation involving these three states today has been characterized by some as a potentially **deadly triangle** in which Afghanistan could become the main source of destabilization.

Region India: Giant of the Realm

A Federation of States and Peoples

That India has endured as a unified state is a politico-geographical miracle. The country contains a cultural mosaic of immense ethnic, religious, linguistic, and economic diversity and contrast; it is truly a state of many nations. Upon independence in 1947, India adopted a democratic, secular, federal system of government, giving regions and peoples some autonomy and identity, and allowing others to aspire to such status.

The map of India's political geography shows a federation of 29 States, 6 Union Territories (UTs), and 1 National Capital Territory (NCT) (**Fig. 8-18**). The federal government maintains direct authority over the UTs, all of which are small in both territory and population. The NCT, however, includes most of the Delhi/New Delhi urban region and now contains almost 26 million inhabitants (**Box 8-12**). The country's State boundaries reflect the broad outlines of the country's cultural mosaic: insofar as possible, the system recognizes languages, religions, and cultural traditions. Indians speak 14 major and numerous

minor languages, and although Hindi is the official language it is by no means universally used (see Fig. 8-5).

As Figure 8-18 shows, the territorially largest States lie in the heart of the country as well as on the massive southward-pointing peninsula. Uttar Pradesh (200 million) and Bihar (104 million) together cover much of the Ganges Basin and also form the core area of modern India. Maharashtra (112 million), anchored by the great coastal city of Mumbai (called Bombay before 1996), also has a population larger than that of most countries. West Bengal, the State that adjoins Bangladesh, contains more than 91 million residents, 15 million of whom live in its urban focus, Kolkata (known as Calcutta before 2000).

These are staggering numbers, and they do not decline much toward the south. Southern India consists of four States linked by a discrete history and by their distinct Dravidian languages. Facing the Bay of Bengal are Tamil Nadu (72 million) and Andhra Pradesh (50 million), both part of the hinterland of the city of Chennai (formerly Madras) that is located on the coast near their joint border. Facing the Arabian Sea are Karnataka (61 million) and Kerala (33 million).

India's smaller States lie mainly in the northeast, on the far side of Bangladesh, and in the northwest, toward Jammu and Kashmir (Fig. 8-18). The map becomes even more complicated in the remote northeast, beyond the narrow corridor between Bhutan and Bangladesh. The dominant State here is Asom (the old spelling is Assam), home to just over 31 million, famed for its tea plantations, and important because its petroleum and gas production amounts to more than 40 percent of India's total. In the Brahmaputra Valley, the axis of Asom, this State resembles the India of the Ganges. But in almost every direction from Asom, things change.

North of Asom, in sparsely populated Arunachal Pradesh (1.4 million), we are in the Himalayan offshoots once again. To the east, in Nagaland (2.0 million), Manipur (2.7 million), and Mizoram (1.1 million), lie the forested and terraced hills that separate India from Myanmar (Burma). This is an area of myriad ethnic groups (more than a dozen in Nagaland alone) and of frequent rebellion against the central government. And to the south, the States of Meghalaya (3.0 million) and Tripura (3.7 million), hilly and still wooded, border the teeming floodplains of Bangladesh. Here in the country's northeast, India confronts one of its strongest regional challenges.

India's Changing Map

Devolutionary pressures have continued throughout India's existence as an independent state. In some of the cases, the federal government and the military have come down hard on the insurgents, but in others the government has been more inclined to negotiate. In 2000, for instance, three more new States were recognized: Jharkhand, carved out of southern Bihar State on behalf of 18 poverty-stricken districts there; Chhattisgarh, where tribal peoples had been agitating since the 1930s for separation from the State of Madhya Pradesh; and

Box 8-11 Technology & Geography

Drones of War

On May 23, 2016, the Afghani Taliban leader Mullah Mansoor was targeted and killed by a missile fired from a U.S. military drone. **Drone warfare [24]** today has become a standard element of high-technology military strategy in the arsenal of the United States and other powers.

Unmanned aerial vehicles (UAVs), or drones, are aircraft that fly without an on-board pilot. Advances in aircraft design as well as imaging, navigational, and communications technology now permit those who control drones to be located halfway around the world from the field of operations. For instance, American military drones in embattled Pakistan, Syria, Yemen, and Iraq are guided by pilots based in the United States who watch live-streaming video feeds via satellite uplinks from cameras mounted on each active drone. Sensors, computers, and on-board instruments enable drones to be flown remotely, similar to a flight simulation game on a computer. But make no mistake: when armed with weaponry such as air-to-ground missiles, drones become deadly instruments of war.

Figure 8-19 maps communications relays and satellite links that enable the remote piloting of drones by military personnel stationed on their home territory. Creech Air Force Base near Las Vegas, Nevada is a major U.S. intelligence and coordination center. From Creech, commands are sent via high-speed transatlantic fiber-optic cable to the U.S. Air Base in Ramstein, Germany, and from there the command is directly relayed via satellite to drones deployed in Afghanistan and elsewhere. Drone take-offs and landings are controlled locally (see photo).

There are several tactical advantages to using drones in war: they are difficult to detect from the ground, flight times

STAFF/REUTERS/Newscom

An airman guides a U.S. Air Force MQ-9 Reaper drone as it taxis toward the runway at Kandahar Airfield, Afghanistan in March 2016.

in excess of twelve hours are possible without refueling, and advanced imaging, thermal, and night-time-vision capabilities enable drones to 'see' what is otherwise invisible to on-board pilots.

As such, drone technology influences where and how wars are fought, but it also raises issues of international law. That Taliban leader was eliminated while driving in a convoy on a road in Balochistan, the southwestern province of Pakistan. As noted earlier, Afghanistan is a transition zone (Fig. 8-15), and its conflicts often spill across the border with Pakistan. The United States deployed drone technology to reach into Pakistani territory—not surprisingly, Pakistan's government strongly condemned the American incursion.

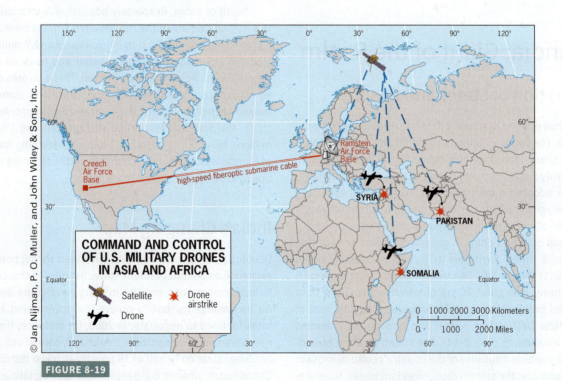

© Jan Nijman, P. O. Muller, and John Wiley & Sons, Inc.

COMMAND AND CONTROL OF U.S. MILITARY DRONES IN ASIA AND AFRICA

- Satellite
- Drone
- Drone airstrike

FIGURE 8-19

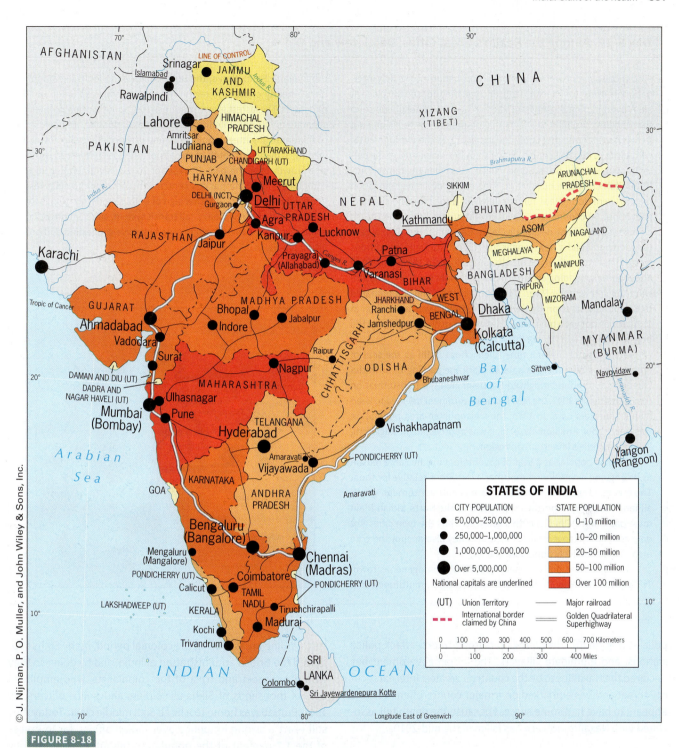

FIGURE 8-18

© J. Nijman, P. O. Muller, and John Wiley & Sons, Inc.

Uttarakhand (originally named Uttaranchal), which split from Uttar Pradesh. Most recently, in 2014, the new State of Telangana (India's 29th) was created out of the northwestern sector of Andhra Pradesh. Telangana is an inland area that long saw itself marginalized by coastal elites. The new State contains more than 35 million people, covers roughly one-third of pre-2014 Andhra Pradesh, and its capital is Hyderabad—one of the country's leading technopoles and its sixth-largest city—which will serve jointly as the capital of the new State of Andhra Pradesh until 2024.

During the past several years, there have been troubling signs of yet another challenge to India's federal government: communist-inspired rebellions that seem to be transforming into a coordinated revolutionary campaign. It is known in India as the ***Naxalite*** movement, named after a village in the State of West Bengal where it was founded in the 1960s. Mainly active in India's poorest and most disaffected States—such as Bihar, Jharkhand, and what is now Telangana—the Naxalites appeal to the poor and other minorities (especially tribal people), whose plight is blamed on India's elites and their neoliberal economic

---

Box 8-12   Among the Realm's Great Cities: Delhi New and Old

Fly directly over the Delhi–New Delhi conurbation into its new international airport (opened in 2010), and you may not see the place at all. A combination of smog and dust creates an atmospheric soup that can limit visibility to a few hundred meters for weeks on end. Delhi was rated the most polluted major city in the world in 2015, with air toxicity levels much higher than in Beijing and other better known polluted cities. Some relief comes when the rains arrive, but Delhi's climate is mostly dry. The tail-end of the wet monsoon reaches here during late June or July, but altogether the city only gets about 60 centimeters (25 in) of rain a year. When the British colonial government decided to leave Calcutta more than a century ago and build a new capital city adjacent to Delhi, conditions were different. South of the old city lay a hill about 15 meters (50 ft) above the surrounding countryside, on the right bank of the southward-flowing Yamuna River, a tributary of the Ganges. Compared to Calcutta's hot, swampy environment, Delhi's was agreeable. In 1912, it was not yet a megacity. Skies were mostly clear. Raisina Hill became the site of a New Delhi.

This was not the first time rulers chose Delhi as the seat of empire. Ruins of numerous palaces mark the passing of powerful kingdoms. But none brought to the Delhi area the transformation the British did. In 1947, the new Indian government decided to keep its headquarters here. In 1970, the metropolitan-area population was only about 4 million. By 2016, it was approaching a staggering 26 million, India's largest urban region.

Delhi is popular as a seat of government for the same reason as its ongoing expansion: the city has a fortuitous relative location. The regional topography creates a narrow corridor through which all land routes from northwestern India to the North Indian Plain must pass, and Delhi lies in this gateway. Thus the twin cities not only contain the government functions; they also anchor the core area of this massively populated country.

Old Delhi was once a small, traditional, homogeneous town. Today, Old and New Delhi form a multicultural, multifunctional urban giant. From above, the Delhi conurbation looks like an ink-blot, a nearly concentric region that has steadily expanded in all directions. The fastest growth has been to the south, where formerly separate towns and satellite cities such as Faridabad are now part of this sprawling megalopolis. Gurgaon is especially well known, a leading activity hub south of the airport that has witnessed explosive growth over the past 15 years or so, an agglomeration of IT companies, international call centers, and new middle-class residential developments. Another sign of Delhi's modernization is the construction of a new, heavy-rail transit system, aimed at providing some relief to the metropolitan area's massive congestion and pollution problems.

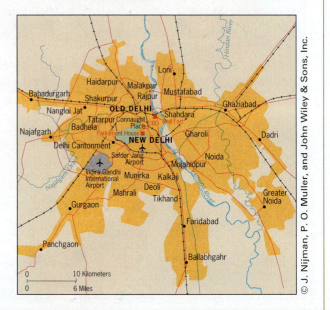

© J. Nijman, P. O. Muller, and John Wiley & Sons, Inc.

---

policies (**Fig. 8-20**). They blow up railroad tracks, attack police stations, and intimidate villagers. The Indian government has declared the Naxalites to be the country's greatest internal security threat and has embarked on a major counteroffensive. This appears to have had some success because the death toll associated with Naxalite violence has fallen in the mid-2010s.

Communal Tensions

The term **communal tension [25]** (communal disharmony) refers to the several different categories of conflicts that recur among India's highly diverse sociocultural groups. Most commonly, these conflicts have a base in (politicized) religion, but they can also be caste-based as we shall see.

The **Sikhs** (the word means "disciples") adhere to a religion that was created about five centuries ago to unite warring Hindus and Muslims into a single faith. The Sikhs rejected what they considered to be the negative aspects of Hinduism and Islam, and Sikhism gained millions of followers in Punjab and surrounding areas. During the colonial period, many Sikhs supported the British administration of India, and by doing so they won the respect and trust of the colonialists, who employed tens of thousands of Sikhs as soldiers and police officers. By 1947, Punjab was home to a large Sikh middle class. Today they still exert a strong influence over Indian affairs, far in excess of the 1.7 percent of the population (about 23 million) they constitute.

For a period after India's independence, Sikh extremists created India's most serious separatist problem, demanding the formation of an independent state they wanted to call *Khalistan*. The conflict intensified in the 1970s and culminated in an attack by federal military forces on Sikh rebels barricaded in the Golden Temple in the city of Amritsar, Sikhdom's holiest site (see photo). Shortly after this 1984 attack, then Prime Minister Indira Gandhi was assassinated by her own Sikh bodyguards, unleashing a lengthy period of anti-Sikh violence across much of India. Since these dramatic events of the 1980s, both sides have worked to gain better understanding.

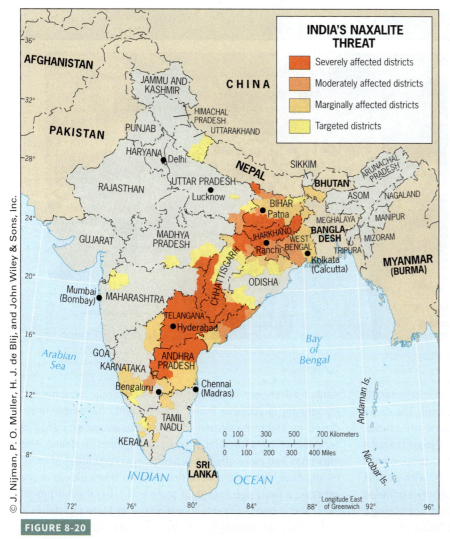

INDIA'S NAXALITE THREAT
- Severely affected districts
- Moderately affected districts
- Marginally affected districts
- Targeted districts

FIGURE 8-20

© J. Nijman, P. O. Muller, H. J. de Blij, and John Wiley & Sons, Inc.

confrontation in Kashmir, and from acts of Islamic terrorism that have bedeviled India in recent years. As a result, tensions between Muslims and Hindus have been rising. The most serious threat to the better integration of Muslims into Indian society arises from their comparatively low level of education and inferior economic standing. Official statistics show that less than 4 percent of the country's Muslims have completed secondary school. And according to one recent government report, Muslims overall have now become as poor as the members of India's lowest-ranked Hindu caste (**Box 8-13**).

Hindu extremism or fundamentalism may seem to be a contradiction in terms, but a movement has come to the fore in this century that seeks to remake India as a society in which Hindu principles prevail. *Hindutva* **[26]**, or Hinduness, is variously expressed as Hindu nationalism, Hindu heritage, or Hindu patriotism. It has been the guiding agenda for the Bharatiya Janata Party (BJP), a powerful force in national politics as well as in large States like Maharashtra, Gujarat, and Madhya Pradesh. *Hindutva* fanatics want to impose a Hindu curriculum on schools, change the flexible family law in ways that would make it unacceptable to Muslims, constrain the activities of non-Hindu religious proselytizers, and forge a new India in which non-Hindus are essentially outsiders.

It is easy to see how Hindu hardliners and Islamic revivalists can fuel each other's agendas and maintain a vicious cycle of conflict. This naturally worries Muslims as well as other minorities, but it also concerns those who understand that India's secularism—its separation of religion and state—is indispensable to the survival of democracy. Moderate Hindus and non-Hindus in India oppose this strengthening of *Hindutva*, whose notions are as divisive as any India has faced.

The Caste System

Hinduism's sprawling and accommodating properties (there are few orthodoxies in Hinduism, and there is no central authority)

Its **Muslim** minority may constitute less than 15 percent of India's population, but they number just under 190 million, nearly the total population of Pakistan. Muslims are in the majority in the States of Jammu and Kashmir, Asom, and West Bengal, and their absolute numbers are substantial in the big States of Uttar Pradesh, Bihar, and Maharashtra. To India's great advantage, the Muslim minority is not geographically concentrated, avoiding what could otherwise lead to a dangerous regional secessionist movement.

The status of Muslims within Indian society is not that easy to separate from India's relations with Pakistan, from the

© Jan Nijman

One of India's most famous landmarks, the Golden Temple in the city of Amritsar in Punjab State, which dates back to the late sixteenth century. Sikh places of worship are called *gurdwaras*, and this is the holiest one of all, known to the faithful as *Harmandir Sahib* ("The Temple of God"). More than 100,000 Sikhs pray here every day of the year. Adherents of all religions are welcome, but visitors must dress appropriately, remove their footwear, and cover their heads.

Box 8-13 From the field Notes...

"It is a Friday afternoon as we make our way through a main thoroughfare in one of Mumbai's biggest slums. Dharavi 'houses' approximately half a million people on less than 2 square kilometers! About 20 percent are Muslims, and the overwhelming majority of all others are dalits. Increasingly, the Muslims live spatially segregated in their own tightly knit neighborhoods. There are several mosques but little space inside them. So the men, taking a break from work nearby, place their mats alongside the road and prepare for prayer outdoors. Having a faith is important to people living and working in Dharavi's filthy and impoverished environment, especially for the Muslim minority."

© Jan Nijman, Photo by Zach Woodward

combine with a system of social stratification that is generally derided in the West and by some Indians as well. The **caste system** has its origins in the early social divisions into priests, warriors, merchants, and farmers, and it is also thought to have had a racial basis (the Sanskrit term for caste, *varna*, means color). More specifically, caste became associated with specific occupations, and over the centuries its complexity expanded until India had thousands of castes, some of them containing a few hundred members and others consisting of millions.

Hindus believe in reincarnation, and a person is believed to be born into a particular caste based on his or her actions in a previous life. Hence, it would not be appropriate to counter such ordained caste assignments by permitting mobility (or even contact) from a lower caste to a higher one. Persons of a particular caste could perform only certain jobs and worship only in prescribed ways at particular places. They and their children could not eat, play, or even walk with people of a higher social status.

The *untouchables* occupying the lowest tier of all were the most debased, wretched members of this rigidly structured social system. Indeed, the term "untouchable" acquired such negative connotations that it was replaced on several occasions. Mahatma Gandhi, a powerful critic of the system, introduced the term *harijans*, meaning children of God. But more recently this label was perceived to have a condescending connotation, and it gave way to the term **dalits** (the oppressed), indicating a greater sense of awareness and assertiveness among them. Today, dalits are estimated to comprise about one-sixth of all Hindus; Brahmins, the highest-ranking caste, account for a similar proportion. The rest of the population are members of a wide range of in-between castes.

Untouchability was officially abolished at independence, but the caste system has proven difficult to dismantle. Many dalits have chosen to convert to other religions, most notably Buddhism. Successive Indian governments have now introduced an elaborate system of affirmative action on behalf of *Scheduled Castes* (the official government label for dalits). This effort has had more effect in urban than in rural areas of India. In any case, dalits now have reserved for them places in the schools, a fixed percentage of State and federal government jobs, and a quota of seats in national and State legislatures. These jobs are often highly desirable because they tend to be white-collar and much better paid than those generally available to dalits.

Just how far India's political pendulum can swing was demonstrated in the 2007 provincial legislative election in Uttar Pradesh State, where a dalit party won an absolute majority and where a woman named Mayawati Kumari was the first dalit to become the chief minister of India's most populous State. It was a stunning victory that made headlines throughout the country and shook up the political establishment, revealing the growing power of the lowest castes and marking a turning point for India's representative government.

Most lower-caste Indians are still faced with limited opportunities, widespread discrimination, and the most extreme forms of poverty, particularly in less accessible rural areas. In traditional India, caste provided stability and continuity; in contemporary India, it constitutes an often painful and difficult legacy.

Economic Geography

The most commonly cited, and most clearly evident, regional division of India is between north and south. The north, and especially the Ganges Valley, is India's heartland, where Hinduism was born (as was Buddhism), and where Indo-European

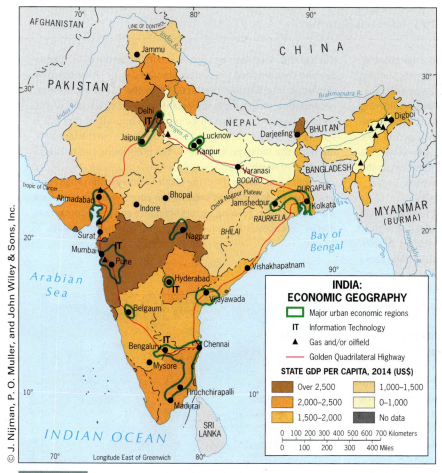

J. Nijman, P. O. Muller, and John Wiley & Sons, Inc.

FIGURE 8-21

INDIA: ECONOMIC GEOGRAPHY

- ▭ Major urban economic regions
- **IT** Information Technology
- ▲ Gas and/or oilfield
- — Golden Quadrilateral Highway

STATE GDP PER CAPITA, 2014 (US$)

- Over 2,500
- 2,000–2,500
- 1,500–2,000
- 1,000–1,500
- 0–1,000
- No data

0 100 200 300 400 500 600 700 Kilometers
0   100   200   300   400 Miles

Persian Gulf countries. Hundreds of thousands of workers from western India have found jobs there, sending remittances back to families from Punjab to Kerala. More importantly, several have used their foreign incomes to establish service industries back home. Outward-looking western India, in ever sharper contrast to the inward-looking east, is well in the lead in strengthening linkages to the global economy. For instance, satellite and fiber-optic cable links have driven Bengaluru (formerly Bangalore) to become the center of a burgeoning software-producing complex that reaches world markets.

Manufacturing and IT Industries

The industrial geography of India is in part a legacy of colonial times, with coastal Mumbai, Kolkata, and Chennai anchoring major industrial zones, and textiles—the entry-level industry of disadvantaged countries—dominating the manufacturing scene (Fig. 8-21). A number of industries, such as steel, machinery, and building materials, have become much stronger, and some leading companies have already become major global players. For example, the Tata Steel Group ranked tenth in size among the world's steel-producing companies in 2015.

India's information-technology (IT) industries, clustered in metropolitan Bengaluru, Delhi, Hyderabad, and Mumbai, draw much international attention. Leading Indian software companies such as Infosys, Wipro (see Box 8-4), and Cognizant are now household names all across urban India. The growth of software and IT services has been spectacular and now accounts for about 8 percent of GDP and no less than one-quarter of merchandise exports. But this sector employs only about 2 percent of the workforce.

What India needs far more, in terms of employment growth, are manufacturing industries that competitively sell goods in the global marketplace, putting tens of millions to work and transforming the economy. Clearly, India is very different from China. The two may have comparably sized populations, but the Chinese economy employs about 10 times as many workers in manufacturing and has a much larger urban middle class. India may have experienced spectacular recent growth, but too much of it is concentrated in the IT sector and benefits only a small, highly educated urban elite.

Life Is in the Harvest

Agriculture provides more jobs in India than any other employment sector, and India's fortunes (and misfortunes) remain strongly tied to farming. More than two-thirds of the population still lives on (and from) the land, spread in and around the country's hundreds of thousands of villages. There, traditional farming

---

languages (mostly variations of Hindi) are spoken. In the Dravidian south, an entirely different language family prevails, and as a second language, southerners prefer English over Hindi. In the south, too, people tend to have darker skin, the food gets spicier, and the climate is more tropical.

But there is another, potentially more significant, divide across India. In **Figure 8-21**, draw a line from Lucknow on the Ganges River south to Chennai on the southeastern coast. To the west of this line, India's economy is stronger and more modern, and the map shows that incomes are higher. To the east, India has more in common with the less promising economies of Bangladesh and Myanmar (Burma) that also face the Bay of Bengal. India's heavy industries built by the state near Kolkata in the 1950s are increasingly outdated, uncompetitive, and in steady decline. The State of Bihar represents the stagnation that afflicts much of India east of our line: by several measures, it ranks among the poorest of the 29 States.

Now compare this situation to western India's. The State of Maharashtra, the hinterland of Mumbai (**Box 8-14**), leads India in many categories, and Mumbai leads Maharashtra. Many smaller, private industries have taken root here, manufacturing goods that range from umbrellas to satellite dishes and from toys to textiles. Across the Arabian Sea lie the oil-rich economies of the

Box 8-14 Among the Realm's Great Cities: Mumbai (Bombay)

In some ways, Mumbai is a microcosm of India, a burgeoning, crowded, chaotic, fast-moving agglomeration of humanity. Shrines, mosques, temples, and churches evince the pervasive power of religion in this multicultural society. Street signs come in a bewildering variety of scripts and alphabets. The Victorian-Gothic architecture of the city center is a legacy of the British colonial period (see Box 8-2). Creaking double-decker buses compete with ox-wagons and handcarts on the congested roadways. The throngs on the sidewalks—businesspeople, holy men, sari-clad women, beggars, clerks, homeless wanderers—spill over into the streets.

In precolonial times, fishing folk living on the seven small islands at the entrance to this harbor named the place after their local Hindu goddess, Mumbai Devi. The Portuguese, first to colonize it, called it Bom Bahia, "Beautiful Bay." The British, who came next, corrupted both names to form *Bombay*, and so it remained for more than three centuries. In 1996, the government of the State of Maharashtra, with support from the federal authorities, changed the city's name back to Mumbai.

With almost 23 million people shoehorned into its peninsula and spilling over onto the mainland (**Navi Mumbai**), this is India's second-largest urban region. Think of Mumbai as a combination of New York and Los Angeles—India-style. This is the country's commercial and financial center, with most of the banks, the leading stock exchange, and most foreign multinationals. And it is also India's most glamorous city, owing in no small part to the presence of Bollywood, center of the Hindi film industry that produces more motion pictures (for more viewers) than anywhere else on Earth.

And then there are the approximately 8 million slum dwellers who struggle to make a living alongside the movie stars and the burgeoning middle classes. In part because its peninsular geography imposes strict limits, Mumbai is an extremely crowded city; its overall density is about seven times that of major U.S. central cities, and its slums exhibit much higher densities than that. Anyone visiting the city is bound to be overwhelmed by the sheer scale of its poverty, which stands in sharp contrast to the lavish, hugely expensive homes of the rich.

Mumbai is very much a product of colonial times. During the Industrial Revolution, England needed raw materials at low prices and India was a leading supplier, especially of cotton.

Much of the raw cotton was processed in the textile mills of Bombay and then shipped to London, Manchester, or Liverpool to be converted into finished products. From the 1860s on, Bombay became one of the most important colonial cities and a vital node in the far-flung British Empire. The imprint of that era is still visible in the architecture of the Fort District in the far south, the port, the railways, and the textile mills lining the rail corridors farther north.

Today not much is left of the textile industry in the urban economy. Instead, Mumbai has become the country's most important world-city with its growing sectors of finance and producer services that include accounting, advertising, and consulting companies.

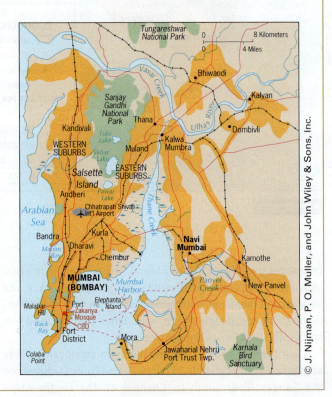

© J. Nijman, P. O. Muller, and John Wiley & Sons, Inc.

---

methods persist, yields per unit area remain among the world's lowest, and hunger and malnutrition still continue to afflict millions even as grain surpluses accumulate. The relatively few areas of modernization, such as the productive wheatlands of the Punjab, are islands in an ocean of agricultural stagnation.

As everywhere, food production is closely tied to climate and physiography, and farm output varies significantly across the country. In India, the monsoon is absolutely critical, its abundance and timing a reliable predictor of the harvest. **Figure 8-22** maps India's agricultural geography and reflects the rainfall patterns and monsoonal cycle depicted in Figure 8-3. Rice dominates all along the Arabian Sea-facing southwestern coast (on the rainy side of the Western Ghats) and in the monsoon-drenched

peninsular northeast; where drier conditions develop, wheat and other grains prevail.

Even though the country produces an ample variety of crops, yields usually are disappointingly low as a result of inefficient land ownership and small farmers' lack of access to inputs such as fertilizer, irrigation equipment, and machinery. If there is any to sell, getting produce to market is yet another struggle for millions of farmers. Almost half of India's 600,000-plus villages cannot be accessed by truck, let alone automobile, and in this era of modern transportation animal-drawn carts still outnumber motor vehicles nationwide.

To meet these challenges, India is now engaged in an unprecedented series of infrastructural improvement projects, ranging

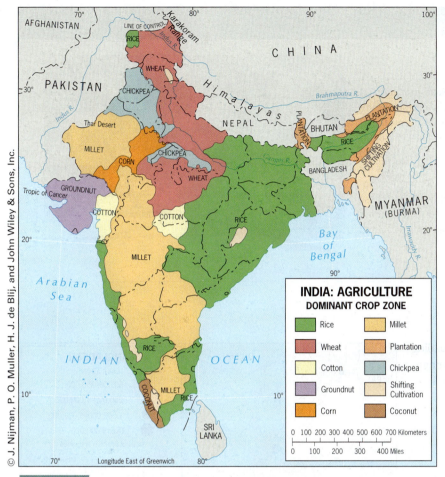

INDIA: AGRICULTURE
DOMINANT CROP ZONE

Rice	Millet
Wheat	Plantation
Cotton	Chickpea
Groundnut	Shifting Cultivation
Corn	Coconut

0 100 200 300 400 500 600 700 Kilometers
0   100   200   300   400 Miles

FIGURE 8-22

do everywhere, but many villagers are also driven off the land by desperate conditions in the countryside. As these migrants manage to establish themselves in Mumbai or Kolkata or Chennai, they help their relatives and friends to join them in squatter settlements that often are populated by newcomers from the same locality, bringing their language and customs with them and cushioning the stress of relocation.

Figure 8-18 displays the distribution of major urban centers in India. Except for Delhi–New Delhi, the biggest cities have coastal locations: Kolkata dominates the east, Mumbai the west, and Chennai the south. The overriding influence of these coastal cities is a colonial legacy. But urbanization also has expanded in the interior, notably within the core area (symbolized by Delhi overtaking Mumbai in 2011 to become India's largest city). In 2016, the country had more than 50 metropolitan areas containing populations of at least one million—and an urban population exceeding 400 million (well above the U.S. population total).

When you arrive in any Indian city, you are struck by the abundance of small shops everywhere—tiny businesses wedged into every available space in virtually every non-public building along every street. Even the upper, walk-up floors are occupied by shops, their advertising signs suspended from windows and balconies. What keeps all these small stores in business? Most of them earn very little and can afford to stay open only because they are part of the **informal sector [27]**—they are essentially unregistered, pay no taxes, utilize family labor, have been handed down through generations, and keep going because of the strong social and communal networks that characterize India's societal fabric. Moreover, high densities and the relatively low level of spatial mobility of many urbanites combine to create substantial demand for the local availability of daily goods and services.

This bustling retail scene notwithstanding, change is now under way throughout urban India. The country's growing consumerism, propelled by a rapidly expanding middle class estimated to number almost 300 million in 2015, has led to the introduction of a new feature on the landscape of major cities—the shopping mall (**Box 8-15**). As recently as 2000, India did not have a single shopping center. But only five years later the 100th had opened its doors, and by 2016 the total was closing in on a thousand. You will see American fast-food restaurants among the establishments in these malls as well as the brand names of numerous other foreign companies, proving that the tide of globalization has already washed up on India's shores.

Nonetheless, if the 25 percent of the Indian population that makes up the middle class can afford to shop at a mall, the

from highways to state-of-the-art airports. The most ambitious of these mega-projects is the construction of a 5900-kilometer (3700-mi)-long superhighway known as the Golden Quadrilateral. Completed in 2012, this fifth-longest highway on Earth interconnects the four anchors of the Indian urban system (Delhi, Mumbai, Chennai, and Kolkata) in a gigantic circuit that also passes through 15 other major cities (Fig. 8-21). The impacts of this all-important artery are multiple: it is expanding urban hinterlands; commuters are using it to travel farther to work than ever before; several once-remote rural areas now have a link to markets; and it is accelerating the rural-to-urban migration flow that will continue to transform India well into the future.

Urbanization

India is very well known for its enormous and teeming cities, but the country is not yet an urbanized society. Only about 32 percent of the population currently lives in cities and towns—compared to an average of around 80 percent across the developed world. In absolute terms, however, urbanization in India is a massive phenomenon. People by the hundreds of thousands are arriving in the already overcrowded cities, swelling urban India by about 3 percent annually—twice as fast as the country's overall population growth. Not only do the cities attract as they

"As I set foot into this spacious (and partially air-conditioned) mall from the steamy and crowded Andheri Link Road in Mumbai, I thought of the implications—is this the future in a land of small family-owned shops where 'retail' has a different and historic meaning? Infinity Mall, an upscale shopping center opened in 2005 in suburban Mumbai, caters to the megacity's new middle class. My Indian friend Pankaj thinks this is all for the better. 'On the weekends my wife and I often come here. There is a nice food court and the shopping is great. These are the things you Americans want, and these are the things we Indians want.' Malls have sprung up across India's major cities in the 2010s, transforming both urban landscapes and consumer behavior."

remaining one billion people cannot—an astounding inequality that widens by the day. Can India's ongoing economic transformation run its course without severe social dislocation?

Region Bangladesh

On the map of South Asia, Bangladesh looks like another State of India: the country occupies the area of the **double delta [28]** of India's great Ganges and Brahmaputra rivers, and India almost completely surrounds it on its landward side (**Fig. 8-23**). But Bangladesh is an independent country, born in 1971 after its brief war for independence against Pakistan, with a territory about the size of Wisconsin. Today it remains one of the poorest and least developed countries on Earth, with a population of 163 million that is growing at an annual rate of 1.2 percent.

Bangladesh remains largely a nation of subsistence farmers. Only a third of the population lives in urban settlements, and more than half of the workforce is engaged in agriculture. Dhaka, the megacity capital (**Box 8-16**), is home to over 16 million; the cities of Chittagong, Rangpur, Khulna, and Rajshahi are the only other urban centers of consequence. Moreover, Bangladesh has one of the highest physiologic densities (people per unit area of arable land) in the world: 1946 people per square kilometer/5040 per square mile.

A Vulnerable Territory

Not only is Bangladesh a poor country; it also is highly susceptible to damage from natural hazards. During

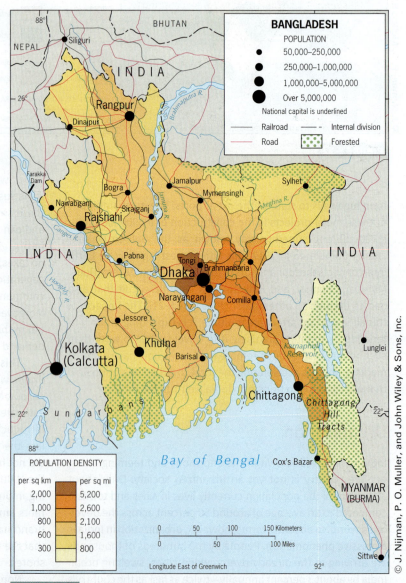

FIGURE 8-23

Box 8-16 Among the Realm's Great Cities: Dhaka

Dhaka is situated in the heart of Bangladesh in the northern sector of the double delta formed by the Ganges and Brahmaputra. It is the economic, political, and cultural focus of the country, and with a population of 16.2 million it is by the far the biggest city.

Known as the "city of mosques and muslin" (a type of cotton), Dhaka began its modern history in the early seventeenth century when the area came under the rule of the Islamic Mughals, along with what is today northern India. Dhaka occupied an advantageous position on the main waterways, and to this day the local Buriganga River is at the center of the city's activities. The oldest section of Dhaka lines the north bank of the waterfront. The British turned the place into a provincial capital of sorts, and their legacy can still be seen in some of the colonial architecture near the old center alongside the remaining Mughal structures. At independence in 1947, the city became the administrative capital of East Pakistan, and in 1971 it became the seat of government for the newly formed state of Bangladesh.

As part of the great double delta, Dhaka lies close to sea level and is prone to flooding during the summer monsoon. Every so often, major cyclones barrel in from the Bay of Bengal and cause massive human and material devastation. The greatest disaster in modern times was triggered by Cyclone Bhola in 1970, which killed as many as 500,000 of the country's inhabitants, destroyed hundreds of thousands of homes, and flooded most of the city.

Dhaka has grown rapidly in recent decades as it has attracted large numbers of migrants from rural areas, and the city has gradually expanded northward. At the time of partition, many Hindus left and Muslims arrived, and now fully 90 percent of the population adheres to Islam. Religious fanaticism is not as widespread as it is in Pakistan. The main schism here is between the tiny rich minority and the enormous poverty-stricken majority. Dhaka is also known as the bicycle-riksha capital of the world: some 400,000 colorfully painted rikshas criss-cross the thoroughfares of this city on a daily basis—surely the cheapest mode of urban transportation on the planet.

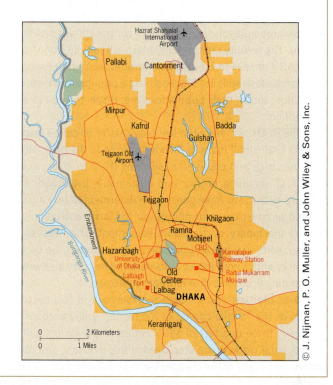

© J. Nijman, P. O. Muller, and John Wiley & Sons, Inc.

the twentieth century, two of the deadliest weather disasters in the world struck this small country. The most recent of these occurred in 1991 when a cyclone (as hurricanes are called in this corner of the world) killed more than 150,000 people. Smaller cyclones strike several times every year and almost routinely kill anywhere between dozens and thousands of people.

The reasons for Bangladesh's vulnerability can be deduced from Figures 8-23 and 8-1. Southern Bangladesh consists of the vast deltaic plain of the combined Ganges–Brahmaputra river system, integrating extremely fertile alluvial (river-deposited silt) soils that attract farmers with the low elevations that endanger them when the water rises. Moreover, the shape of the Bay of Bengal forms a funnel that steers cyclones and their storm surges of wind-driven water to barrel into the double delta's coast. Without money to build seawalls, floodgates, elevated shelters in sufficient numbers, or even adequate escape routes, hundreds of thousands of people are at continuous risk—with deadly consequences.

The country's relations with neighboring India have at times been strained over water resources (India controls the Ganges River, which is the lifeline of Bangladesh), cross-border migration (about 8 percent of the population is Hindu), and transit between parts of India across Bangladesh's north.

Limited Opportunities, Creative Development Strategies

Even though geography seems to offer Bangladesh limited options and the country overall remains very poor, some remarkable successes have been achieved here in recent years. First, since 1990, life expectancy in Bangladesh has increased by ten years and now exceeds that of India, despite India's considerably higher incomes. Second, primary school enrollment among girls today is double that of 2010, and the female literacy rate is steadily improving. And third, since 1990 infant (< 1 year) mortality rates have been more than halved, and child (< 5 years) mortality rates as well as maternal mortality rates have dropped by about three-fourths. According to these indicators, Bangladesh outperforms India and also fares much better than Pakistan.

Interestingly, these improvements were achieved without significant economic growth. Indeed, incomes in Bangladesh are considerably lower than those in India, and even Pakistan does better in this regard. Bangladesh is an overwhelmingly Muslim society, but in some ways it has also been a very progressive one, particularly when it comes to the role of women. For example, 30 seats in the national legislature are reserved for women. Much of this success is attributed to the role of **non-governmental organizations (NGOs) [29]** that, independently of (but supported by) the national government, have promoted programs to improve the quality of life. An important example concerns family planning programs in which women play the lead role (free contraceptives and education). The fertility rate in Bangladesh fell rapidly from 7.0 in 1970 to 3.4 in 1994 to 2.2 in 2016. Consider this: when Pakistan and Bangladesh split in 1971, each had a population of around 65 million; today, their populations are, respectively, 193 and 163 million.

The NGOs also were the key providers of so-called **micro-credits [30]**, small loans at favorable terms to the poor, along with guidance and counseling, allowing them to invest in the means to secure a proper and sustainable livelihood. Frequently, such loans have gone to small farmers to buy a piece of land or construct a farmhouse, or to starter entrepreneurs to buy machinery or transport equipment. The success of micro-credits has spread from Bangladesh around the Global Periphery and is now widely considered to be a key element in formulating economic development strategies.

Over the past several decades, Bangladesh has become known as the "textile capital of the world," today accounting for nearly three-quarters of all export earnings. Almost half of all industrial workers are employed in this industry, and most are women. Many of the cheaper items of clothing that are sold worldwide under popular labels from big-name marketers are made in factories in Bangladesh that are unsafe and where the minimum wage is little more than one U.S. dollar a day. Western companies impose strict production quotas and deadlines, and factory managers force their employees to work long hours under often dangerous conditions. There are no unions to protect workers; disastrous fires and building failures often make the international news (e.g., the 2013 collapse of a factory building in the Dhaka suburb of Savar that killed more than 1300 laborers). Globalization has made Bangladesh's garment industry a valuable contributor to the nation's commercial economy, but those low prices you see advertised in Western stores raise painful questions about what constitutes "fair trade."

While Bangladesh is praised by many for its progressive social and economic programs, it is at the same time confronted by the rising militancy of Islamic fundamentalists. In the first half of 2016 alone, there were two dozen assassinations of liberal politicians, gay activists, academicians, and journalists who had been vocal about democratic freedoms and what could be perceived as criticism of fundamentalist Islamic organizations. Most of the victims were brutally hacked to death. It is not clear whether the perpetrators are associated with al-Qaeda or ISIS, or whether they belong to isolated local terrorist cells. Some observers, and many liberal Bangladeshis along with them, worry that the country's progressive course is

at risk and that Bangladesh may be going down the same violent path as Pakistan.

Region The Northern Mountain States

As Figures 8-1 and 8-15 show, two landlocked countries lie in the central sector of the mountainous zone that walls India off from China: Nepal and Bhutan. Together they constitute a distinct region of South Asia, but both are relatively small (Nepal is roughly the size of Illinois, and Bhutan only one-quarter of that). Importantly, these countries function geopolitically as buffer states between the two Asian giants.

Nepal

Culturally and (until recently) economically, Nepal lies within India's sphere of influence, but China's presence here is being felt ever more strongly. Maoist-communist groups are supported by China but loathed by India (whose Naxalites espouse a similar ideology), resulting in fractious and contentious governance. China, as it does in some other parts of the world, is using its growing economic clout in Nepal as leverage to gain greater political influence.

Nepal, located just to the northeast of India's Hindu core area, has a population of 29 million, of whom 82 percent are Hindu. Nepal's Hinduism, however, is a unique blend of Hindu and Buddhist ideals. Thousands of temples and pagodas, ranging from the simple to the ornate, grace the cultural landscape, especially in the Vale of Kathmandu, the country's core area. The Nepalese are a people of many origins, including India, Tibet, and interior Asia. Even though well over a dozen languages are spoken, most of the people also speak Nepali, a language related to Indian Hindi.

Closer examination of Figure 8-1 shows that Nepal contains three geographic zones: a southern, subtropical, fertile lowland called the Terai; a central belt of Himalayan foothills with swiftly flowing streams and deep canyons; and the spectacular high Himalaya itself (topped by Mount Everest) in the north. The capital, Kathmandu, lies in the east-central part of the country in an open valley of the central hill zone.

The country's soaring Himalayan peaks are a worldrenowned tourist attraction, but visitor expenditures are relatively modest and tourism is often disrupted by flare-ups of political instability. This is a troubled country suffering from severe underdevelopment, recording the lowest per capita income in the realm, even below that of Afghanistan. Many young people go abroad to find work, and Nepal depends heavily on remittances. It also faces strong centrifugal social and political forces. Environmental degradation, overused farmlands, soil erosion, and the blight of deforestation scar the highly corrugated countryside. In the spring of 2015, Nepal's dysfunction became painfully clear when the government proved to be

An earthquake registering a magnitude of 7.8 struck central Nepal on April 25, 2015, the country's most severe quake since 1934. It produced widespread destruction and the loss of over 8600 lives as well as thousands of injuries. More than 70,000 dwellings were destroyed in the initial quake and subsequent aftershocks, leaving 10 percent of Nepal's population homeless and without immediate access to food and clean water. These survivors are searching (by hand) for victims trapped in the rubble of damaged buildings in Bhaktapur, an eastern suburb of the capital, Kathmandu.

useless in the effort (taken over by foreign relief agencies) to assist the victims of a massive earthquake that killed more than 8600 people, injured thousands more, and left nearly three million homeless (see photo).

Bhutan

Mountainous Bhutan, with a population of less than 800,000, lies wedged, fortress-like, between India and China's Tibet (Fig. 8-1). Bhutan has a long history as a constitutional monarchy, ruled by a king whose absolute power was unquestioned by his subjects. But in 2007, the newly crowned monarch, who had just succeeded his father—and perhaps with an eye on the recent overthrow of the royal rulers in nearby Nepal—decided to order his subjects to vote for a political party in an abruptly created democracy. Thus Bhutan overnight went from an absolute monarchy to a multiparty democracy, and Thimphu, the capital, became the seat of a newly elected National Assembly. Throughout the mountainous countryside, the symbols of Buddhism, the state religion, dominate the cultural landscape, and the government's development policies emphasize the importance of spiritual fulfillment alongside the satisfaction of material needs.

Forestry, hydroelectric power, and tourism all have much potential here, and Bhutan also possesses ample mineral resources. The country did not open up to foreign visitors until the 1970s, and even today tourism is kept within strict limits. Although isolation and inaccessibility continue to preserve traditional ways of life, Bhutan is not immune to the turbulent social forces that permeate this realm. Tensions between the Buddhist Bhutanese majority and a Nepalese Hindu minority

have led many Nepalese Hindus to flee across the sliver of India that separates the two mountain states (Fig. 8-1). To date, the governments of Bhutan and Nepal have been unable to work out a solution to this problem.

Region The Southern Islands

As Figure 8-15 shows, South Asia's subcontinental landmass is flanked by several sets of islands. The largest, Sri Lanka, lies just to the southeast of the southern tip of India. The Maldives form an archipelago farther out in the Indian Ocean to the southwest of southernmost India. And the Andaman and Nicobar Islands (which will not be discussed here), a Union Territory of India, mark the southeastern edge of the Bay of Bengal.

The Maldives

The Maldives consists of some 1200 tiny islands whose combined area is just 300 square kilometers (115 sq mi) and whose highest elevation is less than 2 meters (6 ft) above sea level. The Maldives' low elevation has been singled out repeatedly in assessments of the future impact of **rising sea level [31]** caused by global warming (see **Box 8-17**). The population of 375,000 drawn from Dravidian and Sri Lankan sources is now virtually 100 percent Muslim, with one-quarter of it concentrated on the island of the capital named Maale. This country exhibits the realm's highest GNI per capita (see Fig. 8-8), thanks to the Maldivians having translated their palm-studded, beach-fringed island chain into a tourist mecca that annually attracts tens of thousands of mainly European visitors. The Maldives, however, also faces social and political challenges—its nascent democracy was dealt a blow in 2012 when the first freely-elected president was deposed in a *coup* engineered by the country's former dictator and radical Islamists.

Sri Lanka: Paradise Lost and Regained?

Sri Lanka (known as Ceylon before 1972), the compact, pear-shaped island located just 35 kilometers (22 mi) across the Palk Strait from southernmost India (see Figs. 8-1 and **8-24**), became independent from Britain in 1948. There were good reasons to create a separate sovereignty for Sri Lanka. This is neither a Hindu nor a Muslim country: just over 70 percent of its population of 21 million are adherents of Buddhism.

The majority of the Sri Lankan population is descended from migrants who arrived at the island from northwestern India beginning about 2500 years ago. Most of them probably walked across to the island, which was connected to southernmost India by a land bridge (now submerged beneath the Palk Strait) known as *Rama's Bridge*. If you look closely at Figure 8-1, you can see the Sri Lanka-end remains of that land bridge, which was further eroded by a powerful cyclone in the fifteenth century. The migrants brought with them the advanced culture of their source area, building towns and irrigation

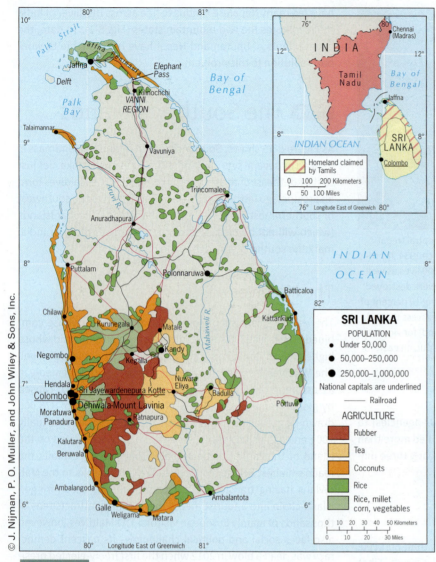

FIGURE 8-24

SRI LANKA

POPULATION
- Under 50,000
- 50,000–250,000
- 250,000–1,000,000

National capitals are underlined
— Railroad

AGRICULTURE
- Rubber
- Tea
- Coconuts
- Rice
- Rice, millet corn, vegetables

systems, and introducing Buddhism. Today, their descendants are known as the Sinhalese and speak Sinhala, which belongs to the Indo-European language family of northern India (Fig. 8-5).

The Tamil-speaking Dravidians, who lived on the mainland side of the Palk Strait, arrived much later. During the nineteenth century, the British colonizers brought hundreds of thousands of Tamils to the island to labor on their tea plantations, and they soon became a substantial component of then-Ceylonese society. Not only did the Tamils bring their Dravidian language to the island—they also introduced their Hindu faith. At the time of independence, one year after the India-Pakistan partition, this minority constituted more than 15 percent of the island's population; in the mid-2010s, they comprise just over 11 percent.

When Ceylon became independent, it was one of the great hopes of the postcolonial world. The country had a sound economy as well as a democratic government, and it was renowned for its tropical beauty. Its reputation soared when a massive effort succeeded in eradicating malaria, which was followed by a family-planning campaign that reduced the birth rate while the rest of South Asia was experiencing a substantial population explosion. Rivers from the cooler, forested, interior highlands fed the paddies that provided ample rice; commercial crops from the moist southwest paid most of the bills,

Box 8-17 From the Field Notes...

"We have just taken off from the Gan airstrip on Addu Atoll, in the southern Maldives. Within seconds, the neighboring island of Villingi comes into full view. It is a resort island, one of many that are vital to the economy of this 'tropical paradise.' Consisting of about 1200 islands with an average elevation of less than 1 meter (3.3 ft), the Maldives have also been labeled 'ground zero' in debates about global warming and rising sea levels. At the moment it is low tide, but you can see how far inland the water reaches at high tide. With stormy weather, flooding is guaranteed. I just spent three days here, and people told me that they face growing problems of beach erosion and soil salinity, and that fish are steadily migrating to greater depths farther from the islands. Given the ominous forecasts of global sea-level rise, this may all be gone (or at least depopulated) by the end of the century."

© Jan Nijman

Sinhalese Buddhist monks visit the famous ruins of the twelfth-century Hatadag Temple in the ancient city of Polonnaruwa in north-central Sri Lanka. The temple that once stood here was also known as the "tooth relic temple" because it was said to have treasured an actual tooth of Prince Siddhartha—the Buddha.

and the capital, Colombo, grew to reflect the optimism that prevailed.

In the midst of this glowing scenario, the seeds of conflict had already been sown. Sri Lanka's Tamil minority soon began proclaiming its sense of exclusion, demanding better treatment from the Sinhalese majority. Tensions steadily escalated until a full-scale civil war broke out in 1983. Militant activist leaders in the Tamil community now demanded a separate Tamil state encompassing the north and east of the country, coinciding with the Tamil-speaking areas mapped in stripes in the inset map of Fig. 8-24. This violent, devastating conflict endured for 26 years (with a death toll estimated as high as 100,000), finally ending in 2009 when the Sinhalese-dominated government declared victory. In the still bitter aftermath, northernmost Sri Lanka was permitted to have a provincial government dominated by the leading Tamil party—but the central government in Colombo maintains very tight sovereign control.

Points to Ponder

- The most skewed sex ratios in the population tend to occur in the most prosperous parts of South Asia.
- According to population growth projections, by 2030 India will be home to the largest national population of Muslims in the world.
- Pakistan, crucially important to the stability of both Southwest and South Asia, is becoming increasingly difficult to govern and faces growing religious fanaticism as well as political turmoil.
- India's IT sector accounts for one-quarter of all exports but employs barely 2 percent of the national workforce.
- Does the United States have the right to conduct drone strikes against the Afghan Taliban on Pakistan's territory, without consent of the Pakistani government?

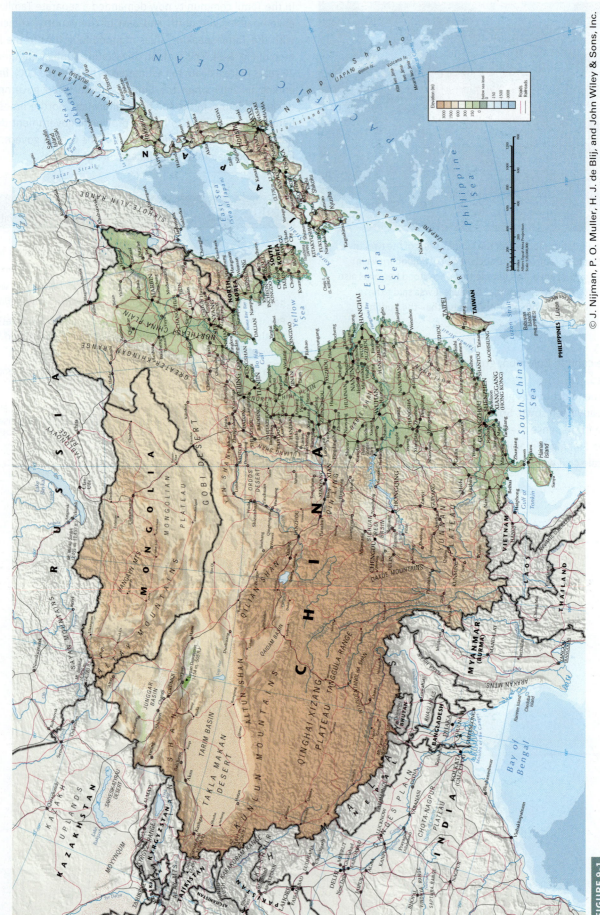

FIGURE 9-1

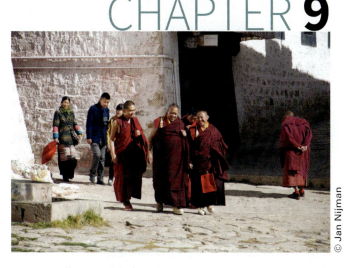

CHAPTER 9

Where were these pictures taken?

The East Asian Realm

East Asia is a geographic realm like no other. At its heart lies the world's most populous country, the product of what may be the world's oldest continuous civilization. On its Pacific mainland shores an economic transformation has taken shape with no parallel in world history. Its offshore islands witnessed the first use of atomic weapons on civilian populations and the postwar emergence of one of the world's most powerful economies. Few lives in this world were left unaffected, directly or indirectly, by the momentous events that have occurred in East Asia over the past two generations. Just look around you. Chinese-assembled smartphones (and so much more), Japanese-made automobiles, South Korean televisions, Taiwanese computers—from toys to textiles and from semiconductors to software—East Asian products fill streets and stores, homes and hotels.

Defining the Realm

It has all happened with such astonishing speed. Not that long ago, you were no more likely to find anything useful made in China than you were to buy anything from Russia. But about 50 years ago, things began to change rapidly. Japan led the way, turning its 1945 World War II defeat into postwar economic triumph two decades later. By the mid-1970s, that country's remarkable economic growth— compared to China's seemingly total stagnation—appeared to justify Japan's recognition as a discrete geographic realm, an economic engine for the world, and a strong competitor for the top-ranked U.S. economy. On a smaller scale, Hong Kong, South Korea, and autonomous Taiwan soon followed, showing the world what other peoples in East Asia were capable of.

But all this was prologue to the main event—the unprecedented, remarkably swift rise of China. Less than five years after the 1978 implementation of its game-changing open-door and market reform policies, this gigantic country had taken center stage in the realm's economy. Since the mid-1980s, led by burgeoning China, East Asia has steadily emerged as the most dynamic realm in the global economy. In the 2010s, it continues to reshape the world—and in the process is itself undergoing a profound transformation.

Geographic Dimensions

As **Figure 9-1** indicates, the East Asian geographic realm forms a roughly triangular wedge between the vast expanses of eastern Russia to the north and the populous countries of South and Southeast Asia to the south, its edges often marked by high mountain ranges or remote deserts. The darker brown on the map designates the highest mountains and plateaus, which create a vast arc north of the Himalaya before bending southward and becoming lower (tan shading) toward Myanmar, Laos, and Vietnam in Southeast Asia. Here in the southwest, where Tibet is located, mountains and plateaus alike are covered by permanent ice and snow, the soaring ranges crumpled up like the folds of an accordion. Three major rivers, their valleys parallel for hundreds of kilometers, disclose the orientation of this high-relief topography. Northward, note how rapidly the

---

Box 9-1 Major Geographic Features of East Asia

1. East Asia is encircled by snowcapped mountains, vast deserts, and Pacific waters.

2. East Asia was one of the world's earliest culture hearths, and China is one of the world's oldest continuous civilizations.

3. East Asia is the second most populous geographic realm after South Asia; its population remains heavily concentrated in its eastern regions.

4. The People's Republic of China (PRC), the world's largest state demographically, is the current rendition of an empire that has expanded and contracted, fragmented and unified, many times during its long existence.

5. Key components of China's sparsely peopled western regions are now being rapidly developed—not only because they are strategically important to the state but also because they lie exposed to minority pressures and Islamic influences.

6. The post-Mao economic transformation launched four decades ago along China's eastern seaboard is today steadily expanding westward, accompanied by a major in-migration of ethnic (Han) Chinese.

7. Widening regional disparities—exacerbated by breakneck, massive urbanization—are straining Chinese society.

8. Japan's economy continues to be ranked among the world's wealthiest, even though it is mired in a quarter-century-long economic slump, largely the result of the continuing inability of its business culture to adapt to changing regional and global circumstances.

9. Geopolitically, this realm is home to the world's newest superpower as China's economic and political influence is increasingly projected beyond East Asia.

10. The political geography of East Asia contains a growing number of flashpoints capable of generating conflict, including North Korea, Taiwan, and several island groups in the seas adjoining the realm.

mountains give way to broad, flat deserts whose names appear prominently: the Takla Makan in the far west, the Gobi where China meets Mongolia, the Ordos in the embrace of what looks like a huge bend of a river we will learn more about shortly, the Huang He (Yellow River).

In this environment of mountains and deserts, living space is at a premium. And speaking of living space, the green areas on the map, which have the lowest relief and (often) the most fertile soils, are home to the vast majority of this realm's population. Here the great rivers that come from the melting ice and snow in the interior highlands have been depositing their sediment load for eons, and when humans domesticated plants and started to grow crops, this was the place to be. That was thousands of years ago—perhaps as long as 10,000 years—and ever since, this has been the largest human cluster on Earth.

But the East Asian realm is not confined to the mainland of mountains and river basins. You can imagine how its off-shore islands were populated: the Korean Peninsula seems to form part of a bridge pointing toward the southernmost large island (Kyushu) of what is today Japan, and from there it seems likely that the early migrants moved farther north until they reached Hokkaido. In warmer times, they may even have ventured beyond, onto the Kurile Islands. And in the south, Taiwan lies even closer to mainland China than Japan does to Korea, whereas near-tropical Hainan Island, the realm's southernmost extremity, is almost—but not quite—connected to the small peninsula that stretches toward it.

In terms of total land area, though, East Asia is mostly mainland—but the islands and their peoples have played leading roles in forging this realm's regional geography. The waters between mainland and islands (the Taiwan Strait, the South China Sea, the East China Sea, the Yellow Sea, the Korea Strait, among others) also figure prominently in the geographic evolution of this realm. Today, the Japanese and the Chinese are arguing over the ownership of small islands in these waters, specks of land surrounded by possible oil reserves and fishing grounds claimed by both sides. So this realm's map is considerably more complicated than that wedge-shaped triangle in Figure 9-1 initially suggests—including even the spellings of many of its contents (**Box 9-1**).

Political Geography

It is all too easy to refer to China when you mean East Asia because China is East Asia's dominant country, contains more than 85 percent of the realm's population, and has commanded an increasingly prominent role on the global stage. But there are five other political entities on East Asia's map: Japan, South Korea, North Korea, Mongolia, and Taiwan. Note that we refer here to *political entities* rather than *states*. In this realm, the distinction is important. Taiwan refers to itself as the Republic of China (ROC), but it is not recognized as a sovereign state by most members of the international community; the communist administration in Beijing, capital of the People's Republic

of China (PRC), regards Taiwan as part of China and as a temporarily wayward province. And North Korea is widely viewed as a rogue state, as an archaic dictatorship that has profoundly failed its people. Having compiled one of the world's most dreadful human rights records, North Korea is not even a fully functional member of the United Nations.

Nevertheless, China is the realm's predominant entity—demographically, economically, and politically. It is important to keep in mind, as you read this chapter, that portions of what we map today as regional components of China were not part of the country in the past, and that other areas now lying outside China are regarded by many Chinese as China's property (for example, a large sector of the Indian State of Arunachal Pradesh and portions of the Russian Far East). China's imperial past saw the state expand, contract, and then expand again, accumulating unsettled border issues on land as well as at sea. On issues like these, Chinese national sentiments can run quite deeply.

Environment and Population

To understand the complex physical geography of East Asia mapped in Figure 9-1, it is useful to refer back to Figure G-4 in the introductory chapter. The high snowcapped mountain ranges of the realm's southwestern interior result from the gigantic collision of the Indian and Eurasian tectonic plates (see Fig. 8-2), pushing the Earth's crust upward and creating not only the mountain ranges of which the Himalaya is the most famous, but also popping up the enormous, domelike Qinghai-Xizang (Tibetan) Plateau. In Figure G-4, note the high incidence of earthquakes associated with this collision, converging on a narrow zone of instability that crosses southwestern China and

DigitalGlobe via / Getty Images, Inc.

Satellite image of the heavily damaged Fukushima Daiichi nuclear power complex, three days after the catastrophic Tohoku earthquake and tsunamis struck northeastern Japan in 2011. This strongest earthquake ever recorded in the country triggered the meltdown of three of the facility's six nuclear reactors. Four years later, radiation was still being released into the air and ocean, continuing to threaten distant as well as local areas. Perhaps the longest-lasting impact of this disaster is the opposition to nuclear energy that continues to grow in Japan, Germany, and other key countries around the world.

stretches into Southeast Asia. The calamitous 2008 earthquake (magnitude 7.9) in China's Sichuan Province that killed almost 90,000 people originated in this danger zone, but was only one in an endless series that will continue to take its toll (the latest occurring here in April 2013—a 7.0 temblor that killed at least 200 and injured more than 10,000).

From Figure G-4 it is also obvious that the Pacific Ring of Fire, with its lethal combination of volcanism and earthquakes, endangers Japan far more than it does China. This vulnerability became a horrific reality on March 11, 2011, when a monstrous 9.0-magnitude earthquake struck off Japan's northeastern coast near the city of Sendai. **Figure 9-2** displays the configuration of the underlying tectonic plates that converge on Japan. The Pacific Plate moves westward at an average of 7–10 centimeters (3–4 in) per year, and the only way it can do so is by pressing forward (or subducting) *beneath* the North American Plate (much of Japan sits atop the western tip of the latter). It was the most powerful Japanese earthquake ever recorded and one of the five biggest in the world since 1900. Even worse, this massive temblor and its torrent of violent aftershocks triggered a series of devastating **tsunamis [1]** (seismic sea waves) that swept across the narrow, densely populated coastal plains that hug the shoreline of Honshu (Japan's largest island) north of Tokyo. And as if the destruction and death left behind was not enough, the leakage of radioactivity from a heavily damaged nuclear power complex near Fukushima south of Sendai (see photo and Fig. 9-2) may have caused serious future health problems in the local population. The final death toll of this event, now officially known as the Tohoku earthquake, has been reckoned to be about 21,000.

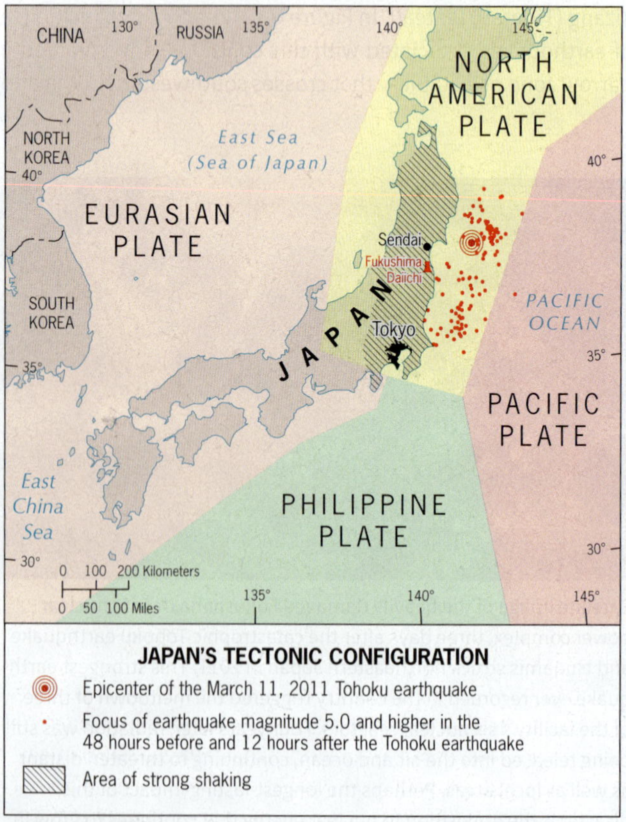

Japan's vulnerability to such disasters results from a dangerous combination of circumstances: it is located in a particularly active tectonic-plate collision zone, and many of the country's habitable (and most densely populated) areas are confined, low-lying plains on the islands' eastern coasts open to flooding by Pacific tsunamis. Not surprisingly, the location of most Japanese nuclear power plants along these susceptible shorelines has now become a hotly debated issue.

Looking at the layout of East Asian climates (**Fig. 9-3**, left map), we should not be surprised that the western and northern sectors of this realm are dominated by conditions that do not favor substantial population clusters. Permanent snow and ice cover much of the area mapped as *H* (highland climates), including Tibet (Xizang) and Qinghai. Northward, the dry *B* climates (desert and steppe) prevail because this vast expanse lies about as far from maritime influences (and moist air masses) as you can get in Asia. Mongolia is one of the driest countries in the world, but even here—and also in frosty Tibet—there are places where people manage to eke out a living. But the map leaves no doubt as to why most East Asians congregate in the eastern segment of this realm.

When we compare the climates prevailing in the East Asian realm to those familiar to us in North America (Fig. 9-3, right map), it is immediately obvious that the *C* or humid-temperate climates are more extensive in the United States than in East Asia. Note especially the comparative location of the milder *Cfa* climate, which in the United States extends beyond 40° North latitude up to New England, but which in China yields to colder *D* climates at a latitude equivalent to Virginia's. Thus the capital, Beijing, has the warm summers but long, bitterly cold winters characteristic of *D* climates. Take a closer look at Figure 9-3 and it is clear that both the Korean Peninsula and the Japanese islands lie astride this transition from *C* to *D* climates. South Korea is significantly milder and moister than North Korea, and Japan's largest island (Honshu) is temperate in the south but cold in the north. Not unexpectedly, the least densely populated major island of Japan is northernmost Hokkaido, where the climate is quite similar to that of northern Wisconsin.

Comparing just the United States and China, observe that whereas *C* or *D* climates prevail over more than half of the United States, these climates predominate over less than one-third of China— even though the United States has only 324 million people compared to China's 1.4 billion. That is what makes the population distribution map (**Fig. 9-4**) so noteworthy: the overwhelming majority of East Asia's inhabitants are located in the easternmost one-third of the realm's territory, creating the largest and most densely settled population cluster in the world, which is mainly associated with the limited green-colored area in Figure 9-1.

Several times in earlier chapters of this book, we have noted the capacity of humans to live under virtually all environmental circumstances; technological developments enable the survival of year-round communities in Antarctica, on oil platforms at sea, in the driest of deserts, and on the highest of plateaus. But the world population distribution map (Fig. G-7) still reminds us

JAPAN'S TECTONIC CONFIGURATION

◎ Epicenter of the March 11, 2011 Tohoku earthquake

· Focus of earthquake magnitude 5.0 and higher in the 48 hours before and 12 hours after the Tohoku earthquake

▨ Area of strong shaking

FIGURE 9-2

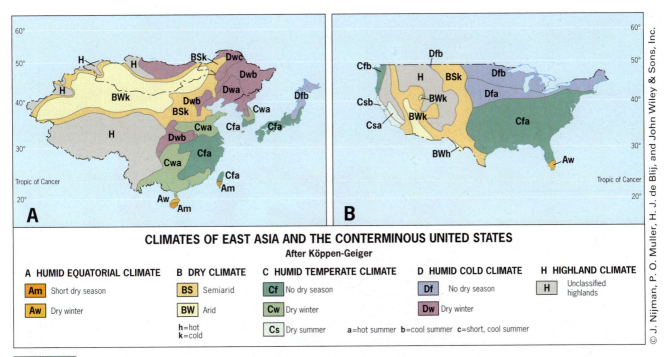

CLIMATES OF EAST ASIA AND THE CONTERMINOUS UNITED STATES
After Köppen-Geiger

A HUMID EQUATORIAL CLIMATE

Am	Short dry season
Aw	Dry winter

B DRY CLIMATE

BS	Semiarid
BW	Arid

h=hot
k=cold

C HUMID TEMPERATE CLIMATE

Cf	No dry season
Cw	Dry winter
Cs	Dry summer

D HUMID COLD CLIMATE

Df	No dry season
Dw	Dry winter

H HIGHLAND CLIMATE

H	Unclassified highlands

a=hot summer   b=cool summer   c=short, cool summer

© J. Nijman, P. O. Muller, H. J. de Blij, and John Wiley & Sons, Inc.

FIGURE 9-3

how we got our modern start—through crop-raising and herding. The fertile river basins and coastal plains of East Asia supported ever-larger farming populations, whose descendants still live on that same land: for all its ongoing industrialization and urbanization, more than 40 percent of the population of China remains rural to this day. Environment, in the form of terrain, water supply, soil fertility, and climate played the crucial role in the evolution of the population distribution displayed in Figure 9-4—and will continue to do so for centuries to come.

The Great Rivers

China is in many respects the product of four great river systems and their basins, valleys, deltas, and estuaries. These rivers and their tributaries are visible but do not stand out as clearly in Figure 9-1 as they do on the physiographic map (**Fig. 9-5**). So important are they as shapers of the realm's regional geography that we should acquaint ourselves with them here. Of the four, the two in the middle are in many ways the most important of all: the **Huang He** (Yellow River) that makes a huge loop around the Ordos Desert and then flows across the North China Plain into the Bohai Gulf, and the **Yangzi** (often still spelled Yangtze), probably the most famous river in China's historical geography, known as the **Chang Jiang** (Long River) upstream.

As the map illustrates, the Yellow River and its tributaries form the sources of water essential to the historic core area of China, the North China Plain, where the capital, Beijing, is located. The Yangzi River is the primary artery of the Lower Chang Basin; at its mouth lies China's largest city, Shanghai, and in its middle course the water flow is controlled by the world's biggest dam (see photo). Both the Huang and the Yangzi originate in the snowy highlands of the Qinghai-Xizang (Tibetan)

Plateau, a reminder that these remote environments are critical to hundreds of millions of people who live thousands of kilometers away.

The other two rivers have much shorter courses, but the **Pearl** River outlet of the one in the south, the **Xi Jiang** (West River), forms an estuary that has become China's (and East Asia's) foremost hub of globalization. This is where you find Hong Kong and, right next to it, the fastest-growing major city in the history of humankind—Shenzhen.

Finally, the northernmost of China's four major rivers, the **Liao** River, originates near the margin of the Gobi Desert and then forms an elbow as it crosses the Northeast China Plain to reach the Bohai Gulf flowing southward. Here the climate is

Three Gorges Dam is emblematic of China's contemporary "era of the mega-project." The dam wall rises 180 meters (600 ft) above the inundated valley floor of the Chang/Yangzi River. It is 2 kilometers (1.2 mi) wide and creates a reservoir—largest of its kind in the world—that extends more than 600 kilometers (400 mi) upstream.

VCG/VCG via Getty Images

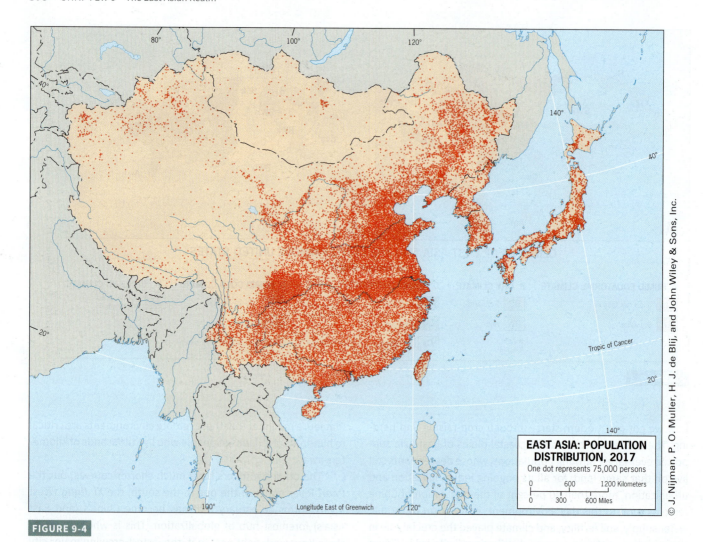

FIGURE 9-4

EAST ASIA: POPULATION
DISTRIBUTION, 2017
One dot represents 75,000 persons

© J. Nijman, P. O. Muller, H. J. de Blij, and John Wiley & Sons, Inc.

much colder, lowlands scarcer, agriculture lagging, and population smaller than in the more southerly river basins, but mineral resources create opportunities for mining and industry. Each of East Asia's river-based population clusters has its own combination of potentials and problems.

But even these major river systems, together with China's lesser ones, are increasingly pressed to satisfy the country's skyrocketing demands for water. In order to meet these urgent needs, China is now proceeding with the massive *South-to-North Water Transfer Project* to bring new supplies to its thirsty northeastern core area, especially metropolitan Beijing and Tianjin (**Box 9-2** and Fig. 9-5).

Along the Coast

East Asia's Pacific margin is a jumble of peninsulas and islands. The Korean Peninsula looks like a near-bridge linking Asia to Japan, and indeed it has served as such in the past. The Liaodong and Shandong peninsulas protrude into the Yellow Sea, which continues to silt up from the sediments of the Huang and Liao rivers. Off the mainland lie the islands that have played such a crucial role in the modern human geography of Asia and,

indeed, the world: Japan, Taiwan, and China's Hainan. Japan's environmental range is expressed by cold northern Hokkaido and warm southern Kyushu, but Japan's core area lies on its main island, Honshu. As Figure 9-1 shows, myriad smaller islands flank the mainland and dot the East and South China seas. As we will discover, some of these smaller islands increasingly affect the political geography of this realm.

Natural Resources

Given that the East Asian realm contains nearly one-fourth of the world's population, it is not difficult to imagine the magnitude of the demand for natural resources here. The world first received notice of this about 125 years ago when Japan's imperial expansion was in part driven by the need for raw materials to feed its rapidly expanding industrial base. Today, East Asia's demand for natural resources is expressed not in imperialism but in the global marketplace: it is a hugely important driver of natural resource exploitation around the world, from Russia to Subsaharan Africa to Brazil to Australia.

While China was moribund under its early Mao-led communist administration, and before post–World War II Japan

FIGURE 9-5

embarked on its headlong rush to become a world economic power, East Asia's requirements remained modest by global standards. But then Japan's economic success, followed by China's swift adoption of market economics, created unimagined and unprecedented needs.

Japan, about the size of Montana but with a population of more than 126 million, showed what lay ahead. With limited domestic resources to support large-scale manufacturing, the Japanese set up global networks through which flowed commodities ranging from oil and natural gas to iron ore and chemicals. Urbanizing and modernizing populations demand ever more consumer goods, and Japanese products poured onto domestic as well as foreign markets. Japanese-owned fleets of freighters now plied the oceans, and for a while Japan even became remote Australia's top-ranked customer for commodities.

When China took off in the 1980s, economic geographers cast a wary eye on the geologic map. Until then, China's biggest resource had been its fertile, river-deposited alluvium (silt): despite the communist regime's best efforts, state-run industries planned to satisfy domestic needs were hardly a match for globalizing Japan. Most Chinese were farmers, and staving off famine was an unrelenting preoccupation. However,

when China opened its doors to the world, and its industries and cities mushroomed, its needs—for oil, natural gas, metals, food, electricity, and water—multiplied. Before long, China had replaced Japan as Australia's leading customer. Moreover, Chinese manufacturers and suppliers were now searching for commodities from Indonesia to Iraq and from Tanzania to Chile. If China holds an advantage, it is in the so-called *rare earth elements* that are not commonly known, such as thulium (used in lasers), praseodymium (aircraft parts), lanthanum (electric automobiles), and promethium (X-ray equipment). Nowadays, China's deposits account for roughly 90 percent of the global production of these minerals, which also are increasingly used in missile technology and "green" energy applications.

But China decidedly needs the world because East Asia's storehouse of other known resources is not particularly favorable (Fig. 9-5). Widely dispersed coal reserves can satisfy the country's expanding coal-fired energy system—but only at the cost of thousands of miners' lives every year. Oil reserves are also widespread, but they tend to be rather modest and diminishing. Deposits in northeastern and far western China are the largest, and exploration is proceeding offshore. Yet nothing in China compares with the abundant iron ore deposits and plentiful ferroalloys available to Russia when it industrialized

Box 9-2 Regional Planning Case

China's South-to-North Water Diversion Project

China's **South-to-North Water Diversion Project [2]** (SNWDP) is the biggest inter-basin water transfer scheme in the world. The project was about two-thirds complete in late 2016, and the cost at that benchmark had surpassed U.S. $80 billion. Its purpose is to deliver approximately 40 billion cubic meters of fresh water annually to northern Chinese cities facing chronic and increasingly severe shortages, exacerbated by surface water pollution, steady depletion of underground supplies, and frequent droughts. (By comparison, New York City consumes about 1.5 billion cubic meters of water each year.)

Using the three diversion routes mapped in Figure 9-5, these aqueducts originate from the Huang He (Yellow) and Chang Jiang (Yangzi) rivers that emanate from the snowpacks of the icy highlands of the Tibetan Plateau. The **Eastern Route**, which opened in 2013, follows the watercourse of China's ancient Grand Canal which dates back to around 500 BC. It has the capacity to supply up to 14.8 billion cubic meters of water per year to the coastal-area provinces of Jiangsu, Anhui, Shandong, Hebei, and the megacity of Tianjin. The **Central Route** was completed in 2014 and can provide up to 9.5 billion cubic meters of water per year from the Danjiangkou Reservoir on the Han River (a major tributary of the Yangzi). Water delivery to residents of Beijing via this route is now well underway (see photo). The far more challenging **Western Route** is just entering its 30-year construction stage. By mid-century, it is expected to begin diverting about 20 billion cubic meters annually from three tributaries of the upper Chang Jiang around Qinghai's Bayankala Mountains and into the Huang He not far below its source.

Although the SNWDP represents an amazing feat of regional planning, calls are rising for more sustainable local water management solutions in northern China. It is feared that the artificial provision of massive water supplies from beyond the North China Plain will obviate the need to address the underlying causes of water shortages, These include widespread pollution, lack of recycling, and inefficient water usage for agricultural, industrial, and urban activities. Moreover, some critics argue that in the longer run the Central Route will not be able to supply all this water because of the growing vulnerability of the crucial Han River to recurring drought.

© Xinhua/Li Bo via Getty Images

An open canal segment of the South-to-North Water Diversion Project's Central Route flowing past Zhengzhou, capital of Henan Province, on its way north to metropolitan Beijing. This smog-choked industrial city of nearly 6 million lies at the midpoint of the SNWDP's 1400-kilometer (865-mi)-long middle route.

---

to meet the Nazi challenge. Similarly, China cannot begin to match the immense gas and oil reserves of contemporary Russia, let alone the gargantuan deposits that surround Southwest Asia's Persian Gulf. Hence, China competes with Japan for energy shipments from Russia and with other sources for its industrial raw material imports. So in just a few short years, China has become both the world's biggest consumer and its leading exporter.

Peoples of the East Asian Realm

Among the many peoples encircling the Chinese are not only the Koreans and the Japanese in the northeast, but also the Mongols and Tatars to the north and northwest; the Kazakhs, Kyrgyz, Tajiks, and Uyghurs to the west; the Tibetans and Nepalese to the southwest; and peoples too numerous to identify individually to the south, including both majorities (Burmans, Thais, Vietnamese) and minorities in states that were still forming in South and Southeast Asia.

China ranks among the world's oldest continuous civilizations, and the Chinese imperial state can be traced back to at least 1766 BC when the first ruling **dynasty [3]** came to power. Most important of all was the formative Han Dynasty (206 BC–AD 220), whose legacy was the molding of the Chinese nation and the consolidation of its territorial base. In fact, to this day the Chinese still define themselves ethnically as the *People of Han* (**Box 9-3**; **Box 9-4**). Like all empires, China expanded and contracted over time. But during its final dynasty, that of the Qing (1644–1911), many of the states and peoples mentioned in the previous paragraph fell under Chinese rule, and the effects of **Sinicization [4]**—also known as **Hanification [5]**—were considerable even if Chinese domination was resented. This expansion, however, would prove to be the emperors' last hurrah: during the dynasty's final decades, European and Japanese imperialists challenged Qing rule, took control over most of the state's core area, ousted the Chinese from much of the periphery, and left the country in chaos, bringing about the end of nearly 3700 years of dynastic rule in 1911.

By that time, the ethnic geography of the East Asian realm had become highly complicated. Even after losing control over peoples from Korea to Vietnam and from Mongolia to Burma (today called Myanmar), the successor Chinese state of the past century still continues to govern numerous minorities. Thus East Asia remains an extremely complex mosaic of ethnicities and languages.

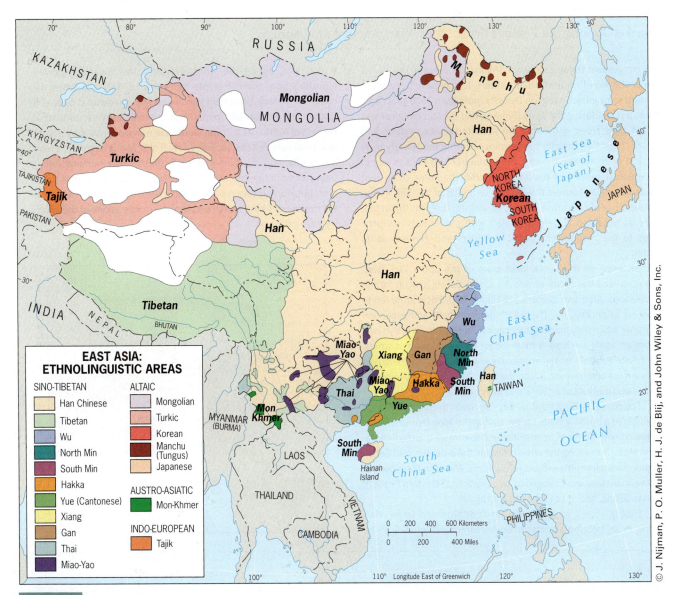

FIGURE 9-6

EAST ASIA:
ETHNOLINGUISTIC AREAS

SINO-TIBETAN
- Han Chinese
- Tibetan
- Wu
- North Min
- South Min
- Hakka
- Yue (Cantonese)
- Xiang
- Gan
- Thai
- Miao-Yao

ALTAIC
- Mongolian
- Turkic
- Korean
- Manchu (Tungus)
- Japanese

AUSTRO-ASIATIC
- Mon-Khmer

INDO-EUROPEAN
- Tajik

© J. Nijman, P. O. Muller, H. J. de Blij, and John Wiley & Sons, Inc.

Box 9-4 From the Field Notes...

© You-tien Hsing

"Xian was the capital city of several imperial dynasties, including those of the prosperous western Han (206 BC–AD 9) and the Tang (AD 618–907). In recent times, Xian has been trying to capitalize on that legacy, asserting its status as the 'ancient Rome' of China and investing heavily in historic preservation. But like any major Chinese city today, Xian is also motivated to modernize and expand. In the winter of 2014, I visited the city's Daming Palace complex, a newly designated World Heritage property. The preserved ruins of the palace complex, built sometime during the Tang Dynasty of the 7th–10th centuries AD, lie in a spacious heritage park covering more than 3 square kilometers (1.2 sq mi). Nonetheless, this open space has now become completely encircled by a ring of ultramodern, fortress-like apartment towers, which conveys the impression of a continuous mountain range surrounding a basin. Perceptive visitors to this heritage park soon come to realize the irony of the nostalgia for the past juxtaposed against this present-day expression of hopes for future development."

Box 9-3   Among the Realm's Great Cities: Xian, Ancient And Modern

The city known today as Xian in Shaanxi Province is the site of one of the world's oldest urban centers. It may have been a settlement during the Shang-Yin Dynasty more than 3000 years ago; it was a town during the Zhou Dynasty, and the Qin emperor was buried here along with 6000 life-sized terracotta soldiers and horses, reflecting the city's importance. During the Han Dynasty (ca. 200 BC to AD 200) the city, then called Chang'an, was one of the greatest centers of the ancient world, the Rome of ancient China. Chang'an formed the eastern terminus of the Silk Route, a storehouse of enormous wealth. Its architecture was unrivaled, from its ornamental defensive wall with elaborately sculpted gates to the magnificent public buildings and gardens at its center.

Situated on the fertile loess plain of the upper Wei River, Chang'an was the focus of ancient China during crucial formative periods. After two centuries of Han rule, political strife led to a period of decline, but the Sui emperors rebuilt and expanded Chang'an when they made it their capital. During the Tang Dynasty, Chang'an again became a magnificent city with three districts: the ornate Palace City; the impressive Imperial City, which housed the national administration; and the busy Outer City containing the homes and markets of artisans and merchants.

Following its Tang heyday, this city again declined, although it remained a bustling trade center. During the Ming Dynasty, it was endowed with some of its architectural landmarks, including the Great Mosque marking the arrival of Islam; the older Big Wild Goose Pagoda dates from the influx of Buddhism. After the Ming period, Chang'an's name was changed to Xian (meaning "Western Peace"), then to Siking, and in 1943 back to Xian again.

Having served as a gateway for Buddhism and Islam, Xian in the 1920s became a center of Soviet communist ideology. The Nationalists, during the struggle against the Japanese, moved industries from the vulnerable east coast to Xian, and when the communists took power they enlarged Xian's industrial base still further. The present city (population: 5.6 million) lies southwest of the famed tombs, its cultural landscape now dominated by a large industrial complex that includes a steel mill, textile factories, chemical plants, and machine-making facilities. Little remains (other than some prominent historic landmarks) of the splendor of times past, but Xian's location on the main railroad line to the vast western frontier of China sustains its long-term role as one of the country's key gateways.

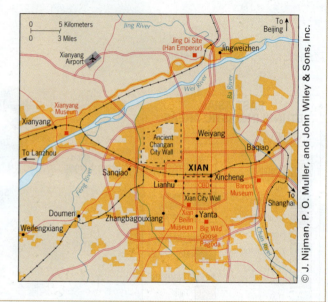

© J. Nijman, P. O. Muller, and John Wiley & Sons, Inc.

Figure 9-6 highlights the cultural diversity of a realm in which China dominates numerically but infusions from elsewhere are evident. We already know of the Mongols, the Muslim Uyghurs, and the Buddhist Tibetans; but the most varied and most numerous minority groups inhabit the southeastern corner of this realm, from Hainan Island northward to the mouth of the Yangzi River. For example, the Yue language (middle green on the map—it used to be called Cantonese) is the common language in the pivotal Pearl River Estuary. Many of the other minority tongues shown here have links to Southeast Asian languages.

East Asia's Post–World War II Development

The 1911 fall of the 267-year-old Qing Dynasty was followed by four decades of revolutionary upheaval. Much of this chaotic period was marked by the ultimately futile effort of the Nationalists to gain control over all of China. They successfully expelled the Europeans by 1927 and in 1934 drove their chief domestic enemy—the fighters of the strengthening communist movement—into the remote western interior. But the Nationalists were unable to resist Japan's expanding imperial thrust into China during the 1930s, and they bore the brunt of the devastating full-scale war that broke out in 1937—soon to be absorbed into the much wider conflict of the Asia-Pacific theater of World War II. Although the Japanese lost that war in 1945, much of China now lay in ruins. The well-armed communists chose that moment to burst out of their far western stronghold and launched widespread attacks on the exhausted Nationalist forces.

Communist China

With the Japanese finally out of the way, civil war resumed with a vengeance between the Nationalists (commanded by Chiang

Kai-shek) and the Red Army (led by the chair of the Communist Party, Mao Zedong). After a four-year losing struggle, Chiang's forces were driven off the mainland, with no other option than to regroup on nearby Taiwan as the newly formed Republic of China. Meanwhile, on October 1, 1949, standing in front of the assembled masses at the Gate of Heavenly Peace in Beijing's Tiananmen Square, Mao proclaimed the birth of the People's Republic of China.

Under communism, Chinese society was completely overhauled. The communist regime, dictatorial and brutal though it was, attacked China's weaknesses on several fronts, mobilizing virtually every able-bodied citizen in the process. Land was confiscated from the wealthy; farms were collectivized; dams and levees were built with the hands of thousands; the threat of hunger for millions receded; health conditions improved; child labor was reduced; literacy was broadened.

At the same time, the regime committed colossal errors and systematically repressed its citizens at an unheard of scale. The so-called **Great Leap Forward** (the propaganda term for what this aimed to accomplish) was a failed attempt at the end of the 1950s to accelerate the growth of industrial and agricultural productivity; it ended as perhaps the worst human-engineered catastrophe in world history, causing between 30 and 45 million deaths, mostly from starvation.

Another calamitous episode of Mao's rule was the so-called **Great Proletarian Cultural Revolution**, launched during his final decade in power (1966–1976). Fearful that Maoist communism was becoming contaminated by Soviet "deviationism" and concerned about his own stature as its revolutionary architect, Mao unleashed a vicious campaign against what he viewed as emerging elitism in society (most likely, it was a strategy of power consolidation in the name of culture). He mobilized young people living in cities and towns into cadres known as Red Guards and commanded them to attack "bourgeois" and "revisionist" elements throughout China, criticize Party officials, and violently root out "opponents" of the system. As a result, thousands of China's leading intellectuals died; moderate leaders and older revolutionaries, including some of Mao's closest allies, were purged and imprisoned; the economy was staggered, especially agriculture, and by the end at least another 30 million had perished. Although the full economic impact of the decade-long Cultural Revolution is still debated, the repetitive political campaigns, purges, and violence created a culture of interpersonal distrust that would take much time to repair. One of those who survived the political upheaval was a Communist Party leader who himself had been purged and thereafter reinstated—Deng Xiaoping. Deng was destined to lead the country into the new era of economic reform following Mao's death in 1976.

Japan's Defeat and Recovery

After Japan's defeat in 1945, the United States forced it to accept a new constitution as well as minor territorial adjustments. The new constitution stipulated that the country could not spend more than 1 percent of its GDP on the military, and it had to accept the permanent stationing of American troops on its soil. Most importantly, this changed postwar condition induced Japan to reset its focus from military might to economic prowess.

East Asia's Economic Transformation

Japan's Pacesetting Achievements

Japan's accelerated economic recovery during the postwar era that led to its rise as a world economic superpower, was one of the greatest success stories of the second half of the twentieth century. Even though Japan had lost a war, only four decades later it had completely reinvented itself as an industrial behemoth, a technological pacesetter, a fully urbanized society, a political power in the new global arena, and one of the most affluent nations on Earth.

By 1980, the U.S. automobile industry had effectively been competed out of the market by the likes of Toyota and Honda. The very same thing happened in the domain of consumer electronics and other high-technology products. Japan did it better and cheaper than anyone else. Now cities everywhere had reliable Japanese cars on their streets; people across the globe adopted Japanese video and portable audio products; laboratories the world over utilized Japanese precision equipment. Japan today continues to be one of the world's richest countries, but its trading dominance—particularly in the sphere of consumer electronics—has decidedly moderated since 2000. Persistent domestic economic problems have contributed to this leveling off, but an even greater challenge was posed by the steady rise of competitors from within the realm.

The Asian Tigers

Japan's sensational development success did not go unnoticed, particularly in the Asia Pacific's smaller-scale, dynamic, upwardly bound economies that were soon being labeled the four **Asian Tigers [6]**: Hong Kong, South Korea, Taiwan, and Singapore. (The first three belong to the East Asian realm and the last to adjoining Southeast Asia.) In the 1960s and 1970s, all four embarked on similar strategies that resulted in rapid industrialization propelled by the attraction of foreign direct investment and the creation of export processing zones for manufacturing **high-value-added goods [7]**, including computers, mobile phones, kitchen appliances, and a plethora of electronic devices. The highly competitive Tigers (nicknamed "The Four New Japans") quickly became formidable exporters to the most affluent markets of North America and Europe, taking full advantage of their large, ultramodern, fortuitously located ports.

Japan with the Tigers in tow shared an important attribute in their export-driven trajectories of industrial upgrading—the strong intervention of the state. National political leaders worked closely with industrial corporations, targeting such strategic sectors as

China is the world's biggest exporter and home to three of the five busiest container ports in the world: Shanghai, Shenzhen, and Hong Kong (shown here). The other two—Singapore and South Korea's Busan—are also located in the Asian Pacific Rim.

automaking, microelectronics, and shipbuilding; nationalized financial institutions were tapped to supply the necessary funding that was soon augmented by the infusion of foreign investment capital. These national governments played an important part in the stimulation of new urban industrial regions that became engines of the national economies—in some ways a model that would be replicated, at a much larger scale, by China (**Fig. 9-7**).

China's Economic Miracle

After Mao's death in 1976, China began a historic metamorphosis that was to have the widest global impact. The essence of China's transformation—at a much grander scale—was comparable to what the Asian Tigers had achieved earlier: the creation of a favorable environment for foreign investment to support the growth

EAST ASIA: ECONOMIC GEOGRAPHY

- China's original Special Economic Zones
- Major urban-industrial region
- Nuclear power plant
- Major highway
- Railroad

National capitals are underlined

0   300   600 Kilometers
0 100 200 300 Miles

FIGURE 9-7

© J. Nijman, P. O. Muller, and John Wiley & Sons, Inc.

Employees on the assembly line at the massive Foxconn plant in Shenzhen. If the United States has long represented "big" to the rest of the world—whether in terms of houses, automobiles, skyscrapers, or the size of soft drinks—China today increasingly stands for "huge." Foxconn, also known as Hon Hai, is a giant electronics contract manufacturer. Think of Foxconn as a huge manufacturing subcontractor for corporations such as Apple, Dell, Hewlett-Packard, and PlayStation. This company was established in 1978 in Taiwan and opened its first gigantic factory on the mainland ten years later in the Longhua Subdistrict of Shenzhen. In 2013, Foxconn was operating 28 plants across the PRC. At Longhua by then, the company employed 240,000 workers on a sprawling "campus" complete with dormitories, stores, banks, and sports facilities as well as a food court that daily dispenses three tons of meat. Worldwide, Foxconn now employs about 1.5 million workers (ten times as many as a decade ago). You are quite likely to own something made by Foxconn, the leading maker of iPhones, iPads, Kindles, myriad video games, and much more.

of a state-of-the-art manufacturing sector geared mostly toward exports. Chinese wages, at least at the outset, were kept low, and technology transfers in the form of training programs were aimed at constantly upgrading the skills of the local workforce. At the same time, political conditions remained stable because in China, more than anywhere else, the (Party-dominated) government maintains very tight control. This was, and remains, a communist state—but one that proved to be extremely adept at understanding how global capitalism operates and how best to put it to use. These days, pragmatism has overwhelmingly become the hallmark of Chinese policies (**Box 9-5**).

Over the past quarter-century, China has emerged as the most dynamic and fastest-growing component of the global economy. To be sure, Japan and the Tigers had also experienced double-digit growth rates, but it is a far different story when the developmental centerpiece is a country of nearly one and a half billion. In 2010, China surpassed Japan to become the world's second-largest national economy.

Geopolitics in East Asia

Sino-Japanese Relations

Understandably, relations between China and Japan have been complex and plagued by problems during the century following the end of Chinese imperial rule. Japan proudly proclaimed to stand for "pan-Asian" ideals, and that is how it legitimized its invasion (1931) and occupation of China (through 1945). But the Japanese committed atrocities during their campaign in China. Millions of Chinese citizens were shot, burned, drowned, subjected to gruesome chemical and biological experiments, and otherwise wantonly victimized.

Box 9-5 From the Field Notes…

"I could not quite believe my eyes when I came upon this Starbucks in the middle of Beijing's Forbidden City: this icon of American consumerism, wrapped in ancient Chinese architecture, in the heart of this revered complex, in a country that calls itself communist. It was back in 2004 that the Chinese authorities had agreed to allow the Starbucks café on the premises in return for a hefty financial contribution to ongoing restoration efforts. But not everybody agreed this was a good idea. Several Chinese groups protested against this 'erosion' of Chinese culture, and by 2007 the government felt compelled to reverse its decision. The deal was rescinded, and Starbucks was evicted from the Forbidden City. It was a splendid illustration of the pragmatism that marks the culture of the Chinese and permeates their economic policies. Call it capitalism or authoritarianism, but the Chinese prefer to do whatever works given the circumstances, and they seem to have little time for inconvenient ideological principles."

Decades later, when China's economic reforms of the 1980s and 1990s led to a renewed Japanese presence in China, the Chinese public and its leaders called for Japan to acknowledge and apologize for these wartime crimes against humanity. The unqualified apology the Chinese desire, in word and deed, has not been forthcoming. Some Japanese history textbooks still avoid acknowledging what happened to the Chinese, and surveys indicate strong public sentiment on this still-sensitive issue.

But now that China has surpassed Japan in terms of economic prowess; now that China has become not only the biggest exporter but also the largest economy in the world; and now that Japan's economic stagnation has entered its third decade—this time it is China that is on its front foot and brimming with confidence. Although the two countries continue to have close economic ties, their diplomatic relations are strained by historical memory, cultural friction, and clashing interests.

One of the new flashpoints to arise in the 2010s involves the Sino-Japanese-Taiwanese dispute over the Senkaku Islands in the East China Sea (**Fig. 9-8**). These tiny uninhabited islands were seized by Japan in 1895, yet are claimed not just by China, which calls them the Diaoyu Islands but also by Taiwan. It is not so much the intrinsic value of the islands that is at stake here: it is far more a matter of national pride and entitlement. Moreover, some oil and gas deposits have recently been discovered in this sector, and ownership of the islands grants rights to their surrounding territorial waters and what lies beneath them. At any rate, China's heightened belligerence over these "rocks" since 2013 has brought Chinese-Japanese relations to their lowest level in decades. It may well be that China has the most compelling claims to these islands because eighteenth- and nineteenth-century maps (emanating from both China and Japan) seem to corroborate that they were in Chinese hands before the Japanese takeover.

The Korea Factor

Throughout history, the Korean Peninsula has repeatedly been divided, partitioned, colonized, and occupied. After Japan—which had annexed Korea as its colony in 1910—was defeated in World War II, the Allied powers divided the peninsula for "administrative" purposes. In 1945, the territory north of the 38th parallel was placed under the control of the forces of the Soviet Union; south of this latitude, the United States was in control. But only five years later, communist armed forces from North Korea invaded the South in a forced-unification drive, unleashing a devastating three-year-long conflict that first swept southward across the peninsula, then back northward across the 38th parallel, and finally drew in the Chinese military who pushed the front southward again. By 1953, a military stalemate halted the hostilities, and the Korean War ended at the cease-fire line not far from where the 38th parallel had marked the original 1945 boundary. Ever since, a heavily armed demilitarized zone (DMZ) has more or less hermetically sealed North from South—with the two Koreas, having grown apart, still in danger of renewed conflict.

But this is not just a bilateral issue. The Korean conflict has long had realmwide and even global implications. A major reason is that North Korea's nuclear capability remains in the hands of a tightly closed, unpredictable, and repressive regime: its nuclear missiles may be aimed at South Korea, but (theoretically at least) they can also reach China, Russia, and Japan—and perhaps even North America. The North Korea issue has also deepened divisiveness within the realm. South Korea and Japan are diametrically opposed to the regime based in Pyongyang, whereas China takes a decidedly more neutral position. The Chinese also appear to be using North Korea in their dealings with Japan and the United States, since it is widely thought that China is crucial to containing the North Korean threat.

Ever since the end of World War II, Japan has adhered to a constitution that essentially forbids its rearmament and commits it to a relationship with the United States that includes the stationing of tens of thousands of American armed forces on Japanese soil. But Japan's 2009 election campaign brought to the fore an unusually forceful reappraisal of these issues, and public opinion has been shifting toward a stronger military posture and the departure of U.S. troops. One result of the international community's failure to constrain North Korea's nuclear aspirations may be Japan's military revival. The country has gradually increased its military posture, and vast resources have been invested in a sophisticated missile defense system to protect against potential North Korean (or Chinese) attacks.

The 2011 demise of the North's leader, Kim Jong-il, and the accession of his son, Kim Jong-un, coincided with one of the many critical food shortages that have afflicted the North Korean people since 1945. With the country desperately in need

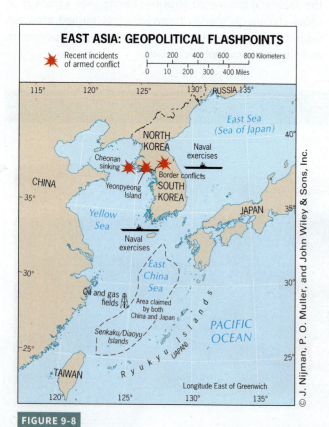

EAST ASIA: GEOPOLITICAL FLASHPOINTS

★ Recent incidents of armed conflict

0 200 400 600 800 Kilometers
0 10 200 300 400 Miles

RUSSIA

East Sea (Sea of Japan)

NORTH KOREA

Naval exercises

Cheonan sinking

Border conflicts

SOUTH KOREA

Yeonpyeong Island

CHINA

Yellow Sea

Naval exercises

JAPAN

East China Sea

Oil and gas fields

Area claimed by both China and Japan

Senkaku/Diaoyu Islands

Ryukyu Islands (JAPAN)

PACIFIC OCEAN

TAIWAN

Longitude East of Greenwich

© J. Nijman, P. O. Muller, and John Wiley & Sons, Inc.

FIGURE 9-8

of outside aid, in early 2012 it appeared that an agreement had been reached between the new regime and the United States: the United States would supply emergency food shipments, and in return, the American government extracted a promise to end the testing of nuclear missiles. However, the deal collapsed just a few months later, following North Korea's launch of a rocket capable of delivering a ballistic missile. North Korean threats of war against South Korea and even the United States have become part of the standard playbook since 2011 and usually peak at the time of the annual joint military exercises of South Korean and U.S. forces. Given the unpredictability of the North Korean leadership, these threats continue to raise alarms throughout the realm and well beyond.

Figure 9-8 displays some of these international geopolitical flashpoints in East Asia. They include Chinese-Japanese conflict in the East China Sea but also military encounters between North and South Korea on- and offshore. The "Cheonan" was a South Korean naval vessel patrolling off the country's west coast when it was torpedoed and sunk by a North Korean submarine in 2010. In the same year, North Korean forces fired artillery shells and rockets at South Korea's Yeonpyeong Island, hitting both military and civilian targets. In the last few years, the North Koreans have employed overland drones sent south across the border for intelligence purposes. Several have been shot down by the South Korean military.

Taiwan: The Other China

Mention the island of Taiwan in mainland China, and you are likely to be greeted with a frown and a headshake. Taiwan, your host may tell you, is a problem foreigners do not understand. Virtually all of the 23.4 million people of Taiwan are Chinese. Taiwan was part of China during the Qing Dynasty. Taiwan was stolen from China by Japanese imperialists in 1895, when it was known as Formosa. Then, when communists and Nationalists were fighting each other for control of mainland China right after World War II, and the communists were about to win, the Nationalists in 1949 fled by plane and boat to Taiwan, where they overpowered the locals. Even as Mao Zedong was proclaiming the birth of the People's Republic of China (PRC) in Beijing, the loser, Chiang Kai-shek, named his regime in Taiwan

the Republic of China (ROC)—and told the world that he headed China's "legitimate" government.

The PRC, of course, ridiculed this assertion, but the ROC had powerful friends, especially the United States. Chiang's regime was soon installed at the United Nations in China's seat. Washington sent massive aid to support the island's economic recovery and weapons to ensure its security. While the PRC languished under communist rule, Taiwan—the name commonly employed for the ROC—advanced economically, and over time its political system matured into a functioning (if turbulent) democracy.

But to the PRC, Taiwan is regarded as a "wayward province" that must be reunited with the motherland. When U.S. President Richard Nixon arrived in Beijing in 1972 for a historic visit, Taiwan was a bargaining chip. Soon, the ROC's United Nations delegation was dismissed, and representatives of the Beijing government were seated in its place. Many countries all over the world that had recognized Taiwan as the legitimate heir to China's leadership now swiftly changed sides. Meanwhile, Beijing's leaders set about trying to isolate the ROC, and to a considerable extent they succeeded.

Geography, however, was to intervene. With billions of U.S. dollars in reserves, fruitful connections to Overseas Chinese in Southeast Asia, and its emergence as an Asian Tiger economic powerhouse, Taiwan had some significant cards to play—and in the late 1970s the Beijing regime discovered it could not afford to deny Taiwanese companies permission to exploit opportunities in the PRC's new development zones. And so, via the "back door" of Hong Kong, Taiwanese entrepreneurs built thousands of factories in mainland China, many of them located directly across the Taiwan Strait. Taiwanese businesspeople now pumped hundreds of millions of dollars into China's development boom, and the economies of Taiwan and the PRC found themselves on a path toward ever tighter integration. In 2015, an estimated 42 percent of Taiwan's exports went to China along with substantial Taiwanese foreign direct investment, and nearly one million Taiwanese were living and working (temporarily) on the mainland.

Today, per capita annual income in Taiwan exceeds U.S. $22,000—more than triple that of China and on a par with South Korea. And even though a sizeable Taiwanese majority opposes reunification with the PRC, few would want to see their economic well-being imperiled by political adventures.

Regions of East Asia

East Asia presents us with a good opportunity to illustrate the changeable nature of regional geography. Our regional classification is based on current circumstances. It is anything but static, and we discuss how this framework may change in the foreseeable future. Today, we can identify seven geographic regions in the East Asian realm (**Fig. 9-9**), the first three of which comprise the People's Republic of China:

1. ***China's Coastal Core***. Anchored by the country's eastern coastal provinces, this is not only China's core area but also East Asia's

most influential region as well as the most dynamic spatial component of the global economy. The development success of this region since 1980 is the biggest miracle of all—the conversion of a repressed economic backwater of more than a billion people into an ultramodern urban and industrial colossus. In this decade, the seaboard component of the Coastal Core has fully consolidated, and its transformative energy is now being redirected westward; the leading edge of this inland penetration of China's core area follows upgraded transportation corridors that lead to the key interior cities of Chengdu, Lanzhou, and Kunming.

FIGURE 9-9

2. ***China's Interior***. This vast crescent of territory centered on Lanzhou, for centuries the gateway to far western China, is the country's fastest growing region today. Propelled by new government policies to reverse the massive and deepening economic divide between the booming Pacific seaboard and the stagnant, poverty-plagued, still-rural Interior, huge investments are pouring into this region to implement a high-priority "Go West" program to counter this uneven development before it triggers protest movements that could threaten the political stability of the PRC.

3. ***China's Western Periphery***. Two large Autonomous Regions, remnants of Han and Qing Dynasty imperialism, constitute China's western tier. One is Xizang—better known as Tibet—whose small Buddhist population is dispersed across the enormous, high-altitude Tibetan Plateau that is framed on its south and west by the world's mightiest mountain range. The other is even larger Xinjiang, whose vast desert basins and encircling mountains comprise the homeland of the Muslim Uyghurs, a primary ethnic component in the human geography of the vital frontier borderland where China meets Islamic Central Asia.

4. ***Mongolia***. The landlocked desert state of Mongolia is vast, sparsely peopled, and poverty-ridden—but in the past decade it experienced a development boom thanks to rich mineral deposits and substantial Chinese investment.

5. ***The Korean Peninsula***. The Korean people form a single nation, but they have been partitioned by ideology since the end of World War II in 1945. Hermetically sealed, China-supported North Korea is a vicious communist dictatorship that relentlessly persecutes its citizens. In the starkest possible contrast, Western-backed South Korea has evolved into a vibrant democracy as well as an Asian Tiger economic powerhouse.

6. ***Japan***. The postwar economic transformation of this longtime rival of China is one of modern history's most successful development stories. But an inability to adapt to changing circumstances over the past quarter-century has all but ended the country's growth trajectory. Nonetheless, Japanese society still ranks among the world's most affluent, and the restoration of Japan's late-twentieth-century regional and global position is not impossible if certain social, economic, and political challenges can be surmounted.

7. **Taiwan**. This thriving Asian Tiger continues to maintain its own regional identity within the East Asian realm, even as the PRC views the island as a "wayward province" that must be reunited with the mainland state.

Before embarking on our survey of East Asia's regions, we first need to set the stage with a general overview of China, as well as a profile of its political-administrative structure, because this structure is critical to the way this huge country is governed.

The People's Republic of China (PRC)

Citizens of the PRC have no doubt: not only does their country still have the largest population of any state in the world ("still" because India is closing in), but theirs is the oldest continuous civilization on Earth. Yet China's territorial size does not match its demography: even with its 1.4 billion people, the PRC is only slightly larger than the conterminous United States (**Fig. 9-10**). Distances in China are comparable to those in the Lower-48: it is about the same distance from Beijing to Shanghai as it is from New York to Chicago. And as the map shows, east-west distances are similar as well. Latitudinally, however, China extends farther north as well as south of the conterminous United States, creating a greater range of natural environments. In its far northeast, China comes very close to Siberian cold. The extreme south exhibits the Caribbean's tropical warmth—with typhoons (as hurricanes are called in Pacific Asia) to match.

Political-Administrative Divisions

China and the United States may be about the same size, but they exhibit very different governance structures: the United States is a federation, whereas China is a highly centralized unitary state that is ruled from the capital Beijing (**Box 9-6**). For administrative purposes, China is divided into the following units (**Fig. 9-11**):

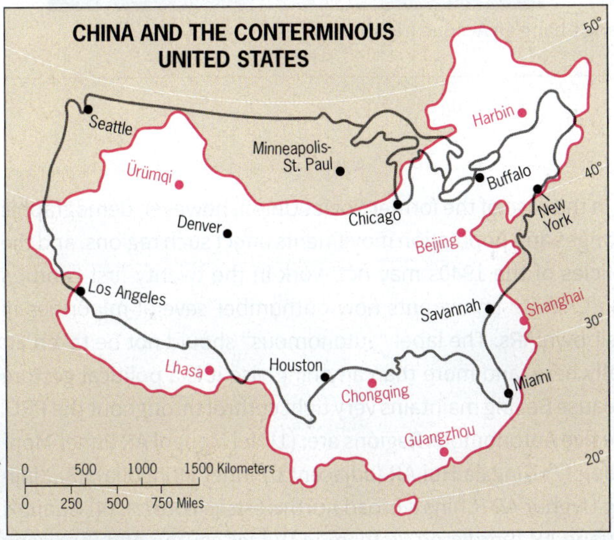

4 Central-government-controlled Municipalities (each is known as a *zhi-xia-shi,* or *shi* for short)

22 Provinces

2 Special Administrative Regions (SARs)

5 Autonomous Regions (ARs)

The four central-government-controlled **Municipalities** (**shi's**) are the capital, Beijing; the nearby port city of Tianjin; China's largest metropolis, Shanghai; and the upper Chang River port of Chongqing in the Sichuan Basin. These megacity *shi's* anchor China's largest internal population clusters (see Fig. 9-4), and direct control over them from Beijing entrenches the PRC government's power.

We should keep in mind that the administrative map of China continues to change—and to pose problems for geographers. The city of Chongqing was made a *shi* in 1996, and its municipal territory was enlarged to incorporate not only the central urban area but an enormous hinterland covering all of eastern Sichuan Province. As a result, the urban population of Chongqing is now officially 30 million, making this the world's "second-metropolis"—but, in truth, the central urban area contains only 7.4 million inhabitants. And because Chongqing's population is officially not part of the province that borders it to the west (Fig. 9-11), the official population of Sichuan dropped by 30 million when the Chongqing *shi* was created.

China's 22 **Provinces**, like U.S. States, tend to be smallest in the east and largest toward the west. The territorially smallest are the three easternmost provinces on China's coastal bulge: Zhejiang, Jiangsu, and Fujian. The two largest are Qinghai, flanked by Xizang (Tibet), and Sichuan, China's Midwest. As with all large countries, some provinces are more important than others. Hebei Province nearly surrounds Beijing. Shaanxi Province is centered on the great ancient city of Xian. In the southeast, momentous economic development is occurring in Guangdong Province, whose urban focus is Guangzhou (now China's third-largest city).

In 1997, the British dependency of Hong Kong was taken over by the PRC and became the country's inaugural **Special Administrative Region (SAR)** (Macau followed in 1999, when it was handed over to China by Portugal). Hong Kong is now supposedly in a transitional administrative phase: China has sovereignty, yet the island retains an 'independent' government and even its own currency (the Hong Kong dollar), and it is exempt from China's notorious "great firewall" that blocks major Western Internet websites. After this transitional stage of 50 years (until 2047), Hong Kong is supposed to be fully integrated into the political and administrative structures of the PRC. That seems a long way off, but even today there is trepidation among the Hong Kong residents about China's long shadow. It is telling that two new infrastructural projects that are under construction to more efficiently connect Hong Kong Island to

CHINA AND THE CONTERMINOUS UNITED STATES

Seattle · Minneapolis-St. Paul · Ürümqi · Denver · Chicago · Los Angeles · Beijing · Harbin · Buffalo · New York · Savannah · Shanghai · Lhasa · Houston · Chongqing · Miami · Guangzhou

0   500   1000   1500 Kilometers
0   250   500   750 Miles

© J. Nijman, P. O. Muller, H. J. de Blij, and John Wiley & Sons, Inc.

FIGURE 9-10

Box 9-6   Among the Realm's Great Cities: Beijing

Beijing (population 20.4 million), capital of China, lies at the northern apex of the roughly triangular North China Plain, just over 160 kilometers (100 mi) from its megacity port, Tianjin, on the Bohai Gulf. Urban sprawl has now reached the hills and mountains that bound the Plain on the north, a defensible natural barrier fortified by the builders of the Great Wall. Today you can drive to the Great Wall from central Beijing in about an hour.

Although settlement on the site of Beijing began thousands of years ago, the city's rise to national prominence began during the Mongol (Yuan) Dynasty, more than seven centuries ago. The Mongols, preferring a capital close to their historic heartland, endowed the city with walls and palaces. Following the Mongols, later rulers at times moved their capital southward, but the government always returned to Beijing (whose name means "northern capital"). From the third Ming emperor onward, Beijing was China's focus; it was ideally situated for the Qing ("Manchus") of the northeast when they took over in 1644. During the twentieth century, China's Nationalists again chose a southern capital (Nanjing), but when the communists prevailed in 1949 they immediately reestablished Beijing as their (and China's) headquarters.

Ruthless destruction of historic monuments, carried out from the time of the Mongols to the communists, has diminished but not totally destroyed Beijing's heritage. With fifteenth-century monuments such as the Forbidden City of the Qing emperors (left photo) and the Temple of Heaven, Beijing remains an outdoor museum and the cultural focus of China. Successive emperors and aristocrats, moreover, bequeathed the city numerous parks and additional recreational spaces that other cities lack. Over the past two decades, Beijing has been transformed

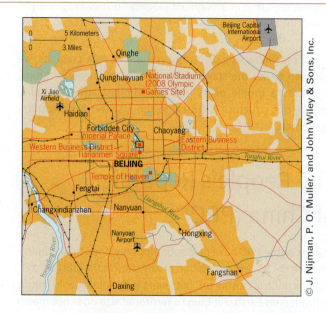

by China's new economic policies. A forest of ultramodern high-rises now towers above the retreating traditional cityscape, avenues have been widened, and expressways have been built. Unmistakably, a new era has dawned in this time-honored capital (right photo).

Beijing, old and new: inside the Forbidden City (left), and traffic moving along Chang'an Avenue (right).

the homeland—a 45-kilometer (28-mi) bridge to Zhuhai and a high-speed rail line to Shenzhen and Guangzhou (see Fig. 9-13)—are causing anxiety in Hong Kong. They are perceived, in part at least, as the PRC coming closer and tightening its grip on the SAR.

In 1999, Portugal similarly transferred Macau to Chinese control, and this former colony, situated opposite Hong Kong on the Pearl River Estuary, became the second SAR under Beijing's administration.

The five **Autonomous Regions (ARs)** were created in order to recognize the non-Han ethnic minorities living there. Some laws that apply to the Han Chinese do not apply to certain minorities.

As in the case of the former Soviet Union, however, demographic changes and population movements affect such regions, and the policies of the 1940s may not work in the twenty-first century. Han Chinese immigrants now outnumber several minorities in their own ARs. The label "autonomous" should not be taken literally here, and more than anything else it is a political gesture because Beijing maintains very tight control throughout the PRC. The five Autonomous Regions are: (1) Nei Mongol AR (Inner Mongolia); (2) Ningxia Hui AR (adjacent to Inner Mongolia); (3) Xinjiang Uyghur AR (China's broad northwestern corner); (4) Guangxi Zhuang AR (bordering Vietnam in the far south); and (5) Xizang AR (Tibet, cornerstone of the southwest).

FIGURE 9-11

China's "Capitalist" Turn

Although China made considerable progress in developing heavy industry (particularly defense industries) during the 27 years of Mao's regime, this advance cannot begin to be compared with the spectacular achievements of Japan and the trailing Asian Tigers. As the post-Maoist era dawned in the late 1970s, it was clear that the old-fashioned socialist production system had to change because now China was falling ever farther behind, especially in terms of technology.

Taking their cues from the accelerated development of Asia's offshore economic powerhouses, the new leadership saw opportunities that could not even be whispered about under Mao's rule. For example, the Lower Chang Basin—focused on the port of Shanghai (China's largest city) and interconnected by the Yangzi and its tributaries—still retained some of its historical identity and energy (**Box 9-7**). Drab, teeming, and decrepit though it had become, Shanghai had not fully lost its intellectual vigor, artistic individuality, risk-taking entrepreneurs, or even its opponents of communist dogma. And, as always, the city's geography continued

to offer immense possibilities. Thus from day one, the new regime in Beijing, led by Deng Xiaoping and his pragmatists, saw in Shanghai what they could not yet foresee in their own backyard. Here was a vibrant hub on the Pacific coast right at the mouth of the realm's greatest river, loaded with talent and accustomed to taking chances. Moreover, Shanghai's vast hinterland contained much more than a farmscape of flood-prone paddies and wheatfields: the Sichuan Basin alone had a population of more than 100 million, growing everything from rice to tea and from fruits to spices.

If China's planners needed any further encouragement, all they had to do was look to southern China's Pearl River Estuary, the main outlet of the Xi River Basin. There, situated at the estuary's mouth, was the burgeoning market economy of Hong Kong, the Asian Tiger still ruled by the British. When Deng took charge, Hong Kong was not only a successful port city importing raw materials by the shipload, but also disgorged manufactured products that sold on markets all around the world. Hong Kong may have been a British crown colony, but it was Chinese managers and Chinese workers who propelled its beehive economy. If you were a visitor to Hong Kong in the 1970s, you would

Box 9-7 Among the Realm's Great Cities: Shanghai

Sail into the mouth of the great Yangzi River, and you see little to prepare you for your encounter with China's largest city (population: 22.7 million). For that, you turn left into the narrow Huangpu River, and for the next several hours you will be spellbound. To starboard lies a fleet of Chinese warships. On the port side you pass oil refineries, factories, and crowded neighborhoods. Next you see an ultramodern container facility, white buildings, and rust-free cranes, built by Singapore. Soon rusty tankers and freighters line both sides of the stream. High-rise tenements tower behind the cluttered, chaotic waterfront where cranes, sheds, boatyards, piles of rusting scrap iron, mounds of coal, and stacks of cargo vie for space. In the river, your boat competes with ferries, barge trains, and cargo ships. Large vessels, anchored in midstream, are being offloaded by dozens of lighters tied up to them in rows. The air is acrid with pollution. The noise—bells, horns, whistles—is deafening.

What strikes you is the vastness and sameness of Shanghai's cityscape, until you pass beneath the first of two gigantic suspension bridges. Suddenly, everything changes. To the left, or east, lies *Pudong*, an ultramodern district with the space-age Oriental Pearl Television Tower rising like a rocket on its launchpad surrounded by a forest of leading-edge glass-and-chrome skyscrapers that make the Huangpu look like Hong Kong's famous harbor. To the right, along the waterfront Bund (Embankment), stand the last remnants of Victorian buildings, monuments to the British colonialists who made Shanghai a treaty port and started the city on its way to greatness. Everywhere, construction cranes rise above the cityscape, and Shanghai now boasts more skyscrapers over 500 meters (1640 ft) than any other city on Earth. The biggest of all opened in 2015—Shanghai Tower, at 824 meters (2703 ft) the world's second-tallest building after Dubai's Burj Khalifa.

The Chinese spent heavily to improve infrastructure in metropolitan Shanghai, including the new Pudong International Airport, connected to the city by the world's first maglev (magnetic levitation) train—the fastest anywhere with a top speed of 430 kph (267 mph).

Pudong itself has become a magnet for foreign investment, attracting as much as 10 percent of the country's annual total. Shanghai's income is rising much faster than China's, and the Yangzi River Delta is becoming a counterweight to the massive Pearl River Estuary development in China's South. Among other things, Shanghai is becoming China's "motor city," complete with a Formula One racetrack seating 200,000 at the center of an automobile complex where all components of the industry, from manufacturing to sales, are being concentrated.

In 2010, Shanghai hosted World Expo on the banks of the Huangpu River (the maglev route was extended to the site), and the city's planners, who have already transformed the place into a quintessential symbol of the New China, took advantage of the opportunity to show the world what has been accomplished here in just a single generation.

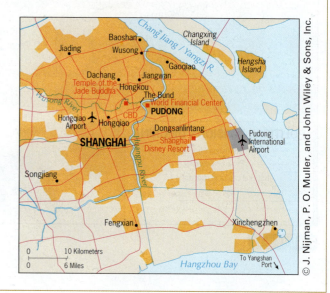

© J. Nijman, P. O. Muller, and John Wiley & Sons, Inc.

---

be taken to a place on a nearby hillside from where you could peer across the fortified border into "Red China"—and observe some villages with duck ponds and rice-producing paddies as well as wooden fishing boats. The contrast was crystal clear.

China's newly formulated policies of the late 1970s, however, did not entail an official departure from communism. This was (and is) still the People's Republic of China: the Communist Party remains firmly in control, the Politburo still calls the shots, Mao continues to be officially revered like a deity, and freedom of speech remains anything but. Yet the Party leadership fundamentally reorganized the way the economy operates in China—beginning with certain vital parts of it. China created highly favorable conditions for foreign investors from around the world with a docile, skilled, hardworking labor force; increasingly efficient facilities; a steadily improving infrastructure; fiscal advantages; and outstanding accessibility vis-à-vis the rest of East Asia, the most rapidly developing regional-scale component in the global economy.

Population Issues

China has long been the world's most populous country, and during the population explosion of the twentieth century it was also one of the fastest growing. Throughout Mao's rule, when China still was a dominantly agricultural society, families were encouraged to produce numerous children, and China grew at a high annual rate of about 3 percent. That was the policy inherited by Deng and his reformers, who immediately grasped that curtailing China's rate of growth was critical to its economic future. Accordingly, the new regime in 1979 embarked on a severe population-control program that imposed on Han Chinese families (but not minorities) a one-child limit. Enforced by sometimes draconian methods, this policy achieved the desired outcome: by the mid-1980s China's annual population growth had been reduced to 1.2 percent, and by 2016 the rate had dipped to below 0.5 percent (less than half the worldwide average of 1.1 percent and two-thirds the U.S. rate of 0.7 percent).

The **One-Child Policy [8]** had its desired economic impact, but its other results were less salutary. China's is a dominantly patriarchal society, where male children are much preferred; rates of female fetus abortion, infanticide, and abandonment skyrocketed, resulting in a gender imbalance that raises alarms for the future. China's government admits that the PRC has the most serious and prolonged **gender imbalance [9]** (sex ratio) on Earth: data for 2015 showed that only 100 girls were born for every 116 boys (105 is the world norm). Demographers also calculated that if the policy was not modified, Chinese society would be short some 30 million brides. That is already increasing the trafficking of women both in China and from neighboring countries, where widespread resentment of Chinese males pursuing (and sometimes abducting) local females is on the rise.

The One-Child Policy had other outcomes as well. An important one was that China's became an aging population as its proportion of youngsters shrank while the older age cohorts mushroomed. That raised the specter of a population implosion (discussed in Chapters 4 and 5): would there be a sufficient number of workers in the younger age groups to support the ever-expanding older population? For China, this concern was magnified by the prospect that, unlike the situation in European countries and Japan, China might grow old before it grew rich, thereby seriously increasing its **dependency ratio [10]**. By the mid-2010s, debate over the unpopular One-Child Policy (**Box 9-8**) spilled out into the open, and the government decided to officially abandon it after 36 years on January 1, 2016. But even though families can add another child, they are required to obtain a permit from the government—or face penalties that include forced abortion or even sterilization.

China's Urban Transformation

As recently as 1980, barely one-fifth of the Chinese people were urban dwellers; today, well over half live in towns and cities. China's urban transformation is on a scale the world has never seen before. The urban population increased from 18 percent in 1978 to 26 percent in 1990 to 31 percent in 2000. Once into the twenty-first century, however, that proportion skyrocketed to 57 percent in 2016. Much of this growth has been planned and controlled by the government, and that alone is an incredible feat. But it has brought profound changes to Chinese society, uprooting tens of millions, infusing awareness even in remote rural areas (where a quarter of China's villages have been abandoned since 2000), creating inequality both within cities and between city and countryside, and spawning an enormous **floating population [11]** consisting of temporary urban dwellers with restricted residency rights.

The so-called ***hukou* system [12]** is based on residency permits that indicate where individuals are from and where they may exercise such rights as education, health care, and housing. The *hukou* tradition dates back to ancient China and is not uncommon in Japan, Vietnam, and other parts of Asia.

During Mao's rule, this residency-permit system became far more rigid and was widely used as a tool to control migration, manipulate labor supplies in the urban-industrial as well as the rural-agricultural sector, and minimize government investment in urban services.

As the market reform movement has intensified over the past three decades, so have the liabilities of the *hukou* system. In Shanghai, for instance, millions of migrants continue to pour into China's largest metropolis, their numbers having tripled in the last ten years alone. They now account for over 10 million of the 23 million people who inhabit this mushrooming urban region. They also account for nearly 70 percent of Shanghai's 20–34 age cohort, the critical life-cycle stage during which many couples start a family. Not surprisingly, in 2013 the metropolis contained an estimated 400,000 children below the age of 6 who were officially categorized as migrants lacking a Shanghai *hukou*. With the total of migrant workers now exceeding 300 million, the *hukou* tradition is becoming an ever greater social problem throughout China. These workers maintain that they are major contributors to the expanding economies of the metropolitan centers in which they are employed, yet they are treated as second-class citizens who receive minimal recognition and are denied access to basic local urban services.

China now counts more than 170 cities containing at least one million people, heavily dominating the metropolitan landscape of this realm, and three times as many as the United States

A crowd of Chinese passengers lined up at an entrance to Beijing Railway Station, one of the capital's two main terminals, to return home for the Spring Festival (Chinese New Year). For many migrants, this official week-long holiday is the only one long enough for such a visit; in recent years, this event has become internationally known as the biggest single movement of people on Earth. In 2013, for many this Festival (centered on the lunar-new-year date of February 10th) lasted as long as 40 days and produced an astounding 3.4 billion trips, an increase of more than 8 percent from the year before. Of that trip total, 3.1 billion were by road, 225 million by rail, and some 35 million by air. In no small way, this army of annual migrants serves to maintain what is left of the ties between China's rapidly developing cities and the widely dispersed rural hinterlands.

Box 9-8 Regional Issue: The One-Child Policy

Population Control is Key to Our Future.

"I was born here in Shanghai in 1963, the youngest in a family of five children. My father and mother worked very hard all their lives, but there was very little they could give us. We lived in a small tenement, sharing sanitation facilities with everyone else on our block. My siblings and I all got a basic education and basic healthcare, but food was scarce and luxury did not exist. Of course, in the countryside things were much worse, and the stories about the terrible famines and grinding poverty there are well known.

"There can be no doubt that, if our government had not enforced family planning, China would never have taken off as a great economic power. And, more importantly, it would have been impossible for the Chinese people to improve their lives as they have done. Why do you think we are so far ahead of India in terms of life expectancy, literacy, income, child mortality rates, and so on? Indeed, within ten years or something like that, they say that India will surpass China as the biggest country in the world in terms of population. My condolences to them!

"Our policy works with incentives meant to discourage couples from having more than one child. Taxes can go up for those having more than one, or their youngest children will not get certain benefits from the state such as a free education. Don't forget, our state provides lots of things to the people, so it seems to me that it is only reasonable that they prevent people from abusing the system.

"Now let's look to the future. I run a factory here on the outskirts of the city and we employ 600 people, but because of rising wages we get more and more into automation, using robots. I know there have been temporary shortages in past years, but most likely we will need less labor in the future. Agriculture is more and more mechanized; our economy moves from labor-intensive manufacturing to high-tech, IT, and services; and China is increasingly producing things abroad. Will we really need all that labor in the future? I don't think so.

"And I have not even said anything about the huge environmental burden of such an enormous population: the air in Beijing and Shanghai is hard to breathe, and there aren't enough parking spaces for all the cars. It is hard to imagine the environmental cost of every Chinese family driving a car, isn't it? We must stick with tight family planning, for the good of China and its prosperity."

The One-Child Policy is Wrongheaded and a Disgrace.

"Rather than getting into these abstract debates about what is good for China's future—and nobody can accurately predict the future

anyway—let me tell you a real-life story that recently circulated on the Internet all over China. And I think you will understand that I would rather remain anonymous when expounding my views. There was a young couple in a small village in the north-central province of Shaanxi. They had one child and the woman was 7 months pregnant with their second. The local authorities found out and slapped them with a huge fine of 40,000 yuan (about U.S. $5,500). If she paid the fine, she could keep the child and obtain *hukou* privileges. Well, the husband earned just 4,000 yuan a month working at a local hydroelectric power station, and so they did not have the money. On May 30, 2012, the husband set out for the coal mines of Inner Mongolia for a month or so to earn extra money. Then the local officials (thugs, better put) go to their home, basically abduct the woman, and force her to sign a consent form to abort the baby. They restrained her and gave her an injection in her belly. The next day she gave birth to a dead 7-month old baby.

"This is the daily reality of the One-Child Policy for lots of people, especially the poor. The reason that more and more Chinese are protesting the policy is that the stories get out on the Internet and people are now daring to raise their voices. It was the family of the woman who posted the story online. You see, this is about basic human rights! How can such practices ever be approved on the basis of abstract economic planning? The so-called incentives are measures of force, that is what they are, and they deprive people of fundamental human choices.

"Moreover, this policy is now completely outdated. First, China's population will begin to shrink by 2026, and the working population is already declining as we speak. We need more, not fewer, people who can provide for the old and the young. When will our government understand that we actually have, as a result of this One-Child Policy, a shortage of working-age people as well as a shortage of women?

"Finally, most people in China are already deciding for themselves that they don't want a second child, let alone a third. The cost of living is too high and living quarters are too small. Today in China, people want to improve their material circumstances and quality of life. If a smaller number of couples, mostly living in rural areas, desire more than one child, let it be so because, overall, our population is not going to keep growing anyway. The One-Child Policy was immoral and unethical from the start. Today it has become a counterproductive approach as well. The sooner it is suspended, the better."

[Authors' note: as indicated in the text, China's One-Child Policy was terminated in 2016.]

(**Box 9-9**). By 2025, there could be dozens more. And rapid urbanization has been closely tied to equally rapid economic growth, as will be explained in the next section (also see **Box 9-10**, **Fig. 9-12**). But the future is uncertain. Will there continue to be enough jobs to employ all these rural-urban migrants? What will be the consequences of China's accelerated program to expand its urban housing stock by millions of new units every year? Can cities that grow at such a frenzied pace be managed in an environmentally

sustainable way? Will the Communist Party be able to maintain control over the new concentrations of urbanites with their changing lifestyles, identities, and expectations? For how much longer can the (officially communist) government justify the widening income inequalities within the major cities, and can it meet the steadily increasing demands of industrial labor? Could these enormous economic and social changes at some point force a political transformation—and where might that lead?

Box 9-9 Major Cities of the Realm, 2016

Urban Area	Population est. (millions)
Tokyo, Japan	37.8
Seoul-Incheon, South Korea	23.6
Shanghai (Shi), China	22.7
Beijing (Shi), China	20.4
Guangzhou, China	18.8
Osaka, Japan	17.0
Shenzhen, China	12.2
Tianjin (Shi), China	11.3
Chengdu, China	10.7
Taipei, Taiwan	8.5
Wuhan, China	7.6
Chongqing (Shi), China	7.4
Hong Kong SAR, China	7.3
Nanjing, China	6.4
Xian, China	6.2
Shenyang, China	6.2
Harbin, China	4.9
Pyongyang, North Korea	2.9

Region China's Coastal Core

Turning now to our survey of the seven East Asian regions mapped in Figure 9-9, we begin with the realm's cornerstone—China's Coastal Core, the crucible of the PRC's post-Mao economic transformation. This revolutionary shift began quietly and was officially launched in 1980 at a few carefully chosen places. By establishing new economic rules that would initially apply only to certain urban areas on the Pacific coast, most of the rest of the country and the majority of the population would remain comparatively unaffected—at first.

Special Economic Zones

The government introduced a three-tier system of **Special Economic Zones (SEZs) [13]**, which would attract technologies and investments from abroad and reshape the economic geography of China's seaboard (see Fig. 9-7). Within these economic zones, investors were offered numerous incentives that included low taxes, eased import-export regulations, and simplified land leases. The hiring of labor under contract was permitted. Products made in the economic zones could be sold on foreign markets and, under some restrictions, in China as

well. Even Taiwanese enterprises could operate here. And profits earned were allowed to be sent back to the investors' home countries.

When the government made the decisions that would reorient China's economic geography, location was a primary consideration. Beijing wanted China to participate in the global market economy, but it also wanted to produce as little impact on interior China as possible—at least in the beginning stages. The obvious answer was to position the Special Economic Zones along the coast. Accordingly, four SEZs were established in 1980, all with particular geographical properties (Fig. 9-7): **Shenzhen** (adjacent to booming, then-British Hong Kong on the Pearl River Estuary in Guangdong Province); **Zhuhai** (next to Macau); Shantou (opposite southern Taiwan); and Xiamen (on the Taiwan Strait). Interestingly, the last three of these areas were former colonial port cities from which many Chinese had fled the country during the Chinese takeover, to settle elsewhere in East and Southeast Asia. Whether or not this was part of the plan, these **Overseas Chinese [14]** were going to play a major part in China's remarkable economic revolution (also see Chapter 10).

Three more SEZs were proclaimed, in 1988, 1990, and 2006, respectively: **Hainan Island**, with a forward location on the doorstep of Southeast Asia; **Pudong**, on the east bank of the Huangpu River directly facing central Shanghai; and **Binhai New Area**, the coastal zone of the northern port city of Tianjin (about 100 kilometers [70 mi] southeast of Beijing). On the outskirts of Tianjin, the Chinese have in the last five years been building what is supposed to become the biggest financial district in the world, *Yujiapu* (see photo). It should be noted that the distinctions between the SEZs and other areas and regions in China

Have you ever seen a construction project of this size? China is studded with these big bets that the country's headlong economic growth will continue apace, but few are more grandiose than the Yujiapu financial district (shown here in the fall of 2012). Already nicknamed "China's new Manhattan," this massive complex aspires to be a world-class, highest-order activity hub. It encompasses more than 50 skyscrapers and is being built within a tight bend of the Hai River on the coastal flats 50 kilometers (30 mi) from central Tianjin. Even though Yujiapu is on track to be completed in 2019, it is not at all certain that it will be able to live up to expectations. In fact, recent reports have warned about substantial overcapacity and suggested that many financial firms may well choose to stay in such established hubs as Hong Kong, Shanghai, or nearby Beijing.

Wei ta - Imaginechina / AP Images

Box 9-10 MAP ANALYSIS

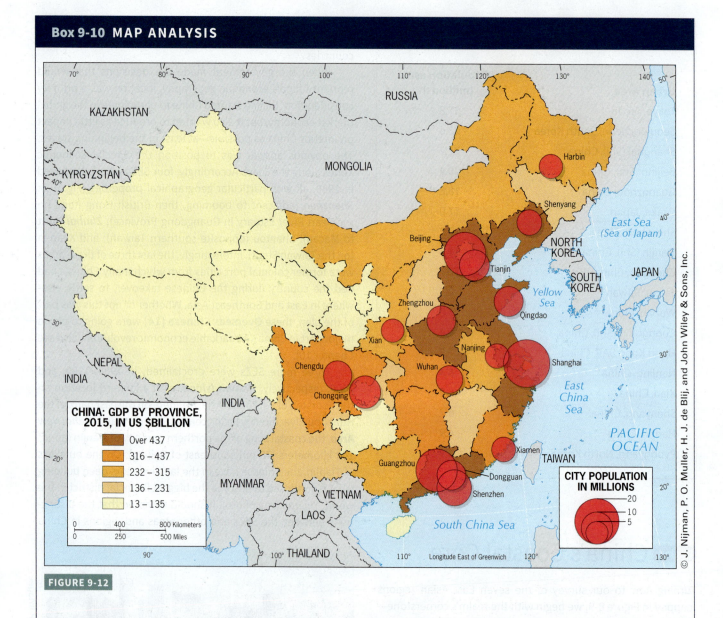

China: GDP by Province, 2015, in US $Billion

	Over 437
	316 – 437
	232 – 315
	136 – 231
	13 – 135

CITY POPULATION IN MILLIONS
20
10
5

© J. Nijman, P. O. Muller, H. J. de Blij, and John Wiley & Sons, Inc.

FIGURE 9-12

Underscoring the prominence of China's Coastal Core region, this map shows the distribution of Chinese cities whose populations exceed 5 million as well as the size of the economy in each province and autonomous region. China's rate of urbanization since 1980 has been so phenomenal that more than 170 cities now contain at least one million people. That is three times as many as in the United States. Simultaneously, the Chinese economy steadily expanded to dominate global trade like no other. Generally speaking, what is the geographic relationship between urban growth and economic development in China? How would you explain that relationship?

Access your Wiley PLUS Learning Space course to interact with a dynamic version of this map and to engage with online map exercises and questions.

have over time decreased and become blurred, because so much of the country has been opened up, to various degrees, to market activity and foreign investment, even though government design and control of the economy has remained intense.

The grand design of China's economic planners, therefore, was to stimulate economic development in the coastal provinces and to capitalize on the exchange opportunities created by location: (1) the availability of funding; (2) the proximity of foreign investors in Southeast Asia, Taiwan, Japan, and—importantly—still-British Hong Kong; (3) the presence of abundant cheap labor; and (4) the promise of world markets eager for low-cost Chinese products.

To date, the poster child of China's economic planners' grand design is Shenzhen (Fig. 9-13). In the 1970s, the name "Shenzhen" was hardly known, but within a decade this small fishing village had become the fastest-growing city in human history—from about 30,000 (mainly fishing folk and farmers) in 1980 to 3 million in the early 1990s to more than 12 million today. Virtually everyone who lives in Shenzhen comes from somewhere else, and because all are really outsiders they

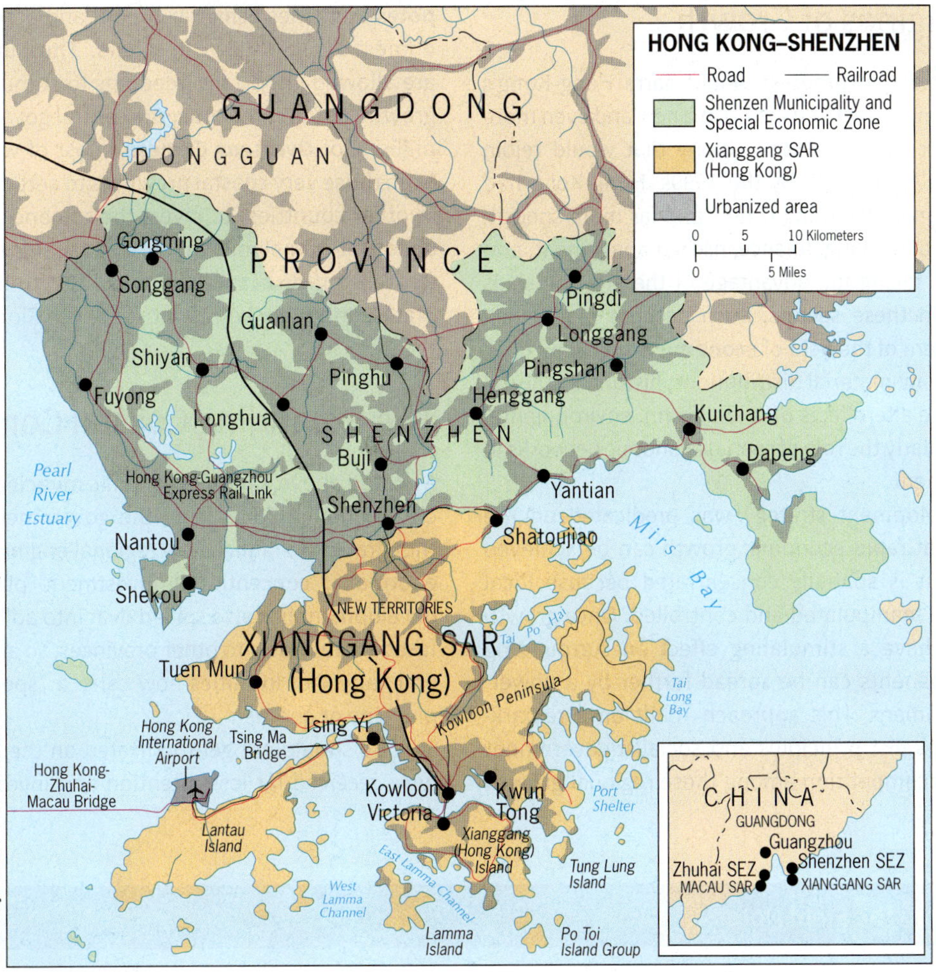

© J. Nijman, P. O. Muller, and John Wiley & Sons, Inc.

FIGURE 9-13

speak to each other in Standard ("Mandarin") Chinese rather than the Cantonese dialect that prevails all around them.

Accompanying this unprecedented growth was the headlong conversion of the surrounding Pearl River Estuary's conurbation into a world-class industrial powerhouse. What caused that to happen? When China opened the Shenzhen SEZ, hundreds of companies moved their factories there from neighboring Hong Kong and offshore Taiwan, capitalizing on lower wages and taxes, looser environmental regulation, and weaker official oversight. At the same time, the Chinese government built business-friendly infrastructure in the form of state-of-the-art seaport facilities, airports, highways, and railroads. Practically overnight, Shenzhen became the icon of Deng's Open China policy, to be emulated (inasmuch as possible) by the other SEZs.

Think about it: Shenzhen, an urban center less than 40 years old, now contains 27 percent more residents than New York City, and Guangdong Province has an industrial workforce larger than that of the entire United States! As for the wider Pearl Estuary conurbation, it has become nothing less than the most productive regional economy of its size in the world. Today, it accounts for just about half of all of China's exports, so if you possess a car, television set, laptop computer, smartphone, and/or any similar device, it is certain to contain components that originated in this massive manufacturing complex. Although Shenzhen looks like an ultramodern city from above (see photo), its population is largely composed of first-generation rural-urban migrants. Moreover, its rapid territorial expansion has swallowed up many small villages, and their former residents now often find themselves in a radically transformed environment (**Box 9-11**).

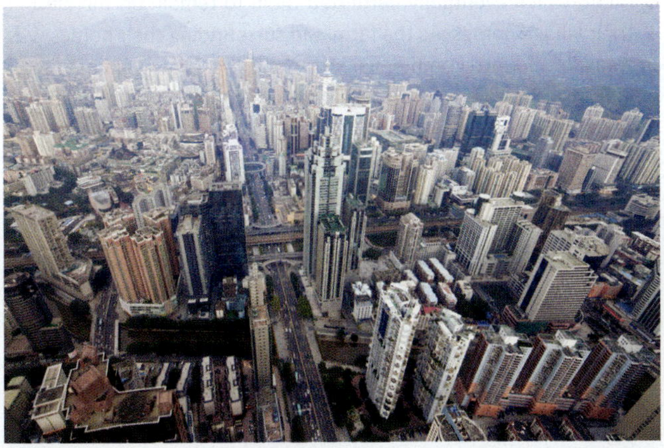

© Nikada / Getty Images, Inc.

A bird's-eye panorama of spectacular Shenzhen. Every structure in this thriving megacity is still less than four decades old.

Regional Engines of Growth

None of this accelerated progress would harm Hong Kong's Asian Tiger economy. In fact, it ushered in a new and even more prosperous era for the still-British colony that would rejoin China as its first SAR in 1997. By the 1990s, Hong Kong had become less of a manufacturing center and far more specialized in international banking, finance, fiscal management, and business services, taking full advantage of the growing skills of its workforce in these sectors. Such occupational transitions are the concern of the field of **economic geography [15]**, which focuses on raw-material distributions, historical-cultural factors (such as the aftereffects of colonialism), environmental issues, and particularly the role of spatial economic networks in regional development.

Chinese development strategy was predicated on the understanding that rapid economic growth can be achieved and managed if it is spatially concentrated because local conditions can be manipulated and controlled. Once growth takes off, it can have a stimulating effect on surrounding places, and the benefits can be spread further by a powerful central government. This approach—built on a remarkable blend of capitalist principles and socialist-government control—includes competition among these regional **growth poles** and cities using tax breaks and other incentives to lure domestic and foreign investors. Thus urban municipalities are allowed important freedoms to stimulate local economic growth. At the same time, the central government encouraged millions of **Overseas Chinese** (most of whom had left China from those very coastal provinces to settle—and often thrive—in other countries) to invest their financial resources in their ancestral homeland, thereby attracting ever more substantial **foreign direct investment [16]** from Japan, the United States, and other key countries of the Global Core.

The Expanding Coastal Core

As noted above, China's economic miracle began with the Special Economic Zones, and the government continues to use this strategy to launch new regional engines of growth through enormous, concentrated investment plans. But economic development has also spilled over into adjoining areas and diffused more widely to other provinces, so an increasing number and variety of localities now carry a "special" designation of one type or another.

We should not get too fixated on the SEZs because some have received far less attention and investment than others.

Box 9-11 From the Field Notes...

"The Pearl River Estuary of Guangdong Province was the first regional growth engine to emerge under the reforms introduced at the outset of China's post-Mao era. As urban expansion swiftly became a leading development priority, huge tracts of farmland were converted into industrial and residential land uses. Although farmers were compensated for these property losses, their forced relocation often sparked serious social conflict, even violence. In the southern city of Guangzhou, located at the apex of the horseshoe-shaped megalopolis lining the Estuary, I encountered something different. Here I met some well-organized villagers who had bargained shrewdly and successfully with the municipal government, which eventually permitted them to remain on the land in their village homes. As the farmland surrounding their homes was transformed into commercial strips and high-rise buildings, their cluster of houses became a "village in the city." Taking advantage of the new premium location of these older homes, the villagers-in-the-city renovated and added several floors to their houses, renting the extra space to migrant workers and vendors. This is one of the largest of these Guangzhou "villages," which claims to have a population density more than six times higher than Manhattan's! But look at the inevitable result of cramming together 5- to 8-story buildings to such an extent: with the compressed streets that separate them ranging from only 0.7 to 3 meters (2.3 to 9.8 ft) in width, so little sunlight can enter (even at midday) that this residential environment has acquired the nickname of 'thread-like sky.'"

© You-tien Hsing

Hainan, for instance, has only experienced minimal increases in manufacturing, mostly in the processing of agricultural products. Furthermore, it does not necessarily require the official status of SEZ to obtain inflows of investment (especially from domestic sources) and to achieve economic growth. Thus the entire tier of seaboard provinces, from Guangdong in the south to Liaoning in the north, now forms the heartland of a highly productive, Pacific Rim-based economy (Fig. 9-9); its relentless growth in recent years has generated the inland expansion of the Coastal Core region as far west as the central Sichuan Basin (**Fig. 9-14**).

Although export industries remain largely clustered along the coast, economic development in this decade is decidedly being steered toward the interior, as shown in the map of GDP growth by province (**Fig. 9-15**). Much of this growth (resulting from the government's "Go West" program referred to earlier) is occurring around such thriving inland manufacturing centers as Wuhan, Chongqing, Zhengzhou, and Xian, all now engulfed by the expanding Core (Fig. 9-9). As for the leading edge of this inland-moving region, by 2017 three extended

lobes had reached the major gateway cities to the western half of China: **Lanzhou** on the Huang River in the north, the jumping-off point for accessing the crucial westernmost ARs of Tibet and Xinjiang; Sichuan's capital, **Chengdu**, in the center; and **Kunming** in the south, a hub for travel into mainland Southeast Asia. This expansion of China's Coastal Core toward the west has posed a challenge in terms of infrastructure and connections between the country's cities and regions. High-speed rail has been hugely important in this regard, and China has swiftly become the preeminent global leader in this technology (**Box 9-12**, **Fig. 9-16**).

We have already noted that the Chinese government is engaged in a race against time to spread the benefits of economic growth from east to west. However, the challenges of mitigating the inequalities produced by this uneven regional development are formidable: an enormous chasm not only persists in the level of prosperity between the high-flying cities and the deprivation-studded rural areas, but also at the national scale between the enlarged Coastal Core and most of the remainder of the country (Fig. 9-14).

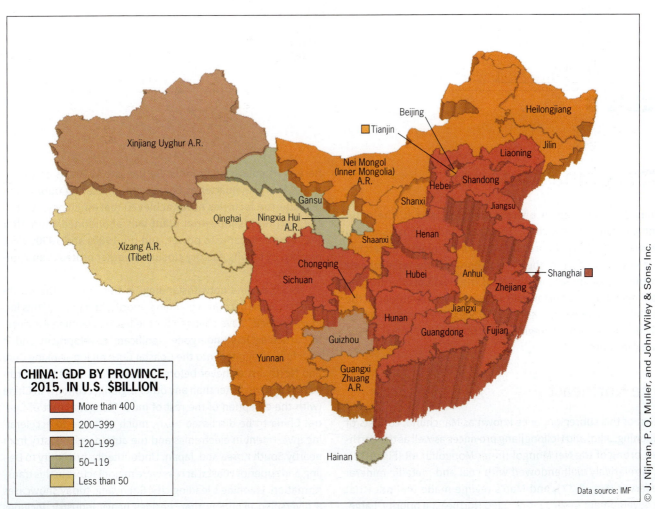

FIGURE 9-14

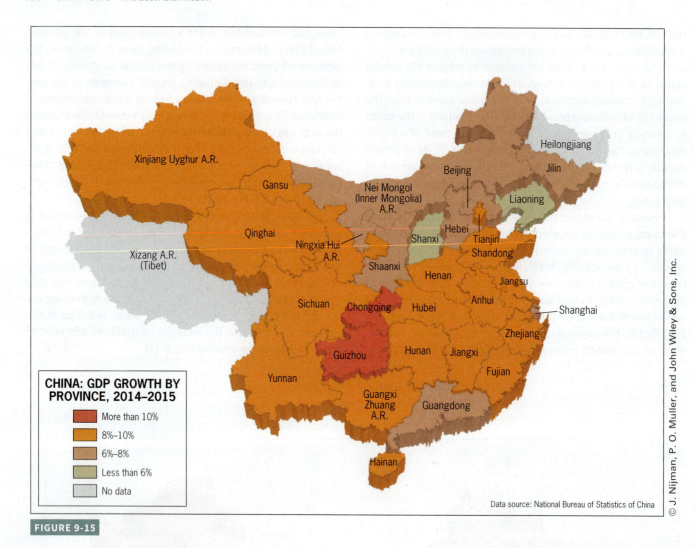

CHINA: GDP GROWTH BY PROVINCE, 2014–2015

- More than 10%
- 8%–10%
- 6%–8%
- Less than 6%
- No data

Data source: National Bureau of Statistics of China

© J. Nijman, P. O. Muller, and John Wiley & Sons, Inc.

FIGURE 9-15

Region China's Interior

China's Interior region offers a stark contrast to the affluent environs of Shenzhen, Shanghai, and Beijing. Situated between the Coastal Core and Western Periphery, it stretches in a vast crescent from the Russian border in the far northeast to the city of Yumen (midpoint of the Hexi Corridor leading to Xinjiang) in the northwest to the Vietnamese border in the farthest reaches of the south (Fig. 9-9). The Interior can be divided into three subregions: the Northeast, Central China, and the South.

The Northeast

Most of this subregion, once known as Manchuria, consists of Liaoning, Jilin, and Heilongjiang provinces as well as the northern prong of the Nei Mongol (Inner Mongolia) AR (Fig. 9-12). It is relatively well endowed with coal and metallic mineral reserves (**Fig. 9-17**), and Mao's regime made the industrial redevelopment of the war-ravaged Northeast a priority. Large, state-owned, heavy industrial enterprises were established here, particularly steel mills. By the late 1950s, it produced more than one-quarter of China's manufacturing output, but its state-run factories never achieved real efficiency. Most of the Northeast's industrial plant was abandoned soon after the reformers came to power, and the 1980s and 1990s were marked by layoffs, factory closures, worker protests, and outmigration.

Such decline notwithstanding, in this decade the Northeast has again become a planning priority, and parts of this former "rustbelt" have changed for the better. Liaoning Province, in particular, has undergone significant development and is now fully integrated into the Coastal Core region—linking Core and Northeast as never before. Indeed, since 2010 the Northeast has grown faster than any other regional economy in China (with the exception of the rest of Inner Mongolia, part of Central China to be discussed next), much of it due to accelerating investment in electronics and the auto parts industry from nearby South Korea and Japan. Undoubtedly, proximity to Beijing and superior coastal access were major factors in this transformation. Liaoning's leading city, Shenyang, today showcases a diversified economy that, besides heavy industry, includes aerospace, electronics, and banking. The Northeast's primary

Box 9-12 Technology & Geography

High-Speed Rail in China

The introduction of railroads to North America and Europe in the nineteenth century sparked a revolution in the circulation of people and goods. Railroads were both critical in the integration of regional economies and propelled the expansion of **urban systems [17]**. In the United States, further improvements in long-distance transportation during the twentieth century saw the railroads supplemented and then surpassed by the rise of new interstate-highway and air-travel networks.

East Asia's economic development over the past half-century evolved during a different historical-technological context, one that offered a new and highly efficient means of overland mobility: **high-speed rail (HSR)**. Today, China, Japan, and South Korea account for well over half of the world's 17,000 kilometers (10,500 mi) of HSR operations. Although there is no global standard at to what exactly constitutes HSR, its trains typically average speeds of 200-300 kilometers per hour (125-185 mph)—at least twice as fast as conventional express trains. Dedicated HSR routes also utilize specially welded rails to minimize track vibrations, and trains are powered electrically via overhead cables. Moreover, aerodynamic designs, engine technologies, and advanced braking systems set them apart from conventional trains. Their dedicated tracks, particularly within built-up areas, are separated from all other rail traffic, including their access to railroad terminals.

Most HSR networks across the East Asian realm are designed for passenger travel between major cities—because speed matters more for moving people than freight. At medium distances of up to about five hours, HSR offers important time-distance advantages over air travel: main stations are located in the city center whereas airports are almost always located in the urban periphery. Furthermore, rail travel takes less check-in time and dispenses with most airport-type security procedures; delays (especially weather-related) are less common; seating is both faster and more comfortable; and ticket prices tend to be cheaper.

For China, HSR is particularly important because it is the best way to connect mushrooming urban regions and provide efficient transport for long-distance commuters, businesspeople, and migrant workers. As Figure 9-16 shows, many of the intercity distances in China's core area are well served by HSR, and its network here constantly expands. The prevalence of HSR in China, however, stands in sharp contrast to its near absence in the United States, whose extensive (but conventional) rail infrastructure is almost exclusively used for freight (see Chapter 1). Americans, of course, are wedded to their cars—a luxury that most Chinese cannot yet afford. But one has to wonder whether HSR could also benefit the U.S., an issue that has been debated for decades. In the meantime, China continues to grow its HSR network at a feverish pace: by 2020, it will connect all 170 cities with a population of more than a million as well as myriad smaller ones located in-between.

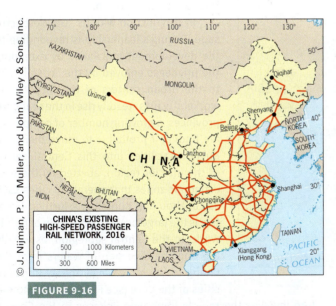

FIGURE 9-16

outlet port—Dalian, located where the Bohai Gulf meets the Yellow Sea—has become a leading export focus for trade with Japan and the Koreas.

Central China

The Central subregion of China's Interior borders Mongolia in the north, Xinjiang and Tibet in the west, and the Chengdu protrusion of the Coastal Core in the south (Fig. 9-9). The very sparsely populated western margins of this subregion are dominated by Tibetan pastoralists. Lanzhou, focal point of the upper basin of the Huang He, is the subregion's leading urban center. This key crossroads city for overland travel to the Western Periphery is also known for its oil refineries and petrochemical industries. Much of the central Interior is arid country that includes parts of the Gobi and Ordos deserts and, farther west, the salt lakes of the Qaidam Basin (Fig. 9-5). But this subregion

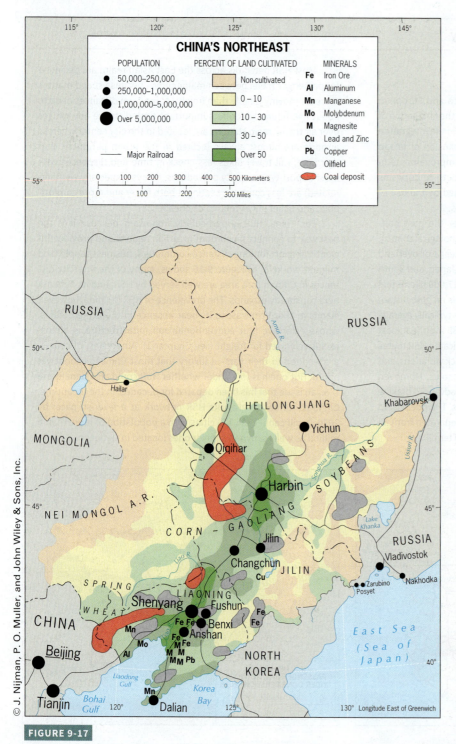

FIGURE 9-17

communities for millennia, and you can readily discern its population cluster in Figure 9-4. The Sichuan Basin, encircled as it is by mountains, is also one of the world's most sharply delineated physiographic regions; the settlement pattern of its roughly 125 million inhabitants reflects that configuration.

Although there are ethnic minorities in the Interior, the overwhelming majority of the people are Han Chinese. Serious regional disparities have long existed between the central interior and coastal China, and the economic gap has widened during the era of market reform. China's Pacific Rim transformation drew tens of millions to the coastal factories and myriad construction sites in what turned out to be the largest internal national migration in human history. Those who stayed behind were too remote from the opportunities introduced by the reform, and now they faced major new problems in the form of water shortages and the negative effects of rapid farmland conversion to nonagricultural uses. Not surprisingly, protests and social conflicts have steadily risen throughout Central China.

In conjunction with its "Go West" crusade, Beijing's leadership is responding to the development imbalance by extending the privileges enjoyed by the coastal SEZs to the urban centers of the Interior. As workers' wages steadily rise in the seaboard provinces, a growing number of companies are finding it profitable to move their production facilities into this Central sector of the Interior region, abetted by the new economic zones that operate in cities such as Lanzhou and Kunming. A further incentive is the large-scale construction of public works in the Central subregion, especially the ongoing diversion of waters from the Chang to the increasingly thirsty northern cities of the Core (see Fig. 9-5). Yet another lure involves the rapidly improving linkages to the Coastal Core in the form of thousands of kilometers of new expressways, a steadily expanding high-speed rail network (Box 9-12; Fig. 9-16), and dozens of new or expanded airports.

The South

The South subregion of the Interior, like its two counterparts, is dominated by Han Chinese, but it also includes sizeable concentrations of minorities, especially near the borders with Myanmar, Laos, and Vietnam (Fig. 9-6). Indeed, adjoining Vietnam

also includes the middle course of the Huang with its agriculturally productive Loess Plateau (**loess** consists of fertile, windblown deposits of glacier-pulverized rock). Add the water of this great river plus an adequate growing season, and a sizeable population can be supported.

To the south, Central China includes the northern half of the Sichuan Basin, another highly fertile area, crossed by China's other great waterway, the mighty Chang Jiang. Despite its vulnerability to devastating earthquakes (most recently in 2013), this huge natural amphitheater has supported human

in the southeast, we find the Guangxi Zhuang AR, which was specially created with these minorities in mind (Fig. 9-11).

The South is largely covered by the Yunnan Plateau, source of the streams that feed the Xi Basin and its Pearl Estuary outlet. Much of southeastern China is quite hilly and in places even mountainous. Among other things, this challenging terrain has inhibited overland contact between China and Southeast Asia—an obstacle soon to be overcome by new routes extending outward from Kunming (see Chapter 10). Yunnan Province itself, in addition to its geopolitical significance, possesses important mineral deposits as well.

Region China's Western Periphery

Historically, empires have expanded and contracted, acquiring and losing territory as well as subjects, and leaving their imprints where they once ruled. Russians in Central Asia, British in East Africa, French in Indochina, Japanese in Taiwan all came and went, leaving behind languages, religions, infrastructures, and traditions.

And so it was with China. The Qing Dynasty had forged a sprawling empire covering eastern Asia but lost most of its periphery after 1850. Nonetheless, Figure 9-9 reveals that China is still an empire today. The PRC's beyond-the-Han domain includes a pair of large Autonomous Regions (Xizang and Xinjiang) as well as a third one lining the Mongolian border (Nei Mongol) that is undergoing rapid integration into Han China itself. Even though they are called autonomous regions, the governance structures of China's far western minority entities are fully integrated into the nationwide Party and government hierarchies. Ethnically, the bureaucracies of these ARs may be a joint force of minority and Han administrators, but almost always under the top leadership of a Han.

The megacity of Chengdu (population: 10.7 million), capital of Sichuan Province, has grown and modernized at a staggering pace since 2000, part of China's unbelievably rapid urbanization. Not long ago, this was part of the Interior; now the city has not only been drawn into China's westward-advancing Core region but also forms its spearhead. Traditionally known as one of the country's most laid-back and liveable cities, Chengdu has quickly acquired a new reputation as a pace-setting economic center that specializes in information technology, finance, and biopharmaceutical industries.

China's urban transformation began nearly 40 years ago in its eastern coastal provinces. Today, just one generation later, it has spread not only to the surrounding provinces of the interior but also to the remote autonomous regions of the far west. This photo, taken in 2015, shows the large-scale construction of apartment buildings on the outskirts of Tibet's capital, Lhasa. The official Chinese presence in Tibet is symbolized by the red flags near the right margin of the photo. The developers and construction companies are all said to be Chinese, and so will the future owners of these apartments. The Han Chinese influx into this urban area continues and within a decade Lhasa's Tibetan population is likely to fall below the 50-percent level.

The Western Periphery has been gaining in importance for China not only because of its natural resources but also its strategic access to Central Asia (especially Kazakhstan; see Chapter 5) and South Asia (especially Pakistan; see Chapter 8).

Xizang (Tibet)

Tibet (called *Xizang* by the Han) is the icebound heartland of Tibetan-Buddhist culture that extends northeastward into Qinghai Province and, most importantly, into a corner of far northeastern India in the Indian State of Arunachal Pradesh. Tibet shook off Chinese domination at the close of the nineteenth century, but in 1950, almost immediately after the communist regime took control in Beijing, the Red Army was ordered to go into Xizang to recapture the territory.

Tibetan society had been organized around the fortress-like monasteries of Buddhist monks who paid allegiance to their supreme leader, the Dalai Lama (see **Box 9-13** and the photo pair at the beginning of this chapter). The Chinese government wanted to modernize this feudal system, but the resistant Tibetans clung to their traditions, and in 1959 the army had to crush an uprising. During the Cultural Revolution, the Chinese destroyed much of Tibet's cultural heritage, looting its religious treasures and works of art.

Under Deng's rule, part of this heritage was returned to Xizang, but Beijing also encouraged Han Chinese to move to Tibet and built "the world's highest railroad" (reaching an altitude

Box 9-13 From the Field Notes...

"It is a cool, crisp, early February morning in 2015 at *Jokhang Temple*, Lhasa. Hundreds of pilgrims arrive here every day, seeking spiritual energy and good *karma* at Tibet's holiest site. Jokhang means 'House of the Buddha,' and the temple's chambers hold some of the most venerated and oldest Buddhist statues in the world. The pilgrims come from all corners of Tibet, and for many it is a long and once-in-a-lifetime journey. I am drawn to some recently arrived pilgrims who are worshipping in front of the temple, next to a colorful collage of prayer flags. Their devotion seems absolute, expressed in the serenity and intensity on their faces, in the endless patterns of standing, kneeling, and prostrating. The entrance of the temple, which was first built in the seventh century, is marked by two dome-structures with poles atop, reaching into the sky. The domes and poles are covered with prayer flags, and it is believed that the wind carries the flags' divine mantras to the world around. Inside the temple, the crowds slowly make their way through the dark labyrinth of incense-rich chambers and narrow passageways, whispering prayers and leaving offerings. To the pilgrims, this is a sacred destination, a place of spiritual fulfillment."

© Jan Nijman

---

of 5072 meters/16,640 ft) across the Tibetan Plateau to the capital of Lhasa (Fig. 9-9). While the Dalai Lama travels the world making the 3-million-plus Tibetans' case and asking not for independence but for genuine autonomy, Hanification continues. A complex set of cultural policies has come to prevail in Tibet: separate school systems, one taught in Chinese Mandarin, the other in Tibetan; dual healthcare systems, one specializing in traditional Tibetan therapeutics, the other in Han (Western) medical practices; and a quota system mandating that a certain percentage of government employees be Tibetan. Nevertheless, driven by the twinned processes of Hanification and modernization, Tibetan culture is existentially threatened. Keep in mind that most Tibetan citizens were born after the PRC seized control, and for them the Chinese presence has become a fact of life.

Tibet today is fiercely controlled by the Chinese government, not only for geopolitical reasons (e.g., the border with India) but also because so many critical parts of China depend on the water provided by the two great eastward-flowing rivers that rise in this soaring highland. The Chinese government refers to the Tibetan Plateau as "China's water tower." Its resources are so abundant that planners are even considering the diversion of some of these waters to serve the expanding populated areas of the Western Periphery's arid northern interior.

Forward economic momentum in Tibet is strong and incomes are climbing. But popular resistance continues, intermittently

accompanied by violence and drama. Since 2009, more than 150 self-immolation suicides have occurred, mainly by Buddhist monks, to protest the Chinese occupation, the marginalization of Tibetan culture, and the callous destruction of the Tibetan environment through mining. Starkly underscoring these grievances is the 2013 landslide disaster at a major copper and gold mining complex near Lhasa that claimed at least 80 lives.

Xinjiang

Xinjiang constitutes the westernmost margin of China's modern empire and is even larger than Xizang (**Fig. 9-18**). With just over 23 million inhabitants, roughly 40 percent of them Han Chinese, the Xinjiang-Uyghur AR is even more important than its Buddhist, Tibetan neighbor. Here China meets the peoples of Central Asia and the faith of Islam; here China has highly significant energy reserves; and here, in the remote, cloudless, desert-dominated far west, China built its original space program. Moreover, across the border to the west lies Kazakhstan, itself endowed with abundant natural resources and providing transit to huge oil and gas reserves in the Caspian Sea Basin. And Kazakhstan is also becoming a key segment of China's **New Silk Road [18]**, connecting western China through Central Asia to Turkey and Europe (see Fig. 5-19).

The modern, tightly controlled, Han-dominated, urbanized component of Xinjiang is located in the northern Junggar

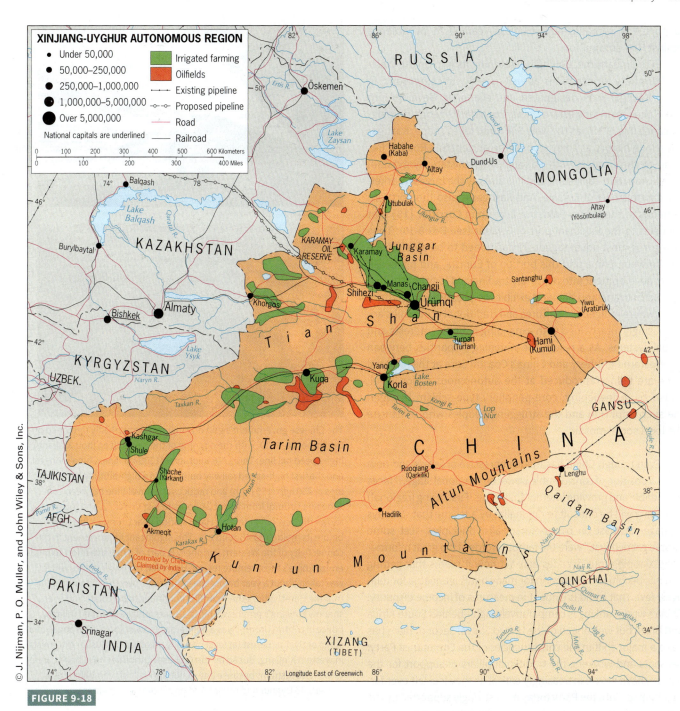

FIGURE 9-18

Basin, focusing on the revitalized capital of Ürümqi and the nearby model city of Shihezi. The traditional, predominantly Islamic, rural component is anchored by the historic town of Kashgar in the extreme western corner of the southern Tarim Basin (Fig. 9-18).

The Xinjiang-Uyghur AR is no more autonomous than Tibet. During the Qing Dynasty, the Muslim peoples here—including Uyghurs, Kazakhs, Tajiks, Kyrgyz, and others—fell under Chinese imperial control and continued their numerical dominance for a half-century after the communist regime came to power in Beijing. But in the twenty-first century, Hanification has shifted into high gear: today around 40 percent of the population are

Han, and Xinjiang has become a two-tiered society. Led by the Uyghurs, who still constitute 46 percent of the AR's population, the peoples of the lower, Islamic tier disdain "Xinjiang," preferring the label "East Turkestan" to signify their inter-realm connection to the Turkic nations of neighboring Central Asia.

When the Uyghurs confront the Han, as often occurs in the factories of Ürümqi and the streets of Kashgar, the results can be tragic. In 2009, a skirmish between Uyghur workers and Han managers at a plant outside Ürümqi spun out of control and led to rioting in which some 200 Han Chinese citizens were killed and hundreds more wounded. Despite the growing risk of central-government retaliation, Uyghur militants continue to pursue a sporadic—and fruitless—campaign of violence against Han domination.

In the meantime, Beijing is tightening its grip on the Western Periphery for economic as well as geopolitical reasons. In 2010, Kashgar was designated an "economic development zone," the first in far western China. Consider its geographic significance (Fig. 9-18): as the **New Silk Road** takes shape, Kashgar has a long history as a key stop on the ancient Silk Road; it lies close to the borders with Pakistan, Afghanistan, Kyrgyzstan, Uzbekistan, and Tajikistan; and it is already the largest marketplace in this part of Eurasia. And Kashgar is also the northern terminus of the **China-Pakistan Economic Corridor [19]** that connects to the Pakistani port of Gwadar (see Box 8-10). Thus Kashgar is most advantageously positioned to be an effective link between China and this portion of Eurasia, and plans call for its development into a primary trading and logistical hub.

Both Xizang and Xinjiang exhibit the properties of peripheries, where indigenous cultures and traditional economies are increasingly intertwined with expanding national interests and global systems. As a result, the widening disparities between the Western Periphery and China's Coastal Core at the national scale are being reproduced at the autonomous-region level, fueled by a deepening, core-periphery-type schism between the successful Han and the struggling peoples who constitute the local social mosaic.

Even though China has made impressive progress and now holds a position of considerable power on the global stage, it faces formidable and growing internal challenges. First, rising (regional) inequality will be difficult to rationalize by a regime that still pays homage—at least on paper—to socialist principles. "Let some people get rich first," were the famous words of the great reformer Deng Xiaoping four decades ago. Yet now it seems that getting rich has become an obsession for many but is attainable for relatively few. That trend is evident in measures of income disparity: China's world ranking is comparatively high—behind South Africa and Brazil but well ahead of the United States. Second, there has been a major shift in the constituencies of the Communist Party: peasants and workers, the traditional pillars of support for the regime, have grown increasingly critical as they feel left ever farther behind. With the Party today most strongly supported by the urban middle class and the *nouveau riche*, how much longer can it maintain the ideological foundation of its policies? And third, the proliferation of the Internet and social media is providing new channels of information flow and opinion expression. Although most online forums are nonpolitical and the "Great Firewalls" were imposed to limit and police Internet communication (you cannot access Google or Facebook in mainland China), this powerful information-transmission technology is bound to become a new platform of political contestation in China.

Region Mongolia

We now turn to the four East Asian regions beyond China and begin by looking to the north, to Mongolia, where the effects of

The Chinese government's grip on Xinjiang was markedly tightened on December 26, 2014, with the opening of the Lanxin (Lanzhou-Xinjiang) high-speed rail line. This 1776-kilometer (1100-mi) line, China's longest and the first high-speed railway built across a high-altitude zone, operates at 250 kph (155 mph) and reduces the travel time between Lanzhou and Ürümqi from about 24 hours to just under 12. The top photo shows the new bullet train at Hami station near Ürümqi a few days before its maiden eastbound trip. This significant improvement in the connectivity of the Western Periphery is part of a massive effort to modernize China's most remote region. However, the lower photo, which was taken about a year before the inaugural Lanxin run, is a reminder that the Han-Uyghur ethnic conflict continues to percolate and overshadow Xinjiang's progress. This 2013 assemblage of armed military police in central Ürümqi was a deliberate show of force in response to a surge of Uyghur unrest (at least partially inflamed by Muslim extremists) in western Xinjiang. But so far the Chinese authorities have been careful to exercise restraint because any major anti-Islamic campaign could exacerbate Uyghur and other budding separatist movements.

the realm's economic transformation have only lately arrived. This immense, landlocked, isolated country wedged between China and Russia, with an area larger than Alaska but a sparse population of just 3 million, suggests a steppe-and-desert-dominated vacuum between two of the world's most powerful countries (Fig. 9-9).

Mongolia used to be the domain of a powerful people who, many centuries ago, swept westward to challenge the Russians and southeastward to rule China. But in more modern times, Mongolia became a weak and vulnerable country whose 800,000-plus herders and their millions of sheep today follow nomadic tracks along the fenceless fringes of the immense Gobi Desert, where Siberian cold periodically causes severe human and livestock losses. During the Soviet era, the location of the capital of Ulaanbaatar symbolized the country's security against Chinese encroachment, and Mongolia functioned as the typical **buffer state [20]** that separated two political adversaries.

In the 2010s, Chinese involvement and investment have been soaring here, and much of this revolves around Mongolia's enormous storehouse of raw materials. Major deposits of gold, copper, and coal have been found, and exploration by large foreign companies ensued. Mongolia was the fastest-growing national economy in the world in 2011 (nearly 18 percent), but it has since declined due to, allegedly, deficient governmental management of the "new economy." The country has become overly dependent on a small number of gigantic foreign mining companies, and negotiations concerning taxes and regulations have proven to be a cause of economic volatility. The main consumer of Mongolia's commodities is China, and Chinese investments continue to pour in, largely for infrastructural projects. It is not clear whether ordinary Mongolians, among the poorest people in Asia, will benefit from this bonanza anytime soon; that uncertainty has not restrained some economists, who speculate that Mongolia is evolving into another Qatar—where a small population grew rich from its enormous natural endowments.

Region The Korean Peninsula

Take another look at Figure 9-9 and you will see one prominent (bicolored) peninsula that seems to reach out from the East Asian mainland toward the islands of Japan. As Koreans will tell you, this isn't just another peninsula. This is a place where human geography took momentous turns as a corridor of migration, an incubator of culture, a cauldron of warfare, and a cradle of economic miracles. Today, this Idaho-sized tongue of land has a population of 76 million (Idaho: 1.7 million) that is divided between two states steeped in the same linguistic and cultural traditions.

The Korean Peninsula has been a most turbulent stage. Thousands of years ago, even before there was a China, lower sea levels made the crossing from Korea to Japan possible and migrants from the mainland began to challenge the even earlier inhabitants of the islands. Later, long-stable states, always influenced by rising China, forged a durable cultural geography marked by distinctly Korean

attributes including ethnicity, language, and customs. Then, more than a century ago, disaster struck as Japan embarked on a campaign of colonial conquest that eventually overran not only Korea, but also large parts of neighboring China. Subsequently, in the aftermath of Japan's World War II defeat, the Korean Peninsula became the scene of a terrible great-power conflict between China-supported communist armies and American-led anticommunist forces. For more than three years (1950–1953) this devastating conflict swept back and forth across the peninsula, claiming more than 3 million lives and ending in an armistice whose cease-fire line can be seen in **Figure 9-19**. Ever

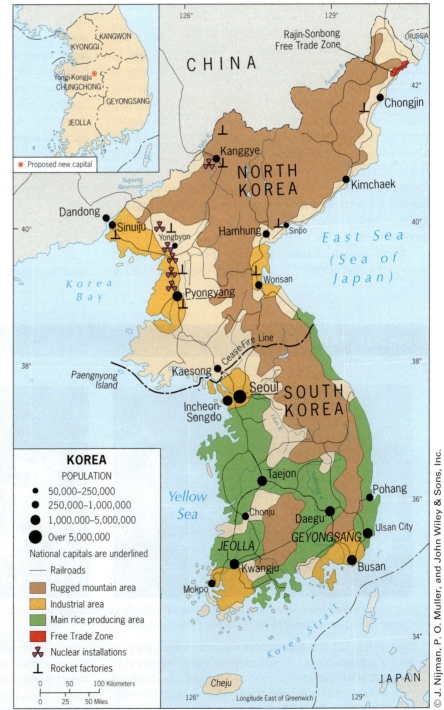

FIGURE 9-19

since, that line has separated—almost hermetically sealed—a vicious communist dictatorship in North Korea from a South Korea that originated as a brutal capitalist dictatorship but then transformed itself into both a free-wheeling democracy and an Asian Tiger. The partition of this region remains at the heart of a politico-geographical tragedy.

North Korea

As the map shows, North Korea is territorially larger than South Korea, but its population of 25 million is half the size of South Korea's. Seven decades of the most barbaric variety of communist rule have turned North Korea into one of the poorest, hungriest, and most regimented countries on Earth (**Box 9-14**). But it also has one of the largest standing armies as well as advanced nuclear weapons and missile technologies. A regime that incarcerates its citizens for the slightest offenses operates a gulag of prison camps reputedly worse than that of the Stalinist-era Soviets, starves its people as punishment, and isolates its subjects from the rest of the world—yet invests heavily in weaponry with which to blackmail its neighbors and infect the world. Refugees' and escapees' stories tell of the most extreme forms of poverty and misery. But China, North Korea's ideological ally, supports the regime based in the capital of Pyongyang in the interest of regional "stability" rather than pressuring it to ameliorate its policies.

North Korean dictators rule with powers unimagined even by the likes of Fidel Castro, Saddam Hussein, and Bashar al-Assad, and the North Koreans have become a society of informers—on each other. Any sign of disaffection risks imprisonment or worse, and mass displays of regimented solidarity reminiscent of Soviet public travesties attend dynastic transfers of power. So it was when the ostensibly "affable" Kim Jong-un, son of the previous dictator Kim Jong-il, took control of the state in 2011. Shortly thereafter he threatened to resume nuclear testing and followed through, he and his ruling elite kept comfortably in power by China's provision of food and energy.

South Korea

The tragedy of the Korean Peninsula is especially painful when put in geographic perspective. As environmental maps reveal, North Korea and South Korea actually need each other. The North has raw materials demanded by South Korean industries. The South produces food the North could urgently use. This condition is known as **regional complementarity [22]**, but in Korea it is precluded by political segregation. This makes the narrow demilitarized zone (DMZ) that separates the two Koreas an ultimate expression of core-periphery demarcation because even as the North stagnated, the South became one of the legendary Asian Tigers.

As noted above, South Korea (50.5 million) emerged from the Korean War in 1953 as an unstable, dictatorial, politically and economically corrupt nation where, as in the North, dissent could cost you your life. But South Korea's rulers also encouraged industrial development, driven by powerful industrial con-

---

Box 9-14 MAP ANALYSIS

Nightlights in East Asia's Pacific Rim, 1992 (left) and 2013 (right). *Source*: Image and Data processing by NOAA's National Geophysical Data Center. DMSP data collected by the US Air Force Weather Agency.

Given widespread, intense electrification in modern times, the presence of nighttime lighting across a country is a good indicator of its economic growth, urbanization, and development. The absence of such artificial light—as in North Korea—reflects a lack of development. **Nightlight maps [21]** have become useful for displaying these patterns, which are spectacularly evident in East Asia. Compare the two images above. Which country shows the most impressive increase in electrification? Which country shows virtually no change at all?

Access your Wiley PLUS Learning Space course to interact with a dynamic version of this map and to engage with online map exercises and questions.

glomerates. This kind of **state capitalism [23]** triggered rapid growth, and the politicians knew that workers would demand better treatment and that the economic and political landscape would have to change. And so it did: dictatorial rule ended, representative government took hold, corruption was confronted, and the economy, once described by early postwar observers as undevelopable, took off like a rocket. South Korea in short order became a boisterous democracy, an economic giant, a nation in universal pursuit of education, the world's leading shipbuilder, a manufacturer of iron and steel as well as chemicals and electronics plus countless other industrial products in addition to producing ample food supplies. South Korean social institutions became ever stronger, and the media were now as free as any on Earth. Freedom of religion prevails, Korean art and culture find a global market, and Korean athletes excel in the Olympic Games and other international sports events. As a key member of the international community, South Korea participates in peacekeeping and security operations worldwide.

The map of South Korea (Fig. 9-19) reveals several of these achievements, most notably the massive urban agglomeration formed by Seoul (**Box 9-15**) and neighboring Incheon, which today constitutes East Asia's second-largest metropolitan complex, and contains almost half of the country's population. Even before democracy took hold, South Korea's rulers decided to further enhance the country's infrastructure by introducing a high-speed rail network to connect Seoul to the traditionally separated subregions of Geyongsang and Jeolla. For a long time, workers in parts of the country not favored

by politicians and manufacturers had trouble finding employment in fast-growing industrial areas, an economic-geographic pattern that has still not completely been erased. As regional integration progressed, the southeastern subregion of Geyongsang, centered on the city of Busan, was always ahead of the southwest, where Jeolla was historically marginalized by the powers in Seoul. Today Geyongsang remains in the lead: here 300,000-ton tankers are built along with smaller freighters and warships; here are the famous Hyundai and Kia automobile plants; and here is the home base of POSCO, headquartered in Pohang, one of the world's largest steel-making corporations.

Despite all this success, and despite the fact that South Korea has become a popular culture trendsetter in East Asia (**Box 9-16**), surveys indicate that the South Koreans are not especially happy as a society. Undoubtedly, a major factor is the stress level associated with the ever-present danger of an attack from bellicose North Korea. But there is another dimension that goes beyond the North-South contrasts. In states that achieve great economic success, women tend to participate in the action even if they are not paid as well as men. Yet South Korea's triumphant story is often recounted as being dominated by males, despite the remarkable contribution of female workers to the economic miracle, especially their labor activism, which resulted in much improved working conditions throughout the country. Nonetheless, the proportion of women gainfully employed in South Korea today is still not representative of a developed economy, and the number of women holding

---

Box 9-15 Among the Realm's Great Cities: Seoul-Incheon

Metropolitan Seoul (population: 23.6 million), located on the Han River, is ideally situated to be the capital of all of Korea, North and South (its name means "capital" in the Korean language). Indeed, it served as such from the late fourteenth century until the early twentieth, but events in that century changed its role. Today the city lies in the northwest corner of South Korea, for which it serves as its capital. Not far to the north lies the tense DMZ (demilitarized zone) that contains the cease-fire line with North Korea. That line cuts across the mouth of the Han River, thereby depriving Seoul of its inland waterborne traffic. Its bustling ocean port, Incheon, emerged as a result, and has now fully coalesced with Seoul.

Seoul's undisciplined growth, attended by a series of major accidents, including the failure of a key bridge over the Han and the collapse of a six-story department store, reflects the unbridled expansion of the South Korean economy as a whole, as well as the political struggles that carried the country from autocracy to democracy. Central Seoul lies in a basin surrounded by hills to an elevation of about 300 meters (1000 ft), and this burgeoning megacity has sprawled outward in all directions, even toward the DMZ. An urban plan designed in the 1960s was subsequently overwhelmed by the steady inflow of immigrants.

During the era of Japanese colonial control, Seoul's surface links to other parts of the Korean Peninsula were improved, and this infrastructure played a role in the city's postwar success. Seoul

is not only the capital but also the leading industrial center of South Korea, exporting sizeable quantities of textiles, clothing, footwear, and electronic goods. South Korea continues to thrive as an Asian Tiger, and Seoul-Incheon is its heart.

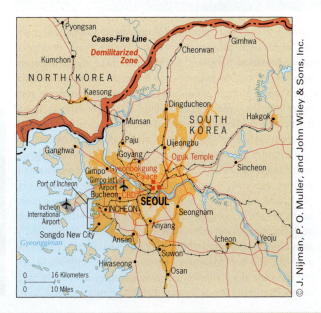

© J. Nijman, P. O. Muller, and John Wiley & Sons, Inc.

S.J. Kim / Getty Images

Songdo New City, South Korea's answer to China's Pacific Rim hub cities. Songdo is the centerpiece of the ultramodern urban complex springing up around the country's second-largest port of Incheon at the southwestern corner of the Seoul-centered conurbation. Modeled after Singapore, Songdo is not only designed to be "the world's gateway to Northeast Asia" (more than 25 percent of the world's population lives within four hours' flying time) but also to be a prototype "Smart City." The gateway function is based on the global accessibility enabled by adjacent, state-of-the-art Incheon International Airport—making Songdo an international magnet or "aerotropolis" serving globalized businesses as well as offering a Las Vegas-style "pleasure carnival" featuring mega-casino/resorts, theme parks, and a Jack Nicklaus-built golf course. At the same time, still-building Songdo's smart-city technology is regarded as a cutting-edge example of sustainable urbanization, one that so highly impressed the UN's Green Climate Fund that it located its headquarters here in 2012.

senior positions remains disturbingly low. Hopes that this serious gender gap may finally be addressed have risen since 2013 when Park Geun-hye, the daughter of a postwar South Korean dictator, took office as the country's first female president.

Region Japan

Japan consists of four main islands—Honshu, the largest; Hokkaido to the north; and Shikoku and Kyushu to the south—in addition to numerous small islands and islets (**Fig. 9-20**). Most of the country is mountainous and steep-sloped, geologically young, earthquake-prone, and studded with volcanoes. A mere 18 percent of the country is considered to be habitable. Japan's high-relief topography has compressed its economic development. Except for the ancient capital of Kyoto, all of Japan's major cities are perched along the coast, and virtually all lie partly on artificial land reclaimed from the sea.

Sailing into Kobe harbor near Osaka (the country's second-largest city), one passes artificial islands designed for high-volume shipping and connected to the mainland by automatic, space-age trains. Enter Tokyo Bay, and the refineries and factories to the east and west stand on huge expanses of landfill that have pushed the bay's shoreline outward.

With just over 126 million people, seven-eighths of whom reside in towns and cities, Japan uses its habitable living space very intensively—and expands it wherever possible. As Figure 9-20 indicates, farmland in Japan is both limited and regionally fragmented. Urban sprawl has invaded much of the cultivable land. In the hinterland of Tokyo, around Osaka, and surrounding Nagoya, major farming zones are under relentless urban pressure. All three of these lowlands lie within Japan's fragmented but well-defined core area (delimited by the red line on the map), the heart of the country's prodigious manufacturing complex.

Coastal Development

Figure 9-20 also tells us a great deal about the nature of Japan's external orientation and its dependence on foreign trade. All

---

Box 9-16 From the Field Notes...

"When I visited Seoul recently, I encountered a large concentration of cosmetic surgery clinics in the trendy, newly developed neighborhood of Gangnam. Significantly, these clinics represent an emerging cultural phenomenon in rapidly changing East Asia. South Korea, formerly an uncool country known mostly for such heavy industrial exports as cargo ships and motor vehicles, has of late become the hottest arbiter and trendsetter of fashion and pop culture in East Asia. Following the Asian financial crisis of 1998, the South Korean government's effort to diversify the economy has scored major successes in both the IT sector and the entertainment industry, especially developing content for TV dramas, films, pop music, and video games. As this new Korean cultural wave rolled across the realm, it triggered the rise of a Korean esthetic. Celebrity K-pop singers and actors have inspired millions of their fans to wear Korean-designed clothing, to use Korean brands of cosmetic products, and even to undergo Korean cosmetic surgeries to look just like their idols! This form of plastic-surgery-dominated medical tourism now attracts at least 200,000 visitors a year to South

Korea, with the largest contingent originating in China. Thus the South Korean development model has expanded from the export of goods to that of tastes. Once again, the state was an indispensable driver in this process."

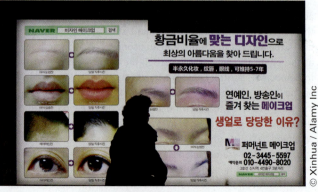

© Xinhua / Alamy Inc

FIGURE 9-20

of the country's primary and secondary regions are found on the coast. Dominant among them is the **Kanto Plain**, the heart of Japan's core area focused on metropolitan Tokyo, which contains about 30 percent of the country's population (see **Box 9-17** and satellite photo). Among its advantages are an unusually extensive area of low relief, a fine natural harbor at Yokohama, a relatively mild and moist climate, and centrality with respect to the country as a whole. (Its principal disadvantage lies in its acute vulnerability to major earthquakes.)

The second-ranking primary economic region within Japan is named the **Kansai District** (Fig. 9-20); it contains the Osaka-Kobe-Kyoto triangle and is located at the eastern end of the Seto Inland Sea between Honshu and Shikoku. Osaka and Kobe are leading industrial centers and busy ports, but the Kansai District also yields large harvests of rice, Japan's staple food. Between the Kanto Plain and the Kansai District lies the **Nobi Plain**, centered on the megacity of Nagoya. And, as the map shows, the Japanese core area is anchored in the west by the conurbation [24] focused on **Kitakyushu**, situated not on Honshu Island but in the northwestern corner of Kyushu, Japan's southernmost major island. This still-expanding edge of the core area is particularly favored by its location relative to South Korea and China.

A Trading Nation

Japan is a remarkable country that despite its poor natural endowments (except for its advantageous location) has done

Yann Arthus-Bertrand/PhotoResearchers, Inc.

Tokyo, centrally located within the world's largest metropolitan region, continues to change. Land filling and bridge building in Tokyo Bay continue; skyscrapers sprout amid low-rise neighborhoods in this earthquake-endangered terrain; traffic congestion steadily worsens. The red-painted Tokyo Tower, a beacon in this district of the city, was modeled after the Eiffel Tower in Paris but, as a billboard at its base proclaims, is an improvement over the original: lighter steel, greater strength, and less weight (i.e., more quake-resistant).

Box 9-17 Among the Realm's Great Cities: Tokyo

Many urban agglomerations are named after the city that lies at their heart, and so it is with the largest of all: Tokyo (37.8 million). Even its longer name—the Tokyo–Yokohama–Kawasaki conurbation—does not begin to describe the congregation of cities and towns that form this massive, crowded metropolis that encircles the head of Tokyo Bay and continues to grow, outward and upward.

Near the waterfront, some of Tokyo's neighborhoods are laid out in a grid pattern. But the urban area has sprawled over hills and valleys, and much of it is a maze of narrow winding streets and alleys. Circulation is slow, and traffic jams are legendary. The train and subway systems, however, are models of efficiency—although during rush hours you must get used to the *shirioshi* pushing you into the cars to get the doors closed.

At the heart of Tokyo lies the Imperial Palace with its moats and private parks. Across the street, buildings retain a respectful low profile, but farther away Tokyo's skyscrapers seem to ignore the peril of earthquakes. Nearby lies one of the world's most famous avenues, the Ginza, lined by department stores and luxury shops. In the distance you can see an edifice that looks like the Eiffel Tower, only taller: this is the Tokyo Tower, a multipurpose structure designed to test lighter Japanese steel, transmit television and cell-phone signals, detect earth tremors, monitor air pollution, and attract tourists (see photo).

Tokyo is the epitome of modernization, but Buddhist temples, Shinto shrines, historic bridges, and serene gardens still grace this burgeoning urban behemoth, a cultural landscape that reflects Japan's successful marriage of the modern and the traditional.

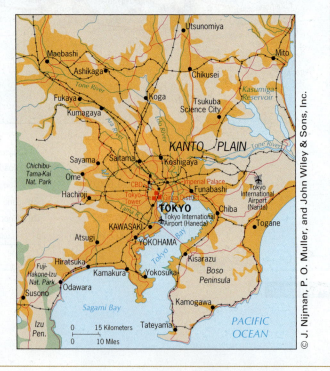

© J. Nijman, P. O. Muller, and John Wiley & Sons, Inc.

extremely well. It became the dominant power of East and Southeast Asia through military force prior to World War II, and its postwar economic miracle astounded the world. In the 1970s and 1980s, people referred to East and Southeast Asia as the "Yen Bloc," one that was dominated by the Japanese economy.

The country defeated the odds by calling on organizational efficiency, massive productivity, dedication to quality, and tight adherence to common goals. Keep in mind that this has always been an exceptionally homogeneous, collectively oriented, and consensus-minded nation with a strong work ethic and a deep-seated conviction that common interests override those of the individual. It is hard to overestimate this cultural context in Japan's outstanding achievements, even if it can become a liability, as when collective rigidity impedes adaptation to new circumstances.

More concretely, Japan dealt with its own limited geographic opportunities by engaging the alternatives. Japan had no choice but to engage other countries, near and far, in order to compensate for what it does not possess itself. Japan, therefore, became wealthy through trade, imported raw materials, accomplished production and processing of high-value-added

goods, and was able to turn these goods into lucrative exports. Look at Figure 9-5 as well as Figure 9-20 and note Japan's substantial past reliance on nuclear energy—a strategy not without major risks in this earthquake-menaced region, as became so painfully clear in the aftermath of the catastrophic Tohoku quake/tsunami in 2011 (see Fig. 9-2). Before the 2011 disaster, about one-third of all consumed energy came from the country's 22 nuclear power plants. In the wake of the disaster, almost all were closed under severe political pressure from the population; in 2016, only one remained operational (the country's southernmost plant, on Kyushu) (Fig. 9-20).

The Challenge of Two Lost Decades

At the end of the 1980s, it seemed that Japan's meteoric rise had been so prodigious that it would only be a matter of time before it would surpass even the United States' economy. But everything came to a rather sudden halt during the early 1990s, and Japan has unexpectedly been stuck in a rut ever since. This now more than 25-year-long economic slowdown had many causes, ranging from government mismanagement and inefficiency to intensifying international competition (such as from South Korean cars and Taiwanese electronic products).

One cause of the slowdown was Japan's lack of adaptability: its culturally embedded economic system seemingly could not cope with rapidly changing conditions. For example, with bankruptcy a near-taboo due to the tight internal networks of firms and their tradition of social responsibility, weak companies were often kept alive. This kind of economic sclerosis inhibits the creation of new businesses (the United States in any given year has three times the number of start-ups as Japan and twice the number of bankruptcies).

Japan's conservative business culture had now become a liability. Whereas women constitute an ever-greater share of Japan's highly educated workforce, only 7 percent of all senior management executives are female (compared to 23 percent in the United States and 30 percent in China). For the first time in decades, Japan suffered a crisis of confidence and, more tangibly, something unfamiliar to the burgeoning Japan of the late twentieth century—rising unemployment. Faced with much leaner economic times, men especially seemed to seek refuge in the past as growing numbers preferred to stay with the same stagnant big companies instead of launching their own businesses.

Japan is a rapidly aging society as well, projected to decline from 126 million today to barely 97 million in 2050 and less than 60 million by the end of this century. It faces an enormous and increasing **demographic burden [25]**: by 2025, it is projected there will be only two workers for every retiree. Ethnically uniform Japan has historically resisted immigration, so when the government tried to recruit ethnic Japanese living in Brazil (and elsewhere) to return home, the experiment was not very successful. Therefore, a shrinking base of qualified workers threatens the government's capacity to sustain social programs for all.

The 2011 nuclear disaster further added to the problems. By 2017, more than one-third of the country's energy needs (that used to be provided through cheap nuclear power) were

This large Pacific inlet is central Honshu's Tokyo Bay, and Japan's capital city is located just to the west of the bay's northernmost corner. But even this Landsat-7 satellite image—measuring roughly 80 by 115 kilometers (50 by 70 mi)—cannot capture the entirety of the massive, Tokyo-centered conurbation. Colored light gray, the far-flung urban landscape covers more than half of the land area in this view and is home to nearly 40 million of the total Japanese population of 126 million. Tokyo Bay also ranks among the world's finest natural harbors, giving rise to Yokohama (located halfway to the ocean from Tokyo on the bay's western shore), widely considered to be the most productive port anywhere.

being supplied through imported oil. Not surprisingly, Japan is feverishly seeking to bring alternative energy online (e.g., wind power), but this too requires major investments.

Still another set of problems has to do with Japan's international relations. Japan never signed a peace treaty with the (then) Soviet Union after World War II because the Russians had occupied and refused to return four small island groups in the Kurile chain northeast of Hokkaido (Fig. 9-20). Failed negotiations for the return of these "Northern Territories," as the Japanese call them, have cost Japan the opportunity to play a key role in the economic development of the Russian Far East, where crucial energy as well as mineral resources abound. Furthermore, relations with both South and North Korea are troubled—with the South over ownership of a small island group in the East Sea (Sea of Japan) and over memories of Japanese misconduct during World War II; and with the North over the kidnapping of Japanese citizens by North Korean agents and, above all else, the threat posed by North Korea's nuclear weapons development. And add to all this a number of lingering issues with China over Japanese actions during the colonial and wartime periods as well as the Senkaku Islands dispute (see Fig. 9-8).

Region Taiwan

When we see where Taiwan is located, it would seem to be inextricably bound up with China because it lies a mere 200 kilometers (125 mi) from the mainland. Not surprisingly, Taiwan has a history of immigration from coastal Chinese provinces. As noted earlier, China's rulers view Taiwan as a "wayward province" that must one day return to the People's Republic. Taiwan lost its seat in the United Nations in 1971 and was replaced by the PRC. The United States at the time also transferred its official recognition of China from Taiwan to the PRC in order to improve relations with the mainland government.

Taiwan's Island Geography

Taiwan, as **Figure 9-21** shows, is not a large island. It is smaller than Switzerland but has a much larger population (23.4 million), most of it concentrated in a crescent lining the western and northern coasts. The Chungyang Mountains, an elongated zone of high elevations (some exceeding 3000 meters [10,000 ft]), steep slopes, and dense forests, dominate the eastern half of the island. Westward, these mountains yield to a zone of hilly topography and, facing the Taiwan Strait, a wide substantial coastal plain. Streams from the mountains irrigate the paddy-fields, and farm production has more than doubled since 1950, even though hundreds of thousands of farmers have left the land for jobs in Taiwan's booming industrial sector.

Today, the lowland urban-industrial corridor lining western Taiwan is anchored by the capital, Taipei, at the island's northern end and by rapidly growing Kaohsiung in the far south. The Japanese developed Chilung, Taipei's outport, to export nearby coal, but now the raw materials flow the other

way. Taiwan imports raw cotton for its textile industry, bauxite (aluminum ore) from Indonesia, oil from Brunei, and iron ore from Africa. Taiwan has a growing steel industry, nuclear power plants, shipyards, a substantial chemical industry, and modern transport networks. Most of all, it has become a leading producer and exporter of high-technology products, especially personal computers, telecommunications equipment, and precision electronic instruments. Taiwan offers a wide range of brainpower resources, and many foreign firms join in the research and development effort. Assisted by the government, a world-class **technopole [26]**, focusing on microelectronics and PC development, has been established at Hsinchu on the northwestern coast (see photo pair). Farther south, the budding

© Bojan Brecelj / Corbis Images

Maurice Tsai/Bloomberg via GettyImages, Inc.

Taiwan's high-tech manufacturers constitute a force to be reckoned with in the global science industries. Domestically, one of the leading contestants for the title of "Taiwanese Silicon Valley" is the burgeoning technopole centered on the city of Hsinchu—a metropolis of nearly 550,000 located on the Taiwan Strait 45 minutes southwest of the capital, Taipei. R&D predominates here, and the top photo shows the Synchrotron Radiation Research Center, focused on its new-generation nuclear particle accelerator. Hsinchu is not without rivals, however, and an hour to its south lies the smaller city of Taichung whose high-tech centerpiece is its thriving Science Park. The lower photo shows one of this complex's state-of-the-art factories, whose owner is one of the world's foremost manufacturers of flat-screen panels. Demand for flat screens has surged in recent years, prompting the AU Optronics Corporation to expand production onto the nearby Chinese mainland.

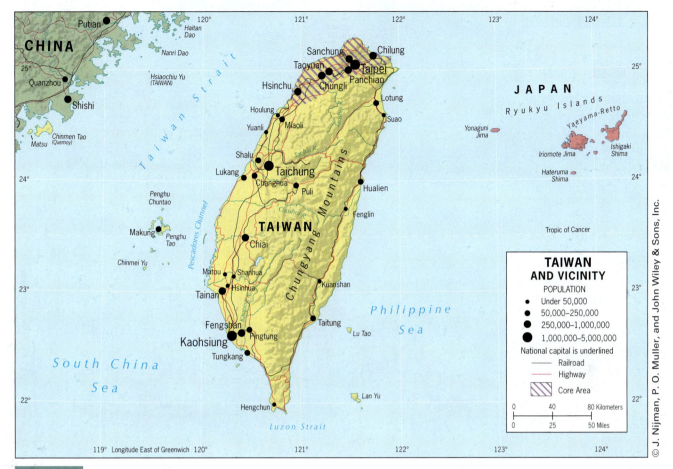

FIGURE 9-21

"science city" of Tainan specializes in microsystems and information-technology innovations.

The creation of China's Special Economic Zones went a long way toward mitigating longstanding problems between Beijing and Taipei. China needed investment capital and Taiwan had it, and in the almost-anything-goes environment of the SEZs, Taiwanese entrepreneurs could buy or build factories just as "real" foreigners were able to do. In fact, the Xiamen SEZ (see Fig. 9-7) was laid out directly across from Taiwan for this purpose (people there will tell you that almost all of the locals have relatives across the Taiwan Strait). Consequently, thousands of Taiwanese-owned factories began operating in the PRC, their owners and investors having a strong interest in avoiding violent confrontation over political issues.

Taiwan's Future

Taiwan's unresolved status entails risks, but there are signs that both parties want to find a solution. Given that the PRC will never allow Taiwan to attain independence, and the growing prospect that the United States, Taiwan's chief guarantor, could not secure Taiwan against Chinese military intervention, many Taiwanese as well as their allies and adversaries are seeking a long-range, negotiated solution. Among the options is one that appears to be working in Hong Kong, where the principle of

One Nation–Two Systems [27] has functioned more successfully than many observers anticipated—despite the sporadic protests led by the political opposition that mark the ongoing debate over the policy's effectiveness.

Points to Ponder

- The fortunes of China and Japan have always seemed to be inversely related: when one was rising, the other was in decline.

- In the past decade, China's defense spending has increased by about 12 percent annually, much of it intended to match and eventually surpass the military strength of the United States in the islands and waters off the Asian mainland.

- Widening social inequalities resulting from China's rapid economic growth are becoming ever more difficult for China's self-proclaimed communist government to justify.

- After 70-plus years of dictatorship, malnutrition, and famine, North Koreans at the age of 21 are an average 6 centimeters (2.4 in) shorter than their South Korean counterparts.

- Since the 2011 nuclear disaster, Japan has closed 21 of its 22 nuclear plants, and the country is now zealously seeking to develop alternative energy sources.

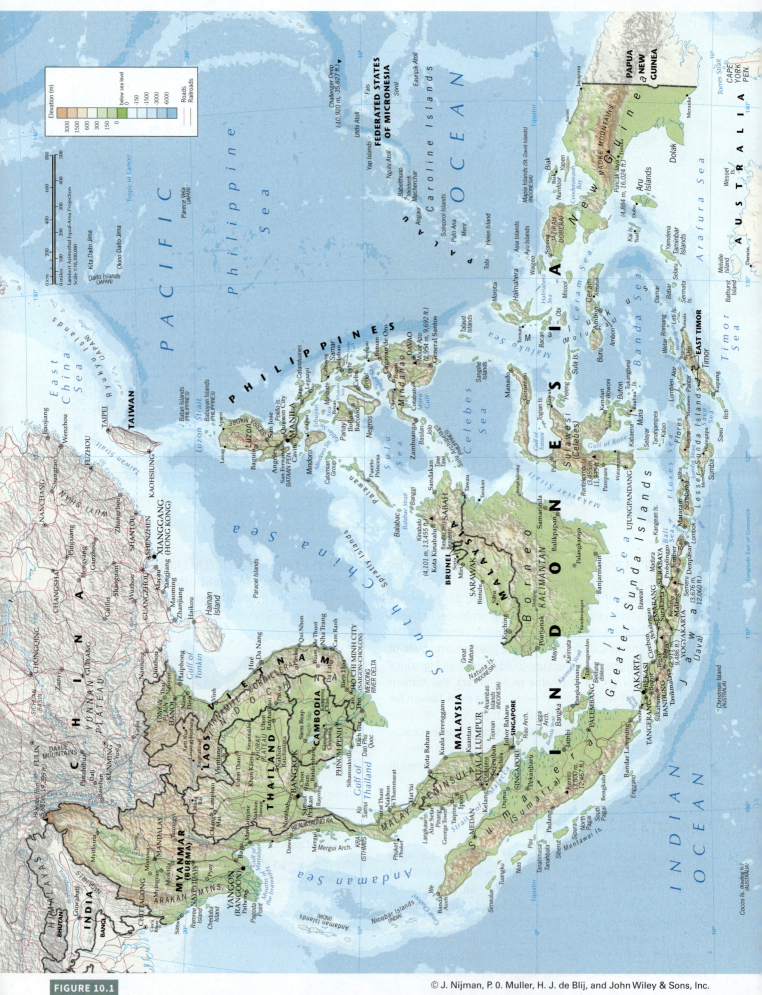

FIGURE 10.1

© J. Nijman, P. O. Muller, H. J. de Blij, and John Wiley & Sons, Inc.

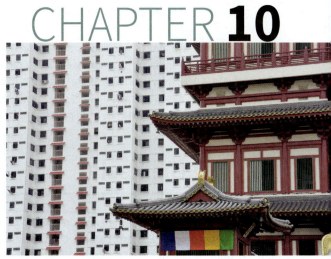

CHAPTER 10

Q Where were these pictures taken?

© Jan Nijman

The Southeast Asian Realm

Southeast Asia is a realm of peninsulas and islands, a corner of Asia bounded by India on the northwest and China to the northeast (**Fig. 10-1**). Its western coasts are washed by the Indian Ocean, and to the east stretches the gigantic Pacific. From all these directions, Southeast Asia has been penetrated by outside forces. From India came traders; from China, settlers; from across the Indian Ocean, Arabs to engage in commerce and Europeans to build empires; and from across the Pacific, Americans. Southeast Asia has been the scene of countless contests for power and primacy—the competitors have come from near and far.

Defining the Realm

Southeast Asia's geography—summarized in **Box 10-1**—in some ways resembles that of Eastern Europe, even if the physiography is quite different. It is a mosaic of smaller countries on the periphery of two of the world's largest states. It has been a **buffer zone [1]** between powerful adversaries. It is a **shatter belt [2]** in which stresses and pressures from without and within have fractured the political geography. Like Eastern Europe, Southeast Asia exhibits great cultural diversity. This is a realm of hundreds of cultures and ethnicities, myriad languages and dialects, global as well as local religions, and diverse national economies ranging from high- to low-income.

A Geographic Overview

Figure 10-1 displays the relative location and dimensions of the Southeast Asian realm, an assemblage of 11 countries situated on the Asian mainland as well as on thousands of islands, large and small, extending from Sumatera in the west to New Guinea in the east, and from Luzon in the north to Timor in the south. We will become familiar with only the largest and most populous of these islands, but if you were sailing the seas of this part of the world, you would pass dozens of smaller ones every day. And if you could stop, you would find even the tiniest populated islands possess an individual character resulting from the cultural sources of their inhabitants, the environmental challenges they face, the opportunities they exploit, their modes of dress, the layout of their dwellings, and even the vivid colors they use to decorate their boats.

The giant of this realm, both territorially and demographically, is the far-flung **archipelago [3]** (island chain) of Indonesia, labeled appropriately in Figure 10-1 by the largest lettering. In its easternmost sector, the Indonesian state extends beyond the Southeast Asian realm into the Pacific realm because it controls the western half of a large island—New Guinea—whose indigenous peoples are not Southeast Asian. We discuss this unusual situation later (and focus on New Guinea as a whole in Chapter 12), but note it here because unusual borders and divided islands are a hallmark of this realm.

On the Asian mainland, as noted above, Southeast Asia is bordered by China and India, both sources of immigrants, cultural infusions, economic initiatives, and other relationships evident in the realm's cultural landscapes. Also originating there are rivers that play a pivotal role in the lives of many millions of people residing to the south of the Chinese border. As we shall discover, the core areas of many of Southeast Asia's most populous countries are located in the basins of major rivers whose sources lie outside the realm.

Before we get started, it is helpful to become acquainted with the key states of this realm, some of which will already be familiar. North of Indonesia lies the Philippines, well known to Americans because this was once an American colony and is still a source of

---

Box 10-1 Major Geographic Features of Southeast Asia

1. Southeast Asia extends from the peninsular mainland to the archipelagos offshore. Because Indonesia controls part of New Guinea, its functional region reaches from the Southeast Asian geographic realm into the neighboring Pacific realm.

2. Southeast Asia has been a shatter belt between powerful adversaries and displays a fractured cultural and a political geography shaped by foreign intervention.

3. Southeast Asia's physiography is dominated by high relief, crustal instability marked by volcanic activity and earthquakes, and tropical climates.

4. A majority of Southeast Asia's 640 million people live on the islands of just two countries: Indonesia, with the world's fourth-largest population, and the Philippines. The rate of population increase in the Insular region of Southeast Asia exceeds that of the Mainland region.

5. Although the great majority of Southeast Asians have the same ancestry, cultural divisions and local traditions abound, which the realm's divisive physiography nurtures.

6. The Mekong River, Southeast Asia's Danube, rises in China and borders or traverses five Southeast Asian countries, sustaining tens of millions of farmers, fishing people, and boat owners.

7. Singapore is the leading world-city in Southeast Asia and lies at the realm's center of trade and business networks.

8. Southeast Asia contains a number of rapidly emerging markets and fast-growing economies that in certain respects imitate development in the neighboring East Asian realm.

9. China's influence in this realm has increased markedly since 2000, frequently triggering local backlash.

many immigrants to the United States. Also lying north of Indonesia is Malaysia, easy to find on the map because its core area is located on the lengthy Malay Peninsula that almost connects mainland Asia with Indonesia. At the tip of that peninsula lies another famous geographic locality: Singapore, the spectacular economic success story of Southeast Asia and a world-city by every measure.

In the realm's mainland component, the name Vietnam still resonates in the United States, which fought a bitter and costly war there in the 1960s and 1970s. As the map shows, Vietnam looks like an elongated strip of land extending from its border with China to the delta of the greatest of all Southeast Asian rivers, the Mekong. Looking westward, neighboring Laos and Cambodia may not be all that familiar, but next comes centrally positioned Thailand, the dream (and reality) of millions of tourists and one of the world's most fascinating countries. And then, on the realm's western margin, we come to Myanmar (formerly Burma): here human potential, natural endowment, and opportunity were thwarted by a half-century of extremely harsh military rule that plunged the country into devastating poverty—but that quite unexpectedly relaxed its vise-like grip in 2012 and reopened Myanmar to the world.

Southeast Asia's Physical Geography

Physiographically, Southeast Asia is in some ways reminiscent of Middle America, a fractured realm of islands and peninsulas flanking a populous mainland studded with high mountains, deep valleys, and stunning landforms (**Box 10-2**). It is useful to look back at Figure G-4 to see why: both Southeast Asia and Middle America contain dangerous terrain, where the Earth's crust is unstable as tectonic plates collide, earthquakes are a constant threat, volcanic eruptions take their toll, tropical cyclones lash sea and land, and floods, landslides, and other natural hazards make life riskier than in most other parts of the world.

In terms of its economic geography Southeast Asia may be part of the Pacific Rim, but physiographically it forms part of the Pacific Ring of Fire—and all the geologic hazards this designation carries with it (Fig. G-4). In 2004, an undersea earthquake off westernmost Indonesia produced a **tsunami [4]** (seismic sea wave) in the Indian Ocean that killed more than 230,000 people along coastlines from Indonesia's Sumatera to eastern Africa's Somalia. Yet this was only the latest in a string of geologic disasters originating in Southeast Asia whose effects were felt far beyond the realm's borders. In 1883, the Krakatau volcano lying between Sumatera and Jawa exploded, resulting in a death toll estimated at more than 36,000. In 1815, the Tambora volcano in the chain of islands east of Jawa known as the Lesser Sunda Islands blew up, darkening skies throughout the world and affecting climates around the planet (the year that followed is still known as the "year without a summer" when crops failed, economies faltered, and people went hungry as far away as Egypt, France, and New England). Research on what may well have been the most calamitous of all such eruptions suggests that, about 73,000 years ago, the Toba volcano on Sumatera exploded with such force that its ash and soot not only darkened skies and affected weather, but altered the global climate for perhaps as long as 20 years and threatened the very survival of the human population, which then was still small in number and widely scattered. In fact, a number of scientists postulate that the widespread impact of this eruption gave rise to so many casualties that human genetic diversity was significantly diminished.

Thus, high relief dominates Southeast Asia, from the Arakan Mountains in western Myanmar to the glaciers (yes, glaciers!) of New Guinea. Take a close look at Figure 10-1 and you can see how many elevations approach or exceed 3000 meters (10,000 ft), and, using the elevation guide in the map legend, how mountainous and hilly much of the realm's topography is. Lengthy ranges form the backbones not only of islands such as Sulawesi and Sumatera, but also of the Malay Peninsula, most of Vietnam, and the border zone between Thailand and Myanmar. In Figure G-4, it is possible to trace these volcano-studded mountain ranges by their earthquake epicenters and volcanic records.

Box 10-2 From the Field Notes...

"After a three-hour drive by bus southward from Hanoi, we reached the Vanlong Nature Reserve in northeastern Vietnam's Red River Delta. It lies in a limestone terrain known as karst, part of the same geology that continues to the north through Halong Bay and into southern China beyond. Karst landscapes are formed through the slow but steady dissolution of limestone by (naturally) acidic water; the harder rock formations are more resistant to erosion and remain as towers, resulting in unique and sometimes whimsical landforms (left photo). This natural reserve is a biologically rich ecotourism site that was established in 2001 and involves local communities. Women from nearby villages guide visitors around in their basket boats, and they welcome the extra income in an area barely touched by Vietnam's emerging economy (right photo)."

© Jan Nijman   © Jan Nijman

Exceptional Borneo

Among the islands, however, there is one major exception. The bulky island named Borneo in Figure 10-1 exhibits high elevations (Mount Kinabalu in the north reaches 4101 meters [13,455 ft]), yet has no volcanoes and negligible Earth tremors. This island has been called a stable "minicontinent" amid a cauldron of volcanic activity, a slab of ancient crust that long ago was uplifted high above sea level by tectonic forces and then subsequently eroded into its present landscapes. Borneo's soils are not nearly as fertile as those of the volcanic islands, so that an equatorial rainforest developed here that long survived the human onslaught, providing sanctuary for countless plant and animal species that include Southeast Asia's great ape, the orangutan. That era is ending today as human encroachment on Borneo's tropical habitat accelerates: logging is destroying vast swaths of the remaining forest, and roads and farms are penetrating the island's interior. Borneo, along with other parts of Indonesia and Malaysia, has also experienced the rapid expansion of palm oil plantations at the expense of tropical woodlands (see **Box 10-3**).

Borneo's tropical forests are a remnant of a much larger stand of equatorial rainforest that once covered most of this realm. Besides Borneo, eastern Sumatera still contains limited expanses of rainforest, including a few orangutan sanctuaries. So does less populous and more remote New Guinea, which was never reached by these great apes. As we noted in the introductory chapter, equatorial rainforests can still be found today in three low-latitude zones: the Amazon Basin of South America, west-equatorial Africa, and here in Southeast Asia. The climatic conditions that sustain these forests—persistently warm temperatures and year-round rainfall—produce the biologically richest and ecologically most complex vegetation regions on Earth. Enormous numbers and varieties of trees and other plants grow in very close proximity, vying for space and sunlight both horizontally and vertically. Yet despite all this luxuriant growth, the soils beneath rainforests are nutrient poor. Most of the surface nutrients are derived from the decaying vegetation on the forest floor, which nurture the growth of the next generation of plants. Long-term indigenous inhabitants have developed methods of shifting cultivation to use these soils, plants, and animals to eke out a living, but when migrant farmers from elsewhere remove the trees to utilize the soil, mistakenly assuming it will be able to support their crops, failure swiftly results. This reality notwithstanding, intensifying population pressure throughout the tropics continues to shrink what remains of the world's rainforests—an issue also discussed in the chapters on Middle and South America (Chapters 2 and 3).

Relative Location and Biodiversity

Look again at Figure 10-1, but in a more general way. The Malay Peninsula, adjacent Sumatera, Jawa, and the Lesser Sunda Islands east of Jawa seem to form a series of stepping stones toward New Guinea, and in the extreme lower-right corner of the

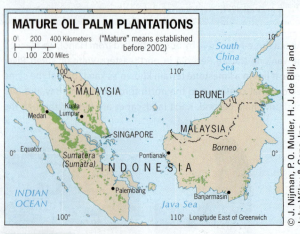

A section of a vast palm oil plantation in Sarawak State in the southern part of Malaysian Borneo. These continually enlarging plantations have proliferated in Malaysia and Indonesia (see **Fig. 10-2**), earning sizeable export revenues but at ever-increasing environmental cost.

Box 10-3 Deforestation and Palm Oil Plantations

Palm oil is used worldwide in processed foods and myriad other products. This form of commercial agriculture is quite lucrative because the yield of oil palms per hectare (which equals 2.5 acres) is much higher than the yield of such competing edible oil crops as soybeans and sunflowers. Moreover, the price of this commodity on global markets has risen steadily over the past several years, triggering successive rounds of production increases. Malaysia and Indonesia have benefited the most, together accounting for 90 percent of the world's palm oil. But because these countries take advantage of economies of scale and produce palm oil on huge, ever-expanding plantations, this crop has become the single biggest threat to tropical forest preservation (**Fig. 10-2;** see photo). Most at risk for survival are such key wildlife species as orangutans, pygmy elephants, Sumatran rhinos, and tigers.

MATURE OIL PALM PLANTATIONS ("Mature" means established before 2002)

FIGURE 10-2

map you can see Australia's Cape York Peninsula also appearing to reach toward New Guinea. Throughout Southeast Asia's geologic history, sea levels have repeatedly dropped, periodically drying up those narrow bodies of water lying between the islands as well as certain peninsulas. During such prehistoric intervals, the orangutans, whose descendants survive in Indonesia today, were able to migrate from the mainland into even warmer equatorial latitudes because islands were temporarily connected by land bridges. Later, early human arrivals here also received assistance from nature: Australia's Aborigines probably managed to cross what remained of the deeper trenches between islands by building rafts, but their epic journey was facilitated by land bridges as well.

Thus time and again, Southeast Asia was a receptacle for migrating species. Combine this with the realm's tropical environments, and it is no surprise that it is known for its **biodiversity [5]**. Scientists calculate that fully 10 percent of the Earth's plant and animal species are found in this comparatively smalll realm. Biogeographers can trace the progress of many of these species from the Asian mainland across the Indonesian archipelago toward Australia, but some were halted by the deeper trenches that remained filled with water even when sea levels dropped. An especially deep trench lies in the Lombok Strait between the islands of Bali (next to Jawa) and Lombok (Fig. 10-1), a key biogeographical boundary (discussed in Chapter 11) first recognized by the naturalist Alfred Russel Wallace (1823–1913), a contemporary of Charles Darwin (1809–1882). As we shall presently observe, Southeast Asia's biodiversity had a fateful impact on its historical geography. Among the realm's specialized plants, it was the spices that attracted outsiders from India, China, and Europe, with consequences still visible on the map today.

Four Major Rivers

Water is the essence of life, and Southeast Asia is comparatively well endowed with moisture (Fig. G-6). Ample, occasionally even excessive, rainfall fills the rice-growing paddies of Indonesia and the Philippines. On the realm's mainland, where annual rainfall averages are somewhat lower and the precipitation is more seasonal, major rivers and their tributaries fill irrigation channels and form fertile deltas. The map of the mainland's population distribution is dominated by this spatial relationship between rivers and people, expressed most clearly in distinct coastal clustering atop the biggest deltas (see Fig. 10-5).

As Figure 10-1 reminds us, rivers that are crucial to life in one realm may sometimes flow from sources in neighboring realms, a geographic issue that can lead to serious regional discord. What right does an upstream state in one realm (or even the same realm) have to dam or otherwise siphon water away from a river whose flow is vital to a state (or states) downstream? In Southeast Asia's case, three of its four principal rivers originate in China: (1) the **Mekong**, which crosses the heart of the mainland from north to south; (2) the **Red** River, which reaches the sea via its course across northern Vietnam; and (3) the **Irrawaddy**, which forms the lifeline of Myanmar. The fourth is an intra-realm river, the **Chao Phraya**, which functions as Thailand's leading artery.

The Mighty, Endangered Mekong From its headwaters high in the frozen uplands of China's Qinghai-Xizang (Tibetan) Plateau, the Mekong River rushes and flows some 4200 kilometers (2600 mi) to its delta in southernmost Vietnam: This "Mother of Rivers" as its name connotes, traverses or borders five of the realm's countries, supporting rice farmers and fishing people, forming a vital transport route where roads are few, and providing electricity from dams upstream. Tens of millions of people depend on the waters of the Mekong, from subsistence farmers in Cambodia to apartment dwellers in China. The Mekong Delta in southern Vietnam is one of the realm's most densely populated areas and produces enormous harvests of rice. The Mekong Basin as a whole (mapped in green in Fig. 10-3) is reputed to be the world's largest and most productive area for freshwater fishing.

But the Mekong is increasingly under threat, and so are the livelihoods of millions in its Basin, as demand for hydroelectric power accelerates all across mainland Southeast Asia. To the north, China has already built seven dams for generating hydropower, and many more are either under construction or planned **(Fig. 10-3)**. And in the Mekong's middle basin, impoverished Laos considers hydropower its most strategic commodity and promising route to development, and is making big plans to sell much of that energy to China, Thailand, and Vietnam. Favored by its hilly terrain, Laos is planning seven new

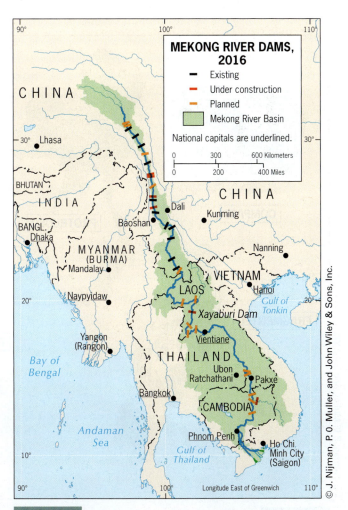

FIGURE 10-3

Construction of the controversial Xayaburi Dam in northern Laos, located on the Mekong River upstream from the capital, Vientiane. For obvious reasons, this site is zealously guarded, and current photos are hard to find (this one dates from late 2012). Opposition groups say this dam will displace and adversely affect countless communities along the Mekong as well as produce significant ecological harm. Meanwhile, the Laotian government is primarily focused on developing the dam's hydropower potential, which is planned to be a main driver of national economic growth through the rising sale of electricity to China as well as other countries. Xayaburi Dam is expected to be in full operation soon after its expected completion in 2019.

dams, and even its lowland downstream neighbor, Cambodia, wants to build another three.

Among these projects, it is the ongoing construction of the Xayaburi Dam in Laos that has by far drawn the most attention and criticism from environmental activists because it sits squarely in the Mekong's main stream (Fig. 10-3; see photo). As such, it will substantially reduce the river's downstream transportation of silt, the alluvial sediment that is highly crucial to the formation of fertile soils on the floodplains of central Cambodia and particularly the massive delta in southern Vietnam. And the sheer number of dams planned for the Mekong Basin is certain to damage ecosystems all along the Southeast Asian courses of the river and its tributaries (starting with the disruption of the migratory routes of freshwater fish). In the long run, therefore, prioritizing energy over environment and even food production may well prove to be a disastrous choice.

Political Geography: Territorial Configurations

Political boundaries define and delimit states. They also create the mosaic of interlocking territories that give individual countries their shape or **state territorial configuration [6]**. A country's shape can affect its condition, and in certain cases even its survival. Vietnam's extreme elongation has influenced its existence since time immemorial. And, as noted later in this chapter, Indonesia has tried to redress its fragmented layout—consisting of thousands of far-flung islands—by promoting unity through the "transmigration" of residents from its most populous, core-area island (Jawa) to many of the others.

Political geographers have identified four major categories of state territorial configuration, all of which we have encountered in our world regional survey but have not classified until now. All of these prototypical shapes can be seen on the political map of Southeast Asia, and **Figure 10-4** displays the terminology and examples:

Compact states [7] have territories shaped somewhere between round and rectangular, without major indentations. This encloses a maximum amount of territory within a minimum length of boundary. Southeast Asian example: Cambodia.

STATE TERRITORIAL CONFIGURATIONS

CAMBODIA
Phnom Penh

COMPACT

THAILAND
Bangkok

PROTRUDED

VIETNAM
Hanoi

Manila

FRAGMENTED

PHILIPPINES

ELONGATED

Ho Chi Minh City
(Saigon-Cholon)

© J. Nijman, P. O. Muller, and John Wiley & Sons, Inc.

FIGURE 10-4

Protruded states [8] (sometimes called *extended*) have a substantial, usually compact territory from which extends a peninsular or other corridor that may be landlocked or coastal. Southeast Asian examples: Thailand and Myanmar.

Elongated states [9] (also known as *attenuated*) have territorial dimensions in which the length is at least six times the average width, creating a state that lies astride environmental and/or cultural transitions. Southeast Asian example: Vietnam.

Fragmented states [10] consist of two or more territorial units separated by foreign territory or a substantial body of water. Subtypes include mainland-mainland, mainland-island, and island-island. Southeast Asian examples: Malaysia, Indonesia, the Philippines, and East Timor.

For so comparatively small a realm, the 11 countries of Southeast Asia display a considerable variety of shapes. But one note of caution: states' territorial configurations do not determine their viability, cohesion, unity, or lack thereof; they can, however, influence these characteristics. Cambodia's compactness, for instance, has not ameliorated its divisive political geography.

Population Geography

Examine the map of Southeast Asia's population distribution (Fig. 10-5), and you are immediately struck by the huge concentration of people on a relatively small island in Indonesia—a cluster larger than any other in the realm, the four mainland river deltas included. This is Jawa, and its population of approximately 150 million not only accounts for well over half of Indonesia's national total but also exceeds that of every other country in the realm.

This population concentration is particularly noteworthy because Indonesia is not yet a highly urbanized country. Today nearly half of all Indonesians still live in rural areas, and although Jawa is the country's most urbanized island, more than 50 million of its inhabitants still live off the land. What makes all this possible is a combination of highly fertile volcanic soil, abundant water, and extremely warm temperatures that enable Jawa's farmers to raise three crops of rice in a single paddy during a single year, helping feed a national population still growing faster than the global average. But we should also be aware that the thickest red clusters on Jawa denote

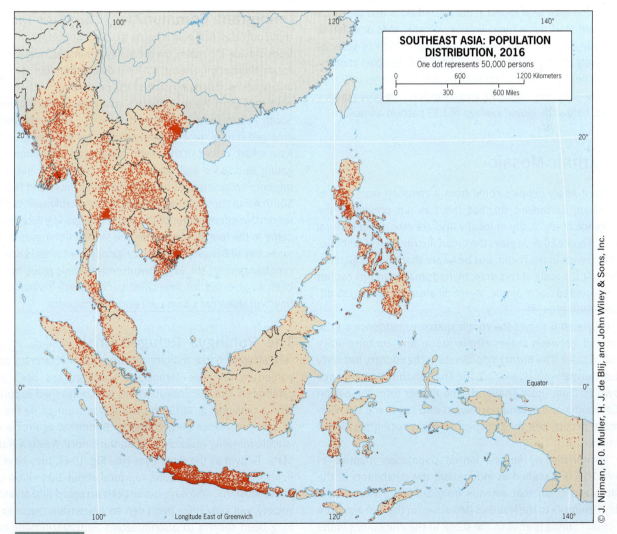

FIGURE 10-5

fast-growing urban areas, cornerstones in the building of a new economy with increasingly global linkages.

Within Indonesia, the contrasts between Jawa and the four other major islands—Sumatera, Borneo (Kalimantan), Sulawesi, and most especially Indonesian New Guinea—reflect the core-periphery relationship between these two sectors of the country. As the map suggests, similar contrasts, less sharply defined, also mark other Southeast Asian countries, and the primate cities of this realm (such as Bangkok, Manila, Jakarta, and Kuala Lumpur) are particularly dominant. In fact, Vietnam even contains two such anchors, which respectively represent its historic northern (Hanoi) and southern (Ho Chi Minh City) core areas (Fig. 10-4).

The states of Southeast Asia are not, by world standards, especially populous. Indonesia today is the world's fourth-ranking country in population size (261 million), but no other Southeast Asian country contains even half that total. Three mainland countries—Vietnam, Thailand, and Myanmar—contain between 50 and 100 million people. But take Laos, quite a large country territorially (about the size of the United Kingdom) and note that its population is less than 7 million; similarly, Cambodia, half the size of Germany, is home to only slightly more than 15 million. In part, such modest numbers on the mainland reflect natural conditions less favorable to farming than those prevailing on volcanic soils or in fertile river valleys; but more generally, this realm did not grow as explosively as neighboring realms did during the past century. Indeed, the populations of three large mainland states of Southeast Asia—Vietnam, Thailand, and Myanmar—are now growing at a rate below the global average of 1.13 percent annually.

The Ethnic Mosaic

Southeast Asia's peoples come from a common stock just as (Caucasian) Europeans do, but this has not prevented the emergence of regionally or locally discrete ethnic and cultural groups. **Figure 10-6** displays the broad distribution of ethnolinguistic groups in the realm, but be aware that this is a generalization. At the scale of this map, myriad smaller groups cannot be represented (for example, Myanmar alone has some 135 different ethnic groups).

Figure 10-6 shows the rough spatial coincidence on the mainland between major ethnic group and contemporary political state. The Burman dominate in the country formerly called Burma (now Myanmar); the Thai inhabit the state once known as Siam (now Thailand); the Khmer form the nation of Cambodia and extend northward into Laos; and the Vietnamese occupy the long strip of territory facing the South China Sea.

Territorially, by far the largest population mapped in Figure 10-6 is classified as Indonesian, the inhabitants of the great island chain that extends from Sumatera west of the Malay Peninsula to the Malukus (Moluccas) in the east and from the Lesser Sunda Islands in the south to the Philippines in the north. Collectively, all of these peoples shown on the map—the

Filipinos, Malays, and Indonesians—are known as Indonesians, but they have been divided by history and politics. And note as well that the Indonesians in Indonesia itself include Javanese, Madurese, Sundanese, Balinese, and other major groups; moreover, hundreds of smaller groups again cannot be mapped at this scale. In the Philippines, too, island isolation and contrasting ways of life are reflected in the cultural mosaic. Also part of this Indonesian ethnic-cultural complex are the Malays, whose heartland lies on the Malay Peninsula but who form minorities in other areas as well. Like most Indonesians, the Malays are Muslims, although Islam is a more powerful force within Malay society than it generally is in Indonesian culture.

In the northern margins of mainland Southeast Asia, numerous minorities inhabit remote corners of the countries in which the Burman (Burmese), Thai, and Vietnamese dominate. Those minorities tend to occupy such peripheral areas because the terrain is mountainous and the forest is dense, thereby hindering the governments of their national states from exerting complete control. This remoteness and sense of detachment give rise to aspirations of secession, or at least resistance to government efforts to establish full authority, often resulting in bitter ethnic conflict.

Immigrant Communities Figure 10-6 reminds us that Southeast Asia also is home to substantial ethnic minorities from outside the realm. On the Malay Peninsula, note the South Asian (Hindustani) cluster. Such Hindu communities with Indian ancestries exist in many parts of the peninsula, but in the southwest they form the majority in a small area. In Singapore, too, South Asians form a significant minority. These communities emerged during the European colonial period. However, South Asians had arrived in this realm many centuries earlier, propagating Buddhism and leaving their architectural and cultural imprints on landscapes as far removed as Jawa and Bali. Most South Asian communities blended in reasonably well, but in the far northwestern corner of the realm one glaring exception has come to the fore in this decade: the Muslim Rohingyas of Myanmar's Bay of Bengal coastland. Originally from what is now adjacent Bangladesh, this long-downtrodden ethnic group has today been singled out for intensified persecution—ironically at the hands of Myanmar's dominant Buddhist majority.

The Rohingya Refugee Crisis Since 2012, the relaxation of harsh military dictatorship in Myanmar, accompanied by tentative democratic openings, has been widely celebrated. But this transition has also unleashed suppressed ethnic tensions: in 2013, the world was stunned by the wave of savage violence perpetrated by Buddhists against a small Islamic minority residing in Myanmar's northwestern Rakhine State. Known as the Rohingyas (see Fig. 10-6), they have lived in Myanmar for generations and total about 1.3 million today. These Muslims, who have never been accepted into Myanmar's society, not only have been denied citizenship but also have long been subject to discrimination and intermittent acts of violence. The Buddhist majority refers to them as "Bengalis," a

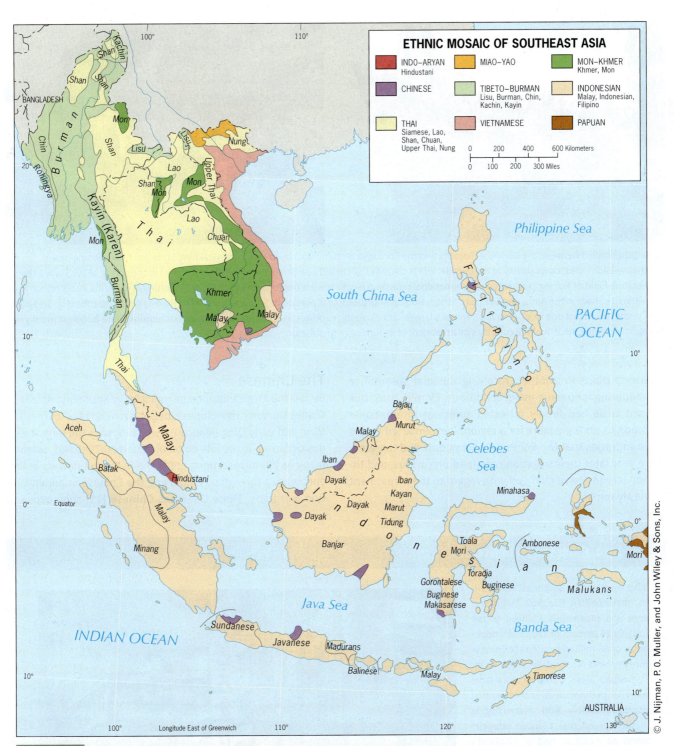

ETHNIC MOSAIC OF SOUTHEAST ASIA

INDO–ARYAN
Hindustani

CHINESE

THAI
Siamese, Lao,
Shan, Chuan,
Upper Thai, Nung

MIAO–YAO

TIBETO–BURMAN
Lisu, Burman, Chin,
Kachin, Kayin

VIETNAMESE

MON–KHMER
Khmer, Mon

INDONESIAN
Malay, Indonesian,
Filipino

PAPUAN

0   200   400   600 Kilometers

0   100   200   300 Miles

FIGURE 10-6

© J. Nijman, P. O. Muller, and John Wiley & Sons, Inc.

reference to their origins in neighboring Bangladesh that borders Rakhine on the north (Fig. 10-6).

Since this upsurge in ethnic violence began in 2013, at least 200,000 Rohingyas have been interned "under protection" in (or fled to) refugee camps operated by the government. Many more fled the country to neighboring Thailand. But when Myanmar's leaders (pandering to heightening anti-Muslim sentiment) closed down these overland escape routes in 2015, Rohingyas by the thousands desperately took to the Indian Ocean. By following coast-paralleling routes southward through the Andaman Sea, they hoped to reach the presumably friendlier shores of Thailand, Malaysia (which now harbors more than 50,000 Rohingyas), or even more distant Indonesia (Fig. 10-1; see left photo). Unfortunately for most, this exodus became yet another horrifying experience: many of their flimsy, poorly equipped, overcrowded boats proved unseaworthy, and hundreds drowned; those who survived the arduous journey beyond Myanmar's waters were often initially denied entry by other countries; and most who finally found safe harbor were forced to move around

The ethnic conflict between the Buddhist majority and the Rohingya Muslim minority in Myanmar's Rakhine State has now become one of Southeast Asia's major news stories of the decade. It first came to light in 2012 and then prominently resurfaced in the spring of 2015 when the accelerating flight of Rohingyas created an international refugee crisis. In the left photo, dated May 20, 2015, Rohingya migrants await rescue from an overloaded boat in the Andaman Sea just offshore from the northernmost portion of Sumatera, one of Indonesia's main islands (Fig. 10-1). The right photo, taken just one week later in Myanmar's largest city (Yangon), captures the protest march of several hundred irate Buddhist monks to denounce foreign criticism of the country's mistreatment of its Rohingya population.

from one place to another, receiving minimal sustenance and enduring inhumane living conditions. By mid-2015, this refugee crisis had finally attracted sufficient international attention to exert pressure for a regional response led by the three countries that the Rohingya boat people had targeted, but a year later their situation remained precarious, and little appeared to have changed with regard to their treatment inside Myanmar.

The Chinese

By far, the largest immigrant minority in Southeast Asia is of Chinese origin. The Chinese began arriving here during the Ming and early Qing (Manchu) dynasties, with the largest exodus occurring in the late colonial period (1870–1940), when as many as 20 million immigrated. The European powers at first encouraged this influx, employing the Chinese in administration and trade. These **Overseas Chinese [11]** soon began to

Box 10-4 From the Field Notes...

"Like most major Southeast Asian cities, metropolitan Bangkok includes a large and prosperous Chinese sector. Roughly 14 percent of Thailand's population of 68 million is of direct Chinese ancestry, and most reside in the country's urban areas. However, this large non-Thai population is well integrated into local society, and intermixing is so widespread that Chinese ethnicity marks as much as half of Thailand's population. Still, Bangkok's 'Chinatown' remains a distinct element in the great city's landscape, and there is no mistaking its limits: Thai commercial signs shift to Chinese, goods offered for sale also change (Chinatown contains many shops selling gold, for example), and the urban atmosphere, from street markets to bookshops, becomes predominantly Chinese. This is a boisterous, noisy, energetic part of multicultural Bangkok, a vivid reminder of the Chinese commercial success in Southeast Asia."

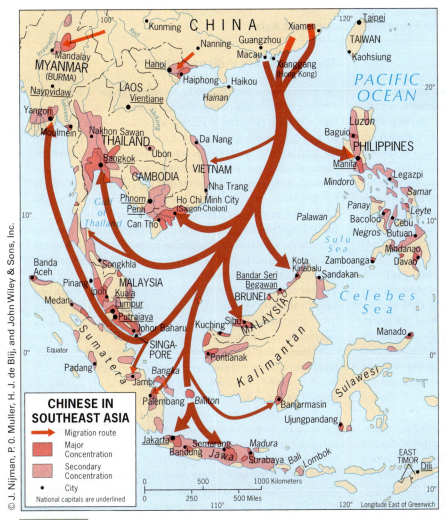

CHINESE IN
SOUTHEAST ASIA
→ Migration route
■ Major
Concentration
■ Secondary
Concentration
• City
National capitals are underlined

FIGURE 10-7

continues to smolder in various parts of Southeast Asia (as discussed in **Box 10-5)**, and episodic flare-ups against the Chinese are likely to continue.

During this decade, Singapore is experiencing its own unique challenges with the latest wave of Overseas Chinese. Unique, because the overwhelming majority of native Singaporeans are themselves of Chinese descent—even though they are third-, fourth-, or fifth-generation, and have to varying degrees blended into the surrounding Malay culture (Singapore and Malaya were both part of the same British colonial entity for more than 130 years before independence in the 1960s). However, that shared Sinic ethnicity has not prevented Singaporeans from becoming ever more critical of the ongoing influx of 'newly rich' mainland Chinese. Resentment continues to rise concerning the role of purportedly arrogant, wealthy immigrants in driving up real-estate prices, commandeering the best jobs, and disrespecting the city-state's strict codes of public behavior. Overall, this clash increasingly erodes social relations because 'old' Chinese and 'new' Chinese today refer to distinctly different identities within Singapore's multicultural mosaic.

Religions of Southeast Asia

Southeast Asia, more than any other realm, is a historic crossroads of religions (see Fig. G-10). With the migrants from the Indian subcontinent came their faiths—first Hinduism and Buddhism, later Islam. The Muslim religion, promoted by the increasing number of Arab traders who appeared on the scene, soon became the predominant religion in Indonesia (where 88 percent of the population adheres to Islam today). But in Myanmar, Thailand, and Cambodia, Buddhism remained supreme, and in all three of these mainland countries at least 90 percent of the people are currently adherents. In culturally diverse Malaysia, the Malays are Muslims (to be a Malay *is* to be a Muslim), and almost all Chinese are Buddhists; but most Malaysians of Indian ancestry remain Hindu.

Although Southeast Asia has generated its own local cultural expressions, most of what remains in tangible form has resulted from the infusion of foreign elements. For instance, take Angkor Wat, the enormous complex of religious structures built in Cambodia during the twelfth century AD and today one of the world's most famous monuments (see photo). It was originally constructed as a Hindu temple, dedicated to the god Vishnu. The carvings on the walls of Angkor Wat tell the stories of the Hindu epics, and the temple's designs are closely

congregate in the realm 's major cities where they established Chinatowns and gained control over much of the local commerce (see **Box 10-4**). World War II was about to break out and the colonial era would end soon thereafter.

Figure 10-7 shows the migration routes and current concentrations of Chinese in Southeast Asia. Most migrants originated in southern China's Fujian and Guangdong provinces, and a large number invested much of their wealth back in China when the country opened up to foreign businesses in the 1980s. Clearly, the Overseas Chinese of Southeast Asia have played a significant role in shaping the People's Republic of China's (PRC's) economic miracle.

Today Southeast Asia is home to almost 30 million Overseas Chinese, more than half of this Sinic diaspora's worldwide total. Their lives here have often been challenging. The conquering Japanese relentlessly persecuted Chinese living in Malaya during World War II. Later, in the 1960s, Chinese in Indonesia were accused of communist sympathies, and hundreds of thousands were killed. In the late 1990s, Indonesian mobs again attacked Chinese and their property because they resented their relative wealth and because many Chinese had converted to Christianity during the colonial era and were now targeted by Islamic throngs. Such hostility

---

Box 10-5   Regional Issues: The Chinese Presence in Southeast Asia

The Chinese Are Too Influential!

"It's hard to imagine that there was a time when we didn't have Chinese minorities in our midst. I think I understand how Latin Americans [sic] feel about their 'giant to the north.' We've got one too, but the difference is that there are a lot more Chinese in Southeast Asia than there are Americans in Latin America. The Chinese even run one country because they are the majority there, Singapore. And if you want to go to a place where you can see what the Chinese would have in mind for this whole region, go there. They're knocking down all the old Malay and Hindu quarters, and they've got more rules and laws than we here in Indonesia could even think of. I'm a doctor here in Bandung and I admire their modernity, but I don't like their philolosophy.

"We've had our problems with the Chinese here. The Dutch colonists brought them in to work for them, they got privileges that made them rich, they joined the Christian churches the Europeans built. I'm not sure that a single one of them ever joined a mosque. And then shortly after Sukarno led us to independence they even tried to collaborate with Mao's communists to take over the country. They failed and many of them were killed, but look at our towns and villages today. The richest merchants and the money lenders tend to be Chinese. And they stay aloof from the rest of us.

"We're not the only ones who had trouble with the Chinese. Ask them about it in Malaysia. There the Chinese started a full-scale revolution in the 1950s that took the combined efforts of the British and the Malays to put down. Or Vietnam, where the Chinese weren't any help against the enemy when the war happened there. Meanwhile they get richer and richer, but you'll see that they never forget where they came from. The Chinese in China boast about their coastal economy, but it's coastal because that's where our Chinese came from and where they sent their money when the opportunity arose.

"As a matter of fact, I don't think that China itself cares much for or about Southeast Asia. Have you heard what's happening in Yunnan Province? They're building a series of dams on the Mekong River, our major river, just across the border from Laos and Burma. They talk about the benefits and they offer to destroy the gorges downstream to facilitate navigation, but they won't join the Mekong River Commission. When the annual floods cease, what will happen to Tonlé Sap, to fish migration, to seasonal mudflat farming? The Chinese do what they want—they're the ones with the power and the money."

The Chinese Are Indispensable!

"Minorities, especially successful minorities, have a hard time these days. The media portray them as exploiters who take advantage of the less fortunate in society, and blame them when things go wrong, even when it's clear that the government representing the majority is at fault. We Chinese arrived here long before the Europeans did and well before some other 'indigenous' groups showed up. True, our ancestors seized the opportunity when the European colonizers introduced their commercial economy, but we weren't the only ones to whom that opportunity was available. We banded together, helped each other, saved and shared our money, and established stable and productive communities. Which, by the way, employed millions of locals. I know: my family has owned this shop in Bangkok for five generations. It started as a shed. Now it's a six-floor department store. My family came from Fujian, and if you added it all up we've probably sent more than a million U.S. dollars back home. We're still in touch with our extended family, and we've invested in the Xiamen SEZ.

"Here in Thailand we Chinese have done very well. It is sometimes said that we Chinese remain aloof from local society, but that depends on the nature of that society. Thailand has a distinguished history and a rich culture, and we see the Thais as equals. Forgive me, but you can't expect the same relationship in East Malaysia. Or, for that matter, in Indonesia, where we are resented because the 3 percent Chinese run about 60 percent of commerce and trade. But we're always prepared to accommodate and adjust. Look at Malaysia, where some of our misguided ancestors started a rebellion but where we're now appreciated as developers of high-tech industries, professionals, and ordinary workers. This in a Muslim country where Chinese are Christians or followers of their traditional religions, and where Chinese regularly vote for the Malay-dominated majority party. So when it comes to aloofness, it 's not us.

"The truth is that the Chinese have made great contributions here, and that without the Chinese this place would resemble parts of South or Southwest Asia. Certainly there would be no Singapore, the richest and most stable of all Southeast Asian countries, where minorities live in peace and security and where incomes are higher than anywhere else. In fact, mainland Chinese officials come to Singapore to learn how so much was achieved there. Imagine a Southeast Asia without Singapore! It's no coincidence that the most Chinese country in Southeast Asia also has the highest standard of living."

---

associated with Hindu cosmology. But during the fourteenth century, Buddhism took over this part of the realm, and the temple complex was transformed into a major place of worship for Buddhists.

Vietnam represents what is surely the ultimate fusion of diverse religious influences in Southeast Asia. In this country, for centuries there has been an almost casual blending of early Hinduism with Buddhism, Daoism, and Confucianism—all mixing comfortably with age-old traditions of ancestor worship.

The Imprint of Colonialism

When the European colonizers arrived in Southeast Asia, they encountered a patchwork of kingdoms, sultanates, principalities, and other traditional political entities whose leaders they tried to co-opt, overpower, or otherwise fold into their imperial schemes. These colonizers forged empires here, often by playing one state off against another; the Europeans divided and ruled. Out of this foreign intervention emerged

Cambodia's Angkor Wat, one of the world's most famous historic religious sites, is a blend of Hindu and Buddhist influences. This treasure-filled architectural triumph embodies the cultural crossroads that is Southeast Asia.

the modem map of Southeast Asia, and only Thailand (formerly Siam) survived the colonial era as an independent entity—because it conveniently served as a buffer between the French sphere to the east and the British to the west.

Figure 10-8 shows the colonial framework in the late nineteenth century (before the United States assumed control over the Philippines in 1898). The colonial powers divided their possessions into administrative units as they did in Africa and elsewhere. Some of these political entities became independent states when the colonial powers withdrew or were ousted by force. The French named their empire **Indochina [12]**. The *Indo* part of Indochina refers to cultural imprints received from South Asia: the Hindu presence; the importance of Buddhism, which came to Southeast Asia via Ceylon (Sri Lanka) and its seafaring merchants; the influences of Indian architecture and art (especially sculpture), writing, and literature as well as social structures. The *China* in the name Indochina signifies the role of the Chinese here: its emperors coveted Southeast Asian lands, and China's power penetrated deeply into this realm. French Indochina consisted of present-day Vietnam, Cambodia, and Laos.

The British ruled a pair of major entities in Southeast Asia—Burma and Malaya—in addition to a large part of northern Borneo and many small islands in the South China Sea. Burma (now Myanmar) became independent in 1948, and the Federation of Malaysia was created in 1963 by the political unification of recently independent mainland Malaya, Singapore, and the former British dependencies on the largely Indonesian island of Borneo. Singapore, however, left this Federation in 1965 to become a sovereign city-state, and the remaining components were later restructured into peninsular Malaysia and, on Borneo, Sarawak and Sabah. Thus the term *Malaya* properly refers to the geographic territory of the Malay Peninsula, including Singapore and other nearby islands; the term *Malaysia*

identifies the politico-geographical entity of which Kuala Lumpur is the capital.

The Dutch had come to Southeast Asia in the late sixteenth century, primarily attracted by the so-called spice trade, and they ended up controlling most of what is now Indonesia. Among the plants domesticated by the local people on these islands of the "East Indies" were black pepper, cloves, cinnamon, nutmeg, ginger, turmeric, and other condiments essential to flavorful meals. In Figure 10-1 you can detect, in eastern Indonesia between Sulawesi and New Guinea, a group of small islands called the Maluku Islands (formerly the Moluccas). The Dutch colonizers called these the **Spice Islands** because of the lucrative commerce in spices long carried on by Arab, Indian, and Chinese traders. Java (Jawa), the most populous and productive island, became the focus of Dutch administration; from its capital at Batavia (now Jakarta), Dutch colonialism threw a girdle around Indonesia's more than 17,000 islands, paving the way for the creation of the realm's largest and most populous state that is now home to more than a quarter-billion people.

Finally, the Philippines had been controlled by Spain since 1571. But these colonizers were confronted with a major uprising here just as the Spanish-American War broke out over the Caribbean island of Cuba in 1898 (there, as well, the locals revolted against Spanish domination). As part of the settlement of that brief war, the victorious United States replaced Spain in Manila as colonial proprietor and eventually guided the Philippines to independence in 1946. Today all 11 of Southeast Asia's states are independent, but centuries of imperial rule have left deeply embedded cultural imprints. In its urban landscapes, education systems, languages, and many other spheres, Southeast Asia will long carry the legacies of its colonial past.

Southeast Asia's Emerging Markets

If Japan and the Asian Tigers set the early example in the 1960s and 1970s, with post-Mao China following suit since the mid-1980s, several Southeast Asian countries have since 2000 joined the ranks of the world's rapidly **emerging markets [13]**. Vietnam, Indonesia, and Malaysia, particularly, have in recent years attracted substantial foreign investment and exhibited decidedly robust economic growth rates.

But this realm is marked by decidedly extreme economic disparities. The tiny oil-rich state of Brunei has an income higher than that of the United States (although this figure is not indicative of the earnings of its relatively large foreign workforce). However, it is comprehensively developed Singapore that records Southeast Asia's highest income—at a level about halfway between those of wealthy Luxembourg and Switzerland. Malaysia comes in a distant third and is followed by Thailand. Populous Indonesia and the Philippines exhibit per-capita incomes less than one-twelfth that of Singapore's. Finally, Vietnam, Laos, Cambodia, Myanmar, and East Timor, in that order,

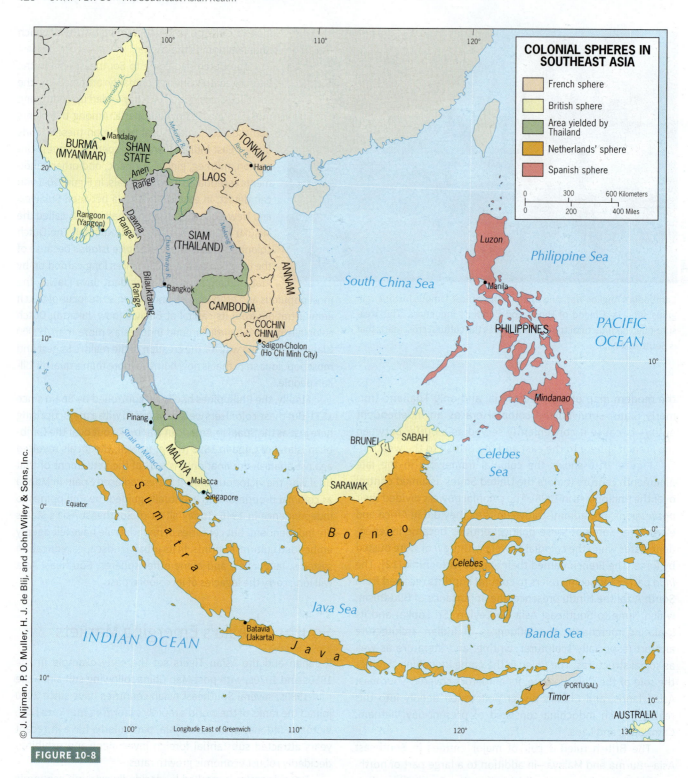

COLONIAL SPHERES IN SOUTHEAST ASIA

- French sphere
- British sphere
- Area yielded by Thailand
- Netherlands' sphere
- Spanish sphere

0   300   600 Kilometers
0   200   400 Miles

FIGURE 10-8

rank at the bottom of the list. Indeed, economically this realm is so internally dissimilar that its countries fall into all four World Bank income categories (see Fig. G-11).

Singapore's Leadership

With its soaring per-capita income, there can be little doubt that Singapore is the economic heart of Southeast Asia. With a mere 5.7 million citizens and only 619 square kilometers (240

sq mi) of territory, we are obviously not referring to size—this is all about the connections and centrality that enable Singapore to function as the predominant **node [14]** in a realmwide economic network. At the same time, geographers also rank Singapore as a top-tier world-city because it has major international linkages and exerts global influence. Singapore's container port is not only the largest in the realm but also the second-biggest in the world, underscoring its key role within and far beyond Southeast Asia (see photo). Moreover, ultramodern Changi

Singapore, Southeast Asia's unmatched world-city, has a state-of-the-art container port whose annual cargo-handling volume is second only to Shanghai's. This thriving city-state simultaneously functions as the nucleus of the realm's spatial economic system as well as its vitally important gateway to the global economy.

Airport Singapore (SIN) is consistently ranked as one of the world's best international airports.

Singapore's commanding regional position results from its exceptional relative location. In the seventeenth century, it was Malacca farther up the Malay Peninsula that was the leading hub for trade and shipping in Southeast Asia. The Strait of Malacca (Fig. 10-8), named after that town, was already the most important sea route, providing access to the realm's waters for ships coming from the west. When the British sought to displace Dutch dominance here during the eighteenth century, they discovered an even better local base of operations: Singapore Island. Because it possessed a larger and deeper natural harbor than Malacca to accommodate the bigger steamships of the time, Singapore swiftly rose to prominence as the British consolidated their power over this key corner of the world.

Most importantly, Singapore today is a symbol of modernity, a model for Southeast Asia's future. Throughout the realm, those who can afford it go there to shop, to connect to intercontinental flights, to transact business, to invest in real estate, or to send their children to one of the city-state's highly ranked universities. Imagine other Southeast Asian localities following in Singapore's footsteps over the next several years and consider how that might impact the realm and the world beyond!

Prospects of Realmwide Integration: ASEAN

The overall development of Southeast Asia still has a long way to go. Political stability and increased regional integration will facilitate the process, and that is the long-term goal of **ASEAN [15]**), the Association of Southeast Asian Nations. Founded during the late 1960s, this supranational organization has primarily been concerned with security. But that has

been a constantly challenging effort because a wide range of conditions mark its ten member-states. With the lone exception of minuscule East Timor (which is now seeking to join), ASEAN encompasses all of the realm's countries. These include one influential city-state (Singapore); an Islamic oil state (Brunei); two impoverished communist regimes (Vietnam and Laos); and a reforming military dictatorship whose population is rising from the ranks of the world's most deprived (Myanmar).

Another problem that ASEAN has failed to resolve, one that literally affects health across much of this realm, is the recurrent air pollution caused by Indonesia's massive, human-ignited forest fires. Depending upon weather conditions and prevailing winds, thick plumes of smoke emanating from Indonesia's Sumatera (where most of the burning occurs, often to enlarge oil palm plantations) stream out toward Jawa, Singapore, Malaysia, and countries farther afield. Repeatedly, Singapore's government has had to advise residents to stay indoors (a new record-high pollution level was set in 2013). Indonesia has been especially slow to address this environmental crisis and even refused to ratify the 2002 ASEAN Agreement on Transboundary Haze Pollution.

In 1992, 25 years after its founding, ASEAN was able to expand into the economic domain through **AFTA [16]**, the ASEAN Free Trade Agreement, and here the payoff has been more substantial. AFTA has both engendered the lowering of tariffs and encouraged the expansion of trade within Southeast Asia. With lower wages than China, certain foreign-investment flows (e.g., in the garment industry) have shifted to Southeast Asia. And intra-realm trade has indeed surged in recent years—a critical forward step to avoid being completely overshadowed by China.

China in Southeast Asia Today

Since the turn of this century, Southeast Asia has increasingly been drawn into the economic (and political) orbit of China. In 2010, ASEAN and China concluded a landmark free-trade agreement, with the Chinese most interested in acquiring raw materials from mainland Southeast Asia as well as accessing the realm's growing export markets. As **Figure 10-9** in **Box 10-6** shows, trade shifts resulting from this agreement have been significant.

East Asia and Southeast Asia are now steadily growing closer economically. This heightened interaction is not only about trade but also about investment flows, industrial development, divisions of labor, and infrastructural integration. With wages in Southeast Asia currently about one-third those of China's, this differential is fueling the relocation of industrial activity from China to Southeast Asian countries.

Those manufactures are then exported directly from those countries to China and/or other parts of the world. Increasingly, it is low-skill jobs that move to, Southeast Asia, whereas the better-paying, higher-skilled jobs remain in China.

Box 10-6 MAP ANALYSIS

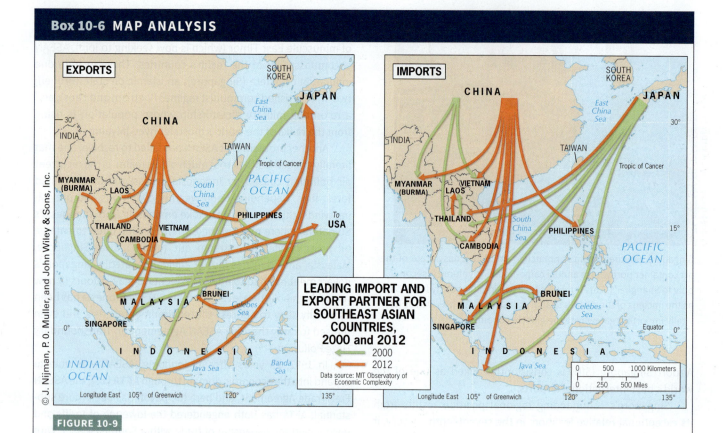

FIGURE 10-9

Figure 10-9 enables you to study major trade shifts between 2000 and 2012. The map on the left shows the largest import origins for Southeast Asian countries in 2000 and in 2012; the right map shows the largest export destinations for those same years. For how many Southeast Asian countries was China the biggest export destination in 2012? For how many countries was China the leading provider of imports? Which two powers outside the realm have lost substantial ground (to China) in Southeast Asian trade since 2000?

Access your WileyPLUS Learning Space course to interact with a dynamic version of this map and to engage with online map exercises and questions.

In order to facilitate the increased production and transportation of goods within as well as to and from China, the Chinese established the Asian Infrastructure Investment Bank (AIIB) in 2015. The AIIB's members include most of the realm's states, and several European countries plus New Zealand have also joined (the United States, for geopolitical reasons, has not). With the Chinese now investing heavily to exploit Southeast Asia's raw materials, infrastructural improvements to upgrade connections to China are a leading priority. Most important of all is the high-speed rail network originating in Kunming, interior southern China's routeway hub for continental Southeast Asia and capital of Yunnan Province. The main rail corridor leads south from Kunming across northern Laos to Vientiane, the Laotian capital on the Mekong River that also forms the border with Thailand. From there, state-of-the-art rail lines will fan out in stages to Thailand, Cambodia, Vietnam, Malaysia, and eventually Singapore. Another high-priority project is the recently completed pair of oil and gas pipelines from Myanmar's Indian Ocean coast to Kunming (discussed later in the section on Myanmar).

As a result of this Chinese regional economic stimulation, Southeast Asia is growing faster than any other developing realm. And since the building of this inter-realm relationship has been fast-tracked, East and Southeast Asia are already more economically integrated than either North America with Middle America or Europe with Africa. Yet even though China's boost to the realm's economic development has been quite successful so far, there is another side to this story. Because the Chinese have not been reticent in asserting themselves, their continuing drive for domination is creating both uneasiness and growing concern in a number of the realm's quarters. Some of this anxiety stems from China's sheer economic prowess, which produces rather asymmetrical relationships; for instance, many infrastructure projects are formidable joint ventures that require comparatively larger investments from these much smaller Southeast Asian states. And, as is usually the case, the benefits are more favorable for China than for the host nation. Moreover, this issue is not entirely about economics: for countries such as Vietnam or Japan, historic sensitivities and/or longstanding rivalries inevitably come into play. Still another issue is a

Box 10-7 MAP ANALYSIS

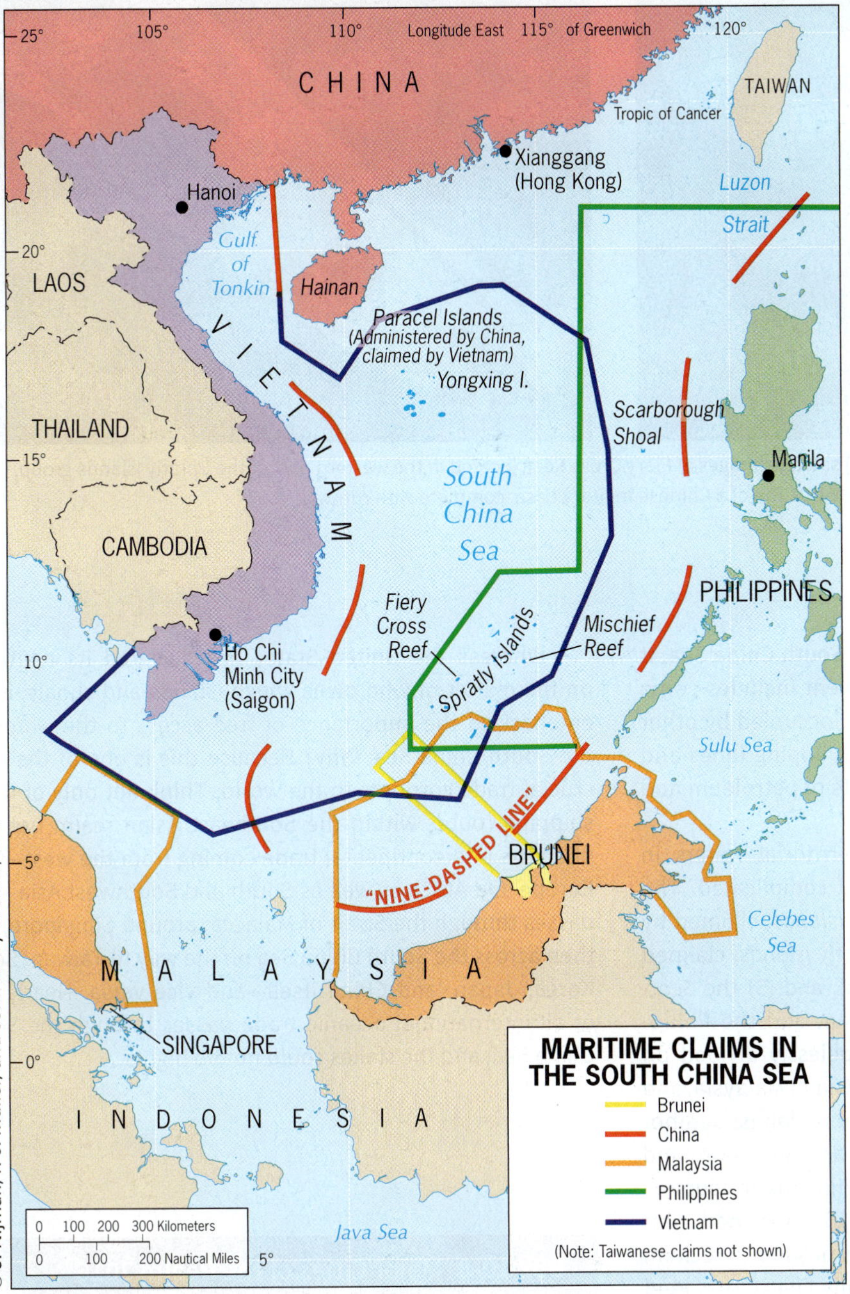

The South China Sea is now the world's most contested maritime geopolitical space. The name South *China* Sea reflects the dominance of China but other countries, such as the Philippines, Malaysia, and Vietnam, have staked their claims as well, mostly on the basis of United Nations maritime conventions (discussed in Chapter 12). The map shows the extent of these conflicting claims. Which pairs of countries are engaged in the greatest territorial conflicts? Are there any conflicts that do not involve China? What would your position be on these conflicts if you were the government of Singapore?

Access your WileyPLUS Learning Space course to interact with a dynamic version of this map and to engage with online map exercises and questions.

MARITIME CLAIMS IN THE SOUTH CHINA SEA

- Brunei
- China
- Malaysia
- Philippines
- Vietnam

(Note: Taiwanese claims not shown)

FIGURE 10-10

© J. Nijman, P. O. Muller, and John Wiley & Sons, Inc.

perceived lack of Chinese diplomacy as well as consideration for regional and national interests. Consider this remark made by China's foreign minister not long ago: "China is a big country and other countries are small countries, and that is just a fact."

Geopolitics in the South China Sea

If China's economic role in Southeast Asia is frequently viewed with mixed feelings, its recent geopolitical and military forays are increasingly met with indignation and opposition. Much of this revolves around China's maritime ambitions and its claims to territorial waters in the South China Sea. Since 2009, the PRC government has circulated its so-called **nine-dashed-line map [17]** displaying Chinese claims in these waters—a map said to date back to 1949, the year the PRC was founded. The 'nine dashes' refer to the delimitation of Chinese maritime claims that effectively cover most of the South China Sea, and are mapped in red in **Figure 10-10** in **Box 10-7**. In the words of one high-ranking Chinese official

Before-and-after (ca. 2013/September 22, 2015) satellite images of Fiery Cross Reef, located in the western part of the Spratly Islands group. The 2015 image shows the nearly completed construction of a Chinese military base, complete with runway.

in 2012, "China does not want all of the South China Sea, it just wants 80 percent." But that 80 percent includes some 40 islands (which China says are illegally occupied by other countries) as well as vital international shipping lanes and seafloor zones potentially rich in deposits of petroleum and natural gas.

The disputes and conflicts over territorial waters in the South China Sea are numerous and complicated. The most important concern: (1) the *Paracel Islands*, claimed by China, Taiwan, and Vietnam; (2) the *Spratly Islands*, claimed by China, Vietnam, Malaysia, and Brunei; and (3) the *Scarborough Shoal*, claimed by China, Taiwan, and the Philippines (Fig. 10-10). ASEAN has been hopelessly divided on these geopolitical controversies, with Vietnam, Malaysia, and the Philippines strongly countering Chinese claims; Cambodia and Laos (tacitly) supporting China; and Singapore and Indonesia trying to steer a more neutral, diplomatic course.

When satellite imagery in mid-2015 confirmed that China was dredging enormous quantities of sand to reclaim and enlarge several major reefs (especially Fiery Cross Reef and Mischief Reef; see photos) in the southern part of the Sea to construct runways and naval bases, Vietnam, Thailand, and the Philippines began to pursue closer ties to the United States, clearly in search of an appropriate counterweight to Chinese leverage. All three quickly concluded military agreements with the United States that include joint military exercises as well as flyovers near Chinese construction sites by the U.S. Air Force. Significantly, the Philippines took its case to the International Court of Justice, which ruled against the PRC in July 2016, saying that China's claims had no legal basis and that it had violated international law. The Chinese quickly declared the ruling "invalid", and with "no binding force." Nevertheless, this ruling is likely to strengthen the non-Chinese claims.

Whereas the United States has declared its neutrality on the matter of who owns which islands and shoals, it has emphasized the importance of free access to the waters of the South China Sea. Why? Because this is one of the most crucial trade corridors in the world. Think not only of major shipping routes within the Southeast Asian realm but also of all the intercontinental trade coming from the west—from Europe and Africa as well as South and Southwest Asia—that passes through the Strait of Malacca, around Singapore, and then across the South China Sea on the way to Taiwan, South Korea, Japan, and China itself—and vice versa. Nearly half of all international oceanic trade passes through the South China Sea, and the stakes could not be higher.

Box 10-8 Major Cities of the Realm 2016

Jakarta, Indonesia	31.3
Manila, Philippines	22.9
Bangkok, Thailand	15.3
Ho Chi Minh City (Saigon), Vietnam	10.1
Hanoi, Vietnam	7.4
Kuala Lumpur, Malaysia	7.4
Singapore, Singapore	5.7
Yangon, Myanmar	5.3
Phnom Penh, Cambodia	1.9
Vientiane, Laos	1.0

Regions of Southeast Asia

Southeast Asia's first-order regionalization recognizes the mainland-island dichotomy that is obvious from any map of this realm. But things are quite a bit more complicated than that because the southern part of the Malay Peninsula, occupied by the states of Malaysia and Singapore, exhibits physiographic, historical, and cultural characteristics to justify its inclusion in the *Insular* (island) rather than the *Mainland* region. Using the political framework as our grid, we map the broadest regional division of Southeast Asia in **Figure 10-11**:

Mainland region: Vietnam, Cambodia, Laos, Thailand, and Myanmar (Burma)

Insular region: Malaysia, Brunei, Singapore, Indonesia, East Timor (Timor-Leste), and the Philippines

Special note should also be taken of the discordance between the political matrix and the cultural-geographic reality along the eastern periphery of the Insular region. Here, even though Indonesia controls the western half of the island of New Guinea, all of New Guinea belongs to the neighboring Pacific realm.

Region Mainland Southeast Asia

Five countries form the Mainland region of Southeast Asia (Fig. 10-11): the three remnants of French Indochina—Vietnam,

Cambodia, and Laos—in the east; Thailand in the center; and Myanmar (formerly Burma) in the west. Although a single religion, Buddhism, dominates cultural landscapes, this is a multicultural, multiethnic region (see Fig. 10-6). This may be one of the less urbanized regions of the world, yet the Mainland contains several major cities and the pace of urbanization is quickening today (see **Box 10-8**). Moreover, as Figure 10-11 shows, two countries (Vietnam and Myanmar) possess dual core areas.

We approach this region from the east, beginning our survey in Indochina where the United States fought and lost a disastrous war that ended in 1975, but whose impact on America is still felt today. After the Indochina War formally began in 1964 (U.S. involvement in Vietnam actually began earlier), some scholars warned that the conflict might spill over from Vietnam into Laos and Cambodia, and from there into Thailand, Malaysia, and even then-Burma. This view was based on the **domino theory [18]**, which holds that destabilization and conflict from any cause in one country can result in the collapse of order in one or more neighboring countries, triggering a falling-domino chain of events that can affect a series of contiguous states in a region. That did not actually happen, and in hindsight, at least, it is clear that some of the rhetoric of the time was infused with alarmist views from the West.

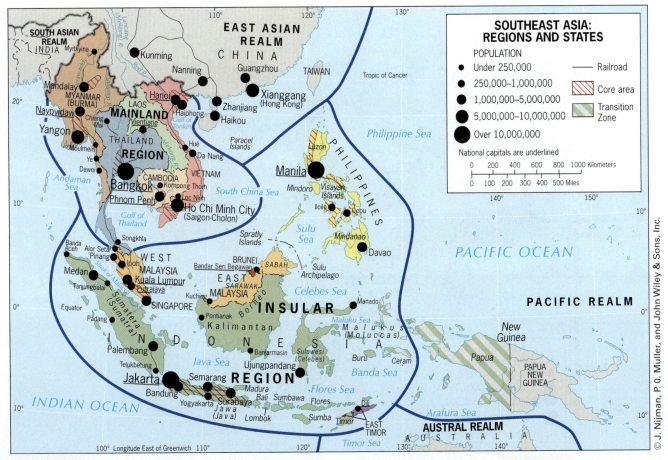

FIGURE 10-11

433

Vietnam: The Legacy of War

Vietnam (population: 94 million) still carries the scars of the Indochina War (1964–1975), even though the overwhelming majority of today's Vietnamese have no personal memory of that terrible conflict. Japan had invaded Vietnam during World War II, after which the French attempted to regain control, fighting Vietnamese nationalists in a war that lasted nine years before they were decisively ousted in 1954. But newly independent Vietnam did not become a unified state. Separate regimes took control: a communist one in the north based in Hanoi and a noncommunist counterpart in the south headquartered at Saigon. Vietnam's pronounced elongation had made things difficult for the French, and now it played a similar role during the postcolonial era. Note, in **Figure 10-12**, that Vietnam is widest in the north and south, with a slender

FIGURE 10-12

"waist" in its middle zone—ensuring that North and South Vietnam were worlds apart in a number of ways. For more than a decade, the United States tried to prop up the Saigon regime that controlled the south, but the communists eventually prevailed; like China, Vietnam is still ruled by a communist government today. After hostilities terminated in 1975, as many as 2 million Vietnamese refugees set out in often-flimsy watercraft onto the South China Sea; of those "boat people" who survived, a majority settled in the United States.

Vietnam in Transition

Vietnam officially became a unified state in 1976, and since then the contrasts between north and south have diminished. The northern capital, Hanoi, has long lagged behind bustling Saigon (now renamed Ho Chi Minh City to honor North Vietnam's revolutionary leader), but today its growing skyline reflects modernization and the overland links to its outport of Haiphong have been upgraded. With 7.5 million residents, Hanoi anchors the northern (Tonkin Plain) core area of Vietnam, the lower basin of the Red River (its agricultural hinterland).

A leading development focus in Vietnam since 2000 has been the expansion of agricultural production. One result is that the country is now the world's second largest coffee producer after Brazil, with most production taking place in the Central Highlands (Fig. 10-12). However, because Vietnam's mass-produced bulk coffee is raised plantation-style in the sun, the diffusion of this crop into the interior highlands has sparked substantial deforestation, worsening the annual monsoon-related floods that plague this now-unstable countryside. Other major Vietnamese crops, led by rice, include rubber, tea, sugar, and spices—all with increasingly favorable export opportunities.

The south has experienced more significant change and faster economic growth (see **Box 10-9**) as the country has opened up to the global economy. Today, Ho Chi Minh City is a burgeoning metropolis of 10.1 million, propelled by the rapid development of the east bank of the Saigon River, which includes a Chinese-style Special Economic Zone as well as the booming New Saigon business/residential district. In fact, the Saigon area now accounts for more than a quarter of Vietnam's industrial output and about a third of its tax revenues.

Somewhat ironically, Vietnam these days has become one of the most pro-American countries in Southeast Asia despite its persistence as a communist state. Undoubtedly, these good feelings toward their former enemy in the "American War" at least partially reflect the growing perception of dominance by China, against whom the United States can provide a counterbalance. As noted above, American-Vietnamese relations have warmed considerably during the 2010s, with Vietnam even opening its strategically located port at Cam Ranh Bay to U.S. naval forces—the original builder of these facilities during the Indochina War a half-century earlier (Fig. 10-12).

Cambodia

Compact Cambodia is heir to the ancient Khmer Empire whose capital was Angkor and whose legacy is a vast landscape of imposing monuments, including the majestic Buddhist-inspired temple complex of Angkor Wat (see **Box 10-10** on Technology & Geography). Today, more than 85 percent of Cambodia's 16 million inhabitants are ethnic Khmers, with most of the remainder divided between Vietnamese and Chinese. The current capital, Phnom Penh, lies on the

---

Box 10-9 From the Field Notes...

© Jan Nijman

"On a walking tour of Ho Chi Minh City (which many here still call Saigon), I encountered two contrasting images of present-day Vietnam, just minutes apart: a statue of the still-revered former communist leader Ho Chi Minh in front of City Hall; and, just three blocks away, the country's first Louis Vuitton store. The latter caters to the expanding base of higher-income Vietnamese consumers—not only a definite sign of national change but also that upscale designer fashion is truly global in scope."

Box 10-10 Technology & Geography

Making the invisible visible—LiDAR at Cambodia's Angkor Wat

Remote sensing refers to the indirect observation and measurement of the Earth with cameras, sensors, and scientific instruments mounted on unmanned aerial vehicles (UAVs or 'drones'), aircraft, or satellites. A special type of remote sensing technology known as 'Light Detection and Ranging' or *LiDAR* is now ushering in a new era of geographic exploration and discovery. Originally developed to measure such atmospheric phenomena as clouds and wind, LiDAR is now used to survey and map the surface of the Earth. LiDAR sensors actively transmit thousands of pulses of light per second with a laser and then record the length of time it takes for that pulse to be reflected back from the surface. When combined with the position, direction, and orientation of the aircraft and sensor, an incredibly rich elevation dataset is produced. What makes LiDAR imagery so special is its high resolution and accuracy. The density of LiDAR data also permits features such as forest and other vegetation cover to be filtered out, thereby making invisible objects on the Earth's surface visible.

LiDAR technology today is used and applied in an ever increasing variety of situations such as agriculture and farming, humanitarian response and disaster preparedness, climate change research, and cultural heritage preservation. Here at the famous site of Angkor Wat in Cambodia, LiDAR is used by archeologists to obtain insights into the Khmer Empire and its preindustrial urban complexes. A recent expedition that mapped the area surrounding Angkor Wat using LiDAR revealed a series of previously unknown roads, canals, and even temple structures that were obscured by dense vegetation— and made even more

difficult to survey due to land mines buried during decades of conflict and war. It turned out that Angkor's urban region was far more extensive than previously thought. LiDAR can play an important role in the preservation and management of such cultural heritage sites by revealing their true geographic extent; providing detailed inventories of structures, temples, and buildings; and even detecting illegal excavations, looting, or encroachment.

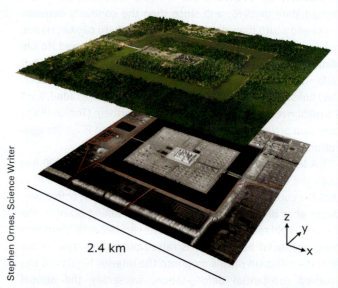

Stephen Ornes, Science Writer

The Angkor Wat temple complex and surroundings. Conventional high-resolution satellite image (top) showing mainly vegetation, compared to the LiDAR image (bottom) that reveals elevation details and terrain features such as roads, mounds, and canals.

Mekong River, which bisects Cambodia before it enters and forms its massive delta in Vietnam (Fig. 10-12).

From 1970 to 1976, Cambodia suffered one of the most brutal and murderous regimes of the twentieth century. The so-called *Khmer Rouge* (*Red Khmer*, the name being a reference to its self-declared communist convictions and to the glorious past of the Khmer people) embarked on a malevolent course of terror and destruction in order to reconstruct *Kampuchea*—as leader Pol Pot referred to Cambodia—as a rural society. They drove townspeople into the countryside where they had no place to live or work, emptied hospitals to send the sick and dying into the streets, and outlawed both religion and family. By the time they were ousted from power by neighboring Vietnamese forces, as many as 2 million Cambodians (out of a population of 8 million at the time) had lost their lives in what forever will be known as the Cambodian Genocide.

Even four decades later, the country's postwar trauma still lingers. Many young children of the 1970s, now in their forties or fifties, became orphans and faced an uphill struggle in life. Almost every family can tell of parents and/or close relatives who were murdered by the regime. The long-lasting effects of

social and economic dislocation, especially in the countryside, have still not been overcome (see **Box 10-11**).

Today, despite Cambodia's continuing status as one of Southeast Asia's poorest countries, it has finally begun to achieve some meaningful economic growth, especially in the garment manufacturing sector. Offshore oil reserves were discovered at the start of this decade, but exploration has yet to gather momentum because lower world petroleum prices after 2015 have done little to attract major oil companies. Other investments from abroad are multiplying, mostly by producers seeking to relocate their operations out of China (where wages are steadily rising) in order to take advantage of Cambodia's cheaper labor. The Cambodian government is also optimistic that tourism, largely focused on the Angkor temple complex in the northwest, will continue to expand and perhaps soon become a mainstay of the economy.

Laos

Landlocked Laos has no fewer than five neighbors, one of which is East Asia's giant, China (Fig. 10-12). The Mekong River forms a long stretch of its western boundary, and the

Box 10-11 From the Field Notes...

"Along one of the access roads leading to the monumental temple complex of Angkor Wat, I encountered a group of middle-aged men playing traditional Cambodian music. When I got closer, I noticed that all were amputees, a legacy of the vicious Pol Pot regime of the 1970s. During that upheaval and the war that followed, which resulted in the regime's ouster, millions (!) of land mines were deployed across the country. Ominously, a large number remain buried in place, and the records indicating their location are poor or nonexistent. Not surprisingly, they continue to inflict casualties to this day, maiming and killing civilian men, women, and children. There are socioeconomic consequences as well: some fertile parts of the country, known to be infested with unexploded mines, are avoided by farmers; and when misfortune strikes, entire families suffer should a parent become disabled and no longer able to work. Most of the men performing here lost their limbs when they were adolescents or young adults. They now earn their livelihood from performing and selling their music—while actively campaigning for

better government efforts to detect and disarm the mines. 'It is 40 years since Pol Pot,' one of the men told me during a performance break, 'but our country is still recovering.'"

© Jan Nijman

important sensitive border with Vietnam to the east lies in mountainous terrain. With only 6.9 million people—more than half of them ethnic Lao who are related to the Thai of Thailand—the country has a woefully inadequate infrastructure, hardly any industry, and less than 5 percent of its land is suitable for agriculture. Laos is only one-third urbanized, but the capital and largest city, Vientiane, has a fast-growing population of just over one million.

Communist Laos has been very slow to follow the lead of its neighbors in opening up its economy, but today momentum is finally beginning to build. The country's first stock market began operating in 2011. Around the same time, a major new hydropower project on the central Laotian Nam Theun River, a tributary of the Mekong, began producing electricity for much of central Laos as well as selling surplus power to Thailand. And we have already discussed the much bigger Xayaburi Dam project on the Mekong itself (see Fig. 10-3), which despite drawing wide regional criticism is scheduled to open before the end of the decade. Moreover, the soon-to-be-completed high-speed rail line from Kunming to Vientiane is certain to pull Laos even more tightly into China's orbit.

Thailand

In virtually every way, centrally positioned Thailand is the leading Mainland state. In contrast to its awakening neighbors, Thailand has been a strong participant in the realm's economic development for decades. This country of 68 million also exhibits some impressive demographic indicators: its annual growth rate is the lowest in the realm (even below Singapore's); it is the only Southeast Asian country whose projected 2050 population is lower than today's total (by some 8 percent); and Thailand leads its region in literacy and life expectancy. Its capital, Bangkok—profiled

in **Box 10-12**—is one of the world's most prominent **primate cities [19]** and the largest urban agglomeration in the Mainland region—a megacity containing more than 15 million residents today.

As Figure 10-11 indicates, Thailand occupies the heart of the Mainland region. Even though it has no Red, Mekong, or Irrawaddy Delta, this country's central lowland is watered by a network of streams that flow off the northern highlands and the Khorat Plateau to the east. One of these waterways, the Chao Phraya, is the Mississippi of Thailand and forms the axis of the country's core area (**Fig. 10-13**). From the head of the Gulf of Thailand to Nakhon Sawan, this river functions as a highway network for boat traffic. Barge trains loaded with rice head toward the coast, ferries move upstream, and freighters transport tin and tungsten (of which Thailand is among the world's leading producers). Bangkok sprawls along both sides of the lower Chao Phraya floodplain, here flanked by skyscrapers, pagodas, modern factories, boatsheds, ferry landings, luxury hotels, and myriad modest dwellings—all crammed together in crowded confusion on the swampy terrain where the central lowland meets the Gulf of Thailand.

At least some of Thailand's internal political instability of recent years can be attributed to the overreaching dominance of Bangkok and its elites as well as this primate city's considerable distance from the substantial electorates of the more peripheral rural areas. Marked by frequent government turnovers and episodes of military intervention, Thailand has yet to face up to the challenge of legitimizing national politics across this sprawling, unevenly developed country. And given Bangkok's location, particularly in the context of Thailand's protruded territorial configuration, the task of national governance is not made any easier by the presence of a large Malaysian—and overwhelmingly Islamic—minority

---

Box 10-12 Among the Realm's Great Cities...

Bangkok

Containing a population of 15.3 million, Bangkok is mainland Southeast Asia's largest city, a sprawling metropolis on the banks of the Chao Phraya River, an urban agglomeration without a true center, and an aggregation of neighborhoods ranging from the immaculate to the impoverished. In this city of great distances and sweltering tropical temperatures, getting around is often difficult because roadways are choked with traffic. A (diminishing) network of waterways affords the easiest way to travel, and life focuses on the busiest waterway of all, the Chao Phraya. Ferries and water taxis carry tens of thousands of commuters and shoppers across and along this bustling riverine artery, flanked by a growing number of high-rise office buildings, luxury hotels, and ultramodern condominiums. Many of these contemporary structures reflect the Thais' fondness for domes, columns, and small-paned windows, creating a skyline unique in Asia.

Gold is Buddhist Thailand's symbol, adorning religious and nonreligious architecture alike (see Box 10-12). From a high vantage point in the city, you can see hundreds of golden spires, pagodas, and façades rising above the townscape. The Grand Palace, where royal, religious, and public buildings are crowded inside a white, crenellated wall 2 kilometers (more than 1 mi) long, is a gold-laminated city within a city embellished by ornate gateways, dragons, and other statuary. Across the mall in front of the Grand

Palace are government buildings sometimes targeted by rioters in Bangkok's volatile political environment. And not far away are Chinatown (see Box 10-4) and myriad markets, all components of Bangkok's throbbing commercial life.

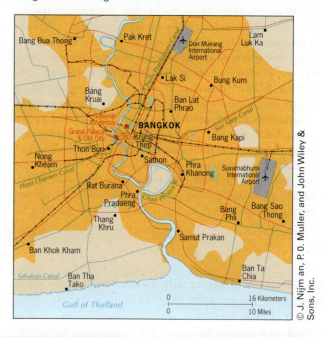

---

on the Malay Peninsula, Thailand's most remote corner that abuts the Malaysian border in the far south (Fig. 10-13).

The Restive Peninsular South

As we have just noted again, Thailand is a classical protruded state, with its southward-pointing land corridor extending away from the more compactly shaped main body of the state. Thailand's protrusion extends for hundreds of kilometers into the upper Malay Peninsula, narrowing to less than 32 kilometers (20 mi) in the vicinity of the Kra Isthmus (Fig. 10-13). A recheck of Figure 10-6 shows the dominance of ethnic Malays in the southernmost periphery of the protrusion—a population that is about 90 percent Muslim (the figure for Thailand as a whole is less than 6 percent). Moreover, the international border between Thailand and Malaysia is quite porous and can be crossed practically at will.

For more than a century, these southernmost provinces of Thailand have had closer ties with Malaysia across the border than with the Thai capital 1000 kilometers (600 mi) to the north. Today, rising Islamic militancy and violence here pose a mounting challenge to the Bangkok government (more than 5000 have been killed and thousands more maimed by local terrorist attacks since the mid-2000s). These events also have economic consequences: Phuket and other coastal resorts facing the Andaman Sea in this area—key components of Thailand's important tourist industry—are located just northwest of the troubled Malaysian border zone and are regarded by many as

increasingly vulnerable to a spillover of continuing hostilities (see Fig. 10-13, inset map).

Myanmar

Thailand's western neighbor, Myanmar (often still referred to as Burma, its former name), has a population of 55 million and is one of Asia's poorest countries. Long oppressed by one of the world's most corrupt and brutal military dictatorships, its government turned a corner in 2011 and soon surprised the world by taking bold steps forward. Grinding poverty, crippling isolation along with sanctions (imposed by much of the world because of human rights violations), and growing dependence on China must finally have played a part in pushing Myanmar's leadership in a new direction. When the West embargoed and worked to isolate Myanmar's military regime in years past, China took the opportunity to invest an estimated U.S. $27 billion here; nonetheless, the Chinese are likely to retain their influence in Myanmar as the country opens up to the rest of the world.

Myanmar's Geographic Challenges

Myanmar's challenges are in some ways comparable to those of Thailand: both are protruded states, and in both cases the primary urban region is quite distant from the outermost rural peripheries (Fig. 10-14). Yangon (previously named Rangoon) is the

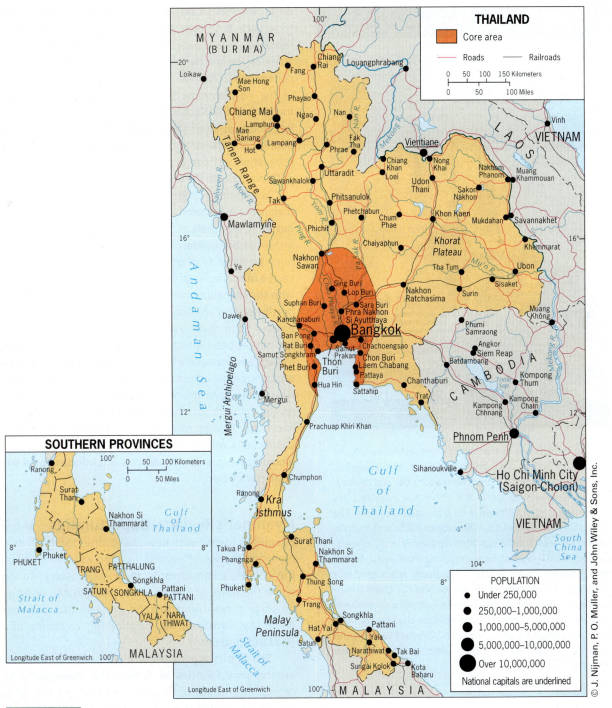

FIGURE 10-13

country's main port, economic center, and primate city, and has a population of about 5.5 million. In 2006, the military junta capriciously relocated the capital from Yangon 300 kilometers (200 mi) northward to a newly constructed headquarters named Naypyidaw, with the government asserting that its more central location would yield greater administrative efficiency as well as better protection against (unspecified) enemies. After a slow start, the new capital city now houses over one million people, not far behind the population size of the second city, the precolonial hub of Mandalay in central Myanmar (Fig. 10-14).

As Figure 10-14 reveals, the peripheral peoples of Myanmar occupy significant parts of the country that are designated as seven different States. The majority Burman-dominated areas, encompassing 68 percent of a national population of 55 million, are designated as Divisions. Among the leading minorities, the Shan of the northeast and far north, who are related to the neighboring Thai, account for 9 percent of the population, or 4.9 million. The Kayin (Karen), who constitute 7 percent (3.8 million), live in the neck of Myanmar's protrusion and have proclaimed their desire to create an autonomous territory within a federal Myanmar. The Rohingyas (2.4 percent; 1.3 million) of

Box 10-13 From the Field Notes...

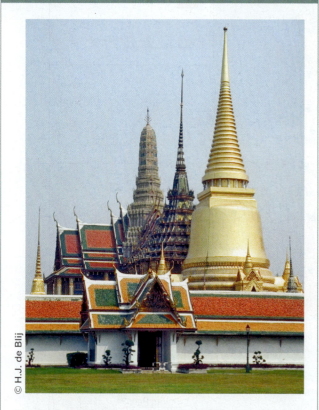

© H.J. de Blij

"Thailand exhibits one of the world's most distinctive cultural landscapes, in both its urban and rural areas. Graceful pagodas, stupas, and spirit houses of Buddhism adorn city and country-side alike. Gold-layered spires rise above and beautify cities and towns that would otherwise be drab and featureless. The architecture of Buddhism has diffused into public architecture, so that many secular buildings, from businesses to skyscrapers, are embellished by something approaching a national style. The grandest expression of the form undoubtedly is a magnificent assemblage of structures within the walls of the Grand Palace in the capital, Bangkok, of which a small sample is shown here. Climb onto the roof of any tall building in the city's center, however, and the urban scene will display hundreds of such graceful structures, interspersed with the modern and the traditional townscape."

press censorship was largely ended, and long-overdue elections took place. But a byproduct of the ending of that lengthy era of repression was the resurfacing of ugly, long-simmering ethnic tensions. Besides the appalling ethnic cleansing of the Rohingya community in Rakhine State, violent clashes have taken place in the northernmost State of Kachin between government forces and the Kachin Independence Army. This strife also has potential international ramifications because the Kachin people—800,000 strong, constituting 1.5 percent of Myanmar's population—are part of a more extensive ethnic community that spills across both the Chinese and Indian borders.

Yet another challenge from the periphery continues to plague the mountainous eastern margins of Shan State, which has long been part of the infamous opium-producing region known as the **Golden Triangle [20]**. That name refers to the huge profitability of opium (the raw material of heroin, harvested from the sap of poppies) on the global market, together with the triangular shape of this territory which encompasses northwestern Laos and northernmost Thailand in addition to eastern Shan State (Fig. 10-14). Although poppy cultivation has almost been wiped out in Thailand, it is as strong as ever in Laos and especially Myanmar. Today, with only Afghanistan producing more opium than Myanmar, it is hard to discern if the Naypyidaw government is truly interested in clamping down on the opium trade: this illicit narcotic remains the source of enormous foreign revenues that, through elusive business connections, tend to find their way into the formal economy—including real estate transactions and infrastructural works in Myanmar's cities.

Myanmar's Regional Opportunities

Myanmar's opening is loaded with new possibilities for itself, its neighbors, and the world beyond—reinforced in no small part by the country's considerable geographic assets. To demonstrate the range of Myanmar's opportunities, we focus here on three major infrastructural projects that are presently underway, involving the neighboring countries of India, Thailand, and China.

The first is a joint venture with India to renovate and expand port facilities in Sittwe, the strategically located coastal city facing the northern end of the Bay of Bengal (Fig. 10-14). India's overriding interest here is to increase the accessibility of its landlocked easternmost States, a deeply impoverished highland periphery difficult to reach via domestic overland routes. A fully developed port at Sittwe would allow the opening of a shorter, easier southern routeway: ships sailing around Bangladeshi waters from nearby Kolkata can offload their goods onto smaller vessels at Sittwe; the smaller vessels could then travel up the Kaladan River to the Indian border, from where trucks using upgraded roads would cover the final leg (Fig. 10-14).

The second project, located in the far south, is a far more significant enterprise from Myanmar's point of view—the so-called Dawei Development Project, a joint initiative with neighboring Thailand to convert the small southern city of

Rakhine State on the northwestern coast comprise the besieged Muslim minority discussed earlier in this chapter, whose plight escalated into a refugee crisis of realmwide proportions in the 2010s. The Mon (2 percent, 1.1 million) were in what is today Myanmar long before the Burmans, and even introduced Buddhism to the national heartland of the Irrawaddy Basin; in fact, they continue to seek the return of ancestral lands from which they were ousted.

When Myanmar finally shook off the excesses of dictatorship earlier in this decade and embraced long-awaited reforms, it was hoped that this change would usher in a period of national progress and a more democratic society. To a certain extent, advances were made—political prisoners were freed,

FIGURE 10-14

Dawei on the Andaman Sea into a world-class port facility and regional trade hub. The key aims and elements of this project are elaborated in **Box 10-14**.

The third infrastructural project has already been mentioned: China's completion of a dual set of oil and gas pipelines from a new port terminal at Kyaukphyu, south of Sittwe, running 1150 kilometers (700 mi) northeastward to Kunming, China's thriving gateway city for Southeast Asia

(Fig. 10-14). These linkages are hugely important to the PRC because they directly supply interior southern China with urgently needed fuels, thereby circumventing the much longer maritime route through the Strait of Malacca and the South China Sea. And Myanmar benefits as well: the pipelines cross the center of the country, providing a series of delivery points in such key urban centers as Mandalay.

Box 10-14 Regional Planning Case

Dawei, Myanmar

Myanmar's stunning reorientation in this decade has produced a range of new possibilities as well as challenges in regional planning. One of the most promising opportunities is focused on Dawei, a small city in the quiet Division of Tanintharyi that forms the country's southernmost extremity (Fig. 10-14). Although this urban center was home to only about 170,000 people in 2016, if current government plans come to fruition Dawei could swiftly be transformed into a major port city. The upgrading of its fine natural harbor would be accompanied by the construction of a new, coast-paralleling land route to link Dawei to Yangon and points north. Nonetheless, an even more important opportunity for hinterland enlargement lies to the east, where another overland route is being built to directly connect Dawei to the nearby heart of neighboring Thailand's burgeoning core area. In fact, from Dawei to Bangkok is only half the distance to Yangon, and the diversion of westward-bound Thai exports to this new Indian Ocean port would eliminate the long and riskier sea journey around the Malay Peninsula.

The Dawei project has three broad aims. First, the government wants to develop the remote, impoverished, and sparsely populated region of Tanintharyi and more tightly integrate it with Myanmar's core. Second, as noted above, the project is expected to transform Dawei into a major international economic node for both Southeast and South Asia. And third, the project will help Myanmar increase its leverage in its relations with Thailand, Japan, and China as well as other Southeast Asian countries.

This U.S. $1.7 billion project has been mired in the planning stage for several years as negotiations have dragged on with foreign partners. Eventually, Japan emerged in 2015 as the main funding source.

The Dawei plan consists of five main components:

- Construction of a deepwater port to accommodate world-class shipping.
- Development of a Dawei Special Economic Zone with industrial complexes offering tax incentives to foreign investors for export processing.
- Construction of a 160-kilometer (100-mi) highway and parallel railroad eastward from Dawei to Thailand's Kanchanaburi (a town with first-rate surface connections to the lower Chao Phraya Basin), located halfway between Dawei and Bangkok (Fig. 10-14).
- Improved rail, road, and sea connections northward from Dawei to Myanmar's Irrawaddy Basin heartland.
- Major new infrastructural facilities to control local flooding, particularly during the wet summer monsoon.

To better grasp the potential impact of this project, take a close look at Figure 10-11 as well as a world map such as Figure G-1. A connection between the Gulf of Thailand and the Andaman Sea arm of the Indian Ocean would create a highly significant transport alternative for an enormous volume of shipping that now passes through the Strait of Malacca. It also explains why Thailand wants to be a strong partner in the project as Dawei comes to anchor a leading new east-west trade corridor across mainland Southeast Asia. Earlier in this chapter we highlighted the role of the South China Sea in global shipping (Map Analysis in Box 10-7), and shortly we will discuss Singapore's dominantly central position in this geographic realm. The Dawei project clearly has the potential to offer a strategic transportation alternative in this politically tenuous corner of the world, and it could well turn out to be faster, cheaper, and safer.

© J. Nijman, P. O. Muller, and John Wiley & Sons, Inc.

DAWEI REGIONAL PLANNING PROJECT

— Existing Main Road
····· Planned Main Road
← Shipping Route

0 250 500 Kilometers
0 100 200 300 Miles

FIGURE 10-15

The site (photo) and situation (map) of Dawei. This 2014 scene at the edge of the city may only show a dirt road, but the coming surge of development is likely to completely transform this landscape within the next few years.

© NYEIN CHAN NAING/epa/Corbis Images

STATES OF WEST MALAYSIA

POPULATION
- Under 250,000
- 250,000–1,000,000
- 1,000,000–5,000,000
- Over 5,000,000

National capitals are underlined

States with greatest Islamic strength

States with largest Chinese and Hindu presence

— Highway
— Main road
— Railroad
✈ Airport

FIGURE 10-16

© J. Nijman, P. O. Muller, and John Wiley & Sons, Inc.

Region Insular Southeast Asia

On the peninsulas and islands of Southeast Asia's southern and eastern periphery lie 6 of the realm's 11 states (Fig. 10-11). Few regions in the world contain so diverse a set of countries. Malaysia, the former British colony, consists of two major territories separated by hundreds of kilometers of South China Sea. The realm's southernmost state, Indonesia, sprawls across thousands of islands from Sumatera in the west to New Guinea in the east. North of the Indonesian archipelago lies the Philippines, another island-chain

country that once was a U.S. colony. Territorially, these are three of the most acutely fragmented states on Earth, and each has faced the challenges that such politico-geographical division presents. This Insular region of Southeast Asia also contains three small but significant sovereign entities: a city-state, a sultanate, and a ministate. The city-state is Singapore, once a part of Malaysia (and one instance in which internal centrifugal forces were too powerful to be overcome). The sultanate is Brunei, an oil-rich Muslim country on the island of Borneo that seems to be transplanted from the Persian Gulf. The ministate is severely impoverished East Timor, which only achieved independence (from Indonesia) in 2002. Few parts of the world are more varied or interesting geographically.

Malaysia

One portion of Malaysia's national territory lies on the Asian mainland and the other on one or more islands. The country is a colonial political artifice that combines a pair of decidedly disparate components into a single state: the southern end of the Malay Peninsula and the northern sector of the island of Borneo. These are referred to, respectively, as **West Malaysia** and **East Malaysia** (Fig. 10-11). The appellation "Malaysia" came into use in 1963, when the original Federation of Malaya, on the Malay Peninsula, was expanded to incorporate the territories of Sarawak and Sabah in Borneo. When the name *Malaya* is used, it refers to the peninsular part of the Federation, whereas *Malaysia* refers to the total political entity.

The Malays of the peninsula constitute 50 percent of the country's population of 30.8 million. They possess a compelling cultural identity expressed in adherence to the Muslim faith, a common language, and a strong sense of territoriality. The Chinese came to the Malay Peninsula as well as to northern Borneo in substantial numbers during the colonial era, and now they account for roughly 23 percent of Malaysia's population. Hindu South Asians were in this area long before the Europeans, and for that matter before the Arabs and Islam arrived on these shores; today they still form a substantial minority of almost 7 percent of the population, clustered, like the Chinese, on the western side of the peninsula (see Figs. 10-6 and 10-16).

Malaysia's ethnic and racial groups have long coexisted in relative harmony, but relations today are marked by rising social tensions. As Malays increasingly assert claims based on 'majority rights,' the Chinese find themselves in much the same minority position that marks the Overseas Chinese across the realm. The Indian population is much smaller and feels increasingly crowded out. Add religious tensions to the mix (that occasionally include Muslim demands for a separate *Sharia* legal system), and the notion of a common Malaysian identity now looks more fragile than ever before. The ruling ethnic Malay government extended its hold on power in the 2013 elections, but did so by the narrowest of margins—a sign of the country's deepening political polarization heightened by growing expressions of dissent among its restive minorities.

West Malaysia The populous peninsular component of Malaysia remains the country's dominant sector, containing 11 of its 13 States and approximately 80 percent of its population. Here, the Malay-dominated government has very tightly controlled economic and social policies as it pursues nation-wide modernization. The Strait of Malacca (Melaka) to the west continues to be one of the world's busiest and most strategic waterways, and forms yet another crucial **choke point** in the sea-borne flow of resources and goods between major geographic realms. A longstanding problem in the Strait, which has markedly resurfaced in the mid-2010s, is piracy. The spatial configuration of this narrow, 1000-kilometer (625-mi)-long waterway makes it highly susceptible to such attacks. Moreover, the Strait is always a most inviting target because the 50,000-plus vessels that annually transit through it constitute fully one-third of the world's shipping and carry one-quarter of its oil.

Malaysia, despite the loss of Singapore and notwithstanding its recurrent ethnic troubles, has become a leading player in contemporary Southeast Asia. Its new **Multimedia Supercorridor** not only anchors the national core area but also showcases the country's commitment to twenty-first-century technology—symbolized by Putrajaya, the recently built high-tech administrative capital. Elsewhere, the strong skills and modest wages of the local workforce have attracted many multinational corporations, and the government has capitalized on its opportunities, such as encouraging the creation of an ultra-modern manufacturing complex on the far northwestern island of Pinang, where Chinese outnumber Malays by two to one.

East Malaysia The decision to combine the 11 sultanates of Malaya with the States of Sabah and Sarawak on Borneo (Fig. 10-11), creating the country we now call Malaysia, had far-reaching consequences. Those two offshore States may constitute 60 percent of Malaysia's territory, but they are home to barely 20 percent of the population. They endowed Malaysia with abundant energy resources, huge stands of timber, and vast tracts of land suitable for palm oil plantations (the country is now the world's leading exporter, with Sarawak and Sabah producing 55 percent of the national crop). But they also complicated Malaysia's ethnic complexion because each State contains more than two dozen indigenous groups (for example,

immigrant Chinese form Sarawak's largest ethnic minority). These locals complain that the federal government in Kuala Lumpur treats Malaysian Borneo as a colony, and politics here are contentious and fractious. It is therefore quite likely that Malaysia will confront intensifying devolutionary forces in East Malaysia.

Brunei

Also located on Borneo—near the midpoint of the north-western coast, adjacent to where Sarawak and Sabah meet—is Brunei (Fig. 10-18 inset map), a rich, oil-exporting, Islamic sultanate far from the Persian Gulf. Brunei, which became independent from the United Kingdom in 1984, is only slightly larger than Delaware and has a population of just 430,000. Thus the sultanate is a mere ministate—but a most prominent one since the discovery of oil and natural gas, which turned it into Southeast Asia's richest country. The Sultan of Brunei rules as an absolute monarch; his palace in the capital, Bandar Seri Begawan, is reputed to be the world's largest. And he will have no difficulty finding customers for his oil in energy-poor eastern Asia. Even though Brunei's oilfields may become exhausted by mid-century, major new supplies have recently been discovered beneath the sea-floor of Brunei's territorial waters. Not taking any chances, and with an eye clearly on the future, the government has been investing heavily in overseas assets such as real estate in London and elsewhere—working to develop alternative revenue sources should oil income begin to decline.

Brunei's expanding population—in which immigration plays a larger role than internal natural increase—is ethnically two-thirds Malay and about 12 percent Chinese. Most inhabitants live and work near the offshore oilfields in the northern district of Brunei Muara that contains the capital city. Evidence of profligate spending is everywhere, ranging from sumptuous palaces to magnificent mosques to the most luxurious of resort hotels. In this respect Brunei offers a stark contrast to surrounding East Malaysia, but even within Brunei there are significant differences. The country's interior, where a small minority of indigenous groups survive, remains an area of subsistence agriculture and rural isolation, its villages a world apart from the modern splendor of Bandar Seri Begawan.

Singapore

In 1965, a fateful event occurred in Southeast Asia. Singapore, the crown jewel of British colonialism in this realm, seceded from the recently independent (1963) Malaysian Federation and became a sovereign state, albeit a tiny city-state (Fig. 10-17). With its unparalleled relative location, its multiethnic and well-educated population, and its no-nonsense government, Singapore then overcame the severe limitations of space and the absence of raw materials to become one of the four original Asian Tigers in the late 1970s.

FIGURE 10-17

With a mere 619 square kilometers (239 sq mi) of territory and 5.7 million people, space is at a premium in Singapore, and this is a constant challenge for the government. The highest priority is to develop cutting-edge but space-conserving manufacturing and service industries. Benefiting from its relative location, the old port of Singapore had become one of the world's busiest even before independence. It thrived as an *entrepôt* [21] between the Malay Peninsula, the rest of Southeast Asia, Japan, and other emerging economic powers on the Asian Pacific Rim. Crude oil from Southeast Asia still is unloaded and refined in Singapore, then shipped to East Asian destinations. Raw rubber from the adjacent Malay Peninsula and from Indonesia's nearby island of Sumatera is shipped to China, Japan, the United States, and many other countries. Timber from Malaysia, rice, spices, and other foodstuffs are processed and forwarded through Singapore. In return, automobiles, machinery, and equipment are imported into Southeast Asia and distributed almost exclusively via Singapore.

Singapore's current development strategy stresses two primary objectives. The first is to focus its industries more tightly on leading-edge information technology, automation, and biotechnology. The second is to forge an ever stronger **Growth Triangle [22]** involving Singapore's developing neighbors, Malaysia and Indonesia. Accordingly, Malaysia and Indonesia would supply the necessary raw materials and relatively inexpensive labor, and Singapore would supply the capital and technological expertise. The ethnic composition of Singapore's population today is just below 75 percent Chinese, 13 percent Malay, and 9 percent South Asian. The government is Chinese-dominated, and its policies have served to sustain ethnic-Chinese control. Indeed, the government of Singapore has encouraged immigration from the People's Republic of China to stabilize the ethnic mosaic of the city-state, where natural–increase rates, especially among its native Chinese, have for some time been well below the replacement level. But, as noted earlier, this may not be a long-term solution because tensions are mounting between the resident and newly arrived immigrant Chinese communities.

Indonesia

The fourth-largest country in the world in terms of human numbers is also the globe's most expansive archipelago. Spread across a chain of more than 17,000 mostly volcanic islands, Indonesia's 261 million people live both separated and clustered—separated by water and clustered on islands large and small. With more than 300 discrete ethnic clusters, over 250 languages, and just about every religion practiced on Earth, Indonesia's survival as a unified state is remarkable. Indeed, Indonesia's national motto is *bhinneka tunggal ika*—diversity in unity. Nevertheless, Indonesia nominally is the largest Muslim country in the world: 88 percent of the people adhere to Islam, and in the cities the silver domes

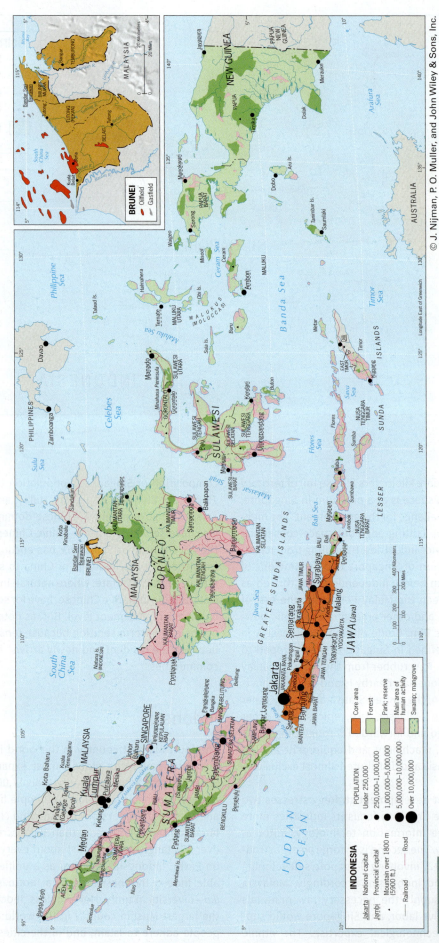

FIGURE 10-18

and minarets of neighborhood mosques rise like a dense forest above the townscape. Although until recently Indonesian Islam has been relatively moderate, as noted earlier, more orthodox Islamization has been on the rise, with new laws now in effect to ban public displays of interpersonal affection and to dictate the kinds of clothing women may wear outside their homes.

The Major Islands

With more than 150 million inhabitants, **Jawa** is one of the world's most densely peopled territories (see Figs. G-7 and 10-5) as well as one of the most agriculturally productive, with its terraced rice-growing paddies rising up the highly fertile flanks of myriad active volcanoes. Jawa also is the most highly urbanized component of a country in which almost half the people still live off the land. The primate city of Jakarta, profiled in **Box 10-15**, has today become the heart of the much larger Greater Jakarta conurbation that now exceeds 30 million, nearly 25 percent of Indonesia's urban population. Thousands of factories, their owners taking advantage of low prevailing wages, have been built in this sprawling conurbation (which goes by the acronym *Jabodetabek*, after its constituent cities), badly straining its infrastructure and overburdening the port of Jakarta.

Sumatera, spelled *Sumatra* in the colonial era, is Indonesia's westernmost island and forms the west shore of the critical Strait of Malacca; Singapore lies across the Strait from approximately the middle of this island. Although much larger than Jawa, Sumatera has only about one-third as many people (roughly 50 million). In colonial times, the island became a base for rubber and palm oil production; its high relief makes possible the cultivation of a wide variety of crops, and neighboring Bangka and Belitung yield petroleum and natural gas. Palembang is the key urban center in the south. In recent years, the northeastern coastlands of Sumatera, along with parts of mainland Malaysia and southern Kalimantan, have been blanketed by huge and continually expanding palm oil plantations to drive up export revenues—but at enormous cost to the regional environment (as noted earlier in the chapter and underscored in the accompanying satellite image of Sumatera's smoke plumes).

Box 10-15 Among the Realm's Great Cities...

Jakarta

Jakarta, capital of Indonesia and the realm's second-largest city, sometimes is called the Kolkata (Calcutta) of Southeast Asia. Stand on the elevated highway linking the port to the city center and see the villages built on top of garbage dumps by scavengers using what they can find in the refuse, and the metaphor seems to fit. There is poverty here unlike that in any other Southeast Asian metropolis.

But there are other sides to Jakarta, which recently attained megacity status (2017 population: 31.3 million). Indonesia's economic progress has made its mark here, and the evidence is everywhere. Television antennas and satellite dishes rise densely from rusted, corrugated-iron rooftops. Cars (almost all, it seems, late-model), mopeds, and bicycles clog the streets, day and night. A meticulously manicured part of the city center contains a cluster of high-rise hotels, office buildings, and luxury-apartment complexes. Billboards advertise planned communities on well-located, freshly cleared land.

Jakarta's population is a cross-section of Indonesia's, and hundreds of mosques dot the cityscape together with Christian churches and Hindu temples. The city always was cosmopolitan, beginning as a cluster of villages at the mouth of the Ciliwung River under Islamic rule, becoming a Portuguese stronghold, and later—as Batavia—the capital of the Dutch East Indies. Advantageously situated on the northwestern coast of Jawa, Indonesia's most populous island, Jakarta is bursting at the seams with growth. Sail into the port, and hundreds of vessels, carrying flags from Russia to Argentina, await berths. Travel to the outskirts, and huge shantytowns are being expanded by a constant stream of new arrivals. So vast is the human agglomeration—nobody really knows how many people have descended on this megacity—that most live without adequate (or any) amenities.

Although Jakarta has long been the biggest metropolis in the world without a rail transit system, two new subway lines are finally nearing completion after endless construction delays; the north-south route will be first to open—perhaps as soon as late 2017. In the meantime, shockingly inadequate public transport continues to rely exclusively on buses, and the city's incessant traffic jams are legendary (transport delays are estimated to cost at least U.S. $3 billion annually), especially when heavy equatorial rains cause frequent flooding in many neighborhoods. The steadily worsening gridlock has even prompted discussions about moving the capital some 50 kilometers (30 mi) to the south; but the government has now distanced itself from that idea, stressing instead the urgent need to improve drainage throughout the metropolis while building a much better highway infrastructure. Most Jakartans, however, remain unimpressed, and none of this talk has made a dent in alleviating their frustrations.

© J. Nijman, P. O. Muller, and John Wiley & Sons, Inc.

In Sumatera's far north lies Aceh, the land area closest to the epicenter of the colossal 2004 undersea earthquake that triggered a disastrous tsunami which claimed about 230,000 lives in the Indian Ocean Basin. Prior to this catastrophe, the province of Aceh was designated a "Special Territory" within Indonesia because of the separatist movement of its indigenous Acehnese population. However, that crisis became defused thanks to the massive post-tsunami relief effort and follow-up of the central government. As a result, relations between the province and Jakarta today are much improved.

Kalimantan is the Indonesian sector that comprises 73 percent of the island of Borneo, a slab of the Earth's crust whose mountainous backbone is of erosional, not plate-tectonic, origin. Larger than Texas, Borneo has a deep, densely forested interior that is a last refuge for some 20,000 orangutans as well as dwindling numbers of Asian elephants, rhinoceroses, and tigers. Along with a number of indigenous peoples, these wildlife species survive even as loggers, plantation builders, and small farmers relentlessly penetrate their shrinking habitat. Borneo is home to a comparatively small human population (roughly 20 million on the Indonesian side, less than 8 percent of the country's total) on poor tropical soils. Aboriginal peoples, principally the Dayak clans, have traditionally had a less negative impact on the natural environment than the Indonesian and Chinese immigrants as well as the multinational corporations that log the rainforests and clear space for new farms. As Figure 10-17 shows, the only towns of any size in Kalimantan lie on or near the coast; access routes into the interior are still few and far between.

Sulawesi consists of a set of intersecting, volcanic mountain ranges rising above sea level; the 800-kilometer (500-mi) Minahasa Peninsula, propelled by active volcanism, continues

© Jeff Schmaltz, LANCE/EOSDIS Rapid Response / NASA Earth Observatory

In June 2013, human-ignited forest fires on Sumatera (center left) generated enormous plumes of thick smoke that drifted eastward to engulf Singapore and surrounding areas of Malaysia and Indonesia. The Singaporean authorities warned people to stay indoors, and the city-state's government loudly insisted to its Indonesian counterpart that a greater effort must be made to prevent plantation owners and smaller-plot farmers from deliberately burning vast stands of trees to clear land for additional palm oil production.

to build itself up from the floor of the Celebes Sea. This northern peninsula, a favorite of the Dutch colonizers, remains the most developed sector of an otherwise rugged and remote island (see **Box 10-16**), with Manado its relatively prosperous urban focus. Seven major ethnic groups inhabit the valleys and basins scattered among the mountains, but the population of about 18 million also includes many immigrants from Jawa,

Box 10-16 From the Field Notes...

"I drove from Manado on the Minahasa Peninsula in northeastern Sulawesi to see the ecological crisis at Lake Tondano, where a fast-growing water hyacinth is clogging the water and endangering the local fishing industry. On the way, in the town of Tomolon, I noticed this side street lined with prefabricated stilt houses in various stages of completion. These, I was told, were not primarily for local sale. They were assembled from wood taken from the forests of Sulawesi's northern peninsula, then taken apart again and shipped from Manado to Japan. 'It's a very profitable business for us,' the foreman told me. 'The wood is nearby, the labor is cheap, and the market in Japan is insatiable. We sell as many as we can build, and we haven't even begun to try marketing these houses in Taiwan or China.' At least, I thought, this wood was being converted into a finished product, unlike the mounds of logs and planks I had seen piled up in the ports of Borneo awaiting shipment to East Asia."

© H. J. de Blij

especially in and around the southern center of Ujungpandang. Subsistence farming is the leading mode of life, although logging, scattered mining, and fishing augment the economy. Clashes between Muslims and Christians occur intermittently in remote areas.

Papua, the Indonesian name for the western half of the island of New Guinea, has become an issue in national politics. Papua is bordered on the east by a superimposed, mostly straight-line boundary that separates it from the state of Papua New Guinea on the eastern half of the island (Fig. 10-18). This territory was seized by Indonesia from the Dutch in 1969, the last remnant of the Netherlands East Indies colony that survived after Indonesia became independent in 1945. Papua contains about 22 percent of Indonesia's territory, but its (fast-growing) population totals only 3.8 million—just under 1.5 percent of the country's total. The indigenous inhabitants of this territory—which is essentially an Indonesian colony—are Papuan, with most living in the remote reaches of this mountainous, densely forested island. Papua is economically important to Indonesia because it contains what is reputed to be the world's richest gold mine as well as its third-largest open-pit copper mine. But political consciousness has now reached the Papuans: the Free Papua Movement has been active for the past quarter-century, mostly holding small demonstrations in the capital (Jayapura), displaying a Papuan flag, and demanding recognition.

Indonesia's Challenges and Prospects

Indonesia's development challenges are complicated by spatial fragmentation and a particularly uneven population distribution (see Fig. 10-5). From 1974 to 2001, Indonesia's central government pursued a policy known as **transmigration [23]**, inducing millions from the densely populated inner islands (especially Jawa, Bali, and Madura) to relocate to such sparsely inhabited, peripheral islands as Kalimantan and Sulawesi. As many as 8 million Jawanese were relocated to outer islands, but a majority of them experienced a decline in their standard of living; many were even reduced to bare subsistence to scratch out a living in the tropical woodlands confiscated from indigenous inhabitants—which turned out to be unsuitable for the farming methods used by the settlers. In the end, cultural conflict, ecological havoc, and rampant deforestation finally led the Indonesian government to admit failure and terminate this grandiose social program in 2001.

Today, Indonesia's stature within Southeast Asia is steadily rising. This is not just the biggest country in the realm: it is also striving to achieve a commensurate level of economic development and prosperity. The economy expanded vigorously during the 2000s; global recession notwithstanding, per capita incomes here now exceed those of Laos, Cambodia, and Myanmar—yet they still rank well below those of Thailand or Malaysia. At the same time, centrifugal forces are intensifying (growing Islamic militancy in Jawa is an overriding concern), and persistent decentralization and devolution throughout this immense archipelago is certain to continue and challenge the Indonesian government well into the future.

East Timor

As Figure 10-18 shows, Timor is the easternmost of the sizeable Lesser Sunda Islands, and throughout the colonial Dutch East Indies period the Portuguese maintained their own colony on the eastern half of it. In 1975, Indonesian forces overran that colony and formally annexed it the following year. Indonesian rule, however, was even less benign than that of the Portuguese, and soon a bitter struggle for independence was under way. Eventually, in 1999, under UN supervision, the people of East Timor voted overwhelmingly for independence, but Indonesia still refused to let go, and unleashed a brutal military occupation. It took another three years of armed conflict and foreign intervention (led by Australia, which previously had supported Indonesia) until independence was finally realized in 2002. By then, the violence had devastated most of this ministate and its infrastructure. East Timor's leaders proclaimed the official name of their country to be *Timor-Leste*—Portuguese for "Timor East." Not surprisingly, a decade and a half later nation-building in this Connecticut-sized country remains a formidable proposition.

As Figure 10-18 indicates, East Timor is a fragmented state, with a dominant east (where the coastal capital, Dili, is located) and a tiny **exclave [24]**, on the northern coast of Indonesian West Timor, named Ocussi. Even though East Timor has always been overwhelmingly agricultural, farming was badly neglected during the turmoil unleashed under the Indonesian occupation. To make matters worse, the country's population of 1.2 million continues to grow explosively, exhibiting a yearly natural increase rate of 2.2 percent (30 percent higher than that of Laos, the realm's second-fastest-growing country); East Timor's 2016 fertility rate of 5.8 children per woman of childbearing age ranked among the world's highest. Most recently, the economic growth rate has been more encouraging, and there is the prospect of billions of dollars to be earned from future oil revenues (the country shares the rights to offshore oil reserves in the Timor Sea with Australia). But the nation-building project here has barely begun to deal with the country's extreme poverty as well as the lingering social devastation of a prolonged conflict in which some 100,000 civilians perished at the hands of the Indonesian military.

The Philippines

North of Indonesia lies a lengthy archipelago of more than 7000 islands inhabited by 102 million people. This island chain constitutes the Philippines, and it can be subdivided into three groups: (1) Luzon, largest island of all, and Mindoro in the north; (2) the Visayan group in the center; and (3) Mindanao, second-largest, located in the south (**Fig. 10-19**). The Philippines' population, concentrated where productive farmlands lie, is densest in three general areas: the northwestern and south-central components of Luzon; the southeastern extension of Luzon; and the islands of the Visayan Sea between Luzon and Mindanao (Fig. 10-5). Luzon is the

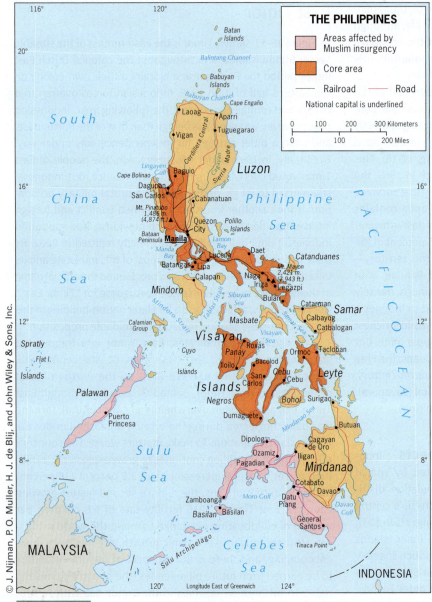

FIGURE 10-19

percent) but dominates numerous local business communities.

The Philippines' small Muslim population, concentrated in the southernmost periphery of the archipelago, and especially on densely forested Basilan Island (Fig. 10-19), has long decried its marginalization in this predominantly Christian country. Over the past generation, a half-dozen Islamic organizations have promoted the Muslim cause here through tactics that have ranged from peaceful negotiation with the government to violent insurgency.

Agriculture continues to dominate the Philippines' economy. Alluvial as well as volcanic soils, together with ample moisture in this mostly humid-tropical environment, enable self-sufficiency in rice and other staples and allow the Philippines to be a net exporter of farm products despite its substantial population growth rate of: 1.6 percent. However, unemployment is rampant, trade linkages are insufficient, and additional land reform is badly needed, as is social restructuring to reduce the controlling influence of a comparatively small group of families over national affairs.

Perhaps more than any other people, Filipinos take jobs in foreign countries in massive numbers, proving their capacity to succeed in jobs they cannot find at home. More than 10 million Filipinos, over one-quarter of the total workforce, are now employed abroad. The global merchant marine would not exist without Filipino sailors, and Filipina nurses and domestic workers can be found from Dubai to Dublin to Dubuque. Funds sent home to family members by the emigrants make the Philippines a world leader in remittances, a monetary inflow that is the basis for at least one-tenth of the country's GDP in any given year.

Overall, the Philippines sustains a lower-middle-income economy, and given a longer period of political stability and success in mitigating the problems outlined above, it should be able to rise to the next level and take its long overdue place among the Asia-Pacific's growth poles.

Since 2000, the Philippines has established itself in a pair of niches in the global economy: international call centers and the outsourcing of digital services. The latter operate through such Internet websites as *oDesk*—one of several that serves as a global marketplace for digital services by freelancers who perform data entry and other routine back-office work. Although such outsourcing is not a promising growth industry and compensation is low by worldwide standards, the **business process outsourcing [25])** sector today employs about one million full-time workers and is the second-largest

site of the capital, Manila (with 22.9 million, 22.5 percent of the entire population of the country), a sprawling megacity (profiled in **Box 10-17**) that faces the South China Sea.

Today, the Philippines, lying adjacent to the world's largest Muslim state (Indonesia), is 81 percent Roman Catholic, 9 percent Protestant, and only 5 percent Muslim. Out of the Philippines melting pot, where Malay, Arab, Chinese, Japanese, Spanish, and American elements met and mixed, has emerged the distinctive Filipino culture. It is not a homogeneous or unified culture, as is reflected by the dozens of Malay languages in use in the islands, but it is in several ways unique. At independence in 1946, the largest of the Malay languages, Tagalog (also known as Pilipino), became the country's official language. But English is widely learned as a second language, and a Tagalog-English hybrid, "Taglish," is increasingly heard today. The Chinese component of the population is small (1.6

Box 10-17 Among the Realm's Great Cities...

Manila

Manila, capital of the Philippines and the realm's largest city, was founded by the Spanish invaders of Luzon more than four centuries ago. The colonial conquerors made a good choice in terms of site and situation. Manila sprawls at the mouth of the Pasig River where it enters one of Asia's finest natural harbors. To the north, east, and south, a crescent of mountains encircles the city, which lies just 1000 kilometers (600 mi) southeast of China's Hong Kong.

Manila, named after a flowering shrub in the local marshlands, is bisected by the Pasig, which is bridged at numerous points. The old walled city, Intramuros, lies to the south. Despite heavy bombardment during World War II, some of the colonial heritage survives in the form of churches, monasteries, and convents. St. Augustine Church, completed in 1599, is one of the city's landmarks.

The CBD of Manila lies on the north side of the river. Although Manila has a well-defined commercial center with several avenues of luxury shops and modern buildings, the skyline does not reflect the high level of energy and activity common in Pacific Rim cities on the opposite side of the South China Sea. Neither is Manila a city of notable architectural achievements. Wide, long, and straight avenues flanked by palm, banyan, and acacia trees give it a look similar to San Juan, Puerto Rico.

In 1948, a newly built city immediately to the northeast of Manila was inaugurated as the *de jure* capital of the Philippines and named Quezon City. The new facilities were eventually to house all government offices, but many functional components of the national government never made the move. In the meantime,

Manila's growth overtook Quezon's, so that it became part of the Greater Manila metropolis (which is now home to 23 million people). Although the proclamation of Quezon City as the Philippines' official capital was never rescinded, Manila remains the *de facto* capital today.

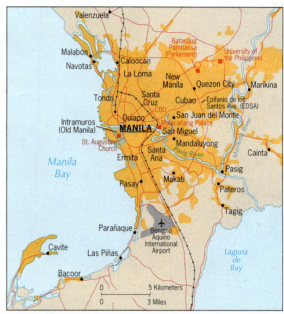

© J. Nijman, P. O. Muller, and John Wiley & Sons, Inc.

contributor (after remittances) to the Philippines' foreign exchange earnings.

In its foreign relations, this country is changing course. Nearly 30 years ago, it terminated the lease for the huge U.S. Subic Bay Naval Base—then the largest of all American foreign military installations. But these days the Philippine government is once again seeking closer relations with the United States. As noted earlier, this is in response to China's growing maritime-territorial assertiveness in Southeast Asia— not only regarding the South China Sea in general but also longstanding specific disputes between China and the Philippines concerning the Spratly Islands and Scarborough Shoal.

Points to Ponder

- The two smallest countries of the realm are also by far the wealthiest.

- Is Southeast Asia becoming a new Chinese sphere of influence, or is the PRC going to face a backlash against its assertive strategies in this realm?

- In your daily life, how many products do you use that contain palm oil? How can you find out? As a consumer, do you thereby contribute to deforestation?

- Under Malaysia's constitution, a person of Malay ancestry (unlike someone with Chinese or South Asian roots) is automatically registered as a Muslim.

- If all current plans come to fruition, China will have built 27 dams in the upper basin of the Mekong River before the waterway enters the Southeast Asian realm in Laos.

- With the deepening involvement of India, China, and Thailand in Myanmar over the past several years, it is hard to imagine that this country could ever return to its isolationist recent past.

- In 2016, the International Court of Justice ruled that China's claims in the South China Sea were unlawful. Beijing retorted that the ruling was invalid and not binding.

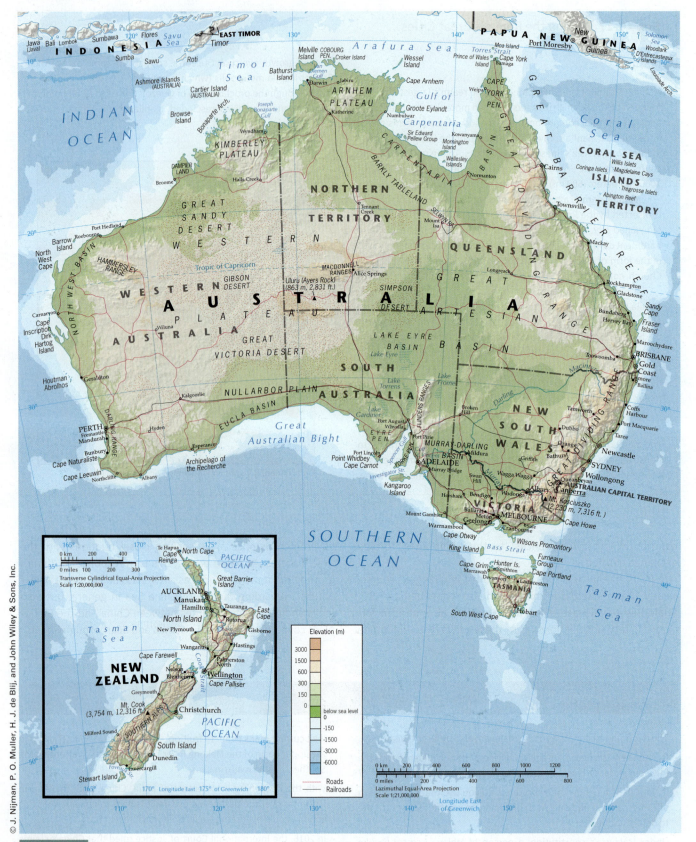

FIGURE 11-1

© Jan Nijman

© Jan Nijman

Where were these pictures taken?

The Austral Realm

IN THIS CHAPTER

- The Austral Realm's Asian turn
- Australia's exceptional biogeography
- China covets Australian commodities
- Aboriginal claims to land and resources
- Earthquake disaster management in New Zealand

CONCEPTS, IDEAS, AND TERMS

Austral	[1]	Outback	[10]
Southern Ocean	[2]	Primary sector	[11]
Subtropical Convergence	[3]	Environmental degradation	[12]
West Wind Drift	[4]	El Niño	[13]
Biogeography	[5]	Desalination	[14]
Wallace's Line	[6]	Aboriginal land issue	[15]
Aboriginal population	[7]	APEC	[16]
Federation	[8]	Peripheral development	[17]
Unitary state	[9]	Disaster management	[18]

The Austral realm is the only one that lies entirely in the Southern Hemisphere. It is also unique in that it has no land link of any kind to a neighboring realm and is thus completely surrounded by oceans and seas. It is second only to the Pacific as the world's least populous geographic realm. Appropriately, its name refers to its location (the word **austral [1]** derives from the Latin for "south")—a location far from the sources of its dominant cultural heritage, but relatively close to its newfound economic partners on the Asian Pacific Rim.

Defining the Realm

Two countries constitute the Austral realm: Australia, in every way the dominant national entity, and New Zealand, physiographically more varied but demographically much smaller than its giant partner (**Fig. 11-2**). Between them lies the South Pacific Ocean's Tasman Sea. To the realm's west lies the Indian Ocean, to its east the Pacific, and to the south the frigid Southern Ocean (**Box 11-1**).

Today, this southern realm is at a crossroads. On the doorstep of populous eastern Asia, the realm's Anglo-European legacies have been diversified accordingly by the infusion of many Asian cultural strains. Markets throughout the Asia Pacific are buying ever larger quantities of raw materials. Chinese and other Asian tourists fill hotels and resorts. The streets of Sydney and Melbourne display a multicultural panorama unimagined just two generations ago. Indigenous minorities—the Polynesian Maori in New Zealand and Aboriginal communities in Australia—are demanding greater rights and wider acknowledgment of their cultural heritage. All these changes have stirred political debate. Issues ranging from immigration quotas to indigenous land rights dominate, exposing social fault lines (city versus Outback in Australia; North Island versus South Island in New Zealand). Aboriginals and Maori first settled this realm, then the Europeans arrived, and now Asians are an increasingly significant economic and cultural element.

Land and Environment

Physiographic contrasts between massive, compact Australia and elongated, fragmented New Zealand are related to their locations with respect to the Earth's tectonic plates (see Fig. G-4). Australia, carrying some of the geologically most ancient rocks on the planet, lies at the center of its own Australian Plate. New Zealand, younger and less stable, lies at the convulsive convergence of the Australian and Pacific plates. Earthquakes are rare in Australia, and volcanic eruptions are unknown; New Zealand has plenty of both. This locational contrast is also reflected by differences in relief (**Fig. 11-1**). Australia's highest relief occurs in what Australians call the Great Dividing Range, the mountains that line the east coast from the Cape York Peninsula in the north to southern Victoria State, with an outlier in Tasmania. The highest point along these old, now eroding mountains is Mount Kosciusko, 2230 meters (7316 ft) tall. In New Zealand, entire ranges are higher—Mount Cook in the Southern Alps, for example, reaches 3754 meters (12,316 ft). West of Australia's Great Dividing Range, the physical landscape mostly exhibits low relief, with some local exceptions such as the Macdonnell Ranges near the continental center; plateaus and plains dominate (Fig. 11-1; **Fig. 11-3**, top map). The Great Artesian Basin is a key physiographic

---

Box 11-1 Major Geographic Features of the Austral Realm

1. Australia and New Zealand constitute a geographic realm by virtue of relative location, territorial dimensions, and dominant cultural landscape.

2. Despite their inclusion in a single geographic realm, Australia and New Zealand differ physiographically. Australia has a vast, dry, low-relief interior; New Zealand is mountainous and has a temperate climate.

3. Australia and New Zealand are marked by peripheral development—Australia because of its aridity, New Zealand because of its topography.

4. The populations of Australia and New Zealand are not only peripherally distributed but also highly clustered in urban centers.

5. The economic geography of Australia and New Zealand is dominated by the exporting of livestock products and specialty goods such as wine. Australia also is a significant producer of wheat and possesses a rich, diverse base of mineral resources.

6. Australia and New Zealand are now integrated into the economic framework of the Asian Pacific Rim, principally as suppliers of raw materials. Mineral-rich Australia's newly discovered and exploited gas reserves are adding to its export base.

7. As immigration from the western Pacific Rim increases, Australia has, in a cultural sense, "returned" to Asia.

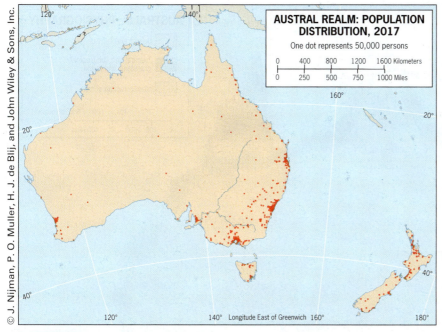

AUSTRAL REALM: POPULATION
DISTRIBUTION, 2017
One dot represents 50,000 persons

FIGURE 11-2

Climates

Figure G-6 reveals the effects of latitudinal position and interior isolation on Australia's climatology. In this respect, Australia is far more varied than New Zealand, its climates ranging from tropical in the far north (**Box 11-2**), where rainforests flourish, to Mediterranean in two corners of the south. The interior is dominated by desert and steppe conditions, the semiarid steppes providing the grasslands that sustain tens of millions of livestock. Only in the extreme east does Australia possess a zone of humid temperate climate, and here lies most of the country's core area. New Zealand, by contrast, is totally under the influence of the Southern and Pacific oceans, creating moderate, moist conditions, temperate in the north and progressively colder in the south.

region, providing underground water sources in what is otherwise desert country. To its south lies the continent's predominant Murray-Darling river system. The area mapped as *Western Plateau and Margins* in the lower map of Figure 11-3 contains much of Australia's mineral wealth.

Ayers Rock, located in the heart of the immense, interior Outback near the geographic center of Australia, is the most famous landform in the Austral Realm. This miniature tableland is all that is left of a much larger surrounding plateau that has been completely eroded away to form the flat plain that frames this iconic formation of red sandstone rock. Geomorphologists use the term inselberg (German for "island mountain") to describe this kind of geologic remnant, which consists of rocks that are far more resistant to the local forces of erosion than those that once surrounded it. Ayers Rock—called Uluru by the Aboriginal people who first settled Australia—stands about 350 meters (1150 ft) above the plain and has a circumference of nearly 10 kilometers (5.8 mi). Despite its extreme remoteness from where most Australians live, about 300,000 visitors annually make the long trek out here, and the tourist infrastructure has kept pace.

The Southern Ocean

Twice now we have referred to the **Southern Ocean [2]**, but try to find this ocean on maps and globes produced by such well-known cartographic organizations as the National Geographic Society. From their maps you would conclude that the Atlantic, Pacific, and Indian oceans reach all the way to the shores of Antarctica. Australians and New Zealanders know better. They experience the frigid waters and persistent winds of this great weather-maker on a daily basis.

For geographers, it is a good exercise to turn the globe upside down now and then. After all, the usual orientation is quite arbitrary. Modern mapmaking started in the Northern Hemisphere, and the cartographers put their hemisphere on top and the other at the bottom. That is now the norm, and it can distort our view of the world. In souvenir shops in the Southern Hemisphere, you see upside-down maps showing Australia and Argentina at the top, and Europe and Canada at the bottom. But this matter has a serious side. An inverted view of the globe shows us how vast the ocean encircling Antarctica is (see Fig. 12-5). The Southern Ocean may be remote, but its existence is real.

Where do the northward limits of the Southern Ocean lie? This ocean is bounded not by land but by a marine transition known as the **Subtropical Convergence [3]**. Here the very cold, extremely dense waters of the Southern Ocean meet the warmer waters of the Atlantic, Pacific, and Indian oceans. It is quite sharply defined by changes in temperature, chemistry, salinity, and marine fauna. Flying over it, you can actually observe it in the changing colors of the water: the Antarctic side is a deep gray, the northern side a greenish blue. Even though the Subtropical Convergence shifts seasonally, its position does not vary far from latitude 40° South, which also is the approximate northern limit of Antarctic icebergs. Defined this way, the great Southern Ocean is a huge body of water that circulates

AUSTRALIA: PHYSIOGRAPHY
Desert and semi-desert
Tropical and sub-tropical grassland and forest
Mediterranean scrub

PHYSICAL LANDSCAPES
Eastern Uplands   Central Lowlands   Western Plateau & Margins

FIGURE 11-3

"The northeastern coastal strip of the Australian State of Queensland contains tropical rainforest. It is not that far from here to Papua New Guinea (Figs. 11-1, 11-3). In 2015, I paid a visit to a unique ethnobotanical garden which calls itself 'The Botanical Ark.' It is run by a couple from the United States, Susan and Alan Carle, who moved here over 25 years ago. They have traveled the tropical habitats of the world and collected an enormous variety of wild plants that yield either fruits or nuts, with the purpose not only of expanding our knowledge of many species but also of preserving them. They grow them on their sprawling estate here, not far from the coastal town of Port Douglas. It is important to remember that plants are crucial to our diets and as medicinals; that myriad species are still unknown or remain poorly researched; and that many are disappearing before we can discover them. Here, standing before a colorful arrangement of edibles, Alan Carle holds up breadfruit, a highly nutritious staple food in many tropical areas, but relatively unknown in the United States and other mid-latitude countries. Cut it in slices, fry the slices briefly, and sprinkle on a little salt, and they make for a delicious snack that goes well with a late afternoon beer—as I found out."

© Jan Nijman

clockwise (from west to east) around Antarctica, which is why we also call it the **West Wind Drift [4]**.

Biogeography

One of the Austral Realm's defining characteristics is its wildlife. Australia is the land of kangaroos and koalas, wallabies and wombats, possums and platypuses. These and numerous other

marsupials (animals whose young are born very early in their development and then are carried in an abdominal pouch) owe their survival to Australia's early isolation during the breakup of Gondwana (see Fig. 7-3). Before more advanced mammals could enter Australia and replace the marsupials, as happened in every other part of the world, this landmass was separated from Antarctica and India, and today it contains the world's largest assemblage of marsupial fauna. Australia's vegetation has distinctive qualities as well, notably the hundreds of species of eucalyptus trees native to this geographic realm. The study of fauna and flora in spatial perspective integrates the disciplines of biology and geography in a field known as **biogeography [5]**, and Australia serves as a gigantic research laboratory for its practitioners.

Biogeographers are especially interested in the distribution of plant and animal species, as well as the relationships between plant and animal communities and their natural environments. (The spatial analysis of plant life is called *phytogeography*, and that of animal life is known as *zoogeography*.) In 1876, one of the founders of biogeography, the great British naturalist Alfred Russel Wallace, posited that the zoogeographic boundary of Australia's fauna was located beyond Australia in the Sunda island chain to the northwest, between Borneo and Sulawesi, and just east of Bali (**Fig. 11-4**).

Wallace's Line [6] soon was challenged by other researchers, who found species Wallace had missed and who visited islands Wallace had not. There was no question that Australia's zoogeographic realm terminated somewhere in Indonesia's Sunda archipelago, but where? Western Indonesia was the habitat of nonmarsupial animals such as tigers, rhinoceroses, and elephants in addition to primates; New Guinea clearly was part of the realm of the marsupials. How far had the more advanced

William Chopart / Age Fotostock America, Inc.

The breathtakingly beautiful Great Barrier Reef, off the coast of the State of Queensland in northeastern Australia (Fig. 11-3, top map). This is the largest coral reef on the planet, sometimes described as the only living thing on Earth that is visible from space. It was designated a World Heritage Site in 1981. Today, it ranks as one of the planet's most prominent ecotourism destinations, despite the looming threat of global warming effects.

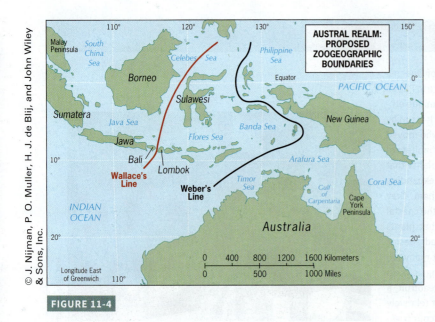

FIGURE 11-4

mammals progressed eastward along the island stepping stones toward New Guinea? The zoogeographer Max Weber discovered evidence that led him to postulate his own *Weber's Line*, which, as Figure 11-4 shows, lay very close to New Guinea.

In Australia, the arrival of the **Aboriginal population [7]** about 50,000 years ago appears to have triggered an ecosystem collapse. Widespread burning of the existing forest, shrub, and grassland vegetation all across Australia probably led to the spread of desert scrub and to the rapid extinction of most of the continent's large mammals soon after the human invasion occurred. Those species that survived faced a second crisis tens of thousands of years later when the European colonizers introduced their livestock, leading to the further destruction of remaining wildlife habitats. Survivors include marsupials such as the koala bear and the wombat, but the list of extinctions is much longer.

Regions of the Austral Realm

Region Australia

Australia is the predominant component of the Austral realm, a continent-scale country in a size category that also includes China, Canada, the United States, and Brazil. For two reasons, however, Australia has fewer regional divisions than these other countries: its comparatively uncomplicated physiography and its small human population.

Positioned on the Pacific Rim, almost as large as the 48 contiguous U.S. States, well endowed with farmlands and vast pastures, rivers, groundwater supplies, mineral deposits, and energy resources, served by good natural harbors, and populated by just 24.3 million mostly well-educated people, Australia is one of the most geographically fortunate countries in the world.

Historical and Political Geographies

From the late eighteenth century onward, the Europeanization of Australia doomed the continent's Aboriginal societies. The first to suffer were those situated in the path of British settlement on the coasts, where penal colonies and free towns were founded. Distance protected the Aboriginal communities of the northern interior longer than elsewhere; but in Tasmania, the indigenous Australians died off in just decades after having lived there for perhaps 45,000 years. Eventually, the major coastal settlements became the centers of seven different colonies, each with its own hinterland; by 1861, Australia was internally delimited by its now-familiar pattern of straight-line State boundaries (**Fig. 11-5**).

On January 1, 1901, following years of difficult negotiations, the Australia we know today finally emerged: the Commonwealth of Australia, consisting of six States and two Federal Territories (Fig. 11-5). The two Federal Territories are the *Northern Territory*, assigned to protect the interests of the substantial Aboriginal population concentrated there and agitating for Statehood, and the *Australian Capital Territory*, carved from southern New South Wales to accommodate the federal capital of Canberra that was completed in 1927.

Australia's six States are New South Wales (capital Sydney), at 7.8 million the most populous and politically powerful; Victoria (Melbourne), small but populous by Australian standards with 6.2 million residents; Queensland (Brisbane), at 4.9 million, with the Great Barrier Reef offshore and tropical rainforests in its north; South Australia (Adelaide), with 1.7 million, where the Murray-Darling river system reaches the sea; Western Australia (Perth) with barely more than 2.6 million people in an area of more than 2.5 million square kilometers (nearly 1 million sq mi); and Tasmania (Hobart), containing 530,000, the island across the Bass Strait from the mainland's southeastern corner that lies in the path of Southern Ocean storms.

In earlier chapters, we referred to the concept of federalism, which is a form of politico-territorial organization. The word "federal" comes from the Latin *foederis*, implying alliance and coexistence, a union of consensus and common interest—a **federation [8]**. It stands in contrast to the idea of a centralized or **unitary state [9]**. For this, too, the ancient Romans devised a term: *unitas*, meaning "unity." Most of Europe's states are unitary, including the United Kingdom. Although most Australians came from that tradition, they managed to overcome their differences and forge a Commonwealth that was, in effect, a federation of States with different viewpoints, economies, and

FIGURE 11-5

objectives, separated by enormous distances along the rim of a particularly remote island-continent.

The Role of Distance

Australians often talk about distance. One of their leading historians, Geoffrey Blainey, labeled it a tyranny—an imposed remoteness from without and a divisive part of life within. Even today, Australia is far from nearly everywhere on Earth.

A trans-Pacific jet flight from Los Angeles to Sydney takes about 14 hours nonstop and is correspondingly expensive. You may have had the fortune of traveling abroad several times, yet never made it to Australia. Freighters carrying products to European markets take ten days to two weeks to get there. Within Australia, distances are of continental proportions as well, and Australians pay the price—literally. Until some upstart private airlines started a price war, Australians paid more per mile for their domestic flights than air passengers anywhere else in the world.

But distance also was an ally, permitting Australians to ignore the obvious. Australia was a British progeny, a European outpost. Once you had arrived as an immigrant from Britain or Ireland, there was a wide range of environments to choose from, magnificent scenery, vast open spaces, and seemingly limitless opportunities. When the Japanese Empire expanded in the 1930s and 1940s, Australia's remoteness saved the day. When boat people by the hundreds of thousands fled Vietnam in the mid-1970s aftermath of the Indochina War, almost none reached the shores of Australia. However, it should be noted that from the beginning in 1901, the "White Australia Policy" operated to prevent immigration from becoming an issue: its self-perceived comforts of isolation led Australia to adopt an all-white admission policy, the last vestiges of which were not officially terminated until 1978.

Core and Periphery

As Figure 11-2 shows, Australia is a large landmass, but its population is almost entirely concentrated in a (discontinuous) core area that lies in the east and southeast, most of which faces the Pacific Ocean's Tasman Sea. Figure 11-5 shows that this crescent-shaped Australian heartland extends from north of the city of Brisbane to the vicinity of Adelaide and includes the largest city, Sydney; the capital, Canberra; and the second-largest city, Melbourne. A secondary core area has developed in the far southwest, centered on Perth. Beyond the core areas lies the enormous periphery that Australians call the **Outback [10]** (see **Box 11-3**).

To better understand the evolution of this spatial arrangement, it helps to refer again to the map of world climates (Fig. G-6). Environmentally, Australia's most favored strips face the Pacific and Southern oceans as well as a sector of the Indian Ocean, and they are not large. We can describe the country as a coastal rimland, with cities, towns, farms, and forested slopes giving way to the immense, arid, interior Outback. On the western flanks of the east coast's Great Dividing Range lie the extensive grassland pastures that catapulted Australia into its first commercial age—and on which still graze one of the largest sheep herds in the world (about 72 million sheep in 2016, producing a major share of all the apparel wool sold in the world). Where it is moister, to the north and east, cattle by the millions graze on ranchlands. This is frontier Australia, over which livestock have ranged for nearly two centuries.

An Urban Culture

Despite the vast open spaces and romantic notions of frontier and Outback, Australia is an urban country, with 89 percent of all Australians living in cities and towns (**Box 11-4**). The large cities lie along the coast, the centers of manufacturing complexes as well as the foci of agricultural areas. For all its vastness and youth, Australia has developed a remarkable cultural identity, a sameness of urban and rural landscapes that persists from one end of the continent to the other.

Sydney, often called the New York of Australia, lies on a spectacular estuarine site, its compact, high-rise central business district overlooking a port bustling with ferry and freighter traffic (**Box 11-5**). Sydney today is a widely spread out metropolis of 4.1 million, with multiple outlying centers dominating its burgeoning suburban ring; brash modernity and reserved

---

Box 11-3 From The Field Notes...

"My most vivid memory from my first visit to Alice Springs in the heart of the Outback is spotting vineyards and a winery in this parched, desert environment as the plane approached the airport. I asked a taxi driver to take me there and got a lesson in economic geography. Drip irrigation from an underground water supply made viticulture possible; the tourist industry made it profitable. None of this, however, is evident from the view seen here: a spur of the Macdonnell Ranges overlooks a town of bare essentials under the hot sun of the Australian desert. What Alice Springs has is centrality: it is the largest settlement in a vast area Australians often call "the centre." Not far from the midpoint on the nearly 3200-kilometer (2000-mi) Stuart Highway from Darwin on the Northern Territory's north coast to Adelaide on the Southern Ocean, Alice Springs also was the northern terminus of "The Ghan," the Central Australia Railway line (before it was extended north to Darwin in 2004), seen in the middle distance. The shipping of cattle and minerals is a major industry here. You need a sense of humor to live here, and the locals have it: the town actually lies on a river, the ephemeral Todd River. An annual boat race is held, and in the absence of water the racers carry their boats along the dry river bed. No exploration of Alice Springs would be complete without a visit to the base of the Royal Flying Doctor Service, which brings medical help to the farthest-flung villages and homesteads."

© H. J. de Blij

Box 11-4 Major Cities of the Realm, 2016

City	Est. Metro Population (millions)
Sydney, Australia	4.1
Melbourne, Australia	4.0
Brisbane, Australia	2.0
Perth, Australia	1.8
Auckland, New Zealand	1.4
Adelaide, Australia	1.2
Canberra, Australia	0.4
Wellington, New Zealand	0.4
Christchurch, New Zealand	0.4

See an urban scene like this, and you realize why Australians refer to their land as 'the lucky country.' This is Sydney Harbour on a sunny, breezy day when sailboats by the hundreds emerge from coves and inlets along the estuary that connects the Pacific Ocean with the booming center of Sydney (whose skyline has grown both vertically and horizontally since this photo was taken around the time of the Australian bicentennial in 1988). Captains of ocean liners plying the world often remark that Australia's biggest city is not only their favorite port but also the grandest harbor entrance of them all. Indeed, this port is built around one of the finest natural harbors on Earth. Overall, it is locally known as Port Jackson, whose irregular shoreline (over 250 kilometers [150 mi] long) provides abundant space for deepwater docking facilities.

British ways somehow blend here. Melbourne (4.0 million), sometimes regarded as the Boston of Australia, prides itself on its more interesting architecture and more cultured ways. Brisbane (2.0 million), the capital of Queensland, which also anchors Australia's Gold Coast and adjoins the Great Barrier

Box 11-5 Among the Realm's Great Cities: Sydney

Well over two centuries ago, Sydney was founded by Captain Arthur Phillip as a British outpost on one of the world's most magnificent natural harbors. The free town and penal colony that struggled to survive evolved into Australia's largest city. Today, metropolitan Sydney (4.1 million) is home to one-sixth of the country's total population. An early start, the safe harbor, fertile nearby farmlands, and productive pastures in its hinterland combined to propel Sydney's growth. Later, as road and railroad links made Sydney the focus of Australia's growing core area, industrial development and political power augmented its primacy.

With its incomparable setting and mild, sunny climate, its many urban beaches, and its easy reach to the cool Blue Mountains of the Great Dividing Range, Sydney is one of the world's most liveable cities. Good public transportation (including an extensive cross-harbor ferry system from the doorstep of the waterfront CBD), fine cultural facilities headed by the multi-theatre Opera House complex, and many public parks and other recreational facilities make Sydney highly attractive to visitors as well (see photo). A healthy tourist trade, much of it from East Asian countries, bolsters the city's economy. Sydney's hosting of the 2000 Olympic Games was further testimony to its rising global visibility.

Increasingly, Sydney is evolving as a multicultural city. Its small Aboriginal sector is being overwhelmed by the arrival of large numbers of Asian immigrants. The Sydney suburb of Cabramatta symbolizes the impact: nearly 75 percent of its more than 23,000 residents were born elsewhere, at least a third of them in Vietnam. Unemployment is high, drug use is a problem, and crime and gang violence persist. Yet, despite the deviant behavior of a small

minority, tens of thousands of Asian immigrants have established themselves in some profession.

These developments herald Sydney's coming of age. The end of Australia's isolation has brought Asia across the country's threshold, and again the leading metropolis is showing the way.

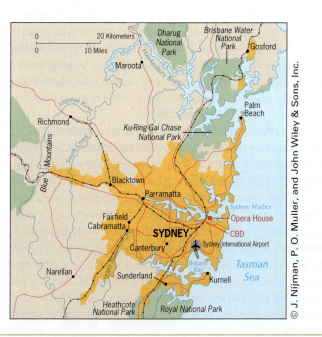

Reef, is the Miami of Australia; unlike Miami, however, its residents can find nearby relief from the summer heat in the mountains of its immediate hinterland (as well as at its beaches). Perth (1.8 million) is separated from its eastern Australian neighbors by three-quarters of a continent, and from Southeast Asia and Africa by thousands of kilometers of ocean—but due to the ever-expanding mining activities of Western Australia it is increasingly drawn into the global economy.

And yet, each of these cities—as well as the capitals of South Australia (Adelaide), Tasmania (Hobart), and, to a lesser extent, the Northern Territory (Darwin)—exhibits an Australian character of unmistakable quality. Life is both orderly and unhurried; streets are clean; slums are uncommon; graffiti rarely seen. Standards of public transport, urban schools, and healthcare provision are high. Spacious parks, pleasing waterfronts, and plentiful sunshine make Australia's urban life more agreeable than just about anywhere else on Earth.

Australia's Economic Geography

Australia today ranks twelfth among the world's countries in GDP, and for the vast majority of Australians life is quite comfortable. In terms of key development indicators, Australia is far ahead of all its western Pacific Rim competitors except Japan and Singapore. The country's initial prosperity was achieved in the mines and on the farms, not in the cities. Australia has material assets that other Asia Pacific countries can only dream of.

Agricultural Abundance

In agriculture, sheep-raising was the earliest commercial venture, but it was the technology of refrigeration that brought world markets within the reach of Australian beef producers. Wool, meat, and wheat have long been the country's big-three income earners; **Figure 11-6** displays the immense pastures in the east, north, and west that constitute the ranges of Australia's huge herds. The zone of commercial grain farming forms a broad crescent extending from northeastern New South Wales through Victoria into South Australia, with a major outlier covering much of the hinterland of Perth.

And keep in mind the scale of this map: Australia is only slightly smaller than the continental United States. Commercial grain farming in Australia is big business. As the climate map (Fig. G-6) suggests, sugarcane grows along most of the humid, subtropical, coastal-lowland strip of Queensland, and Mediterranean crops (including grapes for Australia's highly prosperous wine industry) cluster in the hinterlands of Adelaide and Perth. Mixed horticulture concentrates in the Murray River Basin, including rice, grapes, and citrus fruits, all under irrigation. In addition, as elsewhere across the world, dairying has developed close to the large metropolitan areas. With its considerable range of environments, Australia yields a great variety of crops.

Mineral Wealth, Limited Manufacturing

Australia's mineral resources, as Figure 11-6 shows, also are diverse. Major gold discoveries in Victoria and New South Wales produced a ten-year gold rush starting in 1851 and ushered in a new economic era. By the middle years of that decade, Australia was producing 40 percent of the world's gold. Subsequently, the search for more gold led to the discoveries of additional minerals, including nickel, copper, iron ore, uranium, tungsten, bauxite, and more. Since the 1990s, the enormous demand for raw materials from China as well as emerging India has marked a long-running commodity boom in Australia.

New finds are still being made today, and even oil and natural gas have been discovered both on land and offshore (Fig. 11-6). Most promising are the undersea deposits off the northwest coast, which has sparked an energy boom that is transforming Darwin. This northernmost city of Australia directly faces the fossil-fuel riches of the Timor Sea and is becoming a major processor of natural gas, liquefying it for export to East and Southeast Asia. Another liquid natural gas (LNG) boomtown is the east-coast, central Queensland port of Gladstone: here, an ultramodern processing-plant complex produces LNG from piped-in coal-seam gas, which is extracted by new hydraulic fracturing technologies from coal deposits in interior Queensland's Bowen and Surat Basins (Fig. 11-6).

Australian manufacturing caters mainly to the domestic market and is quite diversified, producing a wide array of industrial goods ranging from machinery made of locally produced steel to chemicals to textiles to paper products. These industries cluster in and near the major urban areas where consumers are located. The domestic market in Australia is not large, but it remains relatively affluent. This makes it attractive to

About 320 kilometers (200 mi) south-southeast of Port Hedland in the Pilbara region of Western Australia, Mount Whaleback near the town of Newman is one of the world's leading sources of high-grade iron ore. China's insatiable demand for this ore has generated the planet's largest "open-cut" iron mine, a huge and growing gash in the natural landscape—just one of many such negative impacts resulting from Australia's role as raw-material supplier to industrializing economies.

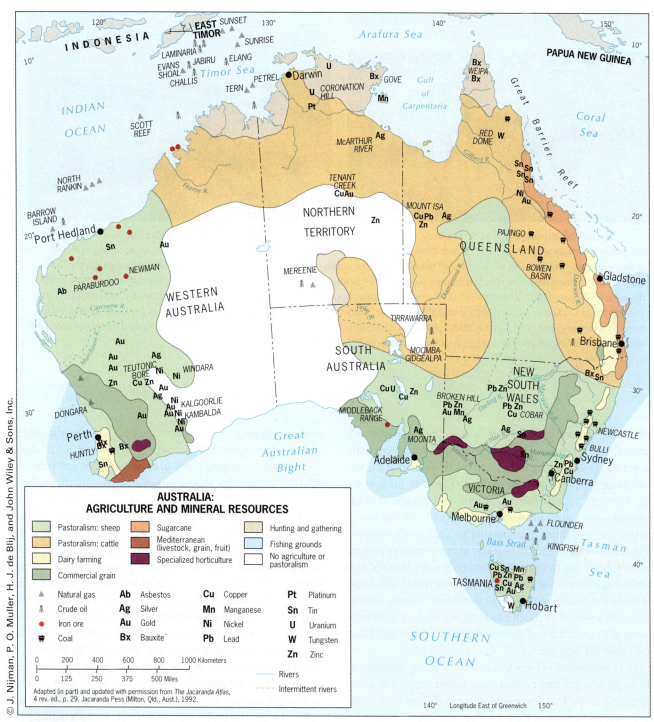

AUSTRALIA: AGRICULTURE AND MINERAL RESOURCES

- Pastoralism: sheep
- Pastoralism: cattle
- Dairy farming
- Commercial grain
- Sugarcane
- Mediterranean (livestock, grain, fruit)
- Specialized horticulture
- Hunting and gathering
- Fishing grounds
- No agriculture or pastoralism

▲	Natural gas	**Ab**	Asbestos	**Cu**	Copper	**Pt**	Platinum
	Crude oil	**Ag**	Silver	**Mn**	Manganese	**Sn**	Tin
●	Iron ore	**Au**	Gold	**Ni**	Nickel	**U**	Uranium
	Coal	**Bx**	Bauxite	**Pb**	Lead	**W**	Tungsten
						Zn	Zinc

0   200   400   600   800   1000 Kilometers
0   125   250   375   500 Miles

— Rivers
--- Intermittent rivers

Adapted (in part) and updated with permission from *The Jacaranda Atlas*, 4 rev. ed., p. 29. Jacaranda Press (Milton, Qld., Aust.), 1992.

FIGURE 11-6

foreign producers, and Australia's retail outlets overflow with high-priced goods from China, Japan, South Korea, the United States, and Europe. Overall, the continuing prominence of the **primary sector [11]** (mining and agriculture) and heavy reliance on raw-material exports indicate that the manufacturing sector is only partially developed, in spite of a long history of protectionist government policies. Whereas exports continue to be dominated by ores, fuels, and other commodities, the leading imports include cars, trucks, and medical equipment

and supplies. In 2015, Australia's top five partners in two-way trade were China, Japan, the United States, South Korea, and Singapore.

Australia's Environmental Problems

Environmental degradation [12], unfortunately, is practically synonymous with Australia. First the Aboriginals, then the Europeans and their livestock, inflicted catastrophic damage

During the first decade of the twenty-first century, Australia lay in the grip of its worst drought on record, with calamitous consequences ranging from deadly wildfires to parched farmlands. One of Australia's most profitable agricultural industries, winemaking, suffered severely as many winegrowers were driven off their land. The left photo shows a once-mature vineyard near Yarrawonga, on the New South Wales-Victoria border north of Melbourne along the Murray River, desiccated and abandoned during the summer of 2009. When the rains finally came (with a vengeance), they caused disastrous floods, washing away topsoil and eroding the heat-baked countryside. The right photo, taken in January 2010, shows the results at Coonamble, New South Wales along a tributary of the Darling River. These extreme-weather cycles persist in the 2010s (see Fig. 11-7). Most recently in mid-2016, a subsiding El Niño unleashed torrential rains and widespread flooding across much of southeastern Australia and Tasmania.

on Australia's natural environments and ecologies. Great stands of magnificent forest were ravaged. In Western Australia, centuries-old trees were simply ringed and left to die so that the sun could penetrate through their leafless crowns to nurture the grass below for pasture for the introduced livestock. In island Tasmania, where Australia's native eucalyptus tree reaches its greatest concentrations (comparable to California's redwood stands), tens of thousands of hectares of this irreplaceable treasure have been lost to chain saws and pulp mills. Many of Australia's unique marsupial species have been driven to extinction, and many more are endangered or threatened. "Never have so few people wreaked so much havoc on the ecology of so large an area in so short a time," observed one prominent Australian geographer not long ago.

More recently, the intensification of mining activities as well as oil and gas exploration have raised ecological stress to the next level. The energy-related development of Gladstone referred to above offers a telling example. Environmental activists are sounding alarms concerning the new hydraulic fracturing techniques for coal-seam gas production because they threaten Queensland's agricultural resources. Moreover, these same groups are vehemently protesting the expansion of port facilities because the coastal zone surrounding Gladstone lies within the supposedly protected confines of the Great Barrier Reef.

Another environmental problem involves Australia's wide and long-term climatic variability. In a predominantly arid continent, droughts in the moister fringes are the worst enemy, and Australia's history is replete with devastating dry spells. Australia is vulnerable to **El Niño [13]** events (see Chapter 3), but global warming may also be playing a role in the

process. **Figure 11-7** shows what are believed to be the effects of global warming on Australia between 1997 and 2011: whereas the northern (tropical) zones have experienced enhanced rainfall, the populated southwestern and southeastern coastal regions received less precipitation and have been more vulnerable to drought. The most serious drought (see photo pair), in fact the worst ever recorded, lasted throughout the 2000s and imperiled the entire Murray-Darling river basin, Australia's breadbasket. Part of this was a water deficit stemming from natural causes—but it was greatly exacerbated by rising demands for water in Australia's burgeoning urban areas in addition to excessive damming, well-drilling, and water diversion in the valleys of upstream tributaries of these two vital rivers. The Australian government responded by initiating the first fully coordinated drainage-basin-control program in the country, but all stakeholders recognize that it will take many years (and delays caused by future dry spells) as well as agricultural adjustments for the full benefits to materialize.

Western Australia, too, faces colossal challenges in the provision of water for its urban, agricultural, and industrial land uses. Here, much like in the western United States, water supplies in this dry environment are increasingly stressed by population growth—and especially all those massive new mining projects. The largest population cluster here is centered on west-coast Perth, and some scientists have warned that it could become the first "ghost metropolis," abandoned because of acute water shortages. To stave off that scenario, Perth's prospects have improved through the use of innovative **desalination [14]** technologies that convert ocean water into a viable source of supply for households and businesses (also see **Box 6-6**). Today, two desalination plants south of Perth provide

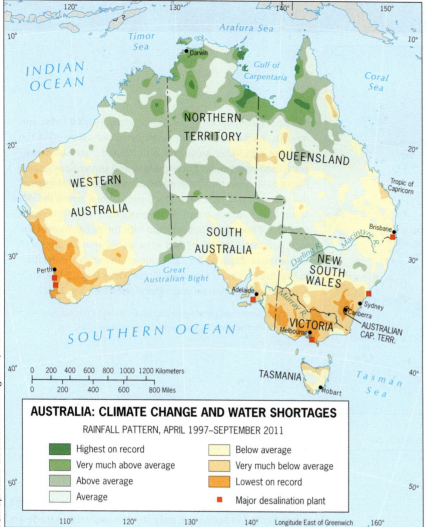

AUSTRALIA: CLIMATE CHANGE AND WATER SHORTAGES

RAINFALL PATTERN, APRIL 1997–SEPTEMBER 2011

- Highest on record
- Very much above average
- Above average
- Average
- Below average
- Very much below average
- Lowest on record
- ■ Major desalination plant

FIGURE 11-7

apologies and reparations; and (2) land ownership. The first question was resolved in 2008 when Prime Minister Kevin Rudd offered the long-awaited formal apology for the historic mistreatment of the Aboriginals. The second question has particular geographic implications. Although comprising less than 3 percent of the total population, the Aboriginal population has been gaining influence in national affairs over the past three decades: since as far back as the 1980s, Aboriginal leaders have been waging a campaign to obstruct mineral exploration on what they regard as ancestral and sacred lands.

Prior to 1992, Australians had taken it for granted that Aboriginals had no right to land ownership, but in that year the Australian High Court rendered the first of a series of landmark rulings in favor of Aboriginal claimants. A subsequent court decision implied that enormous areas (perhaps as much as 78 percent of the entire continent) could potentially be subject to Aboriginal claims. Today, the **Aboriginal land issue [15]** remains mostly (though not exclusively) an Outback issue, but it has the potential to overwhelm Australia's court system and to constrain economic growth (**Box 11-7**).

Immigration and the Changing Cultural Mosaic

Sixty years ago, when Australia had less than half the population it has now, 95 percent of its citizens were of European ancestry, and more than three-quarters of them came from Britain and Ireland. As we know, the White Australia immigration policy maintained this situation until the late 1970s. Today, the picture is dramatically different: of nearly 25 million Australians, only about one-third claim British or Irish origin, and Asian immigrants outnumber those of European ancestry as well as reproducing at a faster annual rate. During the early 1990s, about 150,000 legal immigrants arrived in Australia each year, mainly from Hong Kong, Vietnam, China, the Philippines, India, and Sri Lanka. Annual immigration quotas were subsequently reduced but then were allowed to rise again—set at about 190,000 in 2016—with Asian immigrants continuing to outnumber those from Western sources. Sydney, the leading recipient of the Asian influx, has become a mosaic of ethnic neighborhoods, some of which have gone through periods of gang violence and drug dealing, but have stabilized and even prospered over time (see photo). Still, as the economy has grown, particularly in the mining sector, the country will have to rely on immigration to meet growing skilled-labor demands. Multiculturalism is sure to remain a long-term challenge for Australia.

about half of this metropolitan area's water needs. This technology is now beginning to be applied to Australia's southeastern coast, and a number of plants have already come on line (Fig. 11-7).

Australia's Aboriginal Challenge

Despite Australia's generally positive fortunes, not everyone shares sufficiently in the national wealth. The indigenous (Aboriginal) population, though a small minority today of about 700,000, remains disproportionately disadvantaged in almost every way—shorter life expectancies, elevated unemployment, lower high school graduation rates, and much higher imprisonment rates (Box 11-6).

For the past several decades, the Aboriginal challenge focused on two questions: (1) official acknowledgment, by the government as well as the Australian majority, of mistreatment of the Aboriginal minority accompanied by appropriate

Box 11-6 MAP ANALYSIS

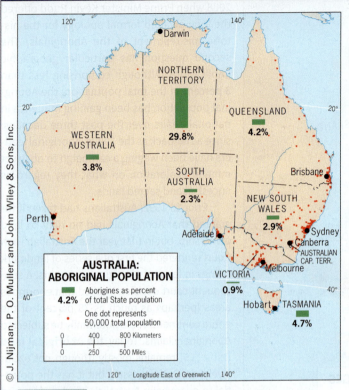

© J. Nijman, P. O. Muller, and John Wiley & Sons, Inc.

FIGURE 11-8

The Aboriginal population totals just under 700,000 today and remains marginalized within Australian society. Not unlike the situation with Native Americans in the United States, land ownership and entitlement are important and contentious issues. Figure 11-8 shows the spatial distribution of both the Aboriginal (bars) and overall Australian (dots) populations across the different States. Where are the Aboriginals most heavily concentrated, and how does that compare with the distribution of the total national population? And what would you say are some of the implications of the clustering of Aboriginals in relation to the distribution of Australia's largest cities (Fig. 11-5) and its geography of agricultural and mineral resources (Fig. 11-6)?

Access your WileyPLUS Learning Space course to interact with a dynamic version of this map and to engage with online map exercises and questions.

Australia's Place in the World

Since World War II, as its historic ties to the United Kingdom have faded, Australia has become ever more embedded in the Asia Pacific—economically, culturally, and politically. The

Robert Francis/Alamy

Cabramatta, one of Sydney's western suburbs, is seen as the "multicultural capital of Australia," a tourist attraction and proof of Australia's capacity to accommodate non-Europeans and especially Asians. This "Freedom Gate" in the Vietnamese community is flanked by a Ming horse and a replica of a Forbidden City lion—all reflecting Australia's growing ties to the Asia Pacific.

country's strengthening orientation toward its Asian neighbors is reflected in economic policies and patterns of trade and investment. Australia is now a key member of the Asia Pacific Economic Cooperation forum, **APEC [16]**; China and Japan, by far, have become its most important trading partners. Australia itself, as the result of changing immigration patterns, is becoming culturally more heterogeneous and increasingly Asian. And, politically, Australia has become more involved with the geographic realms to its north, especially the countries of Indonesia, East Timor, and Papua New Guinea.

Geostrategically, Australia finds itself in a relatively secure location. However, security issues in the South China Sea and the Indian Ocean have become a growing concern precisely because Australia's interdependence with Asia is intensifying. Also of much interest is Australia's intent to forge closer relations with the United States as a counterweight to China's rapidly increasing influence across these realms, especially Southeast Asia. To nobody's surprise, the United States is pleased to reciprocate: since 2012, several thousand American troops have been stationed at Australian military bases, and the U.S. Air Force has been given access to Australian airfields in the Northern Territory that are within easy flying distance of the South China Sea.

Territorial dimensions, relative location, and raw-material wealth have helped determine Australia's place in the world and, more specifically, on the western Pacific Rim. Australia's

Box 11-7 Regional Issue: Aboriginals in Australia

First Australians should come first.

"Australia is becoming a multiracial society. Immigrants from Tajikistan to Thailand are changing the (mostly urban) cultural landscape. Long debates over their rights and privileges roil the political scene. Meanwhile, the Australians who got here first, most of whom live far beyond the burgeoning cities and out of sight, remain the most disadvantaged of minorities. We should remind ourselves of what they went through and why Australia owes them—big time.

"When my European ancestors arrived on these shores, there were more than a million Aboriginal people here, organized into numerous clans and subcultures. Like all peoples, they had their vices and virtues. Among the latter was that they didn't appropriate land. Land was assigned to them by the creator (in what they call The Dreaming), and their relationship to it was spiritual, not commercial. They didn't build fences or walls. Neither had they adopted some bureaucratic religion. Just imagine: a world where land was open and free, and religion was local and personal.

"Then the British showed up and started claiming and fencing off land that, under their European rules, was there for the taking. If the Aboriginal clans got in the way, they were pushed out, and if they resisted, they got killed. You don't even want to *think* about what happened in Tasmania: a campaign of calculated extermination. Between 1800 and 1900, the Aboriginal population dropped from 1 million to about 50,000. When we became a 'nation' in 1901, they weren't even accorded citizenship. They didn't get the vote until 1962.

"None of this stopped certain white Australian men from getting Aboriginal women pregnant. And from 1910 on, church and state managed to make things even worse. They took these young children and put them in institutions, where they would be 'Europeanized' and then married off to white partners, so that they would lose their Aboriginal inheritance. This, if you'll believe it, went on into the 1960s! Think of the scenes, these kids being kidnapped from their mothers by armed officials never to be seen again.

"It's hard to believe that it took nearly another half-century before the Australian government, following a contentious and divisive debate, finally offered a formal apology for this and other misdeeds of the past. But now the question is, do Aboriginal Australians benefit from the country's growing ethnic complexity? I work in a State government office here in Sydney that assists Asian immigrants, and all I can say is that I wish that we'd done for the First Australians what we're doing for the stream of immigrants we admit today."

It's old history: no more special treatment.

"Australia is a nation of immigrants, and we've all gone through rough times. I'm not complaining to the British government for what happened to my ancestors when they were shipped out here as prisoners. Like my father, I was born on this Outback sheep station about 40 years ago, and we employ a dozen Aboriginal workers, most of whom were also born in this area. I had nothing to do with what happened more than a century ago. What I would or might have done is irrelevant, and I can't be blamed for what my great-great-grandparents may have done. All over the world people are born into situations not of their making.

"And I believe that this country has bent over backwards, in my time at least, to undo the alleged wrongs of the past. Look, the Aboriginal minority counts a bit under 700,000 or 2.9 percent of the population. Take a look at this on the map: they've got the whole Northern Territory and other parts of the country too, and that's 15 percent of Australia. And now we're required to give them even more? The Australian High Court keeps awarding Aboriginal claimants more and more land. That affects all of us. Pretty soon all of Australia will be targeted by these Aboriginals and their lawyers. And, by the way, in the old days those people moved around all the time. Who is to say what clans owned what land when it comes to claiming 'native title'? These court cases are going to tie us up in legal knots for generations to come. So, if you're a Japanese or Chinese buyer in search of commodities, are you going to sign contracts when you're not sure who will own the land?

"We should take a lesson from the Kiwis, who agreed to a land deal with the Maori minority in New Zealand, and now look what's happening over there. The place is overrun with Chinese who are making deals for supposedly 'native' land and are converting the whole place to milk-powder production. And the easy residence rules are turning New Zealand into a stepping stone for entering Australia.

"Look, I like the fellows working here, but you've got to realize that no laws or treaties are going to solve all of the problems they have. They're getting all kinds of preferential access to government employment, remedial help in many areas, but still they wind up leaving school, abandoning jobs, winding up in jail. They have to grab the opportunities they now have rather than ask for more. In a lot of countries they would have never gotten them: Aussies are a pretty decent people. It's up to them to make the best of it."

---

population may still be less than 25 million, but the country's importance in the international community greatly exceeds its human numbers.

Region New Zealand

Twenty-four hundred kilometers (1500 mi) east-southeast of Australia, in the Pacific Ocean across the Tasman Sea, lies New Zealand, also known as *Aotearoa* in Maori (meaning

"land of the long white cloud"). In an earlier age, New Zealand would have been part of the Pacific geographic realm because its population was entirely Maori, a people with Polynesian roots (see Chapter 12). But New Zealand, like Australia, was invaded and occupied by Europeans. Today, its population of 4.6 million is about two-thirds European, and the Maori comprise a substantial minority that numbers about 700,000, with many of mixed Euro-Polynesian ancestry (including Pacific Islanders).

New Zealand consists of two large mountainous islands and many scattered smaller islands (**Fig. 11-9**). The two main

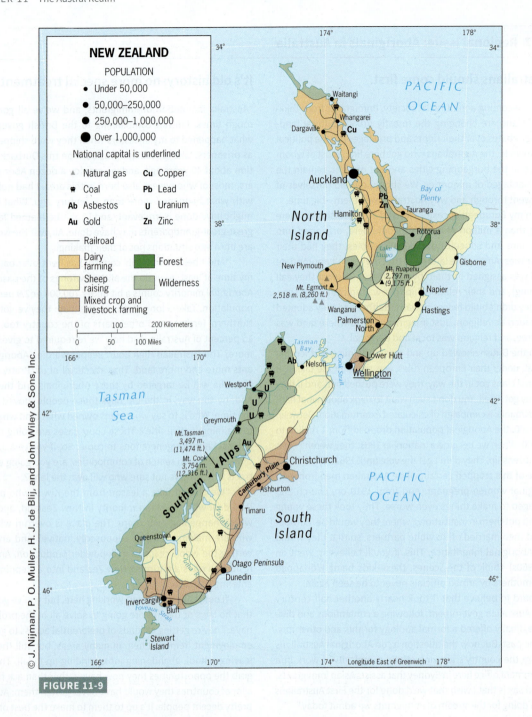

NEW ZEALAND

POPULATION

- Under 50,000
- 50,000–250,000
- 250,000–1,000,000
- Over 1,000,000

National capital is underlined

▲	Natural gas	**Cu**	Copper
⚒	Coal	**Pb**	Lead
Ab	Asbestos	**U**	Uranium
Au	Gold	**Zn**	Zinc

— Railroad

Dairy farming
Sheep raising
Mixed crop and livestock farming
Forest
Wilderness

0   100   200 Kilometers
0   50   100 Miles

© J. Nijman, P. O. Muller, H. J. de Blij, and John Wiley & Sons, Inc.

PACIFIC OCEAN

Waitangi
Whangarei
Dargaville **Cu**
Auckland
Bay of Plenty
Hamilton **Pb Zn**
Tauranga
North Island
Rotorua
Lake Taupo
New Plymouth Mt. Ruapehu 2,797 m. (9,175 ft.) Gisborne
Mt. Egmont 2,518 m. (8,260 ft.) Napier
Wanganui Hastings
Palmerston North
Tasman Bay
Lower Hutt
Ab Nelson
Cook Strait
Wellington
Westport
Tasman Sea
U
U
U
Greymouth
Mt. Tasman 3,497 m. (11,474 ft.)
Mt. Cook 3,754 m. (12,316 ft.) **Au**
Canterbury Plain Christchurch
Ashburton
Timaru
South Island
PACIFIC OCEAN
Southern Alps
Waitaki R.
Queenstown
Clutha R.
Otago Peninsula
Dunedin
Invercargill
Foveaux Strait Bluff
Stewart Island

Longitude East of Greenwich

FIGURE 11-9

islands, with the South Island somewhat larger than the North Island, look diminutive in the great Pacific Basin, but together they are bigger than Britain. In sharp contrast to generally low-relief Australia, the more rugged terrain of the two large islands contains several peaks rising far higher than any on the Australian landmass (**Box 11-8**). Along its western margin, the South Island has a spectacular snowcapped range appropriately called the Southern Alps, with a number of summits reaching beyond 3300 meters (10,000 ft). The smaller North Island has proportionately more land under low relief, but it also has an area of central highlands along whose lower slopes lie the pastures of New Zealand's chief dairying district. The convergence

of the Australian and Pacific tectonic plates underlies much of New Zealand, and it makes the country vulnerable to volcanic eruptions and earthquakes.

Human Spatial Organization

The most promising areas for habitation, therefore, are the lower-lying slopes and lowland fringes on both islands. On the North Island, the largest urban area, Auckland, occupies a comparatively low-lying peninsula. On the South Island, the largest lowland is the agriculture-dominated Canterbury Plain, focused

Box 11-8   From the Field Notes …

"The drive from Christchurch to Arthur's Pass on the South Island of New Zealand was a lesson in physiography and biogeography. Here, on the eastern side of the Southern Alps, you leave the Canterbury Plain and its agriculture and climb into the rugged terrain of the glacier-cut, snowcapped mountains. A vast pasture lies on a patch of flatland in the foreground; in the background is the unmistakable wall of a U-shaped valley sculpted by moving ice. Natural vegetation ranges from pines to ferns, becoming even more luxuriant as you approach the moister western side of the island."

© H.J. de Blij

on the city of Christchurch (still recovering from a months-long swarm of devastating earthquakes in 2010–2011; see **Box 11-9**). What makes these lower-elevation zones so attractive, apart from their availability as cropland, is their magnificent pastures. The range of soils and pasture plants allows both summer and winter grazing. Moreover, the Canterbury Plain, the chief farming region, also produces a wide variety of vegetables, cereals, and fruits. About half of all New Zealand is pasture land, and much of the farming provides fodder for the pastoral industry. About 29.5 million sheep, 6.5 million dairy cattle, and 3.6 million beef cattle dominate these livestock-raising activities, with wool, milk products, and meat providing about 75 percent of the islands' export revenues.

Despite their contrasts in size, shape, physiography, and history, New Zealand and Australia share a number of characteristics. Apart from their joint British heritage, they share a sizeable pastoral economy with growth in specialty goods such as wines, a small local market, the problem of great distances to world markets, and a desire to stimulate (through protectionist policies) domestic manufacturing. The very high degree of urbanization in New Zealand (86 percent of the total population) once again brings up the comparison with Australia: substantial employment in city-based industries—mostly the processing and packing of livestock and farm products—as well as government jobs.

Spatially, New Zealand further shares with Australia its pattern of **peripheral development [17]** and population distribution (Fig. 11-2), imposed not by deserts but by high, rugged mountains and the fragmented layout of the country. The major cities—Auckland and the capital city, Wellington,

on the North Island; Christchurch and Dunedin on the South Island—are all situated on the coast, and the rail and highway networks are therefore dominantly peripheral in their configuration. Furthermore, the two main islands are separated by Cook Strait, a windswept waterway that can only be crossed by ferry or plane (Fig. 11-9). On the South Island, the Southern Alps are the country's most formidable barrier to surface communications.

The Maori Factor and New Zealand's Future

Like Australia, New Zealand has had a history of difficult relations with its indigenous population. The Maori, who today account for 14.6 percent of the country's total population, appear to have reached the islands during the tenth century AD. By the time the European colonists arrived, the Maori had also had a tremendous impact on the islands' ecosystems, especially on the North Island where most of them lived. In 1840, the Maori and the British signed a treaty at Waitangi that granted the colonizers sovereignty over New Zealand but guaranteed the Maori rights covering established tribal lands. Although the British abrogated parts of the treaty in 1862, the Maori still had reason to believe that vast reaches of New Zealand, as well as offshore waters, were theirs in perpetuity.

As in Australia, judicial rulings beginning in the 1990s supported the Maori position, which led to expanded claims and growing demands. Culturally, the declaration of Maori as an

Box 11-9 Regional Planning Case

Rebuilding Christchurch After the 2010–2011 Earthquakes

Earthquakes and other natural (or human-caused) disasters pose major challenges to urban and regional planners. Within the discipline of urban and regional planning, a prominent specialty is **disaster management [18]**, which consists of four phases: hazard mitigation; preparedness; response; and recovery. Mitigation planning seeks to minimize or contain human and material damage if a disaster were to strike. Strategies include land-use regulations, enforcement of building codes, and acquisition of damaged properties in hazard-prone areas. The readiness to respond to a disaster (preparedness), and efforts and actions to save lives and prevent further damage created by a disaster (response), requires coordination with medical and law enforcement professionals. Recovery planning entails efforts to return the community to "normal" or a "new normal", but the pace of recovery may be complicated by the severity of destruction and the degree of population displacement. These four phases can often overlap; mitigation activities are increasingly incorporated during the post-disaster period.

New Zealand, located astride the colliding edges of the Australian and Pacific tectonic plates, is a focus of major earthquake activity (see Fig. G-4). Christchurch, the largest city on South Island (Fig. 11-9), was struck by a swarm of earthquakes between September 2010 and February 2011. The 6.3-magnitude quake on February 22 was the most devastating, resulting in 181 fatalities. In this city of just 385,000 people, thousands were left homeless and sought refuge in surrounding areas. Central Christchurch was the most heavily affected: one out of three buildings was either destroyed or subsequently condemned. As a result, the CBD was designated as a "rebuild zone", and businesses had to relocate to residential and industrial districts of the city as well as nearby towns.

Five years after the earthquake, reconstruction was still ongoing amidst empty lots (many filled with artistic creations) and boarded-up buildings (some in the process of being sold or awaiting demolition). And one key challenge remains: how to deal with the damage to historic, heritage-masonry buildings such as iconic Christchurch Cathedral (see photo).

The threat of earthquakes poses difficult challenges for the future planning of the central city and the broader metropolitan region. The U.S. $27.1 billion Christchurch Rebuild movement, the largest public-private undertaking in New Zealand's history, seeks to create a new Christchurch that is more compact, greener, and internally accessible. The city's blueprint redevelopment plan was launched in 2012 after more than 100,000 ideas were solicited as part of a public consultation process. The plan features 17 anchor projects including state-of-the-art cricket facilities and stadiums in an effort to capitalize on the urban region's sporting heritage.

Questions linger about whether the multi-billion-dollar investment is worthwhile beyond the central city. Such dialogue and debate about using the post-disaster rebuilding period, investments, and timeframes as a window of opportunity to revitalize central cities is nothing new. The bottom line is that cities, especially those vulnerable to disasters, must plan ahead to ensure the future resilience of housing, education, infrastructure, and major economic sectors of their broader urban environments.

Kai Schwoerer/Stringer/GettyImages, Inc.

Christchurch, New Zealand, February 2016: five years after the devastating earthquakes, the city's iconic cathedral is still undergoing reconstruction.

official language in New Zealand and its teaching throughout the school system are seen as significant progress toward an acceptance of the Maori cultural heritage. But the most persistent Maori complaint concerns their slow pace of integration into modern New Zealand society. Although Maori land claims encompass much of rural New Zealand, they also cover prominent sites in the major cities. Today, the Maori question continues to be the foremost domestic issue.

Dominant cultural heritage and prevailing cultural landscape form two criteria on which the delimitation of the Austral Realm is based. But in both Australia and New Zealand, the convergence with neighboring geographic realms is well underway.

Points to Ponder

- In this era of globalization, Australia is becoming more strongly tied to neighboring geographic realms, while its links to faraway Europe steadily erode.
- Australia's advancing desalination technologies provide relief for the country's immediate water needs but are delaying much-needed conservation strategies for the longer-term future.
- New Zealand experiences more than 10,000 earthquakes each year; as many as 150 can be felt, and about 20 are strong enough to cause damage on the human landscape.

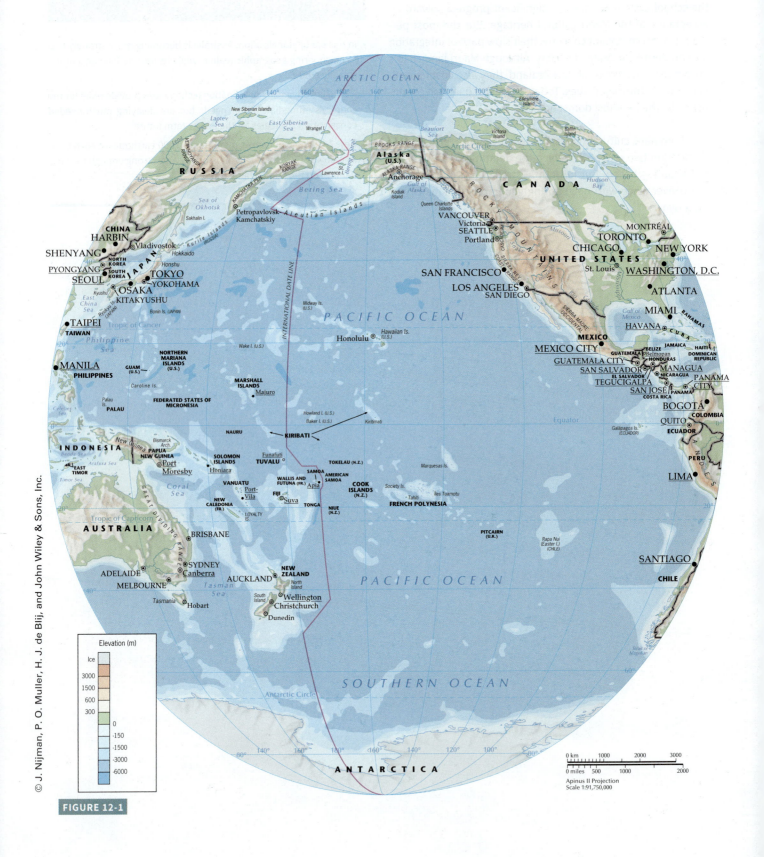

FIGURE 12-1

CHAPTER 12

Where were these pictures taken?

Pacific Realm and Polar Futures

IN THIS CHAPTER

- Water, water everywhere—but who owns it?
- Durable regionalization of Pacific cultures
- Low islands threatened by rising sea level
- The partitioning of Antarctica
- Geopolitics in the warming Arctic Basin

CONCEPTS, IDEAS, AND TERMS

High island	[1]	Exclusive Economic Zone (EEZ)	[10]
Low island	[2]	Maritime boundary	[11]
Atoll	[3]	Median-line boundary	[12]
Tsunami	[4]	Global climate change	[13]
Tsunami Warning System (TWS)	[5]	Rising sea level	[14]
Marine geography	[6]	Antarctic Treaty	[15]
Territorial sea	[7]	Maritime claims	[16]
High seas	[8]	Northeast Passage	[17]
Continental shelf	[9]	Northwest Passage	[18]

In this final chapter we focus on three of the world's gigantic expanses, so different from those discussed in previous chapters that we need an entirely new perspective. Water and ice, not land and soil, dominate the physiography of these areas. Of the utmost importance is that environmental change occurring in all three has consequences for the entire planet.

The largest by far of these expanses, the *Pacific Realm*, covers almost half the globe from the Bering Sea to the Southern Ocean and from Central America to Indonesia (**Fig. 12-1**). The Pacific Ocean, larger than all the world's landmasses combined, is studded with tens of thousands of islands, of which New Guinea is by far the largest. The two polar zones to which it is linked could hardly be more different. The *Arctic* consists entirely of water, frozen at the surface, but you can reach the North Pole in a submarine. The *Antarctic* is mainly land, weighed down by the planet's largest accumulation of permanent ice; the surface journey to the South Pole, first accomplished more than a century ago, has cost many an explorer's life.

Climate change is altering geography everywhere, as is the enhanced technology of mineral (especially oil and gas) exploration and extraction. Rising sea levels would imperil thousands of populated low-elevation islands. Areas of ocean floor once beyond the scope of human intervention are now coming within reach. Melting ice may be clearing new maritime shipping routes. At the same time, as we will see, international rules to govern who owns what in these remote frontiers are subject to dispute. When land boundaries jut outward into the sea, who owns what depends on who draws the maps. We should be aware of the potential for competition, even conflict.

Defining the Realm

Our survey begins with the Pacific Realm, which covers just about an entire hemisphere, the one sometimes called the Sea Hemisphere (Fig. 12-1). This Sea Hemisphere meets the Russian/Central Asian and North American realms in the far north and merges into the Southern Ocean in the south. Despite the preponderance of water, this fragmented, culturally complex realm does possess regional identities. It includes the Hawaiian Islands, Tahiti, Tonga, Fiji, and Samoa—fabled names in a world apart.

In terms of modern cultural and political geography, Indonesia and the Philippines are not part of the Pacific Realm, though Indonesia's political system reaches into it; nor do Australia and New Zealand belong to it. Before the European invasion and colonization, Australia would have been included (because of its Aboriginal population) as well as New Zealand (its Maori population has Polynesian affinities). But the Europeanization of their countries has engulfed indigenous Australians and Maori New Zealanders, and the regional geography of the Austral Realm today is decidedly not Pacific. In New Guinea, by contrast, Pacific peoples, numerically and culturally, remain the dominant element (**Box 12-1**).

In examining the Pacific Realm, it is important to keep in mind the dimensional contrasts this part of the world presents. The realm may be enormous, but its total land area is a mere 975,000 square kilometers (377,000 sq mi), about the size of Texas plus New Mexico, with over 90 percent constituted by the island of New Guinea. Indeed, the realm's inhabitants are so widely and thinly scattered that a population distribution map would be impractical; moreover, the population totals only around 15 million (less than in the compact megacity of Dhaka, Bangladesh—see Chapter 8).

Physiographically, the main landforms of the Pacific can be distinguished in high islands, low islands, and atolls. **High islands [1]** are seafloor-based volcanic peaks and ridges that protrude above the ocean surface and can reach thousands of meters in elevation. **Low islands [2]** are far more common, consist mainly of coral, and sometimes barely rise above sea level.

Box 12-1 **Major Geographic Features of the Pacific Realm**

1. The Pacific Realm's areal extent is the largest of all geographic realms. Its total land area, however, is the smallest, as is its population.

2. The island of New Guinea, with 10.8 million people, alone contains more than three-quarters of the Pacific Realm's population.

3. The Pacific Realm, with its huge expanses of water and numerous islands, has been strongly affected by United Nations Law of the Sea provisions regarding states' rights over economic assets in their adjacent waters.

4. The highly fragmented Pacific Realm consists of three island regions: Melanesia (including New Guinea), Micronesia, and Polynesia.

5. The Pacific Realm's islands and cultures may be divided into volcanic high-island cultures and coral-based low-island cultures.

6. Rising sea level is threatening the low islands in the Pacific, which are desperately trying to adapt to increased flooding.

7. In Micronesia, U.S. influence has been particularly strong and continues to affect local societies.

8. In Polynesia, indigenous culture exhibits remarkable consistency and uniformity throughout the region, its enormous dimensions and dispersal notwithstanding. Yet at the same time, local cultures nearly everywhere are severely stressed by external influences. In Hawai'i, as in New Zealand, indigenous culture has been largely submerged by Westernization.

© Jan Nijman

Aerial view of a circular atoll in the Fijian archipelago, which marks the eastern edge of Melanesia. This atoll is almost completely submerged except for the tiny sliver of sand visible at the extreme left. From the plane, atolls can often be detected first by the clouds clustered above them: the shallow water atop an atoll (for that matter, any low island) heats up more than the deeper ocean around it—causing the air to rise, condense into clouds, and occasionally trigger a local rainstorm.

Livelihoods tend to vary among these different environments: farming prevails on high islands (thanks to fertile soils and orographic rainfall), whereas fishing dominates the economy of low islands (where freshwater supplies are limited). The third landform category is the **atoll [3]**, a ring-shaped coral reef that encircles a lagoon, lying more or less at sea level (see photo). The circular shape results from the reef growing on the rim of an underwater volcano (which may long have eroded away); typically, atolls are uninhabitable.

Whether high or low, the Pacific islands are a reminder of the geological nature of the ocean floor, with its volcanic peaks, ridges, and tectonic activity. Earthquakes in the Pacific, wherever they may occur, can cause huge seismic sea waves known as **tsunamis [4]** (also see chapter 9). Because the low islands and low-lying shores of the Pacific are so vulnerable, an internationally organized **Tsunami Warning System (TWS) [5]** has been established to constantly monitor sea-surface conditions (**Box 12-2**).

Colonization and Independence

The Pacific islands were colonized by the French, British, and Americans—an indigenous Polynesian kingdom in the Hawaiian Islands was annexed by the United States in 1898, and since 1959 has functioned as the fiftieth U.S. State. Because political geography evolves slowly within the Pacific Basin, today's map is still an assemblage of colonial territories and independent microstates (Fig. 12-1). France controls New Caledonia and French Polynesia. The United States administers Guam as well as American Samoa, the Line Islands, Wake Island, the Midway Islands, and several smaller island groups; the United States

also has special relationships with other territories, former dependencies that are now nominally independent. The British, through New Zealand, have responsibility for the Pitcairn group of islands, and New Zealand administers and supports the Cook, Tokelau, and Niue Islands. Easter Island, that storied speck of land in the southeastern Pacific, is part of Chile (see chapter-opening photos). And, as discussed in Chapter 10, Indonesia rules Papua, the western half of the island of New Guinea.

Other island groups have now become independent states. The largest are Fiji, once a British dependency; the Solomon Islands, also formerly British; and Vanuatu, until 1980 ruled jointly by France and Britain. Also on the current map are such impoverished microstates as Tuvalu, Kiribati, Nauru, and Palau, which must depend on foreign aid for their survival. Tuvalu, for instance, has a total area of 26 square kilometers (10 sq mi), a population of barely 10,000, and a per capita GNI of about U.S. $6000—derived from fishing, sales of copra (coconut meat used to make coconut oil), and some tourism. But what really keeps Tuvalu going is an international trust fund set up by Australia, New Zealand, the United Kingdom, Japan, and South Korea. Annual grants from that fund, as well as remittances sent back to families by workers who have left for New Zealand and elsewhere, allow Tuvalu to survive.

The Pacific Realm and Its Marine Geography

Certain land areas may not be part of the Pacific Realm (we mentioned the Philippines and New Zealand), but the Pacific Ocean extends from the shores of North and South America to mainland East and Southeast Asia and from the Bering Sea off Alaska southward to the Subtropical Convergence around latitude 40°S. This means that several seas, including the South China Sea, the East China Sea, the Sea of Japan (East Sea), the Java Sea, and the Tasman Sea, are all part of the Pacific Ocean. As we are about to discover, this relationship matters. Pacific coastal countries, large and small, mainland as well as island, compete for jurisdiction over the waters that bound them.

The ocean-focused Pacific Realm, therefore, provides an ideal setting for introducing some basic principles of **marine geography [6]**. This field encompasses a variety of approaches to the study of oceans and seas; some marine geographers concentrate on the biogeography of coral formations, others on the geomorphology of beaches, and still others on the circulation of ocean currents. A most fascinating specialty within marine geography has to do with the definition and delimitation of political boundaries at sea. Here geography intersects both political science and maritime law.

The State at Sea

Littoral (coastal) states do not end where atlas maps suggest they do. States have claimed various forms of jurisdiction over coastal waters for centuries, closing off bays and estuaries and

Box 12-2 Technology & Geography

Tsunami Warning System in the Pacific

After a major earthquake strikes near or under a large body of water, a series of long, fast-moving waves is sometimes generated. Earthquakes are quite common in the Pacific and Atlantic oceans due to tectonic-plate motion (see Fig. G-4). Such a seismic sea wave is called a *tsunami*. As a tsunami crosses the ocean surface, it can travel at speeds of up to 1000 kilometers per hour (625 mph) and be more than 200 kilometers (125 mi) wide. As the wave approaches shallow water, its forward speed decreases and its height increases. When tsunamis make landfall, at times cresting at over 10 meters (35 ft), low-lying coastal areas and their inhabitants are imperiled. A colossal, magnitude-9.0 quake off Japan in 2011 created a gigantic tsunami that destroyed dozens of coastal settlements, killed more than 20,000 people, and caused disastrous damage at an oceanside nuclear-power generation facility (Fig. 9-2).

During the mid-twentieth century, to warn coastal populations and minimize the loss of life from tsunamis, several Pacific Basin countries joined forces to establish the ***Tsunami Warning System (TWS) in the Pacific***. Consisting of an extensive network of buoys armed with sensors and instruments specifically designed to measure and record the depth of the water column across the Pacific Ocean, the TWS is designed to detect unexpected shifts in water depth and ocean-wave irregularities. Global positioning systems (GPS) satellites record the exact location and movement of these buoys, from which the speed and direction of a tsunami event can be calculated. Should a tsunami be detected, its direction, speed, and potential strength can be estimated, and member-states are swiftly alerted to take appropriate safety measures. **Figure 12-2** shows the location of coastal seismographic stations (that measure both tectonic activity as well as tidal variations), deep-ocean *tsunameters* (that gauge sea-floor pressures and water movement), and average tsunami travel times from Honolulu, Hawai'i.

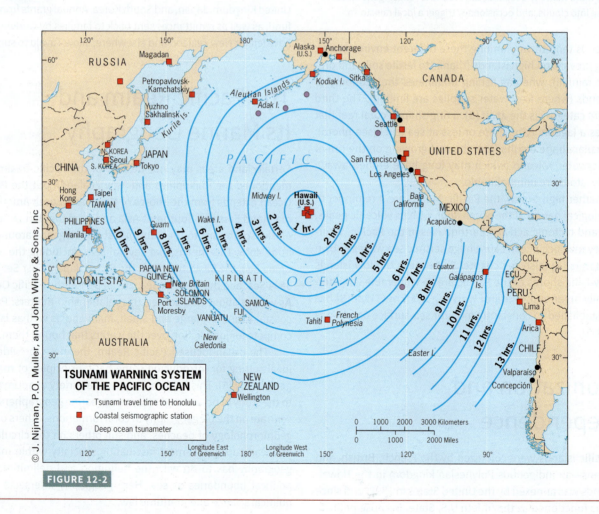

© J. Nijman, P.O. Muller, and John Wiley & Sons, Inc

TSUNAMI WARNING SYSTEM OF THE PACIFIC OCEAN
— Tsunami travel time to Honolulu
■ Coastal seismographic station
● Deep ocean tsunameter

FIGURE 12-2

ordering foreign fishing fleets to stay away from nearby fishing grounds. Thereby arose the notion of the **territorial sea [7]**, where all the rights of a coastal state would prevail. Beyond lay the **high seas [8]**, free, open, and unfettered by national interests—so-called international waters.

It was in the interest of colonizing and/or trading states to keep territorial seas narrow and high seas wide, thereby interfering as little as possible with their commercial fleets. In the seventeenth and eighteenth centuries, the territorial sea was 3, 4, or at most 6 nautical miles wide, and the colonizing powers claimed the identical widths for their colonies (1 nautical mile = 1.85 kilometers or 1.15 statute miles).*

During the twentieth century, these constraints weakened. States without trading fleets saw no reason to limit their territorial seas. States with nearby fishing grounds traditionally exploited by their own fleets wanted to keep the increasing number of foreign trawlers away. States with shallow **continental shelves [9]** (offshore continuations of coastal plains) wished to control their resources on and below the seafloor, made more accessible by improved technology. States also disagreed on the methods by which offshore boundaries, whatever their width, should be defined. Early efforts by the League of Nations (the precursor to the United Nations) during the 1920s to resolve these issues met with only partial success, mainly in the technical sphere of boundary delimitation.

Scramble for the Oceans

In 1945, the United States helped precipitate what has become known as "the scramble for the oceans." President Truman issued a proclamation that claimed U.S. jurisdiction and control over all the resources "in and on" the continental shelf down to its margin, around 100 fathoms (183 meters/600 ft) deep. In some areas, the shallow continental shelf of the United States extends more than 300 kilometers (200 mi) offshore, and Washington did not want foreign countries drilling for oil just beyond the 3-mile territorial sea.

Few observers foresaw the impact the Truman Proclamation would have, not only on U.S. waters but on the oceans everywhere, including the Pacific. It set off a rush of additional claims. In 1952, a group of South American countries, some with little continental shelf to claim, issued the Declaration of Santiago, claiming exclusive fishing rights up to a distance of 200 nautical miles off their coasts. Meanwhile, as part of the Cold War competition, the Soviet Union claimed a 12-mile territorial sea and urged its allies to do likewise.

UNCLOS Intervention

At this point, the United Nations intervened, and a series of UNCLOS (United Nations Conference on the Law of the Sea) meetings began. These meetings addressed issues ranging from the closure of bays to the width and delimitation of the territorial sea, and after three decades of negotiations they achieved a convention that changed the political and economic geography of the world's oceans forever. Among its key provisions were the authorization of a 12-mile territorial sea for all countries as well as the establishment of a 200-mile (230-statute-mi/370-km)-wide **Exclusive Economic Zone (EEZ) [10]** over which a coastal

state would have total economic rights. Resources in and under this EEZ (fish, minerals, oil, and gas) belong to the coastal state, which could either exploit them or lease, sell, or share them as it saw fit. These zones were noted in the discussion of the new competition for the Arctic Ocean in Chapter 5, but as we shall see, EEZs take on even greater importance in the Pacific Realm.

These new provisions had a far-reaching impact on the world's oceans and seas (**Fig. 12-3**), especially the Pacific. Unlike the Atlantic Ocean, the Pacific is studded with myriad islands, so that a microstate consisting of a single small island suddenly acquired an EEZ covering 166,000 square nautical miles. European colonial powers that still controlled minor Pacific possessions (most notably France) saw their maritime jurisdictions vastly expanded. Small, low-income archipelagos could now bargain with large, rich fishing nations over fishing rights in their now huge EEZs. And for all the UNCLOS Convention's provisions for the "right of innocent passage" of shipping through EEZs and via narrow straits, the geographic extent of the world's high seas has been diminished accordingly.

Maritime Boundaries

The extension of the territorial sea to 12 nautical miles and the EEZ to an additional 188 nautical miles created new **maritime boundary [11]** problems. Waters less than 24 nautical miles wide separate many countries all over the world, so that **median-line boundaries [12]**, equidistant from opposite shores, have been delimited to establish their territorial seas. And even more countries lie closer than 400 nautical miles apart, requiring further maritime-boundary delimitation to determine their EEZs. In such maritime arenas as the North Sea, the Caribbean Sea, and the East and South China seas (and more recently the Arctic Ocean), a maze of maritime boundaries emerged, some of them subject to dispute. Furthermore, political changes on land can also lead to concomitant modifications at sea.

EEZ Implications

The UNCLOS provisions created new opportunities for some states to expand their spheres of influence. Wider territorial-sea and EEZ allocations raised the stakes: claiming an island now entailed potential control over a huge maritime area. In Chapter 5, we mentioned the impending need to further carve up the Arctic Basin as the ice melts and untapped resources beckon. In Chapters 9 and 10, we referred to several island disputes off mainland East and Southeast Asia, which involve Japan and Russia, Japan and South Korea, Japan and China, China and Vietnam, and China and the Philippines. Ownership of several offshore Asian islands is uncertain, and small specks of island territory have become large stakes in today's scramble for the oceans. For instance, in the case of the Spratly Islands, several countries claim ownership, including both Taiwan and China (see Fig. 10-10). China's island claims in the South China Sea (within its alleged "Nine-Dashed Line" maritime boundary) support Beijing's contention that this body of water is part and parcel of the Chinese state—a position that increasingly troubles other states with coasts facing it.

---

* Here and in the rest of this section all distances involving maritime boundaries are given in nautical miles only. Throughout the modern history of maritime law, this unit of measurement has been the only one used for such boundary-making. Any distance stated in nautical miles can be converted to its metric equivalent by multiplying it by 1.85.

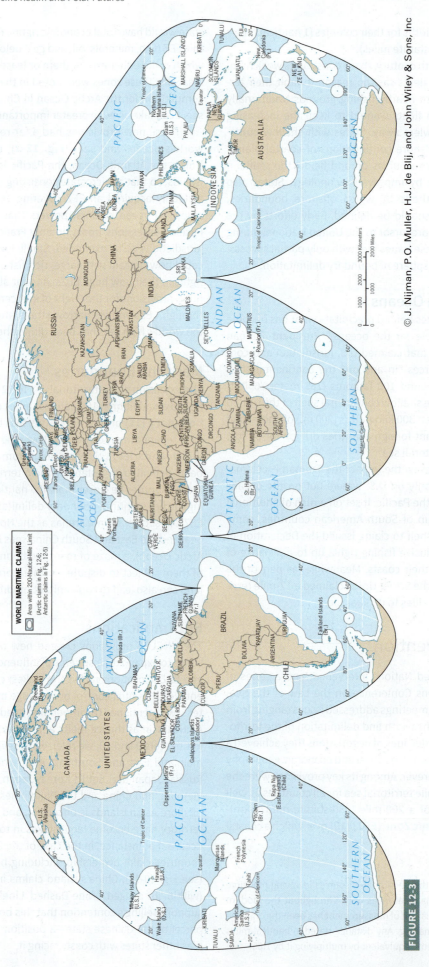

WORLD **MARITIME** CLAIMS

Area within 200-Nautical-Mile Limit
(Arctic claims in Fig. 12-6;
Antarctic claims in Fig. 12-5)

FIGURE 12-3

Figure 12-3 indicates what EEZ regulations have meant to Pacific Realm countries such as Tuvalu, Kiribati, and Fiji, with nearly circular EEZs now surrounding the clusters of islands in this vast oceanic space. Japan, Taiwan, and other nations with fisheries have purchased fishing rights in these EEZs from the island governments. Nonetheless, violations of EEZ rights do occur; not long ago, Vanuatu and the Philippines were at odds over unauthorized Filipino fishing in Vanuatu's EEZ. The process of boundary delimitation continues to evolve, and the Pacific (and world) map of maritime boundaries remains very much a work in progress.

That point is underscored by a development that occurred at the end of the 1990s, when the implications of provisions in Article 76 of the 1982 UNCLOS Convention were reexamined. Whenever a continental shelf extends beyond the 200-mile limit of the EEZ, the Convention apparently permits a coastal state to claim that extension as a natural prolongation of its landmass. Although there is as yet no regulation that will also allow the state to extend its EEZ to the edge of the continental shelf, it is not difficult to foresee such an amendment to the Convention. As it stands, states are now delimiting their proposed natural prolongations under a current deadline,

---

Box 12-3 Regional Issue: Who Should Own the Oceans?

We Who Live Here Should Own the Waters.

"Without exclusive fishing rights in our waters up to 200 nautical miles offshore, our economy would be dead in the water, no pun intended, sir. As it is, it's bad enough. I'm an official in the Fisheries Department of the Ministry of Natural Resources Development here in Tuvalu, and this modest building you're visiting is the most modern in all of the capital, Funafuti. That's because we're the only ones making some money for the state other than our Internet domain (*tv*), which has been the big story around here since we commercialized it. We don't have much in the way of natural resources and we export some products from our coconut and breadfruit trees, but otherwise we need donors to help us and major fishing nations to pay us for the use of our 750,000 square kilometers of ocean.

"Which brings me to one of our big concerns. Three-quarters of a million square kilometers sound like a lot, but look at the map. It's a mere speck in the vast Pacific Ocean. Enormous stretches of our ocean seem to belong to nobody. That is, they're open to fishing by nations that have the technology to exploit them, but we don't have those fast boats and factory ships they use nowadays. Still, the fish they take there might have come to our exclusive waters, but we don't get paid for them. It seems to us that the Pacific Ocean should be divided among the countries that exist here, not just on the basis of nearby Exclusive Economic Zones but on some other grounds. For example, islands that are surrounded by others, like we are, ought to get additional maritime territory someplace else. Islands that aren't surrounded should get much more than just 200 nautical miles. To us the waters of the Pacific mean much more than they do to other countries, not only economically but historically too. I don't like Japanese and American and Norwegian fishing fleets on these nearby 'high seas.' They don't obey international regulations on fishing, and they overfish wherever they go. When there's nothing left, they'll simply end their agreement with us, and we'll be left without income from a natural resource over which we never had control.

"I don't expect anything to change, but we feel that the era of the 'high seas' should end and all waters and seafloor should be assigned to the nearest states. The world doesn't pay much attention to our small countries, I realize. But we've put up with colonialism, world wars, nuclear weapons testing, pollution. It's time we got a break."

The Ocean Should be Open and Unrestricted.

"Ours is the Blue Planet, the planet of oceans and seas, and the world has fared best when its waters were free and open for the use of all. Encroachment from land in the form of 'territorial' seas (what a misnomer!) and other jurisdictions only interfered with international trade and commercial exploitation and caused far more problems than they solved. Already much of the open ocean is assigned to coastal countries that have no maritime tradition and don't know what to do with it other than to lease it. And when coastal countries lie closer than 400 nautical miles to each other, they divide the waters between them for EEZs and don't even leave a channel for international passage. Yes, I realize that there are guarantees of uninhibited passage through EEZs. But mark my words: the time will come when small coastal states will start charging vessels, for example cruise ships, for passage through 'their' waters. That's why, as the first officer of a passenger ship, I am against any further allocation of maritime territory to coastal states. We've gone too far already.

"In the 1960s, when those UNCLOS meetings were going on, a representative from Malta named Arvid Pardo came up with the notion that the whole ocean and seafloor beyond existing jurisdictions should be declared a "common heritage of mankind" to be governed by an international agency that would control all activities there. The idea was that all economic activity on and beneath the high seas would be controlled by the UN, which would ensure that a portion of the proceeds would be given to "Geographically Disadvantaged States" like landlocked countries, which can't claim any seas at all. Fortunately, nothing came of this proposal—can you imagine a UN agency administering half the world for the benefit of humanity?

"So you can expect us mariners to oppose any further extensions of land-based jurisdiction over the seas. The maritime world is complicated enough already, what with choke points, straits, narrows, and various kinds of prohibitions. To assign additional maritime territory to states on historic or economic grounds is to complicate the future unnecessarily and to invite further claims. From time to time, I see references to something called the 'World Lake Concept,' the preliminary assignment of all the waters of the world to the coastal states through the delimitation of median lines. Such an idea may be the natural progression from the EEZ concept, but it gives me nightmares."

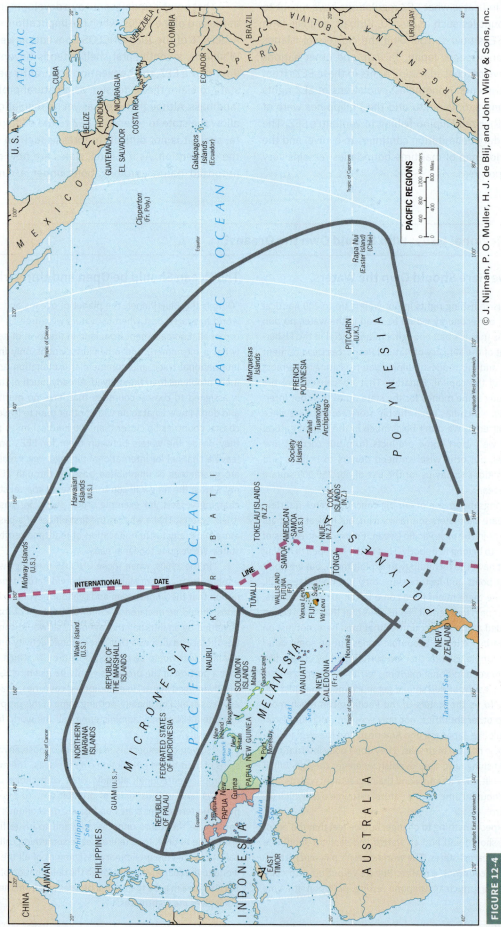

FIGURE 12-4

which puts poorer states at a disadvantage since the required marine surveys are quite expensive. Today, this issue has become quite heated, especially regarding the Arctic Ocean as countries scramble to claim prolongations (most notably Russia, as we shall see at the end of this chapter). The big winners are likely to include Russia, the United States, Canada, Australia, New Zealand, and India—although a total of 60 coastal countries could ultimately benefit to some extent from the provision.

Another complicating factor is the role of new technologies that are accelerating deep-sea exploration and the exploitation of seabed resources. Today, the quest to discover additional mineral deposits is increasingly an underwater endeavor, led by the search for so-called massive sulfides—enormous concentrations of gold, silver, copper, and other metallic minerals formed below the ocean floor in sulfide-rich volcanic environments. The technological breakthroughs entail sensors, advanced robotic devices, and other state-of-the-art gear (such as "sea gliders") developed by the offshore oil and gas industry. The growing use of sea gliders is of special interest because these small, slow-moving, unmanned vessels can track enormous distances as they collect detailed information on everything below—from marine life to seabed resources to the movements of submarines. And so the scramble for the oceans marches ever onward, and with it the constriction of the open waters of the world's high seas (**Box 12-3**).

Regions of the Pacific Realm

Sail across the Pacific Ocean, and one spectacular vista follows another. Dormant and extinct volcanoes, sculpted by erosion into basalt spires draped by luxuriant tropical vegetation and encircled by reefs and lagoons, tower over azure waters. Low atolls with nearly snow-white beaches, crowned by stands of palm trees, seem to float in the water. Pacific islanders, where foreign influences have not overtaken them, appear to take life with enviable ease.

More intensive investigations, however, reveal that such Pacific sameness is more apparent than real. Even the Pacific Realm, with its centuries-old sailing traditions, its still-diffusing populations, and its historic migrations, has a durable regional framework. **Figure 12-4** outlines the three regions of the Pacific that can be distinguished mainly on the basis of culture and language groups:

Melanesia: Papua (province of Indonesia), Papua New Guinea, Solomon Islands, Vanuatu, New Caledonia (France), Fiji

Micronesia: Palau, Federated States of Micronesia, Northern Mariana Islands, Republic of the Marshall Islands, Nauru, western Kiribati, Guam (United States)

Polynesia: Hawaiian Islands (U.S.), Samoa, American Samoa, Tuvalu, Tonga, eastern Kiribati, Cook and other New Zealand-administered islands, French Polynesia, Easter Island (Chile)

Region Melanesia

The large island of New Guinea lies at the western end of a Pacific region that extends eastward to Fiji and includes the Solomon Islands, Vanuatu, and New Caledonia (Fig. 12-4). The human mosaic here is complex, both culturally and ethnically. Most of the 11 million people of New Guinea (divided between the Indonesian province of Papua and the independent state of Papua New Guinea) are Papuans, and a large minority is Melanesian. Altogether there are more than 800 communities speaking different languages; the Papuans are most numerous in the densely forested highland interior and in the lowland south, whereas the Melanesians inhabit the north and east. Minus Papua, Melanesia, contains around 9 million inhabitants, making this the most populous Pacific region by far.

With 7.8 million people today, **Papua New Guinea (PNG)** became a sovereign state in 1975 after nearly a century of British and Australian administration. Almost all of PNG's limited development is taking place along its coasts, whereas most of the interior remains hardly touched by the changes that transformed neighboring Australia. Perhaps four-fifths of the population is part of the self-sufficient subsistence economy, cultivating root crops, raising pigs, hunting wildlife, and gathering forest products. Age-old traditions of the kind lost in Australia continue on here, protected by remoteness and the rugged terrain.

Welding this disparate population into a nation is a task hardly begun, and PNG faces numerous obstacles in addition to its cultural complexity. Not only are hundreds of languages in use, but more than one-third of the population is estimated to be illiterate. English, the official language, is used by the educated minority but is of little use beyond the coastal zone and its towns. The capital, Port Moresby, contains just over 400,000 residents, reflecting the very low level of urbanization (13 percent) in this developing economy and generally across the realm (it is the second biggest city in the realm after Honolulu).

Yet Papua New Guinea is not without economic opportunities. Oil was discovered in the 1980s, and by 2000 crude oil was PNG's largest export by value. More recently, newly found gas deposits have propelled the growth of the energy sector, and Port Moresby's major new PNG-LNG liquid natural gas facility began export operations in 2014 (even larger gas production projects are now being developed). Other important exports include gold, copper, silver, timber, and several agricultural products that include coffee and cocoa, reflecting the country's diversity of environments and resources. And PNG has not been immune to Asia Pacific influences: although the biggest share of exports still go to its nearest neighbor, Australia, Japan ranks second, and fast-rising China is close behind.

Turning eastward, it is a measure of Melanesia's cultural fragmentation that as many as 120 languages are spoken in the approximately 1000 islands that make up the **Solomon Islands** (about 80 of them support almost all the inhabitants, who number almost 600,000). Inter-island, historic animosities among the islanders were worsened by the events of World War II, when U.S. forces moved thousands of Malaitans to Guadalcanal. This started a postwar cycle of violence that finally led Australia to intervene in 2003.

New Caledonia, still under French rule, is in a very different situation. Only around 40 percent of its population of roughly 270,000 are Melanesian; not that far behind are the 29 percent of European ancestry (known as the Caldoches), many of them descended from the inhabitants of the penal colony France established here during the nineteenth century. Nickel mines, based on reserves that rank among the world's largest, dominate New Caledonia's export economy (see **Box 12-4**). The mining industry attracted additional French settlers, and social problems soon arose. Most of the French population lives in or near the capital city of Nouméa, steeped in French cultural landscapes, located in the southeastern quadrant of the island. Melanesian demands for an end to colonial rule have led to violence, and the two communities remain engaged in a long-running process of coming to terms.

On its eastern margins, Melanesia includes one of the Pacific realm's most interesting countries, **Fiji**. On two larger and over 100 smaller islands live just over 900,000 Fijians, of whom about 55 percent are Melanesians and 35 percent South Asians, the latter brought to Fiji from India during the British colonial occupation to work on the sugar plantations. When Fiji achieved independence in 1970, the indigenous Fijians owned most of the land and held political control, whereas the Indians were concentrated in the towns (chiefly Suva, the capital) and dominated commercial life. It was a recipe for trouble, which surfaced as soon as the British had departed. Nearly a half-century of ethnic conflict and political malfunctioning has cast a long shadow across this country. In 2009, after some 40 years of strife and violations of democratic rule, Fiji was suspended from the British Commonwealth. The sugar industry was damaged by the nonrenewal of Indian-held leases by Fijian landowners, which also resulted in a substantial movement of Indians from the countryside to the towns, where unemployment is already high. Only after democratic elections were finally held in 2014 was Commonwealth status reinstated; more importantly, at the same time both Australia and New Zealand restored diplomatic relations. While Fiji was shunned by these other countries, China stepped in to fill the void. Chinese foreign aid to Fiji has increased sevenfold since the last military coup in 2006, and Chinese investments in infrastructural projects have increased commensurately. This, of course, is yet another sign that China's influence is now being projected globally—and not always in a manner reassuring to the international community.

Box 12-4 From the Field Notes …

"Arriving in the capital of New Caledonia, Nouméa, is an experience reminiscent of French Africa in the postwar era. The French tricolor was much in evidence, as were uniformed French soldiers. European French residents [the Caldoches] occupied hillside villas overlooking palm-lined beaches, giving the place a Mediterranean cultural landscape. And New Caledonia, like Africa, is a source of valuable minerals. It is one of the world's major producers of nickel, and from this vantage point you could see the huge treatment plants, complete with concentrate ready to be shipped (left, beneath the conveyor). What you cannot see here is how southern New Caledonia has been ravaged by the mining operations, which have denuded whole mountainsides. Working in the mines and in this facility are the local Kanaks, Melanesians who make up approximately 40 percent of the population of roughly 270,000. Conflict between Kanaks and French Caldoches has obstructed government efforts to change New Caledonia's political status in such a way as to accommodate pressures for independence as well as the continuance of French administration."

© H.J. de Blij

Region Micronesia

North of Melanesia and to the east of the Philippines lie the islands that constitute the second Pacific region, Micronesia (Fig. 12-4). The name refers to the size of the islands: the 2000-plus islands of Micronesia are not only tiny (many of them are no larger than one square kilometer [half a square mile]), but they are also much lower-lying, on average, than those of Melanesia. Some are volcanic high islands, but they are outnumbered by low coral islands and atolls that barely reach above sea level. Guam, with 550 square kilometers (210 sq mi), is Micronesia's largest island, and no elevation anywhere in Micronesia reaches 1000 meters (3300 ft).

Take, for example, the country of **Kiribati**, consisting of three main island groups that lie across Micronesia and Polynesia (Fig. 12-4). The average elevation of Kiribati's 33 islands is 2 meters (6.5 ft), and, like all other low islands in this realm, they are particularly susceptible to rising sea level driven by global warming (**Box 12-5**).

The high-island/low-island dichotomy, introduced in the earlier overview of Pacific physiography, is particularly applicable in Micronesia. The high islands wrest substantial orographic rainfall from the moist maritime air, tend to be well watered, and are usually endowed with productive volcanic soils. As a result, agriculture can thrive, life is reasonably secure, and populations tend to be larger than those of low

---

Box 12-5   Regional Planning Case

Adaptation to Rising Sea Level in Kiribati

Kiribati (pronounced "KEER-uh-bahss"), an elongated archipelago straddling the regional boundary between Micronesia and Polynesia (Fig. 12-4), consists of 33 low-lying coral atolls that are home to just 115,000 people. About half of them reside on the main island of Tarawa Atoll and depend on the surrounding ocean for nutrition and livelihoods. Surprisingly, overcrowded South Tawara has a population density similar to Tokyo or Hong Kong, and its freshwater supply, sanitation facilities, and infrastructure are under immense strain.

Tarawa and the other atolls lie less than 3 meters (10 feet) above sea level, and find themselves directly in the crosshairs of **global climate change [13]** impacts—particularly ocean warming and **rising sea level [14]**. Heightened flooding, coastal erosion, and damage to infrastructure are already scarring the landscape, and

salt-water intrusion into groundwater and soils increasingly endanger the raising of crops. Offshore, ocean warming and acidification are simultaneously degrading coral reefs that serve as feeding grounds for the fish that are critical to the dietary needs of the islanders.

Adaptation planning has now become a key component of the toolkit to mitigate climate change impacts. The Kiribati Adaptation Program (KAP) is an international effort aimed at reducing the country's vulnerability to sea-level rise. The project entails measures such as mangrove planting, construction of seawalls, strengthening laws to reduce coastal erosion, and population settlement planning. It also involves the promotion of skills and capacity building so that local communities can provide safe drinking water and maintain resilient coastal infrastructure.

Unfortunately, such measures can only postpone the inevitable: most studies forecast that Kiribati and certain other Pacific island-countries will become uninhabitable by 2050. Debates continue about the responsibilities of the world's leading emitters of greenhouse gases and the rights of climate-change refugees, and about ways of slowing the effects of global warming on sea-level rise. In the meantime, daily life in Kiribati goes on as usual, with most of the increasingly inconvenienced islanders seeming to accept the higher tides, more frequent floods, and incessant seawall construction (see photos).

Jonas Gratzer / Contributor / GettyImages, Inc.

The AGE / Contributor / Getty Images,Inc.

Kiribati and myriad other low islands in the Pacific Realm face the dire consequences of rising sea level. In the left photo, taken in late 2015, a family wades through the latest invasion of seawater that flooded their village to collect seabed stones. These stones are then used to build seawalls higher in order to deflect future floodwaters away from homes (right).

islands. On those low islands, drought is the rule, and fishing and the coconut palm are the mainstays of life; most of the islands can only support small communities, and over time many have died out. Thus the major precolonial, indigenous migrations, which sent fleets to populate islands from Hawai'i to New Zealand, originated mainly in the high islands.

From 1945 until the mid-1980s, Micronesia was largely a United States Trust Territory (the last of the post–World War II trusteeships supervised by the United Nations). But, as Figure 12-4 shows, Micronesia is now divided into countries bearing the names of independent states. These days the **Marshall Islands**, where the United States used to test nuclear weapons (giving prominence to the name *Bikini*), is a republic in "free association" with the U.S., having the same status as Palau and the **Federated States of Micronesia**. The **Northern Mariana Islands** comprise a commonwealth "in political union" with the United States, which in effect provides billions of dollars in assistance to these countries, in return for which they commit themselves to avoid foreign policy actions that are contrary to American interests.

The Republic of **Palau** is another fascinating example of how some of these small Pacific nations make a living. This group of islands some 800 kilometers (500 mi) east of the Philippines was granted independece in 1994 and has a total population of around 22,000. Palau has no military but entered into an agreement with the United States to provide security in return for a 50-year lease on a military base. Though highly dependent on American foreign assistance, Palau also receives funds and investments from Taiwan—a reward for Palau's official recognition of Taiwan's status as the Republic of China as well as for withholding recognition of the PRC.

Also part of Micronesia is the U.S. territory of **Guam**, where independence is not in sight and where U.S. military installations and tourism provide most of its income. And then there is the remarkable Republic of **Nauru**, with a population of barely 10,000 and just 20 square kilometers (8 sq mi) of land. Nauru got rich by selling its phosphate deposits to Australia and New Zealand, where they are used as fertilizer. At one point, per capita income had risen above U.S. $12,000, making Nauru one of the realm's high-income societies; but the main surface deposits of phosphate have now been depleted, and great uncertainty prevails as the mining of a secondary layer proceeds.

In this region of tiny islands, almost all inhabitants subsist on farming or fishing, and virtually every country needs infusions of foreign aid to survive. The natural economic complementarity between the high-island farming cultures and the low-island fishing communities is all too often negated by distance, spatial as well as cultural. Life here may seem idyllic to the casual visitor, but for too many Micronesians it is an arduous daily struggle.

Region Polynesia

To the east of Micronesia and Melanesia lies the heart of the Pacific, enclosed by a great triangle stretching from the Hawaiian Islands to New Zealand to Chile's Easter Island (see **Box 12-6** and the photos on the opening page of this chapter). This third Pacific region is Polynesia (Fig. 12-4), which consists of numerous islands (*poly* means many) ranging from volcanic mountains rising far above the Pacific's waters (Mauna Kea on Hawai'i reaches over 4200 meters [nearly 13,800 ft]), clothed by luxuriant tropical forests and drenched by well over 250 centimeters

---

Box 12-6 From the Field Notes …

"Easter Island, known as Rapa Nui in the native language, is a typical high island that is constituted by three extinct volcanoes protruding above the ocean surface. In the foreground is the spectacular crater and lake named Rano Kau, its rim reaching about 330 meters (1000 ft) above sea level. We have just taken off from the airport runway aligned left to right behind the crater where the slopes of Rano Kau end. Our pilot executed a 180° right turn to let us have this splendid view of the island. Just beyond the left end of the runway is the town of about 5000 people. Most high islands of the Pacific carry abundant vegetation, and Easter Island was indeed once blanketed by palm trees. However, during the seventeenth century the local population effectively cut down all the trees (for reasons that are still not understood), eliminating an essential resource and triggering massive soil erosion. Today, some scholars consider Easter Island to be a symbol of environmental destruction and societal collapse. The trees now found on Easter Island were all replanted. Most food and supplies for the islanders—who total about 6000—have to be brought in from Chile, 3500 kilometers (2200 mi) away."

© Jan Nijman

(100 in) of rainfall each year, to low coral atolls where, if there is just enough elevation, a few palm trees form the only vegetation and drought is a persistent problem.

Despite its vast extent (it is about twice the areal size of Melanesia and Micronesia combined), Polynesia clearly constitutes a distinct Pacific region. Polynesian culture, though spatially fragmented, exhibits a remarkable consistency and uniformity from one island to the next, from one end of this widely dispersed region to the other. This consistency is particularly expressed in vocabularies, technologies, housing, and art forms. The Polynesians are uniquely adapted to their marine environment, and long before European sailing ships began to arrive in their waters, Polynesian seafarers had learned to navigate their enormous expanses of ocean in huge double canoes as long as 45 meters (150 ft). They traveled hundreds of kilometers to favorite fishing zones and engaged in inter-island barter trade, using maps constructed from bamboo sticks and cowrie shells and navigating by the stars. However, modern descriptions of a Pacific Polynesian paradise of emerald seas, lush landscapes, and gentle people distort harsh realities. Polynesian society was forced to accept much loss of life at sea when storms claimed their boats; families were ripped apart by accidents as well as by migration; hunger and starvation afflicted the inhabitants of smaller islands; and the island communities were often embroiled in violent conflicts and cruel retributions.

The political geography of Polynesia is complex. In 1959, the Hawaiian Islands became the fiftieth State to join the United States. The State's population is now 1.5 million, with more than 70 percent living on the island of Oahu. There, the superimposition of cultures is symbolized by the panorama of Honolulu's seaside skyscrapers framed against the famous extinct volcano at nearby Diamond Head. The Kingdom of Tonga became an independent country in 1970 after seven decades as a British protectorate; the British-administered Ellice Islands were renamed Tuvalu and received independence from the UK in 1978. Other islands continue under French control (including the Marquesas Islands and Tahiti), under New Zealand's administration (Cook Islands), as well as under British, U.S., and Chilean flags.

In the process of politico-geographical fragmentation, Polynesian culture has suffered severe blows. Land developers, hotel builders, and tourist dollars have set Tahiti on a course along which Hawai'i has already traveled far. The Americanization of eastern Samoa has created a new society different from the old. Today, Polynesia has lost much of its ancient cultural consistency, and the region has become a patchwork of new and old—the new often bleak and barren, with the old under ever intensifying pressure.

The countries and cultures of the Pacific Realm lie in an enormous ocean basin along whose rim a great drama of economic and political transformation will play itself out during the twenty-first century. Already, the realm's own former margins—in Hawai'i in the north and in New Zealand in the south—have been so thoroughly reshaped by foreign intervention that very little remains of the kingdoms and cultures that once prevailed. In the 2010s, the world's largest country and next superpower (China) faces the richest and most powerful (the United States) across the Pacific. How will the microstates of the Pacific Realm fare?

Polar Futures

Partitioning the Antarctic

South of the Pacific Realm lies Antarctica and its encircling Southern Ocean (discussed in Chapter 11). The combined area of these two gargantuan geographic expanses constitutes 40 percent of the entire planet—two-fifths of the Earth's surface containing a mere one one-thousandth of the world's population.

Is Antarctica a geographic realm? In exclusively physiographic terms, yes—but not on the basis of the criteria we use in this book. Antarctica is a continent, practically twice the size of Australia, but virtually all of it is covered by a dome-shaped ice sheet nearly 3.2 kilometers (2 mi) thick near its center. The continent frequently is referred to as "the white desert" because, despite all of its ice and snow, annual precipitation is extremely low—averaging less than 15 centimeters (6 in) per year. Temperatures are frigid, with winds so powerful that Antarctica also is called the home of the blizzard. For all its size, no functional regions have developed here, nor have any towns or transport networks except the supply lines of research stations. Antarctica still is a frontier, even a scientific frontier only grudgingly giving up its secrets. Underneath all that ice lie some 70 lakes of which Lake Vostok is the largest at over 14,000 square kilometers (5400 sq mi), and it may be as much as 600 meters (2000 ft) deep (Box 12-7).

Like virtually all frontiers, Antarctica has long attracted explorers and pioneers. Whale and seal hunters destroyed huge populations of Southern Ocean fauna during the eighteenth and nineteenth centuries, and explorers planted the flags of their countries on Antarctic shores. By the first decade of the twentieth century, the quest for the South Pole had become an international obsession; Roald Amundsen, the Norwegian, reached it first in 1911. All this activity soon led to national claims in Antarctica during the interwar era (1918–1939).

The geographic effect was the partitioning of Antarctica into pie-shaped sectors centered on the South Pole (Fig. 12-5). In the least frigid area of the continent, the Antarctic Peninsula, British, Argentinian, and Chilean claims overlapped—and they still do. Note that one sector, Marie Byrd Land (shown in neutral beige on the main map), was never formally claimed by any country.

Box 12-7 From the Field Notes …

"The Antarctic Peninsula is geologically an extension of South America's Andes Mountains, and this vantage point in the Gerlach Strait leaves you in no doubt: the mountains rise straight out of the frigid waters of the Southern Ocean. We have been passing large, flat-topped icebergs, but these do not come from the peninsula's shores; rather, they form on the leading edges of ice shelves or on the margins of the mainland where continental glaciers slide into the sea. Here, along the peninsula, the high-relief topography tends to produce jagged, irregular icebergs. The mountain range that forms the Antarctic Peninsula continues across Antarctica under thousands of feet of ice and is known as the Transantarctic Mountains. Looking at this place, you are reminded that beneath all this ice lies an entire continent, still little known, with fossils, minerals, even lakes yet to be discovered and studied."

© H. J. deBlij

Why should states be interested in territorial claims in so remote and difficult an area? Because these remote areas near the poles contain substantial resources and are attractive for possible future use, despite the formidable challenges of access. Antarctica is no different, as both land and sea contain raw materials that may someday become crucial: proteins in the waters, and fossil fuel as well as mineral deposits beneath the huge ice sheet. As noted earlier, Antarctica at 14.2 million square kilometers (5.5 million sq mi) is nearly twice the size of Australia, and the Southern Ocean is almost as large as the

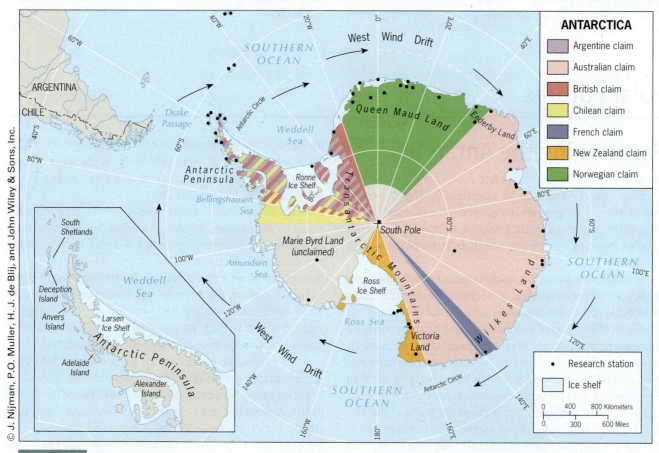

North and South Atlantic combined. However distant actual future exploitation may be, countries want to hang onto their stakes here.

But the states claiming Antarctic territory also recognize the need for cooperation. During the late 1950s, they participated in the International Geophysical Year (IGY) that launched major scientific programs and established a number of permanent research stations across the continent. This spirit of cooperation led to the 1961 signing of the **Antarctic Treaty [15]**, which ensures continued scientific collaboration, prohibits military activities, safeguards the environment, and holds national claims in abeyance. In 1991, when the treaty was extended under the terms of the Wellington Agreement, concerns were raised that it does not do enough to control future resource exploitation.

Notice, in Figure 12-5, that this map shows no **maritime claims [16]** off the pie-shaped sectors that blanket most of Antarctica. In fact, some of the claimant states did draw maps that extended their claims into the Southern Ocean, but the 1991 treaty extension terminated that initiative and restricted the existing claims to the landmass. It is, of course, possible that those claims could be reinstated should the Wellington Agreement disintegrate: national self-interest has abrogated many an international treaty in the past. But delimiting maritime claims off Antarctica is particularly challenging for practical reasons. As Figure 12-5 indicates, Antarctica in a number of places is flanked by *ice shelves* attached to its coast, permanent slabs of floating ice (such as the Ronne and the Ross). These ice shelves are replenished on the landward side and break off (calve) on the seaward side, where huge icebergs float northward into the Southern Ocean to become part of the wide belt of *pack ice* that encircles the continent. From where would any territorial sea or EEZ be measured? From the inner edge of the ice shelf? That would put the territorial "sea" boundary on the ice shelf! From the ever-changing outer edge of the ice shelf? Measuring a territorial sea or EEZ from such an unstable perimeter would not work either. No UNCLOS regulations would be applicable in situations like this, so no matter what the maps showed, such claims are legally unsupportable. That is just as well: the last thing the world needs is a scramble for Antarctica.

In an age of growing national self-interest and increasing raw material consumption, the possibility exists nonetheless that Antarctica and its offshore waters may yet become an arena for international rivalry. Until now, its remoteness and its forbidding environments have saved it from that fate. The entire world benefits from this because evidence is mounting that Antarctica plays a critical role in the global environmental system, so that human modifications are likely to have worldwide (and unpredictable) consequences.

Geopolitics In the Arctic Basin

To observe how different the physiographic as well as the political situation is in the Arctic, consider the implications of **Figure 12-6**. Not only does the North Pole lie on the floor of a relatively small body of water (grandiosely labeled the Arctic Ocean), but the entire Arctic is ringed by countries whose EEZs, delimited under UNCLOS rules, would allocate much of that ocean floor (the *subsoil*, to use its legal designation) to those states. Moreover, again under UNCLOS regulations, states can expand their rights to the seafloor even farther than the 200-mile EEZ if they are able to prove their continental shelves continue beyond that limit—in fact, up to 350 nautical miles offshore.

As the main map shows, the Siberian Continental Shelf is by far the largest in the Arctic; indeed, it is the largest in the world. It extends from Russia's north coast beneath the waters of the Arctic Ocean and under the floating icecap at the ocean's center. Because several island groups off the mainland (such as Franz Josef Land, North Land, and the New Siberian Islands) belong to Russia, the Russians can claim virtually all of Eurasia's northern continental shelf under existing regulations. But even that is not enough: the Russians want to extend their claim all the way to the North Pole itself, where in 2007 they sent a submarine to plant a Russian flag on the seafloor nearly 4000 meters (13,000 ft) below the Pole (see photo).

To bolster their seafloor claim, the Russians assert that the Lomonosov Ridge, a submerged mountain range, is a "natural extension" of their Siberian Continental Shelf. Few, if any, neutral observers might agree that this ridge, or several others rising from the deep floor of the Arctic Ocean, is a continental-shelf landform. But in 2014, Denmark made a new claim on the same grounds: following "ten years of scientific research," their conclusion was that the Lomonosov Ridge is actually connected to Greenland's shelf and therefore lies in Danish territory.

Take another look at Figure 12-6 and observe how far the Danes have extended their claims, which now overlap with those of Canada and, more significantly, Russia (the North Pole is included in the Danish claim as well). The stakes are high: estimates of the quantity of oil and natural gas reserves to be found beneath the Arctic seafloor run as high as 25 percent of the world's remaining total. In the end, whatever method is used to finally resolve the Lomonosov and other claims, Russia will be the biggest winner (**Box 12-8**).

Disputation and Navigation

Even as a dispute over Arctic maritime boundaries looms, another key issue that has emerged in recent years involves the effects of sustained global warming. The Greenland Ice Sheet, in many ways a smaller version of Antarctica's, is now in a melting phase, and the zone of seasonally expanding and contracting Arctic sea ice has also shown significant attrition in both surface extent and thickness. Researchers with the Greenland Ice Core Project have reported in *Science* that during a previous warming episode, more than 450,000 years ago, the southernmost part of Greenland was covered by boreal forest—suggesting that both the ice sheet and the sea ice may disappear entirely if the ongoing global warming trend continues. Indeed, this process seems well underway: the polar icecap is now less than half of what it was in 1979, and the lower-left inset map in Figure 12-6

Box 12-8 MAP ANALYSIS

The Arctic Basin today is a contested geopolitical arena. The main map in Figure 12-6 shows the *EEZs* of the countries involved, including claims that extend to the North Pole itself. Which two countries claim territory beyond their EEZs? Next, compare this map to Figure 12-5: why wouldn't an Antarctic-type, pie-shaped, sectoral partitioning work in the Arctic? Now look closely at the two inset maps

in Figure 12-6 and examine how much, and where, the polar icecap has been melting since 1980. Finally, is the Northeast Passage or the Northwest Passage becoming the more important new oceanic route, and why?

Access your WileyPLUS Learning Space course to interact with a dynamic version of this map and to engage with online map exercises and questions.

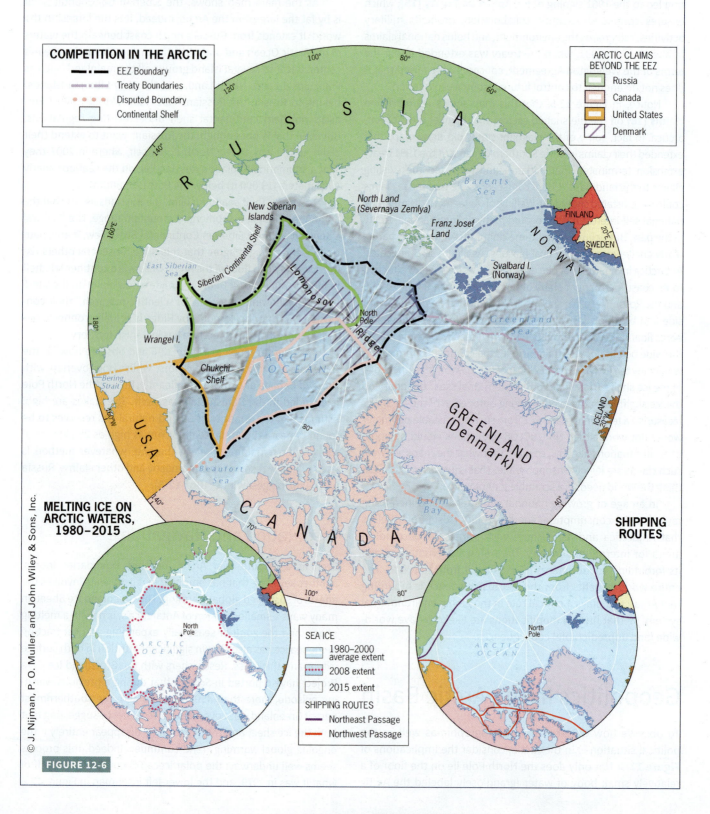

© J. Nijman, P. O. Muller, and John Wiley & Sons, Inc.

FIGURE 12-6

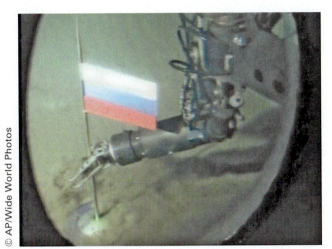

© AP/Wide World Photos

On August 3, 2007, tens of millions of Russian television viewers witnessed live the scene pictured here: the operator of a Russian mini-submarine planting a metal Russian flag at the North Pole, on the seafloor of the Arctic Ocean beneath the polar icecap. As Figure 12-6 shows, the North Pole lies far beyond Russia's continental shelf. Yet in future international negotiations to settle national claims in the Arctic Basin, the Russians will claim that the Lomonosov Ridge actually constitutes an extension of their continental shelf, thereby entitling Russia to draw international boundaries up to, and even beyond, the North Pole. Other countries with actual and pending Arctic claims are making less dramatic but equally assertive moves in the run-up to negotiations that have the potential to transform the map of the Arctic region. With new sea routes as well as newly exploitable oil and gas reserves in prospect, diplomacy will be difficult but crucial.

highlights the current shrinkage phase. The ecological consequences are far-reaching, threatening the habitats of polar bears, whales, walruses, seals, and other species. But the economic- and politico-geographical implications are going to be significant as well.

When the Arctic's sea ice melts in summer and does not fully recover in winter, once-blocked waterways begin to open up and accessibilities can be transformed. The **Northeast Passage [17]**, that legendary high-latitude route between the Far East and Europe, has been the object of hope and despair for centuries, causing much loss of human life in the quest to forge a viable sea route. But in 2007, for the first time in recorded history, it was completely ice-free for a brief period during the late-summer peak of seasonal melting. Since then, the Northeast Passage has been ice-free every year during that August/September window; by 2013, a total of 71 specially equipped vessels—carrying well over a million tons of cargo—were able to successfully navigate the route (mapped in the lower-right inset map of Fig. 12-6). That map also shows the presence of a similar route around the northern perimeter of North America, the **Northwest Passage [18]**. This other icon of (unsuccessful)

pre-twentieth-century polar exploration would connect the Pacific and Atlantic oceans, following an eastward watercourse along Alaska's northern coast, through Canada's massive Arctic Archipelago, and then a final southward traverse via the Davis Strait between Canada and Greenland.

If either or both of these waterways can be activated, the impact on global shipping will be momentous. Consider what it would mean if ships bound for Asia's Pacific Rim from the U.S. eastern seaboard, and too large to fit through the upgraded Panama Canal, could shorten their voyages by several days and thousands of kilometers by rounding North America instead of South America! No wonder countries like China, India, Japan, and even Singapore have applied for "observer status" at the meetings of the Arctic Council, whose regular members include the United States, Russia, Canada, Norway, Sweden, Denmark, Finland, and Iceland.

But who really owns the waterways vacated by the ice? Canada regards the Northwest Passage as a domestic waterway, giving it the right to control all shipping. The United States holds that the Northwest Passage is an international waterway, and American and Russian ships have sailed through it without Canada's permission. The Canadian government has affirmed its position by starting work on a deepwater port at Nanisivik on Nunavut's Baffin Island at the eastern entrance to the Northwest Passage, and by building a military base at Resolute Bay to the west; it has also announced the construction of eight Arctic patrol vessels to police the route. In the meantime, the United States is opening a new Coast Guard station to patrol the narrow Bering Strait between Alaska and Russia. The lines, if not yet battle lines, are already being drawn in the warming Arctic.

Humankind's growing numbers, steadily escalating demands, environmental impacts, and technological capacities are transforming even the most remote recesses of our resilient planet. No international resolution is more critical than the avoidance of destructive conflict in fragile polar environments.

Points to Ponder

- Governments of low-lying island-states in both the Pacific and Indian oceans are demanding international assistance as they are confronted by the threat of the world ocean's rising sea level.

- Fiji receives support from Beijing for officially recognizing the People's Republic of China (PRC) but not Taiwan; Palau gets support from the Taiwanese for officially recognizing Taiwan but not the PRC.

- Singapore, located almost directly on the equator, has requested and been granted "observer status" to monitor the meetings of the Arctic Council because of the impact the possible opening of ice-free, far northern passages will have on global shipping.

Appendix B may be found on the book's website at **www.wiley.com/college/regions17**

METRIC (STANDARD INTERNATIONAL [SI]) AND CUSTOMARY UNITS AND THEIR CONVERSIONS

Length

Metric Measure

1 kilometer (km)	= 1000 meters (m)
1 meter (m)	= 100 centimeters (cm)
1 centimeter (cm)	= 10 millimeters (mm)

Nonmetric Measure

1 mile (mi)	= 5280 feet (ft)
	= 1760 yards (yd)
1 yard (yd)	= 3 feet (ft)
1 foot (ft)	= 12 inches (in)
1 fathom (fath)	= 6 feet (ft)

Conversions

1 kilometer (km)	= 0.6214 mile (mi)
1 meter (m)	= 3.281 feet (ft)
	= 1.094 yards (yd)
1 centimeter (cm)	= 0.3937 inch (in)
1 millimeter (mm)	= 0.0394 inch (in)
1 mile (mi)	= 1.609 kilometers (km)
1 foot (ft)	= 0.3048 meter (m)
1 inch (in)	= 2.54 centimeters (cm)
	= 25.4 millimeters (mm)

Area

Metric Measure

1 square kilometer (km^2)	= 1,000,000 square meters (m^2)
	= 100 hectares (ha)
1 square meter (m^2)	= 10,000 square centimeters (cm^2)
1 hectare (ha)	= 10,000 square meters (m^2)

Nonmetric Measure

1 square mile (mi^2)	= 640 acres (ac)
1 acre (ac)	= 4840 square yards (yd^2)
1 square foot (ft^2)	= 144 square inches (in^2)

Conversions

1 square kilometer (km^2)	= 0.386 square mile (mi^2)
1 hectare (ha)	= 2.471 acres (ac)
1 square meter (m^2)	= 10.764 square feet (ft^2)
	= 1.196 square yards (yd^2)
1 square centimeter (cm^2)	= 0.155 square inch (in^2)
1 square mile (mi^2)	= 2.59 square kilometers (km^2)
1 acre (ac)	= 0.4047 hectare (ha)
1 square foot (ft^2)	= 0.0929 square meter (m^2)
1 square inch (in^2)	= 6.4516 square centimeters (cm^2)

Temperature

To change from Fahrenheit (F) to Celsius (C)

$$^{\circ}C = \frac{^{\circ}F - 32}{1.8}$$

To change from Celsius (C) to Fahrenheit (F)

$$^{\circ}F = ^{\circ}C \times 1.8 + 32$$

Aboriginal land issue   The legal campaign in which Australia's **indigenous peoples** have claimed title to traditional land in several parts of that country. The courts have upheld certain claims, fueling Aboriginal activism that has raised broader issues of indigenous rights.

Aboriginal population   Native or *aboriginal* peoples; often used to designate the inhabitants of areas that were conquered and subsequently colonized by the **imperial** powers of Europe.

Absolute location   The position or place of a certain item on the surface of the Earth as expressed in degrees, minutes, and seconds of **latitude** and **longitude**.

Acculturation   Cultural modification resulting from intercultural borrowing. In **cultural geography**, the term refers to the change that occurs in the **culture** of **indigenous peoples** when contact is made with a society that is technologically superior.

AFTA   The ASEAN Free Trade Agreement that since 1992, through lowered tariffs and other incentives, has fostered increased trade within Southeast Asia. This is the economic centerpiece of the **Association of Southeast Asian Nations (ASEAN)**, a **supranational** organization whose members include 10 of that realm's 11 states (only East Timor does not participate).

Agribusiness   The agricultural operations of large, often multinational, corporations.

al-Qaeda   The terrorist organization that evolved into an expanding global network under the directorship of Usama bin Laden between the mid-1990s and his elimination by the U.S. in 2011. It sought to coordinate the efforts of once loosely allied Muslim revolutionary movements, and unleash a *jihad* aimed at what it perceived to be Islam's enemies in the West.

Altiplano   High-elevation plateau, basin, or valley between even higher mountain ranges, especially in the Andes of South America.

Altitudinal zonation   Vertical regions defined by physical-environmental zones at various elevations (see Fig. 2-4), particularly in the highlands of South and Middle America.

American Manufacturing Belt   North America's near-rectangular **core area**, whose corners are Boston, Milwaukee, St. Louis, and Baltimore.

Antarctic Treaty   International cooperative agreement on the use of Antarctic territory. Originally signed in 1961, it was extended in 1991 under the Wellington Agreement.

Anthropogenic   Environmental impacts emanating from human activities, particularly the production of pollutants associated with atmospheric warming.

Apartheid   Literally, *apartness*. The Afrikaans term for South Africa's pre-1994 policies of racial separation, a system that produced highly segregated socio-geographical patterns.

APEC   The Asia Pacific Economic Cooperation forum. Its 21 Pacific Rim member-economies promote free trade throughout the Asia Pacific region.

Aquifer   An underground reservoir of water contained within a porous, water-bearing rock layer.

Archipelago   A set of islands grouped closely together, usually elongated into a *chain*.

ASEAN   The **Association of Southeast Asian Nations (ASEAN)**, a **supranational** organization whose members include 10 of that realm's 11 states: Brunei, Cambodia, Indonesia, Laos, Malaysia, Myanmar (Burma), the Philippines, Singapore, Thailand, and Vietnam (only minuscule East Timor does not participate, but is currently seeking to join).

Asian Tiger   See **economic tiger**.

Asylum   Legally protected residency status; usually granted by a host country to immigrants fleeing political oppression in their former homeland.

Atoll   A ring-like coral reef surrounding an empty lagoon that probably formed around the rim of a now-completely-eroded volcanic cone standing on the seafloor. They are common in certain tropical areas of the Pacific Ocean where they are classified among that realm's **low islands**.

Austral   South.

Balkanization   The fragmentation of a **region** into smaller, often hostile political units. Named after the historically contentious Balkan Peninsula of southeastern Europe.

Barrio   Term meaning "neighborhood" in Spanish. Usually refers to an urban community in a Middle or South American city.

Basin irrigation   Farming technique devised by ancient Egyptians to trap the Nile's annual floodwaters and their fertile silt by building fields with earthen ridges.

Biodiversity hot spot   A much higher than usual, world-class geographic concentration of natural plant and/or animal species. Tropical rainforest environments have dominated, but their recent ravaging by **deforestation** has had catastrophic results.

Biodiversity   Shorthand for *biological diversity*; the total variety of plant and animal species that exists in a given area.

Biogeography   The study of *flora* (plant life) and *fauna* (animal life) in spatial perspective.

Boko Haram   The violent, jihadist, terrorist organization that is based in—and controls parts of—northeastern Nigeria. It engages in extreme terrorist attacks, increasingly beyond its territory as it seeks to expand into other parts of the country (see Fig. 7-16).

Borderland   General term for a linear zone that parallels a political boundary. The most dynamic of these areas, such as those lining the U.S.-Mexico border, are marked by significant cultural and economic interaction across the boundary that separates them.

Boreal forest   The subarctic, mostly **coniferous** snowforest that blankets Canada south of the **tundra** that lines the Arctic shore; known as the **taiga** in Russia.

Break-of-bulk   A location along a transport route where goods must be transferred from one carrier to another. In a port, the cargoes of oceangoing ships are unloaded and put on trains, trucks, or perhaps smaller river boats for inland distribution. An *entrepôt*.

BRICs   Acronym for the four biggest emerging national markets in the world today—**B**razil, **R**ussia, **I**ndia, and **C**hina.

Buffer state   A country or set of countries separating ideological or political adversaries. In southern Asia, Afghanistan, Nepal, and Bhutan were parts of a buffer zone set up between British and Russian-Chinese imperial spheres. Thailand was a *buffer state* between British and French colonial domains in mainland Southeast Asia.

Buffer zone   A country or set of countries separating ideological or political adversaries. In southern Asia, Afghanistan, Nepal, and Bhutan were parts of a buffer zone set up between British and Russian-Chinese imperial spheres. Thailand was a *buffer state* between British and French colonial domains in mainland Southeast Asia.

Business-process outsourcing   The subcontracting of specific business tasks, such as subscriptions or payroll, to another, specialized, company (sometimes abroad), at lower cost.

Caliphate   An imperial-scale Islamic government led by a caliph, considered a direct successor to the Prophet Muhammad, who rules and exerts moral authority over Muslims worldwide.

Cash economy   An economic system in which all transactions are made in cash.

Caste system   The strict **social stratification** and residential segregation of people—specifically in India's Hindu society—on the basis of ancestry and occupation.

Central business district (CBD) The downtown heart of a central city; marked by high land values, a concentration of business and commerce, and the clustering of the tallest buildings.

Centrality The strength of an urban center in its capacity to attract producers and consumers to its facilities; a city's "reach" into the surrounding region.

Centrifugal forces A term employed to designate forces that tend to divide a country—such as internal religious, linguistic, ethnic, or ideological differences.

Centripetal forces Forces that unite and bind a country together—such as a strong national culture, shared ideological objectives, and a common faith.

Cerrado Regional term referring to the fertile savannas of Brazil's interior Central-West that make it one of the world's most promising agricultural frontiers. Soybeans are the leading crop, and other grains and cotton are expanding. Inadequate transport links to the outside world remain a problem.

China-Pakistan Economic Corridor The 2000-kilometer (1250-mi), northeast-southwest development axis stretching between the westernmost Chinese city of Kashgar and Pakistan's new Indian Ocean port of Gwadar. A major future trade route that aligns with China's **New Silk Road**, its centerpiece is a bundled routeway of ultramodern rail, road, and pipeline connections expected to spawn various development nodes within the corridor, mostly in Pakistan (see Box 8-10).

Choke point A narrowing of an international waterway causing marine-traffic congestion, requiring reduced speeds and/or sharp turns, and increasing the risk of collision as well as vulnerability to attack. When the waterway narrows to a distance of less than 38 kilometers (24 mi), this necessitates the drawing of a **median line (maritime) boundary**.

City-state An independent political entity consisting of a single city with (and sometimes without) an immediate **hinterland**.

Climate The long-term conditions (over at least 30 years) of aggregate **weather** over a region, summarized by averages and measures of variability; a synthesis of the succession of weather events we have learned to expect at any given location.

Cold War The post-World War II standoff between the Soviet Union and the Western alliance led by the United States, which lasted from 1945 to the collapse of the USSR in 1991. Largely coincided with the postcolonial era, in which Soviet-communist influence was aggressively projected in Africa and elsewhere to fill the political vacuum and compete for the hearts and minds of the citizens of newly-independent countries.

Command economy The tightly controlled economic system of the former Soviet Union, whereby central planners in Moscow assigned the production of particular goods to particular places, often guided more by socialist ideology than the principles of **economic geography**.

Commercial agriculture For-profit **agriculture**.

Communal tension Persistent stress among a country's sociocultural groups that can often erupt into communal violence, particularly in India.

Compact state A politico-geographical term to describe a **state** that possesses a roughly circular, oval, or rectangular territory in which the distance from the geometric center to any point on the boundary exhibits little variance.

Complementarity Exists when two regions, through an exchange of raw materials and/or finished products, can specifically satisfy each other's demands.

Connectivity The degree of direct linkage between a particular location and other locations within a regional, national, or global transportation **network**.

Continental drift The slow movement of continents controlled by the processes associated with **plate tectonics**.

Continentality The variation of the continental effect on air temperatures in the interior portions of the world's landmasses. The greater the distance from the moderating influence of an ocean, the greater the extreme in summer and winter temperatures. Continental interiors also tend to be dry when the distance from oceanic moisture sources becomes considerable.

Continental shelf Beyond the coastlines of many landmasses, the ocean floor declines very gently until the depth of about 600 feet (183 m). Beyond the 600-foot line the sea bottom usually drops off sharply, along the *continental slope*, toward the much deeper mid-oceanic basin. The submerged continental margin is called the continental shelf, and it extends from the shoreline to the upper edge of the continental slope.

Conurbation General term used to identify a large multimetropolitan complex formed by the coalescence of two or more major **urban areas**.

Core-periphery relationship The contrasting spatial characteristics of, and linkages between, the *have* (core) and *have-not* (periphery) components of a national, regional, or the global **system**.

Core area In geography, a term with several connotations. *Core* refers to the center, heart, or focus. The core area of a **nation-state** is constituted by the national heartland, the largest population cluster, the most productive region, and the part of the country with the greatest **centrality** and **accessibility** probably containing the capital city as well.

Cultural diffusion The **process** of spreading and adopting a cultural element from its place of origin across a wider area.

Cultural geography The wide-ranging and comprehensive field of geography that studies spatial aspects of human **cultures**.

Cultural landscape The forms and artifacts sequentially placed on the **natural landscape** by the activities of various human occupants. By this progressive imprinting of the human presence, the physical (natural) landscape is modified into the cultural landscape, forming an interacting unity between the two.

Cultural pluralism A society in which two or more population groups, each practicing its own **culture**, live adjacent to one another without mixing inside a single **state**.

Culture hearth Heartland, source area, or innovation center; place of origin of a major **culture.**

Deindustrialization **Process** by which companies relocate manufacturing jobs to other regions or countries with cheaper labor, leaving the newly-deindustrialized region to convert to a service economy while struggling with the accompanying effects of increased unemployment and meeting the retraining needs of its workforce.

Demographic burden The proportion of a national population that is either too old or too young to be productive and that must be cared for by the productive population.

Demographic transition Multi-stage **model**, based on western Europe's experience, of changes in population growth exhibited by countries undergoing industrialization. High **birth rates** and **death rates** are followed by plunging death rates, producing a huge net population gain; birth and death rates then converge at a low overall level.

Dependencia **theory** Originating in South America during the 1960s, it was a new way of thinking about economic development and underdevelopment that explained the persistent poverty of certain countries in terms of their unequal relations with other (i.e., rich) countries.

Dependency ratio An indicator of the pressure on a country's workers, the age-population ratio of (dependent) people who are not in the labor force to those (productive) people who are in the labor force.

Desalination The process of removing dissolved salts from water, thereby producing fresh (drinking) water from seawater or brackish water.

Development The economic, social, and institutional growth of national **states**.

Devolution The **process** whereby regions within a **state** demand and gain political strength and growing autonomy at the expense of the central government.

Digital elevation model A representation of a unit of terrain obtained from **remote sensing** imagery. The technique is discussed in Box 4-2.

Disaster management Subfield of urban and regional planning that consists of four phases: hazard mitigation, preparedness, response, and recovery.

Distance decay The various degenerative effects of distance on human spatial structures and interactions.

Distribution center   A centralized focus of economic activity specializing in the distribution of goods, situated as a major hub on its regional transportation network. Atlanta, Georgia, with its outstanding highway, rail, and air-freight connections to the surrounding southeastern United States, is a classic example.

Domino effect   The belief that political destabilization in one **state** can result in the collapse of order in a neighboring state, triggering a chain of events that, in turn, can affect a series of **contiguous** states.

Domino theory   The belief that political destabilization in one **state** can result in the collapse of order in a neighboring state, triggering a chain of events that, in turn, can affect a series of **contiguous** states.

Double delta   South Asia's combined **delta** formed by the Ganges and Brahmaputra rivers. All of Bangladesh lies on this enormous deltaic plain, which also encompasses surrounding parts of eastern India. Well over 200 million people live here, attracted by the fertility of its soils that are constantly replenished by the **alluvium** transported and deposited by these two of Asia's largest river systems. Natural hazards abound here as well, ranging from the flooding caused by excessive **monsoonal** rains to the intermittent storm surges of powerful cyclones (**hurricanes**) that come from the Bay of Bengal to the south.

Dravidian languages   The language family, indigenous to the South Asian realm, that dominates southern India today; as opposed to the Indo-European languages, whose tongues dominate northern India.

Drone warfare   The use of remote-controlled unmanned aerial vehicles (UAVs) as delivery systems to conduct military attacks. Tactical advantages include highly precise targeting, long flight times without refueling, difficulty of detection from the ground, and advanced imaging capabilities that allow drones to often 'see' what is invisible to human pilots.

Dynasty   A succession of Chinese rulers that came from the same line of male descent, sometimes enduring for centuries. Dynastic rule in China lasted for thousands of years, only coming to an end in 1911.

Economic geography   The field of geography that focuses on the diverse ways in which people earn a living and on how the goods and services they produce are expressed and organized spatially.

Economic integration   The economic benefits of forging **supranational** partnerships among three or more countries. The European Union (EU) is the prototype; NAFTA and CARICOM are examples in the Middle American realm.

Economies of scale   The savings that accrue from large-scale production wherein the unit cost of manufacturing decreases as the level of operation enlarges. Supermarkets operate on this principle and are able to charge lower prices than small grocery stores.

Ejidos   Mexican farmlands redistributed to peasant communities after the Revolution of 1910–1917. The government holds title to the land, but user rights are parceled out to village communities and then to individuals for cultivation.

Electoral geography   The spatial distribution of political preferences as expressed in voting behavior for political parties and/or candidates. The mapping of election results (see Box 1-8) is the foundation of electoral geography.

El Niño   A periodic, abnormal warming of the sea surface in the eastern equatorial Pacific Ocean. Disturbs weather patterns across much of the world, especially in northwestern and northeastern South America.

Elongated state   A **state** whose territory is decidedly long and narrow in that its length is at least six times greater than its average width.

Elongation   In political geography, refers to the territorial configuration of a state that is at least six times longer than its average width. Chile is the most prominent example of this shape on the world map.

Emerging market   The world's fastest growing national market economies as measured by economic growth rates, attraction of **foreign direct investment**, and other key indicators. Led by the **BRICs** (Brazil, Russia, India, and China), but this club is now expanding to include many other countries.

Endemic   Refers to a disease in a host population that affects many people in a kind of equilibrium without causing rapid and widespread deaths.

Entrepôt   A place, usually a port city, where goods are imported, stored, and transshipped; a **break-of-bulk point**.

Environmental degradation   The accumulated human abuse of a region's **natural landscape** that, among other things, can involve air and water pollution, threats to plant and animal **ecosystems**, misuse of **natural resources**, and generally upsetting the balance between people and their habitat.

Environmental determinism   A geographic school of thought, popular in the first quarter of the twentieth century, that maintains the physical environment determines human behavior and/or social outcomes.

Epidemic   A local or regional outbreak of a disease.

Estuary   The widening mouth of a river as it reaches the sea; land subsidence or a rise in sea level has overcome the tendency to form a **delta**.

Ethnicity   The combination of a people's **culture** (traditions, customs, language, and religion) and racial ancestry.

Eurasian Customs Union   A **supranational** organization created by Russia in 2010 to maintain economic ties with the friendliest countries in the **Near Abroad**. The three charter members (Russia, Belarus, and Kazakhstan) were joined in 2015 by the two smaller countries of Armenia and Kyrgyzstan.

European state model   A **state** consisting of a legally defined territory inhabited by a population governed from a capital city by a representative government.

Euro zone   The 19 countries (as of mid-2016) whose official currency is the euro. They are mapped in Figure 4-7.

Exclave   A bounded (non-island) piece of territory that is part of a particular **state** but lies separated from it by the territory of another state.

Exclusive Economic Zone (EEZ)   An oceanic zone extending up to 200 **nautical miles** (370 km) from a shoreline, within which the coastal **state** can control fishing, mineral exploitation, and additional activities by all other countries.

Expansion diffusion   The spreading of an innovation or an idea through a fixed population in such a way that the number of those adopting it grows continuously larger, resulting in an expanding area of dissemination.

Externality effect   A negative consequence of an action, which countries try to minimize. Ethiopia's new dam on the Blue Nile is deliberately situated just upstream from the Sudan border in order to minimize reduced water flow and silt entrapment within Ethiopian territory.

Extreme weather events   Unprecedented, record-breaking departures from the longer-term **weather** patterns of a certain area. In India, such events have tripled since 1980 in the form of severe heat waves, droughts, and non-**monsoonal** rainstorms that trigger massive flooding and widespread landslides.

Failed state   A country whose institutions have collapsed and in which anarchy prevails.

Favela   Shantytown on the outskirts or even well within an urban area in Brazil.

Federation   A country adhering to a political framework wherein a central government represents the various subnational entities within a **nation-state** where they have common interests—defense, foreign affairs, and the like—yet allows these various entities to retain their own identities and to have their own laws, policies, and customs in certain spheres.

Fertility rate   More precisely the Total Fertility Rate, it is the average number of children born to women of childbearing age in a given population.

First Nations   Name given Canada's **indigenous peoples** of American descent, whose U.S. counterparts are called Native Americans.

Floating population   China's huge mass of mobile workers who respond to shifting employment needs within the country. Most are temporary urban dwellers with restricted residency rights, whose movements are controlled by the *hukou* **system**.

Foreign direct investment   A key indicator of the success of an **emerging market** economy, whose growth is accelerated by the infusion of foreign funds to supplement domestic sources of investment capital.

Formal economy   The part of a national economy that is registered with government agencies and complies with laws and regulations, especially taxation. The complement to a country's **informal economy**.

Formal region A type of **region** marked by a certain degree of homogeneity in one or more phenomena; also called *uniform region* or *homogeneous region*.

Forward capital Capital city positioned in actually or potentially contested territory, usually near an international border; it confirms the **state's** determination to maintain its presence in the area of contention.

Fossil fuel The energy resources of **coal**, natural gas, and petroleum (oil), so named collectively because they were formed by the geologic compression and transformation of tiny plant and animal organisms.

Four Motors of Europe *Rhône-Alpes* (France), *Baden-Württemberg* (Germany), *Catalonia* (Spain), and *Lombardy* (Italy). Each is a high-technology-driven region marked by exceptional industrial vitality and economic success not only within Europe but on the global scene as well.

Fragmented modernization A checkerboard-like spatial pattern of **modernization** in an **emerging-market** economy wherein a few localized regions of a country experience most of the development while the rest are largely unaffected.

Fragmented state A **state** whose territory consists of several separated parts, not a **contiguous** whole. The individual parts may be isolated from each other by the land area of other states or by international waters.

Functional region A **region** marked less by its sameness than by its dynamic internal structure; because it usually focuses on a central **node**, also called *nodal region* or *focal region*.

Gender imbalance The demographic imbalance of males outnumbering females resulting from selective birth control. In China, this is an outcome of the **One-Child Policy**.

Gentrification The upgrading of an older residential area through private reinvestment, usually in the downtown area of a central city. Frequently, this involves the displacement of established lower-income residents, who cannot afford the heightened costs of living, and conflicts are not uncommon as such neighborhood change takes place.

Geographic information system (GIS) A form of spatial analysis that integrates computer hardware, mapping software, and such specialized tools as models and algorithms. A versatile technique that is constantly being expanded in its applications (e.g., digital terrain mapping, as shown in Box 4-2).

Geographic realm The basic spatial unit in our world regionalization scheme. Each realm is defined in terms of a synthesis of its total human geography—a composite of its leading cultural, economic, historical, political, and appropriate environmental features.

Geopolitical revanchism Retaliatory policies pursued by a **state** aimed at recovering lost territory. Russia's forcible annexation of the Crimean Peninsula from Ukraine in 2014 is a good example.

Geopolitics Political relations among states or regions that are strongly influenced by their geographical setting, including proximity, **accessibility**, sovereign boundaries, natural resources, **population distribution**, and the like.

Geospatial data Data pertaining to a particular location on or near the Earth's surface.

Gini index A measure of inequality within a given area, ranging from 0 to 100. A value of 0 indicates that income is equally distributed across an area's population; a value of 100 indicates that all income is concentrated in the hands of a single recipient.

Global climate change The shift in the characteristics and spatial distribution of Earth's climates in response to a long-term trend in **atmospheric** warming.

Global core The constellation of countries with the most highly developed and influential economies (anchored by the United States; the 27 member-countries of the European Union; East Asia's China and Japan; and Australia/New Zealand). Although the **global core** (mapped in Fig. G-12) is not a contiguous geographic region, its countries as a whole dominate international trade and investment flows.

Globalization The gradual reduction of regional differences at the world **scale**, resulting from increasing international cultural, economic, and political interaction.

Global periphery All of the countries that lie outside the **global core** (mapped in green in Fig. G-12). Economically, these countries are subordinate to those of the global core in terms of development and international influence. Geographically, the global periphery is constituted by the 8 realms that lie outside North America, Europe, East Asia, and the Austral Realm.

Golden Triangle Opium-producing area encompassing northwestern Laos, northernmost Thailand, and east-central Myanmar's Shan State.

GPS (Global Positioning System) The orbiting-satellite-based navigation system that provides locational and time information, anywhere on or near the Earth's surface where there is an unobstructed line of sight to four or more GPS satellites.

Green Revolution The successful late-twentieth-century development of higher-yield, fast-growing varieties of rice and other cereals in certain developing countries.

Growth-pole concept An urban center with a number of attributes that, if augmented by investment support, will stimulate regional economic **development** in its **hinterland**.

Growth Triangle An increasingly popular economic **development** concept along the western **Pacific Rim**, especially in Southeast Asia. It involves the linking of production in growth centers of three countries to achieve benefits for all.

Hacienda Literally, a large estate in a Spanish-speaking country. Sometimes equated with the **plantation**, but there are important differences between these two types of agricultural enterprise.

Hanification Imparting a cultural imprint by the ethnic Chinese (the "people of Han"). Within China often refers to the steadily increasing migration of Han Chinese into the country's **periphery**, especially Xinjiang and Xizang (Tibet). **Overseas Chinese** imprints, more generally referred to as **Sinicization**, have been significant as well, most importantly in the Southeast Asian realm.

High-value-added goods Products of improved net worth.

High island Cultures associated with volcanic islands of the Pacific Realm that are high enough in elevation to wrest substantial moisture from the tropical ocean air (see **orographic precipitation**). They tend to be well watered, their volcanic soils enable productive agriculture, and they support larger populations than **low islands**.

High seas Areas of the oceans away from land, beyond national jurisdiction, open and free for all to use.

Hindutva "Hinduness" as expressed through Hindu nationalism, Hindu heritage, and/or Hindu patriotism. The cornerstone of a fundamentalist movement that has been gaining strength since the late twentieth century that seeks to remake India as a society dominated by Hindu principles. It has been the guiding agenda of the Bharatiya Janata Party (BJP), which has emerged a powerful force in national politics and in big States like Maharashtra, Gujarat, and Madhya Pradesh.

Hinterland Literally "country behind," a term that applies to a surrounding area served by an urban center. That center is the focus of goods and services produced for its hinterland and is its dominant urban influence as well.

Holocene The current *interglacial* epoch (the warm period of glacial contraction between the glacial expansions of an **ice age**); extends from 10,000 years ago to the present. Also known as the *Recent Epoch*.

Hukou system A longstanding Chinese system whereby all inhabitants must obtain and carry with them residency permits that indicate where an individual is from and where they may exercise particular rights such as education, health care, housing, and the like.

Human Development Index A UN index that is a composite measure of life expectancy, education, and income per capita. It is used to rank countries within a four-level classification under this name.

Human evolution Long-term biological maturation of the human species. Geographically, all evidence points toward East Africa as the source of humankind. Our species, *Homo sapiens*, emigrated from this **hearth** to eventually populate the rest of the habitable world.

Hurricane Alley The most frequent pathway followed by tropical storms and **hurricanes** over the past 150 years in their generally westward movement across the Caribbean Basin. Historically, hurricane tracks have bundled most tightly in the center of this route, most often affecting the Lesser Antilles between Antigua and the Virgin Islands, Puerto Rico, Hispaniola (Haiti/Dominican Republic), Jamaica, Cuba, southernmost Florida, Mexico's Yucatán, and the Gulf of Mexico.

Hydraulic civilization theory The theory that cities which managed to control **irrigated** farming over large **hinterlands** held political power over other cities. Particularly applies to early Asian civilizations based in such river valleys as the Chang (Yangzi), the Indus, and those of Mesopotamia.

Inclusive development The extent of equal economic (and social) development opportunities for different population groups, especially minorities and the poor.

Indigenous Aboriginal or native; an example would be the pre-Columbian inhabitants of the Americas.

Indirect rule British colonial practice that kept indigenous power structures in place, co-opting individual rulers. The purpose was to minimize armed conflict and maximize profits.

Indo-European language family The major world language family that dominates the European **geographic realm**. This language family is also the most widely dispersed globally (Fig. G-8), and about half of humankind speaks one of its languages.

Indo-European languages The major world language family that dominates the European **geographic realm**. This language family is also the most widely dispersed globally (Fig. G-8), and about half of humankind speaks one of its languages.

Indochina Broadly speaking, the region of Southeast Asia that has historically been influenced by both India and China; more narrowly, the part of Southeast Asia under French control during colonial times (Vietnam, Cambodia, and Laos).

Industrial Revolution The term applied to the social and economic changes in agriculture, commerce, and especially manufacturing and urbanization that resulted from technological innovations and greater specialization in late-eighteenth-century Europe.

Informal economy The part of a national economy that is not registered with the government, and for which reliable statistics are rarely available. The complement to a country's **formal economy**.

Informal sector Dominated by unlicensed sellers of homemade goods and services, the primitive form of capitalism found in many developing countries that takes place beyond the control of government. The complement to a country's **formal sector**.

Information economy The new, increasingly dominant, postindustrial economy that is maturing in the most highly advanced countries of North America, Europe, and the Pacific Rim. Here, traditional industry is being eclipsed by a higher-technology productive complex focused on information-related activities.

Insurgent state Territorial embodiment of a successful guerrilla movement. The establishment by antigovernment insurgents of a territorial base in which they exercise full control; thus a **state** within a state.

Intermodal connections Facilities and activities related to the transfer of goods in transit from one transportation mode to another (e.g., the loading of containers from a ship directly onto a truck or railcar).

Intermodal transport system One that smoothly integrates different surface transportation modes. The shipping of cargo containers depends on fast and efficient transfers: they can be stacked on the decks and in the holds of ships as well as attaching to flatbed railcars and trailer trucks.

Irredentism A policy of cultural extension and potential political expansion by a **state** aimed at a community of its nationals living in a neighboring state.

Islamic Front The southern border of the African Transition Zone that marks the religious **frontier** of the **Muslim** faith in its southward penetration of Subsaharan Africa (see Fig. 6-21).

Jihad A doctrine within Islam. Commonly translated as *holy war*, it entails a personal or collective struggle on the part of **Muslims** to live up to the religious standards prescribed by the *Quran* (Koran).

Land alienation One society or culture group taking land from another.

Land bridge A narrow **isthmian** link between two large landmasses. They are temporary features—at least when measured in geologic time—subject to appearance and disappearance as the land or sea level rises and falls.

Land hemisphere The half of the globe containing the greatest amount of land surface, centered on western Europe.

Landlocked location An interior **state** wholly surrounded by land. Without coasts, such a country is disadvantaged in terms of **accessibility** to international trade routes, and in the scramble for possession of areas of the **continental shelf** and control of the **exclusive economic zone** beyond.

Land tenure The way people own, occupy, and use land.

Liberation theology A powerful religious movement that arose in South America during the 1950s, and subsequently gained followers throughout the global **periphery**. At its heart is a belief system, based on a blend of Christian faith and socialist thinking, that interprets the teachings of Christ as a quest to liberate the impoverished masses from oppression.

Lingua franca A "common language" prevalent in a given area; a second language that can be spoken and understood by many peoples, although they speak other languages at home.

Local functional specialization A hallmark of Europe's **economic geography** that later spread to many other parts of the world, whereby particular people in particular places concentrate on the production of particular goods and services.

Low island Cultures associated with low-lying coral islands of the Pacific Realm that cannot wrest sufficient moisture from the tropical maritime air to avoid chronic drought. Thus productive agriculture is impossible, and their modest populations must rely on fishing and the coconut palm for survival. Moreover, they are now threatened by **rising sea level**.

Maquiladora The term given to modern industrial plants in Mexico's U.S. border zone. These foreign-owned factories assemble imported components and/or raw materials, and then export finished manufactures, mainly to the United States. Import duties are disappearing under **NAFTA**, bringing jobs to Mexico and the advantages of low wage rates to the foreign entrepreneurs.

Marine geography The geographic study of oceans and seas. Its practitioners investigate both the physical (e.g., coral-reef **biogeography**, ocean–**atmosphere** interactions, coastal **geomorphology**) as well as human (e.g., **maritime boundary**-making, fisheries, beachside development) aspects of oceanic environments.

Maritime boundary An international boundary that lies in the ocean. Like all boundaries, it is a vertical plane, extending from the seafloor to the upper limit of the air space in the atmosphere above the water.

Maritime claims Intended legal assertions made by **states** concerning **sovereignty** over an oceanic space as parts of their **territorial seas**, or as rights to an **exclusive economic zone** (or as extensions into waters adjacent to land claims, such as in Antarctica [Fig. 12-5]).

Median-line boundary An international **maritime boundary** drawn where the width of a sea is less than 400 **nautical miles**. Because the **states** on either side of that sea claim **exclusive economic zones** of 200 nautical miles, it is necessary to reduce those claims to a (median) distance equidistant from each shoreline. **Delimitation** on the map almost always appears as a set of straight-line segments that reflect the configurations of the coastlines involved.

Medical geography The study of health and disease within a geographic context and from a spatial perspective. Among other things, this geographic field examines the sources, **diffusion** routes, and distributions of diseases.

Megacity Informal term referring to the world's most heavily populated cities; in this book, the term refers to a **metropolis** containing a population of greater than 10 million.

Melting pot   Traditional characterization of American society as a blend of numerous **immigrant ethnic** groups that over time were assimilated into a single societal mainstream. This notion always had its challengers among social scientists, and is now increasingly difficult to sustain given the increasing complexity and sheer scale of the U.S. ethnic mosaic in the twenty-first century.

Mental maps   Maps that individuals carry around in their minds that reflect their constantly evolving perception of how geographic space (ranging from their everyday activity space to the entire world) is organized around them.

Mestizo   Derived from the Latin word for *mixed*, refers to a person of mixed European (white) and Amerindian ancestry.

Metropolis   Urban **agglomeration** consisting of a (central) city and its suburban ring. See also **urban (metropolitan) area**.

Micro-credit   Small loans extended to poverty-stricken borrowers who would not otherwise qualify for them. The aim is to help combat poverty, encourage entrepreneurship, and to empower poor communities—especially their women.

Microstate   A sovereign **state** that contains a minuscule land area and population. They do not have the attributes of "complete" states, but are on the map as tiny yet independent entities nonetheless.

Migration   A change in residence intended to be permanent.

Mobile money   The dominant system of text-messaged money exchange in Subsaharan Africa that is attributable to the realm's general lack of a formal banking infrastructure. Elaborated in Box 7-15.

Monocentric geographic realm   A world geographic realm dominated—territorially and/or demographically—by a single country. Russia in Russia/Central Asia is a prime example; others are the United States (North America), India (South Asia), and China (East Asia).

Monsoon   Refers to the seasonal reversal of wind and moisture flows in certain parts of the subtropics and lower-middle latitudes. The *dry monsoon* occurs during the cool season when dry offshore winds prevail. The *wet monsoon* occurs in the hot summer months, which produce onshore winds that bring large amounts of rainfall. The air-pressure differential over land and sea is the triggering mechanism, with windflows always moving from areas of relatively higher pressure toward areas of relatively lower pressure. Monsoons make their greatest regional impact in the coastal and near-coastal zones of South Asia, Southeast Asia, and East Asia.

Mulatto   A person of mixed African (black) and European (white) ancestry.

Multilingualism   A society marked by a mosaic of local languages. Constitutes a **centrifugal force** because it impedes communication within the larger population. Often a *lingua franca* is used as a "common language," as in many countries of Subsaharan Africa.

NAFTA   The **free-trade area** launched in 1994 involving the United States, Canada, and Mexico.

Nation   Legally a term encompassing all the citizens of a **state**, it also has other connotations. Most definitions now tend to refer to a group of tightly knit people possessing bonds of language, **ethnicity**, religion, and other shared **cultural** attributes. Such homogeneity actually prevails within very few states.

Nation-state   A country whose population possesses a substantial degree of **cultural** homogeneity and unity. The ideal form to which most **nations** and **states** aspire—a political unit wherein the territorial state coincides with the area settled by a certain national group or people.

NATO   Established in 1950 at the height of the Cold War as a U.S.-led **supranational** defense pact to shield postwar Europe against the Soviet military threat. NATO is now in transition, expanding its membership while modifying its objectives in the post-Soviet era. Its 28 member-states (as of mid-2016) are: Albania, Belgium, Bulgaria, Canada, Croatia, Czech Republic, Denmark, Estonia, France, Germany, Greece, Hungary, Iceland, Italy, Latvia, Lithuania, Luxembourg, the Netherlands, Norway, Poland, Portugal, Romania, Slovakia, Slovenia, Spain, Turkey, the United Kingdom, and the United States.

Natural landscape   The array of landforms that constitutes the Earth's surface (mountains, hills, plains, and plateaus) and the physical features that mark them (such as water bodies, soils, and vegetation). Each **geographic realm** has its distinctive combination of natural landscapes.

Near Abroad   The 14 former Soviet republics that, in combination with the dominant Russian Republic, constituted the USSR. Since the 1991 breakup of the Soviet Union, Russia has asserted a sphere of influence in these now-independent countries, based on its proclaimed right to protect the interests of ethnic Russians who were settled there in substantial numbers during Soviet times. These 14 countries include Armenia, Azerbaijan, Belarus, Estonia, Georgia, Kazakhstan, Kyrgyzstan, Latvia, Lithuania, Moldova, Tajikistan, Turkmenistan, Ukraine, and Uzbekistan.

Negative externalities   Undesirable side-effects and/or byproducts of an action. In our case, the downside consequences of dam construction in Brazil's Amazon Basin in the form of further deforestation, other environmental degradation, and the displacement of existing communities.

Neighborhood effect   The impact of one's neighborhood on an individual's outlook, aspirations, socialization, and life chances.

Neoliberal policies   Policies adhering to an ideology or development strategy that advocates the privatization of state-run companies, lowering of international trade tariffs, reduction of government subsidies, cutting of corporate taxes, and overall deregulation of business activity.

Neoliberalism   A national or regional development strategy based on the privatization of state-run companies, lowering of international trade tariffs, reduction of government subsidies, cutting of corporate taxes, and overall deregulation of business activity.

New Silk Road   China's ongoing ambitious project to forge an overland routeway of high-speed railroads to link East Asia to Europe via Central Asia. This new "Eurasian land bridge", mapped in Figure 5-19, follows the general alignment of the fabled ancient Silk Road traversed by Marco Polo from the Mediterranean Basin to medieval China.

Nightlight map   Map that displays the nighttime distribution of artificial lighting in a given area, a good surrogate for the area's level of development (see Box 9-14).

Nine-Dashed-Line map   A map used by Chinese authorities to indicate Chinese claims to the South China Sea.

Node   A center that functions as a point of **connectivity** within a regional **network** or **system**. All urban settlements possess this function, and the higher the position of a settlement in its **urban system** or **hierarchy**, the greater its nodality.

Non-governmental organization (NGO)   A legitimate organization that operates independently from any form of government and does not function as a for-profit business. Mostly seeks to improve social conditions, but is not affiliated with political organizations.

Northeast Passage   The high-latitude sea route of the Arctic Ocean that follows the entire north coast of Eurasia from northern Norway in the west to the northeasternmost corner of Russia where it meets the Bering Strait. Increased seasonal melting of the Arctic ice cap in recent years has begun to open up this waterway as a summer route for shipping between Europe and East Asia.

Northwest Passage   The high-latitude, Arctic Ocean sea route around North America extending from Alaska's Bering Strait in the west to the Davis Strait between Canada and Greenland in the east. Heightened summertime melting of the Arctic ice cap in recent years has increased the likelihood of opening up this waterway for shipping between Asia's Pacific Rim and the Atlantic seaboard of the Americas.

Offshore banking   Term referring to financial havens for foreign companies and individuals, who channel their earnings to accounts in such a country (usually an "offshore" island-state) to avoid paying taxes in their home countries.

One-Child Policy   Chinese population control policy initiated in the late 1970s that proscribed (and enforced) a limit of one child per family of most population groups (mainly urban, **Han** populations). This policy, ended in 2016, had become increasingly controversial over the past decade.

One Nation–Two Systems   The arrangement under which capitalist Hong Kong functions within the PRC's communist economic system. Widely seen as a model for the future reunification of Taiwan with mainland China.

Outback   The name given by Australians to the vast, peripheral, sparsely settled interior of their country.

Outer city   The non-central-city portion of the American **metropolis**; no longer "sub" to the "urb," this outer ring was transformed into a full-fledged city during the late twentieth century.

Overseas Chinese   The more than 50 million ethnic Chinese who live outside China. About two-thirds live in Southeast Asia, and many have become quite successful. A large number maintain links to China and as investors played a major economic role in stimulating the growth of **SEZs** and Open Cities in China's **Pacific Rim**.

Pacific Rim   A far-flung group of countries and components of countries (extending clockwise on the map from New Zealand to Chile) sharing the following criteria: they face the Pacific Ocean; they exhibit relatively high levels of economic development, industrialization, and urbanization; their imports and exports mainly move across Pacific waters.

Pacific Ring of Fire   Zone of crustal instability along **tectonic plate** boundaries, marked by earthquakes and volcanic activity, that ring the Pacific Ocean Basin.

Pandemic   An outbreak of a disease that spreads worldwide.

Partition   The subdivision of the British Indian Empire into India and Pakistan at the end of colonial rule on August 15, 1947.

Peripheral development   Spatial pattern in which a country's or region's development (and population) is most heavily concentrated along its outer edges rather than in its interior.

Permafrost   Permanently frozen water in the near-surface soil and bedrock of cold environments, producing the effect of completely frozen ground. Surface can thaw during brief warm season.

Physiographic region   A **region** within which there prevails substantial **natural-landscape** homogeneity, expressed by a certain degree of uniformity in surface **relief, climate**, vegetation, and soils.

Physiologic density   The number of people per unit area of **arable** land.

Plantation   A large estate owned by an individual, family, or corporation and organized to produce a cash crop. Almost all plantations were established within the tropics; in recent decades, many have been divided into smaller holdings or reorganized as cooperatives.

Population density   The number of people per unit area. Also see **arithmetic density** and **physiologic density** measures.

Population distribution   The way people have arranged themselves in geographic space. One of human geography's most essential expressions because it represents the sum total of the adjustments that a population has made to its natural, cultural, and economic environments. A population distribution map is included in every chapter in this book.

Population geography   The field of geography that focuses on the spatial aspects of **demography** and the influences of demographic change on particular countries and regions.

Population implosion   The opposite of **population explosion**; refers to the declining populations of many European countries and Russia in which the **death rate** exceeds the **birth rate** and **immigration** rate.

Population pyramid   Graphic representation or *profile* of a national population according to age and gender. Such a diagram of age-sex structure typically displays the percentage of each age group (commonly in five-year increments) as a horizontal bar, whose length represents its relationship to the total population (see Fig. 8-13).

Primary sector   Activities engaged in the direct extraction of natural resources from the environment such as mining, fishing, lumbering, and especially agriculture.

Primate city   A country's largest city—ranking atop its urban **hierarchy**—most expressive of the national culture and usually (but not in every case) the capital city as well.

Protruded state   Territorial shape of a **state** that exhibits a narrow, elongated land extension (or *protrusion*) leading away from the main body of territory.

Rain shadow effect   The relative dryness in areas downwind of mountain ranges resulting from **orographic precipitation**, wherein moist air masses are forced to deposit most of their water content as they cross the highlands.

Region   A commonly used term and a geographic concept of paramount importance. An **area** on the Earth's surface marked by specific criteria, which are discussed in the Introduction.

Regional complementarity   Exists when a pair of regions, through an exchange of raw materials and/or finished products, can specifically satisfy each other's demands.

Relative location   The regional position or **situation** of a place relative to the position of other places. Distance, **accessibility**, and **connectivity** affect relative location.

Religious revivalism   Religious movement whose objectives are to return to the foundations of that faith and to influence state policy. Often called *religious fundamentalism*; but in the case of Islam, **Muslims** prefer the term *revivalism*.

Relocation diffusion   Sequential **diffusion process** in which the items being diffused are transmitted by their carrier agents as they relocate to new areas. The most common form of relocation diffusion involves the spreading of innovations by a **migrating** population.

Remittances   Money earned by **emigrants** that is sent back to family and friends in their home country, mostly in cash; forms an important part of the economy in poorer countries.

Remote sensing   The indirect capture of images by specially equipped, Earth-orbiting satellites.

Residential geography   The spatial distribution of a residential population. The term is most often used by urban geographers to describe the clustering of various social groups into the neighborhoods that form the residential fabric of cities and suburbs.

Rift valley   The trough or trench that forms when a thinning strip of the Earth's crust sinks between two parallel faults (surface fractures).

Rising sea level   One of the expected major impacts of **global climate change** on the world ocean resulting from the large-scale melting of Arctic and Antarctic ice. Low-lying coastal settlements and human activities are at greatest risk of inundation. Worldwide, huge urban and rural population concentrations are threatened; in the vast Pacific Basin, its thousands of **low islands** are the most vulnerable.

Rural-to-urban migration   The dominant **migration** flow from countryside to city that continues to transform the world's population, most notably in the less advantaged geographic realms.

Russification   Demographic resettlement policies pursued by the central planners of the Soviet Empire (1922–1991), whereby ethnic Russians were encouraged to **emigrate** from the Russian Republic to the 14 non-Russian republics of the USSR.

Satellite state   The countries of eastern Europe under Soviet **hegemony** between 1945 and 1989. This tier of countries—the "satellites" captured in Moscow's "orbit" following World War II—was bordered on the west by the Iron Curtain and on the east by the USSR. Using the names then in force, they included Bulgaria, Czechoslovakia, East Germany, Hungary, Poland, and Romania.

Scale   Representation of a real-world phenomenon at a certain level of reduction or generalization. In **cartography**, the ratio of map distance to ground distance; indicated on a map as a bar graph, representative fraction, and/or verbal statement. *Macroscale* refers to a large area of national proportions; *microscale* refers to a local area no bigger than a county.

Schengen Area   The territory constituted by most of Europe's countries within which people are free to cross international boundaries without formal border checks. Certain EU members do not fully participate: Bulgaria, Croatia, Cyprus, Ireland, and Romania. Four non-EU countries do participate: Iceland, Norway, Switzerland, and the microstate of Liechtenstein. Another non-participant is the United Kingdom, which voted to leave the EU in 2016.

Sex ratio   A **demographic** indicator showing the ratio of males to females in a given population.

Sharia law The strict criminal code based in Islamic law that prescribes corporal punishment, amputations, stonings, and lashings for both major and minor offenses. Its occurrence today is associated with the spread of **religious revivalism** in **Muslim** societies.

Shatter belt **Region** caught between stronger, colliding external cultural-political forces, under persistent stress, and often fragmented by aggressive rivals. Eastern Europe is a classic example.

Shi'ite Islam The smaller of the two main Islamic sects, comprising about 10 percent of Muslims overall, but in the majority in both Iran and Iraq. The origin of Shi'ism dates back to the death of the Prophet Muhammad, whose adherents believe that only a blood relative of Muhammad (his cousin) could be considered his legitimate successor.

Sinicization Giving a Chinese cultural imprint; Chinese **acculturation**. See also **Hanification**.

Site The internal locational attributes of an urban center, including its local spatial organization and physical setting.

Situation The external locational attributes of an urban center; its **relative location** or regional position with reference to other non-local places.

Small-island developing economies The additional disadvantages faced by lower-income island-states because of their often small territorial size and populations as well as overland **inaccessibility**. Limited resources require expensive importing of many goods and services; the cost of government operations per capita are higher; and local production is unable to benefit from **economies of scale**.

Social stratification In a layered or stratified society, the population is divided into a **hierarchy** of social classes. In an industrialized society, the working class is at the lower end; **elites** that possess capital and control the means of production are at the upper level.

South-to-North Water Diversion Project (SNWDP) The PRC's inter-basin water transfer scheme (the world's largest) to deliver massive quantities of fresh water from the Huang He and Chiang Jiang river systems to the burgeoning urban areas of northern China that face severe water shortages.

Southern Ocean The ocean that surrounds Antarctica (discussed in Chapter 11).

Sovereign Wealth Fund Part of Saudi Arabia's plan for a new economic future as an industrial power rather than an oil producer. This U.S. $2 trillion fund, to be created by selling off state-owned oil assets to private companies, will provide a major part of the financing for the transformation of the economy.

Sovereignty Controlling power and influence over a territory, especially by the government of an autonomous state over the people it rules.

Spatial diffusion The spatial spreading or dissemination of a **culture** element (such as a technological innovation) or some other phenomenon (e.g., a disease outbreak).

Spatial perspective Broadly, the geographic dimension or expression of any phenomenon; more specifically, anything related to the organization of space on the Earth' surface.

Spatial system The components and interactions of a **functional region**, which is defined by the areal extent of those interactions.

Special Economic Zone (SEZ) Manufacturing and export center in China, created since 1980 to attract foreign investment and technology transfers. Seven SEZs—all located on China's Pacific coast—currently operate: Shenzhen, adjacent to Hong Kong; Zhuhai; Shantou; Xiamen; Hainan Island, in the far south; Pudong, across the river from Shanghai; and Binhai New Area, next to the port of Tianjin.

State A politically organized territory that is administered by a sovereign government and is recognized by a significant portion of the international community. A state must also contain a permanent resident population, an organized economy, and a functioning internal circulation system.

State capitalism Government-controlled corporations competing under free-market conditions, usually in a tightly regimented society.

State formation The creation of a **state**, exemplifying traditions of human **territoriality** that go back thousands of years.

State territorial configuration The particular geographic layout of sovereign territory of a state, that influences its internal functioning and external relations.

Stateless nation A national group that aspires to become a an independent **state** but lacks the territorial means to do so.

Subsistence agriculture Farmers who eke out a living on a small plot of land on which they are only able to grow enough food to support their families or at best a small community.

Subtropical Convergence A narrow marine **transition zone**, girdling the globe at approximately latitude 40°S, that marks the equatorward limit of the frigid **Southern Ocean** and the poleward limits of the warmer Atlantic, Pacific, and Indian oceans to the north.

Sunbelt The popular name given to the southern tier of the United States, which is anchored by the mega-States of California, Texas, and Florida. Its warmer climate, superior recreational opportunities, and other amenities have been attracting large numbers of relocating people and activities since the 1960s; broader definitions of the Sunbelt also include much of the western United States, even Colorado and the coastal Pacific Northwest.

Sunni Islam The larger of Islam's two main sects (encompassing) roughly 90 percent of all Muslims) who adhere to the conviction that any devout follower of the Prophet Muhammad is eligible to be his legitimate successor.

Supranationalism A venture involving three or more **states**—political, economic, and/or cultural cooperation to promote shared objectives.

Sustainable development Viable, long-term development that does not deplete resources and/or cause significantly negative side-effects.

Taiga The subarctic, mostly **coniferous** snowforest that blankets northern Russia and Canada south of the **tundra** that lines the Arctic shore. Known as the **boreal forest** in North America.

Taliban The term means "students" or "seekers of religion." Specifically, refers to the Islamist militia group that emerged from *madrassas* in Pakistan and ruled neighboring Afghanistan between 1996 and 2001; it has been trying to regain control of that country in its continuing conflict with U.S.-led NATO troops. Taliban rule, in adherence with an extremist interpretation of *Sharia* law, was marked by one of the most virulent forms of militant Islam ever seen.

Tar sands The main source of oil from non-liquid petroleum reserves. The oil is mixed with sand and requires massive open-pit mining as well as a costly, complicated process to extract it. The largest known deposits are located in the northeast of Canada's province of Alberta, and by most estimates these Athabasca Tar Sands constitute one of the largest oil reserves in the world. The high oil prices of recent years have led to greatly expanded production here, but the accompanying environmental degradation caused by strip-mining and waste disposal has triggered a widening protest movement that may limit the exploitation of this resource.

Technopole A planned techno-industrial complex (such as California's Silicon Valley) that innovates, promotes, and manufactures the products of the postindustrial **information economy**.

Tectonic plates The slabs of heavier rock on which the lighter rocks of the continents rest. The plates are in motion, propelled by gigantic circulation cells in the red-hot, molten rock below. Most earthquakes and volcanic eruptions are associated with collisions of the mobile plates, as is the building of mountain chains.

Territorial sea Zone of seawater adjacent to a country's coast, held to be part of the national territory and treated as a component of the sovereign **state**.

Transculturation Cultural borrowing and two-way exchanges that occur when different **cultures** of approximately equal complexity and technological level come into close contact.

Transferability The capacity to move a good from one place to another at a bearable cost; the ease with which a commodity may be transported.

Transition zone An area of spatial change where the **peripheries** of two adjacent **realms** or **regions** join; marked by a gradual shift (rather than a sharp break) in the characteristics that distinguish these neighboring geographic entities from one another.

Transmigration   The now-ended policy of the Indonesian government to induce residents of the overcrowded, **core-area** island of Jawa to move to the country's peripheral islands.

Triple Frontier   The turbulent and chaotic area in southern South America that surrounds the convergence of Brazil, Argentina, and Paraguay. Lawlessness pervades this haven for criminal elements, which is notorious for money laundering, arms and other smuggling, drug trafficking, and links to terrorist organizations, including money flows to the Middle East.

Tropical deforestation   The clearing and destruction of tropical rainforests in order to make way for expanding settlement frontiers and the exploitation of new economic opportunities.

Tsunami   A seismic (earthquake-generated) sea wave that can attain gigantic proportions and cause coastal devastation. The tsunami of December 26, 2004, centered in the Indian Ocean near the Indonesian island of Sumatera (Sumatra), produced the first great natural disaster of the twenty-first century. Our new century's second major tsunami disaster occurred along the coast of Japan's northeastern Honshu Island on March 11, 2011.

Tsunami Warning System (TWS)   The ocean-basin-scale network of instrumented buoys and other sensors to detect and warn of Pacific **tsunamis** following coastal and seafloor earthquakes (Fig. 12-2). The Pacific TWS is the most advanced and has proven its worth many times. Similar systems are or soon will be in place in other maritime arenas, including the Indian Ocean, the Northeast Atlantic/Mediterranean, and the Caribbean Sea.

Tundra   The treeless plain that lies along the Arctic shore in northernmost Russia and Canada, whose vegetation consists of mosses, lichens, and certain hardy grasses.

Turkish Model   In the wake of the regime changes in the North Africa/Southwest Asia realm brought about by the "Arab Spring" of 2011, moderates have cited Turkey as the best model of democratic governance for this part of the world. Specifically, this involves a multi-party democracy that has a place for, but is not dominated by, Islamic political parties.

Uneven development   The notion that economic development varies spatially, a central tenet of **core-periphery relationships** in realms, regions, and lesser geographic entities.

Unitary state   A **nation-state** that has a centralized government and administration that exercises power equally over all parts of the **state**.

Unity of place   The great German natural scientist Alexander von Humboldt's notion that in a particular locale or region intricate connections exist among climate, geology, biology, and human cultures. This laid the foundation for modern geography as an *integrative discipline* marked by a spatial perspective.

Urbanization   A term with a variety of connotations. The proportion of a country's population living in urban places is its level of urbanization. The **process** of urbanization involves the movement to, and the clustering of, people in towns and cities—a major force in every geographic realm today. Another kind of urbanization occurs when an expanding city absorbs rural countryside and transforms it into suburbs; in the case of cities in disadvantaged countries, this also generates peripheral **shantytowns**.

Urban primacy   Refers to a country's largest city—ranking atop its urban **hierarchy**—most expressive of the national culture and usually (but not in every case) the capital city as well.

Urban system   A **hierarchical** network or grouping of urban areas within a finite geographic area, such as a country.

Viticulture   The growing of grapes for the production of wine.

Wallace's Line   The zoogeographical boundary proposed by Alfred Russel Wallace that separates the marsupial fauna of Australia and New Guinea from the non-marsupial fauna of Indonesia (see Figure 11-4).

West Wind Drift   The clockwise movement of water as a current that circles around Antarctica in the **Southern Ocean**.

World-City   A large city with particularly significant international (economic) linkages that also has a high ranking in the global urban system. Leading world-cities include London, New York, Tokyo, Shanghai, Singapore, and Paris.

Xenophobia   Extreme dislike and fear of foreigners, sometimes fueled by populist politicians.

Index